EOS Annual Meeting (EOSAM 2022)

EPJ Web of Conferences Volume 266 (2022)

Porto, Portugal
12 – 16 September 2022

Editors:

Manuel Filipe Costa
Maite Flores-Arias
Gilles Pauliat
Patricia Segonds

ISBN: 978-1-7138-6266-6

Printed from e-media with permission by:

Curran Associates, Inc.
57 Morehouse Lane
Red Hook, NY 12571

Some format issues inherent in the e-media version may also appear in this print version.

For additional information, please contact EDP Sciences – Web of Conferences at the address below.

EDP Sciences – Web of Conferences
17, Avenue du Hoggar
Parc d'Activité de Courtabœuf
BP 112
F-91944 Les Ulis Cedex A
France

Phone: +33 (0) 1 69 18 75 75
Fax: +33 (0) 1 69 28 84 91

contact-edps@webofconferences.org

Additional copies of this publication are available from:

Curran Associates, Inc.
57 Morehouse Lane
Red Hook, NY 12571 USA
Phone: 845-758-0400
Fax: 845-758-2633
Email: curran@proceedings.com
Web: www.proceedings.com

TABLE OF CONTENTS

Preface

The European Optical Society Annual Meeting, EOSAM, is a major international scientific conference covering all aspects of optics and photonics, covered by the topical meetings and sessions. EOSAM was celebrated this year in Portugal, at the Faculty of Engineering of the University of Porto, FEUP, in co-operation with the Portuguese Society for Research and Development in Optics and Photonics, SPOF, and INESC TEC - Institute for Systems and Computer Engineering, Technology and Science.

EOSAM included high-quality program covering all aspects of optics and photonics, within the 13 topical meetings, posters and special sessions. 360 attendees enjoyed the over 300 presentations during the week. EOSAM this year included these sessions and topical meetings:

Early Stage Researcher Session

EU Project Session

TOM1: Silicon Photonics and Guided-Wave Optics

TOM2: Computational, Adaptive and Freeform Optics - focus on Illumination, AR/VR and information driven systems

TOM3: Optical System Design, Tolerancing and Manufacturing

TOM4: Bio-Medical Optics

TOM5: Resonant Nanophotonics

TOM6: Optical Materials: crystals, glasses, thin films, nanostructures/nanocrystals, metasurfaces, organic molecules and polymers, synthesis...

TOM7: Thermal radiation and energy management

TOM8: Nonlinear and Quantum Optics

TOM9: Opto-electronic Nanotechnologies and Complex Systems

TOM10: Frontiers in Optical Metrology

TOM11: Tapered Optical Fibers, from fundamentals to applications

TOM12: Optofluidics

TOM13: Advances and Applications of Optics and Photonics

In addition, a call was made for submission of selected papers for publication in JEOS:RP, the journal of the European Optical Society, covering the latest advances in both theory and applications of optics and photonics grouped in the included Topical Meetings (TOMs) at the conference.

We would like to express our gratitude to the European Optical Society, the Faculty of Engineering of the University of Porto, FEUP, the Portuguese Society for Research and Development in Optics and Photonics, SPOF, and INESC TEC - Institute for Systems and Computer Engineering, Technology and Science, in addition to the chairs of the different TOMs for their support and help in the organization of the Conference. We finally want to acknowledge all the authors that have submitted their work to be included in this volume and especially thank all those collaborating persons who acted as reviewers of the papers included in this volume for their excellent work.

The editors:

General Chairs: Manuel Filipe Costa, President of SPOF and Maite Flores-Arias, Secretary of EOS

Program Chairs: Gilles Pauliat, President of EOS and Patricia Segonds, President-Elect of EOS

edp sciences

Web of Conferences

Statement of Peer review

In submitting conference proceedings to *Web of Conferences*, the editors of the proceedings certify to the Publisher that

1. They adhere to its **Policy on Publishing Integrity** in order to safeguard good scientific practice in publishing.
2. All articles have been subjected to peer review administered by the proceedings editors.
3. Reviews have been conducted by expert referees, who have been requested to provide unbiased and constructive comments aimed, whenever possible, at improving the work.
4. Proceedings editors have taken all reasonable steps to ensure the quality of the materials they publish and their decision to accept or reject a paper for publication has been based only on the merits of the work and the relevance to the journal.

Title, date and place of the conference: European Optical Society Annual meeting (EOSAM) 2022
12-16 September 2022, Porto, Portugal

Proceedings editor(s): Ralf Bergmann, Bernd Bodermann

Date and editor's signature: 12.09.2022

17, avenue du Hoggar - PA de Courtabœuf – BP 112 - 91944 Les Ulis Cedex A (France)
Tél. : 33 (0)1 69 18 75 75 - Fax : 33(0)1 69 07 45 17 – www.edpsciences.org

Statement of Peer review

In submitting conference proceedings to *Web of Conferences*, the editors of the proceedings certify to the Publisher that

1. They adhere to its **Policy on Publishing Integrity** in order to safeguard good scientific practice in publishing.
2. All articles have been subjected to peer review administered by the proceedings editors.
3. Reviews have been conducted by expert referees, who have been requested to provide unbiased and constructive comments aimed, whenever possible, at improving the work.
4. Proceedings editors have taken all reasonable steps to ensure the quality of the materials they publish and their decision to accept or reject a paper for publication has been based only on the merits of the work and the relevance to the journal.

Title, date and place of the conference

EOSAM2022, September 12-16, 2022, Porto, Portugal

Proceedings editor(s): Manuel Filipe Pereira da Cunha Martins Costa

Date and editor's signature Braga, August 26, 2022

Statement of Peer review

In submitting conference proceedings to *Web of Conferences*, the editors of the proceedings certify to the Publisher that

1. They adhere to its **Policy on Publishing Integrity** in order to safeguard good scientific practice in publishing.
2. All articles have been subjected to peer review administered by the proceedings editors.
3. Reviews have been conducted by expert referees, who have been requested to provide unbiased and constructive comments aimed, whenever possible, at improving the work.
4. Proceedings editors have taken all reasonable steps to ensure the quality of the materials they publish and their decision to accept or reject a paper for publication has been based only on the merits of the work and the relevance to the journal.

Title, date and place of the conference EOSAM 2022, TOM 7, 12-16 September 2022, Porto (Portugal)

Proceedings editor(s): Marco Centini

Date and editor's signature 06/09/2022

Statement of Peer review

In submitting conference proceedings to *Web of Conferences*, the editors of the proceedings certify to the Publisher that

1. They adhere to its **Policy on Publishing Integrity** in order to safeguard good scientific practice in publishing.
2. All articles have been subjected to peer review administered by the proceedings editors.
3. Reviews have been conducted by expert referees, who have been requested to provide unbiased and constructive comments aimed, whenever possible, at improving the work.
4. Proceedings editors have taken all reasonable steps to ensure the quality of the materials they publish and their decision to accept or reject a paper for publication has been based only on the merits of the work and the relevance to the journal.

EOSAM22 TOM3, Porto, 2022

Proceedings editor(s): Oliver Fähnle

220826

Statement of Peer review

In submitting conference proceedings to *Web of Conferences*, the editors of the proceedings certify to the Publisher that

1. They adhere to its **Policy on Publishing Integrity** in order to safeguard good scientific practice in publishing.
2. All articles have been subjected to peer review administered by the proceedings editors.
3. Reviews have been conducted by expert referees, who have been requested to provide unbiased and constructive comments aimed, whenever possible, at improving the work.
4. Proceedings editors have taken all reasonable steps to ensure the quality of the materials they publish and their decision to accept or reject a paper for publication has been based only on the merits of the work and the relevance to the journal.

Title, date and place of the conference

EOSAM 2022, 12-16 sept. 2022, Porto (Portugal)

Proceedings editor(s):

Sylvie Lebrun and Jean-Charles Beugnot

Date and editor's signature

1/09/2022

17, avenue du Hoggar - PA de Courtabœuf – BP 112 - 91944 Les Ulis Cedex A (France)
Tél. : 33 (0)1 69 18 75 75 - Fax : 33(0)1 69 07 45 17 – www.edpsciences.org

Towards Integrated Single Photon Avalanche Detectors for Visible Light (Invited)

Thomas Yong Long Ang[1,*], Jun Rong Ong[1], Ray Jia Hong Ng[1], Soon Thor Lim[1], Trevor Rei Lee[1], Victor Leong[2,*], Salih Yanikgonul[2], Shawn Yohanes Siew[3], Gundlapalli Prithvi[2], Hong Wang[4], Leonid Krivitsky[2], Ching Eng Png[1,*]

1. Institute of High Performance Computing, Agency for Science, Technology and Research, 138632 Singapore
2. Institute of Materials Research and Engineering, Agency for Science, Technology and Research, 138634 Singapore
3. Advanced Micro Foundry Pte Ltd, 117685 Singapore
4. Silicon Technologies – Centre of Excellence, Nanyang Technological University, 639798 Singapore

Abstract. Integrated avalanche photodetectors (APDs) are essential and ubiquitous devices in quantum photonics applications. While free-space APDs are a mature technology, the development of integrated APDs for visible light is still in its infancy. In this invited talk, we review our work on integrated photodetectors – the Germanium photodetector for O band, and the first integrated silicon (Si) APD for visible light. A unique feature of the integrated Si APD system for visible light is the end-fire coupling between the silicon nitride (SiN) waveguide and the Si APD in the same layer. This allows for broadband and high-efficiency coupling of light from the SiN waveguide to the Si APD for light detection, without the drawbacks of conventional interlayer coupling. Our work on integrated Si APDs based on Geiger-mode and our progress towards achieving single photon detection for visible light is discussed in the talk.

1 Introduction

Photodetectors (PDs) are indispensable components of photonic integrated circuits. Various types of PDs designs, such as PIN [1] and avalanche photodetectors (APDs) [2], have been realized on different material platforms, including III–V semiconductors, germanium (Ge), and silicon (Si). For silicon photonics, the common material for PDs is Ge [1, 2]. These are typically designed for optical communication networks that use infrared wavelengths of 1310nm and 1550nm. Integrated silicon APDs have also been demonstrated for short-reach data communications at wavelength of 850nm [3, 4]. However, it is a challenge to realize integrated silicon APDs for visible light. Recently, our group demonstrated the first waveguide-coupled silicon APDs for visible light [5]. This can be used for quantum photonics and other visible light applications. In this invited talk, we review our work on integrated PDs – the Ge PDs for O band [1], and the integrated silicon APD for visible light [5]. Our work [6] and progress towards achieving on-chip single photon detection using silicon APD in the Geiger mode is discussed in the talk.

2 Germanium Photodetectors

Our germanium (Ge) photodetector is based on a P-I-N junction design that consists of p-doped Si slab and n++ Ge, with an intrinsic Ge region for light absorption.

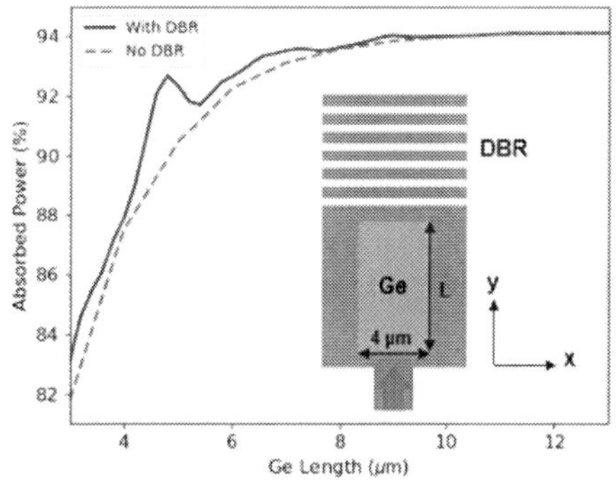

Fig. 1. Absorption of the Ge PD at different lengths. The inclusion of distributed Bragg reflector (DBR) at the exit of the Ge PD enhances the absorption.

To establish ohmic contacts, p++ Si is used. The height of the Si slab height is 220nm, while the thickness of the Ge layer is 500nm. Details regarding the design can be found in our published work [1].

An important design parameter is the absorption length of the Ge PDs. We used 3D finite-difference time-domain (FDTD) simulation to determine the fraction of absorbed power as a function of Ge length at a width of 4μm. As seen in the dotted graph of Fig. 1, the absorption saturates to 94% for length L above 8μm. For short PD with L below 7μm, the absorption is lower than 90%.

* Corresponding authors: thomas-ang@ihpc.a-star.edu.sg, Victor_Leong@imre.a-star.edu.sg, pngce@ihpc.a-star.edu.sg

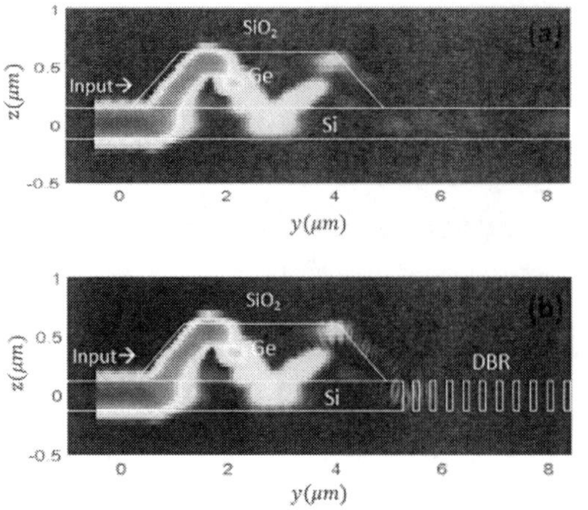

Fig. 2. Simulated electric field distribution in the Ge PD without DBR in **(a)** and with DBR in **(b)**. The use of DBR reflects the unabsorbed light back into the Ge PD for re-absorption.

It is advantageous to use PD with short length for low dark current and high speed. Distributed Bragg reflector (DBR) was therefore placed at the output of the Ge detector. The DBR has a period of 340nm and 50% duty cycle. An additional 3% of absorption was observed for short Ge length of 4.8μm. The use of DBR reflects the unabsorbed light back into the Ge PDs for re-absorption. This phenomenon is illustrated in Fig. 2.

Ge PDs of lengths L = {4.8, 6.8, 11, 17} μm at varying widths W = {4, 4.5, 5, 5.5} μm were fabricated and measured for responsivity, dark current, and O-E 3dB bandwidth performances. DBRs are included in the Ge PDs with length of 4.8μm and 6.8μm. At -2V, dark current below 4nA, responsivity of 0.87A/W, and device bandwidth of 35GHz were observed.

3. Silicon Avalanche Photodetectors

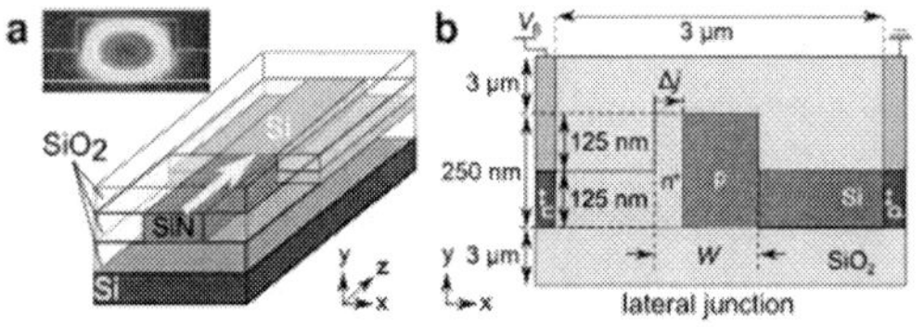

Fig. 3. The on-chip silicon APD designed for visible light. An end-fire coupling scheme in **(a)** is used to couple the input SiN waveguide with the Si rib waveguide that is n+ and p doped to form an integrated silicon photodetector in **(b)**.

Recently, our group designed and experimentally demonstrated the first monolithically integrated silicon avalanche photodetector for visible light [5]. The device structure and doping profile of the proposed on-chip silicon APD are shown in Figure. 3. A unique feature is the end-fire coupling between the SiN waveguide and the and the doped Si rib waveguide in the same layer. This is a straightforward design approach, but difficult for fabrication as both the SiN and Si waveguides are on the same layer. Using 3D FDTD simulation, we analysed the fraction of absorbed power by the Si rib waveguide when visible light is end-fire coupled from the SiN waveguide. In the simulation, both SiN and Si waveguides have the same width W, and a length of 10μm is used for the Si rib waveguide. At visible wavelengths around 680nm, the absorption is above 70% for W above 400nm, as shown in Fig. 4.

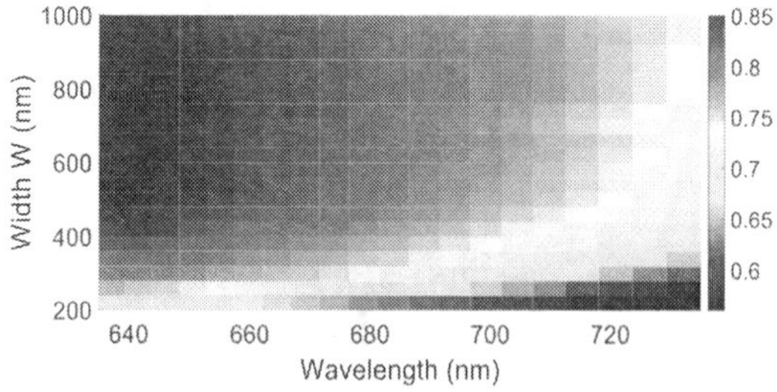

Fig. 4. The fraction of absorbed power by the Si rib waveguide of length 10μm with width W when light is end-fire coupled from the SiN waveguide with width W.

To achieve Si PDs with high absorption, we used W = {750, 900} nm for our fabrication of Si APD based on the end-fire coupling scheme. Two different designs of PN junctions were used on the Si rib waveguide – the lateral and interdigitated PN junctions. Details of the device design, fabrication, and measured results can be found in our work in [5]. A high gain-bandwidth product of 234 ± 25GHz at 20V reverse bias was demonstrated at wavelength 685nm, with low dark current of 0.12μA, and open eye diagrams at up to 56Gbps.

References

1. J. R. Ong, T. Ang, T. X. Guo, S. T. Lim, W. Hong, C. E. Png, 2020 ACP/IPOC, OSA Technical Digest, paper T3D.2.
2. X. Zeng, Z. Huang, B. Wang, D. Liang, M. Fiorentino, and R. G. Beausoleil, Optica 6, 772–777 (2019).
3. M. M. P. Fard, C. Williams, G. Cowan, O. Liboiron-Ladouceur, Opt.Express 25, 5107–5118 (2017).
4. A. Chatterjee, Saumitra, S. K Sikdar, and S. K. Selvaraja. Opt. Lett. 44, 1682–1685 (2019).
5. S. Yanikgonul, V. Leong, J. R. Ong, T. Hu, S. Y. Siew. C. E. Png, L. Krivitsky, Nat Commun 12,1834(2021).
6. S. Yanikgonul, V. Leong, J. R. Ong, C. E. Png, L. Krivitsky, IEEE Journal of Selected Topics in Quantum Electronics, vol. 26, no. 2 (2020).

Integrated optical phased arrays with circular architecture on a silicon platform

Daniel Benedikovic[1,2,3,*], *Qiankun* Liu[1], *Alejandro* Sanchez-Postigo[4], *Ahmad* Atieh[5], *Tom* Smy[1], *Pavel* Cheben[6], and *Winnie* N. Ye[1]

[1]Department of Electronics, Carleton University, Ottawa, Canada
[2]Department of Multimedia and Information-Communication Technologies, University of Zilina, Zilina, Slovakia
[3]University Science Park, Zilina, Slovakia
[4]Telecommunication Research Institute (TELMA), Universidad de Málaga, E.T.S. Ingeniería de Telecomunicación, Málaga, Spain
[5]Optiwave Systems, Inc., Ottawa, Canada
[6]National Research Council Canada, Ottawa, Canada

Abstract. Optical phased arrays (OPAs) are now at the forefront of photonic research as a key beam steering technology for myriad of photonic applications, including in light detection and ranging (LIDAR), communications, and metrology, among others. Integrated OPAs with narrow beam widths and wide-angle steering are in critical need, especially for LIDARs in autonomous vehicle, drone and airplane navigation, or satellites. In this work, we numerically study the performances of OPAs having a circular layout arrangement. Compared to recently available solutions with 1D linear or 2D rectangular arrays, the proposed circular OPAs are poised to deliver effective suppression of the grating sidelobes, while improving beam steering range and obtaining narrower beamwidths. We demonstrate 110-element circular arrays with sidelobe suppression better than 10 dB and an angular beamwidth of 0.5°. Under a monochromatic operation at a 1550 nm wavelength, such array provides a solid angle steering range of 0.21π-sr, with a perspective for performance improvement by using large number of OPA elements and operating under broader spectral range.

1 INTRODUCTION

Chip-scale optical phased arrays (OPAs), which vary the angle of an optical beam by alternating its phase profile, are promising new technology for many applications. Compared to the well-established OPA systems working in fields of imaging and displays, integrated OPAs avoid bulky mechanical mirrors or other spinning and rotating parts. In turn, this helps the system to be more tolerant and insensitive to external vibrations. Thus, on-chip OPAs possess the advantages to be stable, fast, and precise in their functionality, paving the way for next-generation dynamic beam steering and shaping [1,2]. Silicon photonics is a platform of conscious choice for dense integration. Silicon photonics enables realization of compact, low-cost, and high-performance components on a single chip, potentially in large volumes. Leveraging the compatibility with well-mastered tools and processes of complementary metal-oxide-semiconductor (CMOS) fabrication infrastructure, silicon photonics is seen as future technology for scalable product deployment.

Silicon photonics OPAs are realized as one-dimensional (1-D) linear and two-dimensional (2-D) arrays, having either rectangular or circular layouts [1-6]. To steer the light beam, such approaches typically adopt wavelength-controlled antennas or antennas with associated phased shifters [1-3]. OPAs call upon narrow beam widths and large field-of-views, in both azimuthal and elevation angles. Narrow beams can be formed using large number of OPA antennas. On the other hand, large FOV can be assured either by shrinking the element footprint or by reducing the spacing between individual elements within the OPA. It is required to have element separation on a sub-wavelength scale, i.e. separation at a distance that is shorter than a half of a wavelength [1]. Otherwise, grating sidelobes in the far-field become too dominant, which restricts beam steering range and makes phased arrays impractical.

In this work, we report a numerical study on 2-D OPA with circular geometry. Circular OPAs are competitive alternatives to 1-D and 2-D arrays, enabling effective sidelobe suppression, enhanced beam steering spans, and having narrower beamwidths, yet with increased element separation well-beyond the sub-wavelength scale. Under

*Corresponding author: daniel.benedikovic@uniza.sk

a single-wavelength (monochromatic) operation, we design a 110-element phased array with sidelobe suppression larger than 10 dB, narrow angle divergences of only 0.5°, and solid angle steering range of 0.21π-sr.

2 OPA layout and steering performance

Figure 1(a) shows a schematic view of the proposed circular OPA. OPA system is designed for 220 nm / 2 μm (silicon / buried oxide layers) using silicon-on-insulator substrate. Circular array comprises N concentric rings, M grating antennas, with ring radius R_n and element distance dr. Those parameters are optimized and comprehensively studied to obtain reliable performance outputs. For the OPA, antennas are key devices. Optical antennas have subwavelength-assisted L-shaped profile, schematically shown in the inset of Fig. 1(a), enabling superior radiation performance over wide wavelength range [6]. Radiation efficiency as a function of a wavelength is shown in Fig. 1(b).

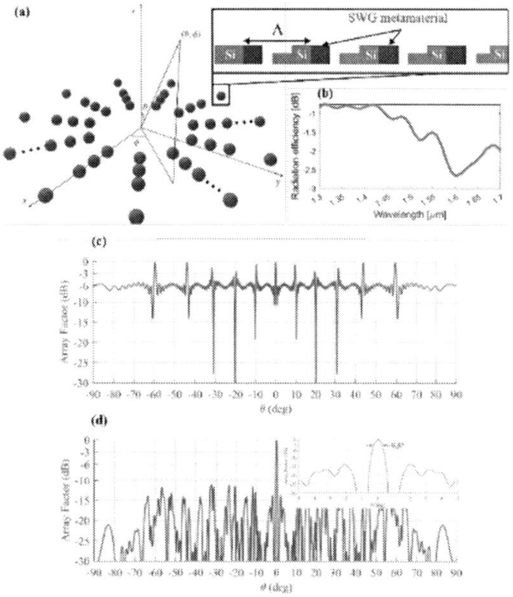

Fig. 1. (a) Schematics of circular array with silicon antennas. Inset: 2D schematics of the L-shaped grating antenna. (b) Antenna radiation efficiency versus wavelength. Elevation cut of array factor amplitudes for (c) $M = 4$ and (d) $M = 11$.

Figures 1(c) and 1(d) show simulated array factor amplitudes for different array configurations, with varying number of antennas. We observed that both sidelobe suppression as well as angular beamwidth are substantially improved. For OPA with $N = 10$; $M = 11$; $R_n = 20$ μm; and $dr = 9$ μm, the main-to-sidelobe grating suppression is -11.1 dB. For this array configuration, the angular beamwidth of about 0.5° was obtained. It is also expected that enhanced performance can be achieved by employing more elements into a circular OPA.

The steering range of proposed circular array was systematically studied by simulating array far-field pattern at different steering angles over the entire hemisphere. The steering range is defined as the set of all steerable $\{\theta, \phi\}$ angles, for which the main-to-sidelobe ratio is larger than 10 dB. All steerable angles are projected on a uv plane. Figure 2 shows the projection of steering range onto a hemisphere under single-wavelength operation at 1550 nm. We estimated a solid angle steering range of 0.21π-sr.

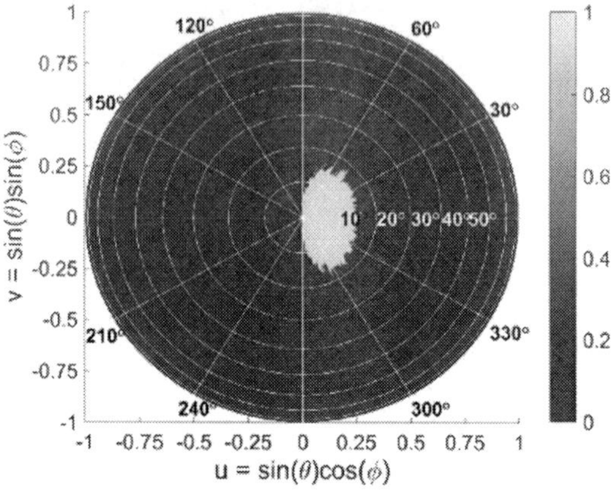

Fig. 2. The steering range projection onto a hemisphere for a circular array operating at 1550 nm wavelength. The bright green area corresponds to the set of all steerable angles with the sidelobe suppression larger than 10 dB, and the dark blue area represents non-steerable angles.

3 Conclusions

We report circular OPAs with efficient antennas that offer effective sidelobe suppression, while improving beam steering range and obtaining narrower beamwidths. We designed a 110-element OPA with sidelobe suppression better than 10 dB and an angular beamwidth of 0.5°. At a 1550 nm wavelength, proposed OPA provided a solid angle steering range of 0.21π-sr. The on-chip circular OPA is a promising technology for practical applications such as LIDARs or communications.

Acknowledgement and Funding

This research was funded by the Natural Sciences and Engineering Research Council of Canada's Collaborative R&D Grant Program by collaborating with Optiwave Systems, Inc., Slovak Grant Agency VEGA 1/0113/22, and Slovak Research and Development Agency under the project APVV-21-0217.

References

1. M.J.R. Heck, *Nanophotonics* **6**, p. 93–107 (2017).
2. C.V. Poulton, et al. *IEEE J. Sel. Top. Quantum Electron.* **25**, p. 7700108 (2019).
3. Y. Guo, et al. *Appl. Sci.* **11**, p. 4017 (2021).
4. J. He, et al. *IEEE Access* **8**, p. 188284–188298 (2020).
5. F. Zhang, et al. *Appl. Opt.* **57**, p. 4977–4984 (2018).
6. Q. Liu, et al. *Nanomaterials* **12**, p. 1938 (2022).

Ultra-dense interferometric chain architecture for datacom and telecom applications

Serge Bidnyk[†,*], *Ksenia* Yadav[†] and *Ashok* Balakrishnan[†]

[†]Enablence Technologies Inc., 390 March Road, Ottawa, Canada K2K 0G7

Abstract. Further increase in the density of integrated planar lightwave circuits (PLCs) depends on the introduction of compact guided-wave layout solutions. We describe a novel architecture for coiling multistage interferometric devices with densities reaching the theoretical limit. Our approach is validated by the design, fabrication, and deployment of state-of-the-art PLCs based on the proposed architecture for use in datacom and telecom applications.

1 Introduction

Planar lightwave circuit (PLC) technology has grown into a powerful platform that is able to meet the challenging demands of today's high-speed optical communication systems [1]. PLCs provide integrated solutions that offer compact form factor, lower cost and higher reliability compared to bulk optics solutions.

The increasing demands of today's communication systems are driving the need to integrate more optical functionality within a smaller footprint. Densification of components can be achieved by using higher refractive index contrast systems, such as silicon-on-insulator (SOI)[2], silicon nitride (Si_3N_4)[3] or III-V photonics [4], which result in tighter optical confinement, allowing smaller radii of curvature and thus higher density of integration. Yet even in these high refractive index contrast systems, the ratio of the active waveguides to the total area of the chip remains very small due to packing limitations of traditional architectures, limiting further progress in the densification of optical components.

Here we present a new approach for achieving highly-optimized, ultra-dense layouts of multistage interferometric devices suitable for both low and high refractive index contract systems. We demonstrate that the devices based on the proposed architecture possess industry-leading performance characteristics resulting in their wide deployment in datacom and telecom applications, including multi-channel LAN-wavelength division multiplexing (WDM).

2 Interferometric coil architecture

In integrated photonics, (de-)multiplexing filters are typically based on arrayed waveguide gratings (AWG) or cascaded interferometric lattice filters. Cascaded lattice filters are constructed as multistage Mach Zehnder interferometer (MZI) chains, and unlike AWG-based systems, they are essentially lossless components. However, in these cascaded systems, each stage of the interferometric chain relies on the optical response from the preceding stage. This serial nature of the interferometric chains usually results in layouts requiring a large area, especially as the number of interferometers increases.

2.1. The generalized coil

The proposed architecture relies on a generalized coil pattern that serves as a basis for the layout of interferometric structures. In order to construct a MZI, one requires an optical delay between two waveguides and two directional couplers. When a ribbon of two or more waveguides is coiled together, an optical delay can be achieved in the bent sections of the coil, where there is a natural physical length difference between adjacent waveguides, while the directional couplers can be implemented in the straight sections of the coil. To construct a multi-channel (de-)multiplexer, the interferometers are chained as a binary tree of cascaded lattice filters [5]. Due to the construction of the coil, the incremental cost of adding additional interferometric stages is very low, both in terms of the added area and the excess insertion loss.

A relatively large radius of curvature is traditionally viewed as a limiting factor to achieving a high density of optical components. However, in our previously reported work with long delay lines, we demonstrated that 500-cm-long waveguides can be folded in a square centimeter even in a relatively low refractive index contrast platform [6]. In practice, even the longest MZI chains require only several centimeters of length and thus do not lead to a substantial increase in the chip footprint. We note that this architecture can be applied to any guided-optics platform.

To validate the proposed interferometric coil architecture, we selected a silica-on-silicon platform with a refractive index contrast of $\Delta n = 2.0\%$ and waveguide

* Corresponding author: serge.bidnyk@enablence.com

dimensions of 3.0×3.0 μm . The advantages of this platform include low waveguide propagation losses of <0.01 dB/cm, efficient fiber-to-waveguide coupling, polarization-invariant performance, and low temperature sensitivity. The length of the waveguides does not significantly affect the device performance due to the very low propagation losses of silica waveguides. However, the relatively low refractive index contrast restricts the maximum radius of curvature to 1 mm.

2.2 8-channel LAN-WDM multiplexer

To demonstrate the applicability of the approach in telecom applications, we have designed and fabricated an 8-channel LAN-WDM multiplexer, governed by the 400GBASE-FR8 and -LR8 standard [7]. An added challenge to the specification is the requirement for a guard band that effectively necessitates eight channels to be arranged in a `4 channels + skip 1 channel + 4 channels` configuration.

An implementation of such a device in buried silica waveguides or silicon nitride has an added advantage of low thermal sensitivity. A silicon nitride-based LAN-WDM based on cascaded Mach-Zehnder interferometers was reported previously [8], albeit for Gaussian-like passbands. The proposed architecture, however, allows us to scale the number of interferometric chains, and individually shape the passbands. We now demonstrate that by adding additional binary-tree filters to a 4-channel LAN-WDM, we can achieve an 8-channel configuration with a square-passband single mode operation.

The layout of the 8-channel LAN-WDM multiplexer is shown in Fig. 1(a). The size of the chip is 0.77×0.49 cm, corresponding to a footprint area of 0.38 cm^2, and it incorporates a total of 11 MZIs. We note that the addition of four additional filters compared to a 4-channel implementation [6] does not substantially increase the required area, highlighting the advantages of the coiled architecture.

It is important to note that the sequencing of the channels using this configuration is 1, 9, 3, 7, 2, 6, 4, and 8. In order to re-order the channels, we have added an additional section onto the coil that uses a braid-like structure shown in the inset of Fig. 1(a). This was done at the expense of additional 300 μm added to the width of the coiled structure.

Fig. 1(b) shows the transmission spectra of the LAN-WDM multiplexer based on the described architecture. The chip was designed for an integrated solution where the coupling of laser diodes occurs directly without using mode converters. As such, the chip does not contain provisions for fiber-to-waveguide couplers on the input side, giving rise to 2.2 dB in perceived insertion loss when measured with an SMF-28 fiber. On the output side, a single mode converter is implemented with an insertion loss of 0.5 dB. Based on these considerations, we estimate the on-chip loss at 0.3-0.5 dB, depending on the channel, with polarization dependent loss of less than 0.2 dB. A single-mode spectral response of each channel is flat with a 1-dB bandwidth of 3.5 nm (80% of the channel pitch).

(a)

(b)

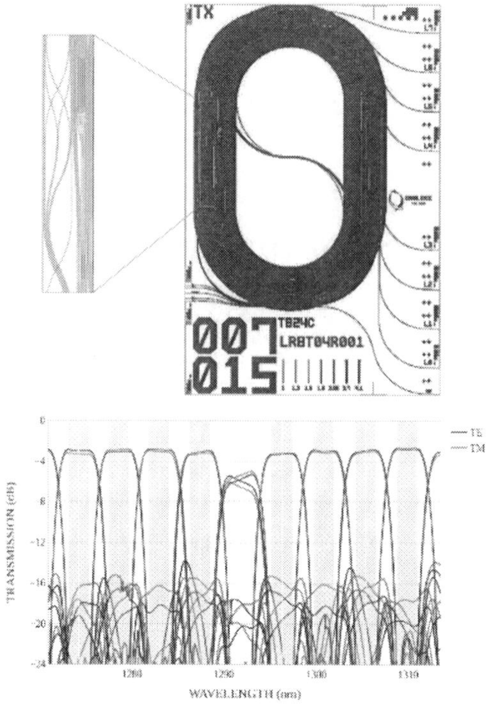

Fig. 1. (a) An 8-channel LAN-WDM multiplexer chip. The inset shows a zoomed-in view of the structure required for the reordering of the multiplexer channels. (b) Measured transmission spectra of the multiplexer, including two fiber couplings.

3 Conclusions

In this work, we have described a new guided-wave layout architecture for coiling multistage interferometric devices. The architecture achieves waveguide densities close to the theoretical limit, and can be used to realize cascaded lattice filters for WDM applications in a small footprint, regardless of the refractive index contrast used. We validated the architecture by demonstrating a production-ready design for a multi-channel LAN-WDM multiplexer chip designed and fabricated in a silica-on-silicon platform with a refractive index contrast of Δn = 2.0%. The resulting devices have a compact footprint and exceptional optical performance that led to their deployment in datacom and telecom applications.

References

1. P. Cheben et al., Nature, **560**, 565 (2018).
2. C. R. Doerr, Frontiers in Physics, **3**, 37 (2015).
3. D. J. Blumenthal et al., *Proceedings of the IEEE*, **106**, 2209 (2018).
4. M. Smit et al., Semicond. Sci. Technol., **29**, 083001 (2014).
5. F. Horst et al., Optics Express, **21**, 11652 (2013).
6. S. Bidnyk et al., Proc. SPIE 12004, 120040F (2022).
7. 200 Gb/s and 400 Gb/s Ethernet Task Force. http://www.ieee802.org/3/bs/ (2021).
8. Z. Zhao et al., Photonics, **9**, 252 (2022).

Scattering parameters of a non-reciprocal magneto-optical integrated coupler used as an isolator

Yann G. Boucher[1,2], *Yuliya S.* Dadoenkova[3,*], and *Florian F. L.* Bentivegna[3]

[1]Institut FOTON, UMR 6082, CNRS, 22305 Lannion Cedex, France
[2]École Nationale d'Ingénieurs de Brest, 29238 Brest Cedex 3, France
[3]Lab-STICC, UMR 6285, CNRS, ENIB, 29238 Brest Cedex 3, France

Abstract. We theoretically investigate a non-reciprocal magneto-optical integrated slab directional coupler. The scattering parameters of the structure are derived in the frame of the Coupled-Mode Theory ~~(CMT)~~. By properly designing the coupler, it is possible to achieve a perfect non-reciprocity (with 100% contrast) between the two directions of propagation. Other operating points can also be defined, especially since modal dispersion in the spectral domain is naturally taken into consideration.

1 Introduction

Couplers are important components of photonic circuits. Using magneto-optical materials as constituents of such systems allows one to design non-reciprocal devices, such as nonreciprocal Mach-Zehnder interferometers or nonreciprocal directional isolators [1, 2]. In this communication, we discuss the principle and the operating conditions of such an isolator based on two coupled magneto-optical waveguides with opposite directions of magnetization.

2 Magneto-optical coupler

The coupler consists of two parallel slab waveguides of equal thickness h made of a bigyrotropic magnetic medium B. Both the spacer of thickness d separating the waveguides and the surrounding space are made of the same isotropic, non-magnetic medium A (Fig. 1). Propagation occurs over a coupling length L along the z-axis. Time dependence is taken as $\exp(+i\omega t)$. Both guides are magnetized at saturation along the y-axis, in an "anti-parallel" transverse configuration.

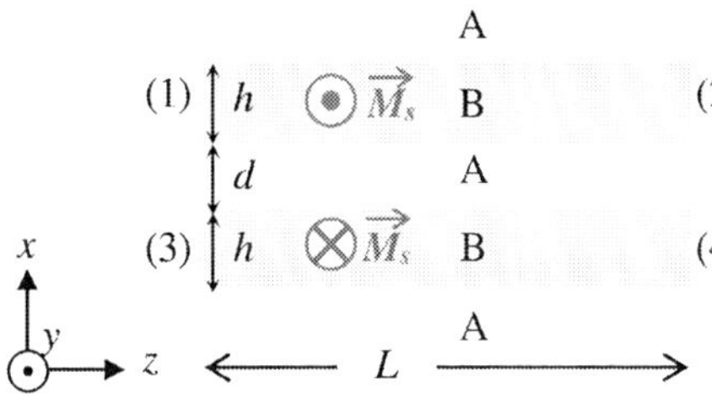

Fig. 1. System of two coupled bigyrotropic slab waveguides (1) in the "anti-parallel" transverse magnetic configuration. The width of both magnetic guiding layers (B) is h, that of the dielectric isotropic spacer (A) is d. L denotes the interaction length. The input-output ports are numbered from (1) to (4).

In the frame of the Coupled-Mode Theory, the coupling constant χ between two phase-matched slab waveguides can be retrieved by evaluating the modal propagation constants β_e and β_o of the *even* (slow) and *odd* (fast) super-modes, since $\chi = (\beta_e - \beta_o)/2$.

According to previous calculations reported in [3], a system made of coupled bigyrotropic slab waveguides in the anti-parallel transverse magnetic configuration becomes non-reciprocal: the two waveguides remain perfectly phase-matched, but the coupling constant depends on the direction of propagation. Let us call χ^+ and χ^- the coupling constants along the z-axis from the left to the right or from the right to the left, respectively. In a non-dissipative system, they are real and positive, with $\chi^+ < \chi^-$.

Besides, considering the spectral effects of modal dispersion, we derive an expression for χ^+ and χ^- that can be linearized around a given operating wavelength.

3 Non-reciprocal *S*-parameters

3.1 *S*-parameters and signal flow-graph

The scattering matrix expresses output fields (b_p) in terms of input fields (a_p). The signal flow graph is the graphical representation of the same connections (Fig. 2).

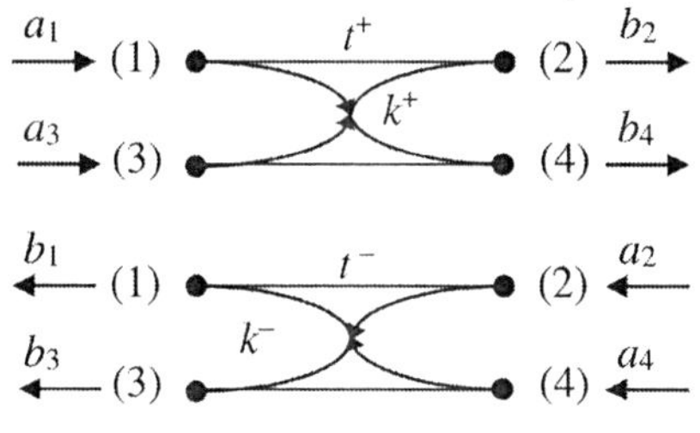

Fig. 2. Signal flow-graph of the directional coupler, in the general case. Input/output fields at port p are denoted (a_p, b_p).

* Corresponding author: dadoenkova@enib.fr

Since the coupler is directional, no reflection (either direct or cross) is experienced:

$$S_{11} = S_{22} = S_{33} = S_{44} = 0, \quad (1a)$$

$$S_{13} = S_{31} = S_{24} = S_{42} = 0. \quad (1b)$$

The remaining *S*-parameters are: from the left to the right, $S_{21} = S_{43} = t^+$, $S_{41} = S_{23} = k^+$; from the right to the left, $S_{12} = S_{34} = t^-$, $S_{14} = S_{32} = k^-$. We get eventually:

$$\begin{pmatrix} b_2 \\ b_4 \end{pmatrix} = \begin{pmatrix} t^+ & k^+ \\ k^+ & t^+ \end{pmatrix} \begin{pmatrix} a_1 \\ a_3 \end{pmatrix} \text{ and } \begin{pmatrix} b_1 \\ b_3 \end{pmatrix} = \begin{pmatrix} t^- & k^- \\ k^- & t^- \end{pmatrix} \begin{pmatrix} a_2 \\ a_4 \end{pmatrix} \quad (2)$$

with

$$t^+ = \cos(\chi^+ L)\exp(-i\beta^+ L), \quad (3a)$$

$$k^+ = -i\sin(\chi^+ L)\exp(-i\beta^+ L), \quad (3b)$$

$$t^- = \cos(\chi^- L)\exp(-i\beta^- L), \quad (3c)$$

$$k^- = -i\sin(\chi^- L)\exp(-i\beta^- L), \quad (3d)$$

where $\beta^+ = (\beta_e^+ + \beta_o^+)/2$ stands for the average propagation constant from the left to the right; similarly, from the right to the left, $\beta^- = (\beta_e^- + \beta_o^-)/2$. Numerical calculations show that the values of β^+ and β^- are almost identical, but anyway, they appear only as a global phase term.

It is possible to design the waveguide width *h* and/or the gap *d* (as well as the interaction length *L*) in order to ensure that, for a given wavelength λ_0, $(\chi^+ L)$ and $(\chi^- L)$ are adjusted at will, so that

$$\frac{\chi^+}{\chi^-} = \frac{2q}{2q+1}, \text{ with } q \in \mathbb{N}, \quad (4)$$

which ensures that the coupler behaves as a perfect directional isolator.

3.2 Perfect directional isolator

For GGG and Bi:YIG as materials A and B, the condition expressed by Eq. (4) can for instance be satisfied (and perfect isolation be achieved) at $\lambda_0 = 1.5$ µm with $h = 0.62$ µm, $d = 0.155$ µm, and $L = 206.5$ µm, so that:

$$(\chi^+ L) \approx 10\pi, (\chi^- L) \approx 10.5\pi. \quad (5)$$

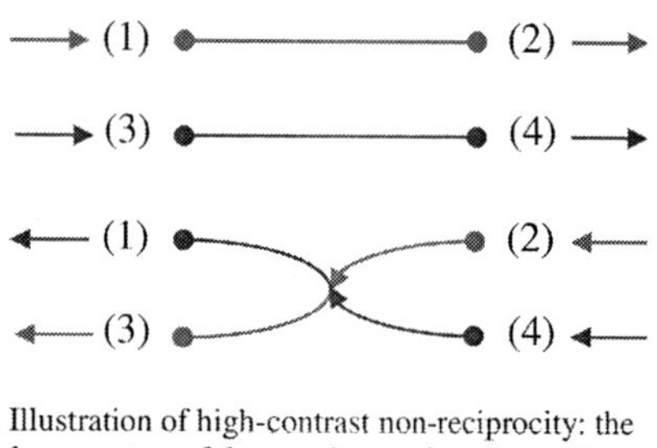

Fig. 3. Illustration of high-contrast non-reciprocity: the material parameters of the coupler can be adjusted in such a way that, for a given operating wavelength, the structure works as a perfect isolator.

As a result, the system becomes highly non-reciprocal, as schematically illustrated in Fig. 3, since

$$|S_{21}|^2 = |S_{43}|^2 = 1, |S_{41}|^2 = |S_{23}|^2 = 0, \quad (6a)$$

$$|S_{12}|^2 = |S_{34}|^2 = 0, |S_{14}|^2 = |S_{32}|^2 = 1. \quad (6b)$$

3.3 Chromatic dispersion and consequences

Once the coupler is designed as a perfect isolator for a given operating wavelength λ_{op}, the knowledge of chromatic dispersion (both modal and material) enables one to assess the bandwidth over which the behaviour remains close to ideal. Around λ_{op}, we write:

$$\chi^{(\pm)} = \chi^{(\pm)}(\lambda_{op}) + (\lambda - \lambda_{op})(\partial\chi/\partial\lambda), \quad (7)$$

since both coupling constants χ^+ and χ^- exhibit roughly the same value of $(\partial\chi/\partial\lambda)$. Let us start from:

$$\chi^+ L = q\pi + \theta \quad (q \in \mathbb{N}), \quad (8a)$$

$$\chi^- L = q\pi + \pi/2 + \theta, \quad (8b)$$

$$\theta = (\lambda - \lambda_{op})(\partial\chi/\partial\lambda)\,L. \quad (8c)$$

Then we get:

$$|S_{21}|^2 = \cos^2\theta, |S_{12}|^2 = \sin^2\theta. \quad (9)$$

If we strive to maintain $\sin^2\theta < 0.01$ (which means $|\theta| < 0.1$), then the corresponding bandwitdth $\Delta\lambda$ is:

$$\Delta\lambda \approx \frac{0.2}{(\partial\chi L/\partial\lambda)} \quad (10)$$

Assuming $(\partial\chi/\partial\lambda) \approx 0.07$ µm^{-2}, $\lambda_{op} = 1.5$ µm and $L \approx 206.5$ µm, we get $\Delta\lambda \approx 14$ nm (or, in frequency terms, $\Delta\nu \approx 1.8 \times 10^{12}$ Hz = 1.8 THz).

The same element can also be seen as a building-block to be inserted into more complex structures, such as a non-reciprocal Mach-Zehnder interferometer.

This work is supported in part by École Nationale d'Ingénieurs de Brest, France, and Collège de France (programme PAUSE).

References

1. Y. Hadj Aissa Fekhar, F. Salah-Belkhodja, D. Vincent, E. Bronchalo, W. Tebboune, R. Naoum. J. Electromagn. Waves Appl. **34**, 623 (2020)
2. N. Bahlmann, M. Lohmeyer, O. Zhuromskyy, H. Dötsch, P. Hertel, Opt. Communications **161**, 330 (1999)
3. Y.G. Boucher, F.F.L. Bentivegna, Y.S. Dadoenkova, IEEE J. Quantum Electron. **54**, 6200208 (2018)

Tantalum pentoxide micro-resonators for frequency comb generation

Jake Daykin[1,*], *Jonathan R C* Woods[2], *Stephen C* Richardson[1], *Oliver J* Trojak[1], *Folly Eli* Ayi-Yovo[3], *Jonathan M* Silver[3], *Nicholas T* Klokkou[1], *Amy S K* Tong[4], *Patrick* Gill[3], *Alberto* Politi[1], *Peter* Horak[4], *James S* Wilkinson[4], and *Vasilis* Apostolopoulos[1]

[1]School of Physics and Astronomy, University of Southampton, Southampton, SO17 1BJ
[2]Aquark Technologies, Unit 2, Abbey Enterprise Centre, Premier Way, Abbey Park Industrial Estate, Romsey, Hampshire, SO51 9AQ
[3]National Physical Laboratory, Hampton Rd, Teddington, TW11 0LW
[4]Optoelectronics Research Centre, University of Southampton, Southampton, SO17 1BJ

Abstract. We present the design, fabrication, simulation and initial characterisation of tantalum pentoxide (Ta_2O_5) optical waveguides and micro-ring resonators for the purpose of supercontinuum and frequency comb generation. Spectral broadening results are presented for linear Ta_2O_5 waveguides for a range of central pump wavelengths between 900 nm and 1500 nm. These results are used as the basis for the dispersion engineering and development of Ta_2O_5 micro-ring resonators. The losses for sputtered and TEOS PECVD deposited SiO_2 top cladded waveguides are characterised using a Fabry-Pérot loss measurement set-up. A solver based on the Lugiato-Lefever equation is presented and used to simulate the expected emission from the Ta_2O_5 micro-ring resonators. Promising initial experimental results show critical coupling and a Q-factor of 3.7×10^4.

1 Introduction

The generation of supercontinuum and frequency combs through spectral broadening in linear waveguides and micro-ring resonators (MRRs) sees many potential applications in fields such as metrology and optical sensing. We present supercontinuum generation in tantalum pentoxide (Ta_2O_5) waveguides and preliminary experimental results towards frequency comb generation in Ta_2O_5 MRRs.

We use Ta_2O_5 sputtered on oxidised silicon wafers as a wave-guiding layer. Ta_2O_5 presents a number of advantages over other commonly used materials in integrated photonics. The linear refractive index of Ta_2O_5 is similar to that of Si_3N_4 at 1550 nm, and the nonlinear refractive index of Ta_2O_5 has been reported to be a factor of 3 greater than that of stoichiometric Si_3N_4 and an order of magnitude greater than of Si, suggesting that Ta_2O_5 is an interesting material for the generation of Kerr frequency combs [1]. Additionally, Ta_2O_5 can be doped with rare-earth elements, such as Er, Nd and Yb, which allows for the creation of waveguide lasers.

2 Linear waveguides for supercontinuum generation

The spectral broadening response of the waveguides was characterised as a function of pump central wavelength. Using the combination of a *Coherent Chameleon* and an optical parametric oscillator (OPO), $\leq$ 200 fs pulses at an 80MHz repetition rate are generated for a range of central pump wavelengths between 900 nm and 1500 nm. The best spectral broadening response is shown in an uncladded, 700×2200 nm linear waveguide for a pump power of 200 mW and central pump wavelength between 900 and 1500 nm and can be seen in Figure 1, showing significant broadening over the majority of pump wavelengths.

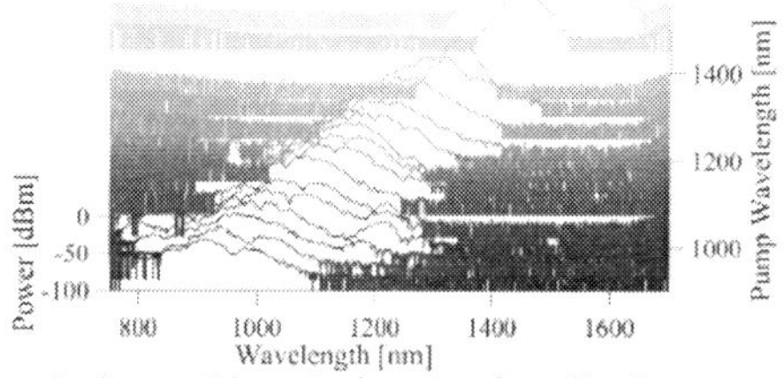

Figure 1: Spectral broadening data for a Ta_2O_5 waveguide for pump wavelengths between 900 and 1500 nm in the TM polarisation [2].

Our simulation solves the generalised nonlinear Schrödinger equation (GNLSE) using the split-step Fourier method. The model [3] allows for the comparison of experimental spectra with simulations for different waveguide modes. Figure 2 shows agreement between the recorded spectrum (blue) for a sputtered SiO_2 cladded 3200×700 nm Ta_2O_5 waveguide and the simulated spectrum for a TM_{02} mode (green) for a pump wavelength of 1050 nm.

A Fabry-Pérot loss measurement set-up was used to characterise the propagation losses of the Ta_2O_5 waveguides for a pump wavelength of 1 µm [4]. Propagation losses in Ta_2O_5 waveguides with RF-sputtered sil-

*e-mail: J.Daykin@soton.ac.uk

ica cladding have been determined to be approximately 3 dB/cm.

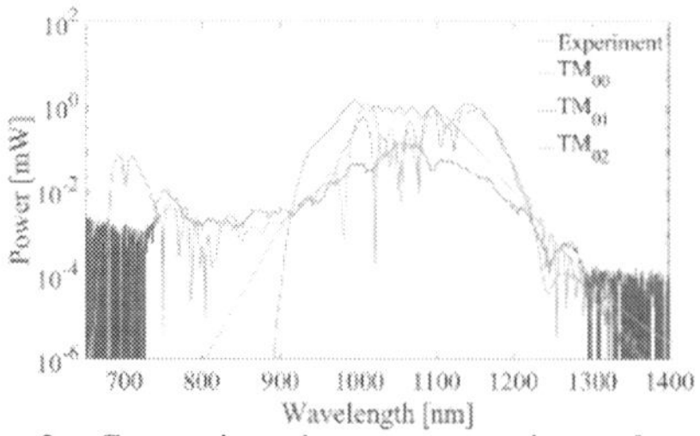

Figure 2: Comparison between experimental measured spectrum and the GNLSE simulated spectra for the TM_{00}, TM_{01}, and TM_{02} modes, showing agreement between the TM_{02} and the measured spectrum.

3 Micro-ring resonators towards frequency comb generation

In order to simulate the expected response from our MRRs, we have developed a Lugiato-Lefever equation (LLE) solver, based on our GNLSE model. The LLE solver uses a continuous wave laser input and periodic boundary conditions in order to simulate MRRs with different radii at different pump wavelengths and powers. As in the case of the GNLSE, the dispersion coefficient, β_2, is calculated for different waveguide geometries and MRR radii by simulating the change in the effective refractive index, n_{eff}, of a chosen mode over a range of frequencies. Variations in waveguide height, waveguide width, etch depth and MRR radius are simulated in order to engineer the waveguides towards broadening around a pump wavelength of 1550 nm. Figure 3 shows the simulated output of a 1500×800 nm Ta_2O_5 MRR with a radius of 180 µm, with a power of 20 mW coupled into the waveguide and a pump wavelength of 1550 nm, resulting in a comb with spectral bandwidth of ≈ 1000 nm.

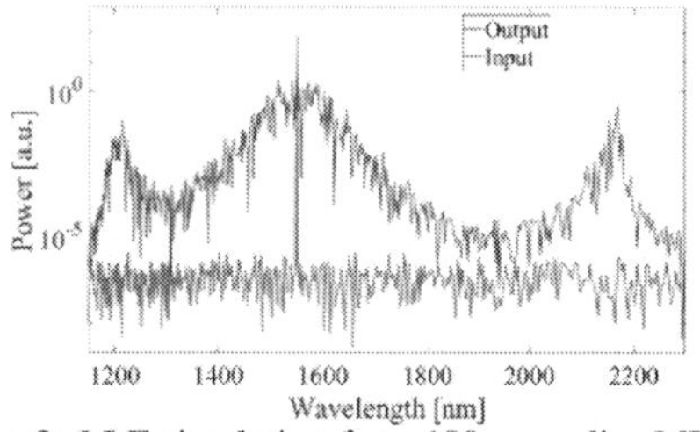

Figure 3: LLE simulation for a 180 µm radius MRR for a pump power of 20 mW and wavelength of 1550 nm.

Figure 4a shows experimental results of Ta_2O_5 MRRs. We show a resonance in a MRR with radius 180 µm at a pump wavelength of 1549.07 nm, yielding a Q-factor of 3.7×10^4 in a chip cladded with RF-sputtered SiO_2.

Figure 4b shows experimentally measured critical coupling, which will be necessary for the generation of a frequency comb, in a Ta_2O_5 MRR with a radius of 200 µm and at pump wavelength of 1548.33 nm. The losses in the system were too great to generate sidebands. In order to decrease our losses and increase our Q-factors further, an investigation into SiO_2 deposited by TEOS PECVD is ongoing, and we have experimentally measured a decrease in the waveguide propagation loss by approximately 1 dB/cm in a linear waveguide (bringing it down to 2 dB/cm).

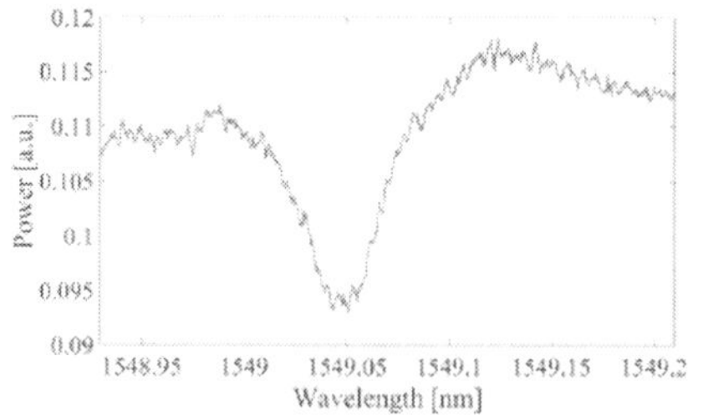

(a) Experimental results from 180 µm radius MRR yielding a Q-factor of 3.7×10^4.

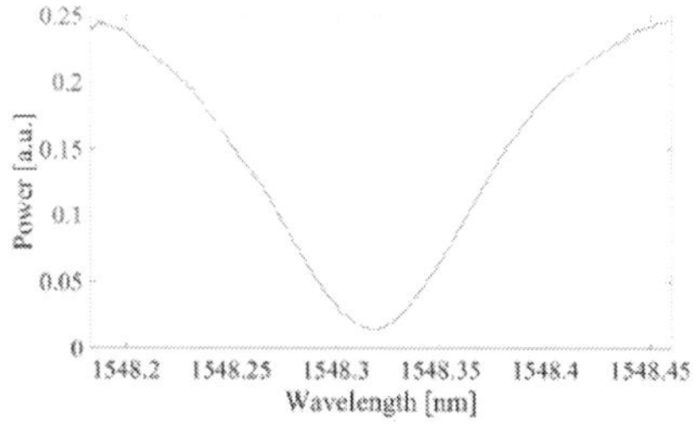

(b) Critical coupling in a MRR with radius 200 µm.

Figure 4: Preliminary experimental results for the output power from Ta_2O_5 MRRs showing a Q-factor of 3.7×10^4 (a) and critical coupling (b).

4 Conclusion

We have previously shown supercontinuum generation in Ta_2O_5 waveguides with excellent agreement with our GNLSE solver. We present preliminary experimental results from Ta_2O_5 MRRs showing critical coupling and Q-factor measurements up to 3.7×10^4 which agrees with our LLE simulations. Simulations of the next generation of Ta_2O_5 MRRs predict Kerr combs with spectral broadening of ≈ 1000 nm, thanks to improved silica cladding deposition and updated designs of racetrack resonators.

Funding

Engineering and Physical Sciences Research Council (EPSRC) (EP/T001046/1, EP/T001062/1).

References

[1] H. Jung, S.P. Yu, D.R. Carlson, T.E. Drake, T.C. Briles, S.B. Papp, *Optica* **8**, 811 (2021)

[2] J.R.C. Woods, J. Daykin, A.S.K. Tong, C. Lacava, P. Petropoulos, A.C. Tropper, P. Horak, J.S. Wilkinson, V. Apostolopoulos, *Optics Express* **28**, 32173 (2020)

[3] F. Poletti, P. Horak, *Journal of the Optical Society of America B* **25**, 1645 (2008)

[4] S. Taebi, M. Khorasaninejad, S.S. Saini, *Applied Optics* **47**, 6625 (2008)

Hybrid polymer-titania waveguides for highly integrated circuits

Isaac Doughan[1,*], *Kehinde* Oyemakinwa[1], *Olli* Ovaskainen[1], and *Matthieu* Roussey[1]

[1]Department of Physics and Mathematics, Institute of Photonics, University of Eastern Finland, P.O. Box 111, 80101 Joensuu, Finland

Abstract. We present an innovative waveguide based on the hybridization of a titanium dioxide nano-waveguide within a polymer strip. Through simulations and design we demonstrate that the waveguide sustains principally the quasi-TM fundamental mode and that even in tight bends (radius smaller than 2 µm) light remains confined in the titania layer. Such a waveguide, in addition of enabling low loss propagation is a way towards efficient evanescent sensing in highly integrated scheme, i.e., small footprint. We also show that the fabrication, here based on electron beam lithography and atomic layer deposition, can be extended easily to large scale manufacturing using nanoimprinting technology.

1 Concept and design

There is a need for efficient sensors with low-cost manufacturing process allowing disposability[1]. Evanescent sensing is based on the probing of the external (to the waveguide) medium by the exponentially decaying tail of the guide mode. The overall interaction of this tail and the outer medium must therefore be maximized. It usually requires either an increase of the optical path length by the implementation of a cavity along the waveguide or a direct increase of the geometrical length of the waveguide. Double spiral waveguides are often used in this case. Unfortunately, it leads to an increase of the footprint of the device in order to minimize the loss due to the smallest radii of the spiral.

In this work we show the development of a nano-waveguide allowing a single mode and low propagation losses together with a high overlap with the outer medium, which can be an analyte for absorption or refractive index sensing. A nano-waveguide can be defined as a strip waveguide with a cross-section of less than 1 × 1 µm². Silicon nano-waveguides usually created on silicon-on-insulator waveguides have dimensions of on the order of 200 × 500 nm², with the largest dimension being the lateral one. This leads to a predominant quasi-TE mode, which may not be the most suitable for evanescent sensing, due to a lower confinement of the field above the waveguide. We propose here to rotate this nano-waveguide by 90° in order to obtain a predominant quasi-TM mode, i.e., the mode of highest effective index.

The cross section of the waveguide is depicted in Fig. 1. The structure is fabricated on oxidized silicon wafer. It consists of a polymer strip in which a trench is performed and further filled with a high refractive index material, intended to be the core of the nano-waveguide. For our purpose, we chose nLOF AZ 2070 electron beam resist as a polymer ($n_P = 1.6$ at $\lambda = 1550$ nm) and titanium dioxide as high refractive index material ($n_H = 2.24$). Fabrication constraints are taken into account during the design process. *Filling* the trench requires to deposit the high refractive index material on top of the polymer which means that a layer of it with lays all over the final sample, if we want to limit the fabrication steps. Using atomic layer deposition, the result is a conformal layer of titanium dioxide covering the waveguide[2], as shown in Fig. 1.

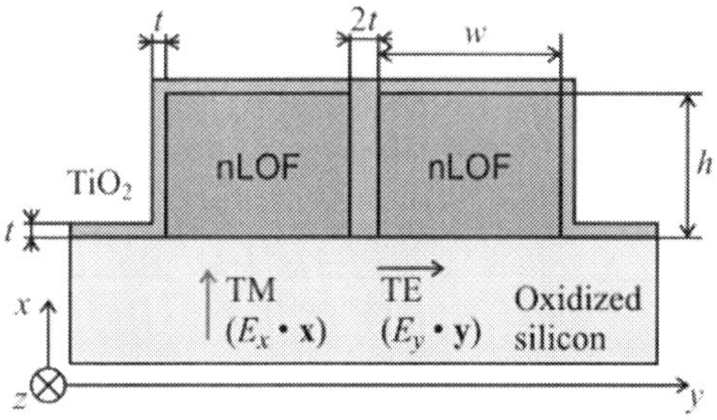

Fig. 1. Cross section of the proposed waveguide

The structure was studied using Lumerical and other mode solvers. The width *w* of each portion of the polymer strip, the thickness *t* and the width of the trench *2t* as well as the thickness *h* of the polymer have been varied. Figure 2a, shows the evolution of the effective index of the TM modes as a function of TiO_2 width *t*. One can see that after 200 nm, the waveguide shows a multi-modal behavior. Similar calculations have been performed to study the light confinement within the waveguide and its interaction with the top medium. The waveguide parameters are t = 200 nm, w = 900 nm, and h = 900 nm. Choosing these parameters allow a single modal operation at $\lambda = 1550$ nm, which is used for the demonstration. However, we want to emphasize that all materials are transparent over a large wavelength range enabling operation over a large portion of the electromagnetic spectrum[3].

Figure 2b is the field distribution of the quasi-TM mode in the optimized waveguide. One can remark that

* Corresponding author: isaac.doughan@uef.fi

light is highly confined within the titania core with still a relatively significant portion (~5%) in the top free region. Moreover, and it is one of the main advantages of the waveguide, the lateral (x-direction, see Fig. 1) tails of the mode are within the polymer strip for limited propagation losses.

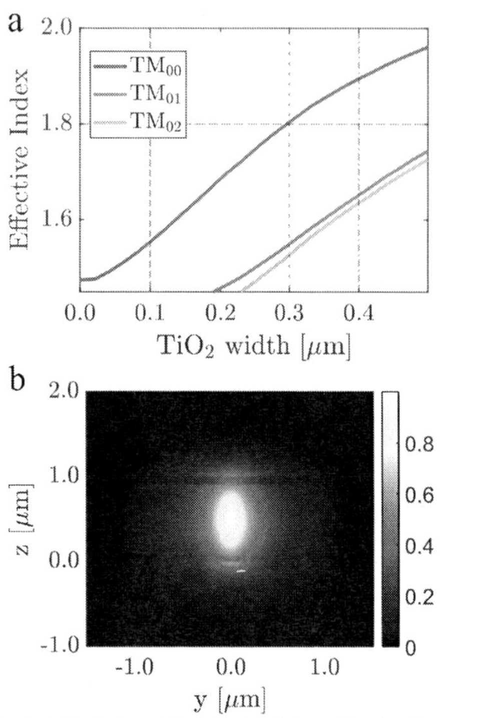

Fig. 2. a) Evolution of the effective index of the few first quasi-TM modes as a function of the titanium dioxide thickness t. **b)** Field distribution of the fundamental quasi TM-mode.

One can remark that the remaining TiO_2 layer has nearly no effect on the lateral tails, and provide an addition increase of the interaction of the field on top of the waveguide.

2 Towards experiments

The fabrication of the waveguide requires a two-steps process. First the polymer waveguide, including the trench, is performed using electron beam lithography by exposure and development of an nLOF layer. Second, a titanium dioxide layer is deposited on top of entire structure. The high conformality of the ALD processes enable the filling of the trench with a smooth and homogenous material. Because we are using a polymer, we set the deposition temperature at 100 °C preventing any degassing or deformation of the polymeric base during the process.

The scanning electron microscope picture in Fig. 3 allows us shows the cross section of the waveguide. One can appreciate the perfect respect of dimensions as well as very horizontal walls despite a slight curvature within the trench. The roughness observable on the top of the waveguide is due to some residues of the conductive layer used for imaging purpose.

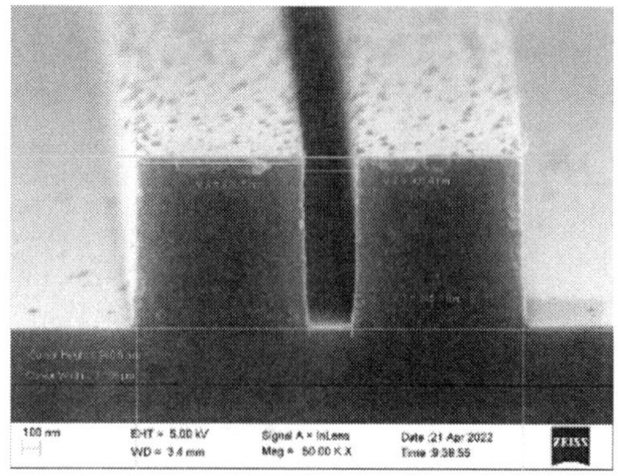

Fig. 3. SEM picture showing a tilted view of the cross section of the waveguide.

The final structure, not shown here, is to be characterized by end-fire coupling method in order to determine the propagation losses (cut-back method). Several patterns have been performed. It includes bends of different radius of curvature as well as long waveguides.

Although we used electron beam lithography in this work, it is to be noted that the polymeric structure can be replicated by means on nanoimprinting. ALD is also a large-scale manufacturing technology[4]. Such a waveguide is therefore one way to enable easier photonics integration with a wider range of materials for a broader field of applications.

Authors acknowledge the financial support of the Flagship on Photonics Research and Innovation PREIN (320166) of the Academy of Finland.

References

1. M. Javaid, A. Haleem, S. Rab, R.P. Singh, R. Suman, Sensors Intern. **2**, 100121 (2021)
2. P. Stenberg, M. Roussey, P. Ryczkowski, G. Genty, S. Honkanen, M. Kuittinen, Opt. Express, **21**, 20 (2013)
3. M. Häyrinen, M. Roussey, A. Säynätjoki, M. Kuittinen, S. Honkanen, App. Opt. **54**, 10 (2015)
4. L. Ahmadi, M. Hiltunen, P. Stenberg, M. Roussey, J. Saarinen and S. Honkanen, Opt. Express **24**, 10, (2016)

 https://doi.org/10.1051/epjconf/202226601007

Bi-directional spectral broadening measurements for accurate characterisation of nonlinear hybrid integrated waveguides

Mikhail Dyatlov[1,2,*], *Philippe* Delaye[3], *Laurent* Vivien[2], and *Nicolas* Dubreuil[1]

[1]Université de Bordeaux, Institut d'Optique Graduate School, CNRS, Laboratoire Photonique, Numérique et Nanosciences, 33400 Talence, France
[2]Université Paris-Saclay, CNRS, Centre de nanosciences et de nanotechnologies (C2N), 91120 Palaiseau, France
[3]Université Paris-Saclay, Institut d'Optique Graduate School, CNRS, Laboratoire Charles Fabry, 91127 Palaiseau, France

Abstract. The emerging interest in integrated optical technologies raises the need for precise characterisation techniques for waveguides presenting nonlinearities. Here we propose a non-interferometric measurement to accurately characterise the Kerr contribution in hybrid waveguides and illustrate its performances using SiN waveguides with a GSS chalcogenide top-layer. The sensitivity of our technique in terms of nonlinear phase reaches 10 mrad and its accuracy makes possible to extract the nonlinear contributions from the top-layer.

1 Introduction

One of the interesting consequences of the strong light confinement in integrated waveguides is the emergence of nonlinear effects, even at low powers. In combination with large capabilities of dispersion engineering in integrated devices, these nonlinear effects lead to observation of various phenomena, such as optical Kerr frequency combs or super continuum generation. Their possible applications extend from spectroscopy to optical communication systems [1].

In order to overcome the limitations imposed by the properties of the most common integrated photonics materials, such as silicon and silicon nitride, hybrid waveguide structures are envisioned. They combine materials selected for their nonlinear properties and their potentialities in terms of hybridisation. The design of such hybrid structures with specific characteristics requires complete simulation tools and, on the other hand, raises the necessity for accurate characterisation methods enabling to verify the real performances of fabricated samples.

Characterisation methods dedicated to 3rd-order nonlinearities of integrated structures are mainly based on the analysis of spectral broadening due to the self-phase modulation (SPM) in pulsed regime or gain amplification of signal beam in the four-wave mixing process. One of the common issue for proper characterisations concerns the systematic measurements of the injected power in the waveguide.

Here we propose a bi-directional method based on the analysis of the spectral broadening of laser pulses with top-hat like spectral shape to improve the sensitivity in terms of nonlinear phase shift. It allows accurate (10 %) and reproducible characterisation of the waveguides in terms of nonlinear index (n_2) and gives access to the comparison between different samples despite the variation in their related coupling efficiencies. Using the calculated field distribution in hybrid waveguides, the specific nonlinear contribution of the deposed layer can be extracted.

2 Proposed method

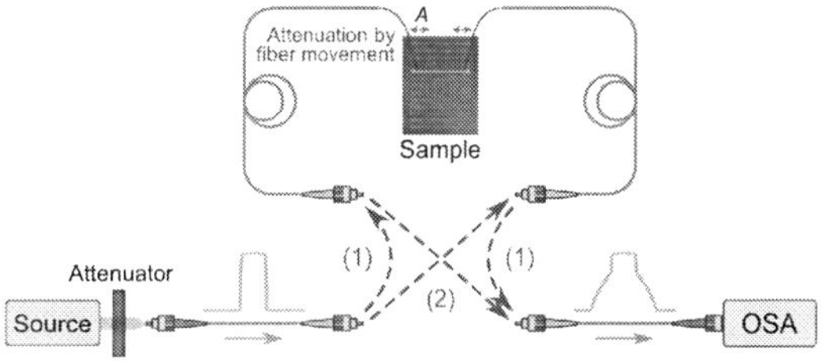

Figure 1. Experimental setup with the typical input and output spectra, and the fibre connections used to perform bi-directional measurements. The source is composed of an erbium-doped fibre laser followed by a grating based pulse shaper.

Our method is based on a previously reported D-Scan method [2]. Here, we present an evolution on the processing of the acquired optical spectra to improve the accuracy of the method in order to reveal the properties of hybrid waveguide structures. The principal scheme of the setup is shown in figure 1. An erbium-doped fiber laser with a grating-based optical filter (pulse shaper) are used as a source that delivers pulses with a top-hat like spectral shape without any residual chirp. Pulses are injected and collected through two cleaved polarisation maintaining (PM) fibres aligned above grating couplers engraved at each waveguide extremity. Self-phase modulation in

*e-mail: mikhail.dyatlov@institutoptique.fr

the waveguide leads to spectral broadening and creates observable 'nonlinear' wings as illustrated in figure 1. Neglecting the chromatic dispersion and other nonlinear effects (verified in practice), the introduced nonlinear phase can be then retrieved by simulating the propagation of the initial pulses through a nonlinear waveguide.

The correct restitution of the results for nonlinear measurements implies the exact peak power in the waveguide to be known. As the coupling efficiencies may not be symmetric at the two waveguide extremities, we perform bi-directional measurements using PM fibres in order to reverse the coupling direction without modifying any position in the injection set-up. Furthermore, one has to take into account for the nonlinear phase shift introduced by the PM fibers. The separation between the nonlinear contributions of the PM fibres and the waveguide is done by varying independently the incident power set by the attenuator (see fig. 1) and the waveguide coupling efficiency achieved by shifting the fibre position along the diffraction grating (introducing an attenuation factor A).

3 Results and Discussion

Our method has been applied to characterise 3 samples of SiN_x waveguides without and with a top-layer cladding of chalcogenide glass (GeSbS, GSS) with a thickness of 100 nm and 150 nm. The waveguides have been designed to support one single mode in quasi-TE configuration (fig. 2). The studied waveguides include straight and spi-

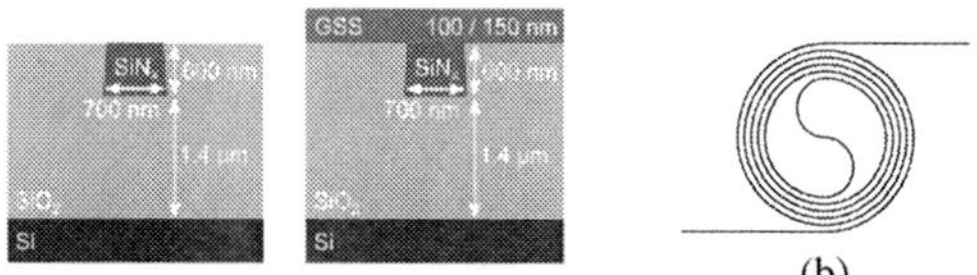

(a) (b)

Figure 2. Cross section of studied waveguides (a). Example of a spiral shaped waveguide used for characterisation (b).

ral geometries with minimal radius of 30 μm and lengths varying from 3.5 mm to 12.9 mm.

The bi-directional measurements have been performed for 7 waveguides on each sample, for 5 different input powers (set with the attenuator) and 6 different coupling efficiencies between the fibre and the waveguide. The contributions of the fibre and the waveguide to the nonlinear phase are extracted using optimisation of the nonlinear phases in a numerical simulation so that it fits the experimental data. The simulation takes into account the SPM effect only, as the chirp introduced by the second-order dispersion effects in the injection fibers and the waveguide are negligible. The spectrum at the minimal power (presenting no wings) is used as a reference spectrum at the input for the simulation and the ratio between the power in the nonlinear wings and the power in the central part serves as the optimisation criteria (see insight in figure 3). The dependency of this ratio with the injected power in waveguides is shown in the figure 3. It demonstrates a good agreement between the numerical simulation and the experimental data. It allows us to retrieve the nonlinear phases accumulated in the fibres and in the waveguides and, using the known linear loss, the effective nonlinear coefficients γ for the waveguides under study. Respect to SiN waveguides, an improvement by more than a factor of 2 of the nonlinear coefficient γ is demonstrated with the hybrbrid SiN-GSS waveguides.

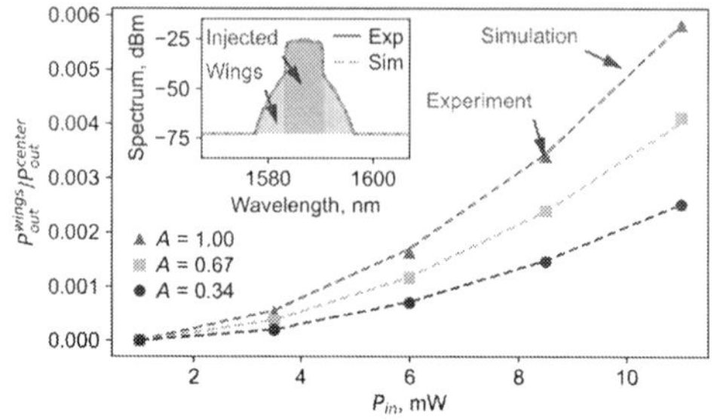

Figure 3. Variation of the ratio between the power in the wings and in the central part of the spectrum with the input power and for three different coupling efficiencies.

Using the effective mode area calculated with Lumerical MODE and the linear loss, estimated from the insertion loss variation with the waveguide lengths, we get access to the effective nonlinear indices for all the samples. The knowledge of the field distribution in the different layers of our samples (also obtained using Lumerical MODE) gives access to the contribution, and though, the nonlinear refractive index of each material, respectively equal to $n_2 = (4.4 \pm 0.6) \cdot 10^{-19}$ m^2/W for SiN_x and $n_2 = (4.1 \pm 1.2) \cdot 10^{-18}$ m^2/W for GSS. These values are in agreement with those previously reported for SiN_x [3] and GSS [4].

4 Conclusion

We have shown the capacity of our method to characterise hybrid waveguides and extract the nonlinear contributions of the different materials. Details about the parameters extraction procedure and their uncertainties will be presented. Our method presents the advantage of a non-interferometric configuration that can be implemented with fully commercialised components and instruments.

References

[1] H. Hu, L.K. Oxenløwe, Nanophotonics **10**, 1367 (2021)

[2] S. Serna, N. Dubreuil, Opt. Lett. **42**, 3072 (2017)

[3] C. Lafforgue, S. Guerber, J.M. Ramirez, G. Marcaud, C. Alonso-Ramos, X.L. Roux, D. Marris-Morini, E. Cassan, C. Baudot, F. Boeuf et al., Photon. Res. **8**, 352 (2020)

[4] S. Serna, H. Lin, C. Alonso-Ramos, A. Yadav, X.L. Roux, K. Richardson, E. Cassan, N. Dubreuil, J. Hu, L. Vivien, Photon. Res. **6**, B37 (2018)

Hybrid integration methodology for quantum cascade lasers with germanium waveguides in mid-IR

Colin J. Mitchell[1,*], *Ahmed* Osman[1], *Ke* Li[1], *Jordi S.* Penadés[1], *Milos* Nedeljkovic[1], *Longqi* Zhou[2], *Kristian M.* Groom[2], *Jon* Heffernan[2], and *Goran* Mashanovich[1]

[1]Optoelectronics Research Centre, University of Southampton, University Road, Southampton, Hampshire SO17 1BJ, UK
[2]Department of Electronic and Electrical Engineering, Sir Frederick Mappin Building, University of Sheffield, Mappin Street, Sheffield, S1 3JD, UK

Abstract. Mid-infrared quantum cascade lasers (QCLs) operating around 5.7 μm have been integrated with germanium waveguides on silicon substrates. QCL bars have been designed and fabricated at the University of Sheffield for the purpose of integration. This hybrid approach uses flip-chip technology that has been successfully transferred from a silicon-on-oxide (SOI) platform working at communication wavelengths, demonstrating the flexibility of this approach. Integration challenges are introduced, and solutions discussed, leading to the next iteration of design presented here.

1 Introduction

Integration of optical sources is a critical technology for Silicon Photonics (SP). The combination of a silicon platform for a CMOS-compatible electro-optical system (or in this case germanium-on-silicon for mid-infrared), and conventional heterojunction compound semiconductor materials for the optical source, provides a 'best of both worlds' solution.

A range of approaches have been demonstrated such as heterogeneous growth of compound heterostructures on SP substrates, wafer bonding or token-printing of unprocessed or partially processed III-V materials, and hybrid integration of fully processed lasers.

Interest in hybrid integration of pre-fabricated III-V lasers on photonic chips has gained interest in recent years as placement tool have improved the potential for reliable placement of laser chips, and hence efficient coupling into waveguides from those chips. This is achieved directly, facet-to-facet, by flip-chip bonders (or novel placement techniques [1]), and may include the use of a spot size converter or coupling lens at the waveguide.

Alternatively, demonstrated hybrid solutions include stand-alone units, such as those produced by Luxtera for placement upon SP chips, that comprise a laser, lens, isolator and mirror for coupling to a grating.

The process described here is based upon the flip-chip bonding of III-V semiconductor laser bars, directly facet-to-facet. This was derived from technology transfer of previous work [2], investigating the integration of commercial telecommunications lasers with an SOI device. An example is shown below (Fig. 1.).

The QCLs utilised in this work were designed at the University of Sheffield National Epitaxy Facility, and were specifically matched to the requirements of the integration methodology presented here. They adopt a double-phonon resonance active region design, based on strain-compensated $In_{0.6}Ga_{0.4}As/Al_{0.58}In_{0.42}As$ material system grown by MOVPE. 4.5 μm wide ridge waveguide Fabry-Perot lasers were fabricated with electroplated top contacts. 2 mm long lasers exhibit CW lasing at room temperature with tens of mW output power.

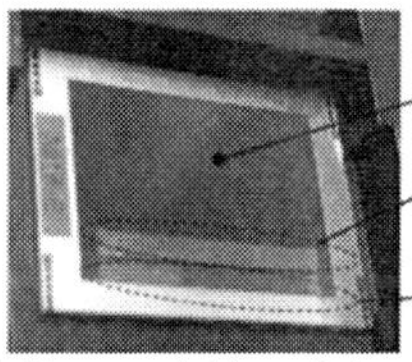

Fig. 1. Commercial laser integrated on SOI at Southampton

2 Flip-chip bonding

Flip-chip bonders produce <0.5 μm positioning accuracy in the plane of the interface, and rotational alignment in that plane is a function of x-alignment (Fig. 2.) at either end of a chip, longer chips increase accuracy. Importantly, systems can be automated for industrial manufacturing. A manual FineTech Fineplacer was used in this work

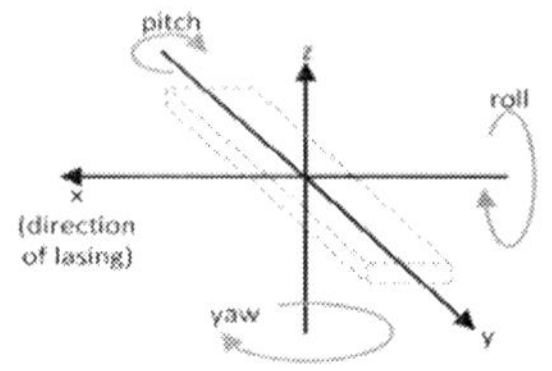

Fig. 2. Degrees of freedom and alignment

However, the vertical and rotational alignments provide challenges. Rotational alignment can be

* Corresponding author: c.j.mitchell@soton.ac.uk

addressed by the use of gimballed placement tools (Fig. 3.), but with mechanical size limitations requiring process design to incorporate both gimbals and high magnification optics for manual alignment in different axes [2], as shown below (Fig. 4.).

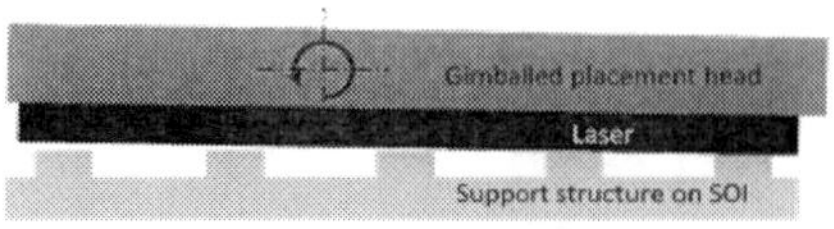

Fig. 3. Rotation alignment through gimbal

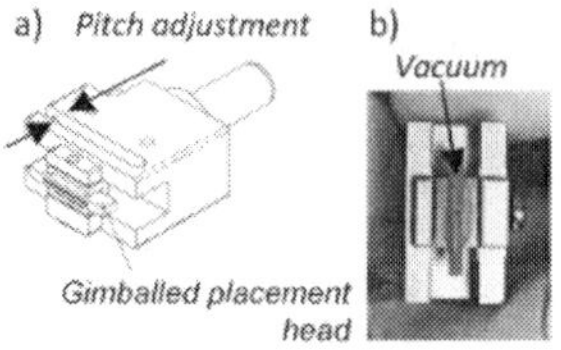

Fig. 4. Rotation alignment through focus a) pitch adjustment schematic, b) placement tool (FineTech GmbH & Co) with gimbal and pitch adjustment

Vertical alignment is most easily addressed by using a support structure fabricated on the SP chip, which is in contact with the unprocessed laser top-epitaxial surface at bonding [1-2], as shown in Fig. 3. in conjunction with a gimballed placement tool. Alternatively, surface tension of a solder material can be used, either with or without a support structure. This requires very high accuracy deposition of the solder and fabrication of wetting areas on the laser. When only a support structure is used, as here, the tolerances on solder material deposition (gold tracks, and gold-tin solder) parameters are reduced.

3 Germanium QCL integration

Germanium waveguides and gratings (3-µm germanium layer on silicon) were formed by e-beam lithography and ICP etch. QCLs were flip-chip bonded to the substrate using AuSn solder - evaporated and patterned by optical lithography. Initial data demonstrated coupling to a germanium waveguide, via a grating, in pulsed operation. Developing experimental setup, particularly the cooling head design yielded 30-fold improvement in intensity and operation up to 50% duty cycle at 10°C. Assessment of both components, and their integration, demonstrated routes for further development.

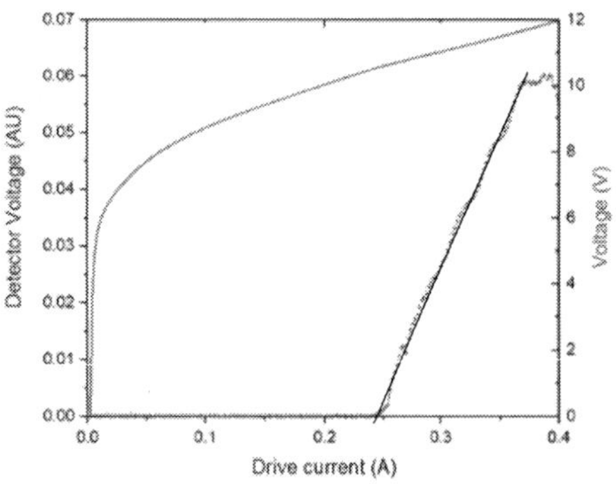

Fig. 5. L-I-V measured (10°C) at germanium grating via fibre

4 Holistic design iteration

Several areas for enhancement were identified on either component individually, and as a single integrated device. A holistic design approach addressing both components simultaneously is critical for such systems, and will be discussed. Processing developments, e.g. vertical ICP etch (Fig. 6.) for the waveguide input facet were also optimised; and original grating designs developed.

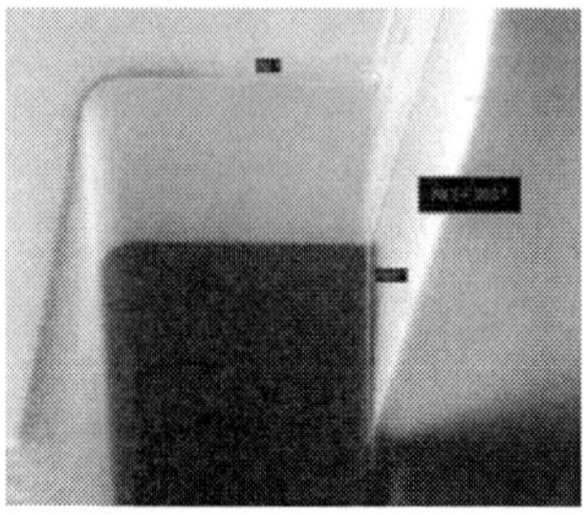

Fig. 6. FIB image of facet showing 90.2° vertical over 8 µm

This intrinsically linked holistic approach has enabled the next generation design (currently in fabrication) to be developed. This includes real-time fibre monitoring from the other side of the QCL (Fig. 7.), whilst monitoring the outputs via gratings and polished facets.

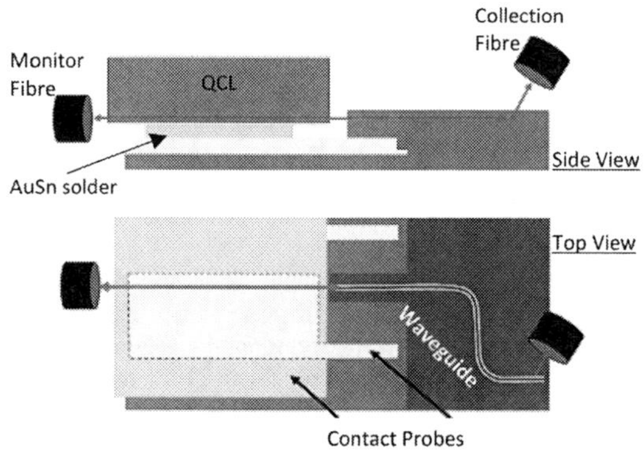

Fig. 7. Latest design with real-time monitor of QCL

The authors would like to acknowledge ESPRC (UK) for funding the Future Photonics Hub programme, and FineTech GmbH & Co for their collaboration.

References

1. Y. Martin, S. Kamlapurkar, N. Marchack, J. Nah and T. Barwicz, IEEE 69th Electronic Components and Technology Conference (ECTC), 2019
2. C.J. Mitchell, K. Li, R. Schachler, A. Khokhar, X. Chen, S. Stanković, X. Wang, N.P. Sessions, K.M. Grabska, D.J. Thompson, C. Daedlow, G.T. Reed, J.S. Wilkinson, IEEE 8th Electronics System-Integration Technology Conference (ESTC), 2020

Polarization independent 2×2 multimode interference coupler with bricked subwavelength metamaterial

Carlos Pérez-Armenta[1,*], *Alejandro* Ortega-Moñux[1], *José Manuel* Luque-González[1], *Robert* Halir[1,2], *Pedro* Reyes-Iglesias[1], *Jens H.* Schmid[3], *Pavel* Cheben[3], *Íñigo* Molina-Fernández[1,2], and *J. Gonzalo* Wangüemert Pérez[1,2]

[1]Telecommunication Research Institute (TELMA) Universidad de Málaga, CEI Andalucía TECH, Louis Pasteur 35, 29010 Málaga
[2]Bionand Center for Nanomedicine and Biotechnology, Parque Tecnológico de Andalucia, Málaga 29590, Spain
[3]National Research Council Canada, Ottawa, Ontario K1A 0R6, Canada

Abstract. The silicon-on-insulator (SOI) platform enables high integration density in photonic integrated circuits while maintaining compatibility with CMOS fabrication processes. Nevertheless, its inherently high modal birefringence hinders the development of polarization-insensitive devices. The dispersion and anisotropy engineering leveraging subwavelength grating (SWG) metamaterials makes possible the development of polarization agnostic waveguide components. In this work we build upon the bricked SWG metamaterial nanostructures to design a polarization independent 2×2 multimode interference (MMI) coupler for the 220 nm SOI platform, operating in the telecom O-band. The designed device exhibits a 160 nm bandwidth with excess loss, polarization dependent loss and imbalance below 1 dB and phase error lower than 5°.

Since their first demonstration in silicon photonics [1,2], SWG metamaterials [3,4] have been used as a powerful engineering tool for overcoming performance limitations of conventional silicon photonic devices [5]. However, achieving polarization-independent and broadband devices is still challenge [6]. In this work we leverage polarization independence provided by bricked SWG waveguides, a recently proposed new subwavelength nano-patterning topology [7,8], to design a polarization-insensitive 2×2 MMI coupler for the 220 nm SOI platform.

The general geometry of bricked SWG waveguide, as well as the MMI device that makes use of it, are shown in Fig. 1. Fig. 2 (solid lines) shows the beat length wavelength dependence of a conventional SWG waveguide for both TE and TM polarizations. By engineering the geometry of the bricked SWG waveguide, it is possible to equalize the beat length for TE and TM polarizations at a given wavelength, as shown in Fig. 2 (dashed lines). This multi-mode waveguide is subsequently implemented in the central region of the MMI coupler, yielding polarization-independent performance.

The performance of the MMI coupler was evaluated by 3D full-vectorial FDTD simulations. The simulated structure included the bricked SWG multimode waveguide and adiabatic tapers for widening the input mode profile and adapting silicon wires to the bricked SWG slab. The simulation results are shown in Fig. 3. Excess loss is below 1 dB in the wavelength range from

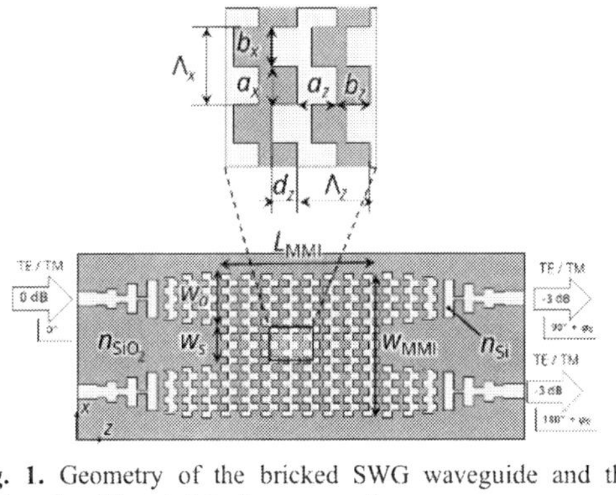

Fig. 1. Geometry of the bricked SWG waveguide and the designed multimode interference coupler.

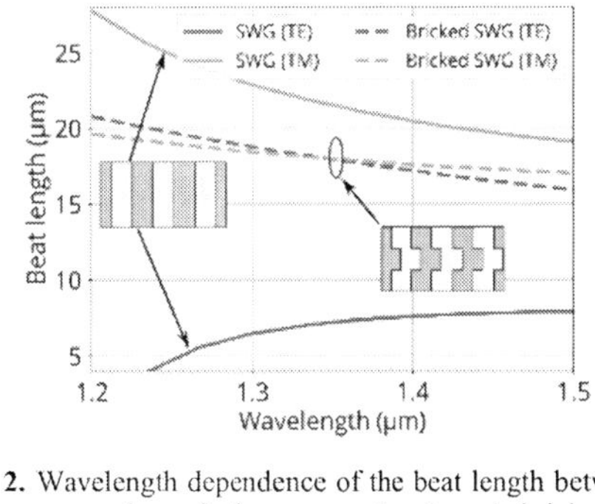

Fig. 2. Wavelength dependence of the beat length between the first two modes of the conventional and bricked SWG waveguides, for the TE and TM polarizations. In both cases, the width of the waveguide is 3 μm.

* Corresponding author: cpa@ic.uma.es

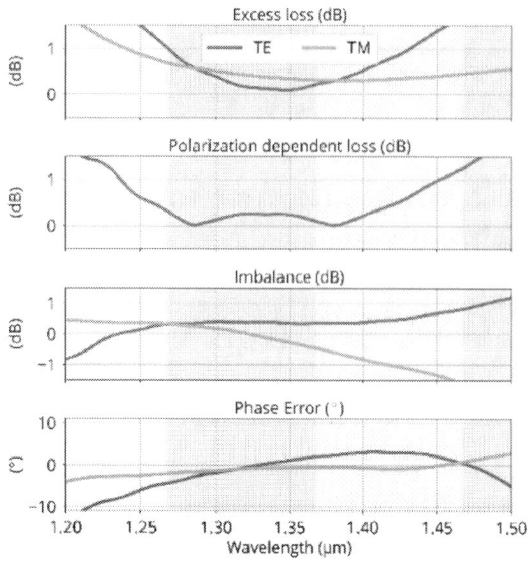

Fig. 3. Performance of the designed device calculated using 3D FDTD simulations.

1260 nm to 1420 nm for both TE and TM polarizations.

Polarization dependent loss is lower than 1 dB from 1240 nm to 1450 nm and power imbalance is less than 1 dB from 1190 nm to 1420 nm. Phase errors of <5° are predicted in the wavelength range 1260 nm – 1500 nm. This is, to the best of our knowledge, the first report of a polarization-insensitive 2×2 MMI for the 220 nm-thick silicon layer on the SOI platform. The results obtained in this work open prospects for the development of polarization-insensitive devices for silicon photonic integrated circuits.

We acknowledge funding from Ministerio de Economía y Competitividad (PID2019-106747RB-I00), Junta de Andalucía (P18-RT-1453, UMA20-FEDERJA-158), Ministerio de Ciencia Innovación y Universidades (FPU16/06762, FPU19/02408) and Universidad de Málaga.

References

1. P. Cheben, D. Xu, S. Janz, A. Densmore, Opt. Express, **14**, 4695-4702 (2006)
2. P. Cheben, P. J. Bock, J. H. Schmid, J. Lapointe, S. Janz, D.-X. Xu, A. Densmore, Opt. Lett. **35**, 2526-2528 (2010)
3. P. Cheben, R. Halir, J. H. Schmid, H. Atwater, D. Smith, Nature, **560**, 565-572 (2018)
4. R. Halir, A. Ortega-Moñux, D. Benedikovic, G. Z. Mashanovich, J. G. Wangüemert-Pérez, J. H. Schmid, Í. Molina-Fernández, P. Cheben, Proceedings of IEEE **106**, 2144-2157 (2018)
5. A. Rahim, T. Spuesens, R. Baets, W. Bogaerts, Proc. of the IEEE, **106**, 2313-2330 (2018)
6. D. Dai, L. Liu, S. Gao, D.-X. Xu, S. He, Laser & Phot. Rev., **7**, 303-328, (2012)
7. J. M. Luque-González, A. Ortega-Moñux, R. Halir, J. H. Schmid, P. Cheben, Í. Molina-Fernández, J. G. Wangüemert-Pérez, Laser & Phot. Rev., **15,** 2000478 (2021)
8. C. Pérez-Armenta, A. Ortega-Moñux, J. M. Luque-González, R. Halir, P. J. Reyes-Iglesias, J. Schmid, Í. Molina-Fernández, J. G. Wangüemert-Pérez, Photon. Res., **10**, A57-A65 (2022)

State-of-the-art and next-generation integrated photonic design

James Pond[1,*], *Xu* Wang[1], *Zeqin* Lu[1], *Federico* Duque Gomez[1], *Ahsan* Alam[1], *Sebastian* Gitt[1], *Dylan* McGuire[1], *Jeff* Young[2], and *Gilles* Lamant[3]

[1]Ansys, Inc., 1700 – 1095 West Pender Street, Vancouver, BC, V6E 2M6, Canada
[2]University of British Columbia, 6224 Agricultural Road, Vancouver, BC, V6T 1Z1,
[3]Cadence Design Systems, Inc., 2655 Seely Avenue, San Jose, CA, 95134, USA

Abstract. The relentless need for higher bandwidth, lower power and lower cost data communications has driven tremendous innovation in integrated photonics in recent years. This innovation has been supported by state-of-the-art electronic-photonic design automation (EPDA) workflows, which enable process design kit (PDK) centred schematic driven design and layout, as well as statistically enabled electro-optical simulation. In addition, custom components can be introduced and optimized for a specific foundry process using advanced methods such as photonic inverse design and machine learning. While much of the innovation has been motivated by data communications, it has enabled a variety of different applications such as sensing, integrated LiDAR and quantum information technologies. We discuss the latest innovations in EPDA workflows and show how a silicon photonic ring-based wavelength demultiplexing (WDM) system can be easily designed, simulated and implemented. In addition, we discuss the extension of these workflows to support the design and simulation of quantum photonic devices, enabling designers to consider the effects of realistic sources and manufacturing imperfections when designing quantum building blocks to meet specific fidelity and fault tolerance thresholds.

1 Electronic-Photonic Workflows

Electronic-Photonic Design Automation (EPDA) combines the best of the highly mature Electronic Design Automation (EDA) tools and workflows with curvilinear layout capability, multiphysics solvers, electro-optical compact model generation and circuit/system simulators to enable the design of complex electro-optical systems such as high-speed optical interconnect transceivers [1,2]. To demonstrate the various aspects of this workflow, we consider the 4-channel wavelength demultiplexing (WDM) circuit shown in Figure 1.

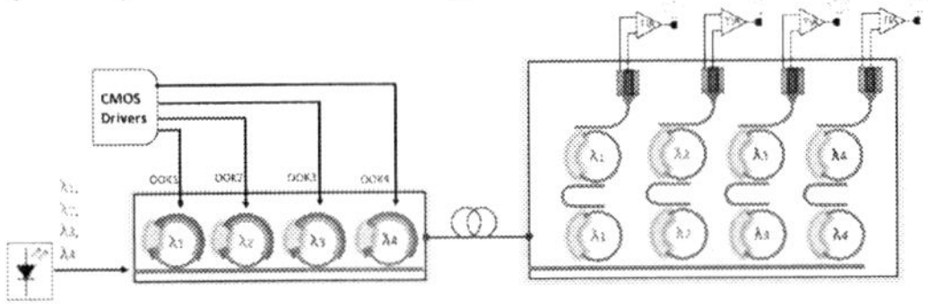

Fig. 1. A 4-channel WDM system using ring modulators and receivers.

1.1 Curvilinear layout and Parameterized Cells

Photonic devices, such as the ring modulator show in Figure 2, are non-Manhattan shapes that require special treatment. We have developed new methods to handle these structures, with a purely mathematical representation of the geometry at its core. This allows for full and accurate parameterization of these curvy geometries using parameterized cells (PCells), within an EDA environment.

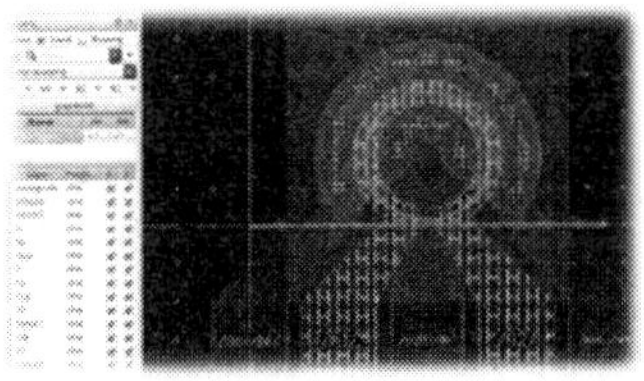

Fig. 2. A ring modulator designed in Virtuoso Layout Suite.

1.2 Physical simulation

Once the layout is complete in Virtuoso Layout Suite, the 3D geometry can be reconstructed in Ansys simulation tools and simulated using various electromagnetic, thermal and charge transport solvers, as shown in Figure 3. The geometry reconstruction is accomplished by a layer builder linked to a GDS file or a direct bridge with Virtuoso Layout Suite. In addition, the layer builder reads a foundry-specific process file with all layer and material information. Physical simulation can be used to optimize components, accounting for the effects of the foundry-specific layers and materials.

1.3 Photonic compact models

Once optimized, the performance of a component can be extracted from the simulation results. Compact models for the design can be created for both photonic Verilog-A and

* Corresponding author: james.pond@ansys.com

for Ansys Lumerical INTERCONNECT using Ansys Lumerical Compact Model Library (CML) Compiler. CML Compiler reads the component data in a specified schema to automate the process of building accurate and reliable model libraries, including the creation of symbols for Cadence Virtuoso. The data for each component is often a combination of simulated, experimental, and statistical information. The models include statistical information to enable corner and Monte-Carlo (MC) analysis of circuits.

Fig. 3. The 3D geometry of the ring constructed with the Virtuoso Layout Bridge (left) and physical simulation results (right) showing, from top to bottom, the electric field intensity in the coupler region, the temperature map under a certain heating bias, the charge density in the diode region under certain bias voltage conditions.

1.4 Schematic design and simulation

Once schematic symbols are available for all components, the schematic design can be created using Cadence Virtuoso. A variety of different circuit and system simulations can be performed. Optical frequency domain analysis can be done to determine bias setpoints for tuning. Electro-optical time domain analysis can be done using either co-simulation between Cadence Spectre and Ansys Lumerical INTERCONNECT, or entirely within Spectre using photonic Verilog-A as shown in Figure 4. In either case, models from foundry PDKs can be mixed with models for custom components.

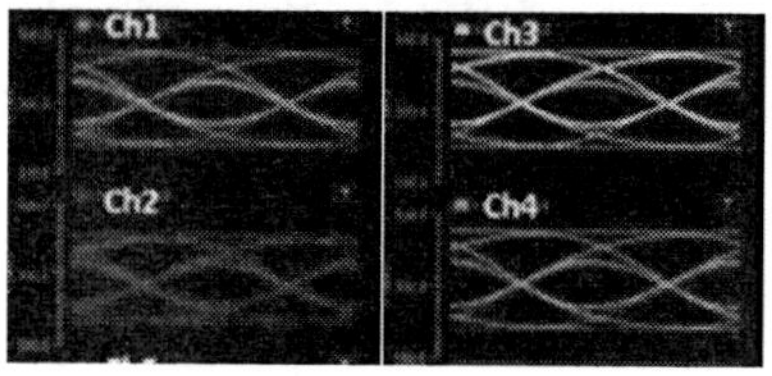

Fig. 4. The simulated eye diagrams for each channel simulated entirely in Spectre using photonic Verilog-A models.

1.5 Schematic driven layout

Schematic driven layout can be accomplished using the highly mature EDA methods in the Cadence Virtuoso environment, including layout vs schematic (LVS) to ensure that the final layout matches the original schematic. The EDA methods are complemented by the CurvyCore technology and the ability to perform accurate optical and electro-optical circuit simulation.

2 Quantum photonics

Various approaches to implement quantum gates and calculations using photonic integrated circuits are under active investigation [3-5]. The necessary components can be designed and optimized using the workflow described above, and the EPDA workflow can be used to create schematics and layout the various gates.

In addition, we have developed new methods to simulate the quantum performance of various sources and gates, for the discrete variable and continuous variable implementations. These methods rely on a classical description of the various components using the compact model libraries described above, which allow us to calculate the quantum performance of these circuits. For example, Figure 5 shows an implementation of a nonlinear sign (NLS) gate [6]. In Figure 6, we show quantum simulation results as a function of the coupling parameter of the directional couplers. The combination of the EPDA workflow, including the statistically enabled compact model libraries, with quantum circuit simulation enables MC simulation of the quantum behaviour of different gates. This allows end-users to understand the impact of manufacturing imperfections on quantities such as fidelity and fault tolerance thresholds, and to optimize quantum gates for specific foundry processes.

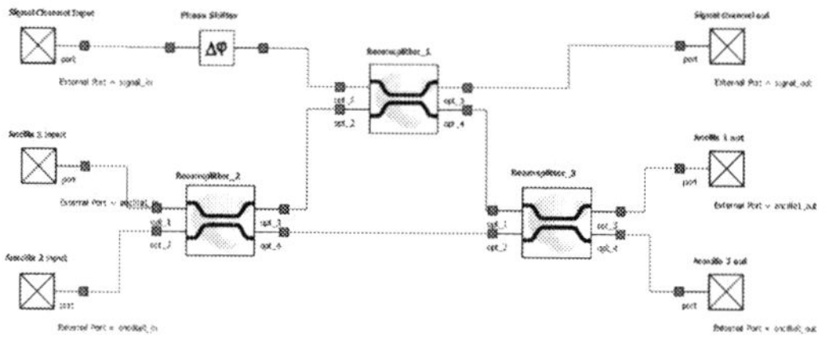

Fig. 5. A schematic in Ansys Lumerical INTERCONNECT of an NLS gate.

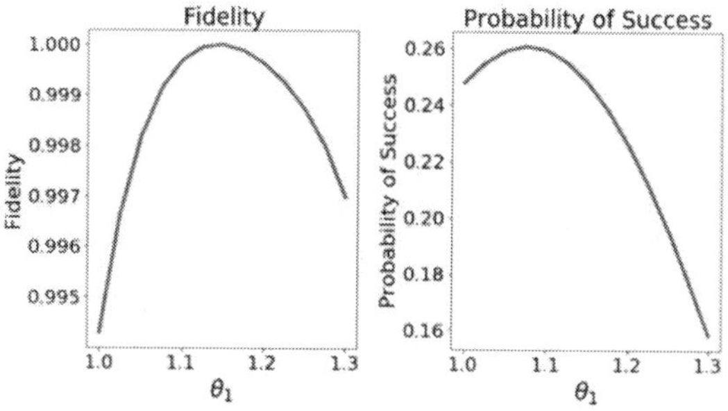

Fig. 6. The NLS gate fidelity and probability of success as a function of the coupling parameter.

References

1. https://www.ansys.com
2. https://www.cadence.com
3. J.M. Arrazola, V. Bergolm, K. Brádler et al, Nature **591**, 54–60 (2021)
4. S. Paesani, A.A. Gentile, R. Santagati et al, Phys. Rev. Lett. **118**, 100503 (2017)
5. H. Wang, Y. He, Y.-H. Li et al, Nature Photonics **11**, 361-365 (2017)
6. E. Knill, R. Laflamme, G. Milburn, Nature **409**, 46–52 (2001)

III-V compound semiconductor membrane quantum well waveguide lasers emitting at 1 μm

Stephen C. Richardson[1,*], *Jonathan R. C.* Woods[2], *Jake* Daykin[1], *Jon* Gorecki[3], *Roman* Bek[4], *Nicholas T.* Klokkou[1], *James S.* Wilkinson[5], *Michael* Jetter[6], and *Vasileios* Apostolopoulos[1]

[1]School of Physics and Astronomy, University of Southampton, Southampton, SO17 1BJ

[2]Aquark Technologies, Abbey Enterprise Centre Premier Way, Abbey Park Industrial Estate, Romsey, SO51 9AQ

[3]School of Engineering and Materials Science, Queen Mary University of London, Mile End Road, London, E1 4NS

[4]Twenty-One Semiconductors GmbH, Kiefernweg 4, 72654 Neckartenzlingen, Germany

[5]Zepler Institute, University of Southampton, Southampton, SO17 1BJ

[6]Institute for Semiconductor Optics and Functional Interfaces, University of Stuttgart, 70569 Stuttgart, Germany

Abstract. We demonstrate epitaxially grown semiconductor membrane quantum well lasers on a SiO_2/Si substrate lasing in a waveguide configuration, for potential uses as coherent light sources compatible with photonic integrated circuits. We study the emission characteristics of $In_{0.13}Ga_{0.87}As/GaAs_{0.94}P_{0.06}$ quantum well lasers, by using real and reciprocal space imaging. The laser cavity length is 424 μm, it emits light at 1 μm, and lasing thresholds as low as 211 mW were recorded. Control over the position and size of the laser spots by the pump was demonstrated.

Introduction

Vertical external-cavity surface-emitting lasers (VECSELs) are semiconductor structures with an active region consisting of multiple quantum wells (QWs) sandwiched between air and a distributed Bragg reflector (DBR). Research on this type of laser began in the mid-90s when their potential capabilities for high power and excellent beam quality were first fully demonstrated [1]. Membrane external-cavity surface-emitting lasers (MECSELs) were recently developed [2] to demonstrate lasing without a DBR. MECSELs can exhibit edge emitting lasing because the excitonic dipole in a semiconductor QW lies on the plane perpendicular to the growth direction and parallel to the surface of the epitaxial substrate [3]. In this work, we investigate the characteristics of this parasitic effect. We present optically pumped membrane quantum well lasers (MQWLs), lasing in-plane as a single laser without the use of an external cavity. The absence of a DBR offers quicker and higher quality growth, flexible wavelength design, and efficient heat extraction when contact bonded to Al_2O_3 or SiC heat spreaders. Finally, high index contrast between the membrane and substrate ensures a high overlap of the guided mode with the QW gain region.

Materials and Methods

In-plane lasing is achieved by pumping non-resonantly at 808 nm on top of a 1.57 μm thick membrane sample, consisting of ten 10 nm thick $In_{0.13}Ga_{0.87}As/GaAs_{0.94}P_{0.06}$ QWs on a SiO_2/Si substrate. The device is similar to the one reported in [4] and emits light at ~1 μm. The experimental setup collects the emitted light and then splits the beam into the real and reciprocal space paths (Fig. 1). A dichroic mirror and long pass filter cut out the pump wavelength. A thermoelectric cooler (TEC) is used to maintain temperature at 14 °C. The frequency output is measured with an optical spectrum analyser (OSA).

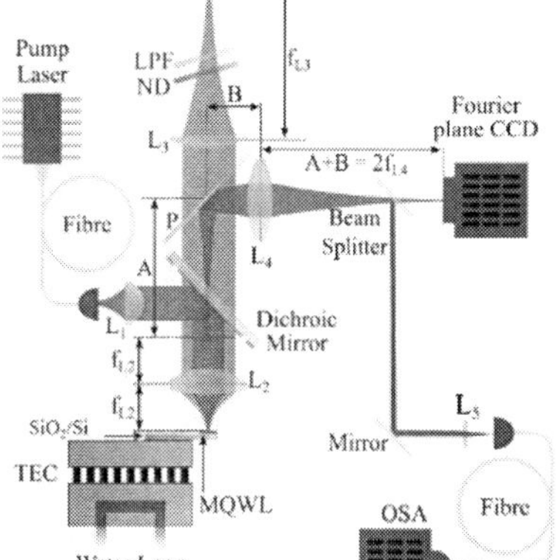

Fig. 1. MQWL experimental setup (modified from [3]).

Results

The real and reciprocal space images of the laser are shown in Figures 2a and 2b respectively. Real space

* Corresponding author: scr1n19@soton.ac.uk

images show the membrane's top view, fluorescence from the pumped region and laser spots at the end facets. Reciprocal space images display the interference patterns formed by the laser spots and strongly indicate laser threshold. From these figures, the length of the cavity can be measured to be 424 μm.

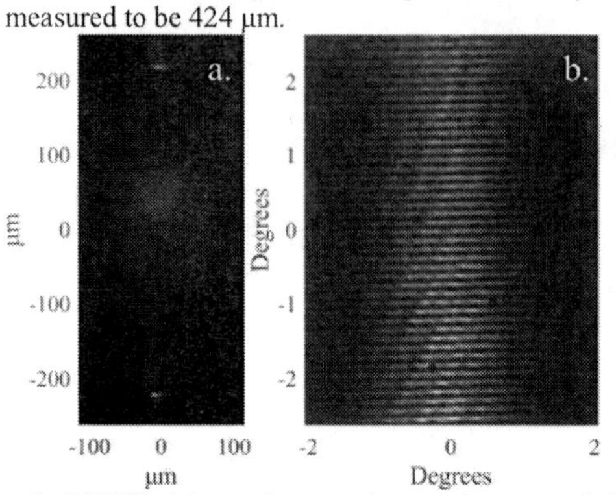

Fig. 2. MQWL (a) top down real space image, and (b) reciprocal space image that shows interference from the laser's end facets.

In order to achieve the lowest pump power, we translate the collimation lens, L_1, of the pump (Fig. 1). Such an action changes the pump spot's full width half maximum (FWHM) which in turn affects threshold pump power as shown in Figure 3. Furthermore, the laser spot width can be increased by broadening the pump spot as it is shown in Figure 4. As expected, the increased laser spot along the end facet is subjected to less diffraction resulting in a narrower interference pattern as it can be seen again in Figure 4.

Threshold values shown in Fig. 4 reach as low as 309 mW, however, 211 mW were achieved on a different area of the sample [3] for a 424 μm long laser cavity. As it is expected from the waveguide configuration, these values are significantly lower than the threshold pump power of ~0.7 W for a MECSEL with the same QW design as our membrane sample [4]. The laser emits at a wavelength of 1013.47 nm at 14 °C and 385.2 mW pump power. As temperature rises, the laser wavelength increases by $(7.92 \pm 0.04) \times 10^{-2}$ nm/°C from 12 °C to 17.7 °C, achieving a 0.43 nm tuning range.

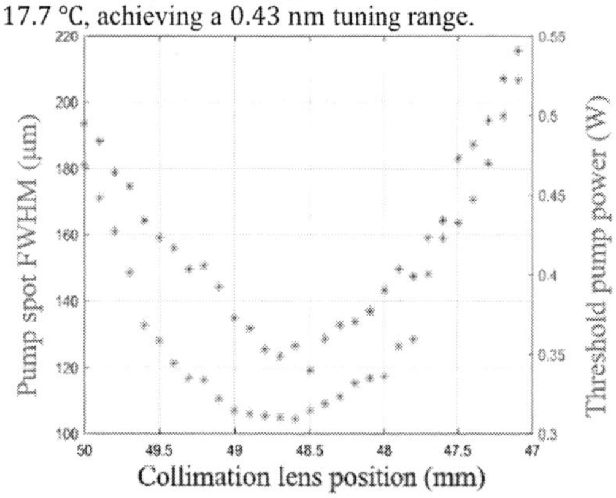

Fig. 3. Evolution of pump spot FWHM and threshold pump power with respect to collimation lens position away from the fibre's end facet (L_1 in Fig. 1).

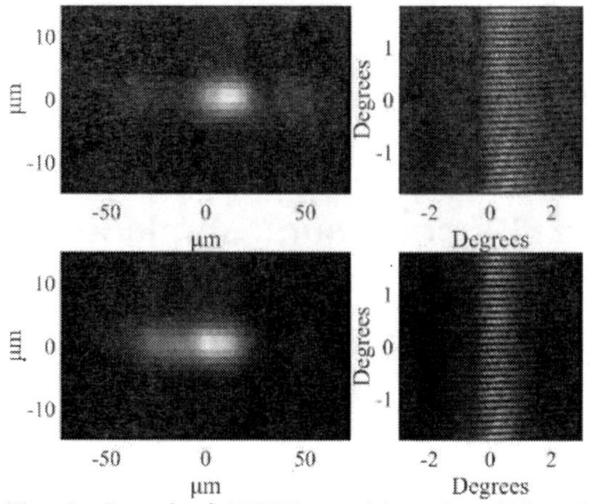

Fig. 4. Control of MQWL spatial mode width and corresponding interference pattern (modified from [3]).

Conclusions

In-plane lasing in waveguide membrane samples by a single laser on a standard SiO_2/Si substrate is realised with 211 mW threshold pump power. The pump spot's location and size allows us to control the shape and position of the waveguide laser's spatial mode, which shows potential for on-chip integration with photonic circuits. Reciprocal imaging displayed the interference between the coherent laser emission of the waveguide cavity's end facets. In [3] we demonstrate coherent laser array operation on a SiC substrate recording thresholds down to 60 mW for 70 μm long cavities. Future work will involve micro structuring of the membranes for novel cavity geometries, and the implementation of a spatial light modulator (SLM) to investigate laser arrays by varying pump illumination patterns. Pump shaping will allow control of the position and coherence of laser arrays and lead to on-chip integration with other photonic devices.

Funding

We would like to thank EPSRC for the grant EP/T001046/1 titled "UK National Quantum Technology Hub in Sensing and Timing".

References

1. M. Jetter, P. Michler, *Vertical External Cavity Surface Emitting Lasers*, Wiley-VCH, 2022.
2. H. Kahle, C. M. N. Mateo, U. Brauch, P. Tatar-Mathes, R. Bek, M. Jetter, T. Graf, P. Michler, Optica, 3, 1506-1512 (2016).
3. J. R. C. Woods, J. Gorecki, R. Beck, S. C. Richardson, J. Daykin, G. Hooper, E. Branagan-Harris, A. C. Tropper, J. S. Wilkinson, M. Jetter, V. Apostolopoulos, arXiv:2201.08674.
4. S. Mirkhanov, A. H. Quarterman, H. Kahle, R. Bek, R. Pecoroni, C. J. C. Smyth, S. Vollmer, S. Swift, P. Michler, M. Jetter and K. G. Wilcox, Electronics Letters, **53**, 1537-1539 (2017).

Fabrication and optical characterization of erbium-doped silicon diode for quantum communication applications

Giulio Tavani[1,*], *Giorgia* Franzò[2], *Michele* Castriotta[3], *Giorgio* Ferrari[4], *Francesco* Picciariello[5], *Giulio* Foletto[5], *Costantino* Agnesi[5], *Paolo* Villoresi[5], *Giuseppe* Vallone[5], *Davide* Rotta[6], *Chiara* Barri[1], *Erfan* Mafakheri[7], *Michele* Celebrano[4], *Marco* Finazzi[4], *Monica* Bollani[7], and *Enrico* Prati[8,9]

[1]L-NESS, Dip. Di Fisica del Politecnico di Milano, I–22100 Como, Italy
[2]CNR-IMM, Via Santa Sofia 64, I–95123 Catania, Italy
[3]Dip. di Elettronica, Informazione e Bioingegneria, Politecnico di Milano, 20133 Milano Italy
[4]Dip. di Fisica, Politecnico di Milano, I–20133 Milano Italy
[5]Dip. di Ingegneria dell'Informazione, Università degli Studi di Padova, via Gradenigo 6B, IT-35131 Padova, Italy
[6]InPhoTec, Integrated Photonic Technologies Foundation, I–56124 Pisa, Italy
[7]IFN-CNR, L-NESS laboratory, 22100 Como, Italy
[8]Istituto di Fotonica e Nanotecnologie, Consiglio Nazionale delle Ricerche, Piazza Leonardo da Vinci 32, I–20133 Milano, Italy
[9]Dip. di Fisica "Aldo Pontremoli", Università degli Studi di Milano, Via Celoria 16, I–20133 Milano, Italy

Abstract. Quantum Key Distribution allows two users to exchange secret keys and it is based on the transmission of single photons or attenuated laser pulses. Recently, sources based on multiple single-photon emitters were demonstrated to be suitable for QKD. Here, we present a CMOS compatible multiple single-photon emitters source realized on a SOI wafer by a standard silicon diode doped with erbium ions. Particular emphasis is placed on the fabrication of such a device enhancing the erbium electroluminescence signal by adopting a proper oxygen co-doping. Finally, electroluminescence characterization at room temperature of the device is presented.

1 Introduction

Quantum communication is expected to cover several market needs in the years to come [1]. For this reason, quantum key distribution (QKD) protocols have been implemented using both single photon sources and attenuated laser pulses, towards integration on photonic chips [2]. Recent works on the security bounds for decoy-state QKD have opened new opportunities for using sources with an arbitrary photon emission statistic, including weak sources based on multiple emitters [3].

In this work, we present the fabrication and a first optical characterization of a silicon diode doped with erbium that behaves like such a weak photon source, compatible with the emission of light at telecom wavelength at 1550 nm directly in a silicon chip [4].

The device consists of a silicon planar junction with a central region, nearby the p-n depletion area, doped with erbium and oxygen atoms. Upon setting a potential difference between the p-n regions, photons are created by the electroluminescence of erbium ions (Fig. 1). The expected advantages of this kind of device are its emission centered in the third telecommunication window (1520 nm - 1550 nm), its working range of temperature between 77 K and 300 K as well as a photon emission rate below 10 μs. To increase the number of optically active Er sites, an oxygen co-doping has been introduced [5].

*e-mail: giulio.tavani@polimi.it

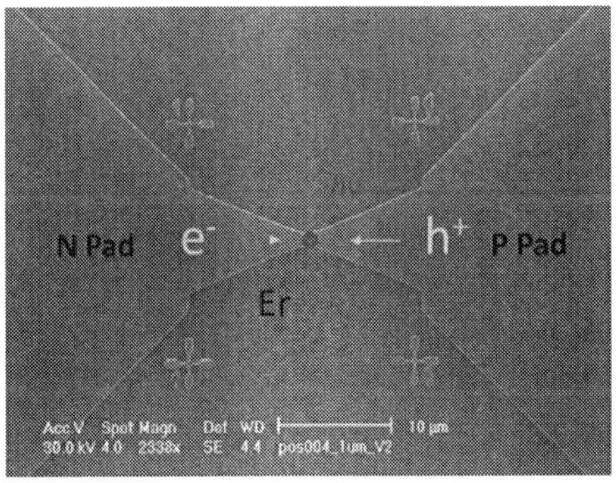

Figure 1. A scanning electron microscope image of one of the fabricated devices. In this device the erbium-doped central area is 1 μm × 1 μm. Since there is no waveguide, photons are emitted in free space.

2 Oxygen co-doping analysis

Since the excitation of erbium ions stems from the same physical mechanism both in electroluminescence and in photoluminescence [6], to determine the proper oxygen co-doping dose, we perform a photoluminescence characterization of three different silicon samples. Specifically, in our analysis the same Er dose is adopted, varying the O one. The erbium dose is equal to 1.0e13 Er/cm^2 whereas the oxygen doses are equal to a) 3.5e13 O/cm^2 b) 7e13

https://doi.org/10.1051/epjconf/202226601012

O/cm^2 and c) 1.4e14 O/cm^2. In Fig.2 the photoluminescence at 11 K of the three samples is present. It is evident that the oxygen co-doping has a dramatic impact on the erbium photoluminescence at cryogenic temperature. Since our devices are supposed to work both at 300 K and at 77 K, we opt for the highest possible oxygen dose in our p-n junctions.

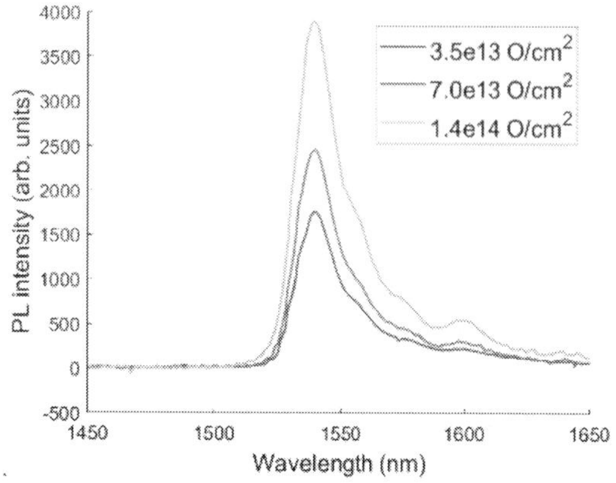

Figure 2. Here are presented the photoluminescence signals at 11K for three flat silicon samples doped with the same erbium dose (1.0e13 Er/cm^2) but with three different oxygen doses. All the samples were annealed for 30 minutes at 900°C.

3 Device fabrication

The devices are fabricated through a top-down process performed by a combination of electron beam lithography, reactive ions etching, electron beam physical vapor deposition and ion implantation starting. The entire process is performed on a 220 nm thick Si layer on a 2 μm thick SiO_2 buried oxide layer . The first step is the definition of the mesa structure, followed by the doping of the n area with phosphorus, of the p area with boron as well as of the optically active zone, in the center of the device, with erbium and oxygen. All the dopants are introduced by ion implantation and activated with a rapid thermal annealing processes. The final step is the definition of the two gold contacts on top of the n and p regions. Three samples are fabricated and characterized, they differ from each other by the size of the erbium doped area: in the first device it is equal to 1 μm × 1μm (Fig. 1), in the second one to 15 μm × 15μm and in the last one to 50 μm × 50 μm.

4 Optical characterization

All the samples are characterized at room temperature by imposing a current and collecting the electroluminescence signal. To reduce the background noise, superconducting nanowires detectors working at cryogenic temperature are employed and the integration time was set equal to one second. The signals from the three different devices are shown in Figure 3. They are approximately comparable in spite of the different erbium doped areas as the detection region is imposed by the collecting fiber which is around 5 μm^2. The highest signals refer to the two smallest junctions, which suggests that the photons in those structures are better directed toward the collecting fiber. A clear hint that the detected signals are due to the erbium ions is that moving away from the erbium doped areas, the signals drop dramatically.

From the previous measurements, it is possible to compute the number of photons emitted by each erbium ion in one second. Considering that we estimate that a tenth of erbium ions is lost after the rapid thermal annealing [7] and that the optical throughput of the setup is approximately 0.1%, a value equals to 50 is obtained in the case of the 1 μm × 1μm device when a current of 50 μA is imposed.

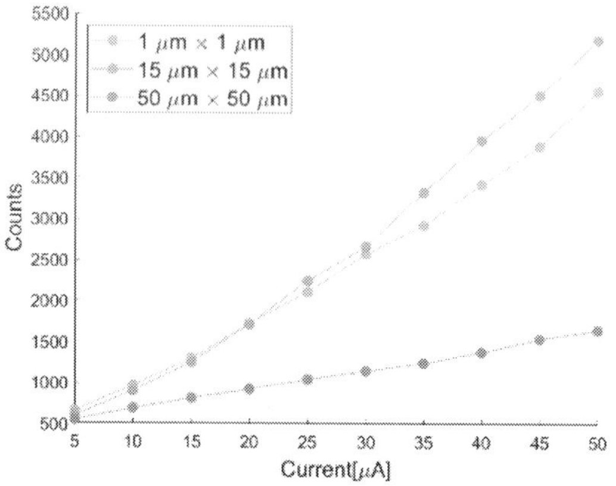

Figure 3. The photon counts obtained for each device imposing a current in the range 5-50 μA. The integration time is equal to one second and the resulting dark counts are around 400.

5 Conclusions

A weak photon source operating at 1550 nm using a silicon diode implanted with erbium and oxygen is presented. We determine the proper oxygen co-doping dose to obtain a significant enhancement of the Er electroluminescence signal at low temperature.

6 Acknowledgement

This research was supported by the QUASIX and PAIDEIA projects.

References

[1] F. Cavaliere, E. Prati, L. Poti, I. Muhammad, T. Catuogno, Quantum Reports **2(1)**, 80-106 (2020)

[2] M. Avesani et al., npj Quantum Information **7(1)**, 1-8 (2021)

[3] G. Foletto, F. Picciariello, C. Agnesi, P. Villoresi, G. Vallone, Physical Review A **105(1)**, 012603 (2022)

[4] M. Di Giancamillo, P. Biagioni, V. Sorianello, E. Prati, EPJ Web of Conferences **255**, EOSAM21 (2021)

[5] F. Priolo, G. Franzò, S. Coffa, A. Carnera, Physical Review B **57(8)**, 4443 (1998)

[6] G. Franzò, F. Priolo, S. Coffa, A. Polman, A. Carnera, Applied Physics Letters **64(17)**, 2235-2237 (1994)

[7] M. Celebrano, L. Ghirardini, M. Finazzi, Y. Shimizu, Y. Tu, K. Inoue, Y. Nagai, T. Shinada, Y. Chiba, A. Abdelghafar, M. Yano, T. Tanii, E.Prati, Optics Letters **42**, 3311-3314 (2017)

https://doi.org/10.1051/epjconf/202226602001

Design and realization of a miniaturized high resolution computed tomography imaging spectrometer

Simon Amann[1,*], *Tobias* Haist[1], *Alexander* Gatto[2], *Markus* Kamm[2], and *Alois* Herkommer[1]

[1]University of Stuttgart, Institut für Technische Optik, 70569 Stuttgart, Germany
[2]Sony Europe B.V, Stuttgart Technology Center, 70327 Stuttgart, Germany

Abstract. The computed tomography imaging spectrometer (CTIS) is a relatively unknown snapshot hyperspectral camera. It utilizes computational imaging approaches to gain the hyperspectral image from a spatio-spectral smeared sensor image. We present a strongly miniaturized system with a dimension of only 36 mm x 40.5 mm x 52.8 mm and a diagonal field of view of 29°. We achieve this using a Galilean beam expander and a combination of off-the-shelf lenses, a highly aspherical imaging system from a commercial smartphone and a 13 MP monochrome smartphone image sensor. The reconstructed hyperspectral image has a spatial resolution of 400 x 300 pixel with 39 spectral channels.

1 Introduction

Hyperspectral imaging is widely used in various fields such as environmental observation, medical diagnosis or food quality control [1]. Most of the used hyperspectral cameras use scanning approaches (spatial, spectral or mixed) to capture the full data cube consisting of a two-dimensional image for each spectral channel. This leads to problems such as motion artifacts or an increased mechanical complexity (moving parts). The computed tomography imaging spectrometer (CTIS) uses a different approach. It computes the hyperspectral information from a single captured sensor image. A typical optical design of such a system is illustrated in Figure 1a.

A diffractive optical element (DOE) is used to create a spatio-spectral smeared two-dimensional sensor image. The zeroth diffraction order is usually placed in the middle of the sensor and is equivalent to a monochrome image of the scene. Several higher diffraction orders, that are similar to an oblique projection of the hyperspectral data cube, are arranged around it. A field stop is used to limit the size of the imaged scene. This is typically done in the intermediate image plane of the first lens. A reconstruction algorithm is required to derive the hyperspectral image from the sensor image. In this paper we use an implementation based on the EM algorithm that utilizes space invariance of the sensor signal [2].

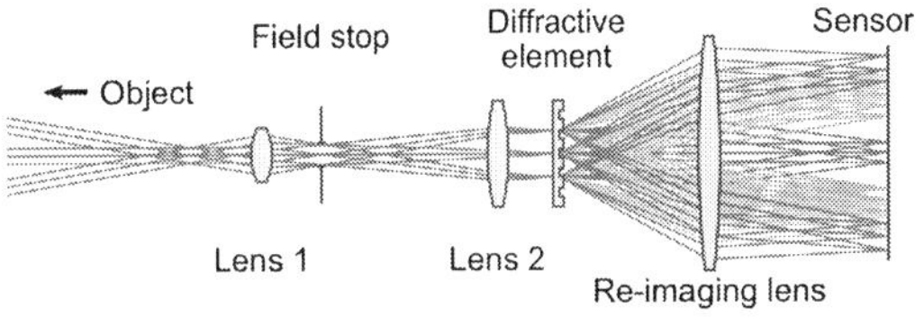

(a) Keplarian (typical) design.

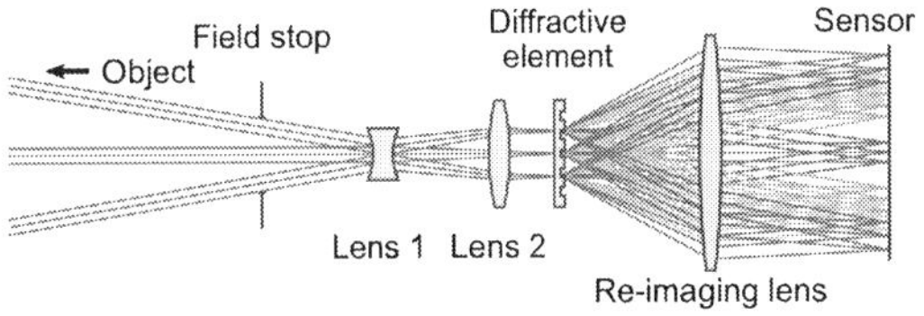

(b) Galilean design.

Figure 1: Different CTIS designs using a Keplarian and a Galilean beam expander.

2 Miniaturization

One major drawback we want to address is the relatively big size of most published CTIS systems, which often hinders their use outside of laboratory environments. The main reason for this is that the focal length of the system has to be very small when a reasonably large field of view (FOV) is used (full scene has the size of the zero order projection in the image plane). The minimal focal length of the re-imaging lens is limited because a distortion effect appears when the angles of the diffracted rays become high. This effect is caused by the non-linear diffraction of light in image space [3]. The first two lenses, therefore, act as a beam expander. They reduce the ray angles and hence the total focal length. The typically used Keplarian type of beam expander does have the big advantage that the intermediate image can be used to crop the imaged scene (field stop). However, we have found that in some cases it

*e-mail: amann@ito.uni-stuttgart.de

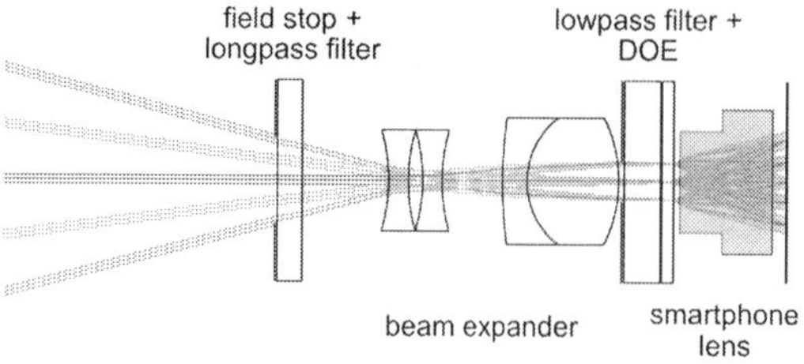

Figure 2: Optic design of the prototype.

Figure 3: Assembled prototype.

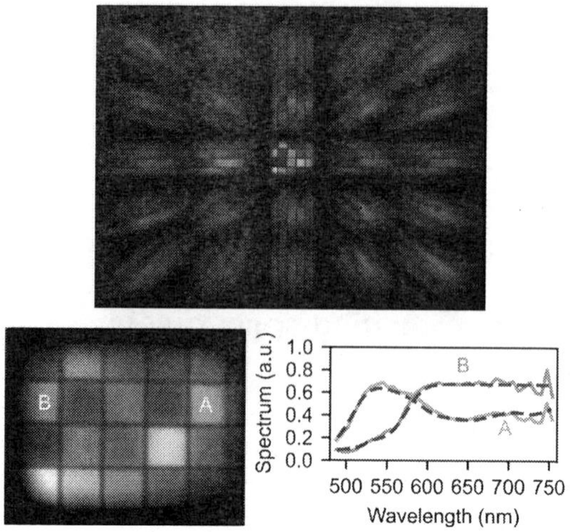

Figure 4: Pre-processed sensor image and reconstruction results. The shown spectra are averaged over a 5 x 5 pixel area. The ground truth is shown in dashed black lines.

is beneficial to use a Galilean type of beam expander instead (Figure 1b). It often achieves a better optical performance in a smaller form factor when the same complexity of lenses is used. The field stop is placed with some distance in front of the first lens. One major drawback of this approach is that the intermediate image, and consequently the reconstructed hyperspectral image, is considerably vignetted when the field stop is placed close to the system or a small f-number is used. The optimal design depends on the requirements of the system.

3 Setup and results

The optical design of our prototype is shown in Figure 2. The first element is the field stop that is made of laser cut cardboard. It is followed by an absorptive longpass filter (Hoya Y48) that is used to block light below 480 nm. Lens 1 is realized using two stacked lenses with a focal length of −4 mm each (SLM-04B-04N from OptoSigma). Lens 2 is an achromatic lens with a focal length of 7.5 mm (#49-275-INK from Edmund Optics). Subsequent to this is a dichroic shortpass filter (#47-817 from Edmund Optics) with a cut-off wavelength of 750 nm. The following DOE is a custom, in-house produced binary computer generated hologram. The lens from a Sony Xperia 10 Plus smartphone (illustrated as black box) is used as the reimaging lens. We chose this because its entrance pupil is close to the front of the lens housing and its big image size (compared to the focal length of approximately 4 mm). The See3CAM CU135M from e-con Systems, with a monochrome 13MP image sensor, is used as the camera body. The vital parts of the mechanics are made of anodized aluminium.

An image of the assembled prototype can be seen in Figure 3. The total optical track length is only 21 mm; the total size including the housing is 36 mm x 40.5 mm x 52.8 mm. The effective focal length of 0.9 mm results in a diagonal FOV of 29°.

Figure 4 shows one exemplary, pre-processed sensor image taken of a ColorChecker. The zero order image is taken with a lower exposure time than the higher diffraction orders and afterwards superimposed. Furthermore we subtracted stray light and corrected the distortion effect caused by the non-linear diffraction. The reconstruction result is illustrated in its RGB representation together with some selected spectra. It has a spatial resolution of 400 x 300 pixel with 39 spectral channels (7 nm steps from 488 nm to 754 nm). The result is in a good agreement with the ground truth values. We assume that remaining inaccuracies originate from noise, stray light and artifacts of the reconstruction algorithm. A custom lens design together with a better reconstruction (for example by using neural networks [4]) can help to improve the result.

4 Conclusion

A strongly miniaturized and portable CTIS has been presented. Despite its size, it has a large resolution and field of view. We achieved this using a Galilean instead of a Keplarian beam expander. It can be used in applications where a small physical size is required.

References

[1] M.J. Khan, H.S. Khan, A. Yousaf, K. Khurshid, A. Abbas, IEEE Access **6**, 14118 (2018)
[2] N. Hagen, E.L. Dereniak, D.T. Sass, SPIE Proceedings **6661** (2007)
[3] J.E. Harvey, R.N. Pfisterer, Optical Engineering **58** (2019)
[4] M. Zimmermann, S. Amann, M. Mel, T. Haist, A. Gatto, Optical Engineering **61** (2022)

Computation of aberration coefficients for plane-symmetric reflective optical systems using Lie algebraic methods

Antonio Barion[1,*], *Martijn J. H.* Anthonissen[1], *Jan H. M.* ten Thije Boonkkamp[1], and *Wilbert L.* IJzerman[1,2]

[1]Eindhoven University of Technology, PO Box 513, 5600 MB Eindhoven, The Netherlands
[2]Signify, High Tech Campus 7, 5656 AE Eindhoven, The Netherlands

Abstract. The Lie algebraic method offers a systematic way to find aberration coefficients of any order for plane-symmetric reflective optical systems. The coefficients derived from the Lie method are in closed form and solely depend on the geometry of the optical system. We investigate and verify the results for a single reflector. The concatenation of multiple mirrors follows from the mathematical framework.

1 Introduction

The calculation of Seidel coefficients of rotationally symmetric optical systems is well known in the optical design literature [1, 2]. Seidel sums provide a clear insight into the contribution of each interface of the system to the third order transverse ray aberrations at the image plane and provide analytic expressions for the aberration coefficients. We aim to derive similar closed form expression for aberration coefficients of plane-symmetric optical systems. To construct the optical mapping for plane-symmetric reflective systems, we employ Lie algebraic methods [3–5]. This mapping systematically produces the analytic aberration expansion up to the desired order in terms of initial position and direction.

2 Analytic Ray-Tracing Equations for Reflecting Interfaces

In classical Hamiltonian optics, the propagation of light rays in an optical medium with refractive index $n(\boldsymbol{q}, z)$ is determined by the Hamiltonian equations, where the Hamiltonian H is given by [6]:

$$H(\boldsymbol{q}, \boldsymbol{p}, z) = -\sqrt{n(\boldsymbol{q}, z)^2 - |\boldsymbol{p}|^2}. \tag{1}$$

Here, $\boldsymbol{q}$ are the space and $\boldsymbol{p}$ are the momentum coordinates of the ray. Since at each z-plane, z = const., a ray is fully determined by $(\boldsymbol{q}, \boldsymbol{p})$, they represent the phase space as in classical Hamiltonian mechanics, with the only restriction that $|\boldsymbol{p}| \leq n(\boldsymbol{q}, z)$.

In order to follow the ray path from object to image plane, we need to be able to reflect the ray at a plane-symmetric mirror, to rotate the coordinate system such that the optical axis ray (OAR) remains aligned with the z-axis and to propagate the reflected ray further with respect to the rotated system. For instance, reflection at an interface $z = \zeta(\boldsymbol{q})$ can be described by its vectorial formulation as:

$$\boldsymbol{p}' = \boldsymbol{p} - 2\frac{\nabla\zeta(\bar{\boldsymbol{q}})}{1 + |\nabla\zeta(\bar{\boldsymbol{q}})|^2}(\boldsymbol{p} \cdot \nabla\zeta(\bar{\boldsymbol{q}}) - p_z), \tag{2}$$

where $\boldsymbol{p}'$ are the momentum coordinates after reflection and $\bar{\boldsymbol{q}}$ are the position coordinates of the point of impact between the ray and the interface. Position coordinates before and after reflection are projected onto the z-plane at the point of impact of the OAR and the coordinate system rotates in order to keep the z-axis aligned with the OAR before and after its reflection, respectively, see Figure 1.

The optical axis ray, with coordinates $\boldsymbol{q} = \boldsymbol{0} = \boldsymbol{p}$, will have reflected coordinates in the rotated system $\boldsymbol{q}^R = \boldsymbol{0} = \boldsymbol{p}^R$ and all the considered rays will lie in a neighbourhood of it. As such, the mapping $\mathcal{M}$ representing reflection with rotation of the coordinate system can be expanded around the OAR.

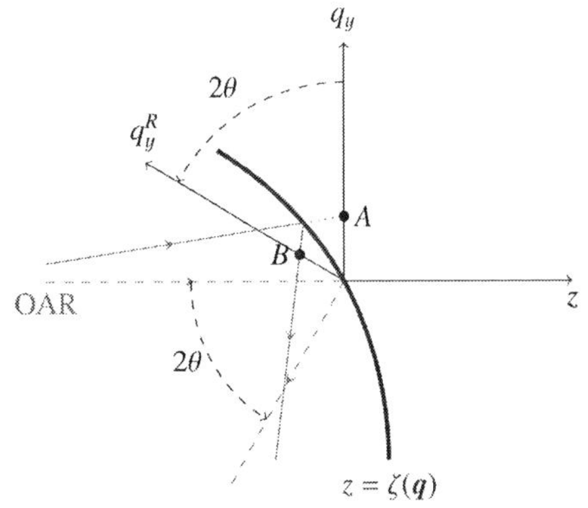

Figure 1. The reflection operator maps the point A of the incoming ray to the point B of the outgoing ray in the rotated coordinate system. The OAR (dashed) is mapped from the origin to itself. The incidence angle of the OAR on the reflector is equal to θ.

*e-mail: a.barion@tue.nl

3 Lie Algebraic Tools

The Poisson bracket is an operator $[\cdot,\cdot]$, which maps any pair of functions f, g in $(\boldsymbol{q}, \boldsymbol{p})$ to a single function of $(\boldsymbol{q}, \boldsymbol{p})$, denoted by $[f, g]$:

$$[f, g] := \frac{\partial f}{\partial \boldsymbol{q}} \cdot \frac{\partial g}{\partial \boldsymbol{p}} - \frac{\partial f}{\partial \boldsymbol{p}} \cdot \frac{\partial g}{\partial \boldsymbol{q}}.$$

The Poisson bracket turns the space of polynomials on phase space into a Lie algebra [3–5]. We define the linear Lie operator $[f, \cdot\,]$ associated with f acting on g as follows:

$$[f, \cdot\,]g = [f, g].$$

If the argument of $[f, \cdot\,]$ is a vector-valued function $\boldsymbol{g}$, then it acts component-wise on $\boldsymbol{g}$. The main Lie algebraic tool we use is the Lie transformation. The Lie transformation $\exp([f, \cdot\,])$ associated with f and generated by $[f, \cdot\,]$ is defined as:

$$\exp([f, \cdot\,]) = \sum_{k=0}^{\infty} \frac{[f, \cdot\,]^k}{k!}. \qquad (3)$$

The powers in (3) follow the recursive definition

$$[f, \cdot\,]^0 = I, \quad [f, \cdot\,]^k = [f, [f, \cdot\,]^{k-1}], \quad k = 1, 2, \dots.$$

It can be proven that the propagation and the reflection with rotation mappings are representable as infinite concatenations of Lie transformations [3, 7] of the form

$$\mathcal{M} = \exp([g_2, \cdot\,])\exp([g_3, \cdot\,])\exp([g_4, \cdot\,])\cdots. \qquad (4)$$

Here, the generators g_2, g_3, etc. are homogeneous polynomials on phase space of degree 2, 3, etc. For instance, the second-order generator h_2 for free propagation in a medium of constant refractive index n over a distance d along the optical axis reads:

$$h_2(\boldsymbol{p}) = -\frac{d}{2n}|\boldsymbol{p}|^2. \qquad (5)$$

The polynomial g_2 associated with the combined action of reflection and rotation to the new coordinate system reads:

$$g_2(\boldsymbol{q}, \boldsymbol{p}) = \frac{\cos(\theta)}{\rho} q_x^2 + \frac{1}{R\cos(\theta)} q_y^2, \qquad (6)$$

where ρ and R are the sagittal and tangential radii of curvature, respectively, and θ is the incidence angle of the OAR on the mirror, see Figure 1.

4 The Fundamental Element and Optical Systems

By representing each mapping in the form (4) and concatenating propagation from an intermediate object plane, reflection by a tilted interface with rotation to the outgoing reference system and, lastly, propagation to an intermediate image plane, we can describe a so-called fundamental optical element by its mapping of the form (4). If we are interested in aberration terms up to third order, then the final mapping $\mathcal{M}_{\text{el}}$ corresponding to one single optical element will be

$$\mathcal{M}_{\text{el}} = \exp([\tilde{g}_2, \cdot\,])\exp([\tilde{g}_3, \cdot\,])\exp([\tilde{g}_4, \cdot\,]), \qquad (7)$$

where all the information about the optical element is contained in the coefficients of $\tilde{g}_2, \tilde{g}_3, \tilde{g}_4$, that are thus dependent on the geometry of the reflector.

By concatenating multiple optical elements, using the mathematical tools described in [3–5], it is possible to derive a mapping $\mathcal{M}_{\text{tot}}$ that describes the complete optical system up to third order aberrations, which has the same form as in Eq. (7). We therefore have

$$\mathcal{M}_{\text{tot}} = \exp([t_2, \cdot\,])\exp([t_3, \cdot\,])\exp([t_4, \cdot\,]), \qquad (8)$$

where the coefficients of the generators t_i are related to the geometry of the complete system.

5 Results

We have been able to recover the results recently presented in [8], where the authors determine the fourth order surface expansion coefficients of a reflector. The requirement for such reflector is that all field-independent aberrations, i.e., $\boldsymbol{p} = \boldsymbol{0}$ for incoming rays, are zero up to third order. By constructing the optical mapping of the form (7) and investigating its action on the initial phase space coordinates, we reproduced exactly the results presented in [8, Table 3].

We proceeded similarly for the case where we required that $\boldsymbol{q} = \boldsymbol{0}$ at the object plane for incoming rays and zero third order aberrations. The spherical ellipsoid is known to be a perfect imager for an object and image point lying at its respective foci [9]. The surface coefficients we calculated from the requirement $\boldsymbol{q} = \boldsymbol{0}$ agree exactly with the surface expansion of the corresponding spherical ellipsoid.

References

[1] W.T. Welford, *Aberrations of Optical Systems* (CRC Press, 1986)

[2] J. Braat, P. Török, *Imaging Optics* (Cambridge University Press, 2019)

[3] A.J. Dragt, J.M. Finn, Journal of Mathematical Physics **17**, 2215 (1976)

[4] A.J. Dragt, E. Forest, K.B. Wolf, in *Lie Methods in Optics* (Springer, 1986), pp. 105–157

[5] K.B. Wolf, *Geometric Optics on Phase Space* (Springer Science & Business Media, 2004)

[6] R.K. Luneburg, *Mathematical Theory of Optics* (University of California Press, 1964)

[7] A.J. Dragt, JOSA **72**, 372 (1982)

[8] J. Caron, S. Bäumer, JOSA A **38**, 80 (2021)

[9] A. Gomez-Vieyra, D. Malacara-Hernández, Appl. Opt. **50**, 66 (2011)

Design of optical surfaces conform the hyperbolic Monge-Ampère equation

M.W.M.C. Bertens[1,*], M.J.H. Anthonissen[1], W.L. IJzerman[1,2], and J.H.M. ten Thije Boonkkamp[1]

[1]CASA, Department of Mathematics and Computer Science, Eindhoven University of Technology, PO Box 513, 5600 MB Eindhoven, The Netherlands
[2]Signify Research, High Tech Campus 7, 5656 AE Eindhoven, The Netherlands

Abstract. We present a method for designing freeform optical surfaces for illumination optics. By the laws of reflection, refraction and conservation of energy, a fully nonlinear PDE, the Monge-Ampère equation, is derived for the optical surface. By the edge ray principle a transport boundary condition is obtained. We solve the hyperbolic variant of the PDE using a least-squares method, resulting in optical saddle surfaces for a parallel source and far-field target.

1 Introduction

Since the advent of LED lighting interest in freeform optics has significantly increased. Forward design tools as ray tracing can take a lot of design time, which can be improved upon by inverse methods. One such inverse method involves solving the fully nonlinear Monge-Ampère (MA) equation with transport boundary conditions. The derivation of the MA equation allows for both hyperbolic and elliptic solutions [1]. It is conjectured that saddle (optical) surfaces may lead to more compact optical systems than convex/concave surfaces, which are solutions to the hyperbolic and elliptic MA equation, respectively. Recently a lot of research has been done on the elliptic variant, and the closely related field of optimal transport. Solution strategies for the hyperbolic MA equation remain scarce. Here, we modify a least-squares solver for the elliptic MA equation to solve the hyperbolic variant.

2 Mathematical model

We aim to construct an algorithm for designing freeform optical saddle surfaces under the geometrical optics limit. We assume a collimated light source with support $\mathcal{X}$ and given emittance $f(x_1, x_2) > 0$, positioned parallel to the x_1x_2-plane, emitting light in the $\hat{\boldsymbol{s}} = (0, 0, 1)^{\mathrm{T}}$ direction. We place a reflector above the light source with the freeform surface at $z = u(x_1, x_2)$. The light is reflected in the direction $\hat{\boldsymbol{t}}(x_1, x_2)$. We prescribe a far-field target intensity $g(y_1, y_2)$ in terms of the stereographic coordinates (y_1, y_2) of $\hat{\boldsymbol{t}}$. By applying conservation of energy it follows that the equation describing u is given by

$$u_{x_1x_1}u_{x_2x_2} - u_{x_1x_2}^2 = \pm\frac{f(x_1, x_2)}{J(u_{x_1}, u_{x_2})g(u_{x_1}, u_{x_2})}, \quad (x_1, x_2) \in \mathcal{X}, \tag{1a}$$

$$\nabla u(\partial\mathcal{X}) = \partial\mathcal{Y}, \tag{1b}$$

where $J > 0$ is the Jacobian of the stereographic projection $\hat{\boldsymbol{t}} \mapsto \boldsymbol{y}$. The above PDE is known as the Monge-Ampère equation. Furthermore, the sign in (1a) may be chosen and determines whether the PDE is elliptic (+) or hyperbolic (-). For the remainder, we replace the right-hand-side of (1a) by $F(x, y, \nabla u) < 0$, yielding the hyperbolic MA-equation.

3 Numerical Method

Let $\boldsymbol{m} = \nabla u$, known as the optical map. The MA equation then reads

$$\det \mathrm{D}\boldsymbol{m} = F, \quad (x_1, x_2) \in \mathcal{X}, \tag{2}$$

with $\mathrm{D}\boldsymbol{m}$ the Jacobi matrix of $\boldsymbol{m}$, which is also the Hessian matrix of u. We formulate a minimization problem for $\boldsymbol{m}$ which we solve numerically. We assume $\boldsymbol{m}$ to be twice continuously differentiable and introduce the auxiliary functions $\boldsymbol{P} : \mathcal{X} \to \mathbb{R}^{2\times2}$ with $\boldsymbol{P} = \boldsymbol{P}^{\mathrm{T}}$, $\det \boldsymbol{P} = F$ and $\boldsymbol{b} : \partial\mathcal{X} \to \partial\mathcal{Y}$, which we use to approximate $\mathrm{D}\boldsymbol{m}$ on the interior domain and $\boldsymbol{m}$ on the boundary, respectively. This is achieved by succesively optimizing three seperate functionals, viz.

$$J_{\mathrm{I}}(\boldsymbol{m}, \boldsymbol{P}) = \frac{1}{2}\iint_{\mathcal{X}} ||\mathrm{D}\boldsymbol{m} - \boldsymbol{P}||^2 \mathrm{d}x\mathrm{d}y, \tag{3a}$$

$$J_{\mathrm{B}}(\boldsymbol{m}, \boldsymbol{b}) = \frac{1}{2}\oint_{\partial\mathcal{X}} |\boldsymbol{m} - \boldsymbol{b}|^2 \mathrm{d}s, \tag{3b}$$

$$J(\boldsymbol{m}, \boldsymbol{P}, \boldsymbol{b}) = \alpha J_{\mathrm{I}}(\boldsymbol{m}, \boldsymbol{P}) + (1 - \alpha)J_{\mathrm{B}}(\boldsymbol{m}, \boldsymbol{b}), \tag{3c}$$

where $||\cdot||$ is the Frobenius norm, $|\cdot|$ the standard 2-norm and $0 < \alpha < 1$ a control parameter to either place more or less weight on the boundary or the interior. Concluding convergence of the iterations, we calculate the optical surface from

$$u = \underset{\psi\in C^1(\mathcal{X})}{\operatorname{argmin}}\, I(\psi), \quad I(\psi) = \frac{1}{2}\iint_{\mathcal{X}} ||\nabla\psi - \boldsymbol{m}||^2 \mathrm{d}x\mathrm{d}y. \tag{4}$$

The minimization of $J_{\mathrm{I}}(\boldsymbol{m}, \boldsymbol{P})$ can be done pointwise as no derivatives of $\boldsymbol{P}$ in the integrand occur and the minimum can be obtained analytically. For the minimization

*e-mail: m.w.m.c.bertens@tue.nl

https://doi.org/10.1051/epjconf/202226602003

of $J(\boldsymbol{m}, \boldsymbol{P}, \boldsymbol{b})$ and $I(\psi)$ we apply calculus of variantions, yielding Poisson equations with Robin boundary conditions for m_1 and m_2, and a Poisson equation with Neumann bounary condition for u. The Poisson problems are solved using standard second order finite differences. For a more details see [2].

For the minimization of $J_{\rm B}(\boldsymbol{m}, \boldsymbol{b})$, we segmentate the boundaries $\partial\mathcal{X}$ and $\partial\mathcal{Y}$, such that $\boldsymbol{m}$ maps a source segment to a target segment. We parametrize both the target segments, and the mapping of the source segments by their arclengths w.r.t. a reference point (in general a corner of $\partial\mathcal{Y}$ and $\boldsymbol{m}(\partial\mathcal{X})$, respectively). Each target segment is then discretized and approximated by piecewise linear functions of the arclength parametrization. Evaluation of the piecewise linear functions using the appropriately scaled arclengths of the mapping, yields $\boldsymbol{b}$.

4 Numerical results

To test the algorithm we consider the MA equation for the example

$$u(x, y) = \cos(x)y^2, \tag{5}$$

$$F(x, y, m_1, m_2) = \frac{m_2^2}{4} + m_1 \sin(x) - 3y^2, \tag{6}$$

on the domain $\mathcal{X} = [-1, 1] \times [1, 1.5]$ and the target domain (and numerical solution of the mapping) as depicted in Figure 1. Figure 2 shows the 1-norm of the errors in u, m_1, m_2 and the residual of the MA equation. Clearly, second order convergence in all relevant variables is obtained and the described numerical method is confirmed to work.

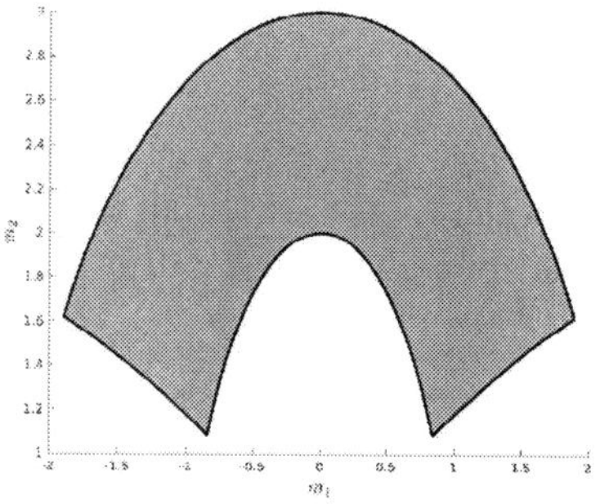

Figure 1: The numerical solution of the mapping $\boldsymbol{m}$ for a 321×321 grid and 10001 boundary discretization points.

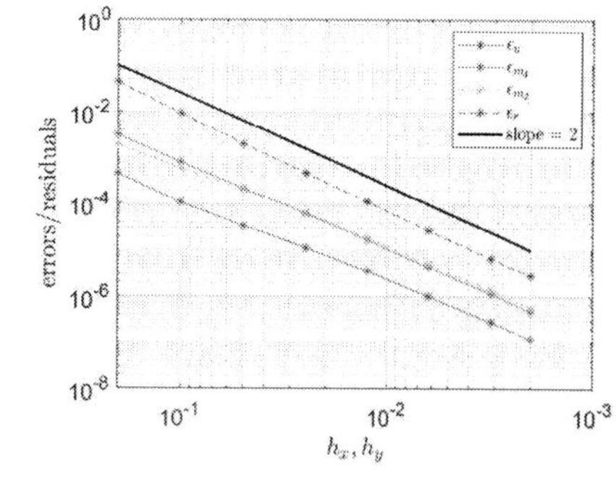

Figure 2: Convergence of the 1-norm of the errors in u, m_1, m_2 and the residual of the MA equation.

An optical example is shown in Figure 3, where the reflector is shown for $\mathcal{X} = \mathcal{Y} = [-1/2, 1/2]^2$ and a source and target intensity given by $f(x_1, x_2) = 1 + x_1$ and $g(y_1, y_2) = 1/2 + 3y_1^2 + 3y_2^2$, respectively. Figure 4 shows the estimated target intensity, obtained by quasi-Monte Carlo raytracing using 101 bins in each direction and 10^6 rays in total. The shown intensity is the approximation of $g(y_1, y_2)$ by binning the stereographic projections of $\hat{\mathbf{t}}$. The expected parabolic structure of g is recovered.

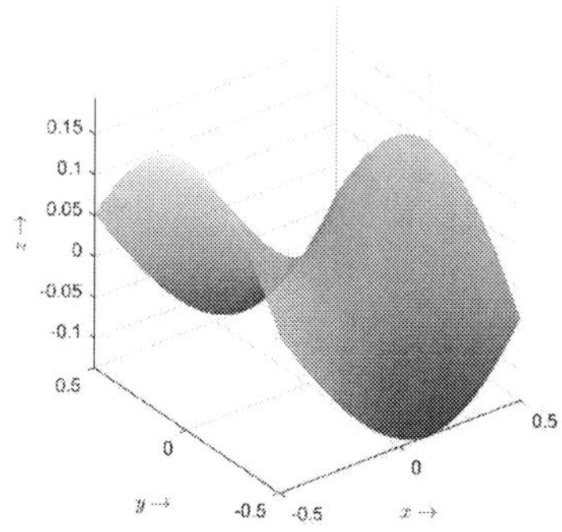

Figure 3: Numerical solution of an optical saddle surface for a 101×101 grid.

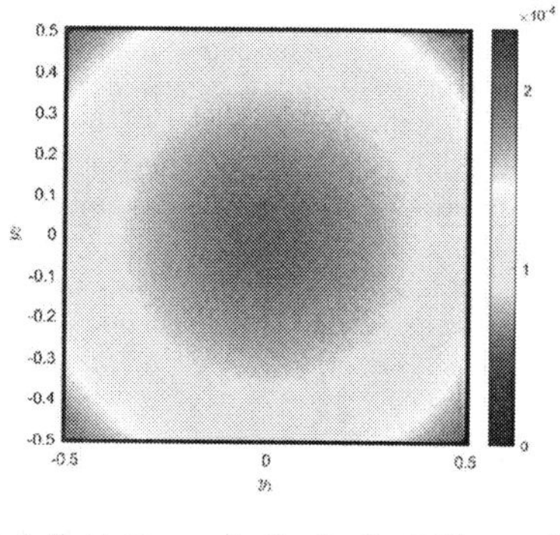

Figure 4: Raytrace results for the far-field target intensity $g(y_1, y_2)$ with the expected radial pattern.

References

[1] C.R. Prins, Ph.D. thesis, Eindhoven University of Technology (2014)

[2] C.R. Prins, R. Beltman, J.H.M. ten Thije Boonkkamp, W.L. IJzerman, T.W. Tukker, SIAM Journal on Scientific Computing **37**, B937 (2015)

https://doi.org/10.1051/epjconf/202226602004

Sparse Mid-Infrared Spectra Enable Real-time and In-vivo Applications in Tissue Discrimination

Felix Fischer[1,*], *Karsten* Frenner[1], *and Alois M.* Herkommer[1]

[1]Institute of Applied Optics, University of Stuttgart, 70569 Stuttgart, Germany

Abstract. Differentiation of malign and benign tissue based on spectral information can be done by only using a small fractional amount of the original mid-infrared spectrum. An optimally selected arrangement of a few narrow-band quantum cascade lasers provides proficient signal-to-noise ratios and can drastically reduce the data acquisition time with constant discriminability, such that real-time applications will be possible in short-term and in-vivo diagnostics in the long-term.

1 Introduction

Most information in analyses of biological structures in human tissue is found in the broad mid-infrared (MIR) region [1]. In order to achieve reasonable signal-to-noise ratios (SNRs) throughout this complete spectrum, either the light source has to be utterly powerful, the detector has to be more efficient or the spectrum has to be averaged over multiple measurements [2]. Whereas the former yields an increased energy consumption and naturally more waste heat that has to be dealt with, the latter entails a longer acquisition time. Both of which is not ideal for preferential real-time and intra-operative tissue discrimination with spectral measurement technologies.

Thus, modern MIR spectrometers tend to use quantum cascade lasers (QCLs), that are more powerful light sources with a higher spectral irradiance than conventionally used thermal SiC sources [3]. When confined to narrower bands, e.g. in gas analyses of particular media [4], a real-time analysis of the spectrum is viable. Since most discriminative information in tissue spectra is also found in these distinct peaks (i.e. biomarkers) [5], the logical conclusion is that a single targeted approach to the relevant areas yields a drastic decline of the spectral acquisition time.

With the help of several different data-driven feature selection (FS) methods, we show that only a small fraction of the original MIR spectrum is necessary to reliably separate benign from malign urothelia. According to that, we introduce an optical design that enables the spatially resolved real-time discrimination of bladder tissue.

2 Wavelength Selection for MIR Spectra

The selection of relevant wavenumbers $\tilde{\nu}$ is based on the data that has been measured with the FT-IR spectrometer Spectrum Two from Perkin Elmer and an attenuated total reflection (ATR) accessory with a diamond crystal. A spectral range from 4,000-450 cm^{-1} is obtained with this combination. The spectra obtained are from multiple patients that have been diagnosed with bladder cancer. Due to non-macroscopic tumors in the resected bladder, it has not been possible to acquire tumorous spectra yet. For the sake of discriminating different bladder wall constituents, we decided to measure the urothelium and the detrusor muscle (DM) of benign bladder samples, see Fig. 1a and b.

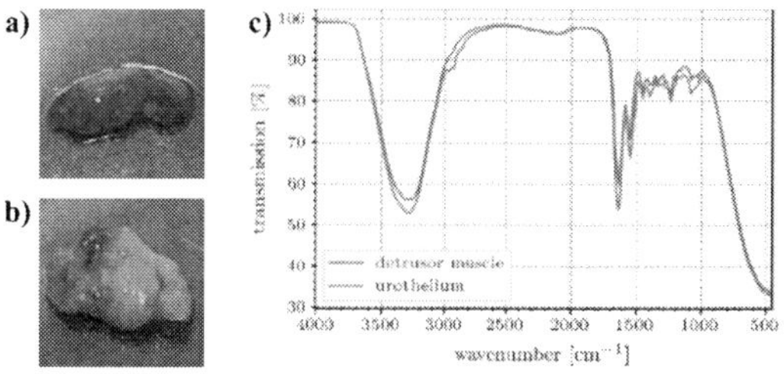

Fig. 1. Investigated samples of **a)** urothelium, **b)** detrusor muscle and **c)** mean spectra of the two investigated spectra. The correlation coefficient between both spectra is 99.75 %.

In order to select the most relevant wavenumbers for an optimal differentiation, we have chosen two different methods. The results of a principal component analysis (PCA) are compared to the results of a recently developed feature selection algorithm called FeaSel-Net [6], that is based on recursive elimination of the input signal in neural networks.

2.1. Feature Selection with PCA

Figure 2a shows the scores plot from PCA of the two different tissue types, where a clear separation between the two clusters is shown along the first component. Except from one sample of the urothelium class, all samples can be clearly detached using the y-axis as border. Even though the correlation between both mean spectra in Fig. 1c is nearly 100 %, it is still possible to

* Corresponding author: felix.fischer@ito.uni-stuttgart.de

transform the data into distinct clusters using PCA. The data has been standardized along the wavenumber axis during the pre-processing.

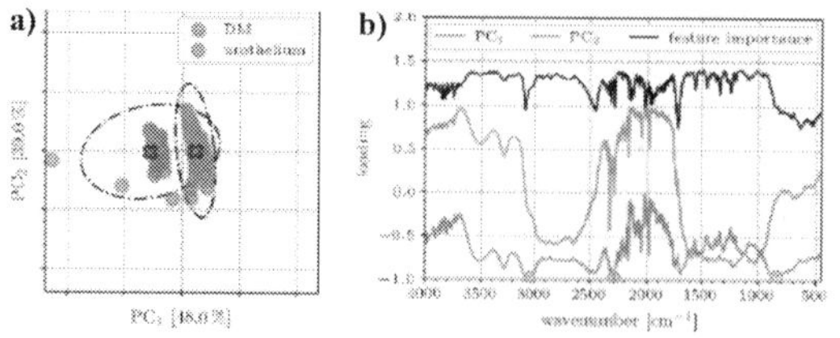

Fig. 2. Scores **a)** and loadings **b)** plot of a PCA showing the first two PCs that describe nearly 90 % of the data. A distinct separation of DM and urothelial tissue can be observed.

When looking at the loadings for the transformation in Fig. 2b, it can be seen that the PC1 and PC2 loading vectors are in the same order of magnitude. When calculating the sum of absolutes of the loadings, it seems that the wavenumber range between 1,600-900 cm^{-1} is slightly more important than the rest of the spectrum. This coincides with the difference between the mean spectra that occurs in Fig. 1c. Since PC1 is attributed with the discriminability though, only the loadings from PC1 are considered. The green circles in Fig. 2b indicate the three most important wavenumbers in this component. The exact wavenumber values are given in Table 1.

2.2. Feature Selection with Neural Networks

As a non-linear evaluation of the wavenumbers, we use FeaSel-Net [6]. During the training process of a classifier neural network, the feature importance

$$\mathcal{I}_f(\mathbf{X}_f) = -\frac{1}{\mathrm{n_s}} \cdot \sum_{i=1}^{n_s} \sum_{\mathbf{x} \in \mathbf{X}_f} \hat{\mathbf{y}}(\mathbf{x}) \cdot \log \mathbf{y}(\mathbf{x}) \qquad (1)$$

is evaluated, where the original dataset $\mathbf{X}$ with $n_f = 888$ and $n_s = 79$ samples is masked at each feature f. Subsequently, the impact on the classification performance is analysed and features are discarded, if the impact of masking is negligible. With a pruning rate of $\pi = 0.2$, a convergence with $n_f' = 3$ distilled features occurs after 25 epochs. Inherent randomness in neural networks make it indispensable to statistically evaluate multiple selections. The wavenumbers that are chosen most often are given in Table 1. We additionally trained another neural network with only the two stated wavenumbers and calculated the mean and standard deviation of the overall classification accuracy (OCA) for ten individual optimization processes. Both methods clearly surpassed random wavenumber selections.

Table 1. Relevant biomarkers and the OCA for both methods.

FS Method	$\tilde{\nu}_1$ [cm^{-1}]	$\tilde{\nu}_2$ [cm^{-1}]	OCA
PCA based	3,048	2,288	97.5 ± 2.0 %
FeaSel-Net	1,696	1,740	96.9 ± 2.2 %
Random	3,592	2,260	78.7 ± 5.9 %

3 Multi-spectral Spectrometer

With the few wavelengths selected in section 2 we propose a multi-spectral approach for the illumination of tissue samples inside an ATR measurement system. As depicted in Fig. 3, an echellette grating is used to guide all distributed feedback (DFB) QCLs onto the optical axis. The QCLs are chosen such that they correspond to our previous findings. With well-timed shutters, the sample is illuminated sequentially.

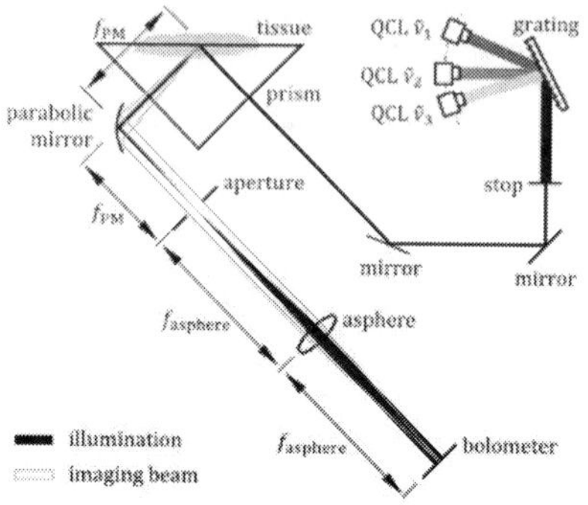

Fig. 3. Possible setup for a multi-spectral ATR spectrometer. The tissue is illuminated by multiple DFB-QCLs and the attenuated reflections are imaged using a microscope.

In this configuration, the lasers are imaged on the focal plane array (FPA) bolometer as a reference at first. As soon as there is close contact with the tissue, a part of the laser light is absorbed depending on the tissue's absorption coefficient at the interface. A magnification of the internal reflective element (IRE) surface is made possible using an $4f$ beam expander.

The work described in this paper was conducted in the framework of the Graduate School 2543/1 "Intraoperative Multisensory Tissue Differentiation in Oncology" (project A2) funded by the German Research Foundation (DFG – Deutsche Forschungsgemeinschaft).

References

1. V.F. Kalasinsky, Appl. Spectrosc. Rev., *Biomedical Applications of Infrared and Raman Microscopy*, **31**, 193 (1996)
2. G.W. Ewing, Rev. Sci. Instrum, *Signal-to-Noise Enhancement in Infrared Absorption Spectrophotometry*, **42**, 169 (1971)
3. M. J. Weida, B. Yee, Eds. Proc. SPIE, *Quantum cascade laser-based replacement for FTIR microscopy* (2011)
4. R. Ghorbani, F.M. Schmidt, Appl. Phys. B, *Real-time breath gas analysis of CO and CO2 using an EC-QCL*, **123**, 144 (2017)
5. D.I. Ellis, W.B. Dunn, J.L. Griffin, J.W. Allwood, R. Goodacre, Pharmacogenomics J., *Metabolic fingerprinting as a diagnostic tool*, **8**, 1243 (2007)
6. F. Fischer, A. Birk, K. Frenner, A.M. Herkommer, *FeaSel-Net: A Recursive Feature Selection Callback in Neural Networks* (2022)

A discontinuous Galerkin method to solve Liouville's equation of geometrical optics

Robert A.M. van Gestel[1,*], *Martijn J.H.* Anthonissen[1], *Jan H.M.* ten Thije Boonkkamp[1], and *Wilbert L.* IJzerman[1,2]

[1]Eindhoven University of Technology, PO Box 513, 5600 MB Eindhoven, the Netherlands
[2]Signify, High Tech Campus 7, 5656 AE, Eindhoven, the Netherlands

Abstract. We present an alternative method to ray tracing that is based on a phase space description of light propagation. Liouville's equation of geometrical optics describes the evolution of the basic luminance on phase space. At an optical interface, the laws of optics describe non-local boundary conditions for the basic luminance. A discontinuous Galerkin method is employed to solve Liouville's equation for a dielectric total internal reflection concentrator.

1 Mathematical model

In the design of optical systems for illumination purposes, ray tracing is typically used to provide insight into the effectiveness of the system. A large number of rays needs to be traced to get enough information. An alternative approach is based on a phase space description of light, where the evolution of a light ray can be described in terms of a Hamiltonian [1]. A light ray is described by its position $(\boldsymbol{q}, z) \in \mathbb{R}^{d+1}$ and momentum vector $(\boldsymbol{p}, p_z) \in \mathbb{R}^{d+1}$ $(d = 1, 2)$, where the momentum vector is simply the unit direction vector multiplied by the refractive index n. Assume that light is travelling in the positive z-direction, such that we can parametrise the light ray in terms of the z-coordinate. Furthermore, this allows us to write $p_z = \sqrt{n^2 - |\boldsymbol{p}|^2}$. The coordinates $\boldsymbol{q}$ and $\boldsymbol{p}$ describe a point in the $2d$-dimensional phase space, and they evolve according to Hamilton's equations [1].

In the absence of Fresnel reflections, scattering and other losses, the energy (per time) of a beam of light remains constant when propagating along the z-axis. This energy is the luminous flux Φ. An infinitesimal element of luminous flux $\mathrm{d}\Phi$ can be related to an energy density ρ on phase space by $\mathrm{d}\Phi = \rho\,\mathrm{d}\mathcal{U}$. Here, $\mathrm{d}\mathcal{U} = \mathrm{d}\boldsymbol{q}\,\mathrm{d}\boldsymbol{p}$ represents an infinitesimal volume element on phase space, which in optics is known as étendue and the energy density ρ is called basic luminance. Conservation of étendue and luminous flux leads to the basic luminance invariance of ρ which can be expressed as

$$\rho\left(z + \Delta z, \boldsymbol{q}(z + \Delta z), \boldsymbol{p}(z + \Delta z)\right) = \rho\left(z, \boldsymbol{q}(z), \boldsymbol{p}(z)\right). \quad (1)$$

The basic luminance invariance even holds when light is reflected or refracted [2, 3]. If we need to consider both forward and backward propagating light, then relation (1) needs to be augmented with the direction of the light, see [4].

From relation (1) one can derive Liouville's equation, which reads

$$\frac{\partial \rho}{\partial z} + \frac{\partial H}{\partial \boldsymbol{p}} \cdot \frac{\partial \rho}{\partial \boldsymbol{q}} - \frac{\partial H}{\partial \boldsymbol{q}} \cdot \frac{\partial \rho}{\partial \boldsymbol{p}} = 0, \quad (2)$$

where $H(z, \boldsymbol{q}, \boldsymbol{p}) = -\sqrt{n(z, \boldsymbol{q})^2 - |\boldsymbol{p}|^2}$ is the Hamiltonian. At an optical interface (that is: a discontinuity in the refractive index field n) the momentum of a light ray changes discontinuously. For forward propagating light and no losses, the basic luminance remains invariant so that at an optical interface the following relation holds

$$\rho\left(z^+, \boldsymbol{q}(z^+), \boldsymbol{p}(z^+)\right) = \rho\left(z^-, \boldsymbol{q}(z^-), \boldsymbol{p}(z^-)\right), \quad (3)$$

where the $\pm$ denote one-sided limits towards the optical interface that correspond to incident and outgoing light for $-$ and $+$, respectively. Moreover, $\boldsymbol{p}(z^+)$ is computed from $\boldsymbol{p}(z^-)$ with the vectorial laws of reflection and refraction. Relation (3) states that at an optical interface different parts of phase space are in contact with each other, i.e., at an optical interface we have non-local boundary conditions. The change in momentum depends on the physical surface unit normal of an optical interface. Consequently, solving Liouville's equation for curved optical interfaces, where the normal changes as a function of the position, is in particular very challenging.

2 Numerical solution to Liouville's equation

In [4] we describe a discontinuous Galerkin method to solve Liouville's equation numerically for two-dimensional optics $(d = 1)$. The method is energy-preserving and allows for control over the accuracy of the numerical solution by choosing discretisation parameters. The method is computationally efficient as it can converge

*e-mail: r.a.m.v.gestel@tue.nl

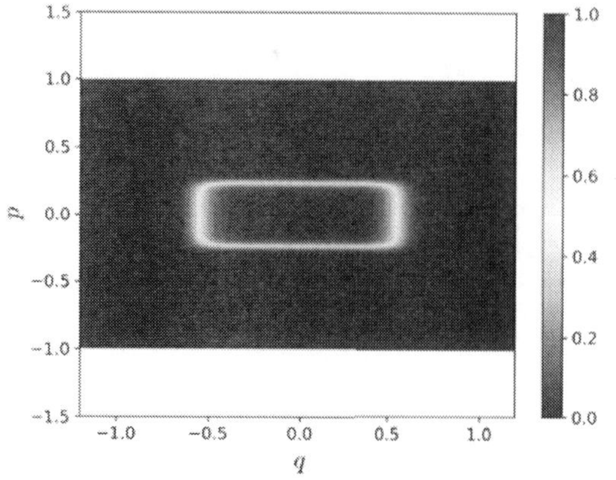

(a) $z = 0$

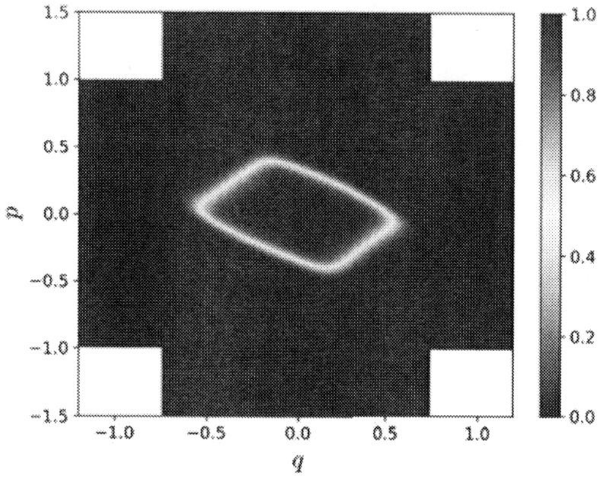

(b) $z = \frac{1}{2}Z_t$

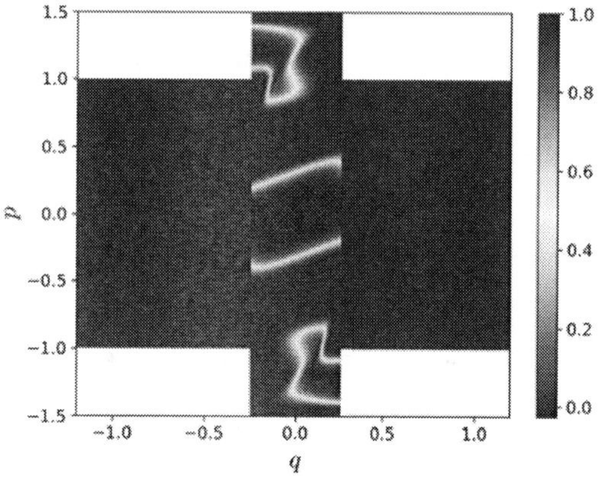

(c) $z = Z_t$

Figure 1: Distributions of ρ for the DTIRC

rapidly to exact solutions for model problems, see [4]. Furthermore, the illuminance and luminous intensity can be computed from ρ.

We apply the method to the dielectric total internal reflection concentrator (DTIRC) shown in Figure 2. For details about the design of such a device, we refer the reader to [2, 5]. The source is located at $z = 0$ and the target/exit of the device is at $z = Z_t = 2.648668$. In Figure 1 we show the numerical solutions, where at $z = \frac{1}{2}Z_t$ one can see that light has been refracted into the medium n_1 and at $z = Z_t$ some of the light has been reflected via total internal reflection yielding the top and bottom patches. The numerical solutions are of high accuracy, as indicated by the good resolution of high gradients present in the solution. Moreover, the numerical solution can capture the discontinuity across the optical interfaces perfectly. In the solution process we naturally compute ρ at certain z-values when stepping towards the target. Thus, in addition to the solution at the target we also know intermediate distributions, hence, providing a complete picture of light propagation through the optical system.

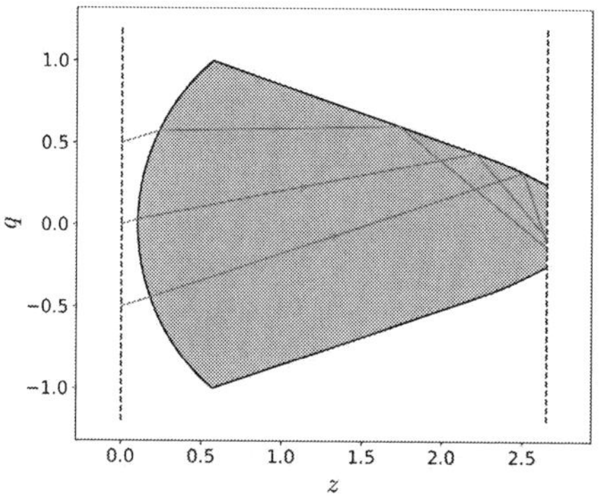

Figure 2: A DTIRC and a couple of light rays. The gray colour represents a refractive index $n_1 = 1.5$ and the white colour the background medium with $n_0 = 1$

References

[1] K.B. Wolf, *Geometric optics on phase space* (Springer Science & Business Media, 2004)

[2] J. Chaves, *Introduction to Nonimaging Optics* (CRC press, 2017)

[3] F.E. Nicodemus, American Journal of Physics **31**, 368 (1963)

[4] R.A.M. van Gestel, M.J.H. Anthonissen, J.H.M. ten Thije Boonkkamp, W.L. IJzerman (in preparation)

[5] X. Ning, R. Winston, J. O'Gallagher, Applied optics **26**, 300 (1987)

Irradiance tailoring with multiple sources using B-spline refinement

Alexander Heemels[1], *Aurèle* Adam[1], and *Paul* Urbach[1]

[1]TU Delft, Applied Physics, Optics Research Group, Lorentzweg 1, 2628 CJ, Delft, The Netherlands

Abstract. To increase the irradiance generated by an illumination system, multiple sub-systems, each generating their own irradiance distribution can be used. We propose a method using B-spline refinement to find the irradiance distribution that a single sub-system produces, so a desired irradiance distribution is obtained using multiple sub-systems.

1 Introduction

To increase the irradiance E_{des} generated by an illumination system consisting of a source and irradiance tailoring optic (freeform, diffuser) one can choose to add more sources to the illumination system increase the amount systems Fig.1. One way to do this, while keeping the desired distribution, is to find a combination of N_s irradiance distributions $E_k, k = 1, \dots N_s$, generated by N_s sub-system, which add up to a total irradiance distribution E_{tot} such that $E_{tot} \approx E_{des}$. Assuming only a single sub-system design which generates an intermediate irradiance distribution E_0, so that $E_k(x, y) = E_0(x + x_k, y + y_k)$. The total irradiance is then given by:

$$E_{tot}(x, y) = \sum_{k=1}^{N_s} E_k(x, y) = \sum_{k=1}^{N_s} E_0(x + x_k, y + y_k). \quad (1)$$

We define an illumination kernel,

$$K(x, y) = \sum_{k=1}^{N_s} \delta(x - x_k, y - y_k), \quad (2)$$

denoting the positions of the sub-systems at (x_k, y_k). Finding an appropriate intermediate irradiance distribution comes down to solving the following deconvolution problem:

$$E_{tot}(x, y) = E_0(x, y) \star K(x, y). \quad (3)$$

2 Method

Under the assumption that the sub-systems are uniformly, equidistantly positioned with separations Δx and Δy the illumination kernel becomes:

$$K(x, y) = \sum_{n=-N/2}^{N/2} \sum_{m=-M/2}^{M/2} \delta(x - m\Delta x, y - n\Delta y). \quad (4)$$

*e-mail: A.N.M.Heemels@tudelft.nl
**e-mail: A.J.L.Adam@tudelft.nl

We show that if a desired irradiance distribution can be approximated by a B-spline [1] with uniform knot vectors, such that:

$$E_{des}(x, y) \approx \sum_i w_i B_i^n(x, y), \quad (5)$$

where B_i^n is the ith basis function of order n and w_i its respective weight. It can be generated by any uniformly distributed set of $N \times M$ sub-systems Fig.2.

References

[1] D. F. Rogers, *An introduction to NURBS: with historical perspective* (Morgan Kaufmann Publishers, San Francisco, 2001) 179-197

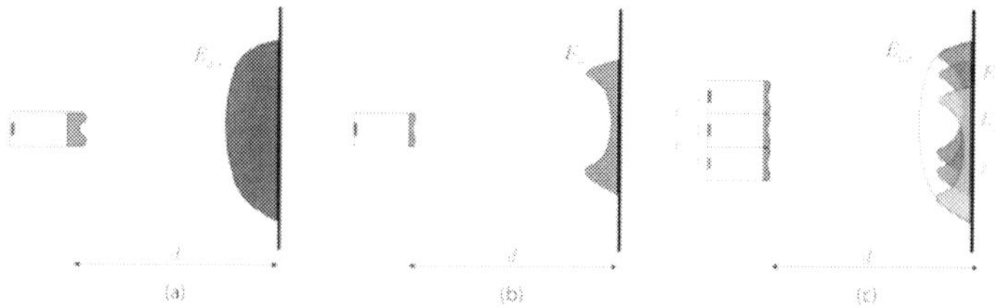

Figure 1. (a) A single illumination system projects a desired irradiance distribution E_{des} onto a plane a distance d away. (b) A single sub-system generates an intermediate irradiance distribution E_0 (c) The combination of multiple illumination systems all projecting a irradiance distribution E_k add up to a total irradiance distribution $E_{tot} \approx E_{des}$.

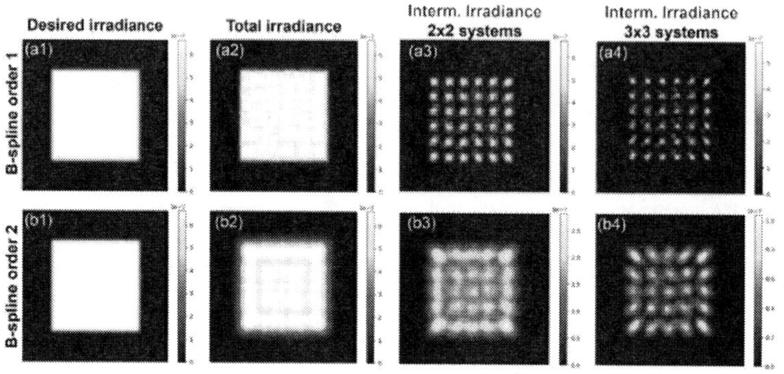

Figure 2. Results of the irradiance refinement: (a1) and (b1) show the desired irradiance distribution, (a2) and (b2) the obtained total irradiance distribution for B-spline of orders 1 and 2, (a3) and (b3) the resulting intermediate irradiance for 2x2 system, (a4) and (b4) intermediate irradiance for 3x3 systems

Trans-scleral Illumination - The Future of Retinal Imaging?

Joel Terry[1], *Daniel* Geddes[1], *Victor* Ochoa-Gutierrez[1], *Zhiyuan* Yang[2], *Kenneth* J. Smith[2], and *Andrew R.* Harvey[1,*]

[1]Imaging Concepts Group, School of Physics and Astronomy, University of Glasgow, Glasgow G128QQ , United Kingdom
[2]Department of Neuroinflammation, UCL Queen Square Institute of Neurology, University College London, London, UK

Abstract. We report how illumination of the retina through the sclera enables the recording of high-contrast reflectance and fluorescence images of the retina, free of the cornea and lens glare and autofluoresence that degrades images recorded using traditional illumination through the pupil.

1 Introduction

High-quality retinal imaging underpins all screening and early detection of retinal disease. The salient objectives are the detection of morphological and structural changes of vascular and neural structure or the chemical abnormalities that may precede them, such as abnormal blood oxygenation or accumulation of chemical waste products. [1–3]. For both reflectance and fluorescence imaging of the retina, both illumination (or excitation) of the retina and imaging of the retina is through the eye pupil. The so-called Gullstrand principle separates the illumination and imaging light paths in the retina to prevent the faint light from the retina from being swamped by the much brighter reflection or fluorescence from the cornea and ocular lens [4]. Typically annular illumination at the pupil is combined with imaging through the centre of the illumination annulus but this is effective only for imaging modest fields of view, and consistent alignment of the annulus with the pupil is difficult to maintain in a handheld device.

Somewhat counter-intuitively the retina may also be efficiently illuminated through the sclera (the white of the eye): in the region of the the pars planar, an annular region a few millimetres outside the iris, a thinning results in light transmission that may be as high as ~ 30% in lightly pigmented eyes, or significantly less for people with high skin pigmentation. It is likely that the impact of the associated stray light at the retina is mitigated by the waveguiding effect of the photoreceptors which tends to limit sensitivity only to light from the centre of the eye pupil.

Here we present retinal imaging systems using trans-scleral illumination for wide-field imaging reflectance and for fluorescence intensity and fluorescence-lifetime imaging.

2 Wide-Field Imaging

An increased field of view of the retina offers the enhanced capabilities for detecting disease but scleral reflections and

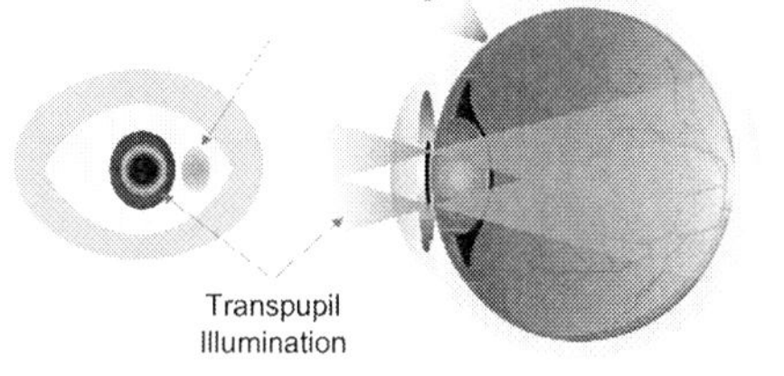

Figure 1: Comparison of annular illumination, where an annulus-shaped field of light is focused onto the cornea and then diverges to both illuminate the retina, and the new technique of trans-scleral illumination where a beam of light is focused onto a spot on the sclera which diffuses and uniformly illuminates the retinal surface.

ocular aberrations increase rapidly with increasing field angle. We have recently reported a proposal to employ multi-camera computational imaging to increase the field of view from a typical 50° to 140° - enabling 80 % of the retina to be imaged [5]. Our multi-camera architecture employs local aberration correction to offer wide-field near-diffraction-limited imaging, but the Gullstrand principle is ineffective for removal of illumination reflections for such wide fields of view. Reflections are however eradicated using trans-scleral illumination and the inherent simplicity also reduces both the optical system complexity and volume. As a step towards realising multi-camera retinal imaging we have demonstrated trans-scleral illumination with a wide-field (80°) single camera, which offers the advantage of a short image acquisition time of < 1 s and of being completely non-contact whilst still being handheld, as shown in Fig. 2 a-c.

3 Oximetry

Oximetry of retinal vascular blood exploits oximetric variations in contrast of narrowband spectral images [6, 7].

*e-mail: Andy.Harvey@glasgow.ac.uk

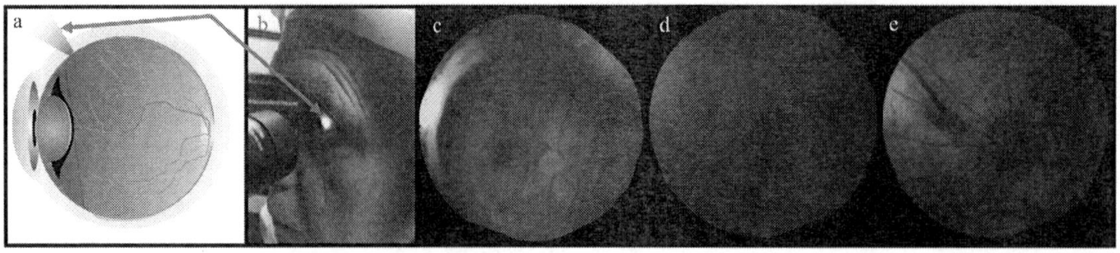

Figure 2: Handheld system utilising trans-slceral illumination. a) Trans-scleral illumination separates illumination and imaging paths b) scleral illumination pattern on a human using a white-light LED c) image human fundus d) red- and green-filtered images of a rat retina for vascular oximetry

Even low-levels of glare from the cornea and lens significantly reduces the contrast, however [8]. We have demonstrated two-wavelength *in vivo* retinal vascular oximetry in rats using scleral illumination, as highlighted in Fig. 2 d-e. This provides promise for anaesthetic-free longitudinal assessment of oximetry in animal models of disease - as well as demonstrating the feasibility of trans-scleral illumination for retinal oximetry in humans

4 Spectral intensity and spectral fluorescence lifetime-imaging

Fluorescence intensity imaging and fluorescence-lifetime imaging (FLIM) of the retina provide additional discrimination for quantification of retinal chemicals associated with metaboloic activity or with residues such as lipofuscin [9]. We record multi-spectral fluorescence images - using a 128 × 192 SPAD array with integrated Time-Correlated Single Photon Counting (TCSPC) module and a co-aligned sCMOS camera (to enable co-registration multiple TCSPC images for enhanced signal-to-noise ratio). We modify the algorithm reported in [10], where phasor analysis and principal component analysis are combined to extract end-member concentrations and lifetimes in photon starved-conditions. However, using trans-pupil illumination, fluorescence images are accompanied by relatively intense lens fluorescence, which degrades accuracy. We have demonstrated the feasibility of trans-scleral fluorescence retinal imaging devoid of contamination by lens fluorescence.

5 Outlook

The transmission of illumination through the sclera, rather than through the pupil, enables almost total eradication of the excitation and reflection of light from the lens and cornea, providing major enhancement of image contrast for quantitative, wide-field reflectance and fluorescence imaging of the retina.

References

[1] D.J. Mordant, I. Al-Abboud, G. Muyo, A. Gorman, A. Sallam, P. Ritchie, A.R. Harvey, A.I. McNaught, Eye **25**, 309 (2011)
[2] C. Dysli, S. Wolf, M.Y. Berezin, L. Sauer, M. Hammer, M.S. Zinkernagel, Progress in retinal and eye research **60**, 120 (2017)
[3] S.N. Patel, A. Shi, T.D. Wibbelsman, M.A. Klufas, Therapeutic advances in ophthalmology **12**, 2515841419899495 (2020)
[4] A.R. Harvey, G. Carles, A. Bradu, A. Podoleanu, in *Computational Retinal Image Analysis* (Elsevier, 2019), pp. 19–57
[5] J. Terry, G. Carles, A.R. Harvey, *Multi-scale Aberration Corrected Imaging of the Retina*, in *Optica Imaging and Applied Optics Congress 2022 (COSI)* (2022), p. ITh5D.2
[6] D.J. Mordant, I. Al-Abboud, G. Muyo, A. Gorman, A. Sallam, P. Rodmell, J. Crowe, S. Morgan, P. Ritchie, A.R. Harvey et al., Investigative Ophthalmology & Visual Science **52**, 2851 (2011)
[7] D.J. Mordant, I. Al-Abboud, G. Muyo, A. Gorman, A.R. Harvey, A.I. McNaught, Eye **28**, 1190 (2014)
[8] G. Carles, M. Preciado, P. Zammit, V. Ochoa-Gutierrez, J. Terry, J.M. Cooper, J. Reboud, A.R. Harvey, *Wide-field, illumination-agile, oximetric retinal imaging with a handheld camera*, in *Computational Optical Sensing and Imaging* (Optical Society of America, 2021), pp. CTh4E–6
[9] D. Geddes, J. Noorbakhsh, , Z. Yang, M. Preciado, M. Normand, K. Smith, , A.R. Harvey, *Quantification of Metabolic Function in the Retina Using Spectral Imaging and Phasor-FLIM*, in *Optica Imaging and Applied Optics Congress 2021 (COSI* (2022), p. JW4C.3
[10] L. Scipioni, A. Rossetta, G. Tedeschi, E. Gratton, Nature Methods **18**, 542 (2021)

Including Fresnel reflection losses in freeform lens design

A.H. van Roosmalen[1,*], *M.J.H.* Anthonissen[1], *W.L.* IJzerman[1,2], and *J.H.M.* ten Thije Boonkkamp[1]

[1]CASA, Department of Mathematics and Computer Science, Eindhoven University of Technology, PO Box 513, 5600 MB Eindhoven, The Netherlands
[2]Signify Research, High Tech Campus 7, 5656 AE Eindhoven, The Netherlands

Abstract. We present an inverse method for optical design that compensates local Fresnel reflections. We elaborate this method for a point source and far-field target. We modify an existing design algorithm based on the least-squares method. This is done in such a way that the shape of the transmitted intensity is as desired.

1 Introduction

With the introduction of LED lighting, a lot of research has been done on freeform optics. Specifically inverse methods are of interest due to their relatively short computation time. Often the geometrical optics approximation is used, ignoring some physical phenomena like Fresnel reflections. However, the arbitrary shapes of freeform lenses can lead to varying angles of incidence and thus varying Fresnel losses. We present an inverse method for designing a freeform lens for shaping a point source to a far-field distribution while compensating for Fresnel losses. For this, we adapt the least-squares solver developed by Romijn et al [1]. A more complete overview is given in [2].

2 Mathematical model

Our goal is to construct a lens with one freeform surface to shape light from a point source to a far-field target. Light is emitted by the point source with each ray defined by its unit direction vector $\hat{\boldsymbol{s}} = (s_1, s_2, s_3)^\mathsf{T}$. We assume that the first surface is spherical, centered around the point source. The second surface is freeform, defined by the radial distance to the point source $u(\hat{\boldsymbol{s}})$. After refraction, the direction of a ray is given by the unit vector $\hat{\boldsymbol{t}} = (t_1, t_2, t_3)^\mathsf{T}$; see Figure 1. At each optical surface, a part of the flux is

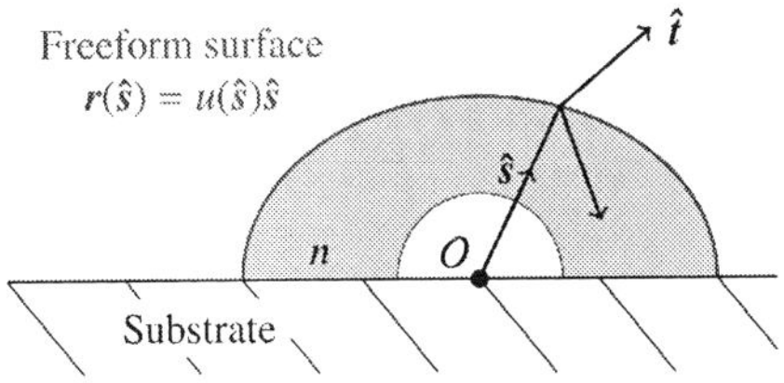

Figure 1: Intersection of a lens with the x,z-plane.

transmitted and a part is reflected, depending on the incident angle [3, Sec. 4.6]. We assume that the reflected light ends up on the substrate and is absorbed there. We have a given source distribution $f = f(\hat{\boldsymbol{s}})$ and a desired target distribution $g = g(\hat{\boldsymbol{t}})$, both with total flux equal to 1. Note that g is hypothetical, since we lose part of the flux from f. The source domain $\mathcal{S}$ is given as the support of f, and similarly $\mathcal{T}$ is the support of g. We parameterize $\hat{\boldsymbol{s}}$ and $\hat{\boldsymbol{t}}$ by the stereographic projections $\boldsymbol{x} = (x_1, x_2)^\mathsf{T}$ and $\boldsymbol{y} = (y_1, y_2)^\mathsf{T}$, respectively, and introduce $\mathcal{X} = \boldsymbol{x}(\mathcal{S})$ and $\mathcal{Y} = \boldsymbol{y}(\mathcal{T})$.

The goal is to compute the surface shape, given by u. For that, we first find a mapping $\boldsymbol{m}$, with $\boldsymbol{y} = \boldsymbol{m}(\boldsymbol{x})$. We can derive an equation of the form

$$\nabla_{\boldsymbol{x}} c(\boldsymbol{x}, \boldsymbol{y}) - \nabla u_1(\boldsymbol{x}) = \boldsymbol{0}; \qquad (1)$$

see [1]. The function u_1 is given by $u_1(\boldsymbol{x}) = \log(\hat{\boldsymbol{s}}(\boldsymbol{x}))$ and c is the so-called cost function, given by $c(\boldsymbol{x}, \boldsymbol{y}) = -\log(n - \hat{\boldsymbol{s}}(\boldsymbol{x}) \cdot \hat{\boldsymbol{t}}(\boldsymbol{y}))$. Substituting $\boldsymbol{y} = \boldsymbol{m}(\boldsymbol{x})$ and differentiating w.r.t. $\boldsymbol{x}$ gives

$$\boldsymbol{C}\mathrm{D}\boldsymbol{m} = \mathrm{D}^2 u_1 - \mathrm{D}_{\boldsymbol{xx}} c =: \boldsymbol{P}, \qquad (2)$$

where $\boldsymbol{C} = \mathrm{D}_{\boldsymbol{xy}} c$, the matrix containing mixed second order derivatives.

We now have a description of the propagation of transmitted rays, but not of the flux at the target. To take that into account, we consider conservation of energy in relation with Fresnel coefficients. These coefficients are functions of the incident and transmitted angles [3]. We can express these in terms of $\hat{\boldsymbol{s}}$ and $\hat{\boldsymbol{t}}$ [2]. We write T_1 and $T_2 = T_2(\hat{\boldsymbol{s}}, \hat{\boldsymbol{t}})$ for the transmittance at the first and second surface, respectively. Note that the transmittance at the spherical first surface is constant. Let g_t be the transmitted intensity, we then have for any $\mathcal{A} \subset \mathcal{S}$

$$\int_{\mathcal{A}} T_1 T_2(\hat{\boldsymbol{s}}, \hat{\boldsymbol{t}}) f(\hat{\boldsymbol{s}})\, \mathrm{d}S(\hat{\boldsymbol{s}}) = \int_{\hat{\boldsymbol{t}}(\mathcal{A})} g_t(\hat{\boldsymbol{t}})\, \mathrm{d}S(\hat{\boldsymbol{t}}). \qquad (3)$$

We can choose g_t to be a scaling of g, i.e., a distribution with the same shape as g, but a total flux equal to the transmitted fraction of the source flux. We write $g_t = \beta g$ for a $\beta \in (0, 1)$. Setting $\mathcal{A} = \mathcal{S}$ in Eq. (3) gives

$$\beta(\boldsymbol{m}) = T_1 \int_{\mathcal{S}} T_2(\hat{\boldsymbol{s}}, \hat{\boldsymbol{t}}(\boldsymbol{m})) f(\hat{\boldsymbol{s}})\, \mathrm{d}S(\hat{\boldsymbol{s}}). \qquad (4)$$

*e-mail: a.h.v.roosmalen@tue.nl

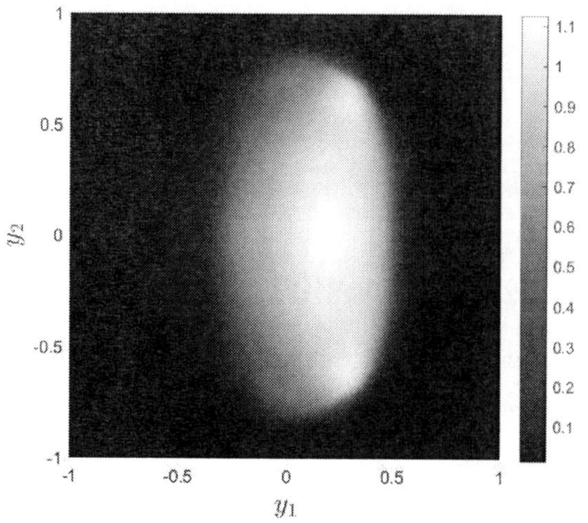

Figure 2: Target distribution in stereographic coordinates

Writing Eq. (3) in differential form in terms of stereographic coordinates gives us the generalized Monge-Ampère equation

$$\det(\mathrm{D}\boldsymbol{m}) = T_1 T_2(\boldsymbol{x}, \boldsymbol{m}(\boldsymbol{x})) \frac{J(\boldsymbol{x})}{J(\boldsymbol{m}(\boldsymbol{x}))} \frac{f(\boldsymbol{x})}{g_\mathrm{t}(\boldsymbol{m}(\boldsymbol{x}))} =: F(\boldsymbol{x}, \boldsymbol{m}(\boldsymbol{x})), \quad (5)$$

where J is the Jacobian resulting from the stereographic coordinate transformation. We slightly abuse notation by writing for example $f(\boldsymbol{x})$ instead of $f(\hat{\boldsymbol{s}}(\boldsymbol{x}))$. The boundary condition for this problem is given by $\boldsymbol{m}(\partial\mathcal{X}) = \partial\mathcal{Y}$.

Summarizing, we compute a mapping $\boldsymbol{m}$ from Eq. (2), where $\boldsymbol{P}$ satisfies $\det(\boldsymbol{P}) = F \det(\boldsymbol{C})$ due to Eq. (5). The mapping is subject to $\boldsymbol{m}(\partial\mathcal{X}) = \partial\mathcal{Y}$. Subsequently, we calculate u_1 from Eq (1) and from this the freeform surface.

We solve both problems iteratively in a least-squares sense. For more details on this method, see [2].

3 Numerical results

We use our method to design a lens for a typical street lamp. We consider a point source with a Gaussian light distribution on the positive half-sphere. The target is an intensity typically used by street lamps [1]; see Fig 2. With the algorithm, we obtain a so-called peanut lens. Calculating the reflectance coefficients on the lens (see Fig. 3) shows us quite large variations, with a loss of up to 20 percent in some places. This indicates that ignoring Fresnel losses in the algorithm would lead to a transmitted intensity significantly different from the hypothetical g. To stress this point, we perform a raytrace on the resulting lens from the algorithm with and without considering Fresnel losses. To compare the results, we scale the raytraced intensities to the same total flux as g. A comparison along the line $y_1 = 0$ is shown in Fig. 4. We see that the result of the algorithm with Fresnel is close to the desired intensity, while the result of the unadapted algorithm is clearly too low in some spots and too high in others. The areas with reduced flux correspond to the peaks in the reflectance as shown in Fig. 3.

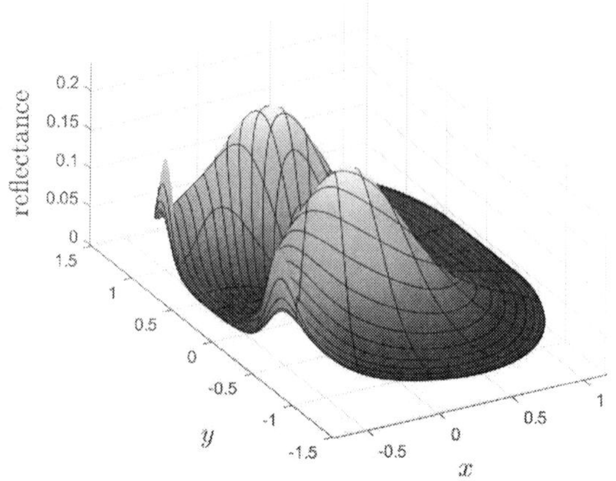

Figure 3: The reflectance on the freeform surface resulting from our algorithm as function of Cartesian x, y-coordinates.

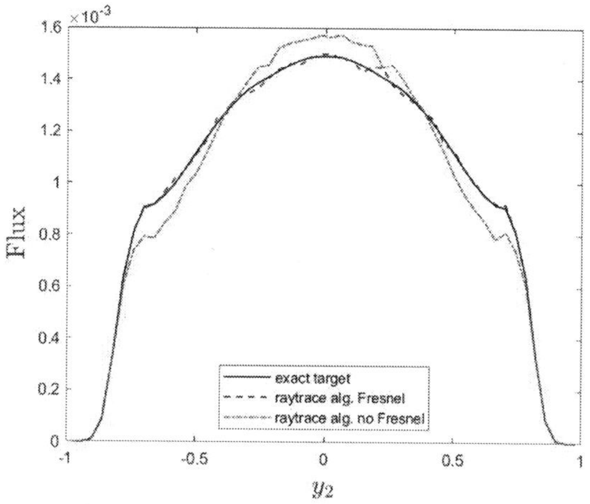

Figure 4: A comparison between the desired output intensity and the results of the algorithm with and without Fresnel reflection scaled to have the same total flux.

References

[1] L.B. Romijn, J.H.M. ten Thije Boonkkamp, W.L. IJzerman, J. Opt. Soc. Am. A **36**, 1926 (2019)

[2] A.H. van Roosmalen, J.H.M. ten Thije Boonkkamp, M.J.H. Anthonissen, W.L. IJzerman (2022), preprint, arXiv:2202.07984

[3] E. Hecht, *Optics* (Pearson, 2016), ISBN 9781292096933

https://doi.org/10.1051/epjconf/202226602009

Extended field-of-view light-sheet microscopy

*Tom Vettenburg**

School of Science and Engineering, University of Dundee, DD1 4HN, Dundee, UK

Abstract. Light-sheet fluorescence microscopy enables rapid 3D imaging of biological samples. Unlike confocal and two-photon microscopes, a light-sheet microscope illuminates the focal plane with an objective orthogonal to the detection axis and images it in a single snapshot. Its combination of high contrast and minimal sample exposure make it ideal to image thick samples with sub-cellular resolution. To uniformly illuminate a wide field-of-view without compromising axial resolution, propagation-invariant light-fields such as Bessel and Airy beams have been put forward. These beams do however irradiate the sample with a relatively broad transversal structure. The fluorescence excited by the side lobes of Bessel beams can be blocked physically during recording; though at the cost of increased sample exposure. In contrast, the Airy beam has a fine transversal structure that is both curved and asymmetric. Its fine structure captures all the high-frequency components that enable high axial resolution without the need to discard useful fluorescence. This advantage does not carry over naturally to two-photon excitation where the fine transversal structure is suppressed. We demonstrate a symmetric and planar Airy light-sheet that can be used with two-photon excitation and that does not rely on deconvolution.

1 Introduction

Light microscopy fulfils a central role in biological research. It can provide direct evidence of processes and cellular or sub-cellular structures, and it can do so non-invasively. Until recently such studies were limited to thin and transparent samples. Beyond a few layers of biological cells, most light is scattered and all contrast is lost due to out-of-focus scattering or fluorescence. Confocal fluorescence microscopy can improve contrast by singling out the light from the focal plane, though at the cost of high irradiation exposure and consequent photo-bleaching and damage. Larger specimen must be chemically fixed and sliced prior to imaging. Naturally, this is an obstacle to the study of the many dynamic processes.

The introduction of planar illumination to fluorescence microscopy has revolutionised how large, dynamic, specimen can be studied. This microscopy technique decouples the illumination and the detection optics so that, optical sections can be imaged rapidly with high resolution and contrast. The orthogonal illumination largely avoids out-of-focus fluorescence. However, while tight focussing of the light-sheet can improve the axial resolution beyond that of wide-field detection optics, diffraction does place a trade-off on our ability to focus the light-sheet throughout the complete field-of-view.

2 Leveraging propagation invariance

To overcome this trade-off, we employed the propagation-invariant Airy beam to form an Airy light-sheet (Fig. 1). Together with Bessel beams and plane waves, Airy beams have a transverse intensity cross section that does not change with propagation distance. In other words, the same image quality can be expected throughout the entire field-of-view. Unfortunately, the axial image-sharpness is severely affected by the large transverse extent of both Bessel and Airy beams. Their width can be tens of times greater than the waist of a Gaussian beam produced with the same numerical aperture, thereby negating the propagation invariance advantage. Moreover, the Airy beam has a parabolically-curved trajectory and thus warped images. Several techniques can be used to recover sub-cellular resolution over the much larger field-of-view. Confocal slit de-scanning or techniques from structured illumination can be used to single out the focal plane fluorescence [1-3]. As with confocal microscopy, this improves contrast in exchange for elevating the sample exposure. It also precludes the use of a simple cylindrical lens to produce the light-sheet [4]. The ability to digitally deconvolve with high contrast is directly related to the characteristic asymmetry of the Airy light-sheet [5-7].

* Corresponding author: t.vettenburg@dundee.ac.uk

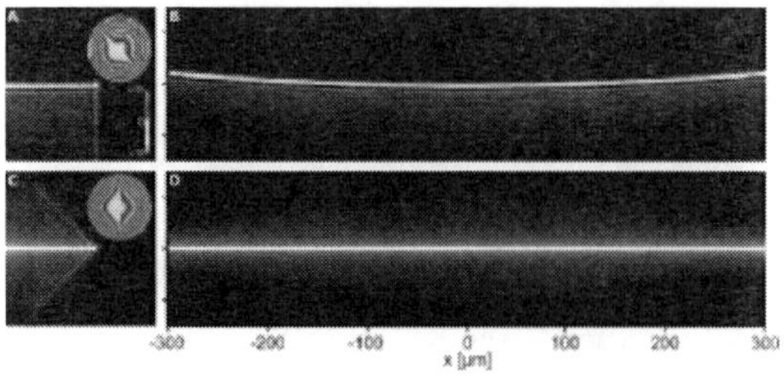

Fig. 1. Scanning an Airy beam to form a virtual light sheet as a curved surface (A-B), or planar illumination (C-D). Adapted from Hosny et al. [9] (CC BY 4.0).

3 Two-photon Airy light-sheet excitation

The impact of light-scattering on image quality grows with our ability to image larger volumes. Two-photon excitation with near-infrared fs-lasers can be used to illuminate deeper into tissue. This requires high peak-powers, hence the light-sheet should be created by scanning a focused beam. However, this non-linearity suppresses the outer lobes of the transversal structure. While this limits the effectiveness of the deconvolution [6,8], it also reduces its need to some extent. Yet, some digital correction is still required to correct for the curved trajectory of the Airy beam. Fig. 1C, shows how digital deconvolution can be completely avoided by rotating the curved Airy beam trajectory 45 degree into the plane, prior to scanning it into a planar virtual light-sheet. Naturally, this light-sheet has the same propagation-invariant profile as the Airy beam that formed it. This blurs the side-lobes into one-another, further reducing the need for deconvolution. We used the two-photon Airy beam light-sheet to image a 0.6mm-thick volume of Wistar rat brain (Fig 2). The high isotropic resolution across the entire field-of-view enabled us to image neuron structures spanning hundreds of microns while simultaneously resolving their microscopic synaptic spines (Fig 2D).

References

1. F. O. Fahrbach & A. Rohrbach *Nat. Commun.* **3**, 632 (2012). https://doi.org/10.1038/ncomms1646
2. T.A. Planchon et al. *Nature Methods* **8**, 417–423 (2011). https://doi.org/10.1038/nmeth.1586
3. B.-C. Chen et al. *Science* **346** (6208): 1257998 (2014). https://doi.org/10.1126/science.1257998
4. Z. Yang et al. *Biomed. Opt. Express* **5** (10), 3434-3442 (2014). https://doi.org/10.1364/BOE.5.003434
5. G. Muyo and A.R. Harvey *Opt. Lett.* **30**, 2715-2717 (2005). https://doi.org/10.1364/OL.30.002715
6. T. Vettenburg et al. *Nature Methods* **11** (5), 541-544 (2014). https://doi.org/10.1038/nmeth.2922
7. Y. Zhou et al. *Phys. Rev. Lett.* **124**, 198104 (2020). https://doi.org/10.1103/PhysRevLett.124.198104
8. P. Piksarv et al. *Scientific reports* **7** (1): 1-8 (2017). https://doi.org/10.1038/s41598-017-01543-4
9. N.A. Hosny et al. *Biomed. Opt. Express* **11** (7), 3927-3935 (2020). https://doi.org/10.1364/BOE.395547

Acknowledgements

T. Vettenburg is a UKRI Future Leaders Fellow supported by grant MR/S034900/1.

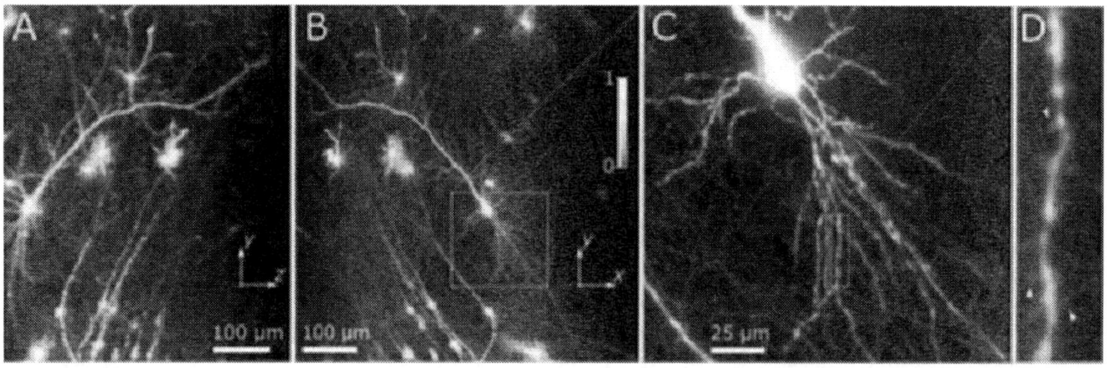

Fig. 2. Projections from a 3D volume of neuronal tissue, reproduced from Hosny et al. [9] (CC BY 4.0). Long-range connections between neurons can be seen in (A-B). A single neuron is highlighted in panel (B) and enlarged in panel (C). A single neuron dendrite is highlighted in panel (C) and enlarged in panel (D). Its dendritic spines are visible in panel (D).

Modelling dynamic 3D heat transfer in laser material processing based on physics informed neural networks

Jorrit Voigt and *Michael Moeckel*[*]

University of Applied Sciences Aschaffenburg, Laboratory for Hybrid Modelling, Wuerzburger Str. 45, 63743 Aschaffenburg, Germany

Abstract. Machine learning algorithms make predictions by fitting highly parameterized nonlinear functions to massive amounts of data. Yet those models are not necessarily consistent with physical laws and offer limited interpretability. Extending machine learning models by introducing scientific knowledge in the optimization problem is known as physics-based and data-driven modelling. A promising development are physics informed neural networks (PINN) which ensure consistency to both physical laws and measured data. The aim of this research is to model the time-dependent temperature profile in bulk materials following the passage of a moving laser focus by a PINN. The results from the PINN agree essentially with finite element simulations, proving the suitability of the approach. New perspectives for applications in laser material processing arise when PINNs are integrated in monitoring systems or used for model predictive control.

1 Introduction

Data-driven machine learning (ML), in particular deep learning, is increasingly applied to analyse measurement data and to model a large variety of phenomena in the physical sciences [1] and technical applications. While it allows for a direct inclusion of empirical data, automatic feature extraction, and promises efficiency gains in modelling complex multi-physics situations, predictions based on purely data-based ML approaches struggle to guarantee consistency with elementary physical laws of nature. Therefore, new paradigms for integrating deterministic, physics-based modelling and data-driven learning techniques are desirable. A plethora of different approaches have emerged, ranging from learning effective input parameters for (e.g., multiscale) deterministic models, posterior correction of simulation results by supplementary data-based models (e.g., for degradation mechanisms), substitutional neural networks trained on synthetic (simulated) data, neural networks with specific, physics adopted topologies e.g., for Lagrangian or Hamiltonian dynamics, and many more.

1.1 Physics informed neural networks

A physics informed neural network (PINN) is a neural network which imposes the validity of governing equations as additional constraints on the loss function. It is capable of seamlessly integrating multimodal experimental data and, hence, to extrapolate between deterministic and empirical modelling. A PINN takes space and time coordinates as neural network inputs and exploits automatic differentiation. As it contains the governing equations it does not require any training data; however, available data can be used to facilitate training. Compared with the finite element method (FEM) it is particularly powerful in solving inverse or ill-posed problems. PINNs have been successfully applied to e.g., fluid dynamics [2], heat transfer [3] or fatigue modelling [4]. This research prepares the way for using PINNS in laser material processing.

[*] Corresponding author: Michael.Moeckel@th-ab.de

1.2 Time dependent heat transport equation

Dynamic heat conduction in bulk materials can be modelled by the time dependent heat conduction equation, a partial differential equation in space ($\vec{x}$) and time (t):

$$c_p \frac{\partial}{\partial t} u(\vec{x},t) = \frac{\kappa}{\rho} \Delta u(\vec{x},t) + \dot{Q}(\vec{x},t). \tag{1}$$

Here c_p describes the heat capacity, κ is the thermal conductivity, ρ the density, $u(\vec{x},t)$ the temperature field and $\dot{Q}(\vec{x},t)$ the (moving) heat source. The temperature field is modelled within a finite cube. To its lower base a Dirichlet boundary condition (BC) was applied and set to 0. For the four lateral faces (F) the Neumann BC was

$$\frac{\partial}{\partial t} u(\vec{x},t) = 0.1 \tag{2}$$

The initial conditions (IC) were set to u = 0 for the whole domain. These conditions are motivated by the typical setup in the laser powder bed fusion process (PBF-LB/M).

2 Implementing PINN for dynamic heat transfer in laser material processing

PINNs have been implemented in a variety of frameworks [5,6]. Here, a dynamic 3D implementation has been realized in the deep learning library PyTorch.

2.1 Heat source modelling and model parameters

For the heat source a Gaussian beam profile with a laser power P_L=1 W and a velocity of 1 mm/s was used. The heat source is modelled by:

$$\dot{Q} = \frac{2 \cdot P_L}{\pi \cdot r_{spot}^3} \cdot \exp\left(-2 \cdot \frac{r_{focus}^2}{r_{spot}^2}\right). \tag{3}$$

The size of the laser spot (r_{spot}) was 0.1 mm and the position of the laser focus (r_{focus}) was a function of time.

2.2 Loss function

Constraints imposed by the time dependent heat equation, as well as by boundary and initial conditions have been included in the loss function with applied multipliers (ω):

$$L = \omega_{PDE} L_{PDE} + \omega_{BC} L_{BC} + \omega_{IC} L_{IC}. \quad (4)$$

The governing partial differential equation (PDE) holds at any point in space and time. Hence, its validity should be imposed at any point of a numerical discretization lattice. However, to relieve the numerical effort, errors resulting from the governing equation are not calculated at all points of the lattice but stochastically sampled on a sufficiently dense subset of all lattice points (20% in each epoch). For the calculation 40,000 collocation points were used and sampled by the Latin-Hypercube method. Further details on modelling the losses will be presented in an extended version of the paper.

2.3 Training and model validation

The PINN is trained for 50,000 epochs. The decay of the contributing losses over the epochs can be seen in ***Fig.*** 1.

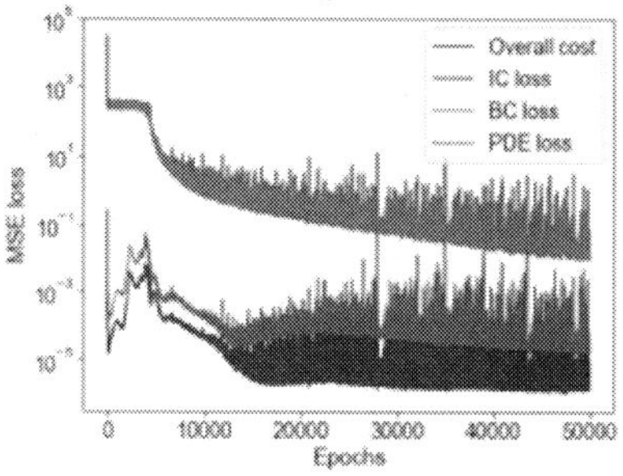

Fig. 1. Convergence of the 3 terms in the loss function. A decrease for the PDE loss can be seen at 5,000 epochs, followed by a smooth convergence of the loss.

The temperatures obtained from the PINN were compared with those from a corresponding FEM simulation, which has been performed in COMSOL Multiphysics environment, using the same parameters as for the PINN.

3 PINN-based modelling results

In *Fig.* 2 a cross-section along the laser path shows the obtained temperature profile $u(\vec{x}, t)$ from the PINN.

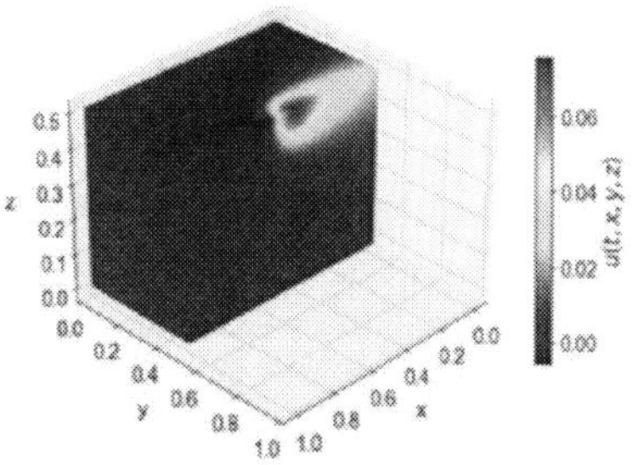

Fig. 2. 3D-temperature profile $u(\vec{x}, t)$ at a given time during and after the passage of a laser focus through model material.

To evaluate the calculations from the PINN, the results are compared to predictions from a FEM simulation, as shown in Fig. 3. The evaluation demonstrates agreement with FEM simulations. However, a deviation can be seen in front of the heat source, which is a result of lack of convergence and a known issue in physics-based modelling.

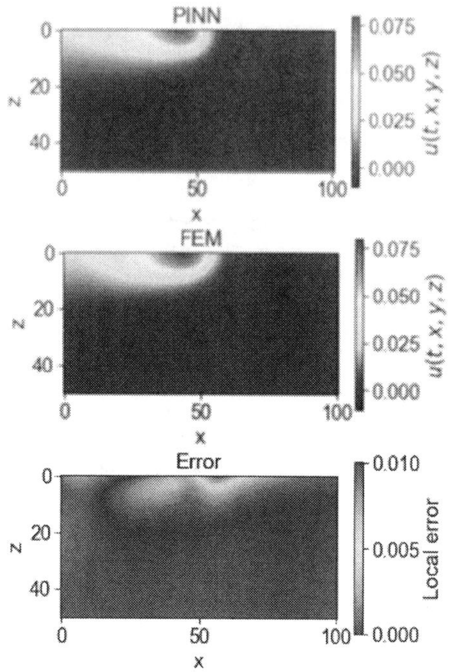

Fig. 3. Comparison of a spatial temperature profile at a given point in time calculated using (i) a PINN (upper panel), (ii) a FEM simulation (middle panel) and (iii) the error measured by the absolute difference between both results (lower panel)

4 Perspectives

The benefit of a PINN approach over finite element modelling becomes visible when additional sensor data is available to fit simulations to experiments or applications. Thus, PINNs can be considered a promising new tool for process monitoring and model predictive control in laser material processing, e.g. as part of a larger digital twin. Technical details on benchmarks and numerical implementations will be presented in an extended research article.

References

[1] G. Carleo, I. Cirac, K. Cranmer, L. Daudet, M. Schuld, N. Tishby, L. Vogt-Maranto et al., Rev. Mod. Phys. 91 (2019).
[2] S. Cai, Z. Mao, Z. Wang, M. Yin, G.E. Karniadakis, Acta Mech. Sin. 37 (2021) 1727–1738.
[3] S. Cai, Z. Wang, S. Wang, P. Perdikaris, G.E. Karniadakis, Journal of Heat Transfer 143 (2021).
[4] A. Dourado, F.A.C. Viana, PHM_CONF 11 (2019).
[5] M. Raissi, P. Perdikaris, G.E. Karniadakis, Journal of Computational Physics 378 (2019) 686–707.
[6] S. Kollmannsberger, D. D'Angella, M. Jokeit, L. Herrmann, Deep Learning in Computational Mechanics, Springer International Publishing, Cham, 2021.

Plasma jet assisted polishing of fused silica freeform optics

Thomas Arnold[1,2,*], *Georg* Boehm[2,3], *Heike* Mueller[2], *Martin* Ehrhardt[2], *Klaus* Zimmer[2]

[1]Technische Universität Dresden, Institut für Fertigungstechnik, 01062 Dresden, Germany
[2]Leibniz Institute of Surface Engineering (IOM), Permoserstr. 15, 04318 Leipzig, Germany
[3]Trionplas Technologies GmbH, Permoserstr. 15, Leipzig, Germany

Abstract. Atmospheric pressure plasma jet machining technology not only provides a flexible and efficient way to generate and correct optical freeform surfaces made of fused silica, it can also be applied as a surface smoothing or polishing technique. Thermal plasma jet treatment leads to softening and redistribution of the material. An accurate temperature regime during the process is inevitable to achieve a uniformly smoothed surface. The possibilities for in-process temperature control are demonstrated. Surface roughness values can be significantly reduced by a factor of 1000 depending on the initial roughness of the ground surface.

1 Introduction

The steadily increasing demand for precision freeform optics e.g. for laser beam shaping or imaging applications call for innovative optical fabrication technologies that allow for highly flexible processes combined with high efficiency. Since freeform surfaces exhibit varying slopes, locally acting sub-aperture tools and computer-controlled machines are employed in the machining processes.
A range of deterministic surfacing methods exist, such as freeform grinding, bonnet polishing, magnetorheological finishing, ion beam finishing, or plasma jet machining (PJM). PJM has been developed during the recent two decades as an alternative process for surface figure generation, figure error correction, as well as for surface smoothing [1,2]. It has been shown that with PJM freeform optics made of fused silica can be produced employing a series of processing steps that rely on the plasma jet technique.
Plasma jet machining technology utilizes a localized plasma dry etching process for precise material removal.
A microwave-driven plasma nozzle generates an Ar/He plasma jet at atmospheric pressure where reactive gases like CF_4 and O_2 are admixed in order to create reactive species such as free fluorine and oxygen atoms by means of dissociation. The fluorine atoms react with the substrate material to form volatile compounds, and material removal from the surface is achieved. For materials like fused silica, silicon, ULE®, or SiC the etch chemistry leads to the formation of gaseous reaction products in the form of SiF_4 and CO_2. When interacting with the substrate surface the plasma jet tool exerts nearly no mechanical forces. Thus, no sub-surface damage is induced. The tool influence function exhibits a rotationally symmetric near-Gaussian shape. Depending on the type of plasma source, the applied microwave power and gas flow rates a lateral removal profile of 0.3 up to 4 mm (full width at half maximum) can be adjusted.

In order to perform a machining process, the plasma nozzle is moved relative to the surface by a CNC motion system. Shaping or surface figure error correction is executed by applying dwell time methods that are usually employed in deterministic surface machining.
Since the dry chemical etching process on fused silica removes the polishing layer, existing residual sub-surface damage will be exposed. Depending on the quality of pre-machining steps like grinding or polishing and the amount of material removal during PJM the resulting surface roughness increases. In order to obtain optical quality with Sq < 1 nm a post-polishing step is often required. In the case of freeform surfaces with relatively small workpiece dimensions of 5 mm up to 50 mm diameter exhibiting steep surface gradients, bonnet polishing may lead to surface form deterioration due to varying tool influence functions. Hence, an alternative smoothing step is required. The application of plasma jets provides the possibility to smooth fused silica surfaces via a thermal interaction similar to laser polishing. In the paper we discuss several aspects of the plasma jet assisted smoothing process, such as process stabilization and investigations to identify optimum process parameters to obtain surface roughness Sq values of less than 1 nm.

2 Plasma jet polishing

A microwave-driven plasma jet source is employed which is fed by inert gases helium and argon with a total gas flow rate of approx. 300 sccm. The effective microwave power is adjusted in the range of 120 W – 170 W.
The smoothing of the optical surface is a thermal process where the substrate is locally heated by the plasma jet. When the jet contacts the substrate, convective local heating of the substrate by the plasma gas near the softening point $T_s(\eta = 10^{7,65}$ dPas$) = 1585°C$ results in softening of the glass material and a viscous flow on the

* Corresponding author: thomas.arnold@iom-leipzig.de

surface. The resulting minimized surface tension leads to a reduction in roughness.

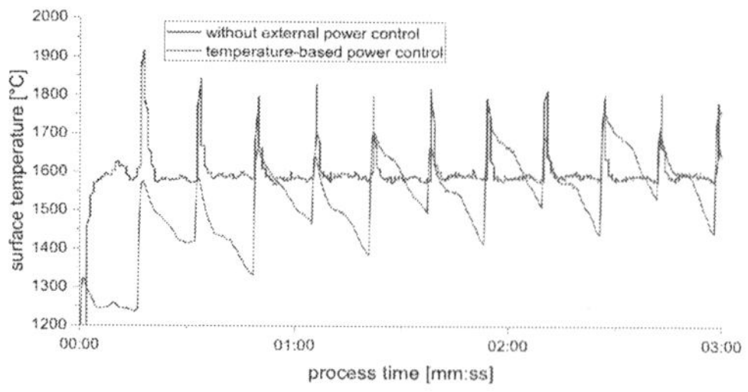

Fig. 1. Plasma jet induced surface temperature with and without PID control system. Temperature peaks occur due to end of scan line deceleration of the motion system.

The resulting maximum temperature distribution while the plasma jet moves in a meander pattern over the surface depends on the applied heat flux as well as on the scan velocity, the thermal properties of the glass, and the thermal coupling of the substrate to the environment. If the microwave power is set constant, a non-uniform temperature profile is observed (Fig.1: black curve). Since a constant temperature is required for uniform processing, a PID-based control system was developed that adjusts the microwave power to reach and hold a preset temperature measured by pyrometer in the center of the plasma jet. The microwave is applied in form of pulses of a frequency of 60 kHz with peak power of approx. 180 W. The control system uses pulse width modulation to adjust the power necessary to reach a maximum temperature of $T_{surf} = 1600°C$ (see Fig. 1). In order to determine the local smoothing capability of the jet, discrete rectangular structures have been produced by ultra-short pulse laser micromachining that were subsequently smoothed by the plasma jet. A fitting algorithm that assumes a Gaussian filter function has been applied to the areal measurement data of the initial and smoothed structures to obtain the 2D-filter size. Fig. 2 shows the original, the plasma smoothed and the filtered original structure. The filter function exhibits a cut-off wavelength of approximately 0.190 mm.

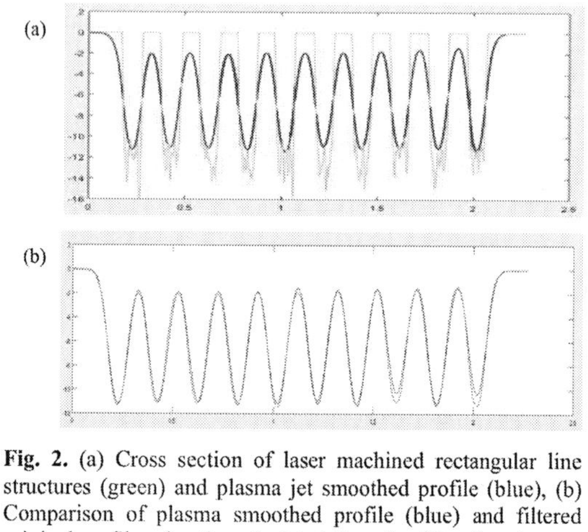

Fig. 2. (a) Cross section of laser machined rectangular line structures (green) and plasma jet smoothed profile (blue), (b) Comparison of plasma smoothed profile (blue) and filtered original profile using the calculated Gaussian filter function.

Areal smoothing of a ground surface with initial roughness of Sq = 330 nm in the spatial wavelength range of 0.8 µm – 40 µm) and waviness of Sq = 230 nm in the spatial range of 8µm – 200 µm was performed with a scan velocity of 2 mm/s and line feed of 0.3 mm. WLI measurement yield a roughness of Sq = 0.3 nm and a waviness of Sq = 9 nm. The PSD functions show that low wavelengths are smoothed more efficiently. Depending on the applied temperature a stronger smoothing effect is observed.

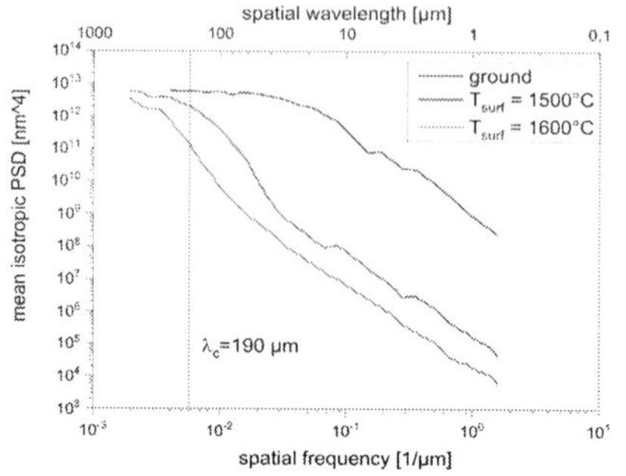

Fig. 3. Isotropic power spectral density (PSD) function of ground and plasma polished surface.

Plasma jet smoothing was applied as final step in a plasma based freeform manufacturing process. The target shape exhibiting a PV of 1300 µm on a diameter of 25 mm is shown in Fig.4 (a). In Fig. 4(b) the process convergence (residual error RMS) is plotted. Starting from a planar surface, shape generation was performed by PJM (step 1). After bonnet polishing (step 2) the form deteriorated. Subsequent PJM error correction steps (step 3,5) with intermediate soft polishing (step 4) could decrease the residual figure error. Finally, plasma polishing (step 5) lead to surface roughness of Sq 0.4 nm, while the figure error was preserved below the specified 1 µm RMS.

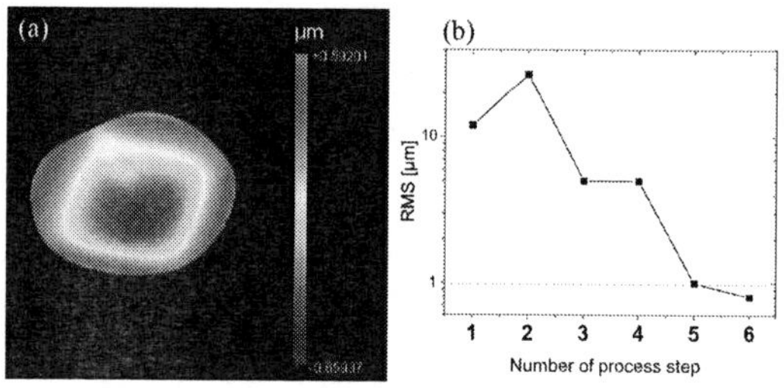

Fig. 4. (a) Freeform target design, (b) process convergence in terms of residual error RMS.

The authors gratefully acknowledge financial support by the German Federal Ministry of Education and Research in the funding program VIP+ (funding reference: 03VP08631, project “ProFreiform”).

References

1. T. Arnold, G. Böhm, H. Paetzelt: Plasma jet polishing of rough fused silica surfaces, Conference proceedings of the 13th International Conference of the EUSPEN V2 19-22 (2013).
2. T. Arnold, G. Boehm, H. Paetzelt, J. Eur. Opt. Soc.-Rapid 11 (2016) 16002

A brief application of material parameters to predict polishing rates for optical glasses

Michael F. Benisch[1,*], *Christian* Trum[1]*O*, *Werner* Bogner[1], and *Oliver* Fähnle[2]

[1]Deggendorf Institute of Technology, Dieter-Görlitz-Platz 1, 94469 Deggendorf, Germany
[2]Eastern Switzerland University of Applied Sciences, Werdenbergstrasse 4, 9471 Buchs, Switzerland

Abstract. This paper investigates the effects of material parameters of optical glasses on the polishing rate of these glasses. For this purpose, the material removal of various glasses was determined in laboratory tests under identical polishing conditions with respect to polishing pad and polishing suspension. The material removal was then evaluated for its dependence on material parameters. The goal of this paper is to derive a rule of thumb which allows an estimation of the material removal and the obtainable surface quality based on certain parameters of the workpiece material under comparable conditions. This rule of thumb can provide an initial insight into the polishability of a material and can be used to estimate polishing times and achievable surface qualities.

1 Introduction

The polishing of optical components is a complicated process step with more than 20 parameters. The single parameters are not independent from each other, which means that they partially interact with other parameters[1]. They depend on the glass to be polished, the polishing slurry, the used polishing pad, and the used machine parameters. Some parameters can be freely adjusted (within certain reasonable process- and machine limits), but many parameters are also defined by the used workpiece and it's mechanical and chemical parameters. For production planning, this represents a major challenge, for example in estimating process times and ensuring quality (roughness and defect sizes).

Existing models are often only applicable under laboratory conditions. The conditions between laboratory and optical shop floor often differ. For example, in the laboratory it is possible to work with small amounts of polishing suspension – in the industrial environment a central polishing suspension tank is usual. This influences the condition of the polishing suspension. For example, the polishing suspension can change during practical operation due to material that has been removed. In the laboratory, this can be solved by using smaller quantities and replacing the polishing suspension; in practice, replacing the polishing suspension too frequently would cause considerable costs.

This makes the use of empirical values in an industrial environment attractive. The disadvantage of this approach is a low flexibility in the case of for example, new types of glass or changed process conditions.

In this publication, we want to demonstrate the effects of certain material parameters of selected glass materials on the material removal rate (MRR) under otherwise identical polishing conditions. These constant conditions lead to a comparability of the polished components and the material removal. This provides the basis for a successful evaluation of the polishing tests.

2 Material and Methods

The experiments were carried out in the laboratories of the Deggendorf Institute of Technology using a synchrospeed polishing machine (Stock RSP 40). The polishing pad was a standard type GR-35 polyurethane material. A Hastilite-PO polishing slurry was used with a densitiy 1.025 g/cm^3. The polishing pressure p was 0.3 *Bar*, the velocity of spindle and pinole was 600 min^{-1}. Assuming the Preston equation[2]

$$dz/dt = K_P \, v_{rel} \, p,$$

it can be assumed that the MRR dz/dt is essentially determined by the Preston constant K_P, since pressure p and relative velocity v_{rel} are identical for all components. The Preston equation, in turn, is a measure of certain process-specific influences, including the influences of the glass material as well as influences from polishing slurry, polishing pad and environmental effects. By using otherwise identical components of the polishing process, influences beyond the glass materials could be minimized. The experiments of the several glass types where repeated several times to ensure a statistically underpinned result.

* Corresponding author: michael.benisch@th-deg.de

Relevant parameters for glass materials are (among others):

- Glass transition temperature Tg,
- Coefficient of thermal expansion α,
- Resistance against water, acid and alkaline attacks,
- Heat capacity c_P

The used material parameters are standard parameters which can easily be found in the glass manufacturer's catalogue. Therefore, no complicated and expensive measuring devices are necessary for a successful use of the presented rule of thumb. Such an approach can also be useful in the optical shop floor for a rapid and successful process planning. Complicated integration of the properties of polishing agent, polishing substrate and machine parameters is (deliberately) omitted in this very basic model. Various glasses with quite different properties were examined. All glass samples have the same diameter of 30 *mm*, which allows the use of identical parameters for all samples and a simple calculation of the amount of removed glass material.

3 Results

The material removal was investigated by weighing the glass components before and after the polishing step. A cleaning step (ultrasonic and acetone C_3H_6O) was used to remove any remaining impurities, such as deposits of polishing agents or any fingerprints. The weight difference can be - under knowledge of the material's density and geometry - used to calculate the removed material and the MRR. Fig. 1 shows the material removal in mm^3.

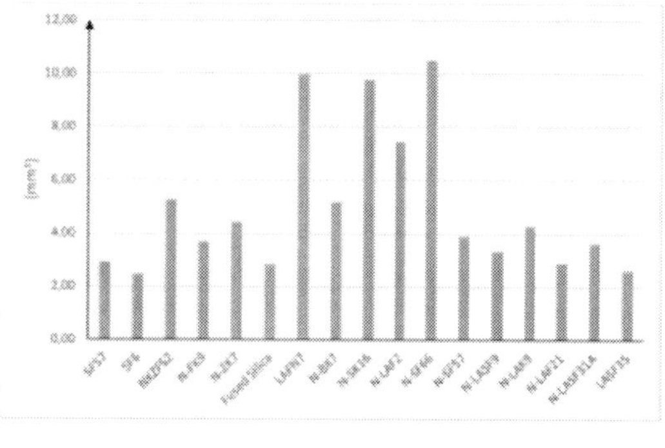

Fig. 1: Removed glass material in mm3. Variations between the removed volume and the removed mass are caused by the different density of the polished glasses. The error bars are not shown here.

The used glass samples have different chemical and mechanical properties. One opportunity is to separate the glasses in types with a low and with a high resistance against acidic attacks. Even tough glass is considered being resistant against acidic attacks (except for HF) acids can change glass surfaces. Over a longer period, this may well lead to a material failure. The acid resistance can be found in datasheets as the SR (1-53,3) value and is used according to standard ISO 8424 (describes the resistance of glass against aqueous acidic solutions at 24 *°C³)*. We have separated the values at a SR of 4. The SR values are not given here in detail but can be found as a standard material parameter in the datasheet.

4 Discussion and Conclusion

The goal of the research activity is the coupling of material removal to material properties. The benefit of this approach is, that a complicated process can be designed more deterministic. Furthermore, it is possible to obtain a faster optimization of the process by using a rule of thumb to find a good starting point for further optimizations.

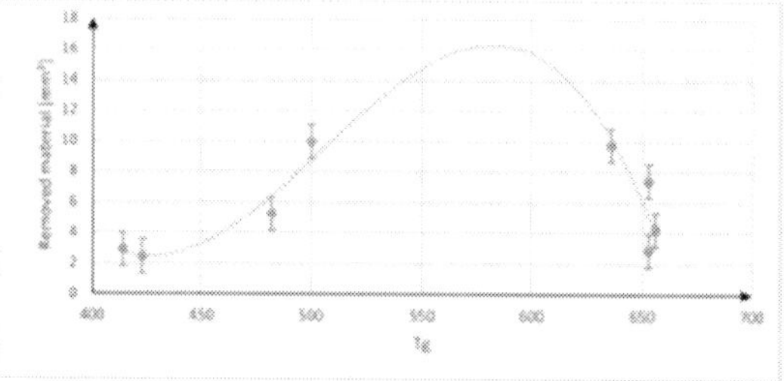

Fig. 2: Removed glass material in mm3 for glasses with low resistance against acid attacks as a function of the glass transition temperature Tg. The function shows a possible, theoretical course for a relationsip between Tg and material removal and has to be verified in further investigations.

Fig. 2 shows a dependency between the individual material parameters of the workpiece and the achievable material removal for glasses with a low resistance against acidic attacks. This relationship between material parameters is the basic for a rule of thumb to forecast material removal and surface quality of unknown or exotic glass types based on datasheet information. The goal of our research work is an industrial rule of thumb that allows an approximate estimation of the achievable material removal rate and the quality to be achieved in high-end polishing based on the glass data provided by the manufacturers. The presented research work is the basis for further research activities in the complicated environment of polishing high quality optical components.

References

1. Dunken H. H.; Bräuer P. and Fink P., *Physikalische Chemie der Glasoberfläche* (VEB Deutscher Verlag für Grundstoffindustrie, Leipzig, 1981)
2. Preston F. W., *The Theory and Design of Plate Glass Polishing Machines* (Journal of the Society of Glass Technology 11,1927)
3. ISO 8424 (International Organization for Standardization, 1996)

Ultra-short pulse laser-based fabrication process for lightweight structures in quartz glass applied for mirrors

David Bischof[1]*, *Michael* Kahl[1], and *Markus* Michler[1]

[1]OST Eastern Switzerland University of Applied Sciences, 9471 Buchs, Switzerland

Abstract. In the following work a manufacturing process for light weight structures in Fused Silica is presented. Such structures can potentially be used for mirrors to decrease the mass by simultaneously ensuring high stiffness. This talk should give the audience the possibility to assess the selective laser etching technology for mechanical structures in the field of optical mirrors.

1 Introduction

Classical manufacturing of glass-based lightweight structures is based on grinding or molding, for example, for large mirrors. In this work we will present a laser-based manufacturing process for lightweight structures (section 1.2). The advantage of the selective laser-induced etching process (SLE) used is that any 3D shape can be inscribed. The disadvantage on the other hand is that the manufacturing time increases linearly with the size of the device. Nevertheless, there are many interesting features that can only be realized with this concept.

The SLE fabrication process is based on two main steps. At first, the glass morphology is modified by the laser treatment and in a second step the glass substrate is shaped by wet chemical etching (Fig. 1).

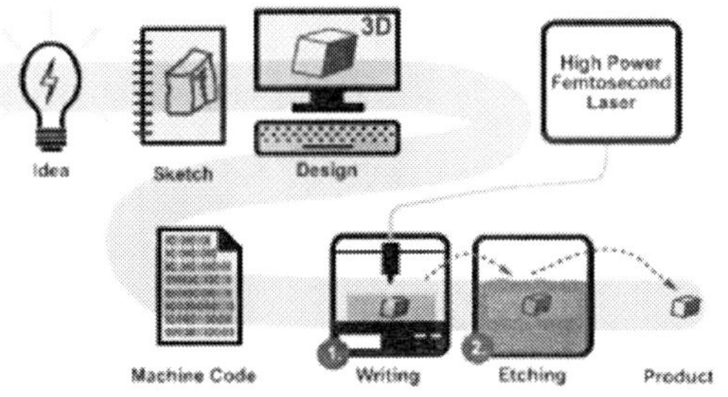

Fig. 1. Illustration of selective laser etching process

For the modification of the synthetic quartz glass a laser with a pulse repetition rate of 750 kHz, a pulse duration of 1 ps, a wavelength of 1030 nm, circular polarization, a pulse energy of about 350 nJ and a focusing objective (20x, NA0.4, -1.1 mm cover glass correction) is used [1]. The focus has a voxel size of about 3 µm, this focus has to be routed with a controlled multi-axis system to form a 3D object. Consequently, the 3D CAD design must be translated into lines and arcs. These curves are finally written into the glass and etched with hot potassium hydroxide (8mol/l, 80°C).

The advantage of this technology is, on the one hand, the design freedom to realize undercuts and freely movable parts, and on the other hand, subtractive manufacturing means that the material quality is defined by the glass and not by the machining process. As a result, of the etching there is an RMS roughness of about 1 to 2 µm observable, but there are no sharp cracks. Compression tests on small etched cylindrical pillars with a size of Ø17 µm and height of 28 µm showed that these structures can withstand a compressive stress up to 8 GPa [2].

The aim of lightweight structures is the reduction of the mass by simultaneously ensuring high stiffness. The stiffness against deflection of a part is inversely proportional to the moment of inertia of the part cross-section. The moment of inertia is defined by the equation $I_y = \int_A z^2 dA$. As a result, the material further away from the center contributes more to the stiffness of the part. For this reason, the material in the center is removed in lightweight structures. [3]

2 SLE applied to generate light weight structures

In this work, SLE technology was applied to create lightweight structures in a fused silica mirror. Two designs with additional features were realized to demonstrate on the one hand the possibility to realize rotating elements (Fig. 2) and on the other hand a linear translation system with flexures (Fig. 4).

The mass of the moveable part of both designs is decreased by a lightweight structure on the backside of the mirror. This structure consists of honeycombs and pillars connected to the front of the sample (Fig. 3). The web width of the honeycomb in Fig. 5 is 250 µm and the structure depth is 5 mm. This corresponds to an aspect ratio of 20.

Fig. 2. Lightweight structure combined with a rotation axis

The rotary axis of the design shown in Fig. 2 and 3 is realized with a plain bearing. Its disadvantage is the clearance fit, which is always greater than about 20 µm due to the etching process. Nevertheless, it can be shown that simple rotary bearings can be integrated with this technology platform.

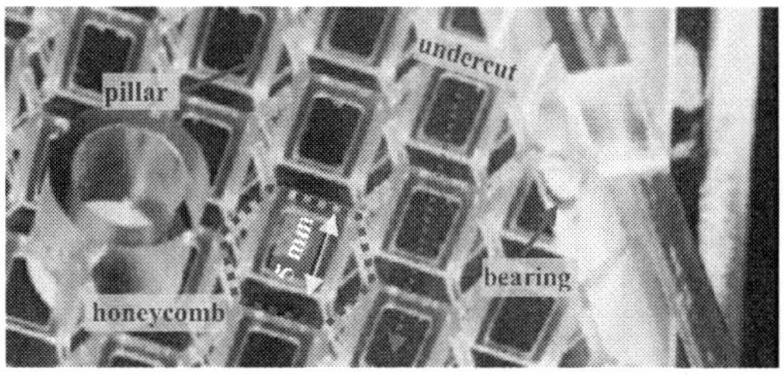

Fig. 3. Detailed view into the lightweight structure. A honeycomb structure is visible at the top and undercuts can be seen in the center, which further reduce the weight

By using two parallel bending beams as shown in Fig. 4 a translation system with two degrees of freedom can be integrated into the quartz glass. In this example the inner lightweight structure can be deflected by ±250 µm in two directions. The advantage of this system is that due to the low thermal expansion coefficient (CTE) of 1 ppm such systems are potentially very stable compared to metal-based solid flexures with a CTE of about 10 ppm.

Fig. 4. Plan Mirror with a lightweight structure on the backside combined with solid flexures. The flexures provide a translation range of ±250 µm in two directions.

Fig. 5. Detail view of the lightweight structure of the mirror with the solid flexure design. On the foreground there is the honeycomb structure with a web width of 250 µm visible

3 Conclusion

Lightweight structures with very small feature sizes and undercuts can be fabricated by using selective laser induced etching. In addition, it was demonstrated that additional features like rotation axes and solid flexures can be integrated into devices using the SLE technology platform.

Acknowledgement

Special thanks go to Jakob Birkhölzer from the OST Eastern Switzerland University of Applied Sciences for photographing the lightweight structures.

References

[1] J. Gottmann, M. Hermans, N. Repiev, and J. Ortmann, "Selective Laser-Induced Etching of 3D Precision Quartz Glass Components for Microfluidic Applications—Up-Scaling of Complexity and Speed," *Micromachines*, vol. 8, no. 4, p. 110, 2017, doi: 10.3390/mi8040110.

[2] R. N. Widmer, D. Bischof, J. Jurczyk, M. Michler, J. Schwiedrzik, and J. Michler, "Smooth or not: Robust fused silica micro-components by femtosecond-laser-assisted etching," *Materials & Design*, vol. 204, p. 109670, 2021, doi: 10.1016/j.matdes.2021.109670.

[3] A. Ahmad, *Handbook of optomechanical engineering*. Boca Raton [Floride]: CRC Press, 1999.

Design of an Optical Parametric Oscillator using a BBO partial cylinder for a continuous tunability between 0.4 µm and 0.9 µm

Baptiste Bruneteau[1,2], *Basile* Faure[1], *Jérôme* Debray[2], *Dazhi* Lu[3], *Grégoire* Souhaité[1], *Jiyang* Wang[4], *Patricia* Segonds[2], *Benoît* Boulanger[2]

[1] *Teem Photonics S.A, 38240 Meylan, France*
[2] *Univ. Grenoble Alpes, CNRS, Institut Néel, 38000 Grenoble, France*
[3] *Shandong University, State Key Lab of Crystal Materials, Jinan 250100, China*
[4] *Tianjin University of Technology, Institute of Functional Crystals, Tianjin 300384, China*

Abstract. This work describes the different steps of the design of a cylindric Optical Parametric Oscillator. It is based on a BBO nonlinear crystal shaped as a partial cylinder to be pumped by a commercial micro-laser at 0.355 µm for an energetic and sub-nanosecond emission continuously tunable between 0.4 µm and 0.9 µm.

1 THE ADVANTAGES OF BBO

The choice of the nonlinear crystal was oriented towards three non-centrosymmetric borates: βBaB_2O_4 (BBO), LiB_3O_5 (LBO) and $CsLiB_6O_{10}$ (CLBO). Indeed, they are transparent down to 0.2 µm allowing a pumping wavelength of $\lambda p = 0.355$ µm emitted by a commercial and compact micro-laser [1]. They can also generate visible light from phase-matched frequency down-conversion [2].

We performed numerical calculations at room temperature using data from [3,4,5,6,7,8] in order to compare the conversion efficiency between these three crystals when pumped at $\lambda p = 0.355$ µm for parametric fluorescence (PF) $\frac{1}{\lambda p} = \frac{1}{\lambda s} + \frac{1}{\lambda i}$. λs and λi stand for signal and idler wavelengths respectively. We considered all the possible phase-matching directions (θ,φ) generating wavelengths between 0.4 µm and 0.9 µm with the three possible polarization schemes labelled Type I, II and III.

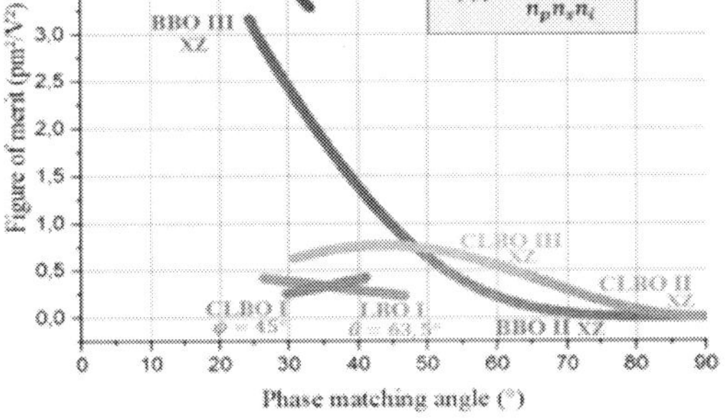

Fig. 1. Figure of merit as a function of phase-matching angles for BBO, LBO and CLBO crystals when pumped at 0.355 µm and generating between 0.4 µm and 0.9 µm from PF.

* Corresponding author: baptiste.bruneteau@grenoble-inp.org

For each phase-matching condition, we calculated the Figure of Merit defined as the ratio of the square value of the associated effective nonlinear coefficient over the product of the refractive indexes of the three interacting waves. It is shown in Fig. 1 as a function of phase-matching angles. It reports that BBO is much more efficient than LBO and CLBO crystals especially for a type I in YZ plane where the pump linear polarization is orthogonal to that of idler and signal.

BBO being a hygroscopic crystal, it is wise to impose to the crystal a working temperature higher than the room temperature in order to prevent any condensation on its surface. It is then necessary to check for any change of the phase-matching conditions as a function of temperature. Indeed, any variation of the principal refractive indices of the order of 10^{-4} can change phase-matching angles by more than 1°.

We performed such a study using our cylinder method [9]. We used a BBO sample cut as a cylinder (Ø10 mm), with a rotation plane corresponding to the YZ plane of the dielectric frame and its curved face polished to the optical quality but uncoated. The cylinder was inserted in an oven allowing a variation of temperature between 20°C and 40°C as shown in inset of Fig.2. With the sample mounted on a rotation stage and by using a commercial laser emitting a fundamental wavelength tunable between 0.7 µm and 1.9 µm, we directly recorded the phase-matching curve for Type I Second Harmonic Generation (SHG) at a given temperature, each data corresponding to a maximum value of the conversion efficiency.

As shown in Fig. 2 for 25 , 30 , 35 and 40°C, there is no influence of temperature on phase-matching conditions between 20°C and 40°C within an accuracy of 1 nm and 0.2° for the phase-matching wavelengths and angles respectively. It also corroborates calculations at room temperature using Sellmeier equations from [3].

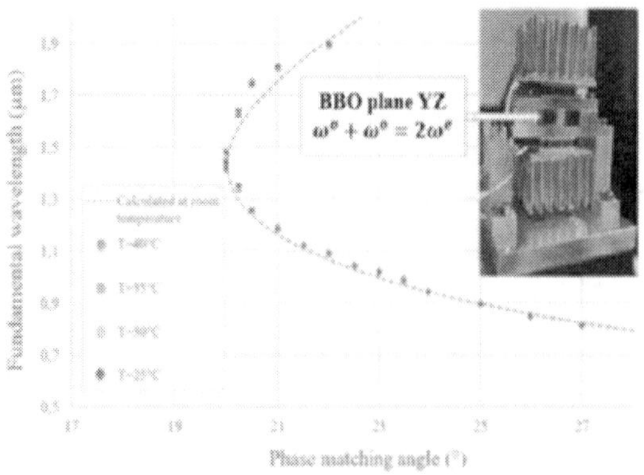

Fig. 2. BBO Type I SHG phase-matching curve for 4 different temperatures in the YZ plane.

2 The OPO cavity

We aimed an Optical Parametric Oscillator (OPO) cavity architecture the simplest and the most compact as possible in a perspective of a commercialization with a pump laser providing 350 ps pulse width at 1 kHz repetition rate. Using our previous work from [10], we selected a linear cavity singly resonant on the signal only (SROPO) with a partial BBO cylinder (Ø27.6 mm) inserted between two plane mirrors for a continuous tunability.

It was inserted in an oven to set the working temperature at 40 °C. Thus, for type I PF in the YZ plane of BBO, our calculations showed (see Fig.1) that the partial cylinder had to be cut oriented to rotate over 11° between 23° and 34° in order to achieve a continuous tunability between 0.4 µm to 0.9 µm. Note that Type I PF has the advantage that in the YZ plane of BBO that belongs to the negative uniaxial optical class, the spatial walk-off applies on the pump beam only so that it is subject to refraction at the exit of the partial cylinder contrarily to generated signal and idler beams.

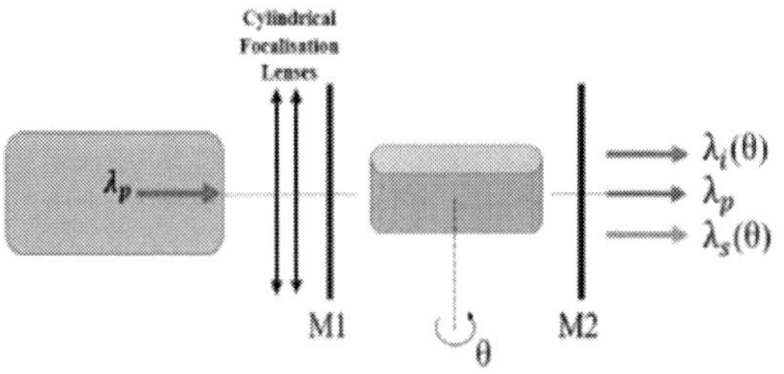

Fig. 3. OPO schematic layout. M stand for plane mirrors.

Among all the possible OPO cavity lengths from the stability domain provided by the partial cylinder, we selected the one for which the waist radius of the signal beam was located at the middle of the partial cylinder (symmetrical OPO) and corresponding to a Rayleigh length larger than its diameter. It provided small waist radii of the signal beam on the two plane mirrors that can be damaged but it had the advantage to reduce the spatial walk-off attenuation of the pump beam on the OPO efficiency. Figure 3 shows the schematic layout of the OPO we implemented.

Obviously, the OPO performances rely in the balance between the intensity of BBO parametric fluorescence and the threshold of our cavity. We recorded the phase-matched parametric fluorescence with a CCD camera imaging the output surface of the BBO partial cylinder oriented at 27° and with the output coupler M2 removed. It is depicted in Fig. 4 with an horizontal astigmatism due to the pump beam spatial walk-off when generating parametric fluorescence along the crystal.

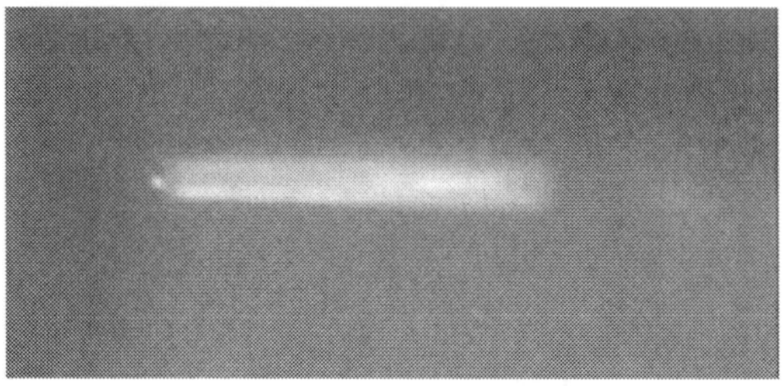

Fig. 4. Parametric fluorescence beam recorded at the exit of the BBO partial cylinder with M2 of the OPO cavity removed.

Measurement of the intensity is in progress. It will be followed by the optimisation of the OPO performances after adding the output coupler M2.

1. https://www.teemphotonics.com/laser/355-nm-laser-1khz-350ps/
2. R. Arun Kumar, Journal of Chemistry (2013).
3. K. Kato, N. Umemura, T. Mikami, Proc. SPIE **7582** (2010).
4. I. Shoji, H. Nakamura, K. Ohdaira, T. Kondo, R. Ito, T. Okamoto, K. Tatsuki, S. Kubota, J. Opt. Soc. Am. B **16**, *620-624* (1999).
5. Nikogosyan, Appl. Phys. A **58**, *181–190* (1994).
6. S. P. Velsko, M. Webb, L. Davis, C. Huang, IEEE J. Quant. Elec. **27**, *2182-2192* (1991).
7. N. Umemura, K. Yoshida, T. Kamimura, Y. Mori, T. Sasaki, K. Kato, ASSL **26** (1999).
8. I. Shoji, H. Nakamura, R. Ito, T. Kondo, M. Yoshimura, Y. Mori, T. Sasaki, J. Opt. Soc. Am. B **18**, *302-307* (2001).
9. B. Ménaert, J. Debray, J. Zaccaro, P. Segonds, B. Boulanger, Opt. Mat. Express **7**, *3017-3022* (2017).
10. V. Kemlin, D. Jegouso, J. Debray, P. Segonds, B. Boulanger, B. Menaert, H. Ishizuki, T. Taira, Opt. Lett. **38**, *860-862* (2013).

https://doi.org/10.1051/epjconf/202226603005

Neural Network for Optical Performance Estimation and Advanced Lens Combination

Robert Brüning[1,*], *Michael Verhoek*[1], and *Uwe Lippmann*[1]

[1]Fraunhofer Institute of Applied Optics and Precision Engineering IOF, Albert-Einstein-Str. 7, 07745 Jena, Germany

Abstract. We developed an algorithm to estimate the performance of an optical system based on the errors of its individual components. After a short training period with classical simulated systems, the performance evaluation for tolerancing could be accelerated by a factor of about three million. Additionally, we propose a probability-based sorting algorithm to combine individual, erroneous components in order to compensate for the tolerance budget within the system and increase the overall yield.

1 Introduction

In the last years machine learning approaches have started to be investigated for optical design tasks. Especially, for the generation of good initial systems [1] or starting points for freeform optimization [2] deep learning algorithm have shown promising results.

In this work we explore the potential of neural network learning for system performance evaluation. In addition, an automated sorting approach was investigated, which combines individual, erroneous lenses to compensate for each other's manufacturing deviations to compensate for individual errors and maximize tolerance budgets.

First, we trained a four-layer neural network to estimate the optical performance of the final system based on the individual deviations of each optical surface with respect to the ideal one. Afterwards, we used this algorithm to evaluate the mutual matching performance of the single lenses with each other to estimate the potential yield within a given tolerance budget.

In a second step, we aim to maximize the yield of performing systems within increased tolerance budgets by matching individual lenses whose tolerances are compensating each other. Starting from a heuristic approach, we developed a probability base algorithm for finite lens batch sizes, which enables its applicability even on a continuous and limited lens flow during manufacturing.

2 Neural network for performance estimation

The modulation transfer function (MTF) value for a system of two lenses L_1 and L_2 was learned using a Multi-Layer Perceptron (MLP) neural network [3]. MLPs can efficiently approximate non-linear functions [4]. The MLP consisted of four layers: an input layer with 6 nodes, two hidden layers with 64 and 32 nodes respectively, and an output node. A ReLU (Rectified Linear Unit) activation function was employed after each hidden layer. The MLP was implemented in PyTorch. The Least Absolute Deviations loss function was used. A diagram is shown in Fig. 1. The input layer nodes denote the values of the 6 lens parameters of the two-lens system: the two radii and the thickness for each of the two lenses. The goal is to teach the MLP to output the MTF for the system in the output node. The training dataset consisted of 200+200 instances of L_1 and 200+200 instances of L_2, combining to $200^2 + 200^2 = 80{,}000$ systems. The instances were created by randomly generating L_1 and L_2 lens parameters according to a uniform distribution within a fixed range and computing the MTFs of the combined L_1+L_2 systems using Zemax ray-tracing software. 85% of the dataset, randomly chosen, was used for training, with the remaining 15% used for validation. Various values of the hyperparameters were tested, eventually settling on 500 epochs with a batch size of 50, and an initial learning rate of 10^{-3}, changing to 10^{-4} after 250 epochs.

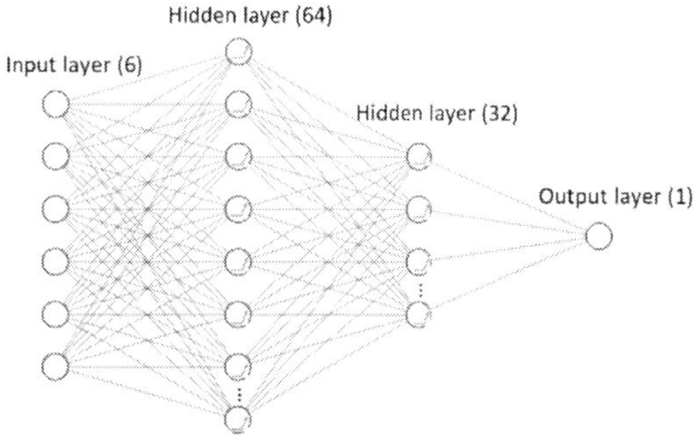

Fig. 1. Diagram of the used four-layer Multi-Layer Perceptron neural network

* Corresponding author: robert.bruening@iof.fraunhofer.de

Evaluating the trained neural network with the remaining 15% of the initial dataset gives a mean relative deviation of about 0.5 percent between the estimated and the simulated MTF value. At the same time calculation of the MTF could be accelerated by a factor of about 3.6 million compared to the classical raytraced based simulation procedure at the same hardware.

3 Lens Sorting Algorithm

The MLP model was used to provide for a method of optimally combining L_1s and L_2s. Suppose these lenses are manufactured with given tolerances of their lens parameters and that these parameters follow a normal distribution independently of each other. The goal is to match up a given L_1 from a pool of L_2s (or vice versa) without explicit knowledge of L_2s produced in the future. To this end, the known probability distributions were utilised. The three-dimensional lens parameter spaces of L_1 and of L_2 were divided into 33 × 33 × 33 bins each. The probability of each bin occurring was calculated using the CDF (cumulative distribution function) for the normal distribution. For each specific L_1, the MTF was calculated using the learned MLP, between the L_1 lens parameters and the central parameters of each L_2 bin. For a given MTF threshold, the bins were selected that satisfied the threshold, and their probabilities were added up. This gave the sum probability that a sufficiently good L_2 can be found for the specific L_1. The algorithm can be repeated with L_1 and L_2 swapped.

Two further datasets were used: one Zemax dataset exactly like the earlier 80,000-system dataset with the exception that the lens parameters were drawn randomly according to a normal distribution (a new MLP was trained on this dataset); and one MLP-generated dataset, in which 10,000 L_1 parameter sets and 10,000 L_2 parameter sets were randomly generated according to the same normal distribution, but the MTF values of the combined 100,000,000 systems were calculated using the trained MLP rather than Zemax.

The number of successful systems, defined by an MTF threshold, that can be combined from the set of L_1s and L_2s was produced according to the probability method as follows: a pool is a subset of the full dataset with N_p L_1 lenses and N_p L_2 lenses. Within this pool, for each L_1, the probability $P(L_1)$ of finding a successful L_2 was calculated according to the above probability algorithm, and vice versa for $P(L_2)$. For each of the N_p^2 combinations, the L_1 and L_2 probability values were multiplied, $P_{12} = P(L_1)P(L_2)$. Also, for each of the N_p^2 combinations, the MTF value was calculated using the trained MLP. The combinations where the MTF values were above the threshold, the P_{12} were compared and the combination with the smallest P_{12} was selected. The rationale is that we want to find matches with good MTF values from unlikely lenses, since such matches are the least likely to be needed for future lenses. The newly matched lenses were removed from the pool, which was refilled with the next in line from the larger dataset. One danger was that the pool fills up with lenses that would never find a match. To avoid this, a removal limit was set: when the number of L_1s or L_2s in the pool that satisfies the removal criteria exceeded this limit, one L_1 or L_2 was discarded from the pool, which was then refilled from the larger dataset. An L_1 satisfies the removal criteria if its MTF is below the MTF threshold for all L_2 in the pool or its P_{12} is zero for all L_2 in the pool (and vice versa).

This method was performed for a range of values for N_p and the removal limit, and the number of successful systems that was yielded, was compared to the number of successful systems according to a heuristic combination algorithm that did have advance knowledge of all the lenses in the dataset. Depending on the pool size N_p, which varies between 0.1% and 1% of complete lens set, our proposed probability method was able to find between 92% and 96% of the successful systems combinations, respectively.

4 Conclusion

We have demonstrated that MLP neural networks can be successfully used to estimate optical performance parameters for erroneous lens simulation as typically during a tolerance analysis. Additionally, the speed advantage for system evaluation algorithm enables to train and prove a probability-based sorting algorithm, which enables the combination of erroneous lenses to a successful optical system in real time.

References

1. G. Cote, J. La Londe, S. Thibault, Opt. Express, Vol. 27, No. 20, 28279-28292 (2019)
2. T. Yang, D. Cheng, Y. Wang, Opt. Express, Vol. 27, No. 12, 17228-17237 (2019)
3. T. Hastie, R. Tibshirani, J. Friedman. The Elements of Statistical Learning: Data Mining, Inference, and Prediction. Springer, New York, NY (2009)
4. G. Cybenko. Math. Control Signals Syst., Vol. 2, No. 4, 303-314 (1989)

Planarization of Lithium Niobate Surface Using a Thin Film Catalyst in Pure Water

Pho Van Bui[1,2*], *Daisetsu* Toh[2], *Masahiko* Kanaoka[1], *Hiromi* Okada[1], *Satoshi* Matsuyama[3], *Kazuto* Yamauchi[2], *Yasuhisa* Sano[2]

[1]JTEC Corporation, Saito-Yamabuki 2-5-38, Ibaraki, Osaka, Japan
[2]Osaka University, Yamada-oka 2-1, Suita, Osaka, Japan
[3]Nagoya University, Furocho, Chikusa, Nagoya, Aichi, Japan

Abstract. A catalytically assisted etching method, named Catalyst-Referred Etching (CARE) was applied to the planarization of Lithium Niobate (LN) surface, which is widely used for optical waveguides, optical modulators, piezoelectric applications. The study demonstrates that an atomically smooth surface with less than 0.1 nm root-mean-square roughness could be achieved on a LN substrate using a thin metal film and pure water as the catalyst and etching solution, respectively. All residual stress and surface damage could be removed completely thanks to the removal mechanism of CARE.

1 Introduction

There is a growing demand for ultraprecision optical components for scientific and industrial applications, especially in extreme ultraviolet (EUV) and X-ray regimes. X-ray microscopy is one of the ultimately non-invasive observation techniques. Using short-wavelength light, scientific imaging of cutting-edge materials /biological samples has become possible thanks to the ultra-precision optical components [1,2]. Although a low figure error of several nm is required for the optics, a highly ordered surface with a root-mean-square (RMS) roughness at the level of several tens of picometer is greatly desired for the highest reflectivity and the lowest unwanted scattering [3].
Recently, Matsuyama proposed an X-ray mirror element for X-ray microscopy using piezoelectric material as a mirror substrate [4]. Thanks to the non-linear optical, electro-optical, piezo-electrical, and pyro-electrical characteristics, single crystal Lithium Niobate ($LiNbO_3$: LN) is the material of choice for the required mirrors [5]. This novel deformable mirror could correct the low spatial frequency mirror shape with nanometer precision with high stability. However, it is challenging to polish the mirror surface to a highly ordered surface.
Chemical mechanical polishing (CMP) is widely used as a final polishing method. In CMP, the state-of-the-art balancing between chemical modification and mechanical removal needs to be determined before a practical polishing. A recent study reported an effect of the chemical-mechanical balance on the polished SiC surface morphology [6]. Several types of step-and-terrace structures were produced depending on the roles of chemical modification and mechanical removal. Thanks to the high hardness of SiC, the mechanical factor is calibrated to avoid any serious subsurface damage and residual stress on the polished surface. However, it is difficult to completely avoid all mechanical damage, especially for a low-hardness material such as silica glass and LN. For such materials, chemical-based polishing methods have advantages in producing a smooth and mechanical damage-free surface.
A new polishing method using a catalyst has recently been developed named Catalyst-referred Etching (CARE) [7]. CARE has been successfully applied to the polishing of oxides and functional material surfaces, such as silica glass, SiC, GaN. Unlike CMP, chemical polishing is sensitive to the electronic and geometrical factors of the polished surface. Thus, chemical attacks at a step edge and a terrace site of a crystalline material would be different. The chemical etching prefers processing at a step edge that leads to the step-flow etching type, resulting a single step-and-terrace structure observed on the polished surface [8]. The removal mechanism of CARE was confirmed to be hydrolysis catalyzed by forming chemical bonds with O/OH at the interface, stabilizing the hypervalent state and promoting the chemical reaction at the interface [9].
In this study, we apply catalyst-referred etching (CARE), an abrasive-free polishing method, to the planarization of a crystalline LN surface, using Pt as the catalyst and pure water as the etching solution. An atomically smooth surface with sub-Angstrom root-mean-square roughness is achieved on the LN substrate.

2 Experimental procedure

To demonstrate the feasibility of CARE for planarization of a LN substrate, a 2-inch single crystal LN with Y-off-

* Corresponding author: buivanpho@j-tec.co.jp, buivanpho@up.prec.eng.osaka-u.ac.jp

cut 36 degree was used in this study. The schematic of the apparatus used in this study is shown in Fig. 1.

A thin film of catalyst was deposited on a polishing pad, which has narrow grooves on its surface to supply etchant to the sample surface. The sample is put in a holder and pressed onto the catalytic pad with a pressure of 20 kPa by an airbag placed behind the sample. The sample and the pad are immersed in pure water during polishing. The planarization is promoted by rotating the sample and the pad on two parallel axes at approximately the same speed of ca. 10 rpm.

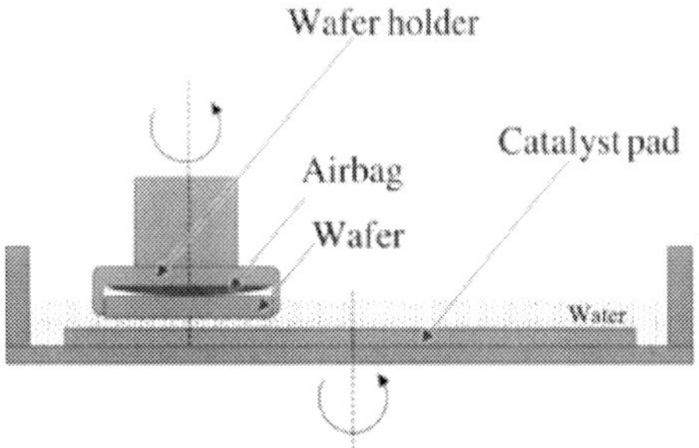

Fig. 1. Schematic of the apparatus

The experimental conditions are summarized in Table 1.

Table 1. The experimental conditions.

	Conditions
Applied pressure	20 kPa
Rotational speed	10 rpm
Catalyst	Pt
Etchant	Pure water

The surface before and after CARE were examined by an atomic force microscope (AFM; Scanning Probe Microscope, SPM-9700HT), phase-shift interference microscope (ZYGO NewView 9000).

3 Results and discussion

The phase shift interference microscopy images of the as-received (left) and CARE-processed (right) LN surfaces are shown in Fig. 2. The size of each image is 170 μm × 110 μm. After CARE processing, the surface roughness was improved from 0.15 nm RMS (as-received surface) to 0.1 nm RMS (CARE-processed surface).

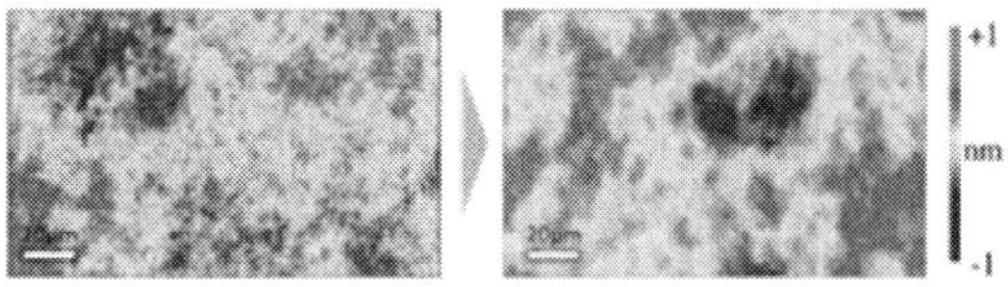

Fig. 2. Phase shift interference microscopy images of the as-received (left, 0.15 nm RMS) and CARE-processed (right, 0.1 nm RMS) LN surface.

AFM images of the as-received (left) and CARE-processed (right) LN surfaces are shown in Fig. 3. Size of each surface is 2μm × 2μm. The surface roughness is improved from 0.175 nm RMS (as-received (left)) to 0.064 nm RMS (CARE-processed (right)).

The surface roughness of the as-received surface is equivalent to previously reported the surface prepared by CMP [10]. After CARE processing, the surface is smoothened, evidenced in low surface roughness of 0.064 nm RMS.

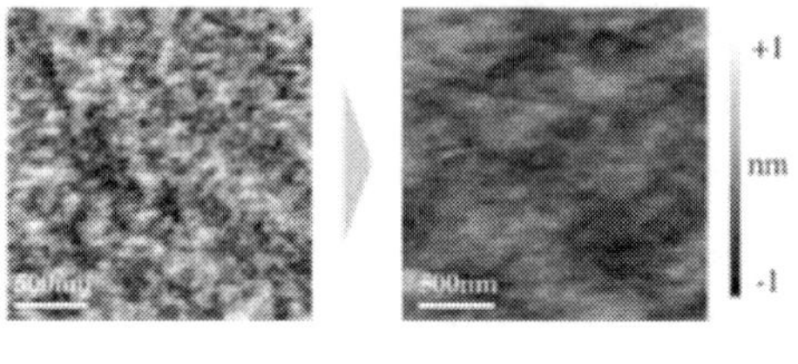

Fig. 3. AFM images (2μm × 2μm) of the as-received (left) and CARE-processed (right) LN surfaces

In CARE processing, the material removal is based on a chemical reaction that occurred at the interface of the catalyst and the substrate. Thanks to its removal mechanism, there is no subsurface damage or residual stress that could introduce on the processed surface. Thus, CARE method is a good choice as a final surfacing method to remove high spatial frequency error and surface damage.

4 Summary

The study demonstrated the feasibility of CARE as a final polishing method for LN substrates. The CARE processed surface is atomically smooth with RMS of less than 0.1 nm RMS over the whole surface. Thanks to its removal mechanism, a geometrically and crystallographically highly ordered surface can be produced.

References

1. C. Wagner, N. Harned, Nat. Photonics **4**, 24 (2010)
2. A. Sakdinawat, D. Attwood, Nat. Photonics **4**, 840 (2010)
3. R. J. Noll, Opt. Eng. **19**, 192249 (1980)
4. S. Nakabayashi, K. Uematsu, T. Inoue, S. Matsuyama, XOPT conference, XOPTp-07 61 (2022)
5. R. Takigawa, T. Tomimatsu, E. Higurashi, and T. Asano, Micromachines **10**, 136 (2019)
6. H. Deng, K. Endo, K. Yamamura, Sci. Report **5**, 8947 (2015)
7. H. Hara, Y. Sano, H. Murata, K. Arima, A. Kubota, K. Yagi, J. Murata, andK. Yamauchi, J. Electron. Mater. **35**, L11 (2006)
8. T. Okamoto, Y. Sano, K. Tachibana, K. Arima, A. N. Hattori, K. Yagi, J. Murata,S. Sadakuni, and K. Yamauchi, J. Nanosci. Nanotechnol. **11**, 2928 (2011)
9. P. V. Bui, D. Toh, A. Isohashi, S. Matsuyama, K. Inagaki, Y. Sano, K. Yamauchi,and Y. Morikawa, Jpn. J. Appl. Phys. **57**, 055703 (2018)
10. S. Wang, Y. Liu, Z. Li, Proc. of SPIE **6722** 67223L-1 (2007)

https://doi.org/10.1051/epjconf/202226603007

Development of a methodology for evaluating the process window of ductile machining for brittle-hard materials

Oliver Fähnle[1], Yannik Steimer[1],Henrik Surberg[1],
Niklas Sass[2], Marco Buhmann[2] and Thomas Liebrich[2]

[1]Eastern Switzerland University of Applied Sciences, Buchs, Switzerland,
[2] Rhysearch Institute, Buchs, Switzerland

This paper presents a standardized methodology for determining the process window for ductile machining of brittle materials. Its application for CaF2 is reported, identifying an optimized process window for single-point diamond turning on UPM machines by determining optimized process parameters.

Introduction

In contrary to the fabrication of mechanical elements, optical fabrication has been dealing with the generation of nanometer range accuracies for millenniums. Demand in industry is growing constantly for components made of brittle-hard materials like ceramics, glass or Silicon, e.g. for dental implants, optical lenses, WC-molds for PGM (precision glass molding) or sensors. Accordingly, it is characteristic that optical-functional surfaces must be produced which have shape deviations from the nominal shape of less than 100 nm and center roughness values of less than 5 nm rms. These features allow, among other things, the direct production of optical surfaces for applications in the infrared wavelength range without requiring subsequent polishing processes.
These are hard to machine materials, which usually results in high tool wear as well as in rough surfaces.
In general, only a greatly reduced process window is available compared to metals, which allows the required surface characteristics to be achieved. This is referred to as ductile machining with chip thicknesses below the critical chip thickness.

Motivation

Machining in ductile regime to improve surface qualities and to reduce subsequent processes like polishing. Therefore, using CaF2 as an example, a methodology for evaluating the machining of brittle-hard machining is developed and presented. For the characterization of the machining sub-surface-damages as well as surface roughness is measured

Procedure

One way of determining the critical chip thickness of brittle-hard materials is to carry out scratch tests with varying chip thickness, as described for example in [1] and [2]. The varying chip thickness makes it possible to identify the transition from ductile to brittle machining behavior with optical measurement methods and to determine the critical chip thickness at this point. That way and applying the three waggons method for optimization of optical fabrication processes [3], the impact of various process determining parameters onto the transition depth between brittle and ductile material removal regime can be studied. This enables the determination of optimized process windows for the ductile machining of materials to be tested following a standardized approach.

The approach implemented in RhySearch / OST is based on a circular indentation that is made in the surface of the material under investigation. The chip thickness is continuously varied along the angular coordinate of the groove. Figure 1 explains the tool path of this methodology. The marked points with the same scratch depth show an exemplary critical chip thickness, which can be identified four times on one tool revolution with the depicted tool path in view of a statistical evaluation.

https://doi.org/10.1051/epjconf/202226603007

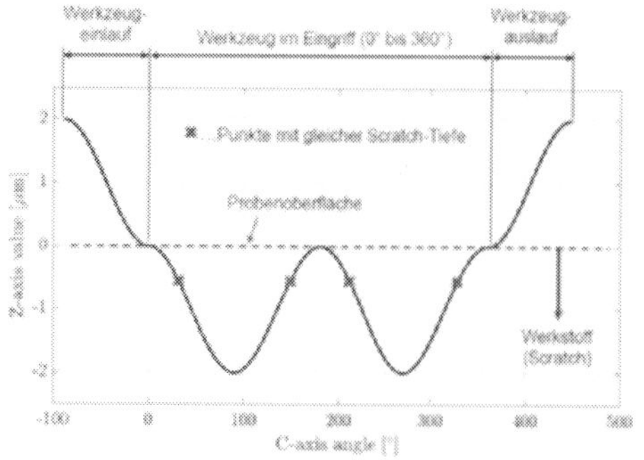

Figure 1: Standardized procedure for determining the critical chip thickness by placing a rotary indentation with varying chip thickness into the specimen surface.

Results and interpretation

Figure 2 shows the microscopy image as well as the photograph of a flat surface of a sample of calcium fluoride. The surface of a flat calcium fluoride sample has been machined with a turning process using a monocrystalline diamond cutting edge. The areas of brittle and ductile machining, which are characterized by different surface qualities, can be seen and are also marked. The area of brittle machining leads, as is also visually apparent, to a significant impairment of the optical transmission behavior.

Conducted analyses of this machining reveal that an unrecognized variation in the material thickness of the sample component on part of the surface has led to the critical chip thickness being exceeded and thus to brittle machining behavior. Therefore, in addition to the knowledge of the material behavior, a more elaborate assurance of the chip thickness with a tolerance in the range of one micrometer or even below is necessary for brittle-hard materials compared to ultra-precision machining of metals in order to keep within the limits of the specified process window. This can be made possible, for example, by tighter tolerances of the component pre-machining or the use of measuring technology integrated on the processing machine.

Repeating this methodology with different sets of process determining paramerets such as rake angle, ground velocity, pH value of the coolant applied and added ultrasound vibrations, the transition point rom brittle to ductile mode machining can be shifted higher values, optimizing the ductile machining process window for the given materia: in this case CaF2.

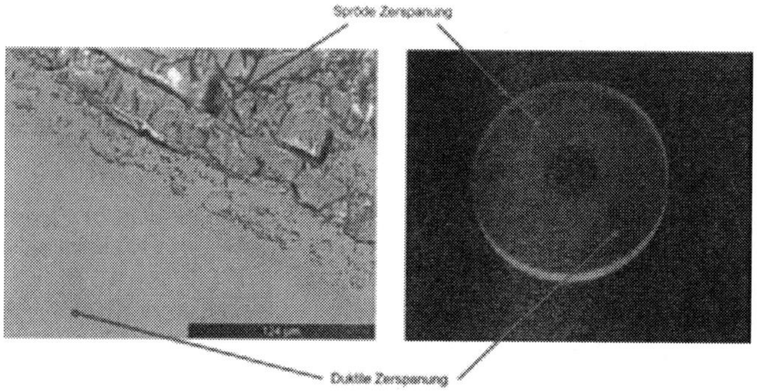

Figure 2: Brittle and ductile machining behavior of calcium fluoride (CaF2) and associated visual appearance of the machined surface (visible impairment of transmission in the region of brittle machining behavior).

References

[1] T. G. Bifano, T. A. Dow, R. O. Scattergood, Ductile-Regime Grinding: A New Technology for Machining Brittle Materials, Journal of Engineering for Industry, vol. 113, 1991.

[2] M. Doetz, O. Dambon, F. Klocke, J. Lee, O. Fähnle, Increasing critical depth of cut in ductile mode machining of tungsten carbide by process parameter controlling, Proc. SPIE 10829, Fifth European Seminar on Precision Optics Manufacturing, 2018.

[3] Oliver Faehnle, "Process optimization in optical fabrication", SPIE Journal on Optical Engineering 0001;55(3):035106. doi:10.1117/1.OE.55.3.035106., 2016

Optical methods for measuring the feature size of optical diffraction gratings with nano-meter accuracy and implementation of suitable feedback control loops.

Thomas Flügel-Paul[1,*], *Martin* Heusinger[1], *Kristin* Gerold[1], *Adriana* Szeghalmi[1], and *Uwe* Zeitner[1,2]

[1]Fraunhofer Institute for Applied Optics and Precision Engineering, Albert-Einstein-Straße 7, 07745 Jena, Germany

[2]Institute of Applied Physics, Friedrich-Schiller-Universität Jena, Max-Wien-Platz 1, 07743 Jena, Germany

Abstract. Surface relief diffraction gratings offer a high flexibility in their design and thus allow to synchronize their optical performance with the specific requirements of the underlying application. However, the accuracy and the specific control of the manufacturing processes are of vital importance. In this contribution, we present optical methods relying on white-light ellipsometry and how they can be exploited for the measurement of the critical dimensions of manufactured surface relief grating structures. We will furthermore present suitable processes (relying on atomic layer deposition) and how they are used in a feedback loop to control the grating's feature sizes on the nanometer scale.

1 Introduction

Modern applications in high-performance spectroscopy or chirped pulse amplification require diffraction gratings (used as dispersive elements) meeting challenging performances in terms of, e.g., diffraction efficiency, bandwidth, spectral dispersion, and stray light. Binary surface relief gratings are well suited to do that job especially when taking into account more complex descendants, e.g., backfilled gratings achieved by atomic layer deposition (ALD) [1,2,3]. Typically, such gratings exhibit periods slightly lower than the wavelength and the lateral feature size scales down to the 100nm region. Aspect ratios might achieve values up to 1:15. The grating geometry, i.e., groove width and depth, must be manufactured and controlled with very high accuracy on dimensions down to few nanometers.

Electron Beam Lithography (EBL) has been shown to be a versatile method, which allows manufacturing of patterns with very high resolution and accuracy. Reactive dry etching processes allow to transfer the exposed lithographic pattern into the underlying substrate [4].

When approaching the limits of the manufacturing processes, the real grating geometry shows inevitable deviations from the ideal structure, i.e., in general non-trivial shape deformations of the trenches. It is thus of crucial importance to be able to accurately characterize such deformations and to feed this information into a control loop to achieve diffraction gratings with unprecedented optical performance.

Here we present an indirect optical characterization method relying on white-light ellipsometry and how it can be exploited for the measurement of shape deformations of manufactured surface relief grating structures. Second, we outline how ALD can be applied to tune the grating's critical dimensions, thus allowing to push the grating's performance to its maximum.

2 White light ellipsometry and shape retrieval

Our measurement setup relies on a commercial white light ellipsometer SE850 from Sentech (typically used for the characterization of thin optical films and stacks).

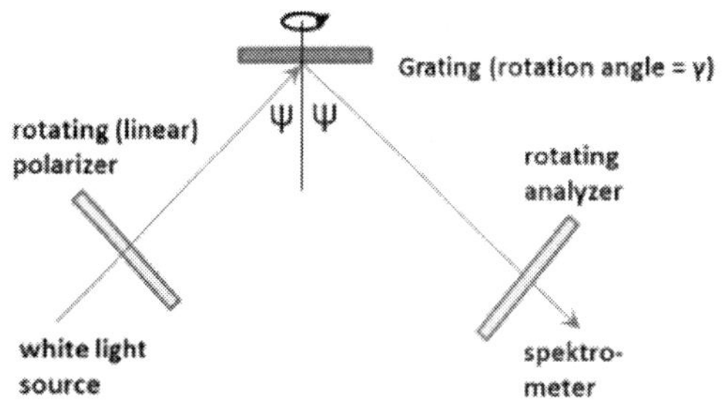

Fig. 1. Scheme of the basic measurement setup.

The basic measurement scheme is depicted in Fig. 1. The grating sample (mounted on a rotation stage) is illuminated with a collimated beam and the specular order in reflection is analysed. Additionally, rotating polarizers/analysers in the input and reflected beam path allow for dedicated selection of linear polarization states. Essentially, the setup allows for the detection of Stokes vector components S1 and S2 of the reflected beam for any arbitrary configuration. Using the rotation angle α of

* Corresponding author: author@e-mail.org

the input polarizer, the incident angle Ψ, and the grating's rotation angle γ as independent parameters, a series of Stokes-vector spectra S1(**p**=**p₀**,λ,α,Ψ,γ) and S2(**p**=**p₀**,λ,α,Ψ,γ) can be recorded ($\mathbf{p_0}$ denotes the sum of all structural parameters, e.g. groove width, depth, etc., which describe the grating's actual micro-structure). It is important to understand that $\mathbf{p_0}$ is assumed to be unknown and its determination is exactly the aim of the outlined method.

For the structure retrieval, the S1/S2-spectra are then compared to full vectorial numerical modelling results of the very same quantities relying on (initially inaccurate) parametrization **p**. Assuming that S1/S2 are sufficiently sensitive to variations in **p**, then a least square fitting procedure between modelled and measured spectra allows to determine the grating's actual microstructure or the corresponding set of parameters, i.e., $\mathbf{p_0}$.

3 Two practical use cases

The first example focuses on a binary transmission grating which is realized via etching into a fused silica substrate. The nominal grating is designed for maximum diffraction efficiency in the -1st order at wavelengths between 545nm – 680nm and an AOI = 41°.

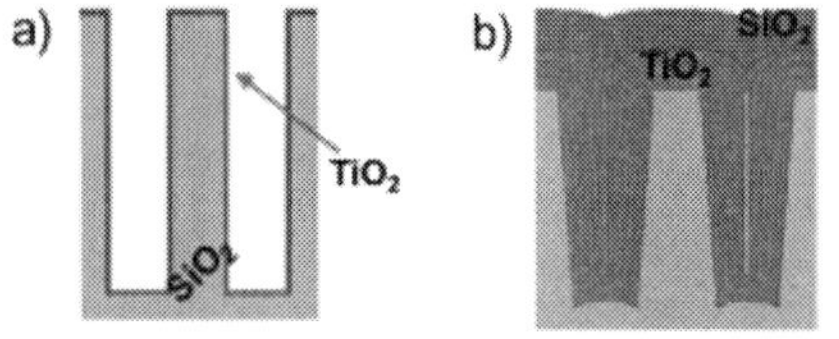

Fig. 2. a) Sketch of an ideal binary surface relief grating with conformal coating of TiO_2. b) backfilled grating with characteristic trapezoidal shape.

The associated grating profile is finished by a thin layer of titanium dioxide (TiO_2) of only 10nm to 15nm to improve performance (see Fig. 2a). The second example is a grating designed for operation at wavelengths between 745nm and 770nm with the AOI being 64°. Maximum diffraction efficiency of larger than 90% can be achieved applying a complete backfilling of the surface relief structure with TiO_2. ALD is used in both cases for deposition and the precise layer thickness intimately depends on the grating's groove width (or better: the actual trench profile) achieved in the previous dry-etching process. Hence, the knowledge of the exact dimensions of the grating's trench profile after etching is a prerequisite for the correct determination of the TiO_2 layer thickness. Figure 3 presents a comparison of selected measured and modelled S1(λ) and S2(λ) spectra for a manufactured sample grating of the type shown in Fig. 2a prior to the application of the TiO_2 ALD coating. The modelled spectra are already the optimum solutions which provide the lowest least square error with respect to the measured spectra. Furthermore, two different shape models and thus parametrizations of the microstructure are assumed. The perfect binary shape (left column) is represented by only two parameters, namely groove depth and width. Second, a more realistic trench profile (right column) is assumed which is represented by five (instead of two) parameters.

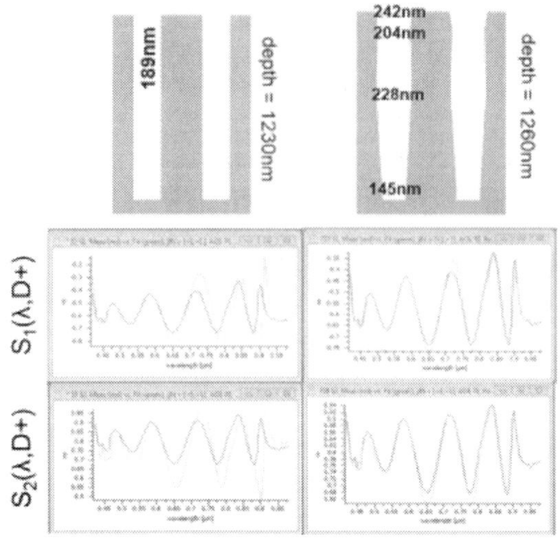

Fig. 3. Modelled (green curves) and measured (red curves) spectra of S1(λ) and S2(λ). D+ denotes diagonal linear illuminating polarization of 45°. Rotation angle is γ = 30° and incident angle Ψ = 60°. The left (right) column represents the best fit for an ideal binary (realistic) microstructure.

Comparing the accuracy of the model fitting for both profiles, it can be clearly seen, that the simple binary shape leads to considerably larger differences to the measured S1/S2 spectra. The mean square error decreases by a factor of 5 which clearly verifies the sensitivity of the method to smallest modifications on the nano-meter scale. Please note, that the left and right columns show the same values, however, on different axis scales.

The details of the second example (Fig. 2b) will be discussed in detail in the full contribution.

4 Conclusion

We have demonstrated that polarization sensitive white light reflection measurements can be used to solve the inverse scattering problem of surface relief grating structures allowing to retrieve the structural shape of the manufactured samples with high precision. Furthermore, we have used this information as a vital input to correctly adjust the precise parameters (layer thicknesses) of subsequently applied ALD coating processes.

References

1. U. Zeitner, H. Dekker, F. Burmeister, et al., Exp. Astron, https://doi.org/10.1007/s10686-022-09840-1 (2022)
2. T. Flügel-Paul, G. Kalkowski, et al., Proc. SPIE **9912**, 99122A (2016)
3. M. Heusinger, T. Flügel-Paul, et al., Optica **9**, 412-418 (2022)
4. U. Zeitner, M. Oliva, F. Fuchs, et al., Appl. Phys. A **109**, 789-796 (2012)

https://doi.org/10.1051/epjconf/202226603009

Removing microdefects on glass surfaces using laser radiation

Kerstin Goetze[1], *Juergen* Bischoff[1], *Jens* Bliedtner[1]
Oliver Faehnle[2], *Michael* Kahl[2]

[1] University of Applied Sciences, 07745 Jena, Germany
[2] OST – Eastern Switzerland University of Applied Sciences, Switzerland

Abstract. Scratches and microdefects on glass surfaces significantly impair the optical and mechanical properties of optical components. They already occur during mechanical processing (shaping) and have to be removed in several specific processing steps. A process is presented with which scratches and microdefects can be removed by means of CO_2 laser radiation.

1 State of the art

Conventional mechanical processing of optical surfaces incurs considerable costs and is time-consuming. In many grinding, lapping and polishing steps, the geometry is created and the surface roughness is reduced bit by bit. This leads to significant material removal and high tool wear, because deep cracks in the surface (sub-surface damage) have to be removed at great expense. [1]
If the surface is sufficiently well mech. polished, scratches can quickly reappear, which negatively affect the function of the optical component and are also the starting point for the mechanical failure of glass under load. To remove them, mech. polishing would have to be performed again or other expensive polishing processes such as ion beam polishing or magnetorheological finishing (MRF) would have to be resorted to. One way to remove microdefects quickly and without additional tools and polishing agents is laser beam polishing (LBP). It has been researched for several years in numerous scientific studies for various materials. In the polishing of metallic surfaces, it has been transferred to industrial production through the development of a special machine technology [2]. In addition, there are numerous publications on studies of surface treatment - especially polishing - of glass (e.g. [3], [4]). In [5] and [6], the method was used to remove scratches on surfaces. In the first example pulsed laser radiation was applied; the scratches could not be completely removed and there was material displacement. The 2nd example is a combination of mechanical and laser processing (partial surface damage removed by LBP, surface then conventionally mech. polished to remove structures from LBP).

2 Proceeding

In the present study, it is shown how microdefects can be smoothed by targeted irradiation of the surface with laser radiation of wavelength 10.6μm. This wavelength is almost completely absorbed by quartz glass surfaces, leads to heating of the surface up to softening. In a flow process, the scratches/micro defects are filled and smoothed. LBP is suitable for polishing glass surfaces in a short time with little effort, without additional tools/polishing agents. On conventionally mech. polished fused silica samples (25 x 25 x 3 mm³), diagonally running scratches and micro defects were selectively introduced and the samples were subsequently smoothed with laser radiation. To demonstrate the effectiveness of LBP, the samples were analyzed before the defects were applied and after smoothing (with confocal sensor (FRT), white light interferometer (WLI), transmission measurement). The first two methods to assess surface quality, transmission measurement to determine how micro defects change the optical properties of the material and that this change is eliminated by LBP.

3 Experiments and Results

A 2 kW CO_2 laser and a 2D scanning system in a polishing chamber are used. The parallel beam diameter is 7.1 mm. To achieve a uniform energy input, the beam is linearly deflected at high speed (1400 mm/s). The resulting polishing line is guided over the sample surface in an additional y-motion ($v = 34$ mm/min). The overlap of the lines is >90%, so that a homogeneous energy input can be assumed, which prevents the formation of a "polishing structure". It is known from previous studies that some energy input is required to heat the surface of amorphous SiO_2 to the point where flow occurs. The calculated energy input resulting from the processing parameters is 500 J/mm², which leads to an average sample heating of about 2100°C (measured with thermal imaging camera for process monitoring). In order not to introduce new defects on the glass surface during LBP, it must be cleaned very well before polishing and the environment must contain only few particles (dust). Residues on the surface and dust from the environment cause defects on laser treated surfaces [9]. Therefore, LBP takes place in the polishing chamber (with filters to reduce particles). During LBP,

https://doi.org/10.1051/epjconf/202226603009

due to the high thermal shock resistance of fused silica (compared to other glasses), no additional heating of the samples is required. After irradiation, the samples are treated in a heating furnace to remove introduced stresses that negatively affect the mechanical durability of the fused silica elements and their optical properties (optical path length difference). The mech. polished samples have a surface roughness of Sq = 18nm. After removal of the microdefects, Sq is reduced to 10 nm. The scratches (detail in Fig. 1) have an average depth of about 5.1 µm, are about 100 µm wide and material upheavals in the range of 1 µm occur in some cases. The roughness along the scratch is Rq = 3.1 µm. This is in the range defined as suitable for LBP in previous studies [7].

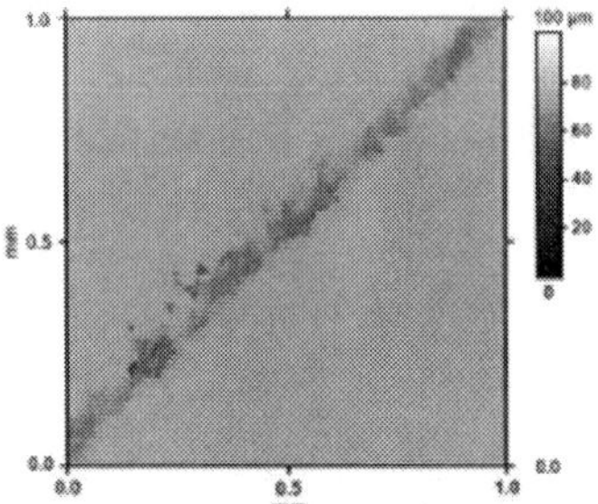

Fig. 1. Detail of a scratch on a mechanically polished surface.

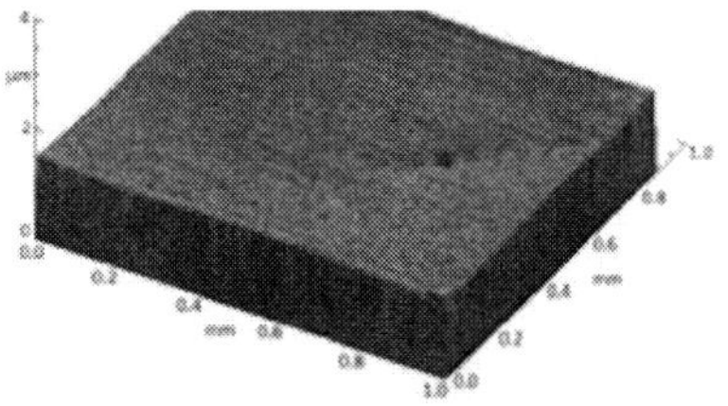

Fig. 2. Scratch, not completely removed (detail).

LBP reduces scratches/microdefects and finally remove them after parameter optimization. Fig. 2 shows the part of a scratch that was not completely removed. The flow of the material has not been sufficient to completely eliminate the defect (tiny pits remain). It could also be observed that if the scratches are not completely leveled, a uniform depression (depth approx. 0.7 µm) remains. In other words: the surface has no clearly defined defects, the roughness remains the same (in relation to the surrounding area), but the shape of the surface is changed.

Table 1. Surface parameters before and after laser polishing.

Sample...	**WLI [nm]**		**FRT [nm]**	
	Sq	**Rq**	**Sq**	**Rq**
mechanically polished	13		18	14
LBP incl./on scratch residue	27	15	16	27
residue-free LBP	12		10	12

Table 1 shows some of the surface characteristics (measurements at WLI, FRT) of the fused silica samples during the tests. Although the measurement conditions (measurement field size, filter settings, etc.) were chosen to be the same, the measured values differ due to the different measurement methods.

Another way to check the quality of optical elements is transmission measurement. A glass component with perfect, defect-free surfaces allows a certain amount of light to pass. If the surfaces have defects, there is a loss of transmission due to absorption or diffusion. This can be used to evaluate the effectiveness of LBP. A Goniospectrophotometer is used to determine the effect of scratches/microdefects on the transmitting light. Since the introduced surface defects account for only about 0.6% of the total surface area, the change in the detected light is also only a few percent. The scratches/microdefects reduce the measured value by 2.1%. After LBP the difference is 1.4% (to the untreated, mechan. polished sample). Even if scratches and micro defects can no longer be detected with FRT or WLI, there are losses whose origin must be further investigated.

4 Conclusion

The results show that it is possible to completely remove scratches and microdefects on quartz glass surfaces by LBP. In future investigations, suitable methods will be examined to check the results better and faster (measurements on FRT: 30 min/mm^2; WLI reaches its limits due to the measuring principle for polished surfaces; transmission measurement not significant enough so far). In the future, it will also be necessary to determine how scratches/microdefects can be categorized on basis of their lateral/vertical extent before machining in order to eliminate them in a targeted manner, since the investigations showed that too much introduced energy leads to deformation of the entire sample.

References

1. J. Bliedtner, *Optiktechnologie*. Carl Hanser Verlag. Munich (2022).
2. E. Willenborg, I. Ross, J. Kumstel and J. Flemmer, *Werkzeugmaschine für das Laserpolieren,* wt Werkstattstechnik online, **102** (2013)
3. J. Hildebrand, K. Hecht, J. Bliedtner and H. Müller, *Laser Beam Polishing of Quartz Glass Surfaces,* Lasers in Manufacturing (2011)
4. C. Weingarten, A. Schmickler, E. Willenborg and K. Wissenbach, *Laser polishing and laser shape correction of optical glass,* Journal of Laser Applications, **29** (2017)
5. P. Cormont, A. Bourgeade, S. Cavaro, T. Donval, T. Doualle, G. Gaborit, L. Gallais, L. Lamaignere and J.-L. Rullier, *Relevance of Carbon Dioxide Laser to Remove Scratches,* ADVANCED ENGINEERING MATERIALS, **17** (2015)
6. C. J. Trum, M. Jung, B. Schmidbauer, S. Sitzberger and R. Rascher, *Hybrid-process-chain for polishing optical glass lenses – HypotO,* Optical Manufacturing and Testing **XIII,** SPIE 11487 (2020)
7. K. Hecht, *Entwicklung eines Laserstrahlpolier-verfahrens für Quarzglasoberflächen,* Universitätsverlag Ilmenau, (2012)

Investigations on a novel process chain for manufacturing of freeform surfaces

Sebastian Henkel[1*], *Christian Schulze*[1], *Samson* Frank[1], *Christoph* Letsch[1], *Jens* Bliedtner[1], *Thomas* Arnold[2], *Heike* Müller[2], *Edda* Rädlein[3]

[1]Ernst-Abbe University of Applied Sciences Jena, Carl-Zeiss-Promenade 2, 07745 Jena, Germany
[2] Leibniz-Institut für Oberflächenmodifizierung e.V. (IOM), Permoserstrasse 15, 04318 Leipzig, Germany
[3]Technical University Ilmenau, Ehrenbergstrasse 29, 98693 Ilmenau, Germany

Abstract. Freeform optical surfaces have become increasingly important in recent years, as they can be used to construct optical assemblies with a reduced number of optical surfaces compared to systems without freeform surfaces, and thus optical systems can get more compact and lighter. However, the flexible and efficient production of precise optical freeform surfaces poses a major problem. This manifests itself in insufficient precision of the optics, long delivery times and high prices. It is shown, that ultrasonic grinding processes, combined with an ultra-fine grinding process and subsequent plasma jet polishing, are very well suited for the production of freeform optics and have a high technical and economic potential. Therefore, the aim is to validate an industrially suitable process chain based on this combination, in order to produce freeform optics of high accuracy (shape deviations <100 nm RMS) that can be manufactured in significantly fewer steps than before.

1 Research approach of a novel process chain

The application of freeform surfaces in optical systems offers many advantages compared to the use of conventional optical elements like spherical lenses. The possibility of optical assemblies with a reduced number of optical surfaces enables more compact and lighter systems, reduces losses of optical radiation power and can even lead to completely new optical functionalities. Unlike spherical or aspherical surfaces, freeform surfaces do not have rotational symmetry, but generally have six degrees of freedom that can be described mathematically with polynomials or splines. To exploit the full potential of freeform optics, both manufacturing and metrology must keep pace with the design. [1]

Therefore, a novel fabrication process chain is currently under development, which combines abrasive grinding processes with a thermal based plasma polishing process as shown in Fig. 1.

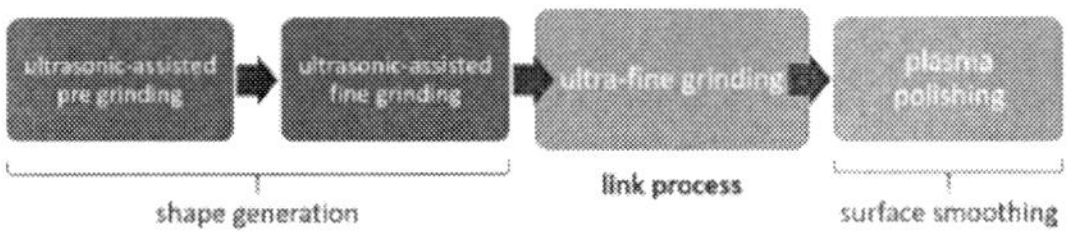

Fig. 1. Schematic representation of novel process chain for freeform fabrication

First, the desired shape of the part geometry made of glass (for example fused silica) is created in a 5-axis CNC-grinding process. It consists of a coarser pre grinding process for fast material removal and a fine grinding process for gradual approximation to the desired surface shape. Both process steps can further be enhanced with ultrasonic (US) assistance, using a high frequency tool oscillation in the micrometre-range. This results in lower grinding forces in the interaction zone and less stress is introduced into the surface, which leads to a reduced area of sub-surface damages. Depending on the glass material and the grinding parameters, force reductions from 10% up to 40% could be determined in the direction of the US oscillation [2]. Tool wear can also be reduced by more than 50% compared to grinding with identical processing parameters without ultrasonic-assistance [3].

Subsequently, an ultra-fine grinding process to further reduce the damage level and the micro-roughness is applied. In this process special resin bond tools from company *Effgen* are used with fine diamond grain size D16. The bond material is highly porous and slightly elastic, which has a damping effect and leads to a reduction in the penetration depth of the abrasive grains into the workpiece material. Thus, it is possible to achieve a surface roughness of ca. Rq=10 nm in fused silica processing [3]. The ultra-fine grinding process has a decisive link function in the process chain, as it is capable of producing already partially transparent surfaces with a particularly favourable roughness and topography for the plasma process.

This plasma polishing process is applied as the final step in the manufacturing chain, to achieve a smooth and ideally defect-free surface of the freeform workpiece. Plasma-beam assisted polishing is based on local heating and softening of the glass surface. Due to the smoothing through thermally induced material redistribution and minimisation of surface tension, no material removal takes place under correct conditions, so that the shape of the before fine grinded surface is preserved. In addition, damage close to the surface is removed, which leads to a significant increase in the laser strength of the surfaces. [4]

2 Freeform definition and digital manufacturing preparation

An important part of the freeform fabrication process is the appropriate definition and digitalisation of the desired geometry. Since its much higher degree of freedom compared to a sphere, the definition of the freeform surface is far more complex. Usually it can be defined using a mathematical function, as an arrow-height table or as a point cloud. In the presented case the freeform is designed according to its desired optical functionality with the optic design software *Zemax* using a mathematical definition of an Alvarez lens. The primary data is then exported as a volumetric body using the STEP format, which is a universal standard for CAD/CAM programs. The resulting file is imported to the program *Creo* for the implementation of CAM-programming. In this, the so-called grinding cycles and processing parameters are chosen and the tool path is planned and simulated (see Fig. 2). Under the use of a post-processor the generated CAM-program is translated into the language and input variables of the specific CNC-machine, which is an Ultrasonic 20 linear from *DMG*.

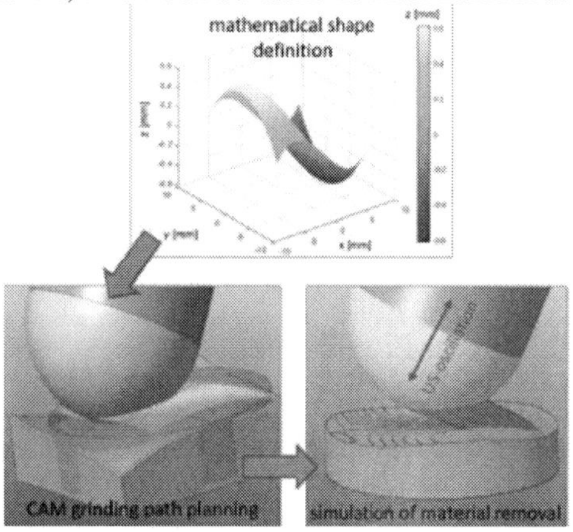

Fig. 2. Digital preparation of freeform grinding process, including geometry definition, optical design and CAD/CAM programming

This approach has proven to be efficient, as well as being able to ensure the necessary digital resolution for a subsequent precise CNC-grinding process of complex geometries.

3 First results of basic grinding experiments and outlook

The individual processes of the manufacturing chain and their interactions are currently under investigation. In addition to fundamental experiments to examine and optimise the processes, the advantageous coupling of the grinding processes with the subsequent plasma polishing is of particular interest. The special kinematic properties of freeform grinding are characterised by a point-shaped tool-surface contact zone and a line-by-line path processing. This usually results in periodic surface structures. For optimal surface smoothing by plasma polishing these structures should ideally be minimised in amplitude and have a short spatial wavelength. Fig. 3 shows an example of the difference topography between the nominal surface geometry and the actual geometry after grinding for an Alvarez freeform before polishing. Local peaks and the described periodic structures can be found.

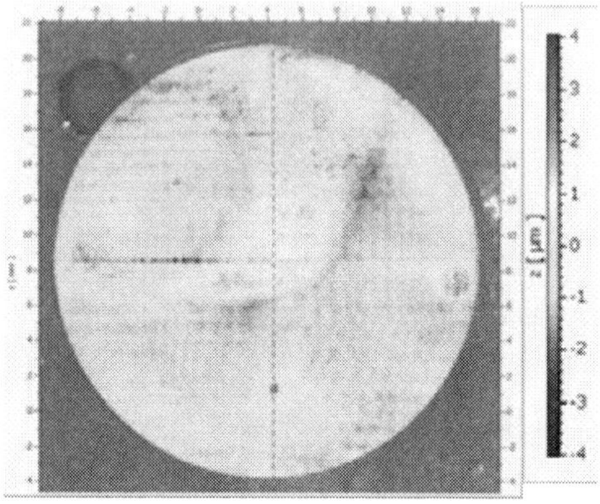

Fig. 3. Difference topography of a grinded fused silica Alvarez freeform (ø 1 inch) measured by white light interferometry

A main approach for the upcoming investigations is therefore the research of the interactions in the freeform grinding process and therefore the control and optimisation of the resulting surface shape and topography.

The authors gratefully acknowledge financial support by the German Federal Ministry of Education and Research in the funding program VIP+ (funding reference: 03VP08632, project “ProFreiform”).

References

1. F.Z. Fang, X.D. Zhang, A. Weckenmann, et al.: Manufacturing and measurement of freeform optics, *CIRP Annals – Manufacturing Technology*, **62(2)**:823-846 (2013).
2. S. Henkel, A. Barz, J. Bliedtner, et al.: Development of adjustable multifunctional optical elements for deflection, splitting and shaping of light beams, *Proc. of SPIE* **Vol. 11171** (2019).
3. S. Henkel, A.-M. Schwager, J. Bliedtner, et al.: New surface smoothing technologies for manufacturing of complex shaped glass components, Rochester, New York: *Proceedings of SPIE* **Vol. 10448** (2017).
4. T. Arnold, G. Böhm, H. Paetzelt: Plasma jet polishing of rough fused silica surfaces, *Conference proceedings of the 13th International Conference of the EUSPEN* V2 19-22 (2013).

Optimization of grinding processes on fused silica components using in-process vibrometry and dynamometer measurements

Sebastian Henkel[1*], *Marcel* Binder[1], *Jens* Bliedtner[1], *Marco* Fritzsche[2], *Abdulla* Huseynov[3], *Franziska* Schöneweck[3], *Sascha* Greiner-Adam[3], *Jörg* Flügge[3], *Edda* Rädlein[4]

[1]Ernst-Abbe University of Applied Sciences Jena, Carl-Zeiss-Promenade 2, 07745 Jena, Germany
[2]Polytec GmbH, Polytecplatz 1-7, 76337 Waldbronn, Germany
[3]Batix Software GmbH, Saalstraße 16, 07318 Saalfeld/Saale, Germany
[4]Technical University Ilmenau, Ehrenbergstrasse 29, 98693 Ilmenau, Germany

Abstract. The presented investigations deal with real-time evaluation and recording of vibrations and forces during a CNC grinding process, as well as the analysis and control of process influences on the surface quality of optical components. The experiments were carried out on a 5-axis CNC machine. Rapid subsequent analysis of the topography resulting from grinding is achieved with the aid of white light interferometry. The aim of the investigations is to reduce the surface deviations (roughness, mid-spatials, waviness) influenced by process factors. It is shown that the vibration data measured during the grinding process correlate to a high degree with the recorded topography data.

1 Motivation

The use of CNC-controlled processing machines for grinding inorganic non-metallic materials allows the production of increasingly complex components for various sectors such as the optical industry, semiconductor industry or medical technology. Using modern 5-axis CNC technology, it is possible, for example, to create freeform surfaces or monolithic components from silicate glasses. Conventionally, the grinding process with bonded grain is used for coarse shape production of a component, which is then followed by several fine machining processes with loose grain, such as lapping and polishing, with gradual reduction of the abrasive grain size. [1] However, the application of grinding processes using bound grain has some advantages regarding better process reproducibility, efficiency and less waste of resources compared with loose grain processes. Lapping and polishing are also very time-consuming and are characterized by high cleanliness requirements. This results in the goal to replace loose grain processes by grinding as far as possible. By application of strategies like ductile regime grinding or usage of novel partial elastic resin bond grinding tools, it is already possible to achieve low roughness values comparable to pre-polishing down to $Rq \approx 10$ nm (planar grinding of fused silica) [2]. While such high frequency surface deviation values can be already reduced, grinding processes of optical materials have to face the problem of the influence of unwanted vibrations and other influencing factors, mainly resulting in low and medium frequency surface deviations. The waviness and the so-called mid spatial frequency, periodic defects or grinding structures are undesirable surface phenomena. These remaining periodic surface structures are particularly hard to remove afterwards, which explains the fact that the processes with loose abrasive grain have so far continued to be indispensable.

Therefore, an essential objective of the investigations is to draw conclusions between chosen grinding parameters, primary process parameters like vibrations and forces and the resulting surface topography and quality. By gaining knowledge of the more precise process correlations, it should be possible in future to influence the described periodic structures in a targeted manner and ideally to prevent them completely.

2 Experimental set-up

All vibrations and small-scale movements that occur during machining in the interaction area between the machine, the tool and the optical component are meant to be considered. This requires highly dynamic and contactless laser vibrometric measuring systems that allow a high amplitude resolution (sub-µm range) and can simultaneously also record a very broad frequency spectrum. Therefore, the application of different laser vibrometry systems from company *Polytec* is considered, including a fibre vibrometer for space-saving installation and the relatively new QTec system for improved signal to noise ratio in difficult measurement environments. One of the investigated preliminary set-ups installed in the grinding machine is shown in Fig. 1. A dynamometer 9129AA from company *Kistler* is also included for in-situ grinding force measurement. The data acquisition of both vibrometer and dynamometer is synchronized in time.

Fig. 1. Preliminary measurement set-up including vibrometer and dynamometer measuring devices

After grinding, the surface topography of produced component surfaces is recorded. Two white light interferometry (WLI) systems from *Polytec* are used, a microscopic measuring instrument (TMS-1400, TopMap Micro.View) and a macroscopic instrument (TMS-500, TopMap Pro.Surf). This enables surface defect analysis from low to high frequencies. On a gained topographical map, the effect of the influencing variables on the grinding process can also be analysed, in analogy to the vibration analysis.

3 Examinations & Results

In order to gain an initial understanding of the process behaviour and its effect on the topography of the workpiece, a simple kinematic strategy was chosen first to detect the influences of the individual parameters. This consists of a linear grinding path using a metal-bonded ring tool of diameter 24 mm and grain size D64. The processed material is fused silica (*Corning* 7980 5F) with sample dimensions of 20 x 20 mm².

In the first preliminary investigations, interesting aspects could already be observed and insights into the processing machine were gained. Noticeable were distinct "steps" on the component surface, which appeared along the grinding path on the entire component but differed in their height. Figure 2 shows the surface topography, recorded with the TMS-500, of a machined workpiece in which periodic structures are visible that are reminiscent of "chatter marks" from the milling process. They were found to be caused by an integrated machine regulation called "temperature compensation". A sensor therefore detects a temperature dependent expansion or contraction of the machine spindle in the micrometre range and a regulation process automatically compensates this by slightly readjusting the z-axis position during the process. Since this should improve dimensional accuracy, the resulting distinctive structures are disadvantageous regarding surface quality.

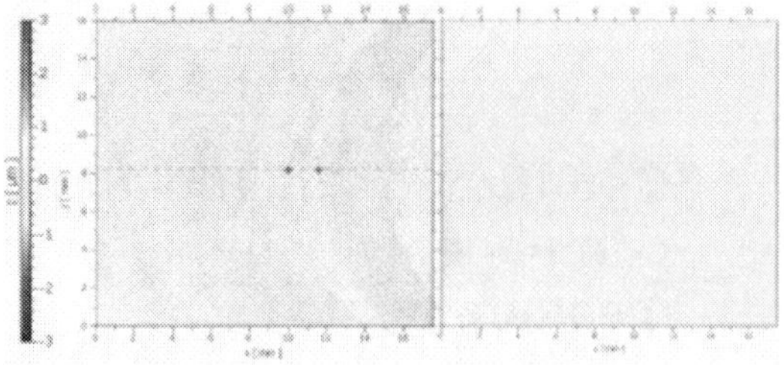

Fig. 2. Sample topographies (area 17,5 x 16 mm²) grinded with temperature compensation on (left) and off (right)

Furthermore, correlations between the in-process-measurements and the resulting sample surface topography were already successfully demonstrated for the first time, as can be seen in Fig. 3. The signal of vibrometry, force measurements and topography have been superimposed in the diagram. The positions of topography steps (blue) and vibrometer distance variation (green) are rather well correlated. The force signal (orange) also shows short peaks on the same positions, which represent the moments the temperature compensation triggers a z-axis regulation.

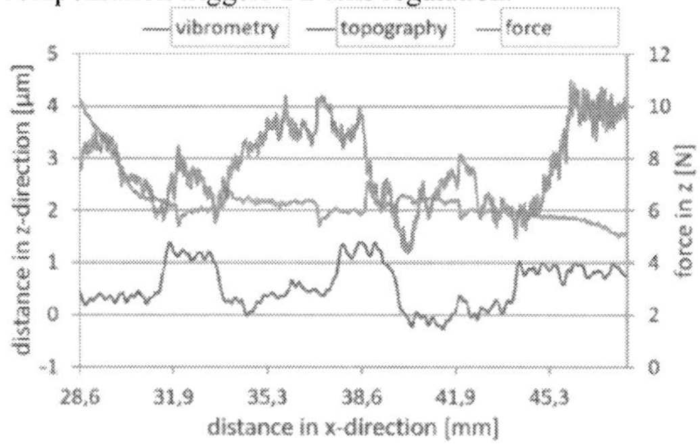

Fig. 3. Synchronized vibrometry, force and topography data for a sample surface (temperature compensation active)

This shows, that it is possible to find the characteristic vibrations dependent on the grinding parameters as surface structures on the components, which makes it possible to predict the surface quality produced in the process. For future investigations a robust AI model will subsequently be created. By means of AI algorithms, possible correlations and thus effects of the error influencing variables on the processing result should be able to be directly concluded. If successful, process regulation could then be carried out with relatively high predictive probability on the basis of the model.

The authors gratefully acknowledge financial support by the German Federal Ministry of Economy and Energy in the funding program ZIM (funding reference: KK5091604KT1, project "VibroKI").

References

1. J. Bliedtner, G. Gräfe, R. Hector: *Optical Technology*, McGraw Hill (2011).
2. S. Henkel, J. Bliedtner, E. Rädlein, et al.: Ultra-fine grinding of silicate materials under the use of new resin bond diamond tools. In: *Proceedings of SPIE* Vol. **10326** (2017).

https://doi.org/10.1051/epjconf/202226603012

Light Scattering from Contamination and Defects – Measurement, Analysis, and Modelling

Tobias Herffurth[1,*], *Alexander* Bergner[1,2], *Sven* Schröder[1], and *Marcus* Trost[1]

[1]Fraunhofer Institute for Applied Optics and Precision Engineering IOF, 07745 Jena, Germany
[2]Institute for Applied Physics, Abbe Center of Photonics, Friedrich-Schiller University, 07745 Jena, Germany

Abstract. Light scattering induced by contamination and defects on optical components can quickly limit the component's performance. Therefore, imperfection analysis and budgeting are mandatory - but also challenging tasks. On the other hand, imperfections can be elegantly characterized using efficient, robust and non-contact light scattering techniques. This will be demonstrated in this contribution for area covering measurement approaches using laboratory instruments with highest sensitivity as well as elaborated sensor systems that are best suited for extended freeform surfaces. Moreover, the measurement results are used to derive practical imperfection scattering data and models that serve as input to model and predict the imperfection induced scattering on optical system level.

1 MOTIVATION

The performance of optical instruments might quickly be limited by particular or molecular organic contamination (PAC, MOC) and surface defects. Beside surface roughness, these imperfections are primary origins of light scattering [1-6] that causes a reduced imaging contrast or a loss of throughput. In particular for super-smooth surfaces or thin film coatings, as those applied in space or lithography industry, even lowest imperfections levels might be performance limiting.

However, imperfections on optical components are unavoidable and, therefore, budgeting and characterization becomes mandatory. On the other hand, the induced light scattering is an ideal, highly sensitive, fast, and flexible characterization tool for imperfection analysis [2-6] even on extended freeform surfaces [5].

Consequently, both aspects light scattering based instruments as well as modelling approaches for the analysis and budgeting of surface imperfections and their scattering are investigated at Fraunhofer IOF. Results ranging from pure angle resolved scattering measurements over full surface characterization regarding roughness, contamination, and defects, to modelling approaches that utilize the measured data for prediction of imaging quality in optical system design are presented.

2 Application examples

2.1 Angle resolved scattering of PAC and MOC

The angle resolved scattering (ARS, [2, 4, 7]) at 532 nm of a contaminated super-smooth silicon wafer with a percentual areal coverage (PAC) of 300 ppm is summarized in the Fig. 1. The upper part shows a full-area mapping into a fixed scattering direction to reveal the local particle distribution. The bottom part shows the ARS curves of selected particles and of the clean Si wafer.

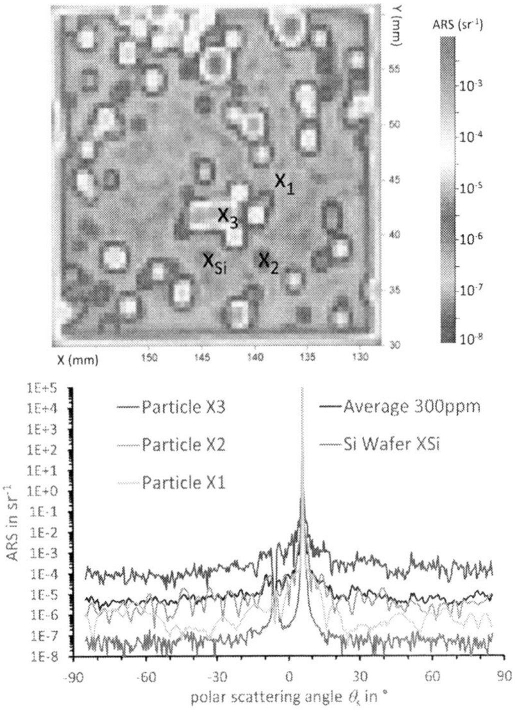

Fig. 1. Light scattering from contamination. Top: scattering mapping obtained by laterally scanning a supersmooth Si-wafer at fixed scattering angle direction; bottom: ARS of this silicon wafer at the indicated particle sites.

* Corresponding author: tobias.herffurth@iof.fraunhofer.de

https://doi.org/10.1051/epjconf/202226603012

The roughness induced scattering level of the silicon wafer is lower than ARS = 10^{-7} sr^{-1} enabling the characterization of PAC levels below 10 ppm. The particle induced scattering significantly increases the ARS level by up to 3 orders of magnitude for singular defects. Moreover, the curves show strong oscillations that correspond to feature size and form [3, 6].

Based on the scattering mapping, single ARS curves of selected postions, and stochastic data analysis, an averaged ARS is derived (blue curve). This curve represents the ARS of the homogeneously illuminated sample and can be utilized as scattering input in optic design software.

2.2 Sensor for freeform surfaces assessment

For a rapid full surface analysis of particles and defects on extended surfaces, a compact hybrid sensor with a CMOS detector matrix for ARS measurement [3] and a microscope for geometrical feature analysis was developed. This sensor is traced over sample surfaces using a robotic handling system [5] that enables the characterization of freeform components. Hence, the resulting mapping data can be analyzed for local rms-roughness, homogeneity, local and averaged 3D-ARS data, or PAC by evaluating the scattering or feature geometry data. Exemplary results and a scheme of the sensor are shown in Fig. 2.

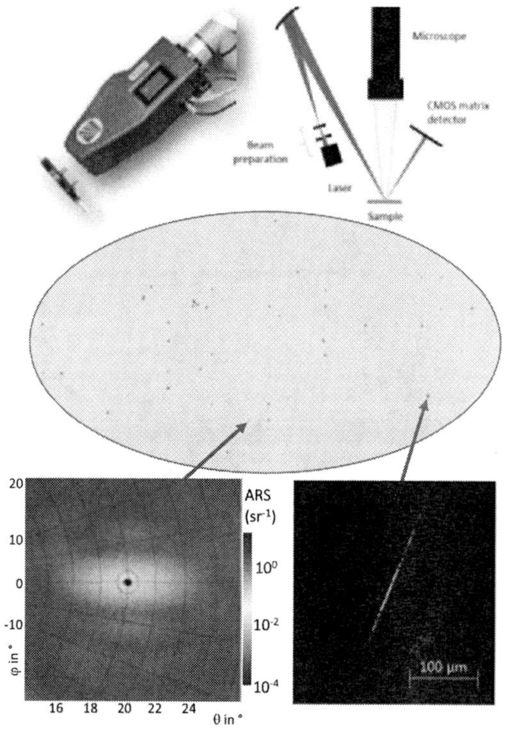

Fig. 2. Hybrid scattering sensor for imperfection assessment. Top: image and scheme of the sensor. Middle: binarized imperfection mapping of an ellipsoidal mirror substrate (120 x 70 mm^2). Bottom: typical 3D-ARS image of a defect (left) and microscopy image of a scratch (right).

2.3 Application in optical system design

Once angle resolved scattering data is obtained that is representative for an entire surface or component, it can be directly transferred into raytracing software (e.g. FRED, ASAP, …) for a realistic prediction of the contamination induced scattering on system level.

For example Fig. 3 shows the possible image degradation for a simple Gaussian double lens induced by a PAC of 0.1 % using transfer function theory [7].

Fig. 3. Image degradation induced by contamination with a PAC of 0.1% for a Gaussian double lens (without aberrations). Left: ideal image, right: degraded image.

Vice versa meaningful and easy to handle scattering models (ARS or BSDF) can be derived that enable the predication of acceptable contamination budgets for the specific optical system application.

References

1. E. Fest, *Stray Light Analysis and Control*, (SPIE Press, Bellingham WA, 2013).
2. S. Schröder, T. Herffurth, H. Blaschke, A. Duparré, Appl. Opt. **50**, C164–C171 (2011).
3. T. Herffurth, S. Schröder, M. Trost, A. Duparré, A. Tünnermann, Appl. Opt. **52**(14), 3279–3287 (2013)
4. S. Schröder, A. von Finck, A. Duparré, Adv. Opt. Techn. **4**(5-6), 361-375 (2015)
5. T. Herffurth, M. Trost, M. Beier, R. Steinkopf, N. Heidler, T. Pertermann, S. Schröder, Opt. Eng. **58**, 092609 (2019)
6. S. Maure, G. Albrand, C. Amra, Appl. Opt. **35**, 5573-5582 (1996)
7. ISO 19986:2020 "*Lasers and laser-related equipment - Test method for angle resolved scattering*", (2020).
8. J.E. Harvey, Proc. SPIE **8841**, *Current Developments in Lens Design and Optical Engineering XIV* 88410W (2013)

Concept, manufacturing and challenges of ultra-compact snapshot multi-spectral multi-aperture imaging systems

Martin Hubold[1,*], *Johanna* Karl[1], *Robert* Leitel[1], *Norbert* Danz[1], and *Robert* Brüning[1]

[1]Fraunhofer Institute for Applied Optics and Precision Engineering, Albert-Einstein-Str. 7, 07745 Jena, Germany

Abstract. Snapshot multispectral imaging is a rising non-invasive and contact-free analysis method and technology to discriminate or identify objects based on their spectral characteristics. We demonstrate a versatile system approach for compact and real-time capable snapshot cameras for the visible (VIS) and the near-infrared (NIR) or the short-wave infrared (SWIR) wavelength range based on a micro-optical multi-aperture system and various spectral filter approaches. In addition, the manufacturing, the calibration, and the limitations of the demonstration systems are described.

1 Introduction

Well-known solutions for multi- and hyperspectral imaging systems are based on one- or two-dimensional scanning approaches suffer from artefacts due to motion blur by non-uniform movements or a reduced signal-to-noise ratio in dynamic scenes. However, several snapshot imaging techniques are reported in Hagen et. al [1], which try to overcome these drawbacks including the TOMBO approach for multispectral imaging [2]. In the last years, a major trend is seen towards miniaturized and real-time capable systems especially adapted for UAV-supported applications for precision farming and environmental monitoring or for handheld devices. There, size, weight and power as well as the costs are crucial parameters.

So, we developed a couple of multispectral snapshot cameras based on a multi-aperture imaging approach using a customized micro-lens array (MLA) and various spectral filters. Each system features different spectral and spatial resolutions in the multispectral data cube depending on the used filter technology as well as the size of the image sensor. The peculiarities and similarities of these are described in this article in excerpts.

2 System design

The object is imaged via multiple regularly (i.e. rectangular) organized miniaturized objective lenses on a common glass substrate (denoted as MLA) onto a commercial image sensor, where the object distance is several times larger than the focal length of each micro objective (aka. channel) shown in Figure 1. The spectral filters are either placed close to the aperture of the system or to the image sensor plane (special image sensor with tiled filter arrays or a customized cover glass is required). Additionally, there are different spectral filter technologies like band-pass filters (BP), linear variable filters (LVF) and Fabry-Pérot filters (FP), which we have applied to our systems (see Sec. 2.1). Furthermore, a baffle array is indispensable for the overall setup serving as a field aperture to suppress cross-talk between neighboring channels and out-of-FoV false light.

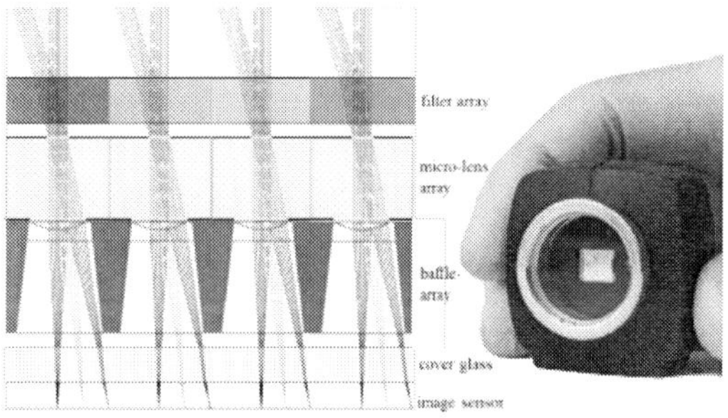

Figure 1. Generalized optical system concept including the main parts of the snapshot multispectral camera (left). Compact size of the ETCHED-VIS demonstration system compared to a hand (right).

Using this approach, a single shot acquisition of the spectral data cube is obtained without an additional bulky objective lens.

2.1 Filter approaches

LVFs are spectral band-pass filters on one glass substrate with a variation of the pass-band along one physical direction. So, it is worthwhile to rotate this filter around the optical axis of the MLA by a certain angle in order to achieve a linear spectral sampling over all channels using a monolithic element (see [3]). However, the limited gradient of such a continuous varying filter requires a relatively

*e-mail: martin.hubold@iof.fraunhofer.de

Table 1. Overview of several parameters of our accomplished systems.

parameter/demo system	LINE-VIS [3]	ETCHED-VIS	BP-VIS	LINE-SWIR	ETCHED-SWIR [4]
filter approach	LVF	FP	BP	LVF	FP
spectral range	VIS-NIR	VIS-NIR	VIS-NIR	SWIR	SWIR
spectral resolution	10 – 16 nm	30 – 40 nm	20 - 40 nm	40 – 50 nm	20 – 30 nm
spectral sampling	continuously (~6 nm)	discrete	discrete (~50 nm)	continuously (~25 nm)	discrete
data cube [X x Y x λ_n]	400x400x66	640x640x12	750x620x9	128x128x20	256x213x6
FoV (diag.)	68°	32°	45°	36°	70°
F#	F/7	F/5.4	F/5.1	F/3.5	F/3.0
pixel size	7.4 µm	2.2 µm (4.4 µm w/ binning)	3.45 µm	15 µm	15 µm
physical size [L x W x H]	60 x 60 x 28 mm³	46 x 47 x 33 mm³	90 x 40 x 30 mm³	80 x 55 x 55 mm³	80 x 55 x 55 mm³
weight	200 g	55 g	110 g	350 g	350 g

large image sensor size and the suppression of the blocking range is lower than for single BP filters. So, BP filters are more feasible for a few spectral channels. Besides BP filters, FP filters are another possibility to apply a spectral filter, which can be narrow-band depending on the complexity of the mirror system at the boundary of the cavity height. The cavity height defines the transmission wavelength which can be tuned via a structuring process and a filter tile array can be fabricated as monolithic element reducing assembly technology costs compared to single BP filters.

2.2 Demonstration systems

An overview of our accomplished systems including several important parameters, i.e. optical and spectral properties, are found in Table 1.

3 Manufacturing and Challenges

All MLAs are fabricated as polymer-on-glass optic via reflow of photoresist or ultra precision tool mastering technologies and a subsequent replication step on wafer-scale inhouse, which offers cost-efficient elements for large quantities. The one- or double-sided MLAs feature also buried black apertures. However, the sag height and shapes of the micro-lenses achieved by those technologies limit the spatial resolution and the f-number of each channel, but there are emerging technologies to overcome these restrictions for multi-aperture imaging systems.

The manufacturing of the filter tile array is performed by a grayscale photolithography process [5] in resist and a subsequent proportional dry-etching process (RIE) into fused silica of the filter's cavity layer. However, the realization of the differences in cavity height between the filter tiles as well as the final absolute height are ambitious due to a varying etch rate factor over time and the selectivity during the dry-etching process. The coating of the filter tile arrays in the ETCHED-VIS and ETCHED-SWIR demonstration system has been realized by magnetron sputtering on Borofloat glass (Schott B33) wafers.

The manufacturability and the performance of the baffle array for the optical system is challenging due to the reflection and scattering coefficients of the materials and coatings under large angle of incidence and the aspect ratio of the separator walls. A thick cover glass and a large gap between cover glass and image sensor reduces the relative illumination behavior of each optical channel due to additional vignetting of outer fields. Removed cover glass image sensors are preferred for very high fill factors. The baffle arrays for the demonstrators were milled out of metal alloys and coated subsequently through plasma chemical oxidation.

The spectral calibration of the individual cameras is performed via measurements using a monochromator setup as presented in [3]. The similarity of all is the wavelength shift of the central transmission wavelength due to the increase of the angle of incidence for outer fields.

4 Conclusion

The versatile approach shows a high potential for capturing spectrally resolved, extended object fields in a snapshot for a wide range of applications in different wavelength ranges, that benefits from a compact architecture. Challenges for further miniaturization, monolithic filter tile array elements, or higher spatial resolution have to be solved in the near future.

References

[1] N.A. Hagen et al., Optical Engineering **52**, 1 (2013)
[2] R. Shogenji et al., Opt. Express **12**, 1643 (2004)
[3] M. Hubold et al., Opt. Express **29**, 7361 (2021)
[4] M. Hubold et al., *Ultra-compact snapshot multispectral camera based on micro-optics and mono lithically fabricated filter array*, in *122. Jahrestagung der DGaO*, edited by G. Häusler, C. Faber (DGaO Proceedings, 2021), Vol. 1, pp. 1–2
[5] H.C. Eckstein et al., *High dynamic grayscale lithography with an LED-based micro-image stepper*, in *Optical Microlithography XXIX*, edited by A. Erdmann, J. Kye, International Society for Optics and Photonics (SPIE, 2016), Vol. 9780, pp. 62 – 68

Nautilus: The advent of large lens-based space telescopes

Daewook Kim[1,2,*], *Tom D.* Milster[1,A], and *Dániel* Apai[2,3,Γ]

[1]James C. Wyant College of Optical Sciences, The University of Arizona, Tucson, AZ 85721, USA
[2]Department of Astronomy and Steward Observatory, The University of Arizona, Tucson, AZ 85721, USA
[3]Lunar and Planetary Laboratory, The University of Arizona, Tucson, AZ 85721, USA

Abstract. One of the most profound and philosophically captivating foci of modern astronomy is studies of Earth-like exoplanets in search of life in the Universe. The paradigm-shifting investigation described here calls for a new type of space telescope that redefines the available light-collecting area in space, far beyond what is currently possible with the 6.5 m diameter James Webb Space Telescope. The Nautilus Space Observatory, which is enabled by multiple-order diffractive optics, is ushering in the advent of large space telescope lenses designed to search for biosignatures on a thousand exo-earths.

1 Nautilus searching a thousand Earths

Nautilus, named after J. Verne's submarine [1], is a revolutionary telescope concept (Figure 1) that replaces the traditional reflective primary telescope mirrors with multiple-order diffractive engineered (MODE) material lens technology [2], offering a scalable and resilient solution for astronomical space telescopes. Its robust optical performance against optical errors/misalignments and fabrication by glass molding process enables significantly lower cost of fabrication, integration-and-testing, assembly, and launch compared to modern reflecting telescopes, which require extremely precise co-phasing and alignment for wavefront control.

Fig. 1. Rendering of Nautilus fleet consisting of multiple unit telescope array during the space deployment phase.

The science goal of Nautilus is to investigate more than 1,000 habitable-zone, Earth-sized exoplanets through transit studies to determine their atmospheric diversity and the occurrence rate of atmospheric biosignatures at a distance of up to 300 pc. [1] A scalable and resilient fleet of multiple Nautilus space telescopes can find, observe, and provide spectra of deeper and fainter exoplanets than any other concept described to date. This statistically meaningful sample of worlds achieves ~2-3 orders of magnitude increase over the direct imaging or exoplanet transit telescope concepts currently envisioned for the next several decades.

2 Ultralightweight large space lenses

A multiple-order diffractive engineered (MODE) lens has been developed [2], in which chromatic focal length variation exhibits both refractive and diffractive characteristics in balance. This hybrid optical component shown in Figure 2 achieves a high Strehl ratio (i.e., diffraction-limited imaging) performance over a wide range of f/# (i.e., focal length and aperture size) design space.

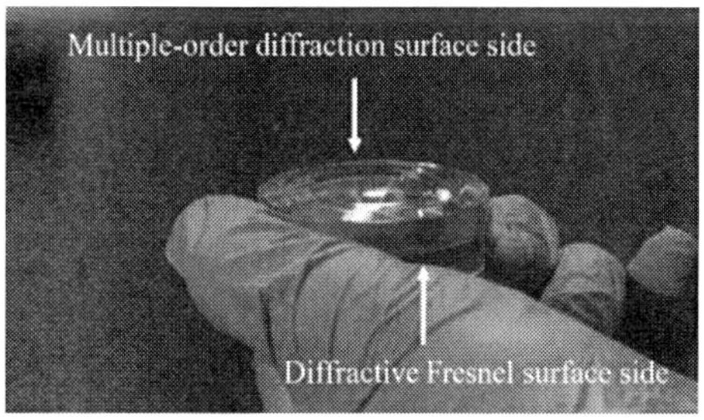

Fig. 2. MODE lens consisting of two engineered optical surfaces, a multiple-order diffraction surface and a diffractive Fresnel surface balancing and compensating chromatic aberrations from refraction and diffraction.

A prototype MODE lens was designed and fabricated for the astronomical R-band wavelength range (i.e., 589 – 727 nm). The measured focal length as a function of wavelength accurately matched design values. Also, an infinite conjugate optical imaging test successfully demonstrated ~2× full-width-half-maximum spot size compared to the ideal Airy disk size despite of all the manufacturing errors and metrology noise with a plastic

https://doi.org/10.1051/epjconf/202226603014

primary lens. Unlike a traditional refractive-only lens, MODE lens technology does not scale thickness as the diameter increases. Thus, large aperture, ultralightweight lens-based space telescopes are enabled.

3 Nautilus Optical Engineering

3.1 Optical design configurations

The Nautilus Observatory consists of many individual unit telescopes in an incoherent array. No precision formation flight is needed. Each unit telescope design can be configured in a one- or two-MODE lens configuration [1, 3] utilizing different MODE lens aperture sizes. The two-MODE-lens configuration is shown in Figure 3.

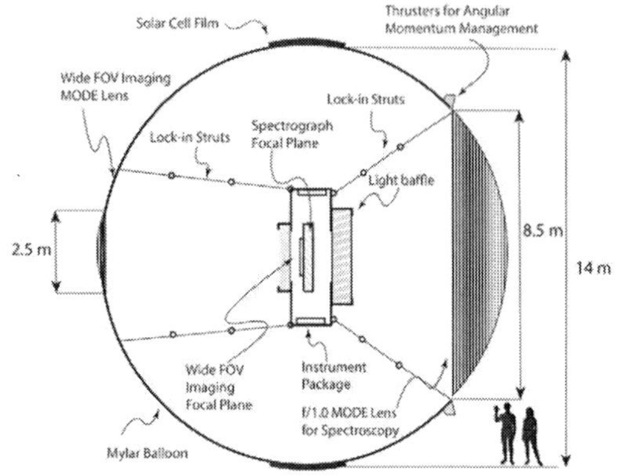

Fig. 3. Nautilus unit telescope concept using 8.5 m MODE lens (on the right side) and 2.5 m MODE lens (on the left side). The 8.5 m MODE lens size is limited by the rocket fairing size. The optional 2.5 m aperture provides parallel imaging capabilities with a wide field of view, optimized for exoplanet transit search or surveys. [1]

The Nautilus unit telescope is packaged in a compact disk-like formfactor and deployed once they are in space, as shown in Figure 1. Once in orbit, a gas canister inflates a Mylar (or any other membrane material) balloon acting mostly as a simple thermal shield and light baffle, and deploys the instrument package along with the MODE lenses. Lock-in linkage struts provide mechanical stability and longevity of the telescope, as shown in Figure 3.

3.2 MODE lens fabrication via glass molding

MODE lenses are fabricated using a diamond-turned mold in a precision glass molding process. The optimal MODE lens manufacturing process going through molding temperatures of ~550°C has been tested and refined via selection of proper mold material, compatible mold fabrication, precision glass molding phase, and stability of the mold itself.

For optimal mold machining and release, as illustrated in Figure 4(top), nickel phosphorous (NiP) plating is applied to the copper-nickel C71500 (CuNi) mold substrate, which has a similar coefficient of thermal expansion (CTE) to the NiP plating.

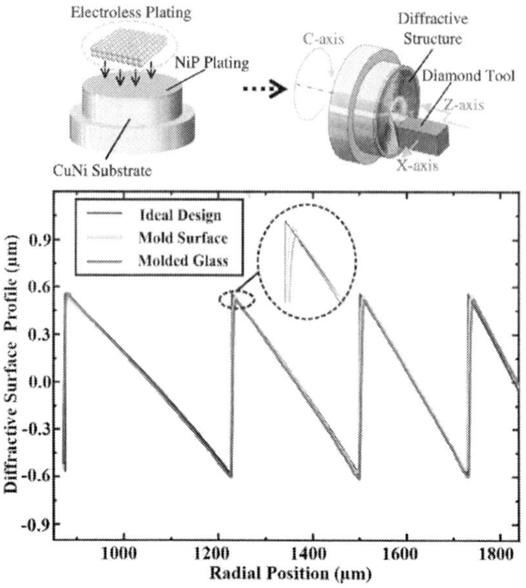

Fig. 4. (top) Schematic diamond turning process of the MODE lens' diffractive Fresnel surface mold. (bottom) Diffractive Fresnel surface profile comparison between the design, the diamond turned mold, and the molded glass surface. [4]

The diffractive Fresnel surface molding capability was demonstrated in terms of surface micro roughness and diffractive optical surface profile accuracy, as presented in Figure 4(bottom). The measured surface roughness values over multiple regions show superb smoothness of Sa = 2 nm. Also, as shown in Figure 4(bottom), the molded diffractive Fresnel surface is accurately molded with <0.05 μm error between the glass and mold. [4] The multi-parameter fabrication process chain optimization with cost modelling is reported to show the scalability of the Nautilus architecture. [5]

4 Conclusions

A very large aperture MODE lens for astronomical space telescopes is ushering in the era of large lens-based space telescopes that provide unprecedented photon collection efficiency in order to probe deep into space for surveying exoplanets for biosignatures.

References

1. D. Apai, et al., The Astronomical Journal, 158:83 (21pp) (2019).
2. T. Milster, et al., Appl. Opt. **59**, 7900-7906 (2020).
3. "Future of Astronomy," *Sky & Telescope*, November Issue, 32-33 (2021).
4. Y. Zhang, et al., Opt. Lett. **45**, 6438-6441 (2020).
5. M. Dupuis, et al., J. Astron. Telesc. Instrum. Syst. 8(2) 027001 (2022).

Ray Transfer Matrix for Onion-Type GRIN Lenses

Veronica Lockett[1], *Rafael* Navarro[*1], and *Jose Luis* López[2]

[1]INMA, Consejo Superior de Investigaciones Científicas and Universidad de Zaragoza, Zaragoza, Spain

[2]Universidad Pública de Navarra, Pamplona, Spain

Abstract. We present the computation of an ABCD matrix for onion-type GRIN lenses. By applying a differential approximation of the layer thickness, the matrix product of a high number of matrices is synthetized into a single matrix where the elements are integrals. The difference between this ABCD matrix and a homogeneous lens matrix is one integration term in element C, which is the GRIN contribution to the lens power. In the case of the crystalline lens, the analytical approximation to the GRIN lens power provides an accurate and concise solution in terms of Gaussian hypergeometric functions.

1 Introduction

Gradient index (GRIN) exists in a wide range of natural and artificial optical systems. There are different approaches to ray tracing computation in GRIN media. Paraxial ray tracing is essential as it determines fundamental properties of optical systems, such as cardinal elements and lens power. Ray transfer matrices are a concise and straightforward way to describe each element of an optical system.

In the particular case of the crystalline lens, there are two main approaches [1] later used to formulate the ABCD matrix. They consider either a slab-type [2] lens or a shell-type lens [3]. The latter involves the product of a high number of matrices.

Our goal is to formulate the ABCD transfer matrix for onion-type GRIN lenses. We apply a differential approximation where the layer thickness tends to zero, and the matrix product becomes a sum; then, each matrix element is computed as an integral. When calculating the power, each layer's contribution is determined by the product of its surface curvature and refractive index increment.

2 Methods

2.1 Ray Transfer Matrix of a GRIN Lens

We formulate the paraxial ray tracing through a GRIN lens with an index distribution $n(r,\theta,z)$ in cylindrical coordinates. The paraxial index distribution depends only on the z coordinate, $n \approx n(z)$. In onion-type GRIN lenses each iso-indicial surface (IIS) can be realized as a refracting optical surface. For a layer k crossing the axis at a distance z from the anterior surface, the transfer matrix is the product of a translation and a reflection matrix.

$$L_k(z_k) = \begin{bmatrix} 1 & -\Delta z \\ 0 & 1 \end{bmatrix} \begin{bmatrix} 1 & 0 \\ \frac{\Delta n}{(n+\Delta n)R} & \frac{\Delta n}{n+\Delta n} \end{bmatrix} \tag{1}$$

Where $\Delta n = n'\Delta z$, $n' = dn/dz$, $n' = n'(z)$, and $R = R(z)$. With some manipulation and simplification, we obtain a matrix which is the addition of the identity matrix, a new matrix A, independent of Δz, and a residue with higher-order terms of $(\Delta z)^2$.

$$L_k(z_k) = I + A \cdot \Delta z + \mathcal{O}(\Delta z)^2 \tag{2}$$

Where

$$A(z_k) = \begin{bmatrix} 0 & -1 \\ \frac{n'(z)}{n(z)R(z)} & -\frac{n'(z)}{n(z)} \end{bmatrix} \tag{3}$$

The whole lens transfer matrix can be computed as the matrix product of all k layers from the back to the front of the lens. We approximate the product to a sum by neglecting higher-order terms of Δz.

$$M = \prod_{k=1}^{K} L(z_k) \approx I + \sum_{k=1}^{K} A(z_k)\Delta z + \mathcal{O}(\Delta z)^2 \tag{4}$$

In the continuous limit, $\Delta z \to 0$, and we convert the sum into an integral and replace Δz with dz.

$$M \approx \begin{bmatrix} 1 & -t \\ \int_0^t \frac{n'(z)}{n(z)R(z)} dz & 1 + \log\left(\frac{n(0)}{n(t)}\right) \end{bmatrix} \tag{5}$$

The anterior lens vertex is at $z = 0$, and t is the lens thickness. Matrix M represents the ray propagation inside the GRIN medium. When this formulation is applied to a biconvex homogeneous lens, the power is equivalent to that of a thin lens where the refractive indices are replaced with their natural logarithms.

The index changes abruptly along the surface between the lens and the external medium, so we formulate the refraction at each external surface independently. The complete lens transfer matrix comprises three matrices:

* Corresponding author: rafaelnb@unizar.es

$L = S_p M S_a$, where S_p and S_a correspond to pure refraction at the posterior and anterior surfaces.

Element L_{21} of matrix L is the most relevant, as it determines the lens power. When comparing L with the transfer of a homogeneous lens, all elements are the same except an extra term in element L_{21}, which accounts for the GRIN contribution to the power.

$$\phi_G = n_0 \int_0^t \frac{n\prime(z)}{n(z)R(z)} dz \qquad (6)$$

2.2 Application to the Crystalline Lens

The human crystalline lens is a paradigmatic example of gradient index (GRIN) optics. Its onion-like inner structure can be pictured as a succession of thin meniscus lenses with their refractive power given by curvature and refractive index increment.

Most advanced lens models replicate the lens index distribution: a greatly inhomogeneous cortex with an increasing refractive index gradient towards the centre and a much more homogeneous nucleus [4-6].

For our formulation [7], we consider a sort of cemented doublet in which the anterior and posterior parts join at a central interface but are treated independently. The axial thickness is the sum of the anterior and posterior thicknesses, $t = t_a + t_p$. The refractive index distributions are [5-7] approximated to power laws of the normalized axial distances to the anterior and posterior external surfaces, z_0.

The external lens geometry–radii of curvature, R_a and R_p, and conic constants, Q_a and Q_p–will determine the shape of the iso-indicial surfaces.

2.2.1 Analytical Transfer Matrix

When we apply Eq. (6) to the crystalline lens geometry and index distribution, the integral cannot be computed in terms of known functions. We worked out a highly accurate approximation for the GRIN contribution to the power using Gaussian hypergeometric functions.

$$\begin{aligned}\phi_{Ga} \simeq &- \frac{n_s \delta n / n_c}{R_a - G(Q_a+1)t_a} {}_2F_1\left(\begin{matrix}1, p\\ p+1\end{matrix}\middle| -\frac{G(Q_a+1)t_a}{R_a - G(Q_a+1)t_a}\right) \\ &+ \frac{n_s(-\delta n/n_c)^2}{2(R_a - G(Q_a+1)t_a)^2} {}_2F_1\left(\begin{matrix}1, 2p\\ 2p+1\end{matrix}\middle| -\frac{G(Q_a+1)t_a}{R_a - G(Q_a+1)t_a}\right)\end{aligned} \qquad (7)$$

Here n_s and n_c are the refractive indices at the lens surface and core, respectively, G is the curvature gradient [7], $\delta n = n_s - n_c$ is the index gradient, and ${}_2F_1$ represents the Gaussian hypergeometric function. There is an equivalent expression for the posterior region of the lens.

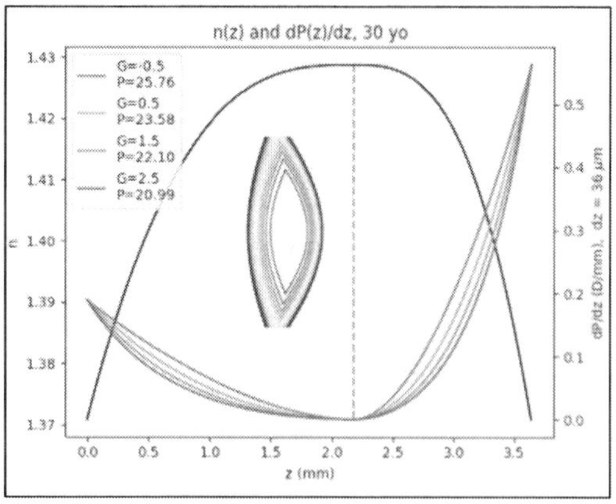

Fig. 1. Axial index distribution n and power differential dP/dz for a 30-year-old model. The label specifies the curvature gradient index G and the resulting refractive power P.

3 Results

To check the validity of our approximation, we compute the crystalline lens power for different values of G both numerically and analytically and compare the results. The average error is 0.14% between analytical and numerical calculations. Compared with a thin lens approximation of the integral, $\int n'(z)dz/R(z)$, the lens power difference with a GRIN lens is 2.13%.

4 Conclusion

The GRIN transfer matrix is like a single-layer transfer matrix, with its elements replaced by integrals, allowing an affordable and straightforward computation.

Matrix L differs from the matrix of a homogeneous lens only in one integration term in element L_{21} that represents the GRIN contribution to the power.

In the case of the crystalline lens, the analytical approximation to the GRIN lens power provides an accurate solution that relies on Gaussian hypergeometric functions, available in both commercial and open-source computing environments.

References

1. D. A. Atchison & G. Smith, Vision Res. **35**(18), 2529-2538 (1995)
2. J.A. Díaz, Appl. Optics **47**(2), 195-205 (2008)
3. S. Giovanzana, T. Evans, and B. Pierscionek, Biomed. Opt. Express **8**(11), 4827-4837 (2017).
4. B. Pierscionek, Exp. Eye Res. **64**, 887–893 (1997)
5. R. Navarro, F. Palos, and L. González, J. Opt. Soc. Am. A **24**(8), 2175–2185 (2007).
6. C. Sheil & A. Goncharov, Biomed. Opt. Express 7(9), 1985-1999 (2016).
7. R. Navarro, S. Baquedano, A. Sánchez-Cano, Opt. Express **29**(20), 30998-31009 (2021)

https://doi.org/10.1051/epjconf/202226603016

Control of thermal emission for thermophotovoltaic systems

Daniela De Luca[1,2,*], *Antonio* Caldarelli[2,3], *Eliana* Gaudino[2,3], *Umar* Farooq[1,2], *Marilena* Musto[2,3], *Emiliano* Di Gennaro[1,2], and *Roberto* Russo[2]

[1]Department of Physics, Università degli Studi di Napoli "Federico II", 80125 Napoli, Italy
[2]Deparment of Industrial Engineering, Università degli Studi di Napoli "Federico II", 80125 Napoli, Italy
[3]Consiglio Nazionale delle Ricerche, Istituto di Scienze Applicate e Sistemi Intelligenti, 80131 Napoli, Italy

Abstract. Thermal emitters play a key role in controlling the thermal radiation emitted in thermophotovoltaic systems and in increasing their energy conversion efficiency. Here, we present different designs of emitters with spectrally selective properties, based on easy-to-fabricate multilayer structures and characterized by a sharp transition from high to low emissivity in the region of interest. Those structures make use of refractory materials to allow working at high operating temperatures and they can be easily customized to maximize the thermal emission in the region of the desired wavelengths.

In recent years, the interest in thermal emissions in the infrared (IR) range is growing fast as the control of thermal emissions in that region is crucial for reaching high efficiency in various energy systems. However, it involves different other fields and applications like thermography [1], thermophotovoltaic (TPV) systems [2], radiative cooling [3, 4], infrared spectroscopy [5, 6], etc.
A TPV system employs a photovoltaic (PV) module and a thermal collector and allows the direct conversion of the thermal radiation emitted from an object heated at temperatures typically higher than 1000 K to electricity. Hence, in those devices, the control of thermal emission is fundamental to improve their thermal efficiency, defined as

$$\eta = \frac{P_{generated}}{P_{absorbed}}, \qquad (1)$$

where $P_{generated}$ and $P_{absorbed}$ are the electrical power generated by the PV cell and the total thermal radiative power absorbed, respectively.
As a result of the well-known PV cell operation, the radiation incident on the PV cell must match its bandgap. Hence, the key role of the thermal emitter, a device that emits thermal radiation and for which selectivity is required: to reduce the amount of wasted heat, it must emit only the photons with energy included in the range of interest to make the PV cell work properly.

Recalling the Kirchhoff's law of thermal radiation, any object in thermal equilibrium radiates a quantity of power that is equal to the power absorbed. Hence, in the case of negligible transmission through the designed structure, the emissivity of an object equals its absorptivity, and we can consider the calculated absorptivity and emissivity as equal:

$$\alpha(\lambda) = \varepsilon(\lambda) \qquad (2)$$

The ideal spectrally selective thermal emitter must have the maximum emissivity for photons with energy in the region of interest (ε=1), while zero emissivity for photons with energy lower than the bandgap energy of the PV cell. In addition, due to the high working temperatures, stability and durability above 1000 K are required.

The structures investigated so far are mainly based on the use of metamaterials [7] and metasurfaces [8,9], photonic crystals [10], gratings [11,12], and plasmonic structures [13-15]. A recent review article [16] highlights the advantages and disadvantages of the various types of emitters, the latter are mostly linked to the thermal and chemical stability at high temperatures. In addition, all of them lack the possibility of being easily reproduced on a large scale. Therefore, in this work, we want to describe and demonstrate the feasibility of easy-to-fabricate selective emitters based on a multilayer structure with a controlled thermal emission.

The materials involved in the metamaterial structure are selected from metals, oxides, carbides, and nitrides, all characterized by good thermal stability and high melting point. In addition, the appropriate choice of a sub-layer allows for ensuring negligible transmissivity in the whole range of investigated wavelengths.
A preliminary result on the simulated absorption spectrum obtained from a multilayer made of Si, W, SiO_2 and Cr_2O_3 is depicted in Figure 1. It has a narrow emission peak with absorptivity close to 1 centered around 1100 nm, reaching a negligible value of emissivity already before reaching 1500 nm.

* Corresponding author: daniela.deluca@na.isasi.cnr.it

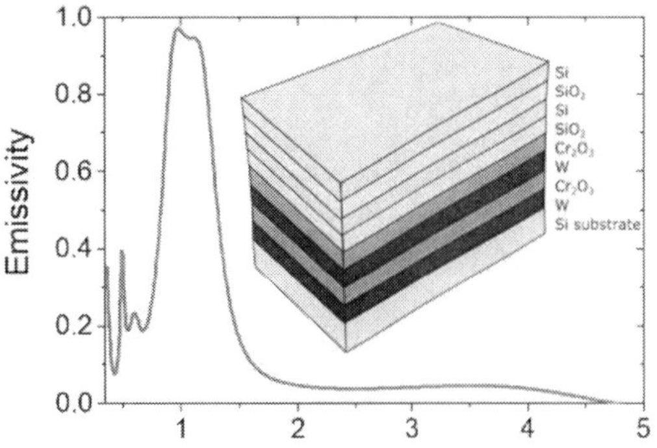

Fig. 1: Absorption spectrum of a multilayer based on Si, W, SiO_2, and Cr_2O_3.

The thermal radiation can be described by the spectral emissivity, $\varepsilon(\lambda)$, and the thermal radiation from a black body, $I_{BB}(\lambda, T)$, according to the Eq. (3):

$$I(\lambda, T) = \varepsilon(\lambda) \cdot I_{BB}(\lambda, T) \quad (3)$$

Figure 2 shows that the emission spectrum of the selective emitters occurs over a narrower region and is always lower than that of the blackbody at the same temperature, highlighting its selective behavior.

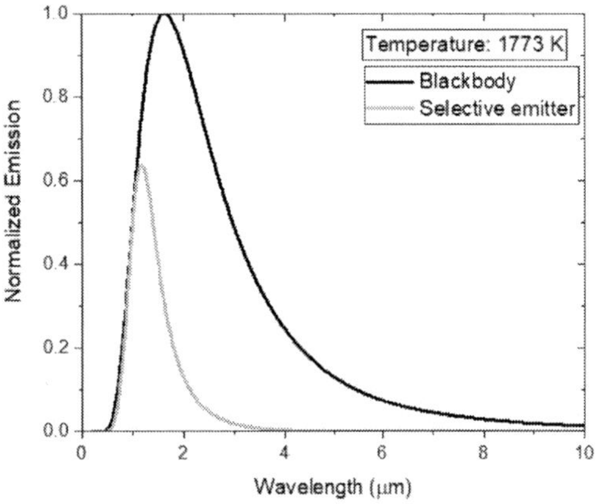

Fig. 2: Normalized spectral radiation emitted by the BB (black line) and the designed selective emitter (green line) calculated at 1773 K.

Further studies are in progress to tune the emission peak at longer wavelengths and to obtain an even more selective behavior in the emissivity curve. In this way, selective emitters could further increase the efficiency of a thermophotovoltaic system compared to what has been demonstrated in [15], and substantially help in thermal energy storage [17].

References

1. C. Meola, G. M. Carlomagno, Meas. Sci. Technol. **15**, R27_R58 (2004).
2. A. Lenert, et al. Nat Nanotechnol **9**, 126–30 (2014).
3. A. P. Raman, M. A. Anoma, L. Zhu, E. Rephaeli, S. Fan, Nature **515**, 540–544 (2014).
4. N. N. Shi, et al. Science **349**, 298–301 (2015).
5. Y. De Wilde, et al. Nature **444**, 740–743 (2006).
6. B. Stuart, Infrared Spectroscopy: Fundamentals and Applications (JohnWiley, 2004).
7. P. N. Dyachenko et al. Nature Communications **7** (2016).
8. D. Costantini, et al. Physical Review Applied **4**, 014023 (2015).
9. C. Luo, A. Narayanaswamy, G. Chen, J. D. Joannopoulos, Physical Review Letters **93**, 213905 (2004).)
10. J.-J. Greffet, et al. Nature **416**, 61–64 (2002).
11. N. Dahan, Y. Gorodetski, K. Frischwasser, V. Kleiner, E. Hasman Physical Review Letters **105**, 136402 (2010).
12. C.-M. Wang et al. Optics Express **15**, 14673–14678 (2007).
13. Y.-H. Ye et al. Applied Physics Letters **93**, 263106 (2008).
14. J. A. Mason, S. Smith, D. Wasserman, Applied Physics Letters **98**, 241105 (2011).
15. Z. Omair et al. PNAS **116**, 15356-15361 (2019).
16. Z. Wang et al. Sol. En. Mat. & Sol. Cells 238 (2022) 111554.
17. Amy, C. et al. Energy & Environmental Science **12**, 334-343 (2018).

Dynamic optimization of optical design process by means of producibility modulations

O.Resnik[1], O.Faehnle[2] and Y.Arazi[1]
[1] JOYA Team, Ramat Yishai, Israel
[2] OST-University of Applied Sciences, Buchs, Switzerland,

In this paper, PanDao is used dynamically in the early stage of the optical design process so that manufacturability aspects can be considered during the process.

Introduction

While optical design translates the application parameters of optical systems such as MTF and image resolution into a set of well-defined technical drawings, their manufacturability can only be assessed retrospectively, e.g., by human analysis or, more recently, by the application of PanDao [1,2]. The logical next step should be the integration of manufacturing analysis already during the creation of the optical design and even incorporation of these features inside the optical design software. This article reports on the first attempts to use PanDao dynamically further upstream, during the early design phase.

Cost impact analysis

For a specific optical design, we define tolerances that are derived from tolerance analysis and then these tolerances are divided to optical elements and to the mechanics – placements of the optical elements and the housing. In this paper, PanDao is applied in the early stage of the optical design process to estimate several levels of optical elements tolerances and their impact on manufacturing processes and costs, knowing the mechanics precision capabilities, and then to assess the need in active alignment and calibration with their potential impact on assembly time and cost. That way, PanDao is applied dynamically during the optical design process; this in contrary to its standard application judging and comparing optical designs after they have been finished.

The following diagram (see figure-2), shows two different options that can be evaluated with PanDao and for which we can extract the production costs of the optical elements. In this way, the costs for machining, coating, centering, tooling, machine setup and maintenance are taken into account in the early design phase and used for design decisions. In other words, we understand the pricing for the different tolerances and can also evaluate pricing of different levels of yield. Based on this information and with additional inputs and knowledge of other assembly steps costs we can ultimately understand the overall cost of each option and use this information to decide which serves better the specific product. In addition, the assembly process can be planned, the assembly tooling, calibration setups and testing setup design can be done in parallel to the optical elements production so that the overall time to build the prototypes can be reduced.

Another benefit of this process is the ability to control the whole process and assess the impact of each decision based on realistic estimates.

The whole optical design to manufacturability process is a complex topic, since many aspects must come into consideration and a lot of information regarding the manufacturing capabilities is not available and transparent to the optical designers.

We can illustrate this complex space with the following diagram (see Figure 2), where the differently colored areas represent the optical and mechanical manufacturing risks; in addition, one can evaluate the different types of assembly processes (sand and marble areas).

Over the entire space lies the constant performances curve and the goal is to reach the point of optimal performances-manufacturability-cost. This curve depends on the overall system complexity, but in general, it passes through different fabrication risk & cost areas, as shown in Figure 2. This is a simplified picture, but it illustrates the different layers of information and various considerations. The earlier in the design stage this picture can be clarified, the easier it is to make the right decisions regarding the whole system approach and ultimately gain the confidence that the designed product can be manufacturable with controlled yield, profitable and scalable.

PanDao helps to cover the "Optical Elements Tolerances" axis of this chart, identifying both the non-manufacturable area and the area with high and low manufacturing costs and risks. Furthermore, using PanDao the optical designer can evaluate manufacturing cost and risk by him/herself without the necessity to consult with dedicated specialists each time a design change is made. Each optics manufacturer has the partial data, but the overall picture is not existent anywhere, except now with PanDao. PanDao output provides the optical elements productions costs and tolerances pricing through more than 340 different manufacturing process types and capabilities, clarifying the whole production costs, that can be expressed using the following formula:

$$\textit{Cost} = \textit{Optics Production} + \textit{Housing Production} + \textit{Assembly Tools} + \textit{Assembly Labor} + \textit{Yield Cost}$$

Each term in 2 different process options costs (cost X and Y in figure-1 mainly depends on the optical elements production costs while other costs change correspondingly, finally generating the overall product cost. Qualitative comparison of the different terms can be done intuitively and easily, but the overall cost point comparison cannot be done until the end of the whole process.

To visualize this, we can picture a cost surface over the producibility space chart (see Figure 2), as shown in figure-3. Using PanDao, we can shift the optical tolerances working point, at the relevant range of mechanical tolerances and the desired assembly process and derive the minimal-possible optical elements cost, which is the overall cost drive. This can be done for each assembly process eventually optimizing the overall producibility working point (located in the "optimal working area" in Figure 2).

Conclusion

In conclusion, first steps have been reported towards the use of PanDao dynamically in an early phase of optical design. That way, the information about fabrication cost of a specific optical design and tolerances can be taken into consideration when making optical design decisions to achieve cost-effective manufacturable product.

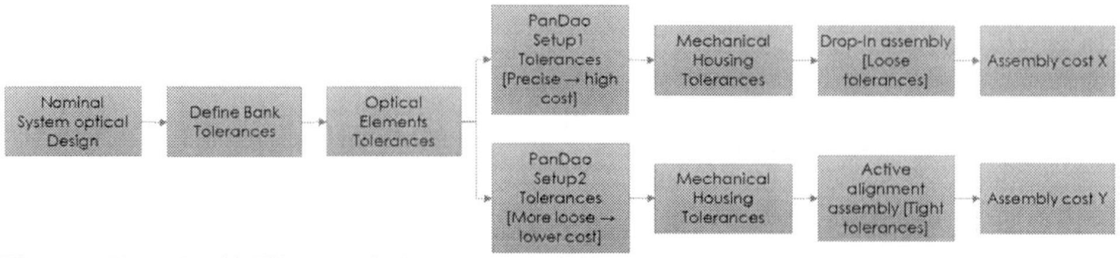

Figure 1 – Example of 2 different optical system producibility options

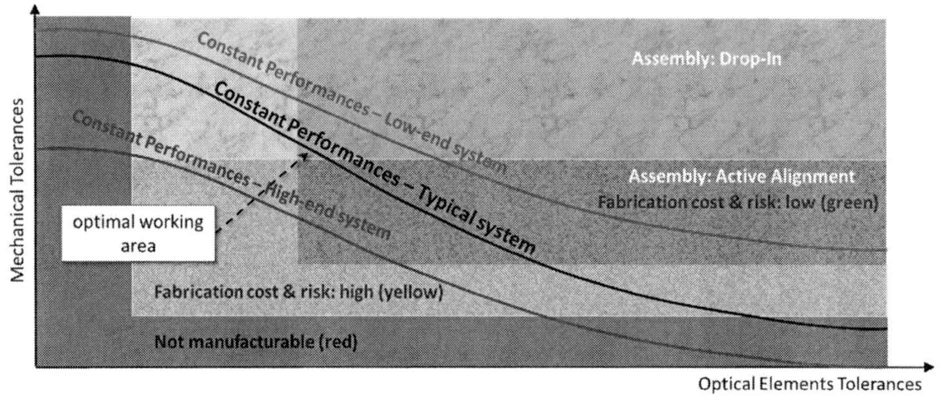

Figure 2 – The producibility space chart: Optical systems fabrication and assembly assessment, where optical tolerances (X axis) and mechanical tolerances (Y axis) increase means that tolerances loosen. In the outlined optimal working area, that can be identified using PanDao, the required performances are met with minimum fabrication cost & risk, as well as drop-in assembly enabled.

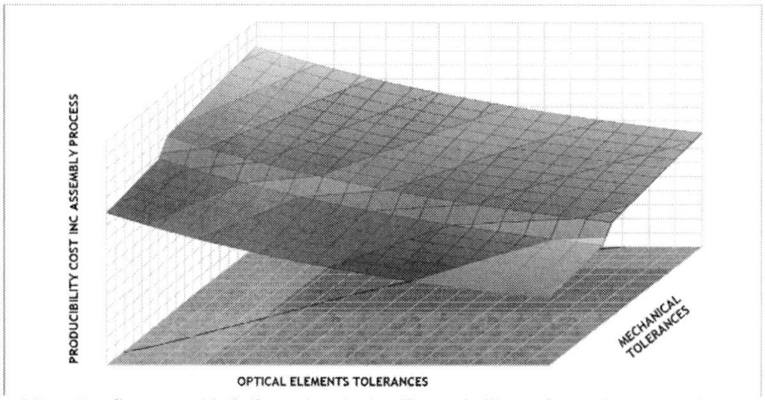

Figure 3 – Adding the Cost as a third dimension to the Figure 2 Chart shows the cost reduction when optical elements tolerances (X axis) and mechanical tolerances (Y axis) increase/loosen; the step-down represents the drop-in assembly process. Optical systems production cost evaluation using PanDao helps to find the most cost-effective point over the optimal working area.

References

[1] www.PanDao.ch
[2] O.Fähnle and I.Livshits, "Optical Fabrication Process Modeling", OSA's Optical Fabrication and Testing conference (OF&T), part of the Optical Design and Fabrication congress (ODF), Rode Island, NY, USA, June 2021

https://doi.org/10.1051/epjconf/202226603018

Improved Production of Large and Multi-Directional Homogeneous Optical Glass: SCHOTT N-BK7® for Challenging Applications

Fabian Rupp[1,*], *Ralf Jedamzik*[1], *Volker Dietrich*[1], *and Uwe Petzold*[1]

[1]SCHOTT AG, Advanced Optics, Hattenbergstr. 10, 55122 Mainz, Germany

Abstract. In many applications, the spatial refractive index variation – called homogeneity – within a measurement aperture either in one or two directions is important. Typical application examples are prisms in ultra-precision metrology with stability in multiple directions. Large lenses are used in artificial laser guide star systems for atmospheric correction in large telescopes. The challenge of enabling highest refractive index homogeneities requires tight control of all production steps from melting to hot forming and fine annealing. Large optical formats can be produced as singular castings in moulds up to 1.2 m in diameter and 250 mm thickness. Smaller formats are available as blocks produced in dimensions of approx. 250 x 250 x 180 mm³. A more economic and ecologic way is the production of continuous strips of glass up to approx. 500 mm width and 120 mm thickness. Recently SCHOTT has improved the homogeneity of these continuously produced jumbo strips significantly. Now homogeneity of up to H4 quality (2 ppm index maximum variation) can be provided on apertures up to approx. 900 mm x 500 mm. This paper gives an insight overview on the latest results and current state of this topic at the optical glass manufacturer SCHOTT.

1 Homogeneity of the refractive index

The absolute refractive index of optical glass is measured with refractometers at SCHOTT. The efficient V-block refractometer provides index control up to the fifth digit, while the spectral goniometer "Ultraviolet to infrared Refractive Index measurement System" (URIS) ensures reliable measurement over a broad spectral range up the sixth digit[1].

Whereas homogeneity is defined as the peak to valley variation of the refractive index at a certain aperture. The measurement is carried out with an optical interferometer. To exclude surface effects from the measurement either oil-on-plate or PHom method are used. The obtained wavefront deformation is a function of the lateral variation of the refractive index integrated over the sample thickness[2].

Corresponding to the variation of the index within the defined aperture the maximum and minimum value define the so-called peak-to-valley (PV) in ppm. Table 1 provides an overview of the homogeneity grades at SCHOTT according to the ISO-norm. The homogeneity grades can be applied in multiple directions for 2D or 3D homogeneity, respectively, depending on customer needs.

Homogeneity is crucial for optical applications with highest performance and tightest tolerances where a low wavefront deformation is mandatory to ensure the proper functionality of the device.

Since homogeneity scales with size/aperture it is important to specify homogeneity properly to the requirements of the application. Typical fields of application are: large lenses in artificial laser guide star systems of extremely large telescopes for atmospheric correction, ultra precision metrology with homogeneity requirements in multi directions for IC lithography, high power laser fusion facilities, or interferometer industry.

Table 1. Homogeneity grades of SCHOTT optical glass according to ISO-norm [3, 4]

SCHOTT homogeneity class indicator	NEW ISO 12123:2018 and ISO 10110-18:2018		OLD ISO 10110-4:1997	General applicable for
	Class indicator	tolerance limits Δn [10^{-6} or ppm]	tolerance limits Δn (+/- notation) [10^{-6} or ppm]	
	NH100	100	± 50	Common application sizes
H1	NH040	40	± 20	
H2	NH010	10	± 5	Partial volumes of the raw glass
H3	NH004	4	± 2	
H4	NH002	2	± 1	Not in all dimensions and not for all glass types
H5	NH001	1	± 0.5	

At SCHOTT we currently run three different interferometers for homogeneity measurements. For large blank and castings a Zygo Verifire MST with 600 mm (24 inch) aperture located in Duryea (USA) and a Zeiss 100 Direct with and 500 mm aperture, located in Mainz (Germany) are available. For larger dimensions multiple

Confidential

* Corresponding author: fabian.rupp@schott.com

measurements of the given aperture are carried out and stitched via software to the full extent of the specimen, as shown in Figure 1. Additionally, SCHOTT operates another Zeiss Direct 100 interferometer with 300 mm aperture in Mainz (Germany).

Fig. 1. Interferometer Zygo Verifire with 600 mm aperture at SCHOTT in Duryea with large casting of SCHOTT N-BK7®.

2 Typical formats of SCHOTT N-BK7®

The optical glass SCHOTT N-BK7® with high homogeneity is available in different formats and geometries of pressings, blocks, jumbo strips, and castings within a broad range of weights from approx. 0.1 – 700 kg and sizes from cm to m, see Figure 2. Beside the format described above individual format are possible.

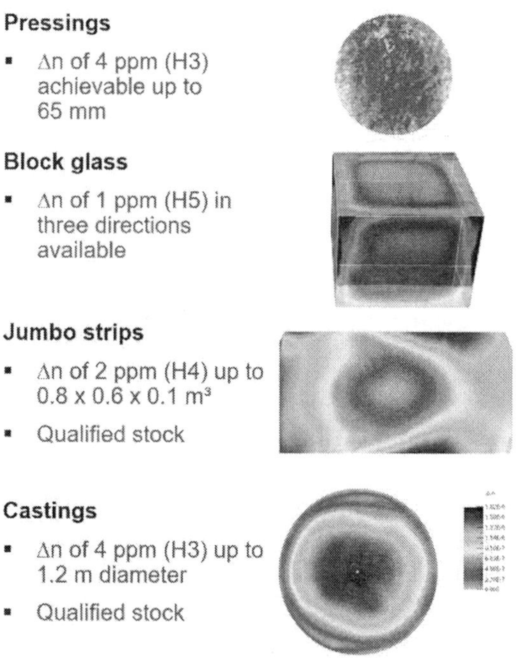

Fig. 2. Overview of different formats and geometries of SCHOTT N-BK7® with typical high homogeneity quality grades in false color code of interferometric measurement. Pressings are available up to 4 ppm (H3), while blocks and jumbo strips can serve up to 1 ppm (H5) and 2 ppm (H4), respectively, even in multiple direction. Large castings achieve highly symmetric homogeneity distributions up to 4 ppm (H3) on a 1-m-scale or even better with reduced diameter.

3 Improved quality of SCHOTT N-BK7®

The production of optical glass with its tight tolerances in the optical position, low stress birefringence and inner quality is extremely challenging. Nevertheless, for optical glass of high homogeneity the requirements are one to two orders of magnitude tighter specified.
To avoid inefficient pick and choose for high homogeneity optical glass affecting reliability and delivery times it was inevitable to re-evaluate each step in the manufacturing process and identify potential levers on homogeneity: raw materials, melting process, refiner, hot forming, as well as coarse annealing and fine annealing. With an interdisciplinary team from SCHOTT it was possible to identify the most relevant levers and to adjust accurate process settings yielding high homogeneity of 2 ppm even on large apertures in the regime of 800 x 400 mm² as depicted in Figure 3.

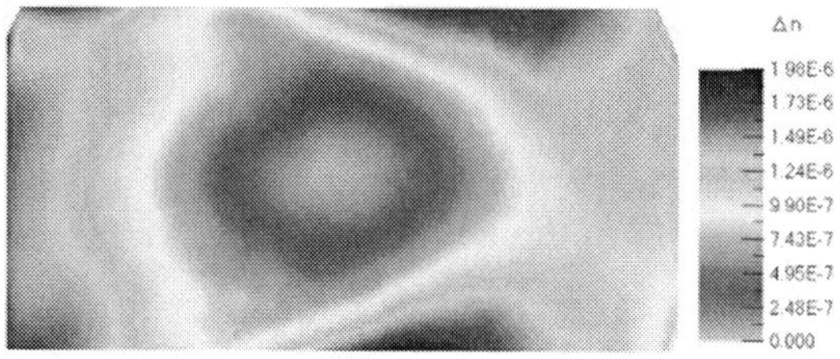

Fig. 3. Representative interferometric measurement of the refractive index of a SCHOTT N-BK7® jumbo strip with an aperture of approx. 800 x 400 mm² in false color code. The low peak-to-valley index variation achieves 1.98 ppm (H4).

Acknowledgemets

We gratefully acknowledge our SCHOTT colleagues from Advanced Optics, RnD and Technical Service for their contributions and fruitful discussions.

References

1. F. Rupp, R. Jedamzik, L. Bartelmeß, U. Petzold, Proc. SPIE 11873 (2021)
2. R. Jedamzik, U. Petzold, Proc. SPIE 10914 (2019)
3. ISO 12123:2018, Optics and photonics – Specification of raw optical glass (2018)
4. ISO 10110-18:2018, Optics and photonics – Preparation of drawings for optical elements and systems – Part 18: Stress birefringence, bubbles and inclusions, homogeneity, and striae (2018)

Confidential

Mechanical Integration of a Prism-Grating-Prism-Assembly for the CO2M Mission

Andreas Kamm[1], *Christian Scheffler*[1*] *and Thomas Peschel* [1]

[1]Fraunhofer Institute of Applied Optics and Precision Engineering IOF, Albert-Einstein-Str. 7, 07745 Jena, Germany

Abstract. We developed and realized a mechanical integration concept of a prism-grating-prism-assembly for the CO2M mission. The mechanical design of the mounts relies on a kinematic mounting of the optical elements. Additional, ultra-light and blackened covers for stray light and disturbance light suppression are included. The complete assembly was investigated by extensive thermo-mechanical simulations to verify the stability of the mechanical design under operational, launch loads and test conditions.

1 Introduction

Climate change has a considerable impact on the future of our planet. Monitoring the amounts of greenhouse gases in our atmosphere from space is a key element for understanding and controlling this process. Earth observation missions like the "Copernicus Anthropogenic Carbon Dioxide Monitoring Mission" (CO2M) represents a considerable part of ESA's contribution to monitor greenhouse gases. Missions like CO2M require ever more efficient spectrometers.

One approach to meet the increased optical requirements for the dispersing assemblies of such spectrometers is the use of prisms in combination with gratings (PG- and PG+P-assemblies). The required complexity, size and mass of such arrangements has increased over the past few years. While the PG-Assembly of the Sentinel 5 mission was still a prism and a grating with an air gap in between and a total mass of the two optical components of 0.34 kg, for the CO2M mission, PG+P-Assemblies are currently being designed and built at the IOF, each of which has a weight of up to 3.5 kg for the optics.

To precisely adjust prismatic components and hold them securely for space missions, a kinematic mount including integration technology for a reflective grating made of quartz glass was developed as part of the "Sentinel 4 Grating Unit" project in 2013. This technology has been further developed for the Sentinel 5 PG and is now also used in the PG+P-Assemblies of the CO2M mission.

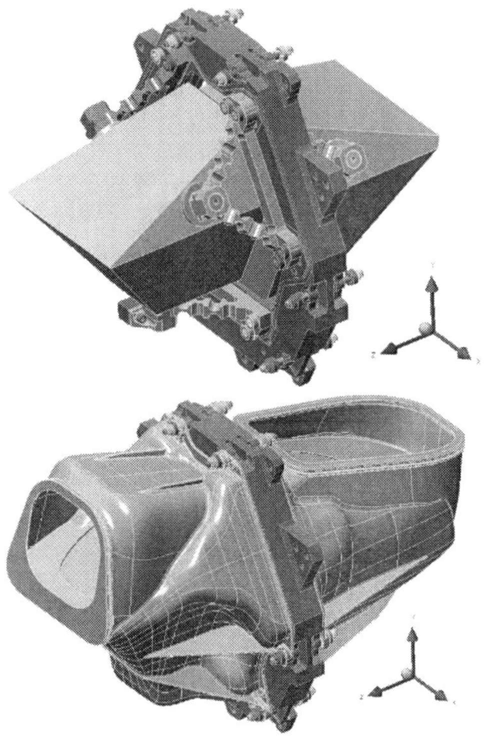

Fig. 1. CAD model of the CO2M-SWIR1-PG+P-Assemblies without covers (top) and complete assembly including the covers (bottom)

2 Mechanical Design

In the PG+P assemblies for the CO2M mission, the grating is already connected to the first prism before the main integration. The glass components are joint using a direct hydrophilic bonding process [1] without any additional material between the components.

The mechanical design of the mount relies on a kinematic mounting of the optical elements using Titanium bipods which form a hexapod-like structure holding each prism (see fig. 1). The connection between the prisms and the bipods is formed by Invar pins which are glued to both partners of the joint.

* Corresponding author: christian.scheffler@iof.fraunhofer.de

To suppress stray light and disturbance light, the prisms are enclosed in two ultra-light and blackened covers, which are additively manufactured at the IOF using Selective Laser Melting (SLM) [2].

The bipods are screwed to the basic structure and the pins are inserted without fixing them. Then the prisms are placed between the pins and adjusted. The gaps required for this are then closed in a two-step gluing process. Since the prisms are only fixed in the final step, all mechanical stresses previously introduced into the assembly are not passed on to the optics. As a result, a very precise, low-stress and long-term stable mounting of the prisms is possible.

3 Results of the thermo-mechanical analysis

FEM analyses are used to verify the stability of the mechanical design of the PG+P assemblies under operational, launch loads and test conditions.

A modal analysis revealed a fundamental frequency of the assembly of 611 Hz, however because of their symmetry properties the lowest three Eigenmodes have negligible impact on the vibration response. The lowest relevant modes are in the range between 882 Hz and 946 Hz, which is compatible with the expected vibration loads.

Similar investigations have been performed with respect to static and temperature loads as well as assembly tolerances. In summary the results of the FEM simulations show that the PG+P assembly will withstand all relevant mechanical and thermal loads.

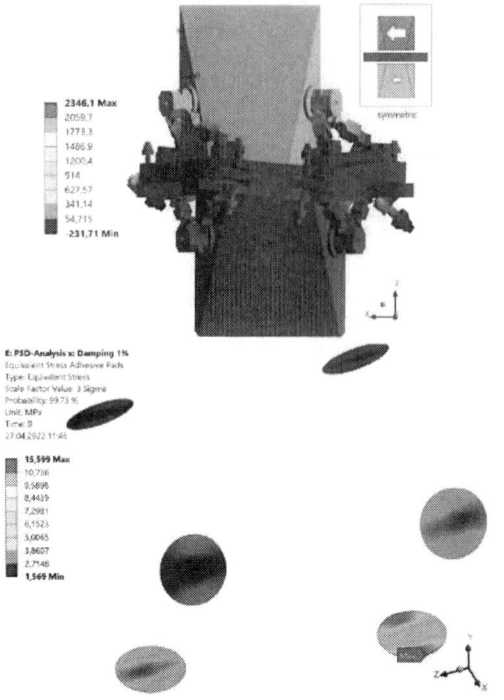

Fig. 2. Lowest relevant mode of the CO2M-SWIR1-PG+P-Assembly (top, covers suppressed for display only) and corresponding stress in the adhesive joints under vibrational excitation in x (bottom).

This work was founded by the Copernicus Program of the European Union, co-founded with ESA.

References

1. T. Flügel-Paul, C. Rothhardt, T. Benkenstein, K. Grabowski, S. Risse, R. Eberhardt, B. Guldimann, U. D. Zeitner, *Proc. SPIE* 11180:1118014 (2019)
2. N. Heidler, H. von Lukowicz, E. Hilpert, S. Risse, L. Alber, J. Klement, F. Heine, R. Bölter and J. M. P. Armengol, EPJ Web Conf. Volume **215** (2019)

Enabling photonic system integration by applying glass based microelectronic packaging approaches

Henning Schröder[1*], *Wojciech Lewoczko-Adamczyk*[1], *Daniel Weber*[2]

[1]Fraunhofer Institute for Reliability and Microintegration (IZM), Gustav-Meyer-Allee 25, 13355 Berlin, Germany
[2]Technical University of Berlin, Gustav-Meyer-Allee 25, 13355 Berlin, Germany

Abstract. Advanced hybrid packaging technologies are used to enhance functionality of glass-based substrates featuring electrical, thermal and optical components including laser diodes, modulators, isolators, photonic integrated circuits, beam-splitters and micro lenses. Such glass-based substrates can be either thin glass layers on large panels containing optical waveguides or more mini-bench-like boards. Optical fiber interconnects, plugs, and electrical-optical integration platforms are used for higher level system integration. We discuss thin glass as a suitable base material for ion exchanged waveguide panels and interposers, precise glass structuring for posts and holders, electrical wiring and the related high precision assembly techniques.

1 Introduction

In this work we address photonic systems in the size range of mm to cm focusing on miniaturization, small module integration, and manufacturing. The need is driven by increasing optoelectronic device diversity, progress in micro-optics and the broad range of materials to adress wavelength specific applications requirements.

Photonic packaging has to follow advances in chip and component manufacturing. Apart from vertical cavity surface emitting laser, edge emitting laser and photo diodes, photonic integrated circuits with integrated active and passive optical functionality require platforms for higher level system integration. It turns out that edge coupling is the most promising approach, e.g. for the development of compact laser sources, but issues stemming from low tolerance to optical misalignments and thermal effects of glued interfaces have not yet been fully resolved.

2 Glass Integration Platform

Thin glass has several advantages over other widely used substrate materials, such as silicon and organic materials [1–3]. The specific dielectric properties of glass provide a high integration potential by combining electrical and optical functionalities in a single substrate. Despite being very thin, panels, as well as the machinery to handle and structure them, are continuing to increase in size and decrease in cost. The composition and tempering of glass, to increase its robustness, e.g. in mobile electronic devices, has also progressed remarkably. Considering the thickness of the used glass substrates we distinguish the thin glass layer concept (sub-mm range) from the glass board concept (mm range) which are described in the following.

Thin glass as interposer material will facilitate thinner consumer products, communication and compute devices with increased bandwidth and performance, and high integration in photonic sensors.

Using thicker glass highly stable and miniaturized optical systems on glass boards can be defined - either standalone or embedded into a PCB environment, with or without integrated optical waveguides. Such glass boards host optical and electrical signal distribution, free-space optical elements, and transparent windows with or without additional beam shaping optics, hermetically sealed capping and interfaces to optical fiber interconnects. To create such complex systems, photonic design for packaging, advanced glass structuring technologies and electronic packaging are merged with high-precision assembly.

2.1 Thin glass layer concept for PCB integration

The combination of electrical circuits and optical waveguides in one circuit board enables high speed signal transmission and a high density of integration. In most cases an optical layer, based on glass is laminated into a standard printed circuit board. However, the excellent properties of

https://doi.org/10.1051/epjconf/202226603020

glass in terms of dimension stability, electrical high frequency and a significantly increasing mechanical strength, also allow so called electro-optical circuit boards to be fully made of glass [4].

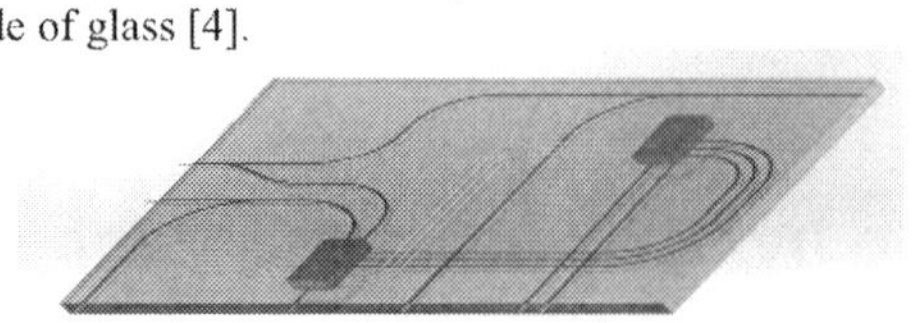

Fig. 1: Schematic drawing of a thin glass layer containing optical waveguides (black) and electrical wiring (yellow) on the top surface.

As shown in Fig. 1, optical waveguides are integrated into the glass, electrical circuits are applied and both optical and electrical components are placed on the board. Such platforms can be realized in single- and double-sided configurations. The application dependent build-up features the flexibility and diversity known from PCB manufacturing. Symmetrical approaches with one central layer only have been developed, but also glass panels with two optical layers on bottom and top side, as well as multilayer boards of stacked or large horizontal glass layer design [4, 5]. The optical waveguides are required to have low propagation loss and can be designed to operate in both single- or multi-mode regime. The process specific gradient index profile is favorable for low signal dispersion communication channels. Waveguides that can fulfill these requirements can be manufactured using an ion-exchange process, in which silver ions replace sodium ions and thus provide a locally increased refractive index.

2.2 Glass board concept and housing

For many photonic applications miniaturized optical benches have to be protected by transparent (hermetic) housings with integrated optical features like lenses, gratings or fiber feed troughs. To configure such micro-optical benches mechanical fixtures made of glass, adapters and posts are structured to not only align various micro-optical or electrical-optical components to the optical axis, but to also include visual and handling features for increased process reliability in automated assembly environments[9]. Especially aspherical or free form optical components often have to be aligned with all six degrees of freedom, often to a precision of less than 500 nm and a few arc seconds.

Fig. 2 shows an example of a rather thicker and stiffer optical bench-like board concept with housing, which includes micro-optic, opto-electronic, and electronic components and connectors. Structured thin metallization on glass that can be electrically bonded enabling co-packaging of surface mounted electronics in the direct vicinity of optical sources or detectors. Component and alignment accuracies are in the 1 to 20 µm range depending on the intended optical performance in free space beams that are split, coupled, combined on the optical board.

Fig. 2: Glass-based electro-optical micro-bench for optical spectrum analysis including micro-optics (filters, isolators, lenses), assembled by sequential active alignment with temporary fixing of components using integrated vacuum-channels, and metallization (solderable, and wire-bondable) to include electro-optical and electronic features (EOM, TIA, PIC).

Stacking of laser structured thin glass has much potential for cost-effective further development: The assemblies as introduced here can be sealed with glass frames and lids, creating compartments with optical windows for thermally isolated or hermetic operation. Here, the inherent low thermal conductivity and transparency of thin glass are advantageous. Thermal and electrical through glass vias TGV can also be created where needed. In Fig. 3 a multi-chip package made of glass using panel level technologies is shown.

Fig. 3: Automated assembly of small (9x9x0.7) mm^3 laser packages with appropriate lenses and optical edge windows on panel before stacking, hermetic sealing, and subsequent singulation by means of laser structuring.

References

1. M. Többer *et al., Proc. ECTC*, Las Vegas, NV, Jun. 2010 - Jun. 2010, pp. 66–73.
2. L. Brusberg *et al., Proc. ECTC*, Las Vegas, NV, Jun. 2010 - Jun. 2010, pp. 269–274.
3. M. Neitz *et al., Proc. ECTC)*, Orlando, FL, USA, May. 2017 - Jun. 2017, pp. 538–544.
4. H. Schröder *et al., Proc. ECTC*, Las Vegas, NV, May. 2016 - Jun. 2016, pp. 468–476.
5. L. Brusberg *et al., J. Lightwave Technol.*, vol. 34, no. 10, pp. 2540–2551, 2016
6. T. J. Cullen *et al., Optics Letters*, vol. 10, no. 4, pp. 134–136, 1984.
7. X. Xu *et al., Photon. Res.*, vol. 8, no. 9, p. 1541, 2020
8. J. Beltran Madrigal *et al., Applied Optics*, vol. 55, no. 36, pp. 10263–10268, 2016,
9. W. Lewoczko-Adamczyk, *et al.*, Proc. *ECTC)*, San Diego, CA, May. 2018 - Jun. 2018, pp. 1136–1139.

https://doi.org/10.1051/epjconf/202226603021

Design and testing of a Kirkpatrick-Baez optics variation

Veronika Stieglitz[1,2,*], *Vadim* Burwitz[1], *Thorsten* Doehring[3], *Rene* Hudec[2], and *Stanislav* Vitek[2]

[1]Max-Planck-Institute for extraterrestrial Physics, Giessenbachstraße 1, D-85748 Garching, Germany
[2] Czech Technical University in Prague, Technicka 2, 166 27, Prague 6, Czech Republic
[3] TH Aschaffenburg – University of Applied Science, Würzburger Str. 45, D-63743 Aschaffenburg, Germany

Abstract. Beside Wolter I X-ray optics, which are used at most in currently operating X-ray space telescopes, there exist also other optical designs and their usability for space observations is still the matter of studies. This article covers preliminary testing results of an optical module which is based on a modified Kirkpatrick-Baez optics. This X-ray optics, consisting of four sub-modules, was assembled in Prague and tested at the PANTER test facility of MPE afterward. The sub-modules use different reflective coatings, in part developed by our research group, on complementary flat mirrors, which approximate the shape of a Kirkpatrick-Baez optical design. In this contribution we summarise the design of the optical modules, the details of applied coating layers, and the X-ray characterisation results at the PANTER test facility.

1 Motivation

By the idea of approximating the concave spherical mirrors, used in the original Kirkpatrick-Baez (KB) design [1], we aimed the simplification of manufacturing. It is therefore an option to produce experimental optical modules more easily. This approach can now be than used for further experiments, mainly performed by university students, as an affordable option in comparison with the demanding production of spherical mirrors. In the presented more simple layout the curved mirrors are replaced by four flat, semiconductor grade silicon wafer substrates with different reflection coatings.

2 Optical design

The used mirror module consists of 4 sets of 17 one-side coated mirrors, housed and oriented as shown in Fig. 1.

Fig. 1. 3D model of tested K-B optic with flat mirrors.

Incoming rays are undergoing reflections only on one surface of each submodule, as is illustrated in Fig. 2. Each incoming ray is reflected two times before reaching the focus, and depending on the trajectory of the photon, the reflection happens on a geometrically available mirror of the module.
Using this layout, a higher intensity of illumination in the focal point is expected due to low grazing incidence angles of the incoming rays, in comparison to a single set of mirrors.

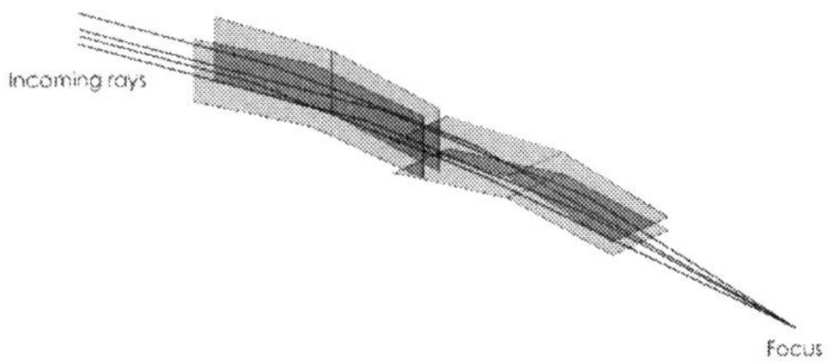

Fig. 2. Schematic of the double reflection in the proposed optical system.

The maximum incident angle is 2.3 degrees, with the total focal length of the system being 2 m and the aperture 80 x 80 mm. The mirrors were mounted into aluminium housings, which provide the correct positioning of the single mirrors.

3 Reflective layers

Because the modular design provides the freedom of different configurations, for the experimental reason of testing different coatings were used two types of reflective layers. Therefore, two sub-modules were coated using pure golden layers, while two sub-modules are equipped with a tri-layer sandwich of chromium and iridium [2].

* Corresponding author: stehlver@mpe.mpg.de

This combination of a main iridium reflective layer and a complementary chromium nano-layer is proven to provide better overall reflectivity in comparison with pure golden layer. This effect can be seen in Fig. 3. All the samples used additional chromium sublayers, which compensate the coating stress and prevent layer degradation over time [3].

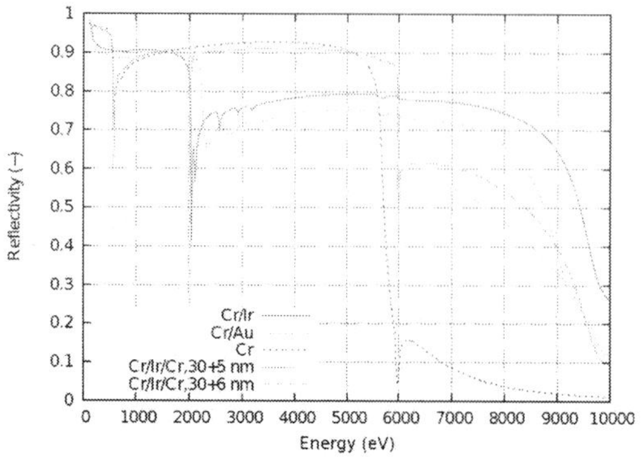

Fig. 3. Comparison of calculated reflectivities for different material combinations.

4 X-ray test results

Optical module was tested at PANTER test facility at different emission lines as well as at low and middle energy band continuum. In Fig. 4, preparation of the module for testing can be seen.

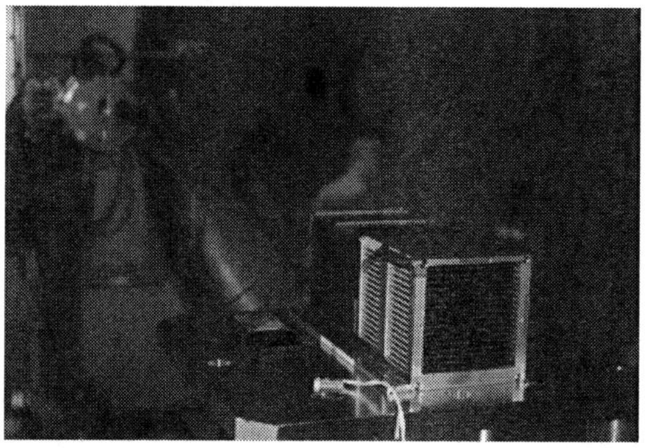

Fig. 4. Optical module during preparation for X-ray tests.

Test campaign consisted of two runs, in which were tried two different configurations of the sub-modules to obtain better understanding of the coating reflective layers behavior [1].
The tests started with focus search to confirm the manufacturing preciseness. Followed imaging of double and, using a mask, single reflections at different angles in focus. For each measurement, statistically sufficient amount of photons needed to be collected, which for the purpose of rather extensive testing campaign meant more focus on softer part of X-ray spectrum (up to 8 keV), although in theory, the optics is suitable even for focusing harder X-rays up to 20 keV.

4.1 Double reflection focus

The double reflection focus represents the complete result of modules focusing ability, as a collection of photons reflected from both pairs of surfaces into the focal point. In Fig. 5, we can see the focussed photons, as captured by the TroPIC detector. The evaluated angular resolution of the optical system corresponds to 1.6 arc min.

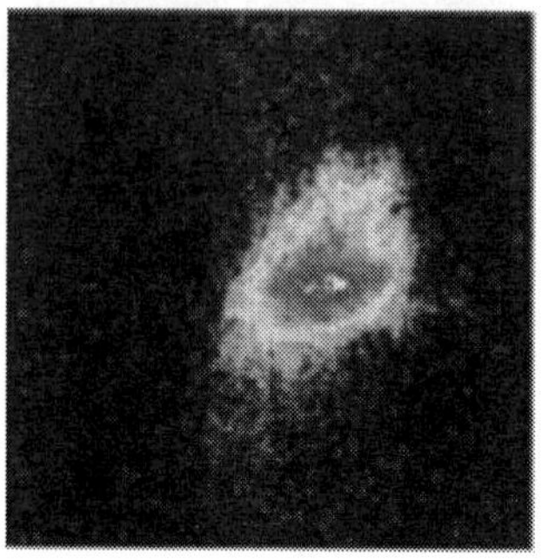

Fig. 5. The double reflection spot at 0.7° at Tungsten-M line.

4.2 Effective area

The effective area of the optics was measured at different angles for double and single reflections. As the optical channels close with rising angles, also the effective area drops. For a 1° tilt of optics, the four-fold reflection resulted in an effective area of 3.42 cm^2, for 2.2° it was only 0.25 cm^2.

5 Conclusion

We built and tested an optical module to show the focusing potential of a simplified, approximated KB type X-ray optics. The obtained results regarding resolution and effective area are inferior to a classic design, but can be improved further by changes of design and material, as the functionality of suggested approach is proven now. This type of an optical system can and will be used in the future for studies and experimental purposes, with the advantage of existing reliable X-ray tests performed by a renown X-ray facility.

This work was supported by an internal grant from the Czech Technical University under the number SGS21/120/OHK3/2T/13 and by collaborative projects funded by the Bavarian-Czech Academic Agency (BTHA). The Cr-Ir-Cr coatings of the telescope have been applied in the coating laboratory of TH Aschaffenburg.

References

1. P. Kirkpatrick and A. V. Baez, J. Opt. Soc. Am. **38**, 766-774 (1948)
2. V. Stehlikova, T. Döhring, M. Stollenwerk et al., Proc. SPIE **11776,** 11776-1 (2021)
3. T. Döhring et al., Proc. SPIE **10399**, 10399-1C (2017)

https://doi.org/10.1051/epjconf/202226603022

Investigations on the production of optical freeforms applying the advanced wheel polishing process

Sebastian Stoebenau[1,*], *Igor* Morozov[1], *Rafael* Hild[1], *Sebastian* Henkel[2], *Christian* Schulze[2], *Christoph* Letsch[2], *Samson* Frank[2], *Jens* Bliedtner[2]

[1]OptoTech Optikmaschine GmbH, Sandusweg 2-4, 35435 Wettenberg, Germany
[2] Ernst-Abbe University of Applied Sciences Jena, Carl-Zeiss-Promenade 2, 07745 Jena, Germany

Abstract. The growing interest in providing additional degrees of freedom to the design of high-end optical systems has led to an increased demand for freeform optical elements. The efficient fabrication of such elements requires a polishing process that provides high removal rates and a stable removal function while working with a relatively small spot size. Taking these constraints into consideration this paper focuses on the successful implementation of polishing processes applying the A-WPT (Advanced Wheel Polishing Tool) technology. Addressing the requirements regarding its removal characteristics as mentioned before, it represents an appropriate choice for providing an efficient pre-polishing as well as corrective polishing technique. In order to maintain perpendicularity towards the freeform surface to be polished, the A-WPT is run on a 5-axis simultaneous machining system. First investigations of the achieved surface accuracy after pre-polishing were carried out as well as an assessment of residual surface features within different spatial frequency regions. In addition, the polished surface is being checked for remaining SSD using an OCT technique.

1 Motivation

Freeform optical elements are of growing importance to provide additional degrees of freedom to the design of optics in the field of beam shaping for laser welding and ablation, off-axis imaging for projection and imaging systems (e.g. head-up-displays), for aberration correction and lens count reduction in optical systems [1]. When reaching into increasingly more complex optical applications the achievable accuracies become more important. Together with the size of the parts this drives the processing times to increase drastically, as only point-like interacting tools are used during grinding and polishing [2]. One approach towards an efficient process chain for the fabrication of optical freeforms therefore is the implementation of polishing processes providing a high removal rate in combination with a temporal stable and highly predictive removal function. This led to our investigations focused on the application of the A-WPT technology to meet these requirements.

2 Freeform, Tool and Machine Set-Up

The Advanced Wheel Polishing Tool (A-WPT – shown on the left hand side picture in Fig. 1) had been introduced by OptoTech back in 2013 as the outcome of an internal R&D project. Wu *et al.* have recently published some closer investigations on the removal behaviour of such tools [3]. The A-WPT provides the specific advantages of a standard wheel polishing tool rotating around only one axis (high removal rates even at small polishing spot sizes, reduced tool wear) while combining them with a homogenization of the resulting micro texture on the polished surfaces by adding a second rotational axis. This second rotational axis is oriented perpendicular to the workpiece surface. To keep the perpendicularity on a freeform surface, the A-WPT needs to be run within a multi-axis machining centre providing at least five CNC-controlled axis.

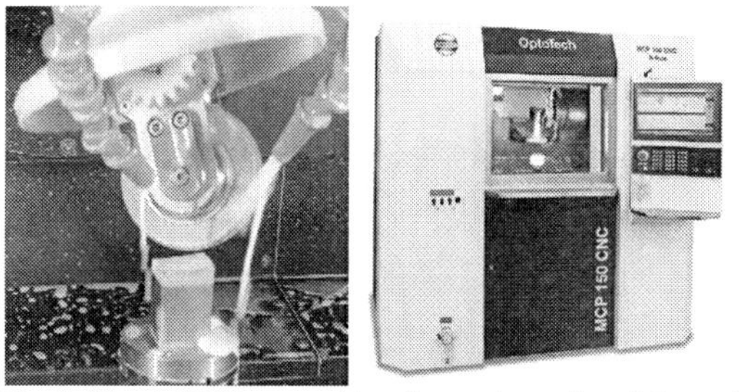

Fig. 1. The A-WPT approaching the starting point of the tool path (left) and 6-axis polishing center MCP 150 CNC (right).

Within our actual investigations a state of the art 6-axis machining centre has been used – the OptoTech MCP 150 CNC as shown on the right hand side picture in Fig. 1. Besides the ability to make use of all different polishing technologies out of the OptoTech MultiTool Concept (spherical polishing using HydroSpeed®; aspherical and freeform polishing using A-FJP, membrane tools, sub-aperture pitch tools) it is equipped with an exchangeable

* Corresponding author: sebastian.stoebenau@optotech.net

workpiece probe. Measuring against predefined reference surfaces it allows for a precise alignment of the freeform within the machine coordinate system.

As a first demonstrator for the proof of the polishing process performance a freeform surface was designed. It is defined by the equation given here below.

$$z = f(x,y) = \frac{(2 \cdot x+25)^2-2.5}{1000} - \frac{(2 \cdot y+5)^2}{380} \qquad (1)$$

A graphical representation of the height profile is shown in Fig. 2. It provides concave as well as convex radii of curvature. For the experimental tests the size of the part was set to 25 x 25 mm^2 with a maximum sag height of 4.87 mm.

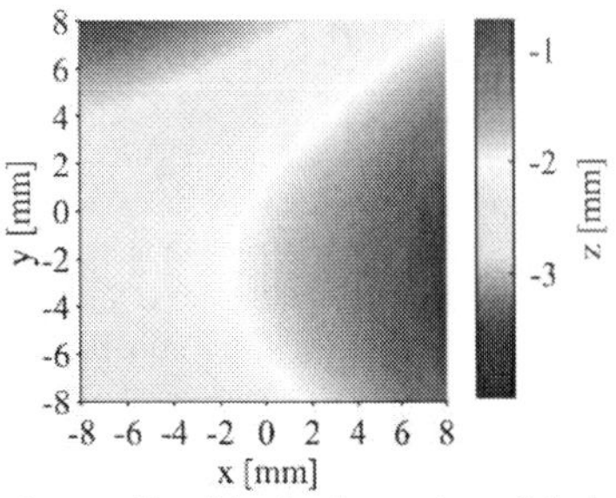

Fig. 2. Surface profile of the freeform to be polished.

3 Experimental Results and Discussions

As mentioned before, for the polishing process tests the A-WPT was used within an MCP 150 CNC machine centre. After alignment of the freeform within the machine using the workpiece probe, the tool was driven across the part in a meandering tool path with a line spacing of 1 mm and a polishing spot size of 6 mm in diameter. After a total polishing time of approx. 50 minutes the pre-polishing was completed and the surface figure was measured using a MarForm MFU200.

Fig. 3 shows the resulting shape accuracy when comparing the polished surface with the designed free-form surface. The P-V error of approx. 13.7 µm partly remains from the grinding process but mainly was introduced during the non-optimized pre-polishing step.

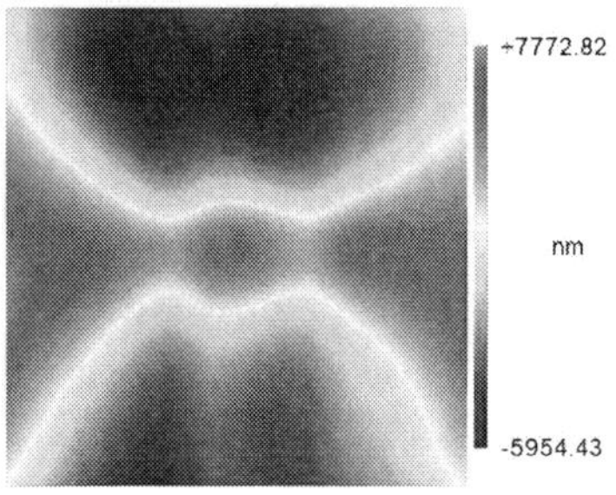

Fig. 3. Shape accuracy measurement of the freeform surface after pre-polishing.

As we first focused on the demonstration of the pre-polishing performance, the result was filtered using a high-pass filter with a cut-off wavelength of 2 mm and was compared to the measurement results achieved from the freeform surface after grinding (Fig. 4). It can clearly be seen that the high spatial frequencies introduced during grinding efficiently could be removed. Residual periodic structures on the polished surface are related to the raster path increment and can be avoided by properly choosing the process parameters in future investigations.

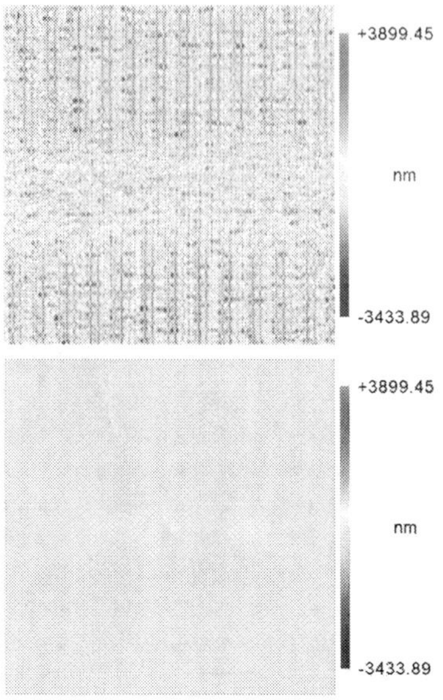

Fig. 4. High-pass filtered measurement results of the freeform surface after grinding (top) and after pre-polishing (bottom).

4 Conclusions and Outlook

The A-WPT technology has proven to be an appropriate choice if an efficient pre-polishing is required. High spatial frequencies as generated during grinding are removed successfully. As the polishing spot size can be varied over a wide range pre-polishing and corrective polishing can be accomplished with the same tool by only changing the process parameters.

Further investigations will focus on the optimisation of the pre-polishing process in order to reduce the PV error occurring in-between as well as on the implementation of correction cycles in order to reduce the total shape error once the pre-polishing has been completed. For an error as shown in Fig. 3 our simulations of the process time needed (polishing spot size of 2 mm in diameter, removal rate of approx. 0.2 mm^2/min) gives an estimated cycle time for the correction of about 20 min.

References

1. J. P. Rolland, M. A. Davies, T. J. Suleski, C. Evans, A. Bauer, J. C. Lambropoulos, K. Falaggis, Optica, **8**, 161-176 (2021)
2. C. Schindler, T. Köhler, E. Roth, Proc. SPIE, **10448**, 1044802 (2017)
3. X. Wu, Z. Huang, Y. Wan, H. Liu, X. Chen, IEEE Access, **8**, 108191-108200 (2020)

EPJ Web of Conferences **266**, 03023 (2022)
EOSAM 2022 https://doi.org/10.1051/epjconf/202226603023

Optical Design at The Age of AI

Simon Thibault[1,*], *Geoffroi Côté*[1], *Julie Buquet*[1], and *Jean-Francois Lalonde*[2]

[1]COPL Université Laval, Québec G1V 0A6, Canada
[2]Université Laval, Québec G1V 0A6, Canada

Abstract. Data-driven methods to assist lens design have recently begun to emerge; in particular, under the form of lens design extrapolation to find starting points (lenses and freeform reflective system). I proposed a trip over the years to better understand why the AI have been applied first to the starting point problems and where we are going in the future.

1 Introduction – Historical Perspective

Early in 80's, D. Dilworth presented his perspective on AI in lens design [1]. He talked about two primary approaches that have been taken at that time in his companie. The first was the 'natural-language interfacing' and the second was the use AI as an expert system. The first is far from our actual conception of AI but the second is somehow a means for the software to 'learn' about optics by studying lenses designed by expert to formulate rules. He recognised that AI was a what he called a 'virgin territory' because no one has investigated the potential of AI in lens design.

In the 90's, the trends in lens design were about global optimization and genetic algorithms [2,3,4]. Dilworth refined his 'expert system' which can be seen today as different as a AI application [5]. The idea of AI in lens design was not anymore on the radar.

At the IODC conference in 2002, Shannon did a talk about 'Fifty years of lens design; What do we know now that we did not know then?' [6]. I peak the following sentence from the manuscript: 'Progress in the future will likely require building more basic knowledge into design programs. Future lens design programs will need to incorporate learning and teaching function. Design programs should become a repository of knowledge as well as a set of tools.'. Shannon saw that design program can do more and more and this is maybe the future.

So the next decade, new applications come in the field, the first one was the computational imaging [7], follow by new type of surface including Zernike and freeform to name a few. These new lens design trends required additional skills that is not fully mastered by lens designer. Consequently to efficiently used it, lens designer need some helps. This is why (maybe) one of the first AI application was about freeform [8].

Over all those years, we can affirm that from a lens design point of view, the AI can be useful if it can:

- Accelerate the development process for optical system
- Enhance productivity of the deisgners
- Reduce costs.

In this presentation we will try to describe three applications that can beneficial from AI in lens design.

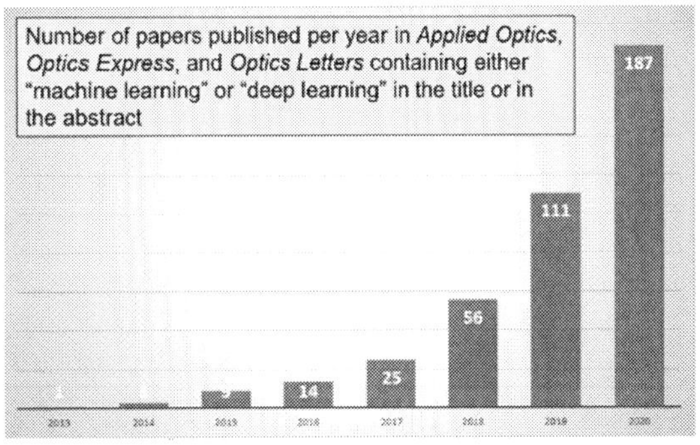

Fig. 1. Number of papers published per year in AI-Optics [9].

2 Few Applications

2.1. Starting Points

The first application is a simple, highly modular deep neural network (DNN) framework to address the problem of automatically inferring lens design starting points tailored to the desired specifications. In contrast to previous work, our model can handle various and complex lens structures suitable for real-world problems such as Cooke Triplets or Double Gauss lenses. Our successfully trained dynamic model can infer lens designs with realistic glass materials whose optical performance compares favorably to reference designs from the literature on 80 different lens structures. Using our trained model as a backbone, we make available to the community a web application that outputs a selection of varied, high-quality starting points directly from the

* Corresponding author: author@e-mail.org

desired specifications, which we believe will complement any lens designer's toolbox [10].

2.2 Diversity

The second example was to build a deep learning-enabled LDE framework with the aim of generating a significant variety of microscope objective lenses (MOLs) that are similar in structure to the reference MOLs, but with varied sequences [11]. As a particular arrangement of glass elements, air gaps, and aperture stop placement. We first formulate LDE as a one-to-many problem—specifically, generating varied lenses for any set of specifications and lens sequence. We then quantify the structure from the slopes of its marginal ray. From only 34 reference MOLs, we generate designs across 7432 lens sequences and show that the inferred designs accurately capture the structural diversity and performance of the dataset. Our contribution answers two current challenges of the LDE framework: incorporating a meaningful one-to-many mapping, and successfully extrapolating to lens sequences unseen in the dataset—a problem much harder than the one of extrapolating to new specifications.

2.3 Challenges using data-driven methods

Finally, as data driven approaches have proven very efficient in many vision tasks, they are now used for optical parameters optimization for application-specific camera design. In this experiment, a neural network is trained to estimate images or image quality indicators from optical parameters. However, the complexity and interconnectivity of optical parameters raises new challenges such as choosing the most relevant inputs with a suitable representation using limited available data. We investigated these challenges in the case of wide-angle systems. We tried to establish a data-driven prediction model to estimate the RMS spot size as a function of the distortion using mathematical or AI-based methods [12].

.

The acknowledgements should be typed in 9-point Times, without title.

References

1. Donald C . Dilworth "Applications Of Artificial Intelligence To Computer-Aided Lens Design", Proc. SPIE 0766, Recent Trends in Optical Systems Design and Computer Lens Design Workshop, (10 June 1987); https://doi.org/10.1117/12.940208M
2. Gregory W. Forbes and Andrew E.W. Jones "Towards global optimization with adaptive simulated annealing", Proc. SPIE 1354, 1990 Intl Lens Design Conf, (1 January 1991); https://doi.org/10.1117/12.47874
3. Doron Sturlesi and Donald C. O'Shea "Invited paper: Future of global optimization in optical design", Proc. SPIE 1354, 1990 Intl Lens Design Conf, (1 January 1991); https://doi.org/10.1117/12.47876
4. Isao Ono, Shigenobu Kobayashi, and Koji Yoshida "Global and multi-objective optimization for lens design by real-coded genetic algorithms", Proc. SPIE 3482, International Optical Design Conference 1998, (21 September 1998); https://doi.org/10.1117/12.321995
5. Donald C. Dilworth "Invited paper: Expert systems in lens design", Proc. SPIE 1354, 1990 Intl Lens Design Conf, (1 January 1991); https://doi.org/10.1117/12.47905
6. Robert R. Shannon "Fifty years of lens design: What do we know now that we did not know then?", Proc. SPIE 4832, International Optical Design Conference 2002, (23 December 2002), https://doi.org/10.1117/12.486483
7. Kenneth Kubala "Computational imaging technologies", Proc. SPIE 7652, International Optical Design Conference 2010, 765202 (31 August 2010); https://doi.org/10.1117/12.868938
8. Tong Yang, Dewen Cheng, and Yongtian Wang, "Direct generation of starting points for freeform off-axis three-mirror imaging system design using neural network based deep-learning," Opt. Express 27, 17228-17238 (2019)
9. Christophe Menke (Zeiss) private communication.
10. Geoffroi Côté, Jean-François Lalonde, and Simon Thibault, "Deep learning-enabled framework for automatic lens design starting point generation," Opt. Express 29, 3841-3854 (2021).
11. Geoffroi Côté, Yueqian Zhang, Christoph Menke, Jean-François Lalonde, and Simon Thibault, "Inferring the solution space of microscope objective lenses using deep learning," Opt. Express 30, 6531-6545 (2022).
12. J. Buquet, J. Parent, J-F Lalonde, and S. Thibault, Challenges using data-driven methods and deep learning in optical engineering, at Current Developments in Lens Design and Optical Engineering XXIII conference (SPIE, August 2022).

https://doi.org/10.1051/epjconf/202226603024

Large Metal Mirrors for Atmospheric Telescopes

Guoyu Yu[1], *David* Walker[1,2,3] Hongyu Li[1], Abdullah Shahjalal[1] and *Trevor Walker*[4]

[1]SciTech Daresbury National Science & Innovation Campus, U-of-Huddersfield Laboratory for Ultra Precision Surfaces, G6, TechSpace One, Keckwick Lane, Daresbury, WA4 4AB, UK
[2]Zeeko Ltd, CoalVille, Leicestershire, LE67 3FW, UK
[3]University College London, Dpt Physics and Astronomy, Gower St. London, WC1E 6BT, UK
[4]Thin Metal Films Ltd, Basingstoke, Hants, RG24 8UG, UK

Abstract. We are developing technologies in processing and metrology to fabricate a 1.4-meter aperture, aspherical surface Aluminium prototype mirror for Cherenkov Telescope Array. The aim is to fast process these large aperture mirrors at low cost. The technical development will ensure the high specifications on the surface quality. Different metrology methods including Swinging Arm Profilometry (SAP) and phase measuring deflectomtry (PMD) are being developed. Recent results have shown very promising progress on these developments. We have excellent record in transferring our research results into industry. These cutting-edge technologies will be transferred to our industry partner to explore further developments.

1 Background

The Cherenkov Telescope Array (CTA) is proposed to be the next generation ground-based observatory for gamma-ray detection at very high energies [1]. The CTA will be the world's largest and most sensitive gamma-ray telescopes. The CTA consists of around 100 telescopes of 4m to 23m primary mirror with segment mirrors.

At the Laboratory for Ultra-Precision Surface, University of Huddersfield, we have long established record in large optical mirror processing. The research and development of European Extremely Large Telescope (E-ELT) [2] that has been successfully delivered is a very challenging project and has similar objective as that of CTA.The R&D on metrology and processing is reported.

2 Research and Development

The first challenge addressed was the metrology of large aperture aspherical surfaces. To facilitate the processing development, a swing arm profilometer (SAP) was developed to enable accurate profile measuring and removal rate calibration. Deflectometry is also under development to enable 3D surface metrology. A further funding is secured to procure a MicroFinish Topographer for surface texture measurement.

2.1 Profilometer

SAP have been developed for different projects [3] having the advantage that the base radius can be cancelled out by setting an appropriate tilting angle of the arm. We have adopted an inverse design in that the component will be tilted instead of the arm. This design suits our development by avoiding the adjustment of delicate probe and arm. Assuming that R is the workpiece's radius of curvature and l the effective arm length, shown in Fig. 1., the tilt angle can be calculated using the equation:

$$\theta = sin^{-1} (\frac{l}{R}) \quad (1)$$

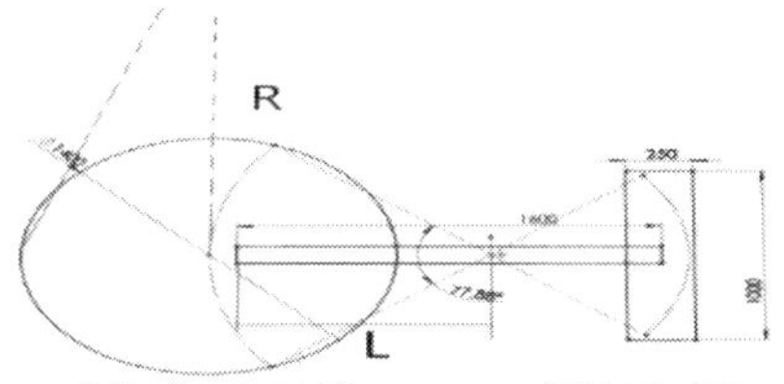

Fig. 1. Design of SAP for segment mirror development.

When implementing this design, (Fig. 2) a tilt stage is employed to provide necessary tilting of the workpiece.

Fig. 2. SAP measuring a full-size mirror.

* Corresponding author: g.yu@hud.ac.uk

2.2 Deflectometry

Deflectometry have been reported to measure large aperture optics [4]. It has the advantage of large dynamic range, flexibility and high accuracy and is especially suitable for the floor level surface metrology. With this technology, a CGH can be avoided for measuring aspherical surfaces so that it is very cost effective. The calibration process is the key step to ensure the accuracy and stereo calibration have been adopted.

2.3 Processing

From previous development work in E-ELT project, we have gained valuable knowledge in process development for such large projects. It is not only to meet the high specifications, but also be able to process the mirrors fast and cost-effectively. Different pads and abrasive grades, shown in Fig. 3., can be used at different process stages to achieve varied removal rates and surface texture. Optimisation is performed to ensure the smooth transition of the process stages and to achieve final specifications.

Fig. 3. Process development and optimisation.

The process development is on a smaller 400mm square sample with the same material. A narrow engraving on the surface is used to calibrate the removal rate of each step with the SAP (Fig. 4 and Fig. 5). Different combination of process steps can be optimised and the total worktime is calculated for the full-size mirror. The surface texture is measured with a MicroFinish Topographer.

Fig. 4. Metrology during process development.

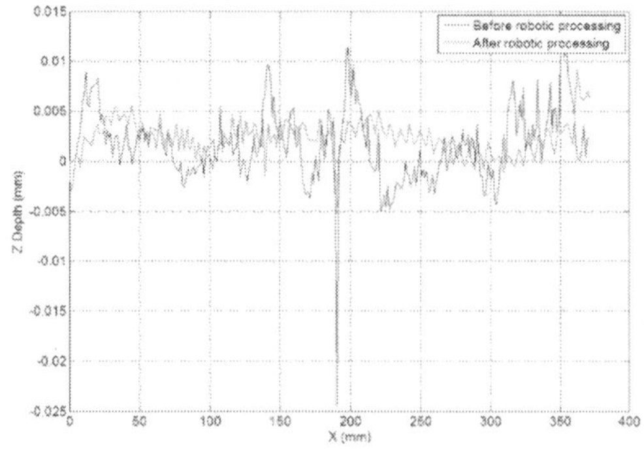

Fig. 5. Calibration of material removal.

3 Results and Discussion

During the process development, Kemet International Ltd, has collaborated on the polishing slurry development. The newly developed slurries have been adopted to compare and superior results on surface texture have been achieved of Sa 4.98nm, as shown in Fig. 6.

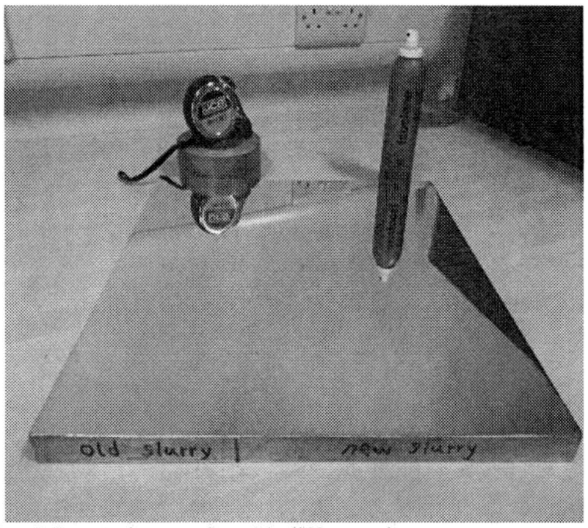

Fig. 6. Processing results with different slurry.

The research and development have been progressed and achieved the defined specification for the full-size mirror processing. The total processing time calculated for this mirror is 220 hours to achieve the final specification.

References

1. The Cherenkov Telescope Array Consortium, https://arxiv.org/abs/1709.07997v1
2. D. Walker, G. Yu, H. Li, W. Messelink, R. Evans, A. Beaucamp, Opt. Express 20(18), (2012)
3. P. Su, R. Parks, J. Burge, Proc. SPIE, 8126, (2015)
4. P. Su, Y. Wang, J. Burge, Opt. Express 20(11), (2012)

A Systematic View of Microscope Objective Design

Yueqian Zhang[1,*], and *Herbert* Gross[2]

[1]Carl Zeiss AG, 07745 Jena, Germany
[2]Fraunhofer Institute for Applied Optics and Precision Engineering IOF, 07745 Jena, Germany

Abstract. The correction of modern microscope objectives is not usually discussed in literature. We have reported a system review and summarized the design principles in a series of papers in 2019 [1-3]. Here we are introducing the systematic view of microscope objective design with an extension of the database till 2021. Furthermore, a systematic synthesis approach aided by AI will also be discussed.

1 Introduction

As the most well-known high numerical aperture (NA) optical system, the light microscope objective has been deeply developed in the past century and utilised in various applications. Owing to a collection of 477 relevant US, JP, DE, and WO patents with purely refractive microscope objectives from 1926 to 2021, the application-oriented development history can be clearly understood. An overview is given in Figure 1.

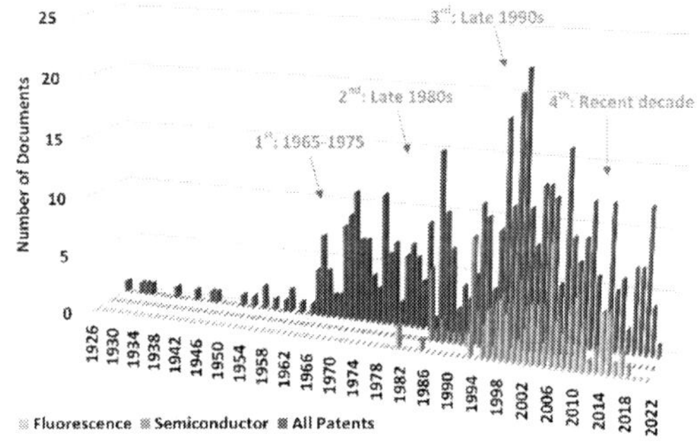

Fig. 1. Number of modern microscope objective patents from 1926 to 2021 with four peaks of patenting activities.

Based on the patent database, an objective database is constructed with 519 entries with different structures. To fairly compare the material selection and system complexity, the constructed systems must have a colour correction covering the visible spectrum (VIS).

The systems cover a full range of magnification from 0.5x to 250x and a full range of NA from 0.025 to 1.70. According to their magnification and NA, the systems can be further classified into six zones corresponding to the etendue and different level of correction, demonstrated in Figure 2:

- Zone 1: typical ach/apochromate two-group systems
- Zone 2: typical Plan-ach/apochromate clear-three-group systems
- Zone 3: novel three-group systems with special correction lens modules
- Zone 4: systems with extremely high NA or etendue, which sacrifice other system parameters, such as parfocal length and immersion liquid type
- Zone 5: very low magnification parfocal telecentric systems
- Zone 6: very high magnification systems

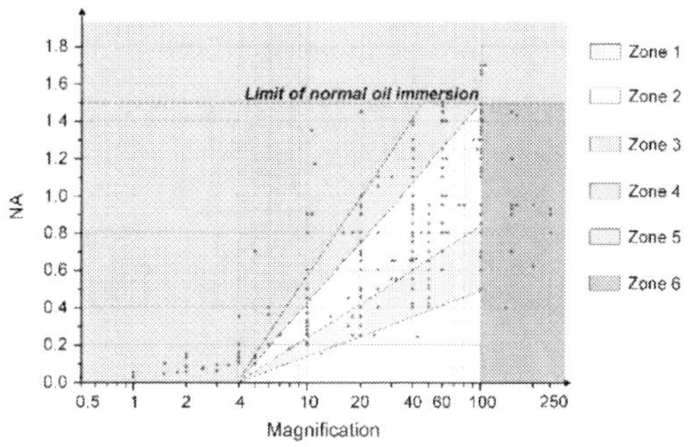

Fig. 2. 6-zone classification of microscope objectives based on etendue.

Focusing on each of the zones, the system structures can be carefully compared by analysing the aberration behaviours of each lens group, thus summarising the lens modules utilised in conventional designs.

2 Conventional Design

We are summarising the conventional design principles of the microscope objective design, which are valid in more than 80% of corresponding systems. They can be classified into three categories: optical power distribution, material selection and lens structures. The combination of

* Corresponding author: yueqian.zhang@zeiss.com

https://doi.org/10.1051/epjconf/202226603025

a material selection and lens structure is regarded as a lens module.

Among the six zones of the microscope objective classification based on etendue, the optical power distribution has four basic types. Selecting one power distribution basically determines one branch of systems in the solution space. The objectives can then be divided into two or three structural/functional groups. 10 lens modules in the front group, 9 lens modules in the middle group and 10 lens modules in the rear group are summarized as the design principles. By inserting the lens modules locally, the microscope objective lens can be evolved from a low-etendue simple structure with poor colour and field correction to a high-etendue sophisticated structure with high correction level. An example is demonstrated in Figure 3.

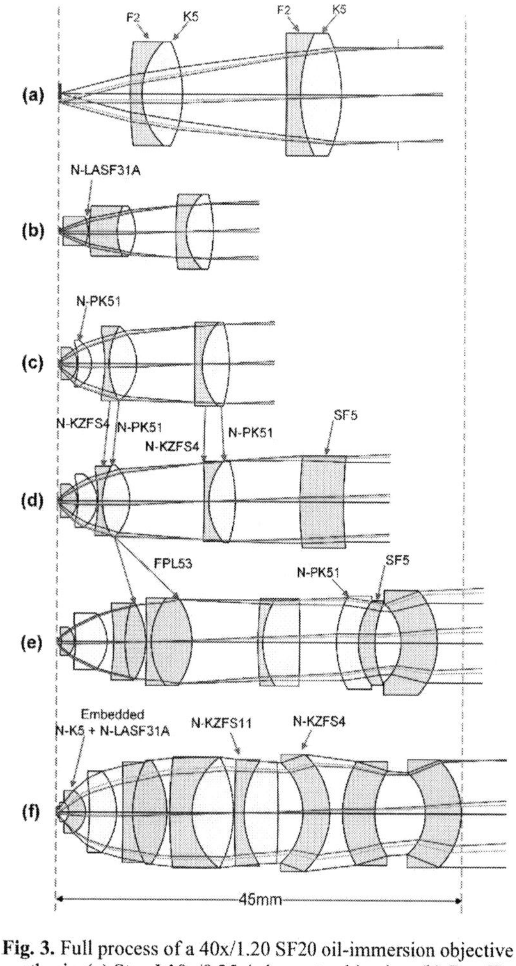

Fig. 3. Full process of a 40x/1.20 SF20 oil-immersion objective synthesis. (a) Step I 10x/0.25 Achromate objective. (b) Step II 40x/0.60 Achromate objective. (c) Step III 40x/0.85 Achromate objective. (d) Step IV 40x/0.85 Plan-semiapo objective. (e) Step V 40x/0.85 Plan-apochromate CORR objective. (f) Final design 40x/1.20 Plan-apochromate objective.

3 Systematic Synthesis with AI

Although dividing the microscope objectives into lens modules provides a systematic view of the lens design, to insert the modules locally for synthesis still requires the experience of optical designers and it is not sufficient to show the overview of the whole solution space.

Since it is difficult to convert these empirical models into mathematical models for direct optimization, a more promising method is to develop a deep learning-enabled lens design extrapolation approach that trains a deep neural network (DNN) model, which takes the objective database as a training dataset.

Recently, selecting 34 systems from the database with object space numerical aperture $0.35 < NA < 0.45$ and half image space field angle $2.5° < HFOV < 4.5°$ as the training domain, we reported such a tool that can both interpolate and extrapolate the solution space of a 20x/0.40 Plan-apochromate solo-corrected infinite-conjugate objective [4]. All the three possible basic structures with distinctive power distributions are reproduced: Lister two-group, double-Gauss and retrofocus. The lens modules are also well exploited in the examples.

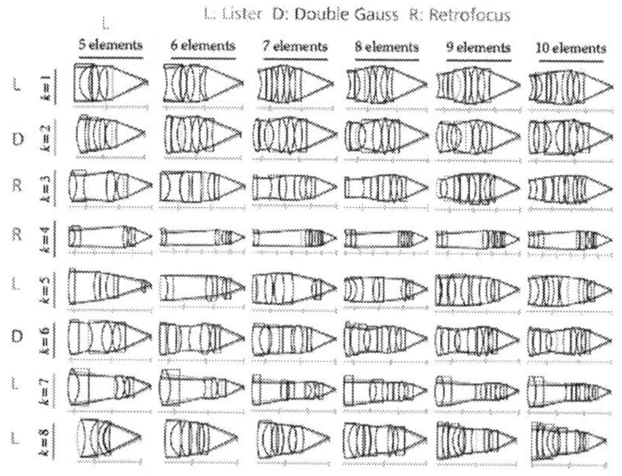

Fig. 4. Subset of designs inferred from all branches of the training model for the 20x/0.40 Plan-apochromate objective.

The tool can generate initial systems for a specific task. With its help, optical designers can have a better view of the solution space to decide whether it is necessary to further increase the number of elements.

References

1. Y. Zhang, H. Gross, Advanced Optical Technology, **8**, 313-347 (2019)
2. Y. Zhang, H. Gross, Advanced Optical Technology, **8**, 349-384 (2019)
3. Y. Zhang, H. Gross, Advanced Optical Technology, **8**, 385-402 (2019)
4. G. Côté, Y. Zhang, C. Menke, J. F. Lalonde, S. Thibault, Optics Express, **30** (2022)

https://doi.org/10.1051/epjconf/202226604001

Polarization Sensitive Digital Holographic Imaging in Biology

Giuseppe Coppola[1], *Maria* Mangini[2], *Gianluigi* Zito[1], *Edoardo* De Tommasi[1], *Anna Chiara* De Luca[2], and *Maria Antonietta* Ferrara[1,*]

[1] National Research Council, Institute of Applied Sciences and Intelligent Systems 'E. Caianiello', Unit of Naples, 80131 Naples, Italy
[2] National Research Council, Institute of Experimental Endocrinology and Oncology 'G. Salvatore', 80131 Naples, Italy

Abstract. A new, simple digital holography-based polarization microscope for quantitative birefringence imaging of biological cells is presented. As a proof of concept, two different class of cells have been characterized by polarization sensitive digital holographic imaging (PSDHI). These two cases study reported are: differentiation of leukaemia cells and identification of reacted sperm cells. Although further experimentation is necessary, the suggested approach could represent a prospective label-free diagnostic tool for use in biological and medical research and diagnosis.

1 Introduction

The main motivation of this work is to fast label-free characterize a birefringent material, i.e. a material with refractive index (RI) dependent on the polarization and propagation direction. There are several paper in literature that demonstrate birefringence in some cells and tissues, where proteins are regularly organized in oriented filaments. Just to cite a few, a large reduction in birefringence in myocardial tissues has been measured in the infarcted regions compared to the healthy tissue values [1]; moreover, birefringence pattern of human breast cancer cells increased when highly invasive cancer cells are considered respect to weakly invasive cells [2]. Therefore, there is a great interest in the study of birefringence distribution in cells and tissue in a label free and non-invasively way.
With this aim, polarization sensitive digital holographic imaging (PSDHI) is an emergent technique based on the study of amplitude and phase of the diffracted beam through two orthogonally polarized reference waves [3]. Thus, not only morphological information is available, such as the classic digital holography imaging, but also intrinsic information about the polarization state of the sample through the phase change quantification. In this work we apply PSDHI to two cases study: leukemia cells differentiation and human sperm cells characterization based on their birefringence and morphological features.

2 Results

2.1. Experimental setup

A polarization sensitive digital holographic imaging system has been implemented by combining two Mach-Zender interferometers obtained by splitting the reference beam in two linearly and orthogonally polarized beams. These two reference beams interfere with the object beam polarized at 45°, producing two different sets of fringe patterns, i.e. two holograms related to the orthogonal components of the object field (Obj_x and Obj_y). By applying the standard reconstruction algorithm on each selected and filtered hologram, amplitude and phase maps for the two components of the object field can be recovered, as detailed descripted in Ref. [4]. Then, to characterize the birefringence pattern of the sample, two characteristic parameters have been evaluated: the amplitude ratio β, associated to the different transmitted intensities for the two orthogonal components, and the phase difference $\Delta\varphi$, connected to the different optical paths due to the anisotropy of the RI [4]:

$$\beta = \arctan(Obj_x/Obj_y) \quad (1)$$

$$\Delta\varphi = \varphi_x - \varphi_y \quad (2)$$

2.2. Acute lymphoblastic leukemia type B (B-ALL) cells differentiation

As first application, we considered the possibility to differentiate acute lymphoblastic leukemia type B (B-ALL) cells based on their birefringence and morphological features. This differentiation process could be very useful for timely and accurate diagnosis. Three different B-leukemia transformed cell lines have been considered: RS4;11, REH and MN60 B-leukemia cells [5,6]. The first two cell lines were both classified as the L2-blast subtype (i.e., B-cell precursor leukemia), while the third cell model was considered as the L3-blast subtype (i.e., B-cell leukemia). Since MN60 cells are much greater in size respect to the other two cell lines and with the aim to perform a birefringence analysis independent on the cells size, a characteristic vector was generated for each cell. It was made up by combining the vectors achieved considering the histograms representing

* Corresponding author: antonella.ferrara@na.isasi.cnr.it

the number of pixel at a specific grey level for both β and Δφ maps, the Δφ maximum value, the total Δφ (birefringent volume), and the birefringent area normalized to the total area of each cell [5,6]. The principal component analysis (PCA) was performed on this set of data to evaluate the discrimination efficiency. The leave-one-out method was used for cross validation of the model. The obtained 3D scatter plot for the first three principal components is reported in Fig.1 as well as the resulting confusion matrix, that identify the correct and incorrect cells classifications.

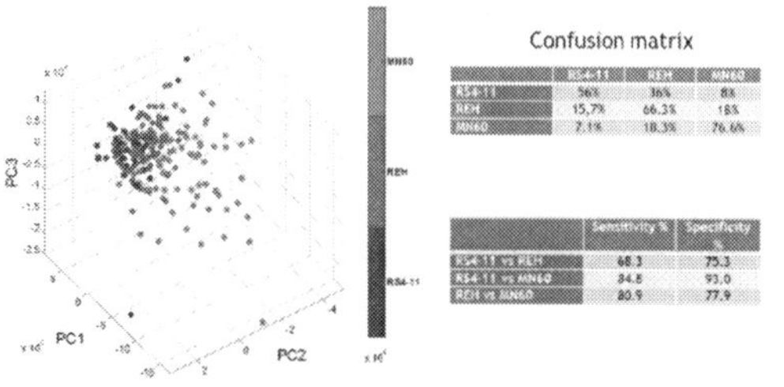

Fig. 1. 3D PCA scatter plots of the first three principal components for datasets obtained by analysing the characteristic vectors and the corresponding confusion matrix. Sensitivity and specificity are reported, too.

2.3 Sperm cells acrosome reaction

As a second example, we studied the birefringence in sperm cells. Indeed, it has been demonstrated that healthy sperm cells show a head totally birefringent due to longitudinally oriented protein filament, while when acrosome reaction occurs, that means that the sperm cell is ripe to enter into the oocyte, birefringence is typically in the post-acrosome area [7,8]. Thus, birefringence could be considered a method to distinguish between reacted and non-reacted spermatozoa and could be used in intracytoplasmic sperm injection (ICSI). Acrosome reaction was induced by exposing the cells to heparin for 4 hours, then, distribution of birefringence patterns of sperm from 3 donors was carried out on control and reacted samples. Analysing the retrieved phase difference maps, we observed a total birefringent head in control semen, while in reacted cells birefringence was present only in the post-acrosomal region [4]. Precisely, when reacted sperm cells are characterized, we note a very high increase in the percentage of analyzed samples that show birefringence localized in post-acrosomal region accompanied by a marked decrease of sperm with the whole birefringent head respect to the control sample (intact acrosome), as showed in the histogram reported in Fig. 2. Additionally, we witnessed a lower number of samples with abnormal birefringence, all of which are associated to the loss of motility and to the existence of abnormal morphology. Moreover, since PSDHI is a quantitative phase microscopy, it allows to retrieve the average phase difference Δφ obtained for control and reacted samples. These values show a decrease of about fourfold of the mean value of Δφ when the acrosome reaction occurs (see Fig. 2).

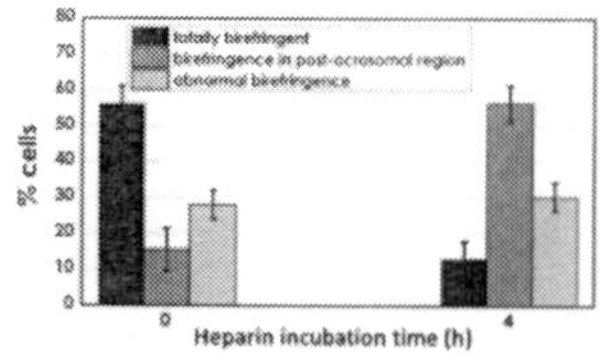

	0h in heparin	4h in heparin
Mean(Δφ) [rad]	1.39	0.36

Fig. 2. Distribution of birefringence patterns of sperm from 3 donors exposed to heparin for 0 h (control sample), and 4 h (reacted sample); evaluated mean phase difference Δφ for the two groups of samples.

3 Conclusions

In conclusion, PSDHI technique has been presented and the results obtained from the birefringence tests performed with the developed holographic system showed a good discrimination between three different lines of leukemia and a good discrimination between intact and reacted sperm cells.

The unique potentialities of holography have allowed to carry out not only the three-dimensional reconstruction of the analyzed samples, allowing an evaluation of their morphological characteristics, as richly reported in the literature, but also birefringence distribution. Results demonstrated that PSDHI can be effectively used to perform a non-invasive and label-free selection on cells and thus could be a valuable label-free tool in biological and medical research.

References

1. M. F. G. Wood, N. Ghosh, M. A. Wallenburg, S.-H. Li, R. D. Weisel, B. C. Wilson, R.-K. Li, I. A. Vitkin, *Journal of Biomedical Optics* **15**, 047009 (2010).
2. I. H. Shin, S.-M. Shin, D. Y. Kim, *Journal of Biomedical Optics* **151**, 016028 (2010).
3. G. Coppola and M. A. Ferrara, *Applied Sciences* **10**, 4520 (2020).
4. A. De Angelis, M. A. Ferrara, G. Coppola, L. Di Matteo, L. Siani, B. Dale, G. Coppola, A. C. De Luca, *Sci. Rep.* **9**, Article number: 4823 (2019).
5. G. Coppola, G. Zito, A. C. De Luca, M. A. Ferrara, *Proc. Digital Holography and Three-Dimensional Imaging 2019*, Article number DH-2019-Th4A.5 (2019).
6. M. Mangini, M. A. Ferrara, G. Zito, S. Managò, G. Coppola, A. C. De Luca, *Proc. of SPIE* **11786**, Article number 117861F (2021).
7. L. Gianaroli, M. C. Magli, A. P. Ferraretti, A. Crippa, M. Lappi, S. Capitani, B. Baccetti, *Fertility and Sterility* **93**, No. 3 (2010).
8. M. C. Magli, A. Crippa, L. Muzii, E. Boudjema, A. Capoti, G. Scaravelli, A. P. Ferraretti, L. Gianaroli, *Reproductive BioMedicine Online* **24**, 352– 359 (2012).

https://doi.org/10.1051/epjconf/202226604002

3D printed FBG based sensor for vital signal monitoring – Influence of the infill printing parameters

Alexandra Ferreira[1*], Cátia Tavares [1,2*], Cátia Leitão [1,2], Daniela Lo Presti [3], M. Fátima Domingues [2], Nélia Alberto [2], Hugo Plácido da Silva [4,5], Paulo Antunes [1,2]

[1] I3N, Department of Physics, University of Aveiro, Aveiro, Portugal.
[2] Instituto de Telecomunicações and University of Aveiro, Aveiro, Portugal.
[3] Unit of Measurements and Biomedical Instrumentation, Departmental Faculty of Engineering, Università Campus Bio-Medico di Roma, Rome, Italy.
[4] Instituto de Telecomunicações, Instituto Superior Técnico, University of Lisbon, Lisbon, Portugal.
[5] PLUX - Wireless Biosignals, S.A, Lisboa, Portugal.

Abstract. The fused deposition modelling technique has been used in the production of strain sensors in which fibre Bragg gratings (FBGs) are encapsulated during the 3D printing process. This paper reports the study of the influence of the FBG position and the material filling, in this case a flexible polymer material, on the sensors' sensitivity and overall performance. In addition, this study preliminarily evaluated the ability of the strain sensor to monitor (heart rate) HR and (respiratory rate) RR as a wearable on the wrist and as a non-intrusive solution on the back of an office chair.

1. Introduction

Fibre Bragg grating (FBGs) sensors have been increasingly used in healthcare applications, in detriment of conventional sensors due to their advantages such as reduced size and weight, biocompatibility, multiplexing capacity, high sensitivity, immunity to electromagnetic interference, and the possibility of operating in environments in the presence of water [1, 2]. These features make FBGs one of the most used optical sensing technologies in the production of biosensors, both in wearable and non-intrusive sensors (i.e., sensors to be placed on objects that are in contact with the body as chairs, cushions, etc.) [1]. The monitoring of vital signs such as respiratory rate (RR) and heart rate (HR) is of great importance in medicine. FBGs have been especially used in detecting HR and RR both embedded into hosting flexible materials and without any encapsulation [1, 2]. Once in contact with the chest surface/skin movements, the FBGs detect the strain changes applied directly to this material during the RR and HR cycles [1, 2]. Different techniques have been reported so far using flexible materials to embed FBGs, such as resins and other plastic composites [2] or 3D printing by fused deposition modelling (FDM) [3]. This last technique allows the creation of sensors with low cost, simplicity, and high reproducibility [3]. FDM has been little explored to produce biosensors [4], and only one manuscript has been published using this technique to produce sensors for vital signs monitoring [3]. This work complements this first publication, testing two different positions of the FBG and three distinct fillings for the sensor's body to verify which one experiences the greatest sensitivity.

2. Sensor's development

2.1 Design and Simulation

This study is a continuation of the work developed by our team, and therefore the material and general dimensions of the sensors have been already reported in [3]: flexible and 30×40×2 mm (Figure 1a). Here, we intend to test the influence of two different factors: the percentage of filling of the sensor body, and the position of the FBG inside the matrix. In detail, three infills' density (10, 20 and 30 %) (Figure 1b) and two positions of FBG (height at which the FBG is on the sensor's body: 1.00 mm and 1.25 mm from the bottom) (Figure 1c) were tested. Therefore, six different sensors (Table 1) were designed in SolidWorks® software exactly with the filling structure obtained by the 3D printer used to print all the sensors, as seen in Figure 1b. Next, the sensors behaviour was simulated using finite element modelling also with SolidWorks®, considering movements in the range of the chest displacement during RR (440 μm) and HR (150 μm) cycles. Figure 2 shows an example of visual result of simulation, and the Table 1 contains the strain values resulting from each simulation.

As it can be seen, in both cases, a lower infill and a higher height (FBG closer to the surface inducing the deformation) are associated with a higher spectral shift, what could indicate that these sensors may have higher sensitivity too. Only the two sensors that presented the highest sensitivities in the simulation were printed (D, E) and after it were subsequently subjected to experimental tests.

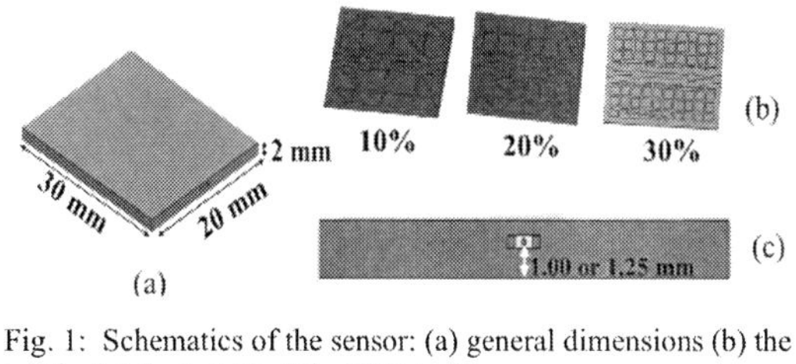

Fig. 1: Schematics of the sensor: (a) general dimensions (b) the infills; (c) the height of the FBG in the sensor.

* Corresponding author: alex.ferreira@ua.pt; catia.tavares@ua.pt

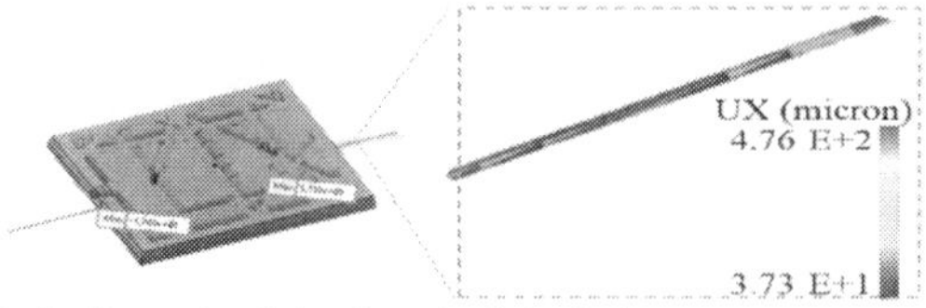

Fig. 2: Example of visual result of simulation.

Table 1. Features of six sensors and their simulations results.

Sensor	*Features*		*ΔL/L (strain)*		*Spectral shift (nm)*	
	FBG height (mm)	Infill (%)	150 μm	440 μm	150 μm	440 μm
A	1.00	10	9.06E-08	1.55E-07	1.09E-01	0.19
B	1.00	20	6.00E-08	9.37E-08	7.20E-02	0.11
C	1.00	30	3.23E-09	6.45E-08	3.87E-03	0.08
D	1.25	10	1.93E-05	6.07E-05	23.20	72.80
E	1.25	20	1.75E-05	2.43E-05	21.00	29.14
F	1.25	30	2.90E-06	3.45E-06	3.48	4.14

2.2. Experimental implementation and characterization

To verify the results of the simulation, sensors D and E were characterized experimentally. The movement of the body induced by the heart beating and breathing was simulated with a waveform generator (33220A, Agilent Technologies, Santa Clara, California). The sensors' response was tested following the protocol in [3].

During the experimental tests, frequencies were varied sequentially between 0.5 Hz and 5.0 Hz for a small amplitude motion (150 μm), and 0.1 Hz and 1.0 Hz for a higher amplitude motion (440 μm). Both sensors were able to respond to all frequencies studied. Then, two frequencies: 1.25 Hz (to simulate HR) and 0.20 Hz (to simulate RR) were considered, varying only the amplitude of motion (from 100 μm to 550 μm in step of 50 μm), first increasing and then decreasing the amplitude. This last test allowed us to retrieve the sensitivity of each sensor. An average sensitivity of $3.56\text{x}10^{-4}$ nm/μm (RR) and $3.98\text{x}10^{-4}$ nm/μm (HR) was obtained for sensor D. To sensor E, it was obtained an average sensitivity of $2.77\text{x}10^{-4}$ nm/μm (RR) and $3.11\text{x}10^{-4}$ nm/μm (HR).

Comparing the results obtained in the finite element simulation and in the experimental test, it was possible conclude that in both cases the most sensitive sensor is the one with the lowest filling, the sensor E. For that reason, this was the sensor chosen for the exploratory tests.

3. Exploratory tests on volunteers

The preliminary assessment of the optical sensor feasibility to monitor RR and HR as a wearable and non-intrusive sensor, was evaluated in one user (a man with 20 years old). The user was invited to place our reference device (BioHarness (BH)- ZEPHYR™ performance systems, Medtronic, Colorado, USA) on their chest and to sit on an office chair. Two tests were carried out: the first with the sensor placed on the wrist, attached to a *Velcro* bracelet; and the second with the sensor glued to the back of the office chair. In both tests, the user was invited to do the following breathing exercise: 20 s apnea (Ap), 60 s of normal breathing (NB), 20 s Ap, 60 s NB and 20 s Ap.

To obtain the vital signals from the exploratory tests, different signal processing steps were applied to highlight each signal component. considering the higher amplitude of breathing-induced chest movements only a smooth filter was applied to detect RR. For HR, since the cardiac vibrations are lower in amplitude and higher in frequency than the ones due to RR, a bandpass filter with cut-off frequencies of 0.8 Hz and 2.0 Hz was applied to discharge the breathing contributions and reveal the masked HR signal. Considering the BH values as a reference, a comparison was made between the number of respiratory and cardiac cycles detected in the same time intervals by proposed sensor and the BH (Figure 3).

From the results of all testes, a root-mean-square deviation (RMSD) value of ~0.00 rpm and 3.99 bpm were obtained. As the mean RR and HR was 10.17 rpm and 68.83 bpm, the RMSD corresponded to an error of ~0.00% to RR and 5.80% to HR.

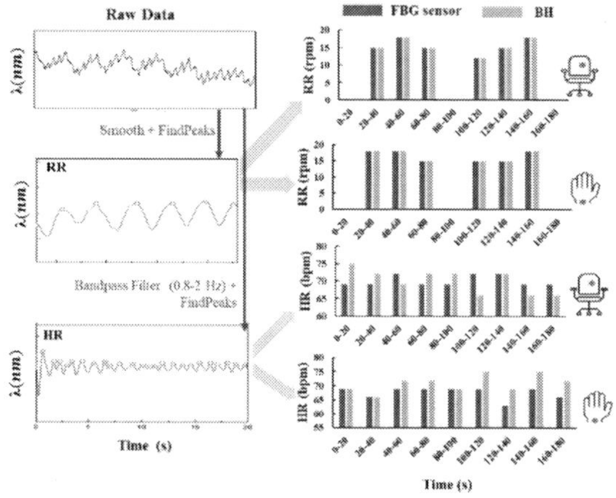

Fig. 3: Example segments for a section of the office chair data acquired (left); RR and HR results represented as bar graphs for two testes (right).

4. Conclusion

This paper presented 3D-printing FBGs, evaluated the influence of both infill density and material on their responses and tested the most sensitive in the monitoring of RR and HR as a wearable (in contact with the wrist) and non-intrusive solution (in contact with an office chair). The sensor showed good performance, with a RMSD value lower than 6%. These promising preliminary results open the possibility of testing the proposed sensor in other wearable and non-intrusive solutions and on more users to confirm the reliability and usability of this device as a sensor for vital signs monitoring.

References

1. D. Presti, C. Massaroni, C. Leitão, M.F. Domingues, M. Sypabekova, et al., IEEE Access, 24, 8 (2020).
2. C. Mayoral, G. Gutiérrez, J. Pérez, M. Treviño, I. Velasco, P. Cruz, et al., Biosensors, 11, 58 (2021).
3. C. Tavares, C. Leitão, D. Presti, M. Domingues, N. Alberto, et al., Biomed. Opt. Express., 13, 4 (2022).
4. Z. Hao, K. Cook, et al. IEEE Sens. Lett., 4 (2020).

Phase contrast imaging to detect transparent cells in the retinal ganglion cells layer

Elena Gofas-Salas [1,2,*], *Nathaniel* Norberg [2], *Céline* Louapre [3], *Ysoline* Beigneux [3], *Catherine* Vignal Clermont [1,4], *Michel* Paques [2], *and Kate Grieve* [1]

[1] Sorbonne Université, INSERM, CNRS, Institut de la Vision, 17 rue Moreau, F-75012 Paris, France
[2] CHNO des Quinze-Vingts, INSERM-DGOS CIC 1423, 28 rue de Charenton, F-75012 Paris, France
[3] Sorbonne Université, Paris Brain Institute - ICM, Assistance Publique Hôpitaux de Paris, Inserm, CNRS, Hôpital de la Pitié Salpêtrière, CIC neurosciences, Paris, France
[4] Hôpital Fondation Rothschild, Paris, France

Abstract. The eye is an optical window giving access to neural networks in a non-invasive way. It is possible to find in the retina biomarkers informing about the pathological state of other parts of the human body, and in particular of the brain. Neurodegenerative diseases could thus be diagnosed early and monitored by high-resolution imaging of the retina. However, a large part of the neurons in the retina are too transparent to be detected by existing techniques. At the Quinze-Vingts hospital, we have a unique retinal imaging platform in which ophthalmologists, neurologists and engineers participate. We implemented a technique based on scanning laser ophthalmoscopy (SLO) to capture the fine variations in refractive index between retinal cells. In this project we aimed at imaging and monitor cellular changes on transparent cells in the retinal ganglion cells layer in vivo on healthy participants and patients with neurodegenerative diseases.

1 Introduction

Adaptive optics technology has been introduced into clinical tools and paved the way for high-resolution imaging of the retina. This technology corrects the wavefront aberrations of the light induced by the eye lens, which cause the lack of resolution in existing clinical systems [1]. These aberrations are calculated by an analyser and then compensated by a deformable mirror allowing the system to reach a resolution of only a few microns. There are two types of clinical cameras corrected by adaptive optics, the full field ophthalmoscope [1], where an extended source is projected over the entire field of the image; and the scanning ophthalmoscope (AOSLO), where a light source is scanned to generate the image [2]. In addition, AOSLO can be equipped with a confocal pinhole that filters out all photons from layers outside the image plane. Using adaptive optics, highly reflective cells, such as photoreceptors, have been revealed [1,2]. However, the confocal configuration turned out to be inefficient in the case transparent cells such as certain neurons or capillaries. To meet this challenge, AOSLO was made sensitive to the fine phase variations of the light passing through these cells by modifying the detection. The AOSLO was optimized by integrating phase contrast imaging techniques from scanning microscopy known as "off-axis" techniques which rely on the displacement of the confocal pinhole of the AOSLO to block the confocal signal reflected by highly reflective structures [3-6]. We can then detect the light backscattered by the deep layers of the retina, which will pass through the transparent cells and therefore illuminate them "by transmission". When crossing these transparent cells, the light is deflected with an angle that depends on the refractive index of the tissue crossed. The displaced confocal pinhole will therefore detect more or less light depending on this angle, generating phase-dependent intensity variations on the detector. A range of modalities based on this off-axis detection already exist. At University of Pittsburgh we developed an optimal detection pattern which consists in a radial combination of multiple offset pinholes allowing for the first time to obtain *in vivo* phase contrast images of microglial and ganglion cells on healthy participants [7]. We implemented this "multi-offset" detection in an AOSLO at the Quinze Vingts hospital in order to image the transparent cells in the retinal ganglion cells layer on healthy subjects as well as patients.

2 Phase contrast Adaptive Optics Scanning Laser Ophthalmoscope (AOSLO)

We modified our AOSLO system previously described [8] at the detection plane to reproduce optimal multi-offset detection configuration determined before [7]. We added a rotation stage to the current detection fiber bundle. The latter is a fiber bundle composed of two opposite fibers that act as offseted pinholes placed at opposite side of a central confocal pinhole. It can be manually rotated to complete the multi-offset pattern as shown in Fig.1. Four acquisitions of 100 frames at 24 Hz need to be recorded at the same region and layer.

The multi-offset technique enables us to convert the phase signal of the photons going through the transparent cells into intensity at the detection plane. The light, which has previously been multiply-scattered by deep layers of the retina, goes through the cells where it is deviated differently depending on the refractive index of the tissue and therefore photons from different regions of the cell do not reach the same offset aperture leading to differences between adjacent tissues with different refractive indexes in transparent cells [7].

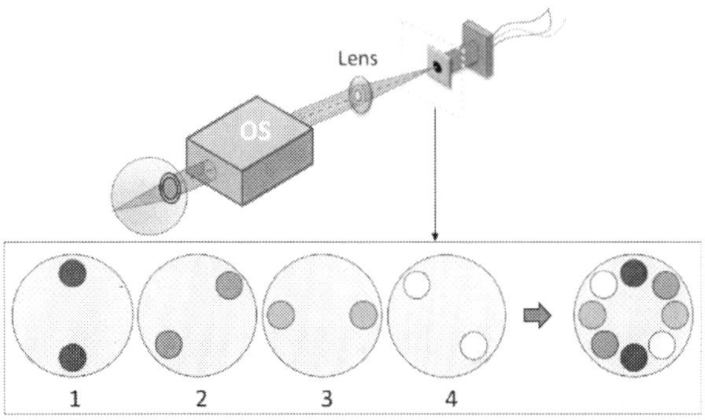

Fig. 1. Schematic drawing of the multi-offset detection configuration implemented on the Quinze-Vingt AOSLO. An arrow indicates the place of the detection fiber bundle which is placed on a rotation stage and rotated to four positions to sequentially obtain the 8 positions multi-offset pattern. OS: Optical system.

By combining the images from the multiple pinholes, we revealed the transparent ganglion cells (Fig.2 d). We also simultaneously acquired confocal images that reveal highly reflective cellular structures such as nerve fibers (Fig.2 c). Image sequences were corrected for eye motion and averaged to increase signal to noise ratio.

3 Detecting and monitoring cells in the retinal ganglion cells layer

In the retina, as in the rest of the central nervous system, immune cells play a key role in homeostasis, neuroprotection and the death of neuronal cells [9]. There is also the potential to observe ganglion cells as well as their axons (Fig.2) which are damaged over the course of neurodegenerative diseases.

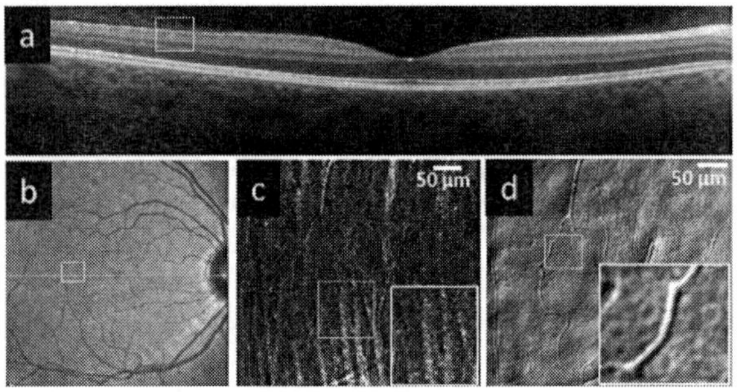

Fig. 2. (a) Optical coherence tomography (OCT) cross-section of healthy volunteer with a rectangle showing were the images were acquired. (b) SLO image of the same volunteer. (c) confocal AOSLO image of the nerve fibers (axons) of the white rectangle region in (a,b). (d) Multi-offset AOSLO images showing the ganglion cells in the same region.

Phase contrast imaging through multi-offset AOSLO has enabled us to image ganglion cells at a cellular resolution in humans in vivo. In the zoom (Fig.2 c) cells with a dark center and surrounded by a bright ring can be distinguished with a size and distribution in accordance with retinal ganglion cells characteristics. Additionally, we were also able to image patients with neurodegenerative pathologies from ongoing clinical studies at the Quinze Vingt hospital and detected putative immune cells in the ganglion cells layer. These cells could become biomarkers informing about the impact of inflammation in these neurodegenerations.

4 Conclusions and Perspectives

Using the phase contrast AOSLO imaging, we were able to detect previously invisible neural network cells as well as putative inflammatory cells specifically in the ganglion cells layer which is targeted by several neurodegenerative diseases. Future work will focus in implementing a fiber bundle detector composed of optical fibers organized according to the optimal multi-offset pattern for simultaneous acquisition of all offset pinholes. This will shorten the total acquisition time and will allow us to detect the phase signal from the same timepoint in all directions.

References

[1] J. Liang, D.R. Williams, D.T. Miller. JOSA A **14**.11, 2884-2892 (1997)

[2] A. Roorda, F. Romero-Borja, W. J. Donnelly III, H. Queener, T.J. Hebert, and M.C.W. Campbell, Opt. Express **10**, 405-412 (2002)

[3] T.Y. P. Chui, D.A. VanNasdale, and S.A. Burns, Biomed. Opt. Express **3**, 2537-2549 (2012)

[4] D. Scoles, Y. N. Sulai, and A. Dubra, Biomed. Opt. Express **4**, 1710-1723 (2013)

[5] D. Scoles, Y.N. Sulai, C.S. Langlo, G.A. Fishman, C.A. Curcio, J. Carroll, A. Dubra, Invest. Ophthalmol. Vis. Sci., **55**(7):4244-4251 (2014)

[6] E. A. Rossi, C. E. Granger, R. Sharma, Q. Yang, K. Saito, C. Schwarz, S. Walters, K. Nozato, J. Zhang, T. Kawakami, W. Fischer, L.R. Latchney, J.J. Hunter, M.M Chung, D.R. Williams, PNAS, **114**(3), 586-591 (2017).

[7] E. Gofas-Salas, Y. Rui, P. Mecê, M. Zhang, V.C. Snyder , K.V. Vienola, D.M.W. Lee, J.A. Sahel, K. Grieve, E.A. Rossi, Biomed Opt Express, **6**;13(1):117-132 (2021)

[8] K. Grieve, E. Gofas-Salas, R. D. Ferguson, J. A. Sahel, M. Paques, E.A. Rossi, Biomed. Opt. Express **9**, 5946-5961 (2018)

[9] G. Rathnasamy, W.S. Foulds, E.A. Ling, C. Kaur, Prog Neurobiol., **173**:18-40 (2019)

https://doi.org/10.1051/epjconf/202226604004

Linearly modulated multi-focal diffractive lens for multi-sheet excitation of flow-driven samples in a light-sheet fluorescence microscope

Meike Hofmann[1,*], *Shima* Gharbi Ghebjagh[1], *Chao* Fan[1], *Yuchao* Feng[1], *Karen* Lemke[2], and *Stefan* Sinzinger[1]

[1]Technische Universität Ilmenau, Fachgebiet Technische Optik, Postfach 100565, 98684 Ilmenau, Germany
[2]Institut für Bioprozess- und Analysenmesstechnik e.V., Rosenhof, 37308 Heilbad Heiligenstadt, Germany

Abstract. Light-sheet fluorescence microscopy (LSFM) with single light-sheet illumination enables rapid 3D-imaging of living cells. In this paper we show the design, fabrication and characterization of a diffractive optical element producing several light sheets along an inclined tube for applications in flow-driven imaging. The element, which is based on a multi-focal Fresnel zone plate and a linear grating, generates in combination with a refractive cylindrical lens five thin light sheets of equal intensity.

1 Introduction

The approach of light-sheet fluorescence microscopy (LSFM) enables 3D-imaging of many types of samples [1–3]. 3D cell cultures are cultivated, stored and characterized in tubes on the droplet based platform of pipe based bioreactors [4], offer the opportunity of using the sample movement for the scanning mechanism providing a z-stack of images in LSFM platforms [5]. The generation of not only one but a series of equal intensity light-sheets enables multi-plane imaging [6, 7], which can effectively reduce the aberrations caused by the mechanical motion of the system and speed. It also enables velocity and correlation measurements.

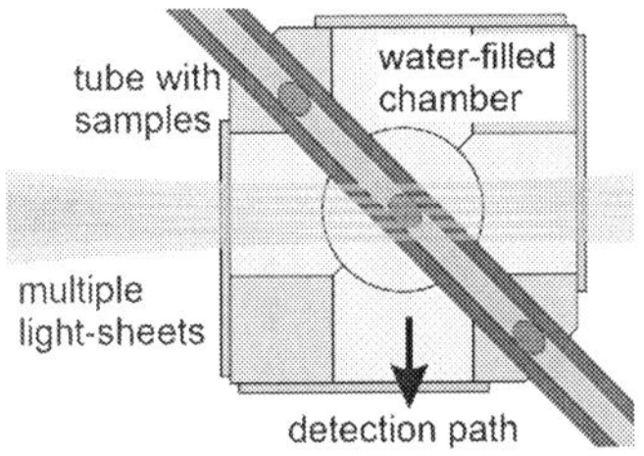

Figure 1. Schematic of the measurement configuration with tube-guided samples illuminated with multiple light sheets. The inclination of the tube of 45° offers the option of a 90°-detection between illumination and detection path.

2 Design, Fabrication and Measurement Results

For the generation of several equal-intensity light-sheets the combination of a multi-focal Fresnel zone plate with a linear grating is optimized. The result is a multi-focal diffractive lens (MFL) which plays the dual role of a multi-focal diffractive lens, as a multi-plane defocuser, and a beam splitting linear grating based on a detour phase approach. The optimized nonlinear phase values in each period affect the energy distribution in the diffraction orders [8, 9] leading to an efficient equal-intensity array of focus spots. Additionally, a refractive cylindrical lens is required to obtain the desired light-sheet distribution.

The transmission function of the MFL is given by

$$T_{\mathrm{MFL}}(\rho, \theta) = \sum_{s=-\infty}^{\infty} c_s \exp\left[is\left(\frac{\pi\rho^2}{\lambda f} + \frac{2\pi\rho\cos\theta}{d} \right)\right], \quad (1)$$

where ρ and θ denote the polar coordinates in the input plane. f and d are the focal distance of the MFL and the period of the linear grating. The coefficients c_s depend on the optimized phase values assigned to each area of the MFL to assure a uniform energy distribution for

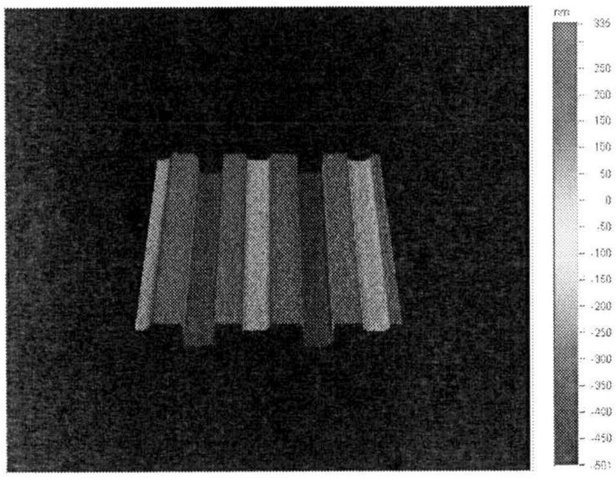

Figure 2. Measurement of the profile of the etched element using white light interference microscopy.

*e-mail: meike.hofmann@tu-ilmenau.de

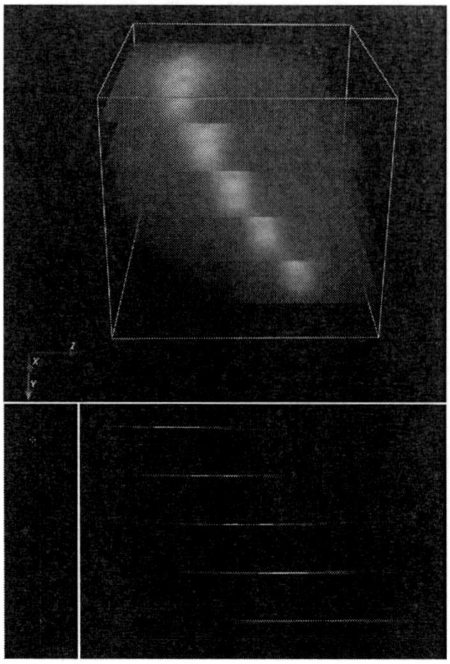

Figure 3. Cube of 6 mm x 6 mm x 6 mm showing the five light-sheets in 3D and cross section.

the generated diffraction orders [8, 9]. The transmission function includes an order-dependent focusing power and a linear phase. This introduces a change of direction leading to an orientation of the focal spots along a desired line coincident with the specimen trajectory within the tube. The general structure of the designed MFL is a slightly curved grating with four phase levels, whereas two levels are equal. Two different elements were designed for the generation of five light-sheets with distances of 1 mm and 0.25 mm to each other. The phase function was adapted for the illumination within a block of water as the aqueous samples are guided in a FEP-tube immersed in a water chamber. The refractive index of FEP is very close to the one of water.

The elements were fabricated in the clean room by means of maskless lithography. Two mask layers were calculated from the phase levels with pixel sizes of 250 nm × 250 nm. The required etching depths were converted from the phase difference values. For the wafer material of fused silica and a wavelength of 532 nm the etching depths were calculated to be 196.3 nm and 617.8 nm, respectively. Reactive ion etching was used for the fabrication process [10]. The achieved etch depths were measured with white light interference microscopy. The measured profile is shown in Fig. 2 With 194 nm and 628 nm they are close to the design values.

The light distribution behind the fabricated MFL in combination with a cylindrical lens was measured plane by plane and reconstructed. Fig. 4 gives the light distribution in 3D and a cross section. The efficiency varies only slightly between the orders. The performance of the element was evaluated by considering the whole optical setup. Figure 4 shows a qualitative impression by showing the tube illuminated with the five light sheets.

3 Conclusion

We successfully designed, fabricated and characterized a DOE which generates in combination with a cylindrical lens five light sheets of equal intensity along a guiding tube for multi-plane LSFM-imaging.

Figure 4. Measurement chamber with inserted FEP-tube filled with water and illuminated by the multiple light sheets.

References

[1] P.A. Santi, S.B. Johnson, M. Hillenbrand, P.Z. GrandPre, T.J. Glass, J.R. Leger, BioTechniques **46**, 287 (2009)

[2] N. Scherf, J. Huisken, Nature Biotechnology **33**, 815 (2015)

[3] M. Hofmann, A. Herrmann, U. Brokmann, tm - Technisches Messen (2022)

[4] K. Lemke, T. Förster, R. Römer, M. Quade, S. Wiedemeier, A. Grodrian, G. Gastrock, Journal of Biotechnology **205**, 59 (2015)

[5] E.J. Gualda, H. Pereira, T. Vale, M.F. Estrada, C. Brito, N. Moreno, Biomedical optics express **6**, 4447 (2015)

[6] P.A. Dalgarno, H.I.C. Dalgarno, A. Putoud, R. Lambert, L. Paterson, D.C. Logan, D.P. Towers, R.J. Warburton, A.H. Greenaway, Optics express **18**, 877 (2010)

[7] K. Mohan, S.B. Purnapatra, P.P. Mondal, I. Georgakoudi, PloS one **9**, e96551 (2014)

[8] S. Gharbi Ghebjagh, D. Fischer, S. Sinzinger, Applied Optics **58**, 8943 (2019)

[9] M. Hofmann, S. Gharbi Ghebjagh, K. Lemke, S. Sinzinger, OSA Imaging and Applied Optics Congress 2021 (3D, COSI, DH, ISA, pcAOP) p. CM1A.7 (2021)

[10] C. Weigel, U. Brokmann, M. Hofmann, A. Behrens, E. Rädlein, M. Hoffmann, S. Strehle, S. Sinzinger, Journal of Optical Microsystems **1** (2021)

Single-shot 3D endoscopic imaging exploiting a diffuser and neural networks

Julian Lich[1,*], *Tom* Glosemeyer[1], *Jürgen* Czarske[1,2] and *Robert* Kuschmierz[1,2]

[1]Laboratory of Measurement and Sensor Systems, TU Dresden, 01062 Dresden, Germany
[2]Competence Center for Biomedical Computational Laser Systems (BIOLAS), TU Dresden, Germany

Abstract. Lens-based endoscopes offer high lateral resolution, but suffer from rigid imaging properties, such as a fixed focal plane. We present a miniaturized 0.5 mm diameter endoscope in which the objective lens is replaced by an optical diffuser. The intensity information of the object space is scattered and passed to a camera via a coherent fibre bundle. The image is reconstructed by a neural network. The field of view and resolution depend on the object distance. 3D-single-shot imaging up to video rate can be enabled. The approach shows great potential for applications like robust 3D fluorescence imaging.

1 Introduction

Endoscopic imaging is an effective and efficient tool for in vivo diagnostics of diseases, such as cancer, and during surgical interventions. Regions that are difficult to reach out, like brain tissue and blood vessels, require ultrathin endoscopes for minimal invasiveness. In combination with the inherently compromised endoscope positioning, applications like brain cancer diagnostics or blood flow measurements require high imaging robustness and resolution at video rate, preferably in 3D.

The smallest endoscope diameters can be achieved with single mode (SMF) and multimode (MMF) fibres. However, necessary scanning movements for SMF [1] and the sensitivity of the MMF transmission matrix towards bending [2] require complex in situ correction to maintain robustness. While coherent fibre bundles (CFB) suffer from deterministic phase noise [3], they offer diameters below 0.5 mm and enable robust and spatially resolved light intensity transmission. Commercially available lens based CFB systems offer high lateral resolution, but the field of view (FoV) is limited to a fixed focal plane. Diffuser-based imaging offers the potential of high lateral resolution [4] as well as single shot 3D-imaging [5]. In contrast to lens-based imaging, diffuser-based imaging requires computational image reconstruction. While iterative convolution algorithms result in high reconstruction times [5], neural networks combine high image reconstruction performance at video rate [6,7].

We present a diffuser-based CFB endoscope in combination with a neural network for 2D image reconstruction at video rate. The setup is also capable of 3D reconstruction [8]. The reconstruction performance of different network architectures is compared regarding correlation coefficient (CC), peak signal-to-noise-ratio (PSNR) and structural similarity (SSIM). An in situ validation with a microscope, is shown for demonstration.

2 Setup and Image Processing

The training data is an augmented version of the Digits MNIST dataset. Up to five digits where randomly rotated, translated and put into one 104x104 ground truth image (example in Fig. 2a, top row). The ground truth is projected from a DLP® LightCrafter Display 4710 EVM-G2, containing a 1920x1080 digital micromirror device and high-power RGB LEDs (see Fig. 1, top right). A validation camera is placed into the virtual image plane of the projection optics. The light field is scattered by the diffuser and propagates through an integrated 0.5 mm glass spacer to the distal facet of the CFB. The CFB is a Fujikura FIGH10-3505 with 10,000 cores and 0.325 mm diameter. The proximal facet of the CFB is imaged onto a Thorlabs Quantalux acquisition camera. The camera image is resized to 104x104 and used as input of a single layer perceptron (SLP) for image reconstruction. Eventually, the SLP reconstruction is denoised by a U-Net.

3 Results

To demonstrate the effectiveness of the combined SLP+U-Net approach, reconstructions are also done with a U-Net and an SLP, individually. The results can be seen in Fig. 2a. The U-Net delivers moderate CC on image data similar to the training data, which however decreases rapidly on other data sets due to overfitting. The SLP reconstructions maintain a higher CC over all datasets but suffer from systematic noise, which keeps the PSNR and especially the SSIM low, as can be seen in Fig. 2b. The SLP+U-Net approach outperforms both individual SLP and U-Net regarding all shown quality parameters. The reconstruction rate is 10 Hz on an office PC (Intel® Core™ i5-4590).

* Corresponding author: julian.lich@tu-dresden.de

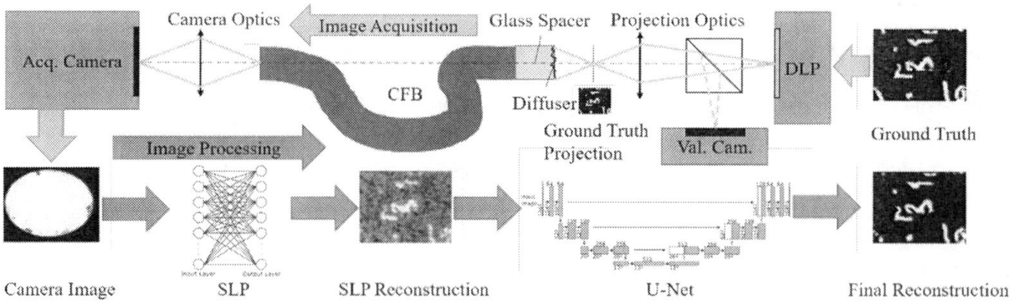

Fig. 1. Sketch of training setup, diffuser endoscope and image reconstruction.

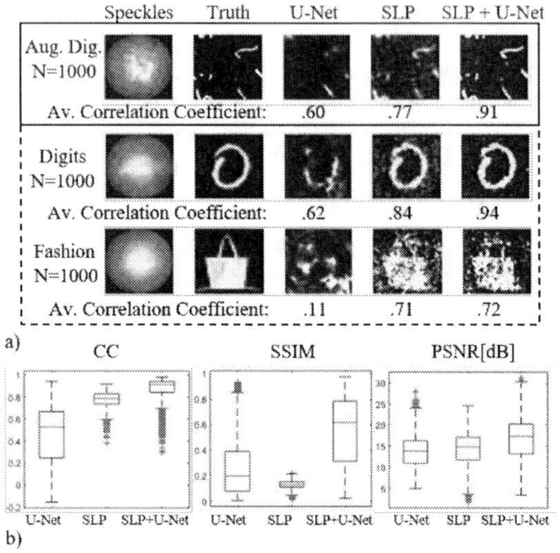

Fig. 2. Comparison between U-Net, SLP and SLP+U-Net after training with Augmented Digits data. a) Reconstruction examples. The CC data set average over N images is displayed below the reconstructions. b) CC, SSIM & PSNR boxplots of all data sets combined.

To demonstrate the feasibility of the method for real life applications, a physical USAF target is placed into the ground truth projection plane at a distance of 800 µm from the diffuser. A square of 400x400 µm is illuminated by the DLP projector. A comparison between validation camera images (reflection) and reconstruction with diffuser endoscope (transmission) can be seen in Fig. 3. The general object structures can be reconstructed and distances down to around 15 µm can be resolved. However, the non-linearity of the network leads to visible distortions of the stripes and numbers.

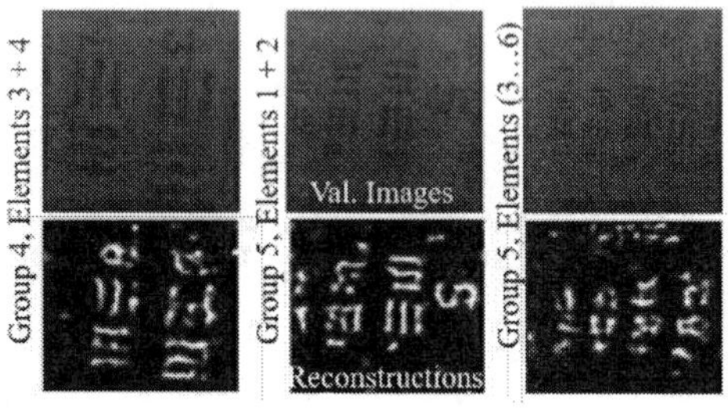

Fig. 3. Reconstruction (bottom) of a physical USAF target and validation images (top).

4 Conclusion

We propose a miniaturized endoscope using an optical diffuser and a neural network for intensity image reconstruction. The reconstruction performance and linearity can be significantly increased by sequentially processing the camera speckle image with a single layer perceptron, followed by a U-Net. Real object 2D imaging is performed on a test target. Video rate reconstruction is possible for 2D reconstruction. The setup can also be used for 3D imaging [8].

References

1. J. Li, S. Thiele, B.C. Quirk *et al.*, *Light Sci Appl* **9**, 124 (2020).
2. S. Rothe, N. Koukourakis, H. Radner, A. Lonnstrom, E. Jorsewieck, J. Czarske, *Sci Rep* **10,** 2740 (2020).
3. E. Scharf, J. Dremel, R. Kuschmierz, J. Czarske, *Opt. Lett.* **45**, 3629-3632 (2020)
4. E.G. van Putten, D. Akbulut, J. Bertolotti *et al.*, *Phys. Rev. Let.* **106**, 193905 (2011)
5. N. Antipa, G. Kuo, R. Heckel *et al.*, *Optica* **5**, 1-9 (2018)
6. S. Li, M. Deng, J. Lee, A. Sinha, G. Barbastathis, *Optica* **5,** 803-813 (2018)
7. K. Yanny, K. Monakhova, R. W. Shuai, L. Waller, *Optica* **9**, 96-99 (2022)
8. R. Kuschmierz, E. Scharf, D. F. Ortegón-González, T. Glosemeyer, J. Czarske, *Light: Adv. Manuf.*, **2(4),** 415-424 (2021)

https://doi.org/10.1051/epjconf/202226604006

Open-view Binocular Double-pass System for the Study of Dynamic Accommodation

Fracisco Ávila [1,*], *Rafael Navarro*[2], *Ebrahim Safarian Baloujeh*[2], *Ana Sánchez-Cano*[1]

[1] Departamento de Física Aplicada, Universidad de Zaragoza, Zaragoza, 50009, Spain.
[2] INMA, Consejo Superior de Investigaciones Cientificas & Universidad de Zaragoza, Zaragoza, 50009, Spain.

Abstract. We present a new open-view dynamic double-pass (DP) system for the study of accommodative response in binocular natural viewing conditions. The DP point spread function (PSF) of the eye is analyzed to compute the Visual Strehl Ratio (VSR) fluctuation as a function of time at 50 Hz frame rate. Preliminary results showed that the proposed method can quantify the temporal dynamic of the optical and visual quality of both eyes simultaneously.

1 Introduction

The interaction of light with the human eye has been widely employed to describe its optical properties [1]. The optical quality of the eye is mainly limited by both scattering and aberrations contributions [2]. In that sense, aberrometers (Hartmann-Shack sensor, Laser Ray Tracing, etc.), and double-pass (DP) systems are commercially available technologies that provide complementary characterization of the optics of the human eye. While aberrometers are used to reconstruct the wavefront aberration through the pupil of the eye, DP technology uses the retinal image from a point source to compute the double-pass *Point Spread Function* (PSF), or the single-pass Modulation Transfer Function (MTF) or the Visual Strehl Ratio (VSR) of the eye [3].
In the last years, accommodation has been one of the most studied functions of the visual system due to the direct relationship to presbyopia progression, myopia development or the restoration of near vision in pseudophakic eyes.

In that sense, DP technology emerged as an excellent objective tool to assess the accommodative response through the information about the optical quality of the eye provided by DP imaging. For the assessment of the dynamics of accommodation it is essential to make binocular measurements [4].
New open-view configurations, have allowed to measure the dynamic response of the optics of the eye in both binocular and natural viewing conditions using DP imaging and aberrometers [5].

The aim of this study was first to introduce a new open-view binocular DP system for the study of accommodation in natural viewing conditions. We present the methodology to obtain optical quality information from DP images. Finally, preliminary testing results from measurements on living human eyes are presented.

2 Methods

2.1 Experimental system

The system incorporates a super luminescent LED (SLED) centred at 840 nm as illumination source. The beam is collimated and split into two identical entrance pupils by means of a 50:50 beam splitter and an adjustable twin mirror system (dual-pupil positioning unit in Fig. 1) to control the spatial separation of the two entrance beams. The entrance pupils are optically conjugated with a refraction correction unit composed of two programmable opto-electronic lenses that compensate for the spherical (defocus) error of each eye, and the natural pupils of the subject itself.

Before reaching the correction unit, the entrance beams are deflected to two optical arms that allow simultaneous control of both eye's pupils and binocular retinal illumination (See Fig. 1). A couple of hot mirrors (HM) separate the infrared illumination from the visual stimuli. One of the HMs was mounted into a micrometric slider to adjust the interpupillary distance of each subject. The DP system was designed in open-view configuration to allow natural visual conditions.

After retinal reflection (i.e. double-passing), aerial DP images are combined within the same optical pathway preserving the original separation at the entrance pupil of the system and are sampled in a single fundus camera. The optical separation of the entrance pupils was stablished to avoid overlapping of the DP images at the (common) sensor plane.

* Corresponding author: avila@unizar.es

https://doi.org/10.1051/epjconf/202226604006

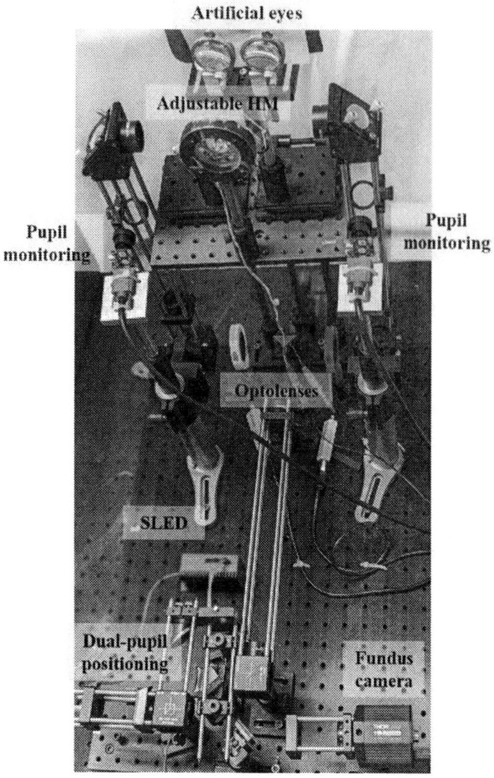

Fig. 1. Top view of the binocular DP system.

2.2 Optical quality analysis

The temporal fluctuations of the optical quality of the eye during accommodation were assessed by computing the VSR from the intensity DP images of each eye, defined as the ratio of the maximum irradiance of the eye's PSF to that of a diffraction limited PSF for the same aperture.

3 Results

Fig. 2 shows the visual Strehl ratios of both eyes computed form the aerial double pass PSF, recorded at frame rate of 50 Hz. We can observe fluctuations of visual quality even for a constant vergence of the fixation target. In this case the VSR was not normalized by that of a perfect system, so that the vertical axis represents relative values. The right eye (Fig.2b) shows roughly double quality than the left eye, as well as higher fluctuations.

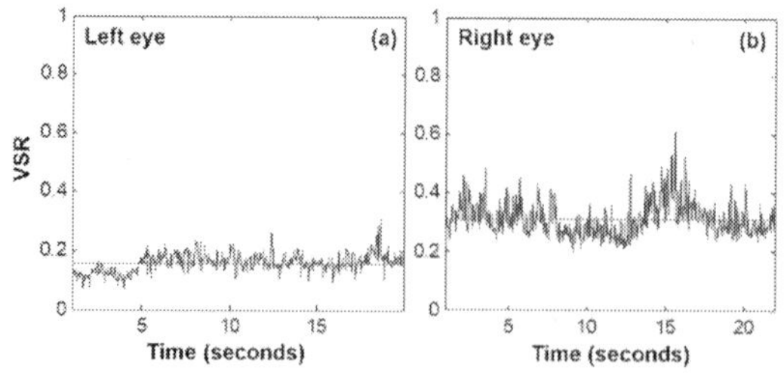

Fig. 2. VSR as a function of time under binocular conditions for one subject. Red line: Averaged VSR value.

4 Conclusions

This work presented an open-view binocular double-pass system to study the temporal dynamics of accommodation in natural viewing conditions. Optical quality fluctuations of the eye were assessed by computing the VSR from intensity DP images and tested in a real eye.

Funding

Funded by the European Union's Horizon 2020 research and innovation programme under the Marie Skłodowska-Curie grant agreement No 956720 and by Ministerio de Ciencia e Innovación y Universidades of Spain (MICIN), grant number PID2019 107058RB-I00

References

1. G. Walsh, W.N. Charman, H.C.Howland. J Opt Soc Am A. 1(9): 987-992 (1984).
2. J. Wanek, M. Mori, M. Sahidi. J Opt Soc Am A Opt Image Sci Vis. 24 : 1296-1304 (2010).
3. F. Diaz-Doutón, A. Benito, J. Pujol, et al. IOVS, 47: 1710-1726 (2006).
4. J. Santamaría, A. Plaza, J. Bescós, Optica Applicata 14(3): 341-347 (1984).
5. C. Garcia-Guerra, M. Aldaba, M. Arjona, J. Pujol. Appl Opt, 1-5 (2016).

Physics-informed machine learning for microscopy

Emmanouil Xypakis[1,2,*], *Valeria deTurris*[1], Fabrizio Gala[3], Giancarlo Ruocco [1] and *Marco Leonetti*[1,2,4]

[1]Center for Life Nano- and Neuro-Science, Istituto Italiano di Tecnologia, Viale Regina Elena 291, 00161, Rome, Italy
[2]D-TAILS srl, Rome, Italy
[3]Crestoptics, S.p.A., Italy
[4]Soft and Living Matter Laboratory, Institute of Nanotechnology, Consiglio Nazionale delle Ricerche, 00185, Rome, Italy

Abstract. We developed a physics-informed deep neural network architecture able to achieve signal to noise ratio improvements starting from low exposure noisy data. Our model is based on the nature of the photon detection process characterized by a Poisson probability distribution which we included in the training loss function. Our approach surpasses previous algorithms performance for microscopy data, moreover, the generality of the physical concepts employed here, makes it readily exportable to any imaging context.

In any optical and non-optical imaging technologies, measurement comes with a noise addition producing a signal that follows a Poisson probability distribution (PPD). Signal enhancement algorithms increase the amount of information by increasing the signal to noise ratio (SNR), making them useful for modeling and visualization of biological data including: microscopy images representing light; medical imaging; computer tomography; positron emission tomography and other in-vivo imaging technologies.

Deep neural networks (DNNs) [1–7] based algorithms achieve the best results in signal enhancement. The performance and the ability to train DNNs, however, depends both on the chosen loss function – a quantity comparing predictions and target that DNN minimizes to learn its internal parameter – but also on the normalization of the network inputs and targets8. The simplest and more frequently used loss functions for denoising, but also in other image enhancement tasks is either the L1-norm or the L2-norm (MSE) [2,4,9,10], where the data are arbitrary normalized. On the other hand, when the desired output comes from a known probability distribution like in semantic segmentation (U-net[11]), or other classification tasks, an entropic loss function, and a probabilistic normalization of data, is of great significance. Although accounting the physics of the camera detection process is known to significantly improve imaging efficiency [12], little research has been done in applying this physical properties in DNNs.

Physics-informed machine learning is a new trend in artificial intelligent [13] and encoding the physics and statistics of light in a DNN is a missing puzzle. Here we report about physically informed DNN which builds on the PPD of signal detections. Our approach aims to provide a general and exportable approach to deal with PPD signal detection: I) we use a non-arbitrary and physics-based normalization process, II) we employ a physically informed loss function, and III) we design DNN architecture which works coherently with the approach. First, we remove any arbitrariness on the normalization just working with images in which each pixel count represents the photon number. Then we design a loss function which considers distance between probability distributions instead of the distance between count number. This enables the algorithm to work with the same efficiency in all dynamic range windows. We employ a custom DNN architecture capable of classifying each pixel on a predicted photon number, thus preserving the photon number encoding and meaning for the output images. We further show that the advantages of the semantic segmentation of U-net and the denoising power of RCAN can be combined in a unique architecture, we call it RESUNET.

Bibliography

1. Belthangady, C. & Royer, L. A. learning for fluorescence image reconstruction. Nat. Methods (2009) doi:10.1038/s41592-019-0458-z.
2. Weigert, M. et al. Content-aware image restoration: pushing the limits of fluorescence microscopy. Nat. Methods 15, 1090–1097 (2018).
3. Laine, R. F., Arganda-Carreras, I., Henriques, R. & Jacquemet, G. Avoiding a replication crisis in deep-learning-based bioimage analysis. Nat. Methods 18, 1136–1144 (2021).
4. Chen, J. et al. Three-dimensional residual channel attention networks denoise and sharpen fluorescence microscopy image volumes. Nat. Methods 18, (2021).
5. Gurrola-ramos, J., Dalmau, O. & Alarcón, T. E. A Residual Dense U-Net Neural Network for Image Denoising. 9, (2021).
6. Byun, J., Cha, S. & Moon, T. FBI-Denoiser: Fast Blind Image Denoiser for Poisson-Gaussian Noise. (2021).
7. Mayorov, A. S. et al. Interaction-Driven Spectrum Reconstruction in Bilayer Graphene. Science (80-.). 333, 860–863 (2011).
8. Shanker, M., Hu, M. Y. & Hung, M. S. Effect of data standardization on neural network training. Omega 24, 385–397 (1996).
9. Qiao, C. et al. Evaluation and development of deep neural networks for image super-resolution in optical microscopy. Nat. Methods (2021) doi:10.1038/s41592-020-01048-5.
10. Xypakis, E. et al. Deep learning for blind structured illumination microscopy. Sci. Rep. 12, 8623 (2022).
11. Falk, T. et al. U-Net: deep learning for cell counting, detection, and morphometry. Nat. Methods 16, 67–70 (2019).
12. Mandracchia, B. et al. Fast and accurate sCMOS noise correction for fluorescence microscopy. Nat. Commun. 1–12 doi:s41467-019-13841-8.
13. Karniadakis, G. E. et al. Physics-informed machine learning. Nat. Rev. Phys. 3, 422–440 (2021).

* Corresponding author: author@e-mail.org

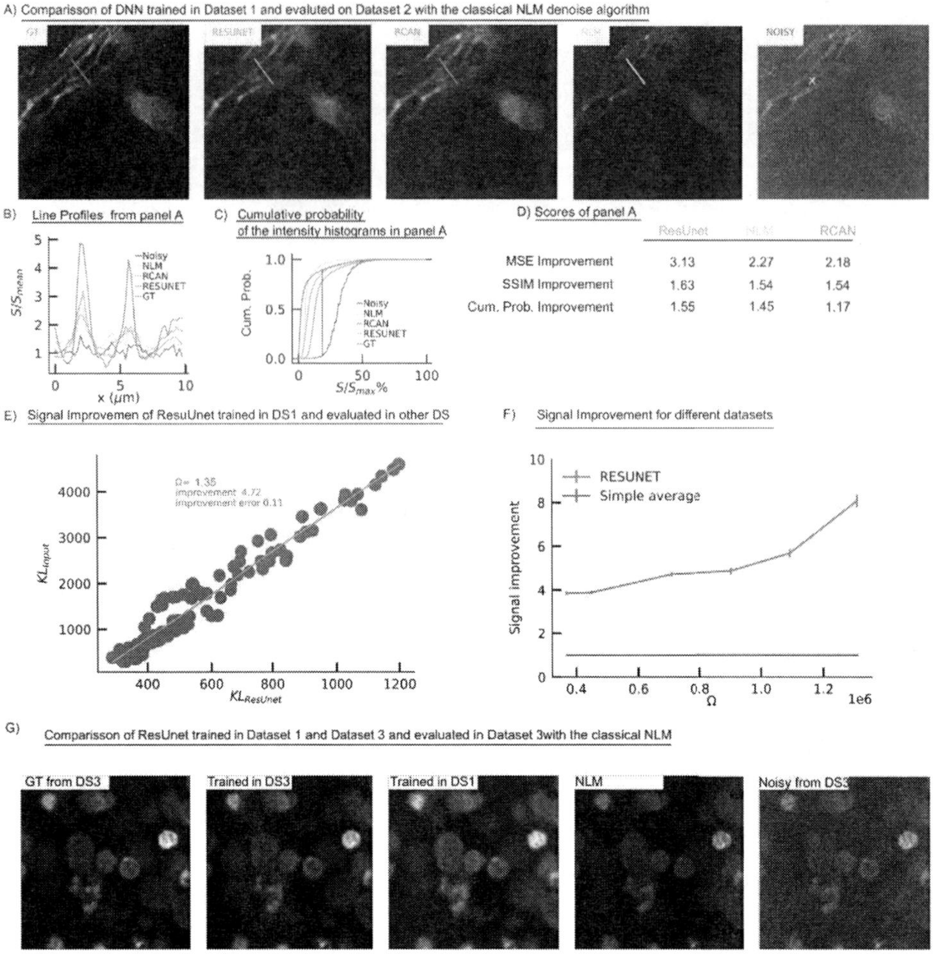

Figure: A) The comparison of the ResUnet with RCAN and NLM algorithm reconstruction of a Noisy input 10msec exposure (Noisy) for a Noci cell image slice from Dataset 2 with the corresponding GT (500msec exposure). Both ResUnet and RCAN have been trained in a different dataset 1. Characterized by different optical systems, fluorophores, and cameras. B) the line profile of the panel A for the line shown above C) The Cumulative probably of the intensity which is characterized by the Kolmogorov distance (vertical line) for panel A. D) The MSE, SSIM, Cumulative probability improvements for Panel A. E) We define the signal Improvement as the slope between the KLinput and the KLResUnet scatter plot linear fit. which we find for each different dataset here shown just for one characterized by the parameter F) The Signal improvements for different datasets with different $\Omega = \frac{V_{input}}{V_{psf}}$. G) Comparison of ResUnet for the same DS trained and different and comparison with the NLM.

EPJ Web of Conferences **266**, 05001 (2022) https://doi.org/10.1051/epjconf/202226605001
EOSAM 2022

Numerical investigation of far-field circular dichroism and local chiral response of pseudo-chiral meta-surface with FEM

Jayeeta Amboli[1,] *, *Guillaume* Demésy[1], *Bruno* Galas[2], and *Nicolas* Bonod[1]

[1]Aix-Marseille Univ, CNRS, Centrale Marseille, Institute Fresnel, 13013 Marseille, France.

[2]Sorbonne Universités, UPMC Univ Paris 06, CNRS, Institut des NanoSciences de Paris, UMR 7588, 75005 Paris, France

Abstract. Circular dichroism spectroscopy is a sensitive and widely applied technique to detect chiral molecules. Recent studies have shown high prospects for plasmonic metasurfaces of pseudo-chiral nano-resonators in enhancing chiral sensitivity. Here we study the far-field circular dichroism for gold U-shaped metasurfaces by calculating Mueller matrix elements with the Finite element method and investigate its response in light of the near field electric energy and optical chiral density.

1 Introduction

Chirality is a key structural property of molecules in biochemistry, that makes them non-superimposable onto their mirror images. Chiral molecules play a widespread role in nature, and their highly sensitive optical detection is a major challenge in life science. The pair of mirror image chiral molecules, called enantiomers, exhibit different chemical and biological properties depending on their handedness, which they unfold by interacting with other chiral objects, such as a circularly polarized light. The enantiomers show preferential absorption to left and right circular polarization, thereby revealing their presence and handedness. Circular Dichroism (CD) spectroscopy is a relevant and well-established method to quantify the electromagnetic chirality of enantiomers by measuring their differential absorption of opposite circularly polarized incident light.

Circular dichroism of a molecule (considered as an electric dipole) can be expressed as [1]

$$CD \propto A^{+}(\mathbf{r}) - A^{-}(\mathbf{r})$$
$$CD \propto \frac{\omega}{\epsilon_0}\alpha''_{ee}[U_e^{+}(\mathbf{r}) - U_e^{-}(\mathbf{r})] + \frac{2}{\epsilon_0}\alpha''_{em}[C^{+}(\mathbf{r}) - C^{-}(\mathbf{r})] \qquad (1)$$

where, $A^{\pm}(\mathbf{r})$, $U_e^{\pm}(\mathbf{r})$, $C^{\pm}(\mathbf{r})$ are the absorptions, electric energy densities and optical chiral densities of a randomly oriented molecule at a position **r** by two opposite circularly polarized beams denoted by the superscript $\pm$, respectively. α''_{ee}, α''_{em}, ω and ϵ_0 denote the imaginary part of the electric polarizability, imaginary part of the chiral polarizability, frequency of the incident light, and the dielectric permittivity of free space, respectively. The second part of equation (1) is attributed to the chiral absorption of the molecule, which is orders of magnitude weaker compared to the background achiral absorption, characterized by the first part of the equation, due to the intrinsic small values of α''_{em}. Therefore, detection of chiral absorption requires high sample concentration and long measurement time. The interaction of chiral molecules with electro-magnetic wave can be characterized by local chiral density C(**r**), expressed as

$$C^{\pm}(\mathbf{r}) = \frac{\omega}{2c^2} Im\{\boldsymbol{E}^{\pm}(\mathbf{r})^{*}\boldsymbol{H}^{\pm}(\mathbf{r})\} \qquad (2)$$

Therefore, the detection sensitivity can be improved by generating strong local electro-magnetic fields. Many chiral and pseudo-chiral plasmonic nanostructures [2] have been designed to enhance the CD response, among which U-shaped scatterers have been of great importance due to their magneto-electric coupling. Recent studies have also established the relevance of dielectric achiral metasurfaces in boosting local C [1]. Previous studies have experimentally quantified the magneto-electric coupling of a single U-shaped scatterer in light of the polarizability tensors [3]. However, far-field CD as a function of the near field electromagnetic responses has not yet been investigated for a metasurface of such split-ring scatterers.

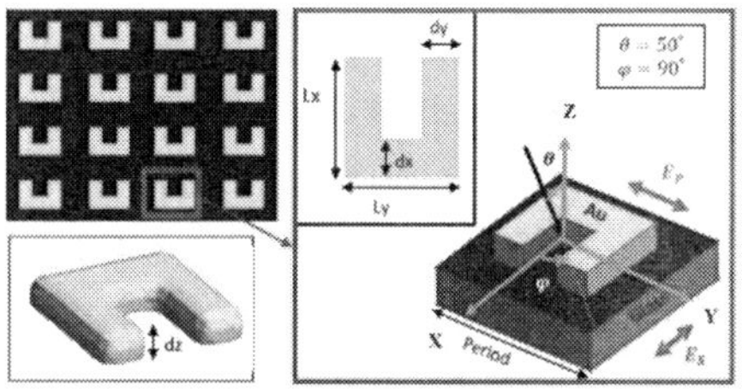

Figure. 1. Schematic of the meta-surface with U-shaped resonators with closer view of it at the inset. The parameters are $L_x = 232\text{nm}, L_y = 210\text{nm}, d_x = 90\text{nm}, d_y = 74\text{nm}, d_z = 40nm$, $Period = 390nm$. The corners of the U have been rounded off to avoid sharp resonances. The resonators are embedded in a

* Corresponding author: jayeeta.amboli@fresnel.fr

glass substrate. The polar and azimuthal angles of incidence are shown with respect to the co-ordinate axes.

In this work, we numerically study the local electric energy density $U_e^{\pm}(\mathbf{r}) = \frac{\epsilon_0}{2}|E^{\pm}(\mathbf{r})|^2$ and chiral desntiy responses of gold U-shaped resonators and investigate the far field CD by computing the Mueller matrix of the metasurface.

2 Numerical analyses

The far-field optical response of a medium can be completely described by its 4x4 Mueller matrix. In this study, we calculate the Mueller matrix elements using a well-suited formulation of the Finite Element Method (FEM) [4]. This formulation rigorously addresses the vector diffraction problems by periodic crossed gratings of U-shaped and nano pillar resonators, and replaces it with an equivalent radiation problem with sources localized inside the resonator itself. Figure 1 shows the structure of the U-shaped nano-resonators. The Mueller matrix elements are calculated in the visible wavelength range (400–1200nm), at polar angle of incidence $\theta_0 = 0°$ and 50° and an azimuthal angle of incidence $\varphi_0 = 90°$, that is, along the lower arm of U. The simulations were performed for electric field components parallel (p) and perpendicular (s) to the plane of incidence containing the lower arm of the U (y-z plane). The amplitude reflection coefficients and Mueller matrix elements are calculated in the far-field by taking only the specular component of reflection into account. The local electric energy density and chiral density are calculated by averaging $\Delta U_e = U_e^+(\mathbf{r}) - U_e^-(\mathbf{r})$ and $\Delta C = C^+(\mathbf{r}) - C^-(\mathbf{r})$ over a 10nm box volume surrounding the U resonator, illuminated by right (+) and left (-) circularly polarized light.

3 Results and discussion

The normalized Mueller matrix is denoted as $[m_{ij}]$ where $m_{ij} = \frac{M_{ij}}{M_{11}}$, for $i,j = 1,2,3,4$. Our study focuses on investigating the element m_{14} as it corresponds to circular dichroism. The m_{14}, ΔU_e, and ΔC spectra of the gold U metasurface for normal and oblique incidence are shown in Figures 2(a), 2(b) and 2(c), respectively. A qualitative comparison of the CD, ΔU_e and ΔC spectra indicates that for normal incidence, the symmetry properties of $U_e^+(\mathbf{r}) = U_e^-(\mathbf{r})$ and $C^+(\mathbf{r}) = -C^-(\mathbf{r})$ hold under parity operation, which, in this case, is equivalent to applying left and right circularly polarized light. Therefore, the resonators behave as achiral objects in normal incidence, without any coupling between electric and magnetic fields, leading to vanishing ΔU_e. In this case, CD can be expressed solely as a function of ΔC, which is significantly small coming only from the chirality of the incident field, thereby leading to nearly vanishing CD. On the other hand, for oblique incidence, ΔU_e becomes the main contributing factor in CD. In this case, as the magneto-electric coupling comes into play, the scattered electric field is enhanced as a function of the incidence angle [3].

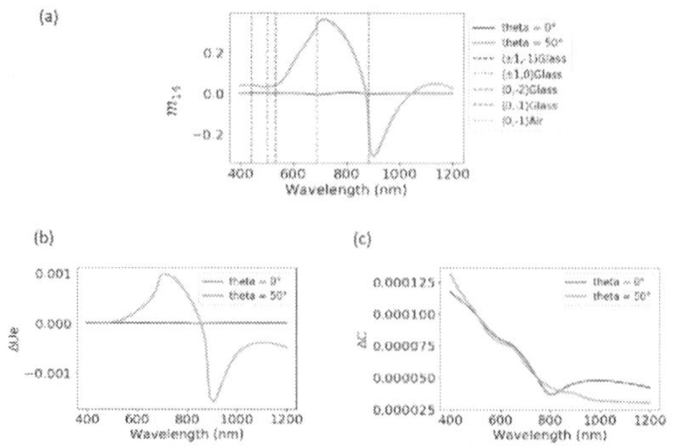

Figure. 2. FEM simulations of (a) far field CD (m_{14}) for the gold U metasurface, (b) near field electric energy density and (c) optical chiral density spectra averaged over a box volume around a unit U resonator for normal (blue) and oblique (yellow) incidence. The vertical lines in (a) represent Rayleigh lines corresponding to the sharp edges of the spectra.

This enhancement of field with angular dependence appears in square in ΔU_e spectra, inducing an extrinsic chirality. However, in presence of a chiral medium, the desired intrinsic chiral signal from the molecules will be dominated by this large extrinsic chiral absorption from the U resonators due to the much higher values of ΔU_e compared to ΔC. The extrinsic chirality would also produce CD signals for achiral molecules with $\alpha''_{em} = 0$ [1]. Therefore, it would be interesting to design the metasurface in a way to remove the contribution from ΔU_e, and still benefit from the enhanced scattered field due to magneto-electric coupling in oblique incidence.

In the next steps, we will investigate new metasurface designs with U resonators in order to describe CD solely as a function of the intrinsic chiral density created by the resonators. The local chiral density will then be optimized by tailoring the geometrical parameters of the resonators. We will also experimentally investigate the polarimetric properties and sensitivity of the designed metasurface to an ambient medium to compare with other relevant metasurfaces.

References

1. F. Graf, J. Feis, X. Garcia-Santiago, M. Wegener, C. Rockstuhl, I. Fernandez-Corbaton, ACS Photonics, 6, 482–491 (2019)
2. J. Mun, M. Kim, Y. Yang, T. Badloe, J. Ni, Y. Chen, C. Qiu, J. Rho, Light: Science & Applications, vol 9, 139 (2020)
3. J. Proust, N. Bonod, J. Grand, B. Gallas, ACS Photonics, 3, 1581–1588 (2016)
4. G. Demésy, F. Zolla, A. Nicolet, M. Commandré, J. Opt. Soc. Am. A/Vol. 27, No. 4 (2010)

https://doi.org/10.1051/epjconf/202226605002

Tailoring Magnetic Dipole Emission by Broken-Symmetry TiO_2 Metasurfaces

Ayesheh Bashiri[1,2,*], *Aleksandr Vaskin*[2], *Katsuya Tanaka*[1,2], *Thomas Pertsch*[2], And *Isabelle Staude*[1,2]
[1]Institute of Solid-State Physics, Friedrich Schiller University Jena, 07743 Jena, Germany
[2]Institute of Applied Physics, Abbe Center of Photonics, Friedrich Schiller University Jena, 07745 Jena, Germany

Abstract. Strong magnetic dipole emission is offered by rare earth ions such as trivalent lanthanides, due to selection rule forbidden electric dipole (ED) transitions. This stimulates the study of optical nanostructures, which efficiently tailor magnetic dipole emission. High refractive index all dielectric nanostructures are promising candidates in this regard due to their strong magnetic response and negligible absorption loss in the visible spectral range. Here, we design and experimentally realize a broken-symmetry titanium dioxide (TiO_2) metasurface supporting an out-of-plane magnetic dipole (MD) resonance at 590 nm wavelength, corresponding to the MD transition of trivalent Europium ions (Eu^{3+}). A strong photoluminescence (PL) enhancement of the MD transition up to a factor of 15.5 is observed.

1 Introduction

High index all-dielectric metasurfaces Offer many important opportunities for the realization of highly efficient flat optical devices [1-3]. In particular, by coupling to their multipolar electric and magnetic Mie-resonances, they can efficiently manipulate spontaneous emission of coupled quantum emitters [4].

A particularly interesting class of all-dielectric metasurfaces for spontaneous emission control are those supporting high quality (Q) factor bound states in the continuum (BIC). Lying inside the continuum while being isolated from the radiation loss, BIC states are characterized by a strong energy confinement and consequently field enhancement localized to the structure supporting them [5]. In a typical symmetric system, the coupling between bound state and continuum is forbidden by symmetry. Through breaking the in-plane symmetry of the nanostructure, a weak coupling between its bright in-plane and dark out-of-plane modes can occur which leads to emergence of high Q- factor sharp resonances since in this case, the radiation loss is suppressed, and the energy is stored inside the nanostructure. These high Q-factor resonances can provide a strong PL enhancement and emission directionality [6].

In optical frequency range, MD transitions in a quantum system are forbidden since their transition rates are usually 4-5 orders of magnitude weaker than allowed ED transition rates [7]. However, MD transitions with a comparable strength as ED transitions exist in lanthanide ions, such as trivalent europium (Eu^{3+}), which makes them a proper candidate for magnetic light matter interaction studies. Eu^{3+} emitters possess a strong MD transition at 590 nm ($^5D_0 \rightarrow {}^7F_1$) and several ED transitions in visible range including the most dominant one at 610 nm ($^5D_0 \rightarrow {}^7F_2$) [8,9].

2 Methods and Results

In this work, we designed Mie-resonant metasurfaces composed of broken symmetry TiO_2 nanoparticles, that support high Q-factor out-of-plane magnetic dipole resonances at 590 nm, corresponding to the MD transition of Eu^{3+}. We fabricated the metasurfaces using electron-beam lithography (EBL) in combination with reactive ion

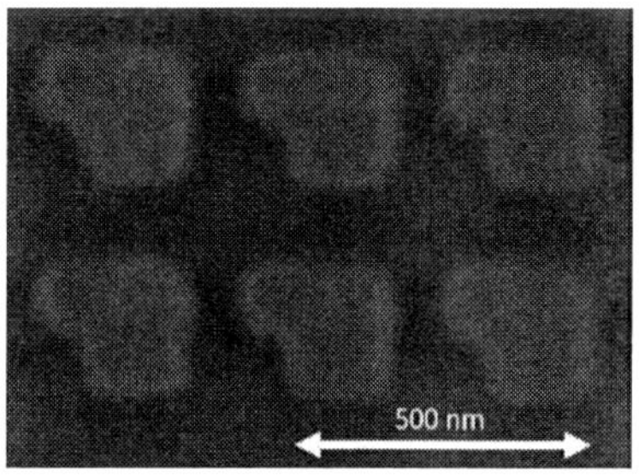

Fig. 1. Top-view SEM image of a fabricated broken-symmetry TiO_2 metasurface.

etching . We then spin-coated the fabricated metasurfaces with a polymethyl methacrylate (PMMA) resist film containing Europium(III) thenoyltrifluoroacetone. Figure 1 shows a top-view scanning electron microscope (SEM) image of a typical fabricated sample.
Next, using a commercially available confocal laser scanning microscope setup (PicoQuant, MicroTime200), we performed PL spectroscopy to investigate PL emission enhancement of Eu^{3+} provided by the broken-symmetry

* Corresponding author: ayesheh.bashiri@uni-jena.de

metasurface. Measured PL emission spectra for both the coated metasurface and an unstructured coated reference area is displayed in the Figure 2.
Using a 4 × /0.1NA collective objective, we achieved a brightness enhancement of up to 15.5 for the MD transition at 590 nm. The ED transition of Eu^{3+} at 610 nm

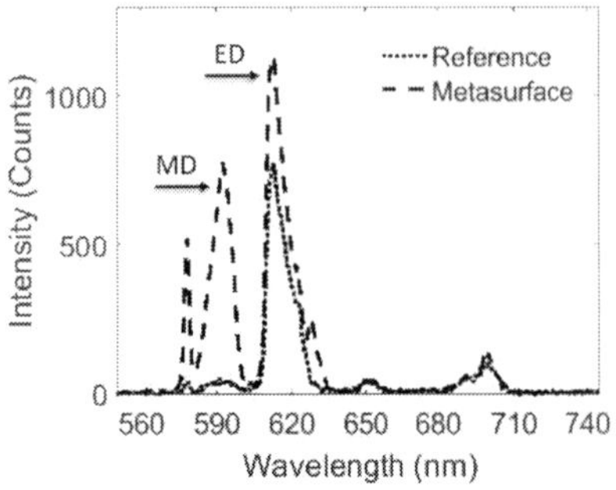

Fig. 2. The PL emission spectra of the Eu^{3+}doped PMMA covered sample.

is also enhanced, but to a much lower degree. The enhancement ratio of the MD transition at 590 nm to the ED transition is ~8.7.

Furthermore, to investigate the spatial emission characteristics of the sample, we performed back focal plane (BFP) imaging of the emission from the coated metasurface at the two wavelengths of 590 nm and 610 nm using a 0.6 NA objective. The result is shown in Figure 3.

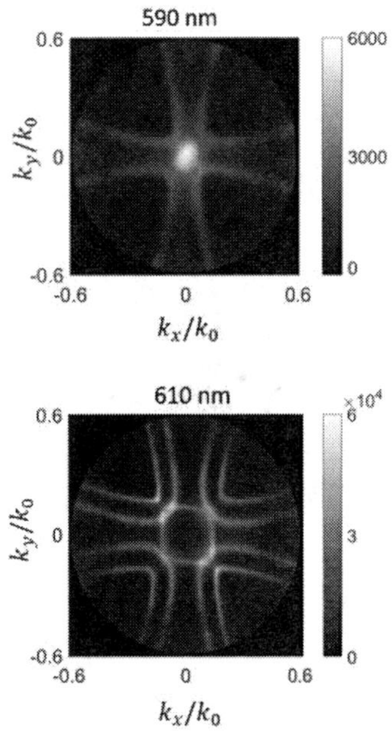

Fig. 3. Measured BFP images of the sample at 590 nm and 610 nm wavelength, corresponding to the electric and magnetic dipole transition, respectively.

The BFP images of the sample show a bright emission for the near-zero polar angles, i.e., a strong forward emission normally out of the sample plane, for the MD at 590. For the ED at 610 nm, instead, forward emission is strongly suppressed. As such, the strong PL emission enhancement for the MD observed while measuring with a low NA collective objective can be explained by the dramatically different emission patterns observed for the two different transitions. From this we can conclude the obtained selective MD enhancement is mainly due to an improved collection efficiency for that transition. Our results offer a novel possibility toward tailoring the directionality of MD-dominated spontaneous emission. It is of a great importance for the research and application exploiting magnetic nature of light. Furthermore, controlling the directionality of the emission is strongly demanded to realize optical devices with high efficiency.

References

1. I. Staude and J. Schilling, Nature Photonics **11**, 274–284 (2017).
2. A. Chen and F. Monticone, ACS Photonics **8** (5), 1439-1447 (2021).
3. A. I. Kuznetsov, A. E. Miroshnichenko, M. L. Brongersma, Y. S. Kivshar, and B. Luk'yanchuk, Science 354 (2016) **10** .1126/science.aag2472.
4. A. Vaskin, R. Kolkowski, A. F. Koenderink, and I. Staude, Nanophotonics **8**, 1151–1198 (2019).
5. K. Koshelev, S. Lepeshov, M. Liu, A. Bogdanov, Y.Kivshar, Phys. Rev. Lett. **121**, 193903 (2018).
6. S. Liu, A. Vaskin, S. Addamane, B. Leung, M.-C. Tsai, Y. Yang, P. P. Vabishchevich, G. A. Keeler, G. Wang, X. He, Y. Kim, N. F. Hartmann, H. Htoon, S. K. Doorn, M. Zilk, T. Pertsch, G. Balakrishnan, M. B. Sinclair, I. Staude, and I. Brener, Nano Lett. **18**, 6906–6914 (2018).
7. T. H. Taminiau, S. Karaveli, N. F. van Hulst, and R. Zia, Nat. Commun. **3** (2012).
8. R. Hussain, S. S. Kruk, C. E. Bonner, M. A. Noginov, I. Staude, Y. S. Kivshar, N. Noginova, and D. N. Neshev, Optics Letters **40**, 1659 (2015).
9. A. Vaskin, S. Mashhadi, M. Steinert, K. E. Chong, D. Keene, S. Nanz, A. Abass, E. Rusak, D.-Y. Choi, I. Fernandez-Corbaton, T. Pertsch, C. Rockstuhl, M. A. Noginov, Y. S. Kivshar, D. N. Neshev, N. Noginova, and I. Staude, Nano Lett. **19**, 1015–1022 (2019).

 https://doi.org/10.1051/epjconf/202226605003

T-Matrix based Scattering Analysis of Photonic Materials with Periodicity in Different Dimensions

Dominik Beutel[1,*], *Carsten* Rockstuhl[1,2], and *Ivan* Fernandez-Corbaton[2]

[1]Institute of Theoretical Solid State Physics, Karlsruhe Institute of Technology (KIT), 76131 Karlsruhe, Germany
[2]Institute of Nanotechnology, Karlsruhe Institute of Technology (KIT), 76344 Eggenstein-Leopoldshafen, Germany

Abstract. Optical devices and artificial photonic materials frequently make use of periodic arrangements of identical scatterers in 3D, 2D, and 1D, e.g. photonic crystals, meta-surfaces, or particle chains. To simplify their analysis, we present here a computational framework based on the T-matrix method that explicitly exploits spherical, cylindrical, and plane waves depending on the geometry and number of dimensions of the lattice. Due to the analytic properties of the chosen basis sets in combination with the use of Ewald's method for the lattice summation, we obtain an efficient framework to simulate such systems. The applicability will be illustrated at the conference by means of selected examples of contemporary interest.

Many designs for optical applications depend on periodic arrangements of individual scatterers. These designs can be a photonic crystal or metamaterial, where the underlying lattice is 3D, or a metasurface, which has a 2D lattice, or a particle chain with a 1D lattice.

Our approach to an efficient simulation method for such systems is the choice of a basis set for Maxwell's equations, that is adapted to the geometrical structure as close as possible. One prominent example for such a suiting basis set are the scattering properties of a sphere that are solved analytically in Mie theory using vector spherical waves (SWs). The approach from Mie theory can be generalized to the T-matrix method, such that calculations for arbitrary scatterers of finite size made from linear materials can be done [1]. If we build a 2D lattice from such scatterers, it is possible to calculate the mutual interaction of the particles with the T-matrix method. However, the effective response for a two-dimensional lattice is more conveniently described by vector plane waves (PWs) [2]. We find that there is a similar connection between a 1D lattice of scatterers described by SWs to an effective description by cylindrical waves (CWs) and a 1D lattice of CWs can be effectively described by PWs (see also fig. 1(a)-(c)). Thus, for each lattice dimension there is a basis set for an efficient description of its total response.

The efficiency of T-matrix based methods is partly due to the fact that the translation coefficients are analytically known [3]. This reduces the computation of the mutual interaction between multiple objects to simple linear algebra operations. For periodic arrangements, however, a notoriously slow converging lattice sum over the translation coefficients is necessary. This challenge can be met by using Ewald's method, where the sum is separated into a short-range and a long-range contribution which – individually – converge exponentially fast in real and reciprocal space, respectively [4]. With Ewald's method, adapted to the different lattice dimensions and the basis sets above, a wide range of optical system can be computed efficiently.

1 Plane, cylindrical, and spherical waves

All three basis sets can be derived by following a common procedure [5]. This solution is denoted by $\boldsymbol{M}_\nu(\boldsymbol{r})$ with the placeholder ν for the indices and it describes a transverse electric (TE) solution. A transverse magnetic (TM) solution is obtained by $\boldsymbol{N}_\nu(\boldsymbol{r}) = \frac{\nabla}{k} \times \boldsymbol{M}_\nu(\boldsymbol{r})$. The PWs are $\boldsymbol{M}_{\boldsymbol{k}}(\boldsymbol{r}) = \mathrm{i}\nabla \times (\hat{\boldsymbol{e}}_z \mathrm{e}^{\mathrm{i}\boldsymbol{k}\boldsymbol{r}}) / \sqrt{k_x^2 + k_y^2}$, the CWs are $\boldsymbol{M}^{(n)}_{k_z,m}(\boldsymbol{r}) = \mathrm{i}\nabla\times(\hat{\boldsymbol{e}}_z \mathrm{e}^{\mathrm{i}m\varphi+\mathrm{i}k_z z} Z^{(n)}_m(\sqrt{k^2 - k_z^2}\rho)) / \sqrt{k^2 - k_z^2}$, and the SWs are $\boldsymbol{M}^{(n)}_{lm}(\boldsymbol{r}) = \mathrm{i}\nabla \times (\boldsymbol{r} Y_{lm}(\theta, \varphi) z^{(n)}_l(kr))$. For the PWs and the CWs the denominators assure a dimensionless basis function. We use $Z^{(n)}_l$ for the Bessel ($n = 1$) and Hankel ($n = 3$) functions of the first kind, $z^{(n)}_l$ for their spherical counterparts, and Y_{lm} for the spherical harmonics. All three solution sets include initially an infinite number of modes: PWs are defined by all wave vectors $\boldsymbol{k}$ that satisfy the dispersion relation, CWs are defined by $k_z \in \mathbb{R}$ and $m \in \mathbb{Z}$, and SWs are defined by $l \in \mathbb{N}$ and $m \in \mathbb{Z}$ with $|m| \leq l$. However, higher multipole or diffraction orders can be neglected by imposing a suitable threshold so that the solution space is restricted to a finite number of basis functions. The transition from the TE/TM basis to the helicity basis by defining $\sqrt{2}\mathbf{A}_{\pm,\nu}(\boldsymbol{r}) = \mathbf{N}_\nu(\boldsymbol{r}) \pm \mathbf{M}_\nu(\boldsymbol{r})$ facilitates the treatment of chiral materials and has also computational benefits over the TE/TM basis [2].

2 T-matrix for CWs and SWs in lattices

While usually defined for SWs, we present the T-matrix method here in a more general way that also applies to

*e-mail: dominik.beutel@kit.edu

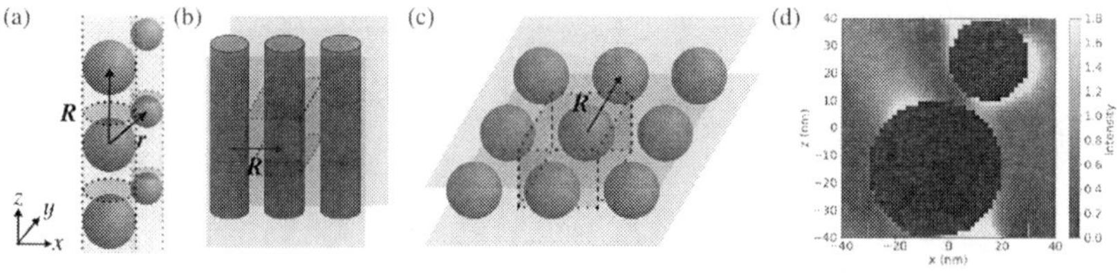

Figure 1. Panels (a) to (c) show the different possible layouts for transitions between SWs, CWs, and PWs which are color-coded in blue, green, and red. In each sketch one unit cell is highlighted. In panel (a) the unit cell contains two scatters. Panel (d) shows the simulated intensities for the arrangement in panel (a) under a PW illumination with along $(\sin\frac{\pi}{3}, 0, \cos\frac{\pi}{3})$ of negative helicity. The electric fields between the red marks are calculated directly with Ewald's method, outside it is much faster to use the efficient CW description. The lack of a discontinuity between these domains shows the accurate match of both approaches.

CWs. These two basis sets allow the electric field to be separated into an incident part $\boldsymbol{E}_{\mathrm{inc}}(\boldsymbol{r}) = \sum_{\lambda,\nu} a_{\lambda,\nu}\boldsymbol{A}^{(1)}_{\lambda,\nu}(\boldsymbol{r})$ and a scattered part $\boldsymbol{E}_{\mathrm{sca}}(\boldsymbol{r}) = \sum_{\lambda,\nu} p_{\lambda,\nu}\boldsymbol{A}^{(3)}_{\lambda,\nu}(\boldsymbol{r})$.

Thus, we can now write down the linear relationship between the incident and scattered mode coefficients, that each are placed in a vector, as $\boldsymbol{p} = \mathbf{T}\boldsymbol{a}$ with the T-matrix $\mathbf{T}$. A particular strength of the T-matrix method are the analytically known translation coefficients $C^{(3)}_{\nu',\nu}(\boldsymbol{r})$. We write the multi-scattering equation within a lattice as

$$\boldsymbol{p}_{i\alpha} = \mathbf{T}_i\left(\boldsymbol{a}_{i\alpha} + \sum_{j=1}^{N}\sum_{\beta}{}' \mathbf{C}^{(3)}(\boldsymbol{r}_{ij} + \boldsymbol{R}_{\alpha\beta})\boldsymbol{p}_{j\beta}\right) \tag{1}$$

where Latin indices i, j refer to N different particles within one unit cell and Greek indices α, β to different positions on the lattice. We now make the assumption that there is a constant phase difference along different positions on the lattice defined by $\boldsymbol{k}_\parallel$. Then, we can find the relation $\boldsymbol{a}_{i\alpha} = \mathrm{e}^{\mathrm{i}\boldsymbol{k}_\parallel\boldsymbol{R}}\boldsymbol{a}_{i0}$ for the incident and, analogously, for the scattered field. After rearranging eq. (1) slightly, we obtain

$$\left(\mathbb{1}\delta_{ij} + \mathbf{T}_i\sum_{j=1}^{N}\sum_{\beta}{}' \mathbf{C}^{(3)}(\boldsymbol{r}_{ij} + \boldsymbol{R}_{0\beta})\mathrm{e}^{\mathrm{i}\boldsymbol{k}_\parallel\boldsymbol{R}_{0\beta}}\right)\boldsymbol{p}_{j0} = \mathbf{T}_i\boldsymbol{a}_{i0} \tag{2}$$

which is the regular multiple scattering equation for T-matrices with the exception of the lattice series $\sum_\beta C_{\nu',\nu}(\boldsymbol{r}_{ij} + \boldsymbol{R}_{0\beta})\mathrm{e}^{\mathrm{i}\boldsymbol{k}_\parallel\boldsymbol{R}_{0\beta}}$. For both the CWs and the SWs these series can be solved for 1D, 2D, and 3D lattices by means of Ewald's method.

3 Transitions between basis sets

Besides the coupling between different cells of the lattice, where the Ewald summation is a key ingredient for an efficient computation, we need the expansion of the incident wave of a particular type into other types, e.g. expanding a CW into PWs. These properties can be summarized by $\boldsymbol{A}_{\pm,\boldsymbol{k}}(\boldsymbol{r}) = \sum_{m=-\infty}^{\infty}\mathrm{i}^m\mathrm{e}^{-\mathrm{i}m\varphi_{\boldsymbol{k}}}\boldsymbol{A}^{(1)}_{\pm,k_zm}(\boldsymbol{r})$ and $\boldsymbol{A}^{(1)}_{\pm,k_zm}(\boldsymbol{r}) = \sum_{l=|m|}^{\infty}\mathrm{i}^{l-m}4\pi\gamma_{lm}f_{\lambda,lm}(\theta_{\boldsymbol{k}})\boldsymbol{A}^{(1)}_{\pm,lm}(\boldsymbol{r})$ using $\gamma_{lm} = \sqrt{\frac{(2l+1)}{4\pi l(l+1)}\frac{(l-m)!}{(l+m)!}}$ and the function $f_{\pm,lm}(\theta) = \frac{\mathrm{d}}{\mathrm{d}\theta}P_l^m(\cos\theta) \pm \frac{m}{\sin\theta}P_l^m(\cos\theta)$ with the Legendre Polynomial P_l^m. The transition from PWs to SWs can be obtained by combining both equations. As a last main ingredient we need the expansion of the scattered field in different basis sets which is more involved because it includes a lattice summation and is necessary, e.g. for the effective description of a chain of (SW-)T-matrices with a CW-T-matrix (fig. 1(a, d)). We make use of the integral representation of the scattered fields in combination with Poisson's formula to arrive at

$$\sum_\beta \mathrm{e}^{\mathrm{i}\boldsymbol{k}_\parallel\boldsymbol{R}_\beta}\boldsymbol{A}^{(3)}_{\lambda,lm}(\boldsymbol{r} - \boldsymbol{R}_\beta) = \frac{\pi\gamma_{lm}}{ak\mathrm{i}^{l-m}}\sum_\beta f_{\lambda,lm}(\theta_{\boldsymbol{k}+\boldsymbol{G}_\beta})\boldsymbol{A}^{(3)}_{\lambda,k_\parallel+G_\beta m}(\boldsymbol{r}) \tag{3}$$

$$\sum_\beta \mathrm{e}^{\mathrm{i}\boldsymbol{k}_\parallel\boldsymbol{R}_\beta}\boldsymbol{A}_{\lambda,k_zm}(\boldsymbol{r} - \boldsymbol{R}_\beta) = \frac{2}{ak\mathrm{i}^m}\sum_\beta \frac{\mathrm{e}^{\mathrm{i}m\varphi_{\boldsymbol{k}+\boldsymbol{G}_\beta}}\boldsymbol{A}_{\boldsymbol{k}+\boldsymbol{G}_\beta}(\boldsymbol{r})}{\sqrt{k^2 - k_z^2 - (\boldsymbol{k}_\parallel + \boldsymbol{G})^2}} \tag{4}$$

where $\boldsymbol{G}_\beta$ are reciprocal lattice vectors and a the lattice pitch. The transitions from SWs to PWs are possible by a combination of the results.

4 Conclusion

With the T-matrix method for lattices and Ewald's method from section 2 and the transitions between different basis sets from section 3 we can efficiently calculate the interaction of objects in a lattice and also have a description of the total response in the most suitable basis set. We implemented the sketched procedure in a program whose open source publication is in preparation.

Acknowledgements

D.B. is funded by the Deutsche Forschungsgemeinschaft (DFG, German Research Foundation) under Germany's Excellence Strategy – 2082/1 – 390761711.

References

[1] M.I. Mishchenko, L.D. Travis, D.W. Mackowski, JQRST **55**, 535 (1996)
[2] D. Beutel, A. Groner, C. Rockstuhl, C. Rockstuhl, I. Fernandez-Corbaton, JOSA B **38**, 1782 (2021)
[3] L. Tsang, J.A. Kong, R.T. Shin, *Theory of Microwave Remote Sensing* (Wiley-Interscience, 1985)
[4] C.M. Linton, SIAM Review **52**, 630 (2010)
[5] P.M. Morse, H. Feshbach, *Methods of Theoretical Physics* (McGraw-Hill, New York, 1953)

https://doi.org/10.1051/epjconf/202226605004

Figure of Merit Comparison Between Surface Plasmon Resonance and Bloch Surface Waves

Bernardo Dias[1,2*], *José* M. M. de Almeida[1,3], and *Luís* C. C. Coelho[1,2]

[1]INESC TEC–Institute for Systems and Computer Engineering, Technology and Science, Faculty of Sciences, University of Porto, 4169-007 Porto, Portugal
[2]Department of Physics and Astronomy, Faculty of Sciences, University of Porto, 4169-007 Porto, Portugal
[3]Department of Physics, School of Science and Technology, University of Trás-os-Montes e Alto Douro, 5001-801 Vila Real, Portugal

Abstract. The sensing performance of two types of electromagnetic surface waves are compared, a Surface Plasmon Polariton, where a gold thin film is used, being a standard material in biosensing applications; and a Bloch Surface Wave, using a photonic crystal made of a stack of silica and titanium dioxide layers. It is verified that the sensing performance (as measured by the Figure of Merit) of the gold film is higher, even though the Bloch Surface Waves can serve specific applications due to its narrow bandwidth. At the same time, it is concluded that further research must be made in order to choose the right set of parameters that maximize the Bloch Surface Wave performance.

1 Introduction

Of the multiple types of electromagnetic surface waves (ESW) reported so far, the Surface Plasmon Polariton (SPP) [1] has dominated the optical sensing research field, and only recently have other types of ESW's been considered for sensing purposes, such as Bloch Surface Waves (BSW's) [2] and Tamm Plasmon Polaritons [3]. In this work, the performance of SPP's and BWS's are analyzed, thus establishing a comparison between the two types of waves.

Created at the interface between a metal and a dielectric (or, more generally, between two media with different permittivity signs), the SPP consists on coupling of light to the free electrons of a metal when the resonance conditions (dictated by the material's properties and thicknesses) are met. In the optical spectrum, this resonance is seen as the absorption of light in a specific wavelength range, creating an absorbance band which can be highly sensitive to refractive index (RI) changes on the surface. This kind of ESW has been used in optical sensing applications for several decades [4] and has developed into an important research field.

On the other hand, BSW's are generated at the interface between a photonic crystal (PhC) and a dielectric (or between two PhCs) [2], producing sharp resonances in the optical spectrum due to the symmetry breaking caused by the PhC – dielectric interface. While the experimental observation of BSW's was made almost five decades ago [5], it is now emerging as a new field for sensing purposes, and a few experimental realizations have already been reported [6].

In this work, a comparison between these two ESW's is made, focusing on the external RI sensitivity and resolution, which are key parameters for optical sensor application in real world scenarios.

2 Experimental Setup

In both SPP's and BSW's, direct excitation of the surface wave is not possible due to a mismatch between the coupling conditions of the wave and the directly incident light. To obtain the wavevector matching for both ESW's, a prism technique is used based on the Kretschmann's configuration [7]. A silica prism was used along with silica substrates, over which the thin films were deposited using an RF Magnetron Sputtering system. Figure 1 illustrates the experimental setup used for the observation of both waves.

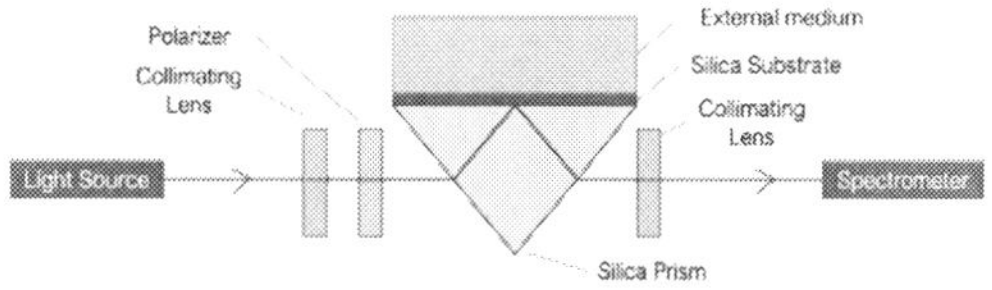

Fig. 1. Experimental setup based on the Kretschmann configuration for ESW excitation.

Using water solutions of sucrose at different ratios, the RI (measured using an Abbe Refractometer – model Atago DR A1 Plus) on the top layer was changed, causing variations in the transmitted spectra which was acquired by a high resolution spectrometer (model AvaSpec-ULS2048CL-EVO). By tracking the band's position and full width at half maximum (FWHM), the performance of each wave as a refractometric sensor can be analyzed, using the Figure of Merit (FOM), defined as [8]

* Corresponding author: bernardo.s.dias@inesctec.pt

$$FOM = \frac{d\lambda/dn}{FWHM} \quad (1)$$

in which $d\lambda/dn$ represents the shift in the wavelength peak due to the change of external RI (commonly referred to as the sensitivity).

For the case of the SPP, a structure composed of Cr (5nm) and Au (50nm) was used. The Cr layer promoted a better adhesion of the Au to the silica substrate, while the Au layer is the active plasmonic component. For the case of the BSW, a photonic crystal made of three bilayers of TiO_2 (135nm) and SiO_2 (180nm) and a final TiO_2 layer (100nm) was fabricated. Both structures were previously simulated using a transfer matrix approach [2], which showed the appearance of ESW's in *p* polarized light for both cases.

3 Results and discussion

For the case of the Au film, the results are shown in Figure 2.

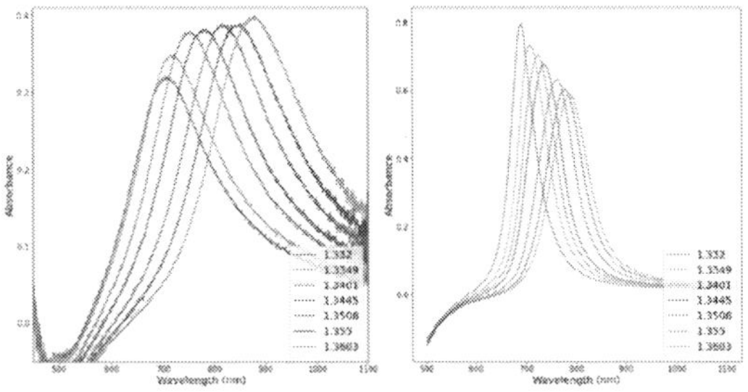

Fig. 2. Absorbance spectra of SPR wave supported on Au thin film: a) Experimental results, b) transfer matrix simulations.

In Figure 2, it is possible to see that the FWHM of the experimental results are considerably larger than the simulations. This is because the simulations employ the permittivity data from bulk Au, which can deviate from thin films. An experimental sensitivity of 6111 nm/RIU was verified, but the FOM was considerably affected by the large FWHM, which was measured to be 203 nm for the pure water RI. Figure 3 illustrates the results of the BSW configuration.

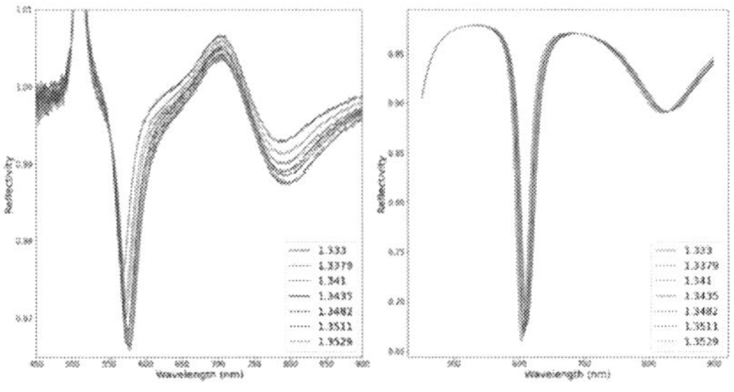

Fig. 3. Reflectivity spectra of BSW wave supported on a photonic crystal: a) Experimental results, b) transfer matrix simulations.

As expected, the BSW produces a much sharper resonance (24.5 nm in pure water), but also a much lower sensitivity, which was measured to be 501 nm/RIU. In this case, the FOM is significantly affected by the low sensitivity inherent to the BSW. The simulations and experimental results seem to agree fairly well.

From the experimental results for both ESW's, the FOM can be calculated, which can establish a comparison between the two structures, as shown in Figure 4.

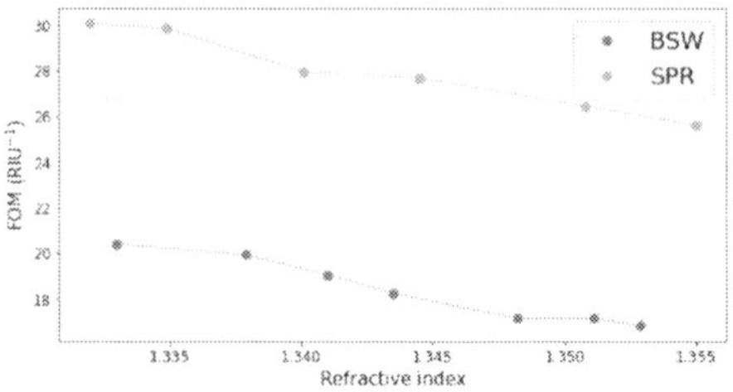

Fig. 4. Experimental values for the Figure of Merit for SPR and BSW.

In Figure 4 it is possible to observe that even with a larger FWHM, the SPR sensor still outperforms the BSW in this particular metric. This is due to the high sensitivity displayed by the surface plasmon, which is one order of magnitude above that of the BSW. Nevertheless, it should be noted that while for the Au SPR there is not much room for improvement (other than improving the optical quality of the gold layer), for the BSW several parameters should be tuned to maximize the FOM, such as the number of bilayers, layer thicknesses and RI ratio (by changing materials). Further research is needed to optimize the performance of BSW's and establish this kind of ESW as a viable alternative to SPR.

References

1. A., Otto, Eur. Phys. J. A **216**, 398-410 (1968)
2. P. Yeh, A. Yariv, and C. Hong, J. Opt. Soc. Am. **67**, 423-438 (1977)
3. M. Kaliteevski, I. Iorsh, S. Brand, R. A. Abram, J. M. Chamberlain, A. V. Kavokin, and I. A. Shelykh, Phys. Rev. B **76**, 165415 (2007)
4. V. Yesudasu, H. S. Pradhan, R. J., Heliyon, 7, 3, 2021
5. P. Yeh, A. Yariv, and A. Y. Cho, Appl. Phys. Lett. 32, 104 (1978)
6. E. Gonzalez-Valencia, I.D. Villar, P. Torres, Sci Rep 11, 11266 (2021)
7. E., Kretschmann, Sens. Actuators B Chem. **35**, 212 (1996)
8. J. Homola, *Surface Plasmon Resonance Based Sensors* (Springer Berlin, Heidelberg 2006)

Engineering high Q/V photonic modes in correlated disordered systems

Nicoletta Granchi[1,*], *Richard* Spalding[2], *Kris* Stokkereit[2], *Matteo* Lodde[3], *Andrea* Fiore[3], *Riccardo* Sapienza[4], *Francesca* Intonti[1], *Marian* Florescu[2] and *Massimo* Gurioli[1]

[1]Department of Physics and Astronomy and LENS, University of Florence, Sesto Fiorentino (FI), Italy
[2]Advanced Technology Institute and Department of Physics, University of Surrey, Surrey, UK
[3]Department of Applied Physics and Institute for Photonic Integration, Eindhoven University of Technology, Eindhoven, NL
[4]The Blackett Laboratory, Department of Physics, Imperial College London, UK

Abstract. Hyperuniform disordered (HuD) photonic materials have recently been shown to display several localized states with relatively high Q factors. However, their spatial position is not predictable a priori. Here we experimentally benchmark through near-field spectroscopy the engineering of high Q/V resonant modes in a defect inside a HuD pattern. These deterministic modes, coexisting with Anderson-localized modes, are a valid candidate for implementations in optoelectronic devices due to the spatial isotropy of the HuD environment upon which they are built.

1 INTRODUCTION

Located in-between random structures and perfectly ordered photonic crystals, there is a special class of disordered photonic heterostructures, called hyperuniform disordered (HuD) photonic structures [1,2]. They have recently been shown to display large isotropic band gaps (BG) as well as optical transparency, to mention two of the most fascinating and promising features. Among the several experimental photonic realizations of HuD structures [3,4], also HuD systems on dielectric slabs have been recently proposed [5] and characterized as photonic materials capable of combining the small spatial footprint typical of random modes with Q/V ratios comparable with photonic crystal cavities for the Anderson-localized modes naturally occurring at the BG edges. These modes with relatively high Q, however do not have predictable spatial locations in the whole structure. Here, we take advantage of the same slab technology with embedded Quantum Dots, that act as internal light source, to engineer a defect inside the HuD luminescent pattern and deterministically control its spatial location in a correlated disordered environment and optimize the light confinement (as it has been recently proposed only theoretically in [6]).

2 STRUCTURE AND EXPERIMENT

The design of the structure is reported in Fig.1a; the central defect is designed in order to support several modes that are labelled accordingly with the Finite Element Methods (FEM) distributions of the magnetic field component Hz: dipole-like (D), hexapole-like (H), quadrupole-like (Q) and octupole-like (O) and so on [6]. By tuning the structural parameters such as the width of the dielectric veins in the HuD network and the radius of the central hole of the defect, we achieve a fine control over every spectral resonance to find the best condition that maximizes the Q factor for every mode. With the theoretical design, a GaAs membrane 180 nm thick, with InGaAs Quantum Dots embedded in the middle, is patterned through Electron Beam Lithography.

We experimentally benchmark this engineering through Scanning Near-field Optical Microscopy (SNOM), used in illumination-collection configuration and capable of subwavelength resolution in the near-IR range (upper right inset of Fig.1a). A room temperature commercial SNOM Twinsnom, OMICRON is used. The sample is excited with light from a diode laser (785 nm) coupled into a chemically etched optical fiber, which allows us to have a direct measurement of the LDOS of the system. Infact photoluminescence (PL) spectra from the sample are collected at each tip position through the same probe.

3 RESULTS AND DISCUSSION

The typical SNOM PL spectrum of the defect modes is shown in the bottom panel of Fig.1b, where D, H, Q and O are detected. The near-field PL map of mode H is reported in Fig.1c, and nicely agree with the FEM electric field intensity map. Thanks to the engineering of the structural parameters, we experimentally detect in a disordered system, modes with Q factors of the order of 6000. All the defect modes, being inside a correlated disordered system, will coexists with all the other localized modes supported by the HuD network.

* Corresponding author granchi@lens.unifi.it

 https://doi.org/10.1051/epjconf/202226605005

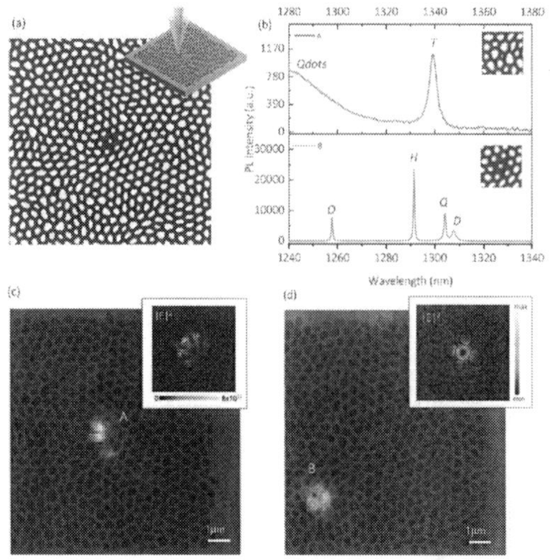

Fig. 1. (a) Design of the engineered defect in a HuD pattern and sketch of the SNOM probe used in illumination-collection configuration. **(b)** Bottom panel: SNOM PL spectrum acquired on the engineered defect displaying four modes D (dipole-like), Q (quadrupole-like), H (hexapole-like) and O (octupole-like). Upper panel: SNOM PL spectrum acquired on the accidental four-fold defect, called topological T. **(c)** SNOM PL map centered around 1326 nm, corresponding to mode H, whose FEM map of the electric field intensity is shown in the inset. **(d)** SNOM PL map centered amount 1339 nm, corresponding to the hyperuniform topological four-fold defect. The inset reports the relative FEM electric field intensity map.

A remarkable example of this innovative feature can be seen in the design (Fig.1a), where the central defect is spatially near to an accidental topological defect (here, a cell of the HuD network with four edges, rather than the average six), typical of stealthy HuD systems [5]. This is an unavoidable defect that depends on the tiling protocol employed to generate the pattern, and it is the first of many tightly localized modes occurring at the PB edge. The SNOM PL spectrum displaying the peak of this mode is shown in the upper panel of Fig.1b, and the correspondent PL map is reported in Fig.1d, again in a fairly good match with the FEM electric field intensity distribution.

The possibility of an experimental deterministic control of a defect inside a correlated disordered environment such as HuD therefore open the way to many applications for which guiding light through modes with different localization properties, a high spectral and spatial density of modes is required as well as a high Q/V ratio. The ability of optically characterizing localized modes of different symmetry and frequency in the same physical cavity can have a great impact on all-optical switching, implementations of linear-optical quantum information processors and single photon sources. Moreover, the average isotropy of HuD pattern in which the defects are placed makes them more adaptable in optoelectronic devices.

References

1. L. S. Froufe-Perez, M. Enge, P.F. Damasceno, N. Muller, J. Haberko, S. C. Glotzer, and F. Scheffold, Phys. Rev. Lett. **117**, 053902, (2016).

2. M. Florescu, S. Torquato, P.J. Steinhardt, PNAS **106**, 20658–20663, (2009).

3. N. Muller, J. Haberko, C. Marichy, and F. Scheffold, Adv. Opt. Mater., **2**, 115-119, (2014).

4. M. Castro-Lopez, M. Gaio, S. Sellers, G. Gkantzounis, M. Florescu, and R. Sapienza, APL Photonics **2**, 061302, (2017).

5. N. Granchi, R. Spalding, M. Lodde, M. Petruzzella, F. W. Otten, A. Fiore, F. Intonti, R. Sapienza, M. Florescu, and M. Gurioli, Adv. Opt. Mat. 2102565, (2022).

6. T. Amoah and M. Florescu. Phys. Rev. B, **91**, 020201, (2015).

Controlling chromaticity by lamellar gratings

Hiroyuki Ichikawa[1,*], *Naoki* Arita[1], *Keigo* Shikimi[1], *Kei* Maeda[1], and *Ryunosuke* Tani

[1]Faculty of Engineering, Ehime University, Matsuyama, 790-8577, Japan

Abstract. Fundamental numerical study on controlling chromaticity with the simplest diffractive structure is carried out. Observed various characteristics on transmission/reflection and dielectric/metal will be useful guidelines for practical optimisation of device structures.

1 Introduction

There have been many reports on generating wide variety of colours by metasurfaces: originally based on metals and then attempted with dielectrics [1, 2]. Noticing range of colour achieved in reported works is relatively limited, we wish to find factors which influence chromaticity. In order to conduct the task through electromagnetic numerical simulation with Fourier modal method, we employ one-dimentional lamellar gratings, because the simplest structure is suitable to concentrate on the least necessary design freedoms.

2 Method

A considered model is shown in Figure 1: a grating is normally illuminated by a plane wave. As grating materials, we assumed two types of optical glasses, fused silica and SF11, and two types of metals, gold and silver. Glasses are illuminated from a substrate side ($\varepsilon_1 > \varepsilon_2 = 1$) and both reflected and transmitted light are taken into account, whereas metals are illuminated from vacuum side ($\varepsilon_1 = 1$) and only reflection is considered. The light source is CIE standard illuminant D65.

Diffraction efficiencies of zeroth-order plane wave either in reflection or transmission is processed to obtain chromaticity in CIE 1931 colour space. In our present study, the spectrum is limited between 380 and 780 nm.

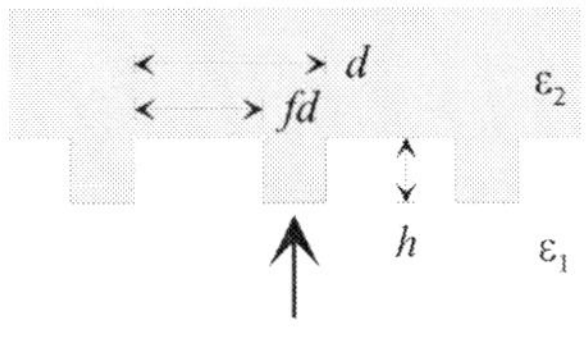

Figure 1. Considered problem.

For the grating structure in Figure 1, there are three design freedoms: period d, depth h and fill factor f. In the present study, we evaluated chromaticity of full combination of $d = 200, 400, \cdots 1800$ nm, $h = 200, 400, \cdots 2000$ nm, and $f = 0.1, 0.2, \cdots 0.9$, in total 810 cases for each grating material. Typical results are shown in Figure 2, where solid and open circles denote all 820 values for gold gratings in reflection and fused silica gratings in transmission, respectively. It is seen that the two materials have noticeably different colour features. Only TE polarisation results are included in this figure. Although TE and TM illumination certainly give different results, only TE results are presented in this report considering our purpose: fundamental study on influencial design freedoms.

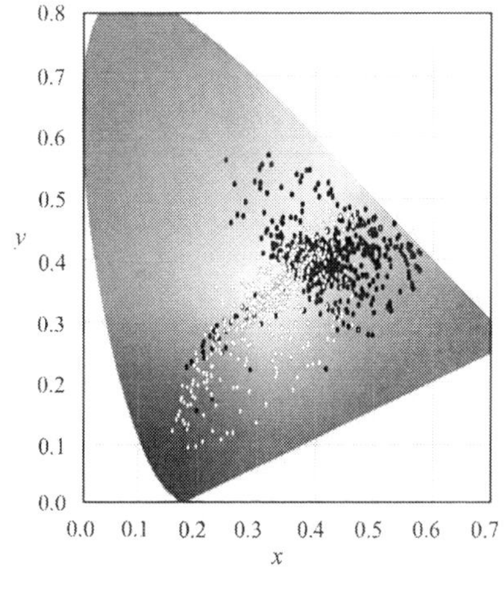

Figure 2. Example chromaticity diagram. •: gold. ○: fused silica in transmission.

All previous study of this topic have stuck to subwavelength structure, because higher-order diffraction is obviously an obstacle when considering practical applications such as colour printing, However, our main motivation here is not finding an optimised grating structure for a particular colour, but investigating influential factors to control chromaticity. Therefore all available design freedoms and their wide combination must be considered.

*e-mail: ichikawa@ehime-u.ac.jp

https://doi.org/10.1051/epjconf/202226605006

In order to handle large number of chromaticity data, directly looking at chromaticity diagram such as Figure 2 is not always convenient. Instead, chromaticity coordinates are converted into relative ones in respect of white point as explained in Figure 3, i.e. from (x, y) to (ρ, ϕ). The white point for D65 is (0.313, 0.329).

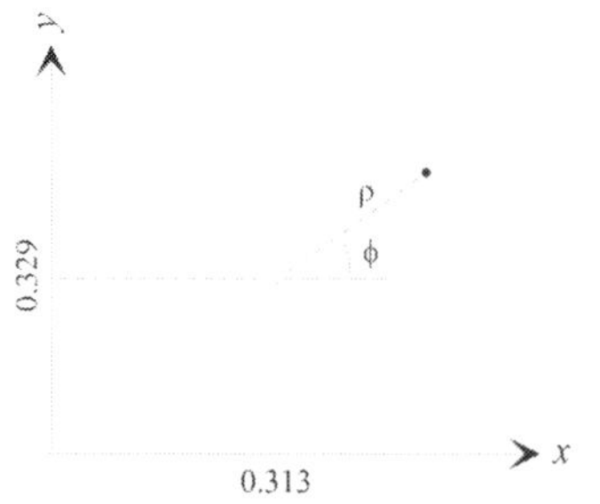

Figure 3. Evaluation of colour diversity. ρ: distance. ϕ; direction. ∘ denotes white point for D65.

3 Results

Relative chromaticity coordinates (ρ, ϕ) defined above are compared among four grating materials: in total six grating modes including reflection/transmission. An example results are shown in Figure 4, where each data point represents maximum value of ρ within 90 combinations of 10 depths × 9 fill factors at each period. This exhibits a degree of saturation, i.e. how far away colour from white can be achieved. Similar graph is obtained for the direction ϕ. This one exhibits a degree of hue, i.e. diversity in colour around the white point. This particular Figure 4 indicates that metal and dielectric gratings in reflection mode show similar tendencies, i.e. maximum distances from the white point are less dependent on periods. On the other hand subwavelength dielectric gratings in transmission-mode perform pretty poorly as colour generators. In more detail, the output light from all 90 structures spanning $200 \leq h \leq 2000$ nm and $0.1 \leq f \leq 0.9$ at $d = 200$ nm present little difference from incident white colour.

We conducted similar observation of ρ and ϕ dependence on depth and fill factor as well. The followings are a list of noticeable findings.

- Subwavelength dielectric gratings in transmission mode do not function as colour generators. This is due to high transmission diffraction efficiencies.
- Range of ϕ is roughly all directions except that ρ of gold grating is limited within fairly narrow range, i.e. well lower than one radian, for $d \leq 600$ nm.
- Dielectric gratings in transmission node need at least 600 nm depth to achieve enough ρ.
- Silver gratings show relatively poor performance in $d \geq 1000$ nm.
- ρ of gold grating is limited within two radian for $d \geq 400$ nm.
- Middle range fill factors, i.e. $0.4 \leq f \leq 0.6$ are preferred in both ρ and ϕ range for all six grating modes.

4 Summary and discussion

Controlling chromaticity with lamellar gratings is investigated in detail n terms of diffractive optics assuming metal and optical glass as grating materials. Some useful information is obtained as presented above, though we have not yet fully analysed the observed data, because the quantity of them is too much to quickly handle.

Many readers may be critical of our inclusion of non-subwavelength gratings. However, gratings in the resonance domain ($d \approx \lambda$) have an attractive feature of rapid change of diffraction efficiency owing to Wood's anomaly. As chromaticity can be regarded as a complicated function of output spectrum, rapid change of efficiency add extra freedom to the spectrum.

In addition to chromaticity, diffraction efficiency is also obtained for each grating structure, though such information is not included here, However, it should be mentioned that high diffraction efficiency over wide spectra is not welcome, but rather disadvantageous. This keeps obtained chromaticity coordinates close to the white point. In this respect, our attempt to employ optical glass gratings in transmission mode may not be a good choice. Instead, non-metal materials with large absorption would be interesting option.

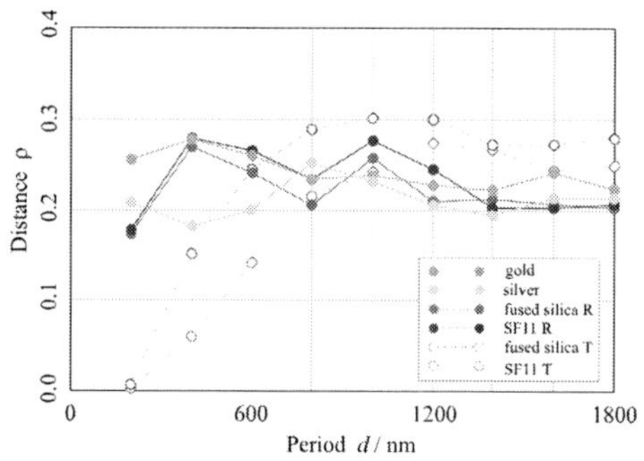

Figure 4. Influence of period on distance ρ. R: reflection. T: transmission.

References

[1] V. Vashistha, et al., ACS Photonics **4**, 1076 (2017)
[2] Y. Nagasaki, et al., ACS Photonics **5**, 1460 (2018)

Dynamics of the optical forces in nanosystems

Andrei Kiselev[1,*], *Karim* Achouri[1,**], and *Olivier J. F.* Martin[1,***]

[1]Nanophotonics and Metrology Laboratory, Department of Microengineering, École Polytechnique Fédérale de Lausanne, Route Cantonale, 1015 Lausanne, Switzerland.

Abstract. We investigate optical forces in the time domain, instead of using the time-average Maxwell stress tensor. We demonstrate first that a plane wave causes on a physical object an optical pressure that fluctuates at optical frequency in the time domain. The analytical formula for the optical force dynamics is presented for this case. The case for two-wave illumination with slightly different frequencies is considered next. It is shown that in this case the optical force acquires a component at the beating frequency. The analytical expression for the transient force is deduced and its relation with average force explained in detail.

1 Introduction

Since Kepler's observation of comet tail bending in the 17th century, the electromagnetic forces created by light have raised a tremendous scientific interest. These forces result from the exchange of momentum between photons and physical objects and are conventionally considered as static pulling [1] or pushing [2–5] entities. The electromagnetic wave theory, on the other hand, predicts that the momentum carried by an electromagnetic wave includes both a static and a time-varying component [6]. The oscillating component plays an important role in practical applications when, for example, optical pulses are utilized as incident illumination for trapping experiments [7, 8]. They also become important when an object is illuminated with two frequency-detuned waves at frequencies ω_0 and $\omega_1 = \omega_0 + \Omega$, $\Omega << \omega_0$. The beating effect in this case gives rise to the force at frequency Ω. The appearance of this force is widely explored in optomechanics at the micro-[9] and macro-scales [10]. Furthermore, several very ambitious ideas for controlling the optical force in the temporal domain by utilizing complex-valued frequency excitation have recently arisen [11].

On looking through the literature, we see that various earlier research publications investigated numerically the optical force in the time domain [12, 13], leaving the theoretical analysis of the transient force quite understudied.

We address this challenge by presenting a generic formalism that serves as a framework for the study of the dynamics of optical forces in the time domain for a single wave illumination and for the illumination with two waves.

*e-mail: andrei.ksielev@epfl.ch
**e-mail: karim.achouri@epfl.ch
***e-mail: olivier.martin@epfl.ch

2 Analytical expressions for the optical force

2.1 Film under monochromatic planewave illumination

Let us consider a planewave having the wavelength λ_0 and impinging at normal incidence on a film with a thickness d, refractive index n_{film} and immersed in a medium with the refractive index equal to 1, see Fig. 1. The reflection (r_f) and transmission (t_f) coefficients in this case can be related to the normalized thickness of the film $\delta = 2\pi n_{\text{film}} d/\lambda_0$ and the refractive index parameter $\Delta_\pm = n_{\text{film}}^{-1} \pm n_{\text{film}}$ as:

$$r_f = \frac{-0.5i \sin(\delta)\Delta_-}{\cos(\delta) - 0.5i\Delta_+ \sin(\delta)}, \quad (1)$$

$$t_f = \frac{1}{\cos(\delta) - 0.5i\Delta_+ \sin(\delta)}. \quad (2)$$

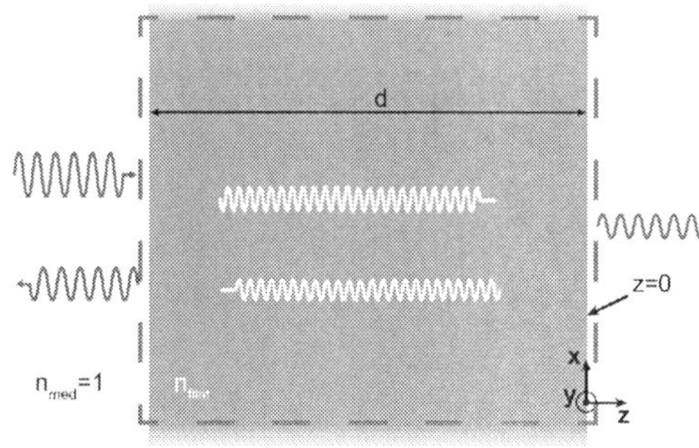

Figure 1. Sketch of an infinite film illuminated with a planewave propagating from left to right. Red dashed box indicates the part of the film, for which the force is calculated.

By analysing the Maxwell stress tensor [14], the average optical force acting on a film can be found in this case. For a segment of film having the total area $2\Sigma_0$ (front+back facets), the force has the form

$$<\mathbf{F}> = \frac{\Sigma_0}{c}\left(\left(1 + r_f r_f^* - t_f t_f^*\right)\right)\mathbf{z}. \tag{3}$$

Here, c is the speed of light in vacuum and $\mathbf{z}$ is a unit vector in the z-direction. Quite interestingly, the film additionally experiences a transient force that oscillates in time. This can be easily understood by recalling that the Poynting vector $\mathbf{S} = \mathbf{E} \times \mathbf{H}$ acquires an oscillating component under a monochromatic plane wave illumination. Hence, as the result of momentum conservation, an oscillating force appears. The magnitude of this force is also linked to the reflection coefficient of the film and has the following form:

$$|\mathbf{F}(t)| = \frac{\Sigma_0}{c}|2e^{-2i\omega_0 t} r_f|\mathbf{z}. \tag{4}$$

The presented formula gives an analytical relation between the average and transient forces acting on a structure. As can be seen, this force in Eq. (4) oscillates at the frequency ω_0, which is typically in the range of PHz for the optical range and is too high for a mechanical oscillator to develop a noticeable response. Quite interestingly, there is a technique that allows downshifting this frequency to the mechanical oscillation frequency.

2.2 Arbitrary object under two-wave illumination

Let us consider an object illuminated with two plane waves having frequencies ω_0 and $\omega_0 + \Omega$, $\Omega << \omega_0$, as shown in Fig. 2(a). Let us also suggest that the intensities of two waves are equal and the waves are propagating along the same direction ($\theta = 0$). In this case, the average force is twice the force produced by each of two waves $\langle \mathbf{F} \rangle$. The oscillating optical force in this case is also linked to the average force and oscillates as

$$|\mathbf{F}(t)| = 2\langle \mathbf{F} \rangle |\cos(\Omega t)|. \tag{5}$$

As can be noted, the force now oscillates at frequency Ω, that can be set within the kHz range, depending on the value of Ω. The dynamics of the force is presented in Fig. 2(b). It can be clearly noted that the peak force achieved in this case is two times stronger than the average one.

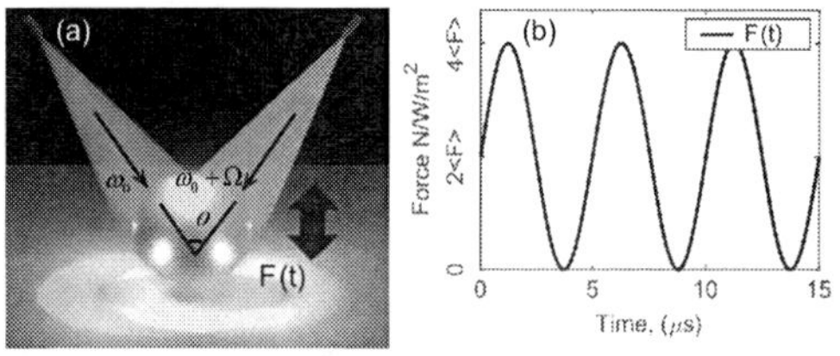

Figure 2. (a) Two optical waves with frequencies ω_0 and $\omega_0 + \Omega$ focus on a particle, generating a beating force at the frequency Ω. (b) The time domain dynamics of the optical force found for two waves of equal intensity and having beating frequency $\Omega/2\pi$ = 200 kHz. Here $<\mathbf{F}>$ denotes the average force generated by each wave.

3 Conclusions

The analytical equations for the optical force in the time domain for a single and two wave illuminations were presented and a link between the average and the transient optical forces was established.

References

[1] A. Ang, A. Karabchevsky, I. Minin, O. Minin, S. Sukhov, A. Shalin, Scientific Reports **8** (2018)
[2] C. Bradac, Advanced Optical Materials **6**, 1800005 (2018)
[3] K. Achouri, A. Kiselev, O.J.F. Martin, Physical Review B **102**, 085107 (2020)
[4] A. Kiselev, K. Achouri, O.J.F. Martin, Optics Express **28**, 27547 (2020)
[5] A. Kiselev, K. Achouri, O.J. Martin, *Multipolar origin of electromagnetic transverse force resulting from TE/TM waves interference*, Vol. 11798 of *SPIE Nanoscience + Engineering* (SPIE, 2021), `https://doi.org/10.1117/12.2593432`
[6] P.W. Milonni, R.W. Boyd, Advances in Optics and Photonics **2**, 519 (2010)
[7] B.J. Roxworthy, K.C. Toussaint, Scientific Reports **2**, 660 (2012)
[8] A. Usman, W.Y. Chiang, H. Masuhara, Science Progress **96**, 1 (2013)
[9] T.J. Kippenberg, K.J. Vahala, Optics Express **15**, 17172 (2007)
[10] M. Partanen, H. Lee, K. Oh, Scientific Reports **10**, 20419 (2020)
[11] S. Lepeshov, A. Krasnok, Optica **7**, 1024 (2020)
[12] M. Partanen, J. Tulkki, Phys. Rev. A **96**, 063834 (2017)
[13] A. Kiselev, K. Achouri, O.J.F. Martin, *Dynamics of optical forces and torques in plasmonic systems: a surface integral equation*, Vol. 11463 of *SPIE Nanoscience + Engineering* (SPIE, 2020), `https://doi.org/10.1117/12.2567651`
[14] J.D. Jackson, *Classical electrodynamics* (John Wiley & Sons, 2007), ISBN 8126510943

Controlling resonant surface modes by arbitrary light induced optical anisotropies

Niccolò Marcucci[1*], *Giorgio* Zambito[2], *Maria Caterina* Giordano[2], *Francesco* Buatier de Mongeot[2], and *Emiliano* Descrovi[1**]

[1]Polytechnic University of Turin, Dept. of Applied Science and Technology, 10129, Turin, Italy
[2]University of Genoa, Dept. of Physics, 16146, Genoa, Italy

Abstract. In this work the sensitivity of Bloch Surface Waves to laser-induced anisotropy of azo-polymeric thin layers is experimentally shown. The nanoscale reshaping of the films via thermal-Scanning Probe Lithography allows to couple light to circular photonic nanocavities, tailoring on-demand resonant BSW confined within the nanocavity.

1. Introduction

Azobenzene-containing polymers are known to exhibit interesting and peculiar properties when irradiated. Thanks to the azo bond present in the molecular structure, azobenzenes can switch between a *trans*- and a *cis*- form, involving a variation in both the steric hindrance and the polarizability of the molecule [1]. Upon a proper choice of the illumination wavelength, the photoisomerization process can be made cyclic, thus leading to macroscopic effects on azopolymeric films such as surface relief grating formation [2-3] and birefringence [4-5].

In this work, we show how such a light-controlled birefringence can be successfully exploited to tune the optical response of a resonant nano-photonic cavity coupled to Bloch Surface Waves (BSW). BSWs are surface defect modes sustained by one-dimensional photonic crystals (or planar multilayers) that are particularly sensitive to variations of the surface optical properties (e.g., refractive index or absorption). Here, we introduce an azo-doped polymer film deposited on top of a dielectric multilayer and use it as an arbitrarily-controlled anisotropic medium to tune some of the BSW properties. The azo-polymeric film birefringence is produced by controlling the polarization state of an external illumination laser beam. Two examples are addressed here, namely the spectral shift of a BSW propagating on a flat multilayer and a resonant BSW confined within a circular photonic nanocavity.

2. Results and discussion

2.1 Tailoring Bloch Surface Waves in flat Azo-doped layers

The dielectric one-dimensional photonic crystal is constituted by alternating layers of SiO_2 and Ta_2O_5 as described elsewhere [6]. Dispersed red 1-methacrylate (DR1M) is mixed with poly-phthalaldehyde (PPHA) (1:3 ratio) and then spun on the multilayer. The thickness of the polymeric layer is about 60 nm. We avoid the use of pure DR1M in order to improve the surface adhesion and the mechanical stability of the azopolymeric blend. In particular, we aim at avoiding strong morphological/topographical deformations that would occur with a pure DR1M structure upon irradiation with the excitation laser. More interestingly, the azopolymer-loaded PPHA layer can be patterned by means of a probe-assisted lithographic technique as will be described in the next section.

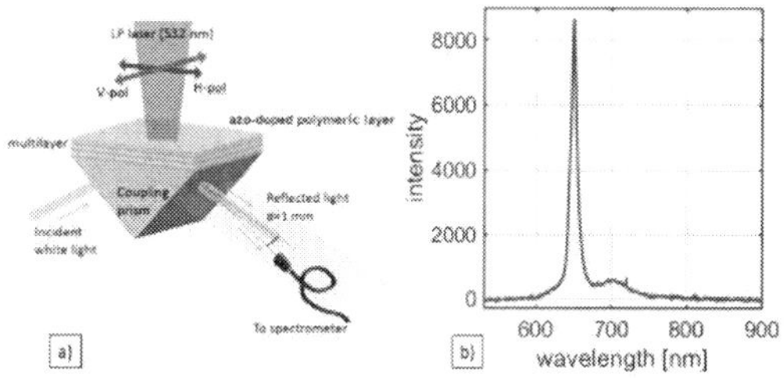

Fig. 1. a) Prism-coupling setup used for the detection of the BSW dispersion shift caused by the birefringence of the azopolymeric layer on top of the one-dimensional photonic crystal. b) Spectrum detected with a fibered spectrometer where the BSW peak is clearly evidenced.

A prism-coupling setup is used to interrogate the planar multilayer. Illumination is provided by a slightly focused white light beam. Differently from the conventional Kretschmann configuration, scattered light leaking in the substrate is detected instead of the reflected light. The zenithal collection angle is kept constant and the BSW signature is represented by a spectral peak

* Corresponding author: niccolo.marcucci@polito.it , ** emiliano.descrovi@polito.it .

Fig.1b). A 532 nm collimated laser beam is used to control the azopolymeric layer birefringence.

An increase/decrease of the BSW effective refractive index is detected as a red/blue-shift of the BSW spectral peak, because of the dielectric loading/unloading effect. In Fig. 2, the red and blue-shift of the BSW coupling wavelength are plotted over time, as the control laser polarization is varied from vertical, to circular and then horizontal (with respect to the BSW electric field orientation, which is oriented as V-pol in Fig. 1). The observations are consistent with the decrease of refractive index produced in DR1M films along the direction of polarization of the illuminating beam.

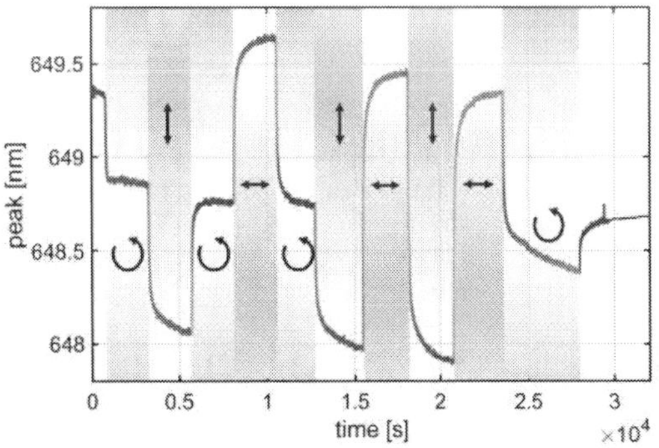

Fig. 2. Time evolution of the BSW peak position detected at a fixed leakage angle, upon variation of the laser polarization state. The overall downward drift of the BSW is likely due to heating effects.

2.2 Tailoring Bloch Surface Waves in photonic nanocavities

Resonant nanophotonic cavities have been engineered at the surface. The use of the novel thermal-Scanning Probe Lithography (t-SPL [7]) has enabled high-resolution nanopatterning of the Azo-doped thin film (PPHA-DR1M) as nanograting cavities. The cavity is constituted by a circular Distributed Bragg reflector (DBR), and it is designed to resonate at a frequency outside of the absorption spectrum of the azopolymer. An additional annular grating encircles the cavity for coupling free-space radiation to the BSW [8], and is shown in Fig.3a-b.

Spectroscopic measurements are based on a modified inverted microscope, with oil-immersion collection optics. The illumination is provided through the condenser. The collected light is Fourier-filtered with a beam blocker in order to let only light propagating at large angles to reach the image plane. In Fig. 3b the optical image of a cavity and outcoupler structure is shown, with the central spot representing the scattered light associated to the cavity resonant mode. The entrance slit of the spectrometer is aligned with the cavity centre, resulting in the spectrum shown in Fig. 3c, where the peaks associated to the cavity modes are highlighted.

A 532 nm laser beam is focused through the condenser over the cavity in order to excite the azopolymer. In Fig.3d the red and blue shift of one of the cavity resonant modes is shown. The shift is similar to the one described in Fig. 2, suggesting an active way to modulate the spectral response of the cavity in the nanometre range.

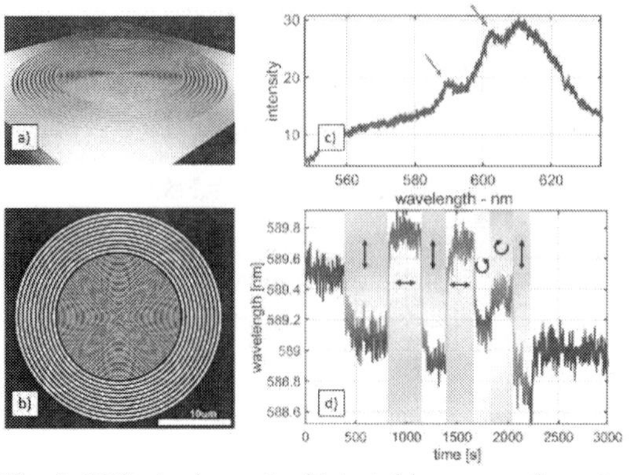

Fig. 3. DBR circular cavity fabricated by means of a probe-assisted indentation technique. a) 3D CAD image of the cavity. b) Image of the cavity obtained by Fourier-filtering the illumination beam. c) Spectrum measured at the centre of the cavity where two resonances are highlighted. d) Time evolution of the left most resonance peak upon variation of the laser polarization state.

3. Concluding remarks

In this work we successfully exploited the photo-induced birefringence of an azopolymer blend in order to tune Bloch Surface Waves. We have shown the capability to optically control the resonant response of photonic nanocavities engineered via thermal nanolithography. In perspective, the proposed approach can be used in luminescent cavities to finely tune resonant emission peaks.

References

1. C. Barrett, A. Natansohn, & P. Rochon, Macromolecules **27**, 4781–4786 (1994).
2. K. Sumaru, T. Yamanaka, T. Fukuda, & H. Matsuda, Appl. Phys. Lett. **75**, 1878–1880 (1999).
3. J. Neumann, K. S. Wieking, & D. Kip, Appl. Opt., AO **38**, 5418–5421 (1999).
4. D. S. Dos Santos et al., Polymer **43**, 4385–4390 (2002).
5. A. Kozanecka-Szmigiel et al., Dyes and Pigments **114**, 151–157 (2015).
6. A. Angelini et al., Sci Rep **4**, 5428 (2015).
7. E. Albisetti et al., E. Riedo, *Nat. Rev. Methods Primer* **2022**, *2*, 32.
8. U. Stella, L. Boarino, N. De Leo, P. Munzert, & E. Descrovi, ACS Photonics **6**, 2073–2082 (2019).

A rigorous computational framework employing coupled-mode theory for assessing lasing with transition metal dichalcogenide bilayers in the nanoscale

Georgios Nousios[1,*], *Thomas* Christopoulos[1], *Dimitrios C.* Zografopoulos[2], and *Emmanouil E.* Kriezis[1]

[1]School of Electrical and Computer Engineering, Aristotle University of Thessaloniki (AUTH), Thessaloniki GR-54124, Greece
[2]Consiglio Nazionale delle Ricerche, Istituto per la Microelettronica e Microsistemi (CNR-IMM), Roma 00133, Italy

Abstract. A whispering-gallery mode nanophotonic laser cavity having as active medium a transition-metal-dichalcogenide (TMD) bilayer is examined. The proposed system is analysed and designed utilizing a strict and rigorous computational framework based on the coupled-mode theory. Our framework is capable of accurately and efficiently handling the gain properties of two-dimensional materials, such as contemporary TMD monolayers, multilayers, and heterostructures. The presented lasing cavity exhibits an adequately low pump threshold and light emission in the order of milliwatts is predicted. Exploiting the capabilities of the developed framework, we were in position to efficiently design the cavity as well as to estimate quantitative lasing parameters such as the pumping threshold and the lasing frequency.

1 Introduction

The study of micro- and nano-cavity lasers has attracted significant attention in the past decade due to their compact size, speed, and power efficiency [1]. The initial focus was on Fabry-Pérot and photonic crystal cavities and nano-cavity lasers exhibited notable performance metrics, which strengthened the research interest and led to other alternatives such as plasmonic or whispering-gallery mode cavities. More recently, two-dimensional (2D) materials were found to have interesting lasing capabilities and researchers started to experiment with their interaction with nanoscale systems. Transition metal dichalcogenides consist such a promising family of 2D materials, being capable of light emission in the deep near-infrared (NIR) and in the visible spectrum [2]. Importantly, each alternative TMD monolayer has a different bandgap giving in general flexibility to the design, which is further expanded when bilayers, multilayers, and TMD heterostructures come into consideration [2].

In this work, we present a rigorous computational framework that allows to systematically estimate the lasing characteristics *and* the output of a resonant system while also providing useful guidelines for designing efficient laser cavities. The framework is appropriatelly developed for considering sheet-type materials, respecting their 2D nature. We utilize this framework to analyse the lasing behaviour of an integrated whispering-gallery mode cavity, having as a gain medium a MoS_2/WSe_2 heterostructure (TMD heterobilayer) [3].

2 Coupled-mode theory framework

The proposed framework is based on the temporal coupled-mode theory (CMT) [4, 5], a powerful tool that allows for the accurate prediction of nonlinear systems response in the time domain without resorting to the solution of Maxwell's equations. Linear full-wave simulations are conducted to extract key parameters of the whispering-gallery cavity (resonance frequency, quality factor), which are then introduced in the simpler, time-dependent ordinary differential equations of CMT. The gain mechanism of the 2D bilayer is carefully treated by modeling both the carrier dynamics through an appropriate number of coupled rate equations, and by connecting them with the electromagnetic response of the system via an induced polarization field [4].

Following [4, 5], the cavity amplitude envelope $a(t)$ (assuming it oscillates at ω) is given by

$$\frac{\mathrm{d}a}{\mathrm{d}t} = -j(\omega-\omega_c)a - \frac{1}{\tau_\ell}a - \xi_1\left(\frac{\mathrm{d}p}{\mathrm{d}t} + j\omega_c p\right), \quad (1)$$

and the induced polarization envelope $p(t)$ by

$$\frac{\mathrm{d}p}{\mathrm{d}t} = -\frac{\omega_m^2 - \omega^2 + j\omega\Gamma_m}{\Gamma_m + j2\omega}p - \frac{\sigma_m}{\Gamma_m + j2\omega}\Delta N a. \quad (2)$$

The cavity (ω_c and $1/\tau_\ell$), gain material (ω_m and Γ_m), and coupling characteristics (ξ_1 and σ_m) are all meticulously included in the framework. The population inversion ΔN emerges generally from a multi-level carrier system, which is also coupled with Eqs. (1) and (2) to correctly encapsulate the various interactions that take place. Our MoS_2/WSe_2 heterobilayer can be modeled by a relatively simple three-level system [3], as schematically depicted in Fig. 1(b). Moreover, in the general case of mismatch between the cavity resonance and the gain peak frequency of the material (i.e., when $\omega_c \neq \omega_m$), the lasing frequency ω_L naturally emerges from the framework by a simple Fourier transformation of the envelope $a(t)$. Additionally, only by making a few simple but rational assump-

*e-mail: gnnousios@ece.auth.gr

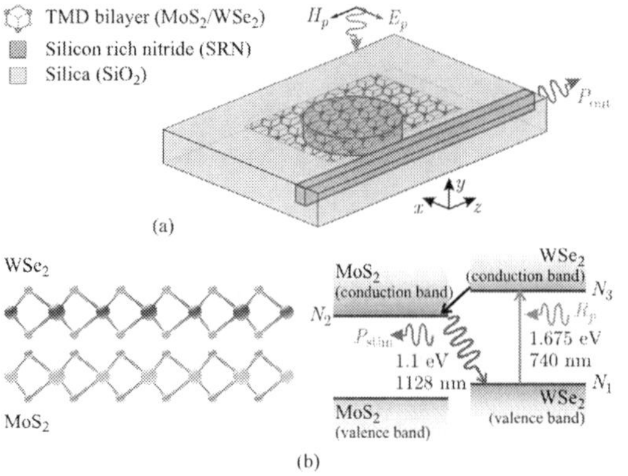

Figure 1. (a) Schematic of the proposed whispering-gallery mode cavity (disk resonator) with the TMD heterobilayer on-top. The vertical pump illumination and the output channel are both clearly marked. (b) Atomic structure of the WSe_2, MoS_2 monolayers and the three-level energy diagram of the heterostructure, which allows for stimulated emission around 1128 nm due to the interlayer excitons formed between MoS_2 conduction and WSe_2 valence bands.

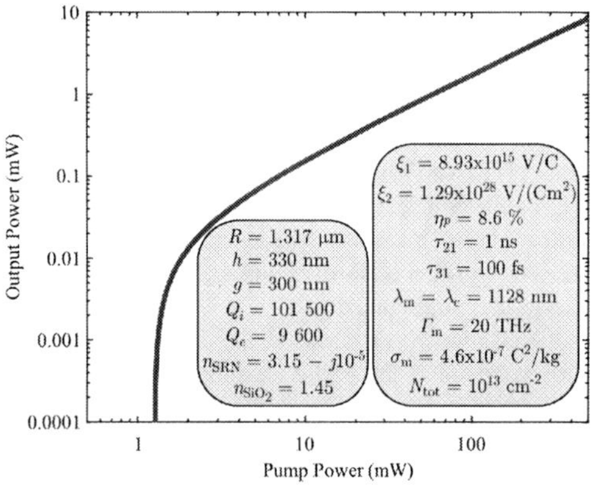

Figure 2. CW lasing output power of the proposed whispering-gallery cavity versus pump power. A low pump threshold of 1.3 mW is achieved and a clear, linear output is observed. The output power saturates for $P_p \sim 10$ W (not shown). Insets: Complete list of the parameters used for the simulation.

tions/simplifications, we are able to get a good estimation of the lasing frequency in a closed-form equation.

3 System design

The practical realization of the envisioned whispering-gallery mode cavity is a silica-cladded, silicon rich nitride (SRN) disk of radius R [Fig. 1(a)]. SRN is chosen for being transparent deep in the NIR where TMDs are able to emit light, in contrast to silicon which starts to exhibit important absorption below 1150 nm. By utilizing the developed framework in the proposed structure, we are able to evaluate that higher order modes with higher quality factors are more beneficial for lasing, despite their weaker interaction with the TMD bilayer. Thus, we opt for a disk with radious $R = 1.317\,\mu$m and a high intrinsic quality factor of $Q_i \sim 101\,500$ (comprising radiation and bulk material ohmic losses). The TMD bilayer is vertically pumped by illuminating with a plane wave at 740 nm, with energy equal to the WSe_2 bandgap. This results in the formation of a vertical Fabry-Pérot cavity and the absorption efficiency is optimized when the height of the resonator is $h = 330$ nm. The structure emits light at 1128 nm due to an interlayer exciton formed between the MoS_2 conduction band and the WSe_2 valence band [Fig. 1(b)]. The emitted light is initially trapped inside the whispering-gallery cavity and then couples to the outside world through a bus waveguide placed near the cavity (300 nm edge-to-edge distance). The respective external quality factor is calculated around $Q_e \sim 9\,600$ since strong coupling between the cavity and the waveguide is desirable.

By solving the system of the ordinary differential equations that consist the developed framework, we are able to calculate efficiently the output power of the laser with respect to the pump excitation. The resulting continuous-wave (CW) output is depicted in Fig. 2. We can readily realize the lasing threshold which is around 1.3 mW. The output power is relatively high, reaching 10 mW for a pumping power of 592 mW. The emission remains linear for a wide pumping range, until starting to saturate for $P_p \sim 10$ W (not shown). The saturated output power is calculated in the vicinity of 120 mW.

4 Conclusion

To conclude, we have developed a rigorous computational framework that is capable of appropriately describing light emission in micro- and nano-cavity lasers with 2D gain materials. Since the framework is well-suited to model the emission from contemporary, sheet-type materials, we have used it to evaluate the performance of a whispering-gallery mode disk-type cavity, combined with a TMD heterobilayer. The proposed system exhibits a lasing threshold of 1.3 mW and emits with a 1.7% efficiency for a wide range of pumping before saturation.

This research work was supported by the Hellenic Foundation for Research and Innovation (HFRI) under the "First Call for HFRI Research Projects to support Faculty members and Researchers and the procurement of high-cost research equipment grant" (Project No. HFRI-FM17-2086).

References

[1] R.-M. Ma and R. F. Oulton, Nature Nanotechnol. **14**, 12 (2019)
[2] S. Manzeli, D. Ovchinnikov, D. Pasquier, O. V. Yazyev, and A. Kis, Nat. Rev. Mater. **2**, 17033 (2017)
[3] Y. Liu, H. Fang, A. Rasmita, Y. Zhou, J. Li, T. Yu, Q. Xiong, N. Zheludev, J. Liu, and W. Gao, Sci. Adv. **5**, eaav4506 (2019)
[4] S.-L. Chua, Y. Chong, A. Douglas Stone, M. Soljačić, and J. Bravo-Abad, Opt. Express **19**, 2 (2011)
[5] T. Christopoulos, O. Tsilipakos, N. Grivas, and E. E. Kriezis, Phys. Rev. E **94**, 062219 (2016)

Towards harmonic generation enhancement on silicon

L. Rodríguez-Suné[1,*], M. Scalora[2], C. Cojocaru[1], N. Akozbek[3], R. Vilaseca[1], and J. Trull[1]

[1]Universitat Politècnica de Catalunya, Physics Department, Rambla Sant Nebridi 22, 08222 Terrassa, Spain
[2]Aviation and Missile Center, US Army CCDC, Redstone Arsenal, Huntsville, AL, United States
[3]US Army Space & Missile Defense Command, Tech Center, Redstone Arsenal, AL 35898, United States

Abstract. Nowadays, nanostructures are routinely fabricated and integrated in different photonic devices for a variety of purposes and applications. For instance, nonlinear silicon photonics is an area of interest due to its high compatibility with CMOS technology, offering structure sizes down to 10nm at low cost. When the nanoscale is reached, light-matter interactions can display new phenomena, conventional approximations may not always be applicable, and new strategies must be sought in order to study and understand light-matter interactions at the nanoscale. In this work, we report a comparative experimental and theoretical study of second and third harmonic generation from silicon with the aim of explaining the nonlinear optical properties of this material at the nanoscale. We measure second and third harmonic efficiencies as a function of angle of incidence, polarization and pump wavelength. We compare these measurements with numerical simulations based on a microscopic hydrodynamic model which accounts for different possible contributions to the nonlinear polarization. This way, we have the ability to explain properly the SH and TH signals arising from different silicon samples. Once we have this knowledge, we are able to design more complex structures, such as silicon nanowires, where higher conversion efficiencies can be achieved.

1 Introduction

Second and third harmonic generation (SHG and THG) are two nonlinear processes in which when light interacts with a nonlinear material, new frequencies are generated. These processes are typically studied in thick materials with high nonlinearities and under phase matching conditions and low material absorption with the aim of achieving high conversion efficiencies. They are described through the nonlinear polarization, which is usually approximated to the electric dipole contribution described by the second and third order susceptibilities $\chi^{(2)}$ and $\chi^{(3)}$. This approximation starts to fail when the nanoscale is reached: SHG and THG conversion efficiencies are drastically reduced and phase matching conditions and absorption no longer play a significant role. In this case, other nonlinear contributions to the nonlinear polarization, such as magnetic dipole and electric quadrupole interactions, must be taken into account.

In this work, we analyse SH and TH fields generated when ps and fs pulses tuned at 1064nm or 800nm, respectively, interact with different silicon samples. This analysis consists on first, measuring the harmonic signals as a function of incident angle, polarization and pump wavelength, and then, comparing them with numerical predictions based on a microscopic hydrodynamic model explained in details in [1]. The study starts by analysing simple structures: an amorphous silicon nanolayer and a crystalline silicon wafer. The comparison of measurements with numerical simulations allows us to obtain information of the material and the light-matter interaction at nanoscale. With this knowledge, we are able proceed the study by designing more complex structures, such as silicon nanowires, where higher conversion efficiencies can be achieved.

*Corresponding author: laura.rodriguez.sune@upc.edu

2 Results

2.1. Experiments

Measurements of SH signals from silicon have been conducted in transmission and in reflection. In both cases, incident pulses tuned at 800nm and 1064nm have been used. The process of THG was conducted either in transmission and reflection using incident pulses tuned at 1064nm. For this purpose, we have developed an experimental set-up capable of analysing the angular dependence of the harmonic signals in both transmission and reflection configurations, and as a function of polarization and pump wavelength. This set-up also allows us to properly separate the harmonic fields from each other and from the fundamental field radiation by means of filters and a dispersive prism.

2.2 Theory

We have performed numerical simulations to model the experimental situation. The theoretical model embraces full-scale time-domain coupling of matter to the macroscopic Maxwell's equations. We use the hydrodynamic model to understand linear and nonlinear optical properties of different materials by accounting for competing surface, magnetic and bulk nonlinear sources, which in this case consist of mostly bound electrons. In semiconductors like silicon, the action of free carriers is neglected, so that the dielectric function of the material can be described through the Lorentz model containing one resonance in the UV range. We measure and estimate the efficiency of the harmonic signal using no assumptions about effective surface or volume nonlinearities other that effective electron masses and densities.

In Fig. 1 we show results of reflected SHG efficiencies as a function of incident angle from a 25nm-thick amorphous silicon nanolayer. Pump pulses tuned at 1064nm were used and both incident and harmonic fields were TM-polarized. Measurements are plotted with full circles while the solid curve corresponds to the predicted efficiencies. As it can be seen, experimental and simulated curves agree remarkably well, showing a maximum predicted and detected efficiency at 50° incident angle.

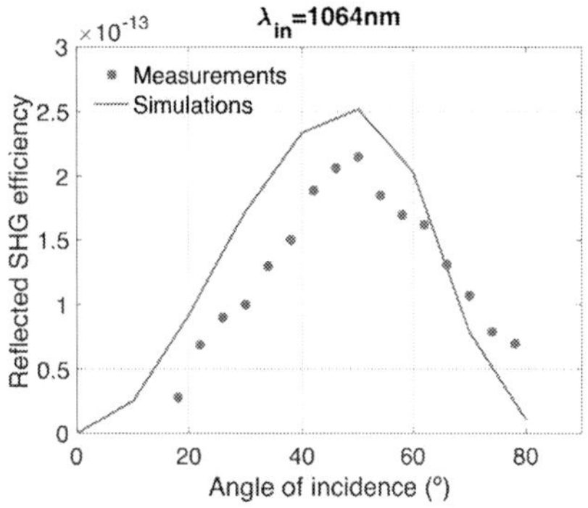

Fig. 1. Measured (full circles) and predicted (solid curve) reflected SHG efficiencies as a function of incident angle from a 25nm-thick amorphous silicon layer when the pump pulse is tuned at 1064nm and both fundamental and SH fields are TM-polarized.

In Fig. 2 we show experimental and predicted results of transmitted and reflected TM-polarized SHG efficiencies as a function of incident angle from a 25nm-thick amorphous silicon layer when an incident field tuned at 800nm and TM-polarized was used. We observe that the reflected experimental and theoretical curves agree well in shape and amplitude, having a maximum around 50° angle of incidence. The results for the transmission case are also satisfactory. Another thing that is reproduced in both experimental and theoretical results is the ratio between transmitted and reflected efficiencies. As a result, we are confident that our theoretical approach suffices to understand and explain most of the relevant aspects of surface SHG.

In Fig. 3 we show the results of the angular dependence of reflected THG efficiencies from a 500μm-thick crystalline silicon wafer. Pump pulses tuned at 1064nm were used. A TM and a TE-polarized TH field was detected for TM and TE-polarized incident light, respectively. Once again, both experimental and theoretical results agree remarkably well in shape and amplitude showing a maximum efficiency at normal incidence. The values of these efficiencies were used to retrieve the nonlinear dispersion of the material $\chi^{(3)}$, as reported in [2].

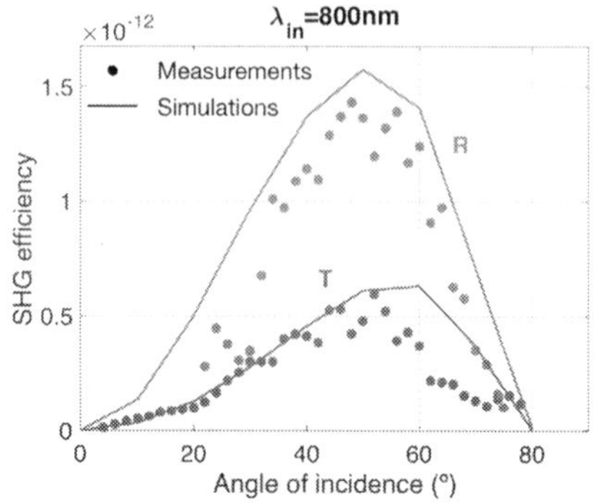

Fig. 2. Transmitted (blue) and reflected (red) measured (full circles) and predicted (solid curves) SHG efficiencies as a function of incident angle from a 25nm-thick amorphous silicon layer when the pump pulse is tuned at 800nm and both harmonic and fundamental fields are TM-polarized.

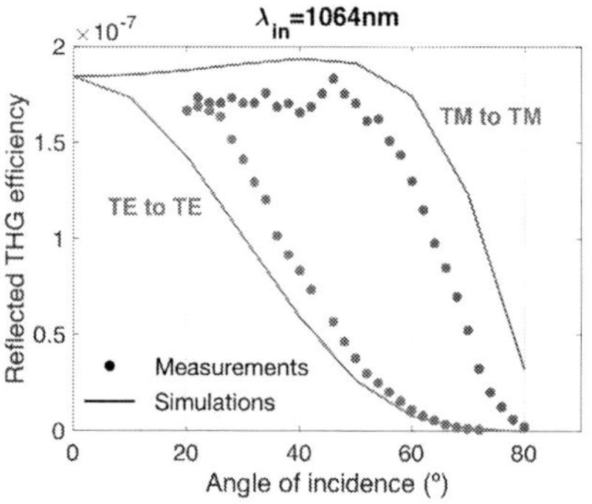

Fig. 3. Transmitted (blue) and reflected (red) measured (full circles) and predicted (solid curves) THG efficiency as a function of incident angle from a crystalline silicon wafer when the incident field is tuned at 1064nm. A TM (blue) and a TE-polarized (red) TH field was detected for TM and TE-polarized incident light, respectively.

LRS, JT and CC acknowledge to US Army Research Laboratory Cooperative Agreement N° W911NF1920279 issued by US ARMY ACC-APG-RTP and Spanish Agencia Estatal de Investigación (project no. PID2019-105089GB-I00/AEI/10.13039750110001 1033).

References

1. M. Scalora, M. A. Vincenti, D. de Ceglia, V. Roppo, M. Centini, N. Akozbek, M. J. Bloemer, Phys. Rev. A **82**, 043828 (2010).
2. L. Rodríguez-Suné, J. Trull, N. Akozbek, D. De Ceglia, M. A. Vincenti, M. Scalora, C. Cojocaru, Front. Photon. **17**, (2021)

https://doi.org/10.1051/epjconf/202226605011

Switchable optics based on guided mode resonance in lithographically patterned vanadium dioxide

Markus Walther[1,*], *Thomas* Siefke[1], *Kristin* Gerold[1], and *Uwe D.* Zeitner[1]

[1]Friedrich Schiller University Jena, Institute of Applied Physics, Albert-Einstein-Str. 15, 07745 Jena, Germany

Abstract. Vanadium dioxide as a phase change material is usually known for its consideration in smart window applications. However, the attention shifts to using it in actively switched optical elements. The main challenges are the deposition of vanadium dioxide with the correct stoichiometry and phase and the patterning of the material. We propose a design with a corresponding manufacturing process for an actively switchable reflector at 1550 nm wavelength with a contrast near 10^5 by using the thermochromic effect of vanadium dioxide. The reflectance of the proposed optical element can be controlled between an ultra-low and a high reflecting state. We elaborate on the proposed optical design, the manufacturing process including deposition, annealing and patterning processes, and discuss already achieved results.

1 Introduction

The reversible semiconductor to metal transition (SMT) of vanadium dioxide (VO_2) at a critical temperature of T_{crit} = 68 °C was first found in 1959 [1], since then many potential applications were described. Including, but not limited to, smart windows [2], adaptive optics [3], and optical switches [4]. The SMT is accompanied by a structural phase transition from the low-temperature monoclinic VO_2(M) to the high-temperature rutile VO_2(R) crystal structure [5], and several changes in physical properties, like the change in electrical resistivity, refractive index, and extinction coefficient, among others.

The hereby proposed design for an actively switching reflector with high switching contrast is based on guided mode resonances. We use a one-dimensional grating with a period of 1 μm on a stack of conductive and nonconductive films (fig. 1). The underlying layer stack serves as a resistive heating element to achieve the phase transition of vanadium dioxide. The heating element consists of a thin film of indium tin oxide (ITO) covered with chromium contacts for the connection of a regulated power supply. On top of the heating element is an electrically non-conductive film of silicon oxide. It isolates the heating element from the vanadium dioxide (VO_2) grating.

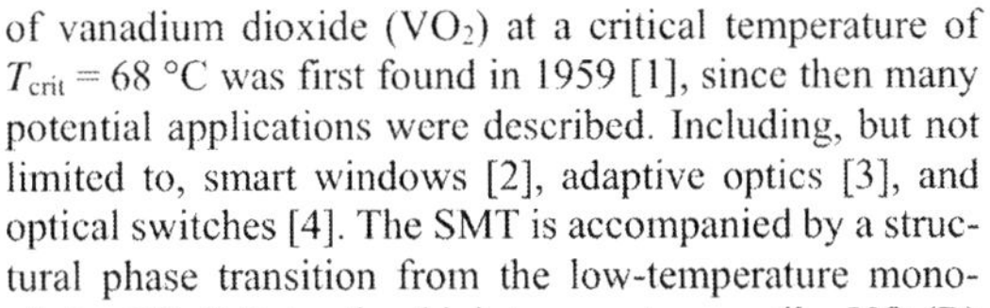

Fig. 1. Design of the resistive heating element underneath and waveguide structure on top. Both parts are electrically isolated by a thin film of SiO_2. The dimensions are not to scale.

The aim of the optical design of the grating is to switch the reflection at the wavelength of λ = 1550 nm. At room temperature (and below T_{crit} = 68 °C), incoming light is coupled into the waveguide structure and absorbed. With rigorous coupled wave analysis (RCWA), the reflectance of this structure can be calculated to be below $R < 0.001\%$. If the temperature rises above T_{crit}, the material changes to the metallic phase and the waveguide grating becomes reflective ($R > 45\%$). This leads to a switching contrast near to 10^5.

2 Manufacturing

A single side polished silicon wafer with <100> orientation is used as a Substrate. ITO was deposited using a sputtering process. The chromium and silicon oxide films were deposited by ion beam deposition and reactive ion beam deposition (IBD/RIBD), respectively. Both layers were patterned by a lift-off process utilizing AZ®nLOF 2070 from MicroChemicals. Afterwards, a vanadium dioxide film was deposited with RIBD. Using X-ray photoelectron spectroscopy (XPS) and Auger electron spectroscopy (AES), we were able to determine the stoichiometric composition and ratio of vanadium to oxygen in the resulting film depending on the oxygen partial pressure during the deposition. The partial pressure of oxygen was controlled by variation of oxygen flow rate. After deposition, an annealing of the layer at 520 °C for 30 min in an atmosphere composed of nitrogen and oxygen in a 100:1 ratio and a pressure of 10 mbar was necessary to form VO_2(M). The formation of this phase could be validated by Raman spectroscopy. We developed a recipe for reproducibly manufacturing VO_2 layers of different thicknesses ranging from at least 100 nm to 300 nm. The resulting layer was patterned by a lift-off process prior to the annealing.

* Corresponding author: walther.markus@uni-jena.de

The VO_2 layer was patterned by inductively coupled plasma (ICP) etching against a chromium hard mask. The latter was patterned using e-beam lithography with positive tone resist FEP-171 (Fujifilm) and reactive ion etching (RIE) directly on the VO_2. After etching the VO_2, the chromium was removed from the top of the ridges with RIE. The presence of the phase change material VO_2(M) in the patterned layer was verified by using Raman spectroscopy.

3 Results

3.1 Complex refractive index of VO_2

The refractive index n (fig. 2 a)) and extinction coefficient k (fig. 2 b)) were measured on samples of VO_2 thin film deposited using RIBD on a silicon substrate by variable angle spectroscopic ellipsometry (VASE). The samples were measured at 25 °C and 90 °C. The temperature was controlled by a regulated hot plate to the specified temperature.

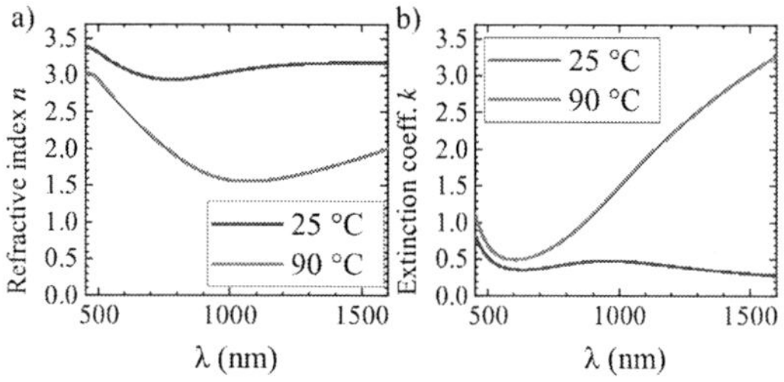

Fig. 2. a) refractive index n and b) extinction coefficient k of prepared VO_2 film below and above the SMT.

3.2 Raman spectra

The Raman spectra (see fig. 3) measured before and after the annealing process show the reconfiguration of the thin film to form crystalline phase of VO_2(M). Most prominent are the two sharp peaks of VO_2 at 194 cm^{-1} and 224 cm^{-1} and the broader peak at 612 cm^{-1}. The peak at 520 cm^{-1} is due to the Si substrate material. The insets show an SEM image of an edge of the respective films on Si substrate.

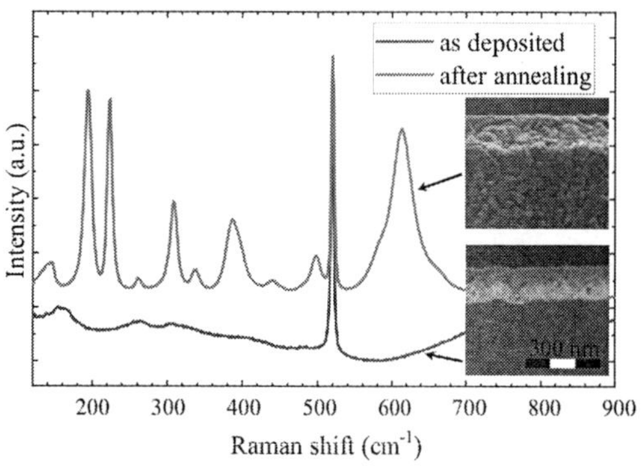

Fig. 3. Raman spectra of VO_2 film on Si: as deposited and after annealing. The insets show SEM images of the respective films.

3.3 Transmission

As an optical performance test, the transmission was measured at $\lambda = 1550$ nm on a non-patterned VO_2 layer with thickness of 265 nm (fig. 4). The sample was heated by applying a heating current to the resistive heating element directly in contact to the VO_2 thin film. Depending on the temperature, the transmission could be switched between $T = 43.6\%$ and $T = 0.17\%$. contrast of $c_T > 250$. According to RCWA, the maximum switching contrast in transmission for the film thickness is $c_T = 375$.

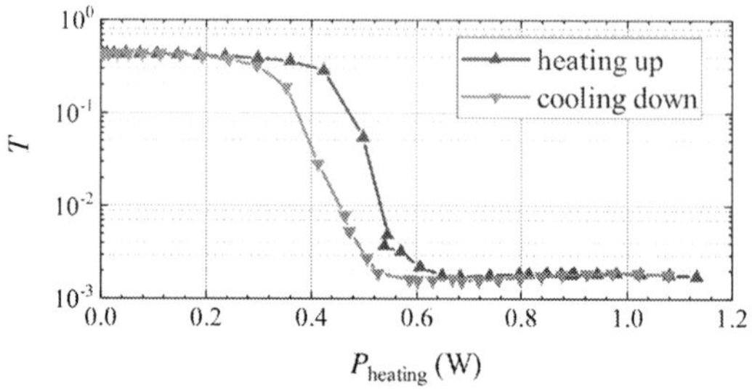

Fig. 4. Transmission change of optical element in relation to heating power inserted in resistive heating film. Each measurement was taken after thermal equilibrium was reached.

4 Conclusion

We presented a design for an actively switching reflector with high switching contrast based on guided mode resonances. Furthermore, we established a manufacturing process for e-beam lithography on VO_2 thin films deposited by RIBD. Characteristic Raman peaks were observed for thin films and patterned layers. These results are very promising for the optical function of such an optical switch. Next, we plan the optical characterisation of the reflector.

Acknowledgement

The authors acknowledge the financial support by the Federal Ministry of Education and Research (BMBF) of Germany and VDI (funding no. 13N15175). This work is funded through the project 20IND04 ATMOC within the Programme EMPIR. The EMPIR initiative is co-founded by the European Unions Horizon 2020 research and innovation program and the EMPIR Participating Countries.

References

1. F.J. Morin. Phys. Rev. Lett. **3**, 34 (1959).
2. X.Q. Gong, J.X. Li, S.Y. Chen, and W.J. Wen, Appl. Phys. Lett. **95**, 251907 (2009).
3. M. Rini, A. Cavalleri, R.W. Schoenlein, R. López, L. C. Feldman, R.F. Haglund, L.A. Boatner, and T.E. Haynes, Opt. Lett. **30**, 558 (2005).
4. T. Ben-Messaoud, G. Landry, J.P. Gariépy, B. Ramamoorthy, P.V. Ashrit, and A. Haché, Opt. Commun., **281**, 6024 (2008)
5. M.F. Becker, A.B. Buckman, R.M. Walser, T. Lépine, P. Georges, and A. Brun, Appl. Phys. Lett. **65**, 1507 (1994).

https://doi.org/10.1051/epjconf/202226605012

Approaches for the RCWA-based non-destructive characterization of subwavelength-structured gratings

Julian Wüster[1,*], *Andreas* Reetz[1], *Rüdiger* Schmidt-Grund[2], *Andrea* Knauer[3], and *Stefan* Sinzinger[1]

[1]Technische Universität Ilmenau, Fachgebiet Technische Optik, Postfach 100565, 98684 Ilmenau, Germany
[2]Technische Universität Ilmenau, Fachgebiet Technische Physik 1, Postfach 100565, 98684 Ilmenau, Germany
[3]Technische Universität Ilmenau, Institut für Mikro- und Nanotechnologien, Postfach 100565, 98684 Ilmenau, Germany

Abstract. Nano-structuring enables us to add additional degrees of freedom to the design of optical elements. Especially the possibility of controlling the polarization is of great interest in the field of nano-structured optics. For being able to exploit the whole range of form-birefringent phase shifts, the aspect ratios of the resulting element are typically much higher than the aspect ratios of conventional diffractive optical elements (DOEs), which does not only pose a challenge on fabrication but also on characterization. We evaluate several well-established approaches for the nondestructive characterization, including Müller-Matrix-Ellipsometry, measurement of the diffraction efficiencies, scattering measurements and calibration with rigorous coupled-wave modelling. The goal is to understand the challenges with all these techniques and combine them to a reliable method for structural reconnaisance of high aspect ratio nanostructures.

1 Introduction

Linear subwavelength gratings, so called "zero-order gratings" show a form birefringence which can be tuned through choice of material and geometric shape. This can be utilized e.g. for the design of anti-reflective layers [1] as well as quarter- or half-waveplates [2–4]. The combination of binary form-birefringent subwavelength gratings and diffractive structures can be exploited for designing polarization-dependent optical elements such as compact polarizing beam splitters [5–7]. To achieve a high difference in phase shift for the different linear polarization states, it is necessary to produce gratings with an extraordinarily high aspect ratio. This poses stress on the fabrication, but also on the characterization of the fabricated grating. Conventional optical measurement techniques such as laser scanning microscopy or white light interferometry are not suitable to measure the depth and exact cross section of subwavelength structures with high aspect ratios. And for scanning electron microscopy (SEM)-imaging, one needs a defined and precisely cut cross-section (e.g. generated through focussed ion beam (FIB) cutting) for directly measuring the geometry (an SEM image of subwavelength structures taken at an opaque angle is shown in Figure 1).

We can circumvent FIB and SEM imaging by measuring the optical properties of the element and compare the results to the simulation results of rigorous coupled-wave analysis models (RCWA). The optical properties include the effects on polarization (Müller-Matrix-Ellipsometry) and the broadband measuring of transmission efficiencies under various angles. Those measurements contain a lot of information about the geometrical properties and fabrication errors, which can be utilized to non-destructively characterize subwavelength gratings.

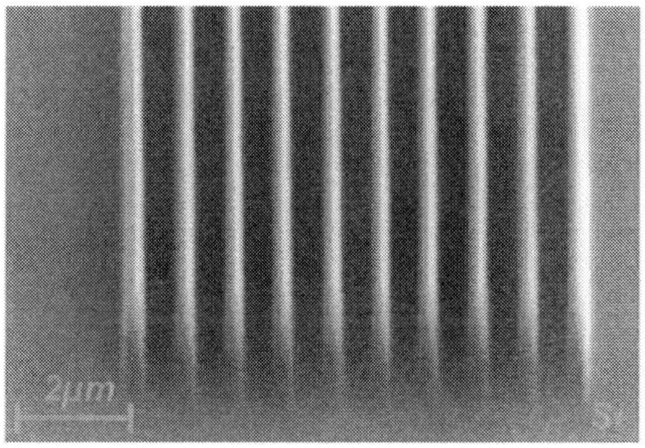

Figure 1. SEM image of subwavelength structures, the full grating layout is shown in Figure 3. Image: Patrick Feßer

2 Grating types

We demonstrate RCWA simulations on three types of binary gratings. Firstly, conventional diffraction gratings with a period of 9 μm with varying etch depths and aspect ratios both in Fused Silica and the Schott glass AF 32 Eco. Secondly, we investigate a linear grating solely consisting of subwavelength structures with the period of 200 nm and positive flank angles of 6°. It is a Silicon Master for Nanoimprint-Lithography (NIL), Eulitha P200L 90d, whose cross-section is shown in Figure 2. Lastly, we in-

*e-mail: julian.wuester@tu-ilmenau.de

vestigate the properties of combined diffractive and subwavelength structures on the example of the aforementioned polarizing beamsplitter shown in Figure 3.

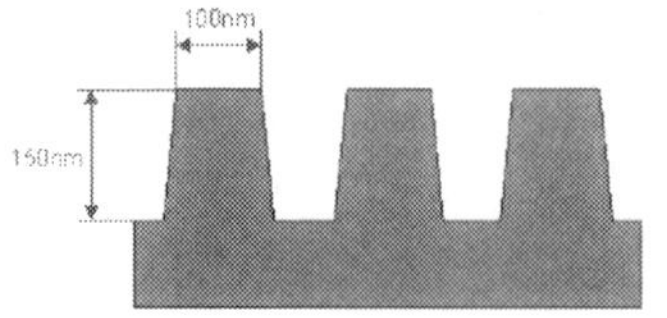

Figure 2. Cross section of Eulitha subwavelength grating (p = 200 nm) in Silicon with positive flank angles of 6°.

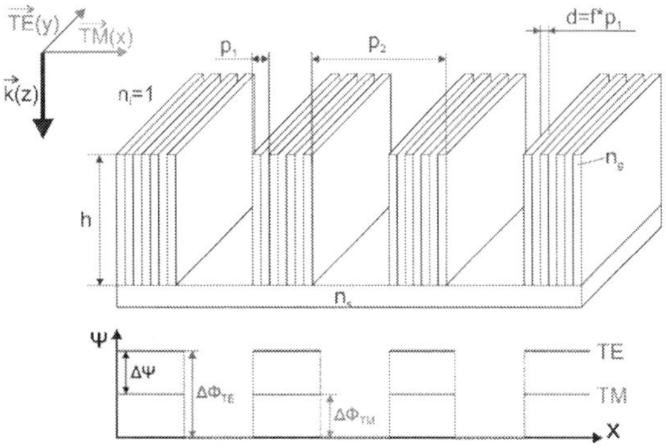

Figure 3. Cross section of subwavelength-structured diffraction grating (p_1 = 500 nm, p_2 = 10 µm, h = 700 nm) in Silicon [7] Below, the relative phase shifts between TE- and TM-polarization are shown.

3 Measurements

The measurements are carried out with three different measurement systems. Mainly we use the Sentech spectroscopic ellipsometer SENresearch 4.0, which offers a spectral range from 190 nm to 2500 nm. We use it for determining the Müller-Matrix as well as transmission curves. We also have our own diffraction efficiency measurement setup including a tunable laser source utilized within the visible range and a scattered light measurement setup wit a fixed wavelength of 532 nm at our disposal (Albatross TT). Not all of the above measuring systems are suitable for each grating type due to the transmittivity of the materials and the possibility to control the polarization. We present an overview which system is most fitting for which grating.

4 Simulation

For modelling with RCWA we use the MC Grating software and for the Müller-Matrix-Ellipsometry the WVASE software of the J.A. Woollam company. For comprehensive simulation of the situation we investigate the effect of the following parameters: etching depth, flank angle, multi layer elements (due to oxidation of the Silicon material), the effects of different etching depths due to varying etch rate on variable size areas, several incidence angles, rounded edges and a broad wavelength range of the light source.

5 Discussion

The measurement techniques we are using are well-known, but the combination of transmission, reflection and polarization measurements leads to new possibilities for characterizing high aspect ratio structures. The RCWA-modelling of gratings enables us to execute extensive parameter studies. This way one can identify the parameter sets (e.g. fabrication errors) which have the main influence on the performance of the optical element. Those are, in descending order, reflectivity (material), etching depth and flank angle. The parameter studies work well for each variable individually but a remaining challenge is the combination of various variables. All the RCWA studies require a sufficient prior knowledge about the geometry of the grating. Nevertheless, we learn both about the charateristics of nanostructures and about the processing deviations, which occurred. Being able to quantize the effect of each parameter is potentially of great importance to utilize those effects for the design of other subwavelength-structured optical elements and optimize the fabrication process.

References

[1] Enger et al., Appl. Opt. **22**, 3220-3228 (1983)
[2] Cescato et al., Appl. Optics **29**, 3286-3290 (1990)
[3] Stock et al., Ilmenau Sc. Coll. **59** (2017)
[4] Isano et al., Jpn. J. Appl. Phys **43**, 5294-5296 (2004)
[5] Xu, F. et al., Opt. Letters **21**, 1513-1515 (1996)
[6] Haidner et al., J. Eur. Opt. Soc. A **7**, 1347-1361 (1998)
[7] Wüster et al., J. Eur. Opt. Soc.-Rapid Publ. **17**, 4 (2021)

Growth and mid-infrared emission properties of "mixed" fluorite-type Er:(Ca,Sr)F_2 and Er:(Ba,Sr)F_2 crystals

Liza Basyrova[1,*], *Pavel* Loiko[1], *Abdelmjid* Benayad[1], *Gurvan* Brasse[1], *Jean-Louis* Doualan[1], *Alain* Braud[1], *Ammar* Hideur[2] and *Patrice* Camy[1]

[1]Centre de Recherche sur les Ions, les Matériaux et la Photonique (CIMAP), UMR 6252 CEA-CNRS-ENSICAEN, Université de Caen Normandie, 6 Boulevard Maréchal Juin, 14050 Caen, France
[2]UMR 6614, CNRS-INSA-Université de Rouen, Normandie Université, Avenue de l'université, BP. 12, 76801 Saint Etienne du Rouvray, France

Abstract. Fluorite-type 5 at.% Er^{3+}:$(M_1,M_2)F_2$ (M_1 = Ca, Ba; M_2 = Sr) crystals were grown by the conventional Bridgman technique and a comparative study of their spectroscopic properties was performed. The vibronic properties of the fluorite-type crystals were studied by Raman spectroscopy. The Er:$(M_1,M_2)F_2$ crystals exhibited a slightly inhomogeneous broadening of mid-infrared luminescence spectra as compared to the 'parent' compound, Er:SrF_2. The luminescence lifetimes of the $^4I_{11/2}$ and $^4I_{13/2}$ manifolds were measured, e.g., for the Er:(Ca,Sr)F_2 crystal, the luminescence lifetimes were estimated to be 8.64 ms and 5.64 ms, respectively, representing a favorable ratio for mid-IR laser operation.

1 Introduction

Calcium fluoride (CaF_2) also called fluorite when appearing in the mineral form is an attractive host matrix for doping with laser-active rare-earth ions (RE^{3+}). This is because of a combination of good thermal properties of the host matrix and a unique spectroscopic behavior of the dopant RE^{3+} ions, namely, very strong inhomogeneous spectral broadening arising from ion clustering observed even at low doping levels [1]. In particular, fluorite-type solid-solution crystals, $(M_1,M_2)F_2$ (where M_1, M_2 = Ca, Sr, and Ba) provide modification of the spectroscopic properties of RE^{3+} ions in the form of inhomogeneous broadening of the absorption and emission bands, which is attractive for broadly tunable and mode-locked lasers [2,3]. Erbium ions are attractive for developing mid-infrared lasers at 2.8 μm according to the $^4I_{11/2} \rightarrow {}^4I_{13/2}$ transition [4]. In the present work, we report on the growth and a comparative spectroscopic study of Er^{3+}-doped (Ca,Sr)F_2 and (Ba,Sr)F_2 crystals.

2 Crystal growth

The crystals of 5 at.% Er^{3+}:$(M_1,M_2)F_2$ (M_1 = Ca, Ba; M_2 = Sr) crystals were grown by the conventional Bridgman technique using MF_2 and ErF_3 powders with high purity (4N) in a graphite crucible. As an example of an as-grown crystal, the Er^{3+}:SrF_2 boule crystal is shown in Fig.1(a). The as-grown crystals were of high optical quality without cracks and inclusions and had a slight rose coloration, Fig. 1(b).

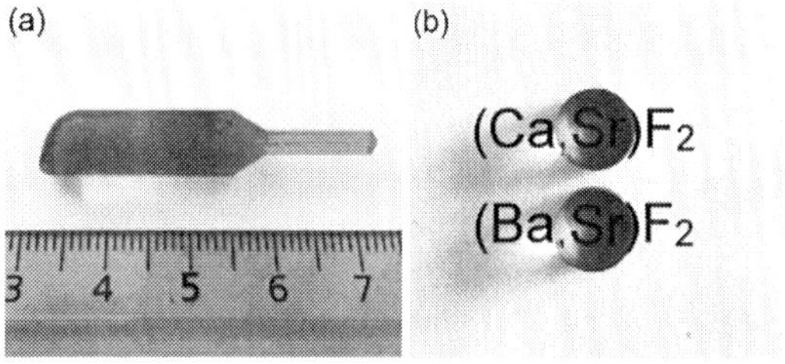

Fig. 1. (a) A photograph of an as-grown Er^{3+}:SrF_2 crystal boule; (b) a photograph of laser-quality polished Er^{3+}:$(M_1,M_2)F_2$ crystals

3 Optical spectroscopy

3.1. Raman spectra

The vibronic properties of the fluorite-type crystals were studied by Raman spectroscopy, Fig. 2. The Raman spectra were measured with an excitation of 514 nm. The factor group analysis predicts the following irreducible representations for the optical and acoustical modes at the Γ-point ($\boldsymbol{k}$ = 0): $\Gamma_{total} = 1F_{2g} + 2F_u$, of which 1 mode ($F_g$) is Raman active and 2 modes (F_u) are IR-active [5]. For the 'parent' fluoride crystals, MF_2, the single modes are observed at 242, 285 and 321 cm^{-1} for BaF_2, SrF_2 and CaF_2, respectively. Raman spectra of the 'mixed' fluorides, $(M_1,M_2)F_2$ are significantly broadened as compared to that of the 'parent' ones. Their Raman peaks are located at 285 and 321 cm^{-1} (for (Ba,Sr)F_2 and (Ca,Sr)F_2, respectively) occupying the intermediate positions between those for 'parent' ones.

* Corresponding author: liza.basyrova@ensicaen.fr

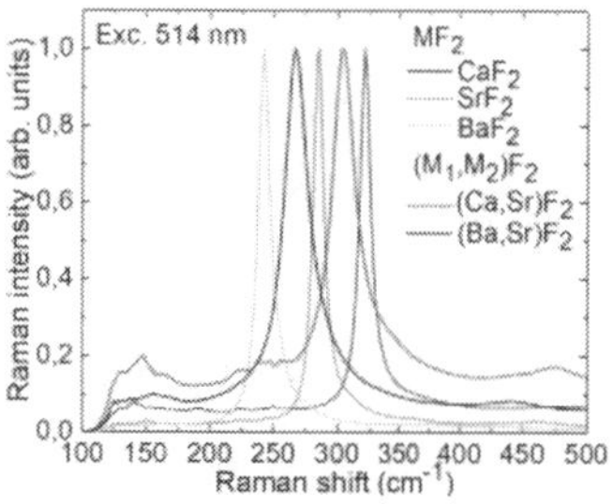

Fig. 2. Unpolarized Raman spectra of the MF_2 and $(M_1,M_2)F_2$ crystals, λ_{exc}= 514 nm

3.2 Absorption

The absorption spectra of the $Er:(M_1,M_2)F_2$ and $Er:SrF_2$ crystals corresponding to the $^4I_{15/2} \rightarrow {}^4I_{11/2}$ Er^{3+} transition are shown in Fig. 3. The spectra of 'mixed' fluorides are slightly broader than that of $Er:SrF_2$, e.g., for $Er:(Ca,Sr)F_2$, the peak absorption is at 969.8 nm and the absorption bandwidth is 16 nm, which is attractive for pumping of these crystals using commercial InGaAs laser diodes.

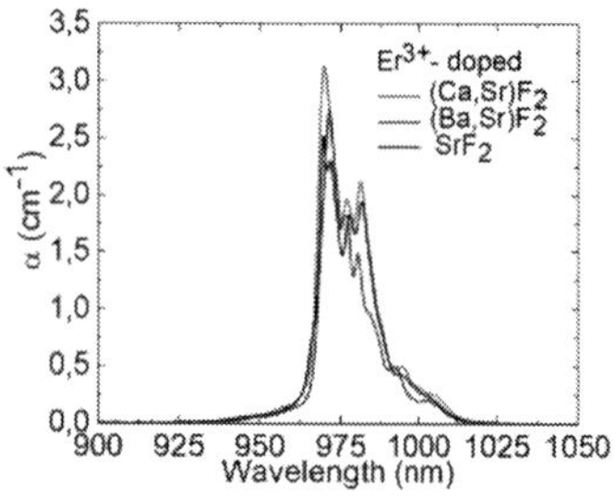

Fig. 3. Absorption spectra for the $^4I_{15/2} \rightarrow {}^4I_{11/2}$ transition of the $Er:(M_1,M_2)F_2$ and $Er:SrF_2$ crystals

3.3 Mid-IR luminescence

The spectra of mid-IR luminescence Er^{3+}-doped $(Ba,Sr)F_2$ and $(Ca,Sr)F_2$ crystals and $Er:SrF_2$ crystal studied for comparison are shown in Fig. 4. The spectra span from 2.5 to 3.05 μm, the luminescence is assigned to the $^4I_{11/2} \rightarrow {}^4I_{13/2}$ Er^{3+} transition. For 'mixed' crystals, the luminescence spectra are slightly broadened as compared to those for the 'parent' compound due to the inhomogeneous spectral broadening arising from the compositional disorder.
The luminescence lifetimes of the upper and lower laser levels, $^4I_{11/2}$ and $^4I_{13/2}$ were measured using finely powdered samples to avoid the effect of reabsorption. For the $Er:(Ca,Sr)F_2$ crystal, they amount to 8.64 ms ($^4I_{11/2}$) and 5.64 ms ($^4I_{13/2}$), and for the $Er:(Ba,Sr)F_2$ crystal, 8.72 ms and 6.32 ms, respectively. In contrast, the lifetime ratio for the $Er:SrF_2$ crystal is 8.96 and 8.87 ms, respectively. Significantly reduced lower laser level lifetime in 'mixed' fluorides allows to achieve low laser thresholds, as well as overcoming the "bottleneck" effect inherent to the $^4I_{11/2} \rightarrow {}^4I_{13/2}$ Er^{3+} transition thus leading to continuous-wave laser operation at 2.8 μm.

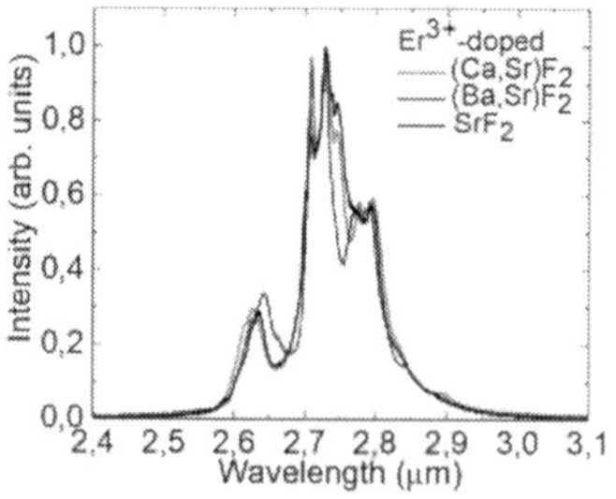

Fig. 4. Mid-infrared luminescence spectra for the $^4I_{11/2} \rightarrow {}^4I_{13/2}$ transition, λ_{exc} = 970 nm

4 Conclusion

To conclude, we report on a comparative spectroscopic study of two Er^{3+}-doped fluorite-type $(M_1,M_2)F_2$ crystals (M_1 = Ca, Ba; M_2 = Sr). They provide inhomogeneously broadened emission bands in the mid-infrared and a favorable ratio of the $^4I_{11/2}$ and $^4I_{13/2}$ level lifetimes making them attractive for 2.8 μm lasers. In particular, $Er:(Ca,Sr)F_2$ appears promising because of its high thermal conductivity compared to CaF_2 and SrF_2 crystals, low phonon energies and long luminescence lifetimes of Er^{3+} ions. Further work will focus on demonstration of laser performances in $Er:(M_1,M_2)F_2$ crystals.

5 Acknowledgement

Agence Nationale de la Recherche (ANR-19-CE08-0028); Région Normandie (Chaire d'excellence "RELANCE").

References

1. V. Petit, P. Camy, J.L. Doualan, X. Portier, R. Moncorgé, Phys. Rev. B. **78**, 8 (2008).
2. K. Veselský, J. Šulc, H. Jelínková, M.E. Doroshenko, V.A. Konyushkin, A.N. Nakladov, Laser Phys. Lett. **17**, 2 (2020).
3. J. Liu, X. Feng, X. Fan, Z. Zhang, B. Zhang, J. Liu, L. Su, Opt. Lett. **43**, (2018).
4. L. Basyrova, P. Loiko, J.-L. Doualan, A. Benayad, A. Braud, B. Viana, P. Camy, Opt. Express **30**, 5 (2022).
5. Bezuidenhout, D. F. *Handbook of optical constants of solids* (Academic Press, 1997).

 https://doi.org/10.1051/epjconf/202226606002

Simultaneous oscillation of dual optical parametric oscillators on monolithic chi(2) nonlinear photonic crystals

Kai-Hsun Chang[1,2,*], *Chia-Chun* Fan[1], *Ming-Shung* Tsai[1], *Fu-Hsiang* Yang[1], *Safia* Mohand Ousaid[2], *Azzedine* Boudrioua[2], and *Lung-Han* Peng[1]

[1]Graduate Institute of Photonics and Optoelectronics, National Taiwan University, Taipei 106, Taiwan, R.O.C.
[2]Laboratoire de Physique de Lasers CNRS UMR 7538, Université Paris 13, Sorbonne Paris Cité,Villetaneuse, France

Abstract. Simultaneous oscillation of two pairs of signals and idlers was demonstrated. We also showed the efficiency of the sum frequency generation of two idlers was comparable with second harmonic generation of each idler. Our design shows the potential of building multi-wavelength laser source from UV to NIR on monolithic crystal.

Introduction

The quasi-phase matching (QPM) technique of nonlinear photonic crystals (NPCs) has great potential to build frequency conversion lasers [1]. Wavelength-tunable lasers based upon parametric down conversion and optical parametric oscillation (OPO) have been reported by using nonlinear optical (NLO) crystals and QPM NPCs [2, 3]. Further, in the regime of multi-wavelengths OPO, Walter et al. had studied the spectral dynamic behaviours on apodized aperiodically poled lithium niobate [4]. However, the latter spectra exhibited unequal power distribution because of gain competition and the cascaded effect in increasing the threshold for the second OPO. To extend the spectral coverage of multi-OPOs, we proposed a simple design of QPM architectures to facilitate multi wavelengths in a linear cavity.

In this paper, we demonstrated the design, and the optical characterization of weaving-like QPM NPCs which can facilitate dual-OPOs in parallel and cascaded NLO processes of second-harmonic generation (SHG) and sum frequency generation (SFG). To deliver our idea, we assumed a confocal cavity and a NPC with two different QPM periodicities transverse to the beam propagation direction, as shown in Fig.1 We also assumed two pairs of (signal, idler) waves (λ_{s1},λ_{i1}) and (λ_{s2},λ_{i2}) which are quasi-matched to the OPOs with QPM periodicity of Λ_1 and Λ_2, respectively. According to the ray optics, the oscillation of (λ_s,λ_i) would follow the optical path marked in Fg.1. As the oscillatory waves propagated along the path of point 1 to 2, they would encounter the QPM period, Λ_1, from which the QPM condition would render the generation of (λ_{s1}, λ_{i1}). Then, from the optical path of point 2 to 3, the NLO waves would encounter both Λ_1 and Λ_2 with a tilt angle, so the phase-matching condition would be lost. Similarly, from the optical path from point 3 to 4, only λ_{s2} and λ_{i2} are phase matched. From the optical path of point 4 to 1, the QPM condition is not fulfilled. As a result, each of these two OPOs will maintain its NLO gain in the respective optical paths, thus to avoid gain competition of them.

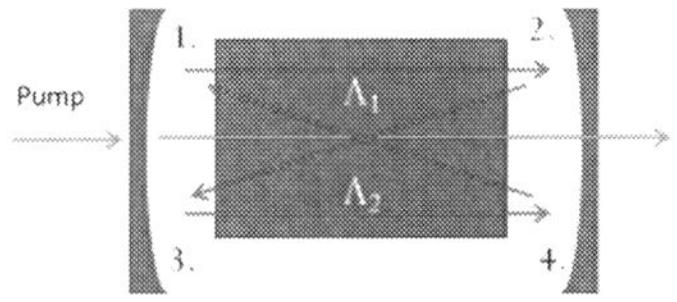

Fig. 1 the design pattern and setup of parallel OPOs

Experimental setup and results

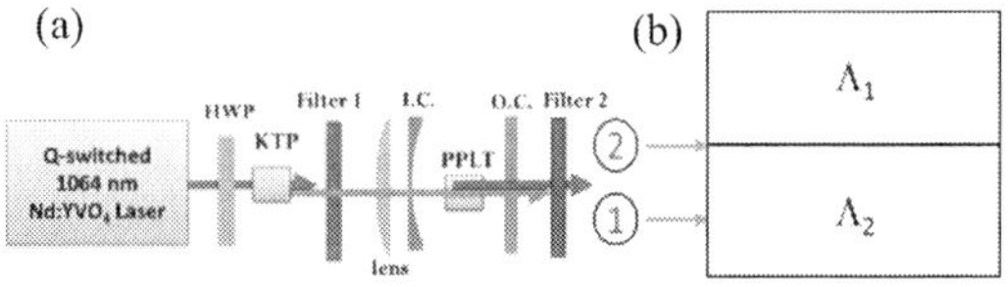

Fig. 2 (a) Experiment setup, filter 1 was to filter out the residual 1064 nm and filter 2 was to filter out the residual 532 nm (b) input beam and its entering position in the experiment

Figure 2(a) showed our experimental setup, a Nd:YVO4 laser with repetition rate of 500 Hz, pulse width of 7 ns and wavelength of 1064 nm. A half-wave plate (HWP) was used to control the polarization of the 1064 nm. A SHG- crystal of KTP was used to convert 1064 nm into 532 nm. The 532nm pump beam was focused to a planar-concave OPO cavity with beam waist radius of 200 μm. The input coupler (I.C.) had high transmittance for pump and signal but high reflectivity for idler. And the output coupler (O.C.) had high transmittance for pump and idler but high reflectivity for signal. Schematically drawn in Fig. 2(b) was the dual-OPO structure, where we chose Λ_1=7.67 μm as and Λ_2=7.66

* Corresponding author: d06941006@ntu.edu.tw

μm. At QPM temperature T=140 °C, their corresponding (signal, idler) wavelengths were (972 nm, 1175 nm) and (980 nm, 1163 nm), respectively. Figure 2 (b) also illustrated a way to enable the operation of either a single- or a dual-OPO by aligning the 532nm pump beam to a proper QPM architecture in the transverse plane. Note that the beam diameter (400μm) was much smaller than the width (~3mm) of each OPO section. So, the setup was a single OPO at position ① but a dual OPO at position ②. In Fig.3 (a), we measured the optical power of the idler waves at these two positions. The corresponding thresholds were about 16 MW/cm^2 for the single OPO and about 28 MW/cm^2 for the parallel OPO, whereas their slope efficiencies were 12.9 % and 12.5 %, respectively. In Fig. 3 (b) and (c) we illustrate the idler wave spectra corresponding to the operation of single and parallel OPO. It can be found that in Fig 3 (c), the two idler waves can exhibit almost an equal optical power distribution while the respective spectral width (~5 nm) was narrower than the case of single OPO. The latter was due to a higher threshold in the operation of dual-OPO, which in turn reduces the NLO gain at the same pump power.

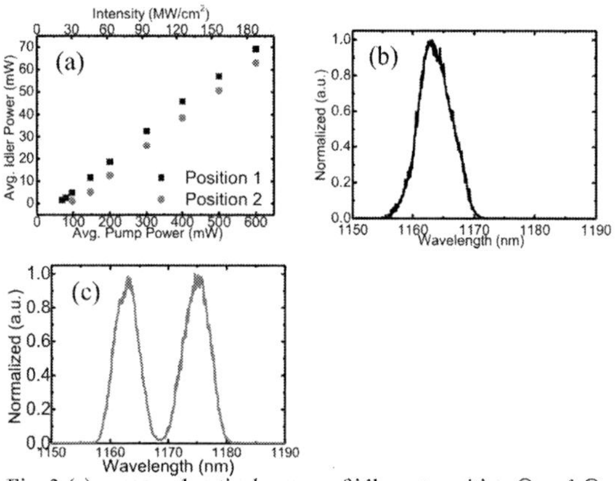

Fig. 3 (a) measured optical power of idlers at position ① and ②, spectra of idler at (b) position ① and (c) position ②

To confirm the spatial mode overlap in the (signal, idler) waves generated from the parallel OPOs, we consider an up-converted structure of QPM periodicity Λ= 10.05μm after the OPO sections in Fig. 4 (a). The formal was chosen to facilitate SFG of two idlers from Λ_1 and Λ_2 at 140 °C and SHG of the idler from Λ_1 and Λ_2 at 137.5 and 142.5°C. As shown in Fig. 4 (b), the respective temperature tuning curves of the idler waves for QPM periodicities Λ_1 and Λ_2 are shown to intersect with the up-conversion curve of Λ. We denote the intersection points marked with solid orange circle as the SHG processes, whereas the open orange circle as the SFG of idler 1 and idler 2. To confirm such NLO design, we changed the output coupling mirror such that idlers were locked inside the OPO cavity. Fig. 4 (c) showed the measured power of intra-cavity up-converter at 137.6 °C and 140 °C. The thresholds at two temperatures were near 100 MW/cm^2. The slope efficiencies were 2% at 137.6 °C and 3% at 140 °C. The experimental results showed that even with parallel OPO design, the two idlers could produce sum frequency generation efficiently.

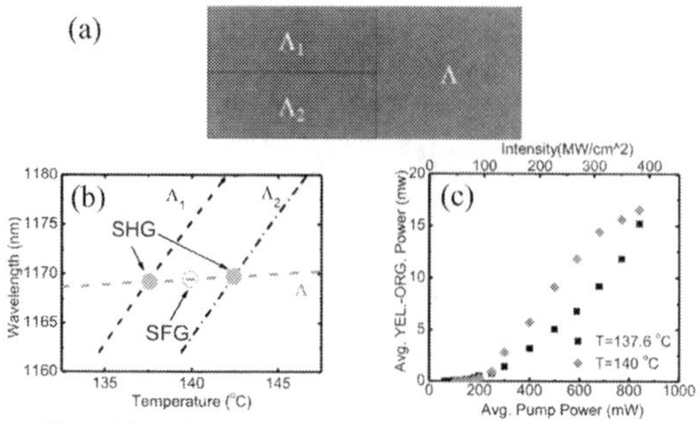

Fig. 4 (a) design pattern with additional cascaded periodicity, Λ (b) phase matching diagram of Λ_1, Λ_2 and Λ versus T (c) experimental results of intra-cavity up converter at two different temperature

Conclusion

Dual-wavelength OPO was demonstrated with periodicities arranged in parallel. Compared with a single OPO, the parallel OPOs had almost same slope efficiency but twice as large as in threshold intensity. The measured spectra and power showed that it had exactly two OPOs which were oscillated in the cavity. Furthermore, with additional cascaded periodicity behind the design, we showed that the sum frequency generation of two idlers was comparable to the second harmonic generation of each idler. This confirmed that the spatial distribution of two OPOs could be overlapped effectively in the cavity. Since our design was simply a combination of two periodicities in parallel, it would preserve the advantage of a single OPO such as wavelength tuning with temperature. Further, our design showed the potential to facilitate multi-wavelength laser from NIR to UV band on monolithic crystal.

References

1. J. A. Armstrong, N. Bloembergen, J. Ducuing, and P. S. Pershan, Phys. Rev. **127**, 1918 (1962).
2. S. E. Harris, *Proceedings of the IEEE*, **57**, no. 12, 2096-2113 (1969)
3. L. E. Myers, R. C. Eckardt, M. M. Fejer, R. L. Byer, W. R. Bosenberg, and J. W. Pierce, J. Opt. Soc. Am. B **12**, 2102-2116 (1995)
4. G. Walter, D. Descloux, J.-B. Dherbecourt, J.-M. Melkonian, M. Raybaut, C. Drag, and A. Godard, J. Opt. Soc. Am. B **37**, 552-563 (2020)

Flexible all-glass planar structured fabricated by RF-sputtering

Alice Carlotto[a,b], *Osman* Sayginer[c], *Anna* Szczurek[a], *Lam* T. N. Tran[a,d,e], *Rossana* Dell'Anna[f], *Stefano* Varas[a], *Bartosz* Babiarczuk[g], *Justyna* Krzak[g], *Oreste* S. Bursi[h,a], *Daniele* Zonta[h,a], *Anna* Lukowiak[i], *Giancarlo* C. Righini[l], *Maurizio* Ferrari[a], *Silvia* M. Pietralunga[b], *Alessandro* Chiasera[a,*]

[a]IFN-CNR, CSMFO Lab. and FBK Photonics Unit, Via alla Cascata 56/C, Povo, 38123 Trento, Italy
[b]IFN-CNR, P.zza Leonardo da Vinci 32, 20133 Milan, Italy
[c]Chair of Biological Imaging and TranslaTUM, Technische Universität München, Ismaninger Str. 22, D-81675 Munich, Germany
[d]Dept. of Physics, Politecnico di Milano, P.zza L. da Vinci 32, 20133 Milan, Italy
[e]Dept. of Materials Technology, Faculty of Applied Sciences, Ho Chi Minh City University of Technology and Education, Vo Van Ngan Str. 1, Thu Duc District, 720214 Ho Chi Minh City, Vietnam
[f]Center for Sensors and Devices, Micro Nano Facility, Fondazione Bruno Kessler, Via Sommarive 18, Povo, 38123, Trento, Italy
[g]Dept. of Mechanics, Materials and Biomedical Engineering, Wroclaw University of Science and Technology, Smoluchowskiego 25, 50-370 Wroclaw, Poland
[h]Dept. of Civil, Environmental and Mechanical Engineering, University of Trento, via Mesiano 77, 38123 Trento, Italy
[i]Institute of Low Temperature and Structure Research, PAS, ul. Okólna 2, 50422, Wroclaw, Poland
[l]IFAC-CNR, MiPLab, Via Madonna del Piano 10, 50019 Sesto Fiorentino, Italy, France

Abstract. Flexible SiO_2/HfO_2 1D photonic crystals and active SiO_2–HfO_2:Er^{3+} all-glass flexible planar waveguides fabricated by radio frequency sputtering, are presented. The 1D photonic crystals show a strong dependence of the optical features on the light incident angle: i) blue-shift of the stopband and ii) narrowing of the reflectance window. Nevertheless, the most interesting result is the experimental evidence that, even after the 1D photonic crystals breakage, where the flexible glass shows naked-eye visible cracks, the multilayer structures generally maintain their integrity, resulting to be promising systems for flexible photonic applications thanks to their optical, thermal and mechanical stability. The flexible planar waveguides, fabricated on ultrathin flexible glass substrate, showed an attenuation coefficient lower than 0.2 dB/cm at 1.54 μm, and exibits emission in the NIR region, resulting particularly suitable as waveguide amplifier in the C band of telecommunications.

1 INTRODUCTION

The benefits obtained in terms of costs and applicability by the development of flexible and stretchable electronics, compared to their rigid counterparts, have fostered the birth of the idea of the photonics analogue. A further step in the development of this kind of systems passes through the all-glass flexible photonics, thus exploiting the typical properties of inorganic glasses (transparency, high thermal and chemical resistance, etc.) to produce flexible systems[1,2]. Here, flexible SiO_2/HfO_2 1D photonic crystals and active SiO_2–HfO_2:Er^{3+} all-glass flexible planar waveguides fabricated by radio frequency sputtering, are presented.

2 Flexible SiO2/HfO2 1D photonic crystals

The photonic crystals, constituted by 41 alternating layers (20 bilayers + 1) of SiO_2 and HfO_2, were deposited on silicon, vitreous silica slides (v-SiO2, 75mm×25mm×1mm) and ultrathin AS 87 eco SCHOTT glass (75mm×25mm×0.175mm) by RF magnetron sputtering. [2]

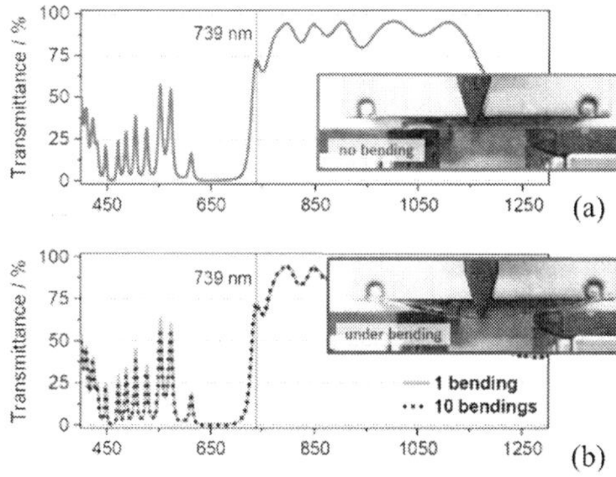

Fig. 1. Picture of the 1D photonic crystal on bending setup a) before and b) during the bending tests (Δd = 2.0 mm). Transmittance spectra of the sample, a) before (grey solid line) and b) after 1 (light blue solid line) and 10 bendings (dark blue dotted line) at a deflection equal to 2.0 mm.

* Corresponding author: alessandro.chiasera@cnr.it

https://doi.org/10.1051/epjconf/202226606003

The photonic crystals deposited on ultrathin SCHOTT glass was subjected to 10 banding cycles using the homemade 3-point bending setup at a deflection (Δd) of 2.0 mm, with a loading speed equal to 2.0 mm/min (Figure 1a and 1b). The mechanical deformation, i.e. bending, does not influence the transmittance properties of the dichroic mirror (Figure 1a and 1b), in fact, both position and intensity of the transmittance spectra are almost perfectly superimposable before and after 10 consecutive bending cycles.

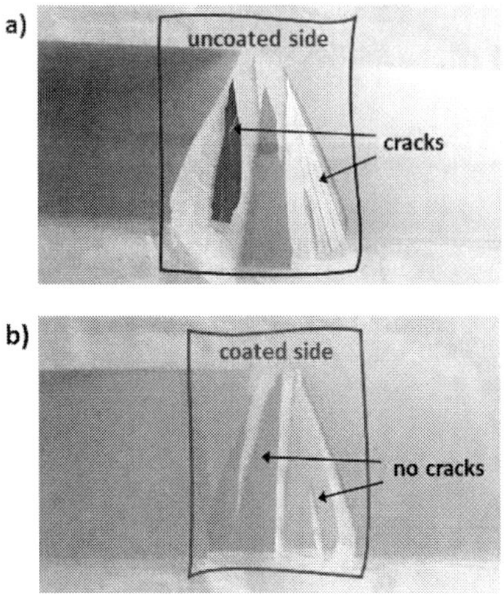

Fig. 2. Illustration photos of 1D photonic crystal after the breakage, showing a) uncoated and b) coated side of AS 87 eco ultrathin SCHOTT glass.

Even where the substrate was intentionally cracked (uncoated side, Figure 2a), the 1D photonic crystal (coated side) can still maintain its integrity (Figure 2b). For this reason, we also exploited Atomic Force Microscopy (AFM) to investigate the sample surface morphology after the ultrathin glass breakage.

3 Active SiO_2–HfO_2:Er^{3+} all-glass flexible planar waveguides

The active SiO_2–HfO_2:Er^{3+} flexible planar waveguide was fabricated via RF-sputtering technique on both ultrathin glass (AS 87 eco SCHOTT glass) and silicon in order to characterize the structure also from the morphological and compositional point of view.

The system, fabricated on ultrathin flexible glass substrate, showed an attenuation coefficient lower than 0.2 dB/cm at 1.54 μm, and the losses do not change after bending the planar waveguide.

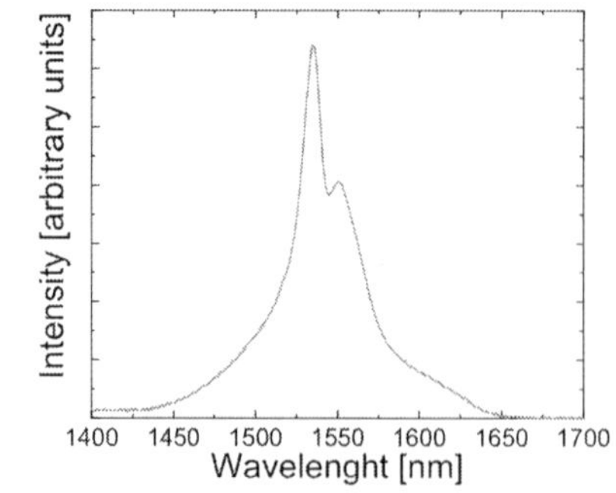

Fig. 3. Normalized photoluminescence spectra obtained exciting at 514.5 nm in prism coupling configuration the flexible SiO_2–HfO_2:Er^{3+} planar waveguide.

To confirm the fabrication of an active waveguide the photoluminescence (PL) properties of Er^{3+} were characterized. In Figure 4, PL emission spectrum in the NIR region is reported. The shape of the NIR PL spectra indicate as the Er^{3+} ions are embedded in a amorphous environment. Further thermal treatment at 400°C does not significantly influence the Er^{3+} spectral features.

This research is supported by the projects CNR-PAS "Flexible Photonics" (2020-2022); the Polish National Agency for Academic Exchange (NAWA) grant no. PPN/IWA/2018/1/00104, Italian Ministry of Education, University and Research (MIUR) in the frame of the 'Departments of Excellence' (grant L 232/2016), ERC-H2020 PAIDEIA GA 816313, NAWA-MAECI Canaletto (2022-2023) and FESR-PON 2014-2020 BEST4U – ARS01_00519.

References

1. L.T.N. Tran, A. Szczurek, S. Varas, C. Armellini, F. Scotognella, A. Chiasera, M. Ferrari, G.C. Righini, A. Lukowiak,Opt. Mat. **124**, 111978 (2022)
2. A. Carlotto, O. Sayginer.c, H. Chen, L. T. N. Tran, R. Dell'Anna, A. Szczurek, S. Varas, B. Babiarczuk, J. Krzak, O. S. Bursi, D. Zonta, A. Lukowiak, S. M. Pietralunga, G. C. Righini, M. Ferrari, A. Chiasera SPIE **2022**, 12142 (2022)

Thermal effects reduction in a diode side-pumped rod-like bonded α-quartz||Nd:glass||α-quartz amplifying medium

Thomas Dubé[1,*], *Patrice* Camy[2], and *Sébastien* Montant[1]

[1]CEA CESTA, 33114 Le Barp, France
[2]CIMAP, ENSICAEN, UMR 6252, 6 boulevard du Maréchal Juin, 14050 Caen Cédex, France

Abstract. We present a numerical study of a square-shape rod-like bonded laser gain medium operating at 1053 nm. The sample is composed of a 5-mm thick Nd-doped phosphate glass bonded to two α-quartz crystals. Owing to this face cooling scheme, the heat generated in the gain medium is effectively reduced, hence resulting in less optical distortions. The simulation of the thermomechanical effects was conducted using COMSOL MultiPhysics® software. Transverse wavefront distortion and birefringence profiles were computed for a given pump mean power. Wavefront and birefringence measurements will be conducted on the bonded sample using a pump-probe setup which is discussed in this paper. These measurements will be presented and weighed with the numerical results.

1 Context of the study

High-power laser facilities, such as the National Ignition Facility (NIF, United States) and the Laser MegaJoule (LMJ, France), rely on Nd-doped phosphate glass for amplification of each laser beam at 1053 nm. The combination of the flash-lamp pumping scheme with the low thermal conductivity of the laser glass leads to strong thermal effects in the amplify medium, resulting in wavefront distortion and stress-induced birefringence [1]. In particular, the large-aperture glass rod of the multi-pass amplifier stage in the LMJ front-end limits its overall performance down to 1 J every 5 minutes.

We investigate the application of bonding to reduce the thermal effects generated in the gain medium. Face-cooling scheme has been demonstrated for heat reduction purposes using sapphire and diamond crystals bonded to Nd:YAG crystals [2,3]. The present work focuses on the reduction of the thermal load in a diode-pumped 5-mm thick LG770 laser glass bonded to two α-quartz crystals.

In what follows, we describe the numerical model conducted using COMSOL MultiPhysics® and the wavefront and birefringence characterisation setup.

2 Numerical model

2.1 Formulation of the problem

The absorbed pump energy contributes to the amplification and the heat source $Q(x, y, z, t)$ due to non-radiative transitions. We assume a temperature continuity at each area of contact between the glass and each quartz crystal. The heat generated during pumping propagates in the gain medium ($i = A$) and the heatsinks ($i = HS$). The Fourier heat equation is expressed as [1,4]:

$$\rho_i C_{p,i} \frac{\partial T_i}{\partial t} - K_{th,i} \nabla T_i = Q_i, \qquad i = A, HS \qquad (1)$$

Where T, ρ, C_p and K_{th} are respectively the temperature the mass density, the specific heat and the thermal conductivity. Q_i is zero in the crystals (HS).

Stress occurs due to local thermal expansion and compression of the material sustaining the temperature rise. Using Hooke's law, mechanical stress is derivated from the temperature field. Temperature and mechanical stress increments give rise to a local change of the refractive index. The refractive index tensor in an isotropic material (e.g. the laser glass) under thermal stress writes [1,4]:

$$\bar{\bar{n}} = \left[n_0 + \frac{dn}{dT}(T - T_0)\right]\bar{\bar{I}} - \frac{n_0^3}{2}\overline{\overline{\Delta B}} \qquad (2)$$

Where n_0 is the unperturbed refractive index at 1053 nm, $dn/_{dT}$ is the thermo-optic coefficient, T_0 is the cooling temperature, $\bar{\bar{I}}$ is the identity matrix and $\overline{\overline{\Delta B}}$ is the second-rank impermeability tensor altered by the photo-elastic effect. It is derivated from the tensor product of the material-dependant piezo-optic tensor with the computed mechanical stress tensor [1]. This expression enables a numerical evaluation of both birefringence and optical path difference (OPD) during pumping.

2.2 Numerical model

We perform a finite element study with COMSOL MultiPhysics® to consecutively solve (i) the heat equation and (ii) the mechanical stresses in the amplifying medium before (iii) evaluating the refractive index.

* Corresponding author: thomas.dube@cea.fr

 https://doi.org/10.1051/epjconf/202226606004

Afterwards, (iv) a wavefront calculation is performed using the ray optics module a single pass through the rod.

The 5-mm thick gain medium is made of Nd-doped phosphate LG770 glass ($5 \times 10 \times 10\ mm^3$). Each transverse surface is bonded to a $\vec{c}$-oriented α-quartz crystal ($10 \times 10 \times 10\ mm^3$) as illustrated in Fig. 1. Optical quartz benefits from a higher thermal conductivity (K_{th}) than LG-770 while matching both Coefficient of Thermal Expansion (CTE) of the glass in the transverse plane (perpendicular to $\vec{c}$-axis) as described in Table 1. The gain medium is side-pumped by diodes emitting at 800 nm with a total mean pump power of 2 W.

Table 1. Key parameters of both LG-770 (supplied by Schott) and α-quartz (supplied by Sawyer Technical materials).

Materials	K_{th} (W.m^{-1}.K^{-1})	CTE (10^{-6} K^{-1})	n_0
LG770	0.57	13.36	1.499
α-quartz ⊥ $\vec{c}$	6.21	13.37	$n_o = 1.534$
α-quartz // $\vec{c}$	10.7	7.97	$n_e = 1.543$

For comparison purposes, we perform the same simulations on two control samples: a single 5-mm thick LG-770 piece and a 25-mm long LG-770 rod with equivalent transverse dimensions. Each sample is illustrated alongside first numerical results in Fig. 1.

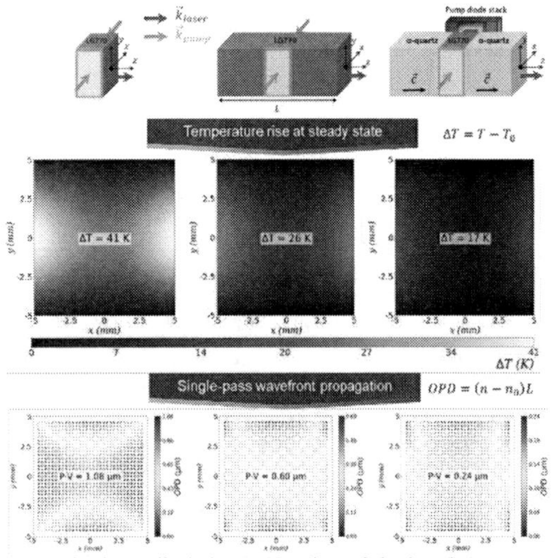

Fig. 1: (Color online) (Top) Sketches of the bonded structure and the control samles. (Middle) Temperature profiles at the center of each gain medium at steady-state. (Bottom) Resulting distorted wavefront after a single-pass.

Owing to the face-cooling scheme, the temperature in the gain medium is respectively decreased by 63% and 35% against both 5-mm and 25-mm long LG-770 rod as shown in Fig 1. (middle). The bottom profiles in Fig. 1 highlight a flattened wavefront with a 5-times lower OPD (left) as compared to the single LG-770 piece (right).

3 Characterization of the bonded sample

The samples are prepared by Onyx Optics Inc. (United States) and Cristal Laser (France) using different methods. The former uses its Adhesive-Free Bonding® method [5] and the latter applies a transparent adhesive layer between the materials [6]. A single-pass pump-probe setup for our measurements is described on Fig. 2.

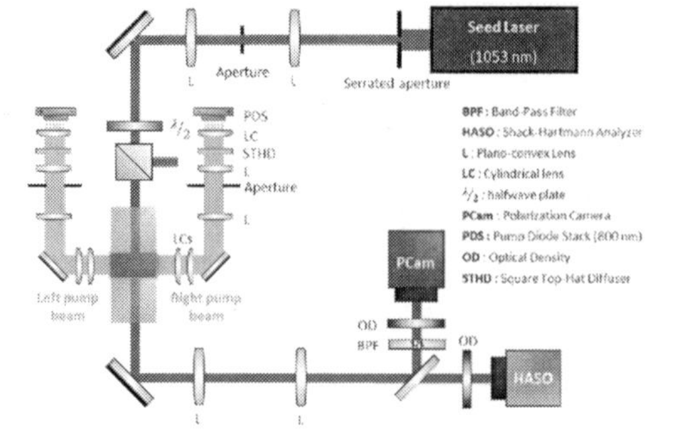

Fig. 2: (Color online) Sketch of the experimental setup. The probe beam is illustrated in red and the pump ones in green.

The sample is maintained with water-cooled copper clamping jaws. We use a regenerative cavity delivering 10 mJ at 1053 nm at 1 Hz repetition rate as the seed laser. The probe beam is shaped into a 8-mm large square profile using a serrated aperture. Then, the beam is transported at the centre of the sample with the help of an afocal system. The medium is pumped on each side by a diode stack using a spatial beam shaping system, which enables the control of both beam size and homogeneity. Its principle hinges on imaging the far field of the pump profile diffused by a square top-hat diffuser (STHD). Both wavefront distortion and birefringence are measured with the help of a Shack-Hartmann wavefront sensor (HASO) and a polarization camera (PCam). Finally, we plan to add a thermal imager for live temperature monitoring.

4 Conclusion

We present a bonded Nd-doped glass||α-quartz rod-like gain structure. First, we conducted a numerical study of both temperature rise and thermal stress in the sample. Then, the wavefront distortion after a single pass is estimated by computing the refractive index variations under thermal stress. The pump-probe characterization setup is also presented. Wavefront and birefringence measurements will be compared to our model.

References

1. W. Koechner, *Solid-State Lasers*, Springer (2003).
2. H. Ichikawa, K. Yamaguchi, T. Katsumata, I. Shoji, Opt. Exp., **25**, 119 (2017).
3. L. Zheng, A. Kausas, T. Taira, Opt. Exp., **7**, 19, (2017).
4. C. Bernerd, B. Da Costa Fernandes, E. Boursier, J. Luce, C. Maunier, A. Braud, C. Meroni, P. Camy, S. Montant, App. Opt., **59**, 31 (2020).
5. H. Lee, O. Meissner, H. Meissner, *Characteristics of AFB interfaces of dissimilar crystal composites as components for solid state lasers,* SPIE Proc. (2005).
6. G. Sraïki, *Sol-Gel method for non-linear optical cristals bonding*, Thesis (2015).

 https://doi.org/10.1051/epjconf/202226606005

DESIGN AND INVESTIGATION OF A LOW-THRESHOLD ORGANIC LASER DIODE USING MIXED-ORDER DFB CAVITIES

Yara El Droubi*, *Quentin* Gaimard, *Jeanne* Solard, *Mahmoud* Chakaroun, and *Azzedine* Boudrioua

Laboratoire Physique des lasers–CNRS UMR 7538, Université Sorbonne Paris Nord, 93430 Villetaneuse, France

Abstract. In this work, we experimentally and theoretically investigate the optical and electrical optimization of an OLED associated to a mixed-order DFB cavity. We, firstly focus on the design and the fabrication of a mixed-order DFB cavity with a high quality factor. We particularly study the impact of the deposition of the organic layers on the topology and the quality factor of the cavity.

1 Introduction

Organic semiconducting materials have a growing interest in the field of photonics. Thanks to their outstanding features in terms of low-cost fabrication, easy processing, chemical versatility, and wavelength tunability across the entire visible range[1], significant research efforts have been made in the last two decades to use them in optically and electrically pumped solid-state laser. However, one of the most important challenge in the field still the realization of organic laser diodes under electrical pumping. Besides, the race actually concerns the development of organic laser diode (OLD) based the utilization of organic light-emitting diode (OLED) with a resonant cavity[3]. For instance, several reports have shown significant interest in the study of new molecules with a high net gain[2]. Sub-nanosecond pulse pumping seems to be emerging as an alternative to avoid the accumulation of non-radiative triplet excitons[4]. Moreover, the realization of a resonant cavity with a high quality factor (Q-factor) compatible with the OLED structure and the electrical pumping scheme is very challenging[6].

Regarding the gain of the standard organic molecules, recent studies predict that the laser threshold would be reached in cavities with Q-factors of the order of a few thousand[5]. Today, the architecture, which has made it possible to have the lowest laser threshold under optical pumping (35nj/cm²) is the mixed-order distributed feedback cavity (DFB)[7]. It allows a strong coupling between the light and the amplifying medium, while providing a large surface area available for radiative recombination. This mixed-order DFB cavity consists of two gratings: the first-order grating allows the confinement of light inside the cavity and the second-order grating is used for extracting light *via* a diffractive mode, orthogonal to the cavity. Recent work has shown a laser effect under electrical pumping by using a promising organic molecule (BSBCz), with a threshold of 600A/cm²[7]. Nevertheless, several questions are still to be considered. In particular, the optimal geometry of such a cavity (topology and dimension of the cavity) and the Q-factor which, could be actually obtained and the optimization of the grating coupling coefficient κ[3] for a homogeneous distribution of the field in the cavity[8]. In addition to that, the deposition of the OLED hetero-structure induces a tilting of the sides of the gratings. Thus, it is important to study the impact of this topological modification on the coupling coefficient.

In this work, we report the theoretical and experimental study of the optical and electrical optimization of an OLED associated to a mixed-order DFB (Distributed FeedBack) cavity. We, particularly, focus on the optimization of the optical resonator in order to obtain the most suitable cavity for our applications.

2 Results and discussion

In this context, we, firstly, performed several simulations by using the FDTD method. For example, our results give us a first approximation of the most optimal cavity length for a homogeneous field distribution. We calculate the quality factor from the resonance wavelength of the cavity shown in figure 1. It gives us a Q ~1000 for a cavity length of 20 μm for order 1 on both sides. This result seems comparable with Adachi's work [7].

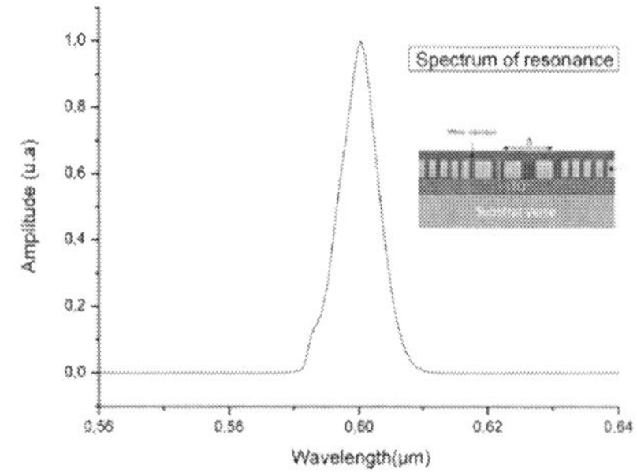

Figure 1: Resonance spectrum of the diffracted wavelength of mixed order DFB cavity using FDTD solution

* Corresponding author: yara.eldroubi@univ-paris13.fr

As an example, the micro-cavity fabricated and studied in this work is shown in Fig 1a. It consists of the implementation of the structure by using electronic lithography in a HSQ resin. The next step will concern the deposition of the organic hetero-structure by thermal evaporation on the top to the obtained DFB structures as displayed on Fig 1b.

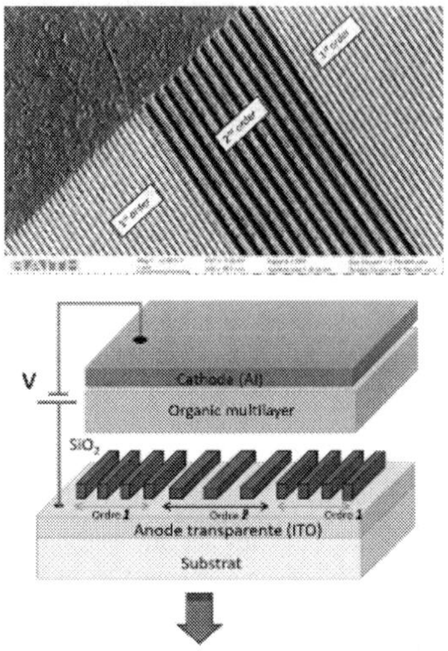

Figure 2: a) SEM image of a mixed-order DFB structure, and b) Schematic representation of the OLED in cavity

Comparing to the mixed-order cavities presented in the literature[6,7], we do not use an etching step which, avoid damaging of the underlying layers. In addition, the use of this process would allow us to substitute the ITO electrode with a graphene layer[9] or a dielectric/metal/dielectric multilayer[10] which is more favorable for the optical confinement of the light in the cavity.

Afterward, we will study the spectral response of our cavities relying on a technique based on a double monochromator experimental setup (Fig 2).

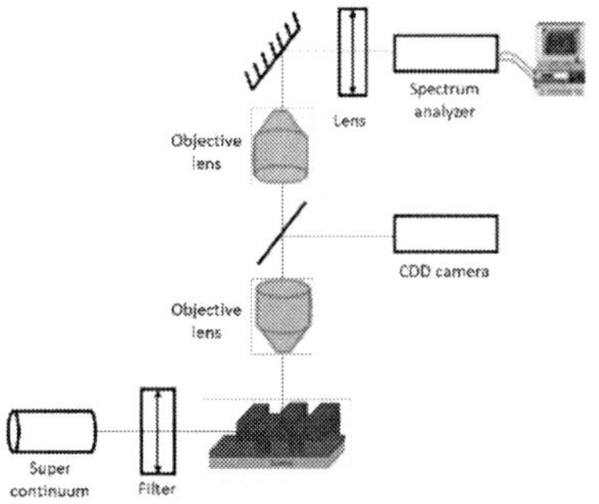

Figure 3 : Experimental setup for the measurement of the spectral response of the cavity

As shown, the spectrum collection system consists of two optical channels, one of which is dedicated to the observation of structures and the other to signal detection. It consists of, pumping our samples with a precise wavelength with a super Continium coupled to a monochromator and, collecting the cavity response by using another monochromator coupled to a CDD camera.

During the conference, we will present the obtained results, compare them to the devices developed by other competing teams[9,12].

3 Conclusion

In this paper, we focus on the optimal design of DFB resonators to be used with organic materials for laser effect. FDTD simulations and the corresponding experimental results will be presented and discussed, in particular through the use of a characterization setup aimed at measuring the spectral response of our cavities.

References

1. Moses, D. High quantum efficiency luminescence from a conducting polymer in solution: A novel polymer laser dye. *Appl. Phys. Lett.* **60**, 3215–3216 (1992).
2. Chenais, S. & Forget, S. Recent advances in Solid-State Organic Lasers. *Polym. Int.* **61**, (2012).
3. Streifer, W., Scifres, D. & Burnham, R. Coupling coefficients for distributed feedback single- and double-heterostructure diode lasers. *IEEE J. Quantum Electron.* **11**, 867–873 (1975).
4. Zeng, L. *et al.* Electrical and Optical Impulse Response of High-Speed Micro-OLEDs under UltraShort Pulse Excitation. *IEEE Trans. Electron Devices* **PP**, 1–7 (2017).
5. Lenstra, D. *et al.* Ultra-short optical pulse generation in micro OLEDs and the perspective of lasing. *J. Opt.* **24**, 034007 (2022).
6. Adachi, C. & Sandanayaka, A. The Leap from Organic Light-Emitting Diodes to Organic Semiconductor Laser Diodes. *CCS Chem.* **2**, 1203–1216 (2020).
7. Sandanayaka, A. S. D. *et al.* Indication of current-injection lasing from an organic semiconductor. *Appl. Phys. Express* **12**, 061010 (2019).
8. Nolting, H.-P. & Ulrich, R. *Integrated Optics: Proceedings of the Third European Conference, ECIO'85, Berlin, Germany, May 6–8, 1985.* (Springer, 2013).
9. Li, N. *et al.* Efficient and bright organic light-emitting diodes on single-layer graphene electrodes. *Nat. Commun.* **4**, 2294 (2013).
10. Cho, H., Yun, C. & Yoo, S. Multilayer transparent electrode for organic light-emitting diodes: Tuning its optical characteristics. *Opt. Express* **18**, 3404–14 (2010).
11. Gourdon, F. *et al.* Optically pumped lasing from organic two-dimensional planar photonic crystal microcavity. *APL: Organic Electronics and Photonics* vol. 5 (2012).
12. Solard, J. Conception et réalisation de cavités à cristaux photoniques à base d'ITO pour des applications dans le domaine de la photonique organique. (2021).

Organophosphorus emitters for OLEDs

Muriel Hissler[1,*], *Pierre-Antoine* Bouit [1], *Denis* Tondelier[2], *Bernard* Geffroy[2]

[1]*Université de Rennes1, CNRS, UMR 6226, Sciences Chimiques de Rennes, Campus de Beaulieu, 35042 Rennes cedex*
[2]*CEA/LPICM, Ecole polytechnique 91128 Palaiseau*

Abstract. In this paper, we present the development of OLEDs using organophosphorus derivatives as emitters. In this study, we were able to show that the structural variations carried out on phospholes, phosphinines and phosphepines (functionalization of the phosphorus atom, nature of the substituents) make it possible to modulate the emission wavelengths and, thus, the emission colour of the diodes. Using this concept, we were able to develop TADF and chiral organophosphorus emitters and hybrids emitting at different wavelengths which can be used for the development of OLEDs. The diode structures used are simple and provide high external quantum yields.

Introduction

The development of technologies based on molecular electronics (field effect transistors (OFETs), light-emitting diodes (OLEDs), solar cells, etc.) has been accompanied by an important development in the synthesis and study of the electronic properties of organic and organometallic compounds with an extended pi-system. [1] Indeed, the development of high-performance "plastic electronic" devices is only possible if the electronic properties and macroscopic organisation of these molecules are controlled. In this perspective, linear oligomers and polymer-conjugates incorporating heteroatoms have received particular attention.[1] Indeed, the introduction of heteroatoms allows the appearance of new geometrical or electronic properties directly related to the nature of the element. The P-rings are used to modify the chemical and physical properties of pi-conjugated systems.[2] The P-rings are also a potential source of chirality that has not been exploited so far in the field of molecular materials. For example, the phosphole has been little studied as a building block for the construction of pi-conjugated systems, although it possesses interesting properties for the construction of pi-conjugated systems[ ] The phosphole possesses (i) a weak aromatic character which should favour electronic delocalisation in a pi-system, and (ii) a reactive heteroatom which offers the possibility of tuning the energetic position of the HO and BV levels by chemical modifications such as the coordination of the phosphorus atom with a metal. These properties indicate that the phosphole may be a good candidate for the construction of extended pi-conjugated systems. However, P-heterocycles are not limited to the phosphole family and during the last few years, other pi-systems based on phosphetenes (4-membered unsaturated P-ring),[3] phosphaphenalenes /phosphinines (6-membered P-ring)[4] or phosphepines (7 membered P-rings)[5] have started to be studied in the context of molecular materials.

P-rings for optoelectronics

Here, we will report on the stereospecific synthesis of helicenoid-based phosphepine as well as a chiral P-containing PAH through an axial to central chirality transfer from readily available 2,2′-bis(diphenylphosphino)-1,1′-binaphthalene (BINAP). [6] In these systems, an axial to central chirality transfer takes place from the BINAP moiety to the P-atom. The impact of the molecular design on the structure, the (chir)optical (including circularly polarized luminescence) and redox properties will be presented.

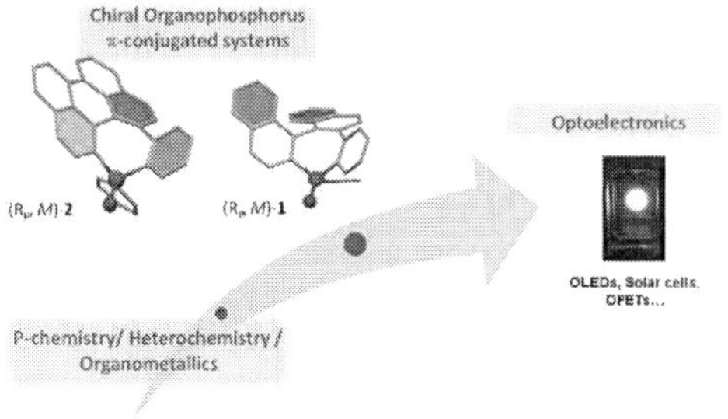

* Corresponding author: author@e-mail.org

We will also present the grafting of a hydroxyoxophosphole-based aggregate induced emitter with a POOH function with ZnO or a chiral cinnabar (-HgS) nanoparticle.[7] The restriction of the intramolecular motions of the organic fluorophore, through either aggregates' formation in solution or processing into thin films, forms highly emissive materials due to a strong aggregation induced emission effect. Theoretical calculations and XPS analyses were performed to analyze the interactions between the organic and inorganic counterparts. Furthermore, this combination allowed us to obtain organic-inorganic hybrid nanoparticles capable of emitting circularly polarised light from achiral fluorophores. The impact of the molecular design on the structure, the (chir)optical and redox properties have been investigated using a joint experimental/theoretical approach, highlighting the potential of these chiral derivatives for further applications in optoelectronics. Preliminary results on the use of these nanohybrids as solution-processed emissive layers in organic light emitting diodes (OLEDs) illustrate their potential for lighting applications.

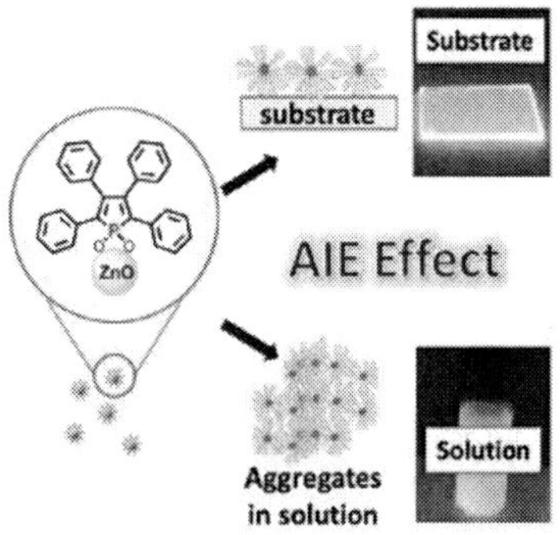

References

1. (a) T. A. Skotheim, R. L. Elsenbaumer, J. R. Reynolds, (2nd edn.) *Handbook of conducting polymers*, Dekker, New York, **1998**. (b) H. S. Nalwa, *Handbook of conductive materials and polymers*. John Wiley and Sons, New York, **1997**. (c) K. Müllen, G. Wegner, (eds.) *Electronic materials: the oligomer approach*. Wiley-VCH, Weinheim, **1998**.
2. (a) D. Joly, D.; P. A. Bouit, M. Hissler, *J. Mater. Chem. C* **4**, 3686 (2016); .(b) M. P. Duffy, W. Delaunay, P.A. Bouit, M. Hissler, *Chem. Soc. Rev.* **45**, 5296 (2016); (c) H. Lauwick, M. P. Duffy, P.A. Bouit, M. Hissler, *Coord. Chem. Rev.* **433**, 213759 (2021),
3. H. Chen, S. Pascal, Z. Wang, P.-A. Bouit, Z. Wang, Y. Zhang, D. Tondelier, B. Geffroy, R. Réau, F. Mathey, Z. Duan, M. Hissler, *Chem. Eur. J.* **20**, 9784 (2014);
4. C. Müller, L. E. E. Broeckx, I. de Krom, J. J. M. Weemers, *Eur. J. Inorg. Chem.* 187 (2013); (b) E. Regulska, C. Romero-Nieto, *Dalton Trans.* **47,** 10344 (2018).
5. a) X. He, J. Borau-Garcia, A. Y. Y. Woo, S. Trudel, T. Baumgartner, *J. Am. Chem. Soc.* **135**, 1137 (**2013**),; (b) Y. Ren, M. Sezen, F. Guo, F. Jäkle, Y.-L. Loo, *Chem. Sci.* **7**, 4211 (2016); (c) W. Winter, *Chem. Ber.* **1976,** 109, 1405-2419.
6. R. Mokrai, A. Mocanu, M. P. Duffy, T. Vives, E. Caytan, V. Dorcet, *T.* Roisnel, *L.* Nyulászi, Z. Benkő, P.-A. Bouit, M. Hissler *Chem Commun* **57**, 7256 (2021)
7. J. Phelipot, N. Ledos, T. Dombray, M. P Duffy, M. Denis, T. Wang, Y. Didane, M. Gaceur, Q. Bao, X. Liu, M.Fahlman, P. Delugas, A.Mattoni, D. Tondelier, B. Geffroy, P.-A. Bouit, O. Margeat, J. Ackermann, M.Hissler, *Advanced Materials Technologies*, 2100876 (2021)

Laser micromachining of diamond: A viable photonic and optofluidic platform

Ottavia Jedrkiewicz[1,*], *Akhil* Kuriakose[1,2], *Argyri N.* Giakoumaki[3], *Giulio* Coccia[3,4], *Monica* Bollani[3], *Roberta* Ramponi[3,4] and *Shane M.* Eaton[3]

[1]Istituto di Fotonica e Nanotecnologie IFN-CNR, Udr di Como, Via Valleggio 11, 22100 Como, Italy
[2]Dipartimento di Scienza e Alta Tecnologia, Università dell'Insubria, Via Valleggio 11, 22100 Como, Italy
[3]Istituto di Fotonica e Nanotecnologie IFN-CNR, Piazza Leonardo Da Vinci 32, 20133 Milano, Italy
[4]Dipartimento di Fisica, Politecnico di Milano, Piazza Leonardo Da Vinci 32, 20133 Milano, Italy

Abstract. We describe how the ultrafast laser micromachining technique applied with different writing methods can be used for the creation of various building blocks essential for the realization of a photonic and optofluidic diamond platform. Waveguides, NV centers, conductive wires, microchannels and microholes can be obtained thanks to laser microfabrication with suitable pulse parameters, making use not only of standard Gaussian laser beams but also of non-diffracting Bessel beams, the latter especially in all those cases where single pass high aspect-ratio microstructures or ablated areas are needed.

1 Introduction

Diamond has attracted great interest as a quantum technology platform thanks to its optically active nitrogen vacancy (NV) center. The NV spin state with the brighter luminescence yield can be exploited for spin readout, exhibiting millisecond spin coherence times at ambient temperature. Since their energy levels are sensitive to external fields, NVs are attractive as a scalable platform for efficient nanoscale resolution sensing and for quantum information systems. Integrated diamond photonics would be beneficial for optical magnetometry, due to the enhanced light–matter interaction and associated collection efficiency provided by waveguides, and for quantum information, by means of the optical linking of NV centers for long-range entanglement. Diamond is also compelling for microfluidic applications due to its outstanding biocompatibility, with sensing functionality provided by NVs. Furthermore, laser written micrographitic modifications could lead to efficient and compact detectors of high energy radiation in diamond. However, it remains a significant challenge to fabricate photonics, NVs, and microfluidics in diamond.

Ultrafast laser micromachining with fs or ps pulses is one of the most efficient and precise techniques for microfabrication of transparent materials thanks to the minimization of heat transfer during the nonlinear absorption process [1]. Here we report the most important fabrication results of quantum emitters, optical waveguides, and microfluidics in diamond, in view of a combination of integrated quantum photonic and microfluidics components that will open the way to lab-on-chip devices enabled with quantum sensing capabilities [2,3].

2 Photonics components in diamond

2.1 Laser writing of waveguides

Femtosecond (fs) laser writing was first applied to glasses to inscribe optical waveguides [4] and has since been extended to other passive and active glasses, polymers, and crystals. Contrarily to glass, in crystals the use of tightly focused pulses leads to a decrease of the refractive index at the focus. In diamond, with suitable laser parameters, one can write two closely spaced modification lines leading to a stressed central region capable of guiding light. We have formed 5 mm long waveguides 50 μm below the surface, having an insertion loss of 6 dB, with a mode field diameter (MFD) of 10 μm at 635-nm wavelength (Fig.1a). Waveguides with larger separation have been laser written for guidance of near-IR, telecom and mid-IR wavelengths.

2.2 Creation of deterministically placed NV centers and NV ensembles

In a different pulse energy regime, fs laser pulses can be used to form single NVs in the bulk of optical and electronic grade diamond, useful for room temperature quantum sensing. When followed by annealing at very high temperatures, the vacancies created by static exposure are mobilized and can be captured by substitutional nitrogen impurities, which are randomly distributed in the diamond [5]. We show that single pulses are required to avoid amorphization and damage of the crystal lattice, and there is a range of pulse energies which

* Corresponding author: ottavia.jedrkiewicz@ifn.cnr.it

can be tuned to control the concentration of laser formed NVs. To increase the sensitivity or the measurement speed, ensemble of NVs have also been written and integrated within a waveguide (Fig.1b).

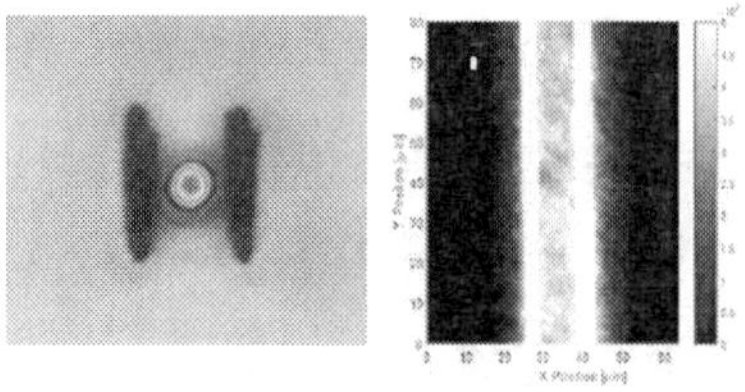

Fig. 1. (a) Transverse view of a waveguide in diamond (b) Cloud of NVs emission along the waveguide mode

2.3 Graphitic wire generation

In diamond, quantum emitters may be electrically controlled, both in their charge state and stimulated emission, with graphitic microchannels. Graphitic modifications can be realized by tailoring the laser processing, and will allow for both electrical interfacing within diamond and post-etching to produce conveniently buried microfluidics.

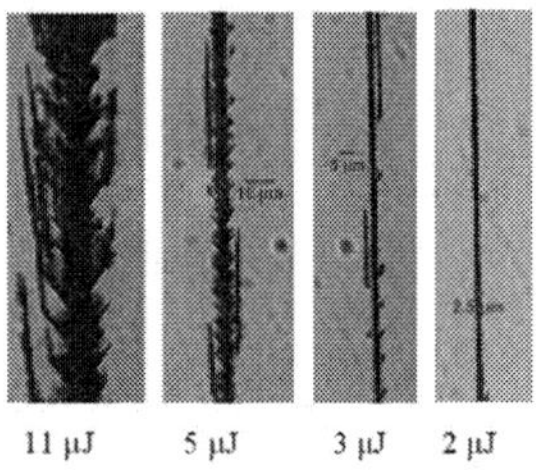

Fig. 2. Bessel beam laser-written graphitic wires in diamond

We report the fabrication of tailored graphitic wires throughout a 500 μm-thick CVD diamond sample, by means of Bessel beams with different geometries and pulse parameters, without any sample translation (Fig.2).

3 Microfluidics components in diamond

3.1 Laser writing of surface microchannels

The laser fabrication technique can be applied for a deep surface ablation of diamond. This can be done in single pass if using Bessel beams, obtaining high-aspect ratio V-shaped trenches featured by a micrometer size central channel, or in multiple pass with a more standard writing technique, leading to regular square shaped channels (Fig.3); the resulting microstructures offer a great potential for microfluidics or biosensing applications.

3.2 Microfabrication of through-holes

Through-holes in thick diamond samples can be realized by means of a micro-drilling process based on the combination of Bessel machining with a trepanning-like technique. The combination of through-holes with surface microchannels allows the implementation of microfluidic lab-on-chip prototypes (Fig.4).

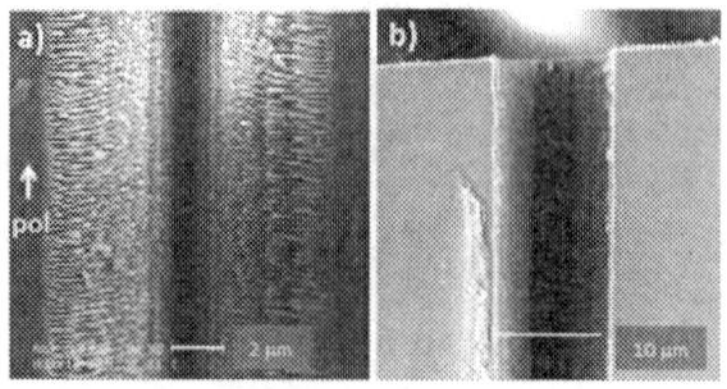

Fig. 3. Microfluidic channels realized by single pass Bessel beam (a) and multiple pass Gaussian (b) writing technique

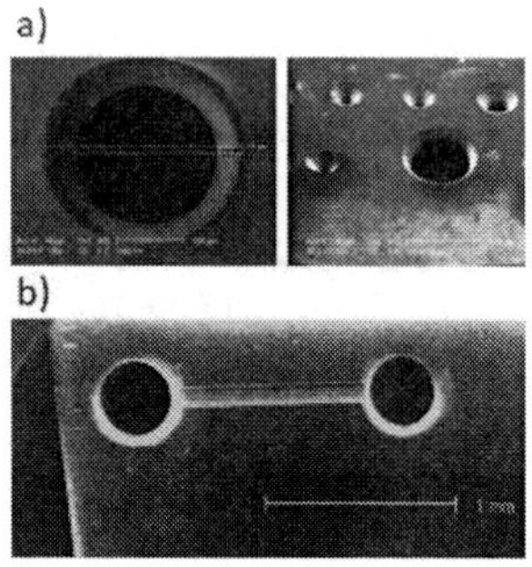

Fig. 4. (a) SEM images of through-holes in diamond (b) lab-on-chip prototype for microfluidics applications

4 CONCLUSIONS

The ultrafast laser writing technique can be applied to fabricate photonic and microfluidic networks in diamond. We have realized different building blocks such as NV centers, waveguides, graphitic wires, microchannels and micro-holes, whose integration will lead to prototype devices for use as quantum-based electric and magnetic field sensors.

References

1. R. R. Gattass, E. Mazur, Nat. Photonics **2**, 219 (2008).
2. V. Bharadwaj, O. Jedrkiewicz, J P Hadden, B. Sotillo, M. Ramos Vázquez, P. Dentella, T. T Fernandez, A. Chiappini, A. N. Giakoumaki, T. Le Phu, M. Bollani, M. Ferrari, R. Ramponi, P. E Barclay and Shane M Eaton, J. Phys. Photonics **1**, 022001 (2019).
3. K. Lim, C. Ropp, B. Shapiro, J. M. Taylor, and E. Waks, Nano Lett. **15**, 1481 (2015).
4. K. M. Davis, K. Miura, N. Sugimoto, and K. Hirao, Opt. Lett. **21**, 1729 (1996)
5. J. R. Rabeau, P. Reichart, G. Tamanyan, D.N. Jamieson, S. Prawer, F. Jelezko, T. Gaebel, I. Popa, M. Domhan and J. Wrachtrup, Appl. Phys. Lett. **88**, 023113 (2006)
6. O. Jedrkiewicz, D. Valetti, P. Di Trapani, SN Applied Sciences **1**, 1267 (2019).

Laser-printed emissive metasurface implemented with a planar thin-film resonator

Myeongkyu Lee*, *Yeongseon* Kim, *Dongkyun* Kang, and *Jaeyong* Kim

Materials Science and Engineering Department, Yonsei University, Seoul 03722, Korea

Abstract. Optical security is a promising application of metasurfaces because light has large degrees of freedom in metasurfaces. Although many different structures/materials have been proposed for this purpose, the fabrication of dynamic metasurfaces in a straightforward and scalable manner while maintaining a high security level remains a significant challenge. Herein, a metasurface consisting of a phase-changing $Ge_2Sb_2Te_5$ (GST) layer and a thin metal back reflector is presented to space-selectively and dynamically control the infrared emission of the surface by a spatially modulated pulsed laser beam. Unlike conventional laser processes using a focused beam, the employed laser printing is an expanded beam-based parallel process that enables the fabrication of wafer-sized emission patterns. Owing to the multispectral responses of GST, mutually independent visible and infrared images can be printed in one region. Grayscale emission patterns can also be obtained by gradually modulating the spatial profile of the laser beam, which makes the replication of laser-printed emission patterns extremely difficult. These encouraging features are experimentally verified using rigid and flexible substrates, indicating that the presented emissive metasurface has the potential for use as an effective platform for anti-counterfeiting.

1 Introduction

A metasurface is an artificial interface with subwavelength thickness that manipulates light by spatially arranged structures called meta-atoms. With the advent of spatial control over the phase and amplitude of electromagnetic waves, metasurfaces have enabled the realization of optical cloaking, super-resolution imaging, and planar optics, which cannot be achieved with the existing bulky optical elements. One of the promising applications of metasurfaces is in the field of optical security [1–5]. The demand for anti-counterfeiting methods with high security levels has continuously increased, particularly in the era of pandemics. The large degrees of freedom (i.e., phase, amplitude, and polarization) in metasurfaces make them suitable for use in optical security technologies against forgery.

Traditional anti-counterfeiting schemes (intaglio, guilloche, and latent image) allow non-experts to easily detect forgeries, but these material-based techniques can be easily replicated. As metasurface-based methods can provide significantly higher security levels, a number of different structures and materials have been investigated, including meta-holograms, polarization-selective metasurfaces, viewing angle-dependent metasurfaces, and Malus metasurfaces. However, in these approaches, as the security level increases, decoding the encoded information becomes more complex and can be performed only by experts with a specific setup. Additionally, the fabrication of metasurfaces operating in the visible range requires a nanofabrication process such as electron-beam lithography, which limits their scalability.

In this study, we present a parallel laser-printing process to independently control the IR and visible responses of a GST metasurface. This enabled the printing of different visible and IR images in one region. Both images can be easily captured using visible and IR cameras by non-experts, providing a simple forgery-detecting scheme. However, the replication of the thermal image becomes extremely difficult as its emission profile is gradually modulated by an expanded laser beam through a grayscale photomask. The printing process is scalable and can produce wafer-sized patterns. Lithography requires photoresist (PR) steps (PR deposition, development, and etching), whereas the fabrication of this GST metasurface involves only thin-film deposition and laser irradiation. All of these provide favorable features for anti-counterfeiting.

2 Results and discussion

The metasurface had a planar cavity structure comprising a GST film on top of a metal back reflector, which was stationed on a substrate. Our approach was to print crystallization patterns inside and on the surface of an amorphous GST film using a spatially modulated pulsed laser beam. The visible reflection of the cavity was determined solely by the top-surface structure of the GST film because of the short penetration depth of visible light into GST, whereas the IR emission depended on the internal structure of the film.

* Corresponding author: myeong@yonsei.ac.kr

Fig. 1 shows the structure of a GST cavity formed on a Si substrate, along with its visible and IR images. Here, *a*-GST represents amorphous GST and *c*-GST refers to crystalline GST. To fabricate the cavity, a 100-nm Cr layer and a 200-nm *a*-GST film were sequentially deposited onto an Al-coated Si substrate using a shadow mask (QR shadow mask). The surface of the *a*-GST film was irradiated by a laser beam passing through the shadow mask. Subsequently, an additional 200-nm *a*-GST film was uniformly coated over the structure. Finally, the surface of the coated GST film was irradiated by a laser beam that was spatially modulated through a photomask (barcode image).

The difference in reflectance between *c*-GST and *a*-GST phases leads to a visual contrast. The visible reflectance of a GST cavity depends solely on the surface structure (i.e., phase) of the GST film owing to the short penetration depth of visible light into GST. Therefore, the visible image of the cavity is independent of its internal structure and is determined by the surface crystallization pattern. In other words, the visible image is defined by the photomask pattern used in the final step. In contrast, the thermal image of a cavity is strongly influenced by its internal structure. The areas with a 400-nm-thick GST film on top of the Cr layer exhibit much stronger emission than those with the 200-nm-thick GST film without Cr. The former has a GST thickness of 400 nm, resulting in an emissivity peak in the detection range (7.5–13 μm) of ordinary IR cameras. The embedded Cr and *c*-GST layers also increase the emissivity; Cr is more absorptive than Al in the IR range, and *c*-GST is also more absorptive than *a*-GST in the same range. The latter has a smaller GST thickness of 200 nm, resulting in an emissivity peak outside the detection range of the IR camera.

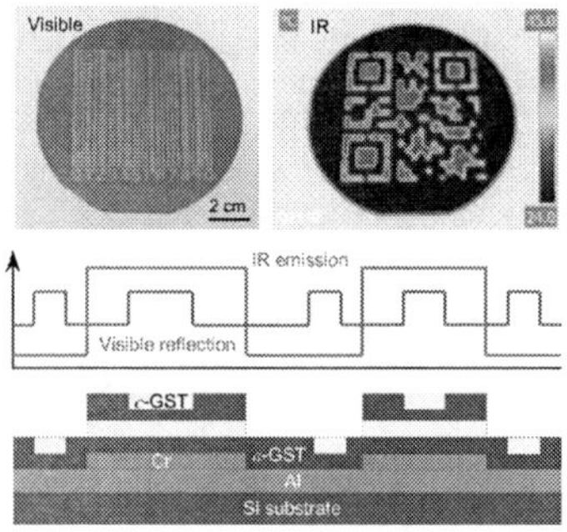

Fig. 1. Structure of a GST cavity fabricated on a Si substrate along with its visible and IR images. *a*-GST represents amorphous GST and *c*-GST refers to crystalline GST.

Another GST emitter was fabricated on a Korean banknote of 1000 won using the same QR shadow mask and a different photomask containing an "Einstein image" (**Fig. 2**); 1000 won is worth approximately one dollar. Unlike the Si substrate, the banknote substrate has a rough surface; nevertheless, the visible image recorded was clearly recognizable. The emitter was also highly flexible; no material delamination or image distortion was observed the substrate was completely folded.

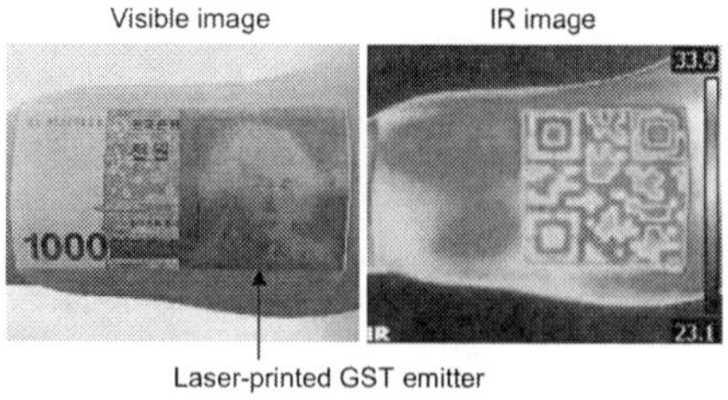

Fig. 2. Visible and IR images of a GST emitter fabricated on a Korean banknote.

3 Conclusion

In this study, emissivity-modulated GST emitters were fabricated on rigid and flexible substrates. The emissivity was spatially controlled by forming crystalline GST phases within the amorphous GST film using a spatially modulated pulsed laser beam and by selectively depositing a Cr layer on the Al back reflector. The multispectral properties of GST enabled different visible and IR images to be simultaneously recorded in the same area. The fact that a single emitter can exhibit different visible and IR images is particularly attractive for optical security applications, including anti-counterfeiting.

This work was supported by National Research Foundation of Korea(NRF) grants funded by the Korean government(MSIT) (2020R1A2C2003575 and 2022M3H4A1A02046445) and a Korea Agency for Infrastructure Technology Advancement(KAIA) grant funded by the Ministry of Land, Infrastructure and Transport (Grant 21CTAP-C163910-01).

References

1. I. Kim, J. Jang, G. Kim, J. Lee, T. Badloe, J. Mun, J. Rho, Nat. Commun. **12**, 3614 (2021)
2. Z. Li, C. Chen, Z. Guan, J. Tao, S. Chang, Q. Dai, Y. Xiao, Y. Cui, Y. Wang, S. Yu, G. Zheng, S. Zhang, Laser Photonics Rev. **14**, 2000032 (2020)
3. J. Deng, L. Deng, Z. Guan, J. Tao, G. Li, Z. Li, Z. Li, S. Yu, G. Zheng, Nano Lett. **20**, 1830 (2020)
4. C. Zhang, F. Dong, Y. Intaravanne, X. Zang, L. Xu, Z. Song, G. Zheng, W. Wang, W. Chu, X. Chen, Phys. Rev. Appl. **12**, 034028 (2019)
5. F. Yue, C. Zhang, X. Zang, D. Wen, B. Gerardot, S. Zhang, X. Chen, Light: Sci. Appl. **7**, 17129 (2018)

One-dimensional photonic crystal for polarization-sensitive surface-enhanced spectroscopy

Erika Mogni[1,*], *Giovanni* Pellegrini[2], *Jorge* Gil-Rostra[3], *Francisco* Yubero[3], *Giuseppina* Simone[1,4], *Stefan* Fossati[5], *Jakub* Dostálek[5,6], *Rebeca* Martinez-Vazquez[7], *Roberto* Osellame[7], *Michele* Celebrano[1], *Marco* Finazzi[1] and *Paolo* Biagioni[1].

[1]Politecnico di Milano, Dipartimento di Fisica, 20133 Milano, Italy
[2]Università degli studi di Pavia, Dipartimento di Fisica, 27100 Pavia, Italy
[3]CSIC-Universidad de Sevilla, Instituto de Ciencia de Materiales de Sevilla, 41092 Sevilla, Spain
[4]Northwestern Polytechnical University, School of Mechanical Engineering, Xi'an Shaanxi, People's Republic of China
[5]AIT Austrian Institute of Technology GmbH, Biosensor Technologies, Tulln an der Donau, Austria
[6]Czech Academy of Sciences, FZU-Institute of Physics, Prague, Czech Republic
[7]Istituto di Fotonica e Nanotecnologie (IFN)-CNR, 20133 Milano, Italy

Abstract. We realize and experimentally characterize a novel platform for surface-enhanced sensing through Bloch Surface Waves (BSWs). We test a one-dimensional photonic crystal, with a high index inclusion in the top layer, that sustains surfaces modes with, in principle, arbitrary polarization. This is achieved through the coherent superposition of TE and TM dispersion relations of BSWs, which can also provide superchiral fields over a wide spectral range (down to the UV). The resulting platform paves the way to the implementation of polarization-resolved surface-enhanced techniques.

1 Introduction

Bloch Surface waves (BSWs) are supported by semi-infinite one-dimensional photonic crystals (1DPCs), where both the transverse electric (TE) and the transverse magnetic (TM) eigenmodes can be sustained. Thanks to the high sensitivity to the local refractive index, platforms supporting BSWs are widely used for sensing analytes which are selectively preconcentrated at the surface of such truncated photonic crystals [1-3]. Moreover, the resonant frequencies of the modes can be tuned by a proper design of the 1DPC, acting on the materials (i.e. the refractive index) and the thicknesses of the layers. A further step in surface-enhanced spectroscopies using BSWs would be the control over the polarization, which can be attained by engineering the overlap between the dispersion relations of the TE and TM modes. For example, the surface-enhanced study of oriented molecules anchored to the 1DPC surface could be realized through differential TE-TM reflectivity measurements, as customarily done in standard attenuated total internal reflection spectroscopy [4]. More specifically, the superposition of the TE and TM modes enables to amplify an optical response in linear and circular dichroism (CD) spectroscopies. CD, in particular, is gaining more and more importance due to the wide applications in chemistry, biology, medicine and pharmacology [5]. In general, CD signals are very weak if compared to those obtained in standard absorption spectroscopies, therefore novel "superchiral" approaches have been recently introduced to enhance the CD signal by engineering the optical chirality of the electromagnetic field with nanostructured systems [6]. Both metallic and dielectric approaches are being used to obtain superchirality through the fabrication of nanostructures generating hot spots of enhanced optical chirality [7,8]. In this framework, we recently proposed a platform [2] that allows for the generation of superchiral surface waves. Here we fabricate a 1DPC that simplifies the previous design by reducing the number of layers while keeping similar polarization performance. We experimentally demonstrate the energy and momentum superposition of the TE and TM modes, paving the way to the employment of such a platform for surface-enhanced polarization-sensitive spectroscopies in the visible-UV range and, more generally, to the implementation of polarization-resolved surface-enhanced techniques.

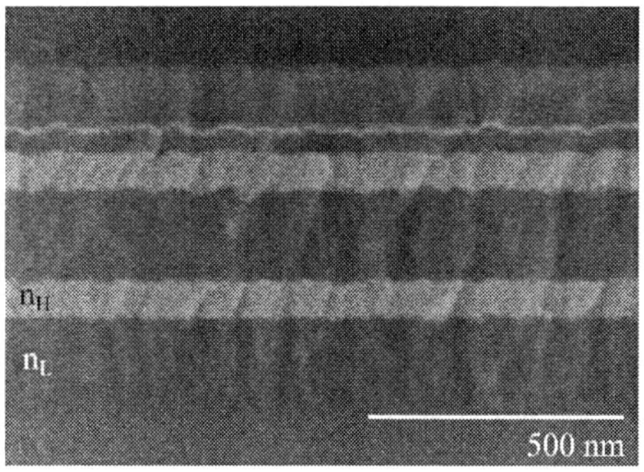

Figure 1. Cross section SEM micrograph of the manufactured 1DPC

* Corresponding author: erika.mogni@polimi.it

2 Materials and methods

The platform consists in a 1DPC grown by reactive magnetron sputtering and made by 2 periods of alternating compact layers of SiO_2 (low refractive index) and Ta_2O_5 (high refractive index) with different thicknesses, terminated with an additional low-index layer incorporating a high refractive index inclusion as shown in the cross-section SEM image in Figure 1. The characterization of the optical response of the sensing platform was done in water using a fluidic cell realized using a polymeric double-side tape with the fluidic chamber cut by femtosecond laser micromachining. The top layer is a fused silica slide that contains the housing to plug the fluidic tubes.

3 Results and discussion

For the experimental demonstration of the superposition of the TE and TM modes, reflectance measurements are performed on the 1DPC with the fluidic chamber filled with water. The resulting dispersion relations are reported in the experimental reflectivity maps of Figure 2 for TE and TM illumination (panels a and b, respectively), together with representative angular plots at specific wavelengths (panels c and d), confirming the close overlap of the two modes in good agreement with simulations. Upon increasing the wavelength, the excitation angle of the modes decreases, approaching the onset of total internal reflection. Moreover, it can be observed that, while the TE mode is narrow and sharp, the TM mode is broader and weaker but still recognizable, allowing us to experimentally confirm the mode superposition. The overlapping of the dispersion relations offers the possibility of facile generating of an arbitrary polarization of the surface modes in order to pave the way to the implementation of polarization-resolved surface-enhanced techniques. This can be achieved by the simultaneous coupling to both TE and TM modes with controlled relative phase and amplitude.

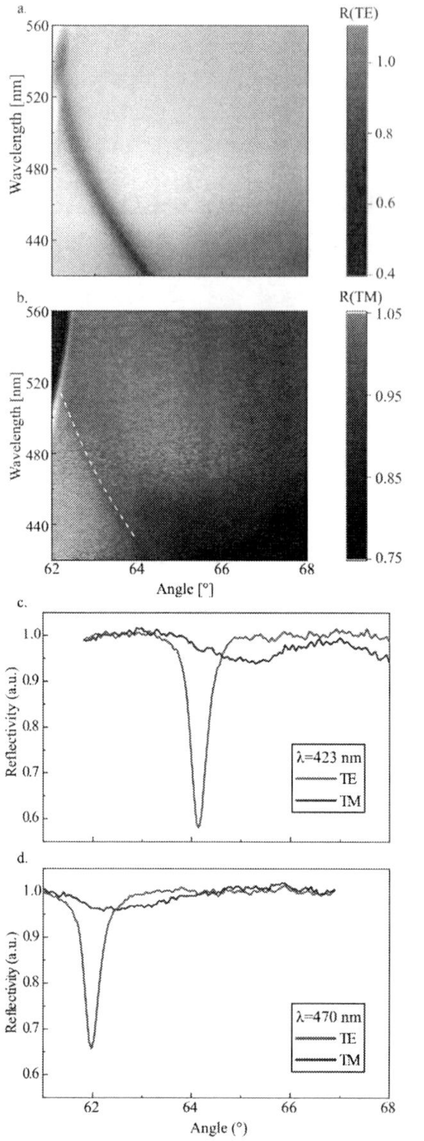

Figure 2. Experimental reflectivity (R) maps for a) TE illumination and b) TM illumination when the surface of the 1DPC is immersed in water. The dispersive TE and TM modes are visible as dips in the reflectivity maps. Experimental angular plot for a wavelength equal to c) 423 nm and d) 470 nm.

F. Yubero acknowledges MCIN for Salvador de Madariaga grant PRX19/00485 and project PID2019-110430GB-C21 funded by MCIN/AEI/10.13039/501100011033 and ERDF. This work was partially supported by the Grant PON ARS01_00906 "TITAN - Nanotecnologie per l'immunoterapia dei tumori", funded by FESR in the framework of PON "Ricerca e Innovazione" 2014 -2020 - Azione II - OS 1.b).

References

1. A. Sinibaldi, N. Danz, E. Descrovi, P.Munzert, U. Schultz, F. Sonntag, L. Dominici, F. Michelotti, Sensors and Actuators B **174,** 292-298 (2012)
2. G. Pellegrini, M. Finazzi, M. Celebrano, L. Duò, P. Biagioni, Phys. Rev. B **95,** 241402 (2017)
3. F. Giorgis, E. Descrovi, C. Summonte, L. Dominici, F. Michelotti, Optics Express **18,** 8087-8093 (2010)
4. L. Fornasari, F. Floris, M. Patrini, G. Canazza, G. Guizzetti, D. Comoretto, F. Mirabelli, Appl. Phys. Lett **105,** 053303 (2014)
5. P. Ferenci, Aliment Pharmacol Ther **19,** 157-165 (2004)
6. Y. Tang, A. E. Cohen, Phys. Rev. Lett. **104,** 163901 (2020)
7. F. Mattioli, G. Mazzeo, G. Longhi, S. Abbate, G. Pellegrini, E. Mogni, M. Celebrano, M. Finazzi, L. Duò, C. G. Zanchi, M. Tommasini, M. Pea, S. Cibella, R. Polito, F. Sciortino, L. Baldassarre, A. Nucara, M. Ortolani, P. Biagioni, ACS Photonics **7,** 2672-2681 (2020)
8. M. L. Solomon, J. Hu, M. Lawrence, A. García-Etxarri, J. A. Dionne, ACS Photonics **6,** 43-49 (2019)

The role of heat treatment in improving photoluminescence and optically stimulated luminescence of HfO_2

Alvaro de Farias Soares[1,2,*], *Sonia* Hatsue Tatumi[1,3], *Clemens* Woda[2] and *Lilia* Coronato Courrol[4]

[1]Escola Politécnica, Universidade de São Paulo, 05508-010, São Paulo/SP – Brasil
[2]Helmholtz Zentrum München, Institute of Radiation Medicine, Ingolstädter Landstraße 1, D-85764, Neuherberg, Germany
[3]Instituto do Mar, Universidade Federal de São Paulo, 11070-100, Santos/SP – Brasil
[4]Instituto de Ciências Ambientais, Químicas e Farmacêuticas, Universidade Federal de São Paulo, 09972-270, Diadema/SP - Brasil

Abstract. HfO_2 is a metal oxide from the IV-B family whose properties have been widely applied in electronics and which displays an important blue emission. Despite this fact, few studies have been dedicated to understanding the role of heat treatment on its luminescence properties. Therefore, this study aimed to investigate the Photoluminescence (PL) and Optically Stimulated Luminescence (OSL) of HfO_2 powder synthesized by the precipitation method and the influence of calcination on its luminescence. PL spectra results showed a broad emission band at about 2.6 eV, which was related to absorption at 4.1 eV. Green and blue stimulated luminescence spectra depicted emission bands in a similar region, varying between 2.44 up to 2.71 eV. The increase in the heat treatment temperature promoted signal enhancement, which could be associated with higher oxygen vacancy concentrations. OSL dose-response curves for the sample calcined at 1600 °C are linear up to 0.6 Gy and a good signal reproducibility is observed, which makes the material suitable for OSL dosimetry.

1 Introduction

Hafnium Oxide (HfO_2) is a dielectric material that has been increasingly used in the microelectronic area since it shows properties such as high thermal stability, a large bandgap of about 5.6-5.8 eV, and chemical stability [1-2]. It can be synthesized in different ways, allowing it to obtain distinct properties and/or crystallographic phases [2]. Due to the presence of several defects in the material, mainly oxygen vacancies, which explains the possibility of charge transport on it, the material displays important electronic and luminescence properties [1]. Applying high-temperature treatment is an option to increase the defect concentrations and improve its luminescence. In our previous study, it was already pointed out that a calcination treatment at 1600 °C intensified the Thermoluminescence (TL) and Optically Stimulated Luminescence (OSL) emissions [3].

While the noticeable luminescence from this crystal, mostly in the blue region, has already been reported [1], the effect of heat treatment on Photoluminescence (PL) and OSL spectra was not in-depth performed. Therefore, this work aimed to contribute to the luminescent characterization of HfO_2 using PL and OSL spectral measurements and to evaluate the use in OSL dosimetry.

2 Methodology

HfO_2 nanoparticles were synthesized using the precipitation method [3,4], with hafnium tetrachloride ($HfCl_4$) and potassium hydroxide (KOH) as precursors. Obtained powders were calcined for 2 hours at different temperatures of 1000, 1200, 1400, and 1600 °C.

PL was analysed using a Horiba Fluorolog fluorimeter, which has a 450-W Xenon arc lamp, coupled with a bifurcated fiber-optic probe, composed of illumination and collecting cables and with a numerical aperture of 0.22. OSL spectra and decay curves were obtained using a Lexsyg research TL/OSL reader, with a built-in β source (dose rate of 0.02 Gy/s), and an Andor Technology iDus 420 Series CCD camera coupled to a Shamrock 163 spectrometer. A Schott-BG 39 filter 3 mm was used for IR stimulated luminescence (IRSL) spectra, and a combination of BG39 and an Edmund Optics 525 nm OD4 short pass filter for green stimulated luminescence (GSL) spectra. For Continuous-wave OSL (CW-OSL), a Hoya-U340 7.5 mm was used for blue stimulated luminescence (BSL), different short pass filters centered at 330, 365, and 410 nm for GSL, and for IRSL, the same as used in the spectrum. X-ray irradiation was performed in an Xstrahl chamber model RS225.

3 Results and discussion

Figure 1 depicts the emission and excitation spectra for HfO_2. When the sample is stimulated in the UV region (300 nm), a broad emission band can be noticed in the blue region. The increase in the heat treatment temperature leads to an increase in signal intensity, which may be associated with a higher concentration of defects.

* Corresponding author: alvarodefarias@usp.br

The excitation spectrum shows that the absorption band is centered at about 300 nm, which coincides with the excitation wavelength used for measuring the emission spectrum.

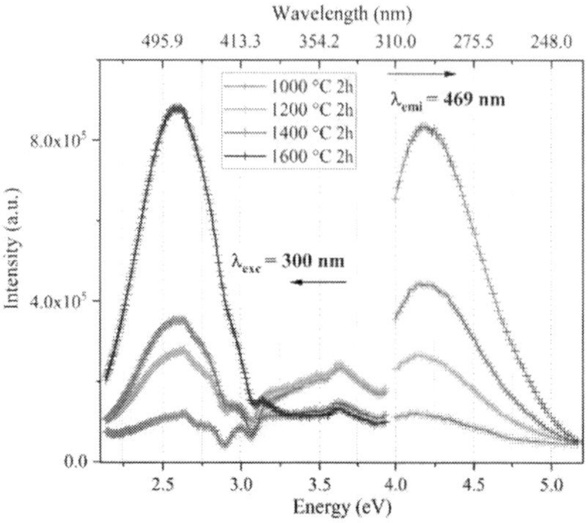

Fig. 1. Emission (λ_{exc}=300 nm) and excitation (λ_{emi}=469 nm) spectra for HfO_2 at different temperatures of calcination.

This blue emission has been previously associated with oxygen vacancies and characterized by an absorption band at 5.2 eV [1]. Here, the emission occurs in a similar energy region, whereas the absorption band has a lower energy value. The Chromaticity Diagram from PL data (Fig. 2) presents that besides the signal increase observed, the emission is slightly shifted to a darker blue region.

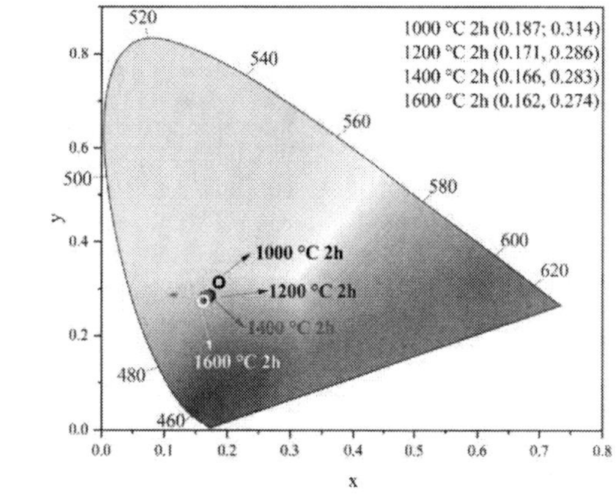

Fig. 2. Chromaticity Diagram for HfO_2 calcined at different temperatures.

The heat treatment was also effective in improving the OSL properties, increasing the intensity for both IRSL and GSL spectra. Figure 3 illustrates the case for IRSL measurements. The emission occurs in the same region as for the PL spectra, which indicates that the same defects could be responsible for both luminescence types.

OSL spectra and PL curves fitted with Gaussians equations indicated central energies that vary between 2.44 to 2.71 eV, depending on the type of stimulation and heat treatment (Table 1). IRSL emission bands have energy values slightly lower than those from GSL, which could be related to the higher stimulation wavelength.

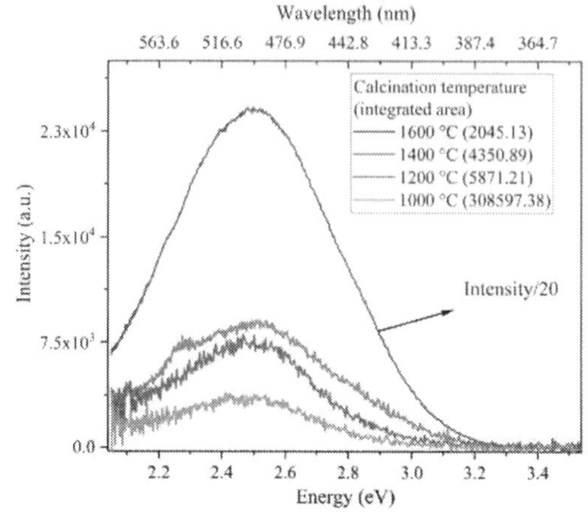

Fig. 3. IRSL spectra for HfO_2 at different temperatures of calcination after 800 Gy of X-rays irradiation.

Table 1. Emission band (eV) for HfO_2 calcined at different temperatures.

Technique	1000 °C	1200 °C	1400 °C	1600 °C
PL (λ_{exc} = 300 nm)	2.56	2.58	2.58	2.59
GSL Spectrum	2.61	2.71	2.69	2.67
IRSL Spectrum	2.44	2.46	2.48	2.49

For HfO_2 calcined at 1600 °C, CW-OSL dose-response curves showed a linear behavior up to 0.6 Gy for green, blue, and IR stimulation. A reproducibility study with 8 cycles demonstrated a variation lower than 10 %.

4 Conclusions

HfO_2 crystals display a notable blue emission, related to oxygen vacancies, which can be improved by heat treatment. OSL properties indicate that the material has the potential to be applied as a dosimeter.

Acknowledgments: this study was supported by grant #2020/15562-8, São Paulo Research Foundation (FAPESP), by grant #2019/20121-3 (FAPESP) and #2020/08592-8 (FAPESP) and financed in part by the Coordenação de Aperfeiçoamento de Pessoal de Nível Superior - Brasil (CAPES) - Finance Code 001.

References

1. V. Gritsenko, T. Perevalov, D. Islamov, Phys. Rep. **613**, 1-20 (2016)
2. J. Choi, Y. Mao, J. Chang, Mater. Sci. Eng. C. **72**, 97-136 (2011)
3. A. Soares, S. Tatumi, R. Rocca, Radiat. Phys. Chem. **180**, 109292 (2021)
4. A. Ramadoss, K. Krishnamoorthy, S. Kim. Mater. Res. Bull. **47**, 2680-2684 (2012)

Polarized light guiding anisotropic deformation and relaxation in photosensitive polymeric substrates

David Urban[1,*], *Dag Roar* Hjelme[1], and *Emiliano* Descrovi[2,**]

[1]Department of Electronic Systems, Norwegian University of Science and Technology, 7034 Trondheim, Norway
[2]Department of Applied Science and Technology, Polytechnic University of Turin, 10129, Torino, Italy

Abstract. Light-responsive polymers offer unique possibilities for anisotropic manipulation of objects on the micron scale. Here we demonstrate the reversible anisotropic stretching of a polymeric surface made out of an azopolymer-elastomer blend, in response to green laser irradiation with varying polarization. We quantify the stretching parameters and the residual strain after relaxation by means of a Fourier-based analysis, which exploits a periodic 2D pattern imprinted onto the surface.

1 Introduction

Light-responsive polymers have recently attracted increasing attention due to their potential applications in areas such as smart micro-structured surfaces, adaptive optics, and biology. For applications necessitating contactless, anisotropic deformations, the subclass of sidechain azopolymers has been found of particular interest, as some of them tend to deform along the polarization axis of linearly polarized light. Application examples using directional deformation of azopolymer pillars range from inducing anisotropic behavior in living cells [1] to tuneable anisotropic surface wettability [2]. However, a major limitation of these types of microstructures is the limited reversibility of the induced plastic deformations. In a recent work carried out by Ryabchun et. al., the reversible deformation of single azopolymer domains sparsely dispersed in a Styrene-Ethylene-Butylene-Styrene (SEBS) thermoplastic elastomer matrix was reported. Pristine domain shapes were reobtained by heating or irradiation with circularly polarized light, making use of the strain stored in the elastomeric matrix surrounding the azopolymer domains [3]. Furthermore, an example of the potential use of this material system in a macroscopic actuator has been provided.

2 Results

Here, we study the deformation and reversibility of a light-responsive polymer slab whose top surface is patterned with a hexagonal array of cylindrical pillars (lattice constant $d = 6\ \mu$m). The sample is prepared by dissolving azopolymer poly[(methyl methacrylate)-co-(Disperse Red 1 methacrylate) (Sigma-Aldrich) and SEBS (Mediprene 500120M, kindly provided by Hexpol AB) in Toluene and casting the solution on a soft PDMS master, reaching a final concentration of 30 %wt azopolymer.

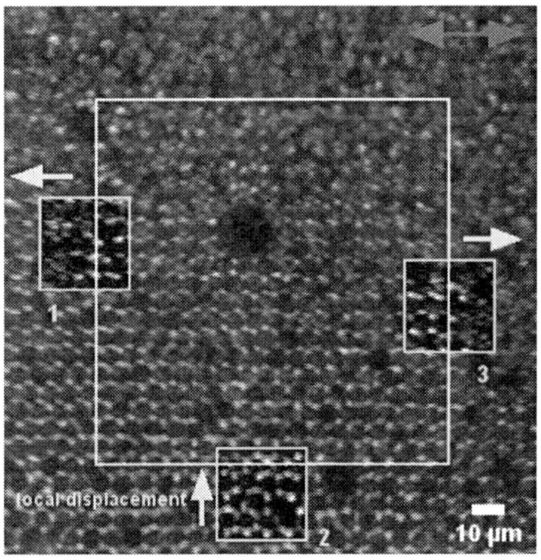

Figure 1. B/W: raw transmission image of the substrate top surface patterned with a hexagonal array of pillars. Magenta: pixels of transmission image after irradiation with horizontal polarization (red double-arrow), initial image subtracted. Yellow: Zone of irradiation. Green: Enhanced contrast regions.

Figure 1 shows the local displacement (magenta) resulting from the irradiation of the central area (yellow square) with a horizontally linearly polarized CW green laser (561 nm, 3 W/cm2, 15 s exposure) in a confocal microscope setup operating in continuous scanning mode. Enhanced contrast insets exemplify the local displacement of brighter spots, which extends also outside the directly irradiated region through strain propagation.

*e-mail: david.urban@ntnu.no
**e-mail: emiliano.descrovi@polito.it

An analysis in the Fourier space of the directly irradiated area allows to estimate the shifts of the peaks associated to the spatial harmonics of the periodic pattern (Figure 2, top). Fourier peaks moving inward/outward indicate real space stretching/compression along the horizontal/vertical axis, which is consistent with horizontally polarized irradiation [4]. Subsequent irradiation of the same area switching to circular polarization partially reverses the strain and the peak shifts associated thereto, as shown in Figure 2, bottom.

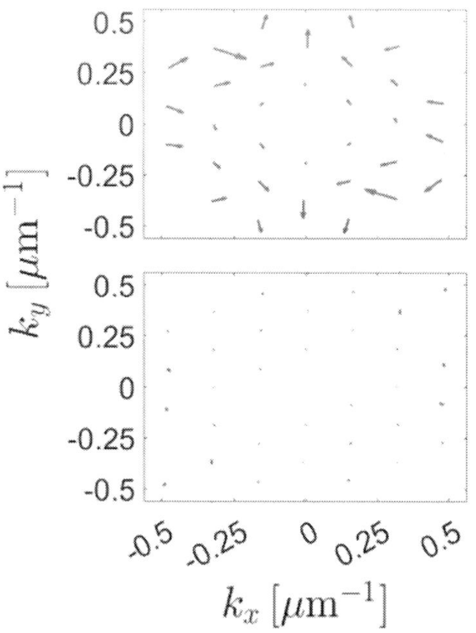

Figure 2. Arrows showing displacement of peaks in Fourier space before and after irradiation with linear polarization (top) and before and after a complete cycle of linear + circular polarization (bottom). Arrow magnification: x10.

A least squares algorithm with parameters θ (angle of strain- with respect to x-axis), ϵ_1 (strain along θ-axis) and ϵ_2 (strain along perpendicular axis) is implemented to fit the strain-associated shifts in Fourier space (Figure 2, top), aiming to quantify the expansive and compressive strain after exposure to linearly polarized light. The resulting parameter values are summarized in table 1, together with the respective sample standard deviation over five subsequent stretching cycles. The computed stretching axis angle θ is close to the horizontal axis, while the associated strains ϵ_1 and ϵ_2 signify stretching by (1.29 ± 0.25) % and compression by (-1.18 ± 0.05) % respectively. R^2-values of the fits are 0.71 ± 0.05, where R^2 has been defined as

$$R^2 = 1 - \frac{\sigma_{res}^2}{\sigma_{tot}^2} \quad (1)$$

with σ_{res}^2 being the squared residual norm of the fit and σ_{tot}^2 the squared sum of arrow magnitudes. Finally, while the shifts after a full cycle of irradiation with linearly and subsequently circularly polarized light are considerably smaller (Figure 2, bottom), it is known that circular polarization can lead to in-plane expansion of azopolymers [3]. Applying the same fitting algorithm to the full deformation cycle the relative area change can be calulated to the first order as:

$$\Delta A_{rel} = \frac{A_{new} - A_{old}}{A_{old}} = \frac{a(1+\epsilon_1)\cdot b(1+\epsilon_2) - a\cdot b}{a\cdot b} \approx \epsilon_1+\epsilon_2 \quad (2)$$

where we have used a rectangular area of size $A = a \cdot b$ as example. The relative area change after deformation and subsequent relaxation is then found to be (0.17 ± 0.02) %, meaning there is a slight effect of in-plane area expansion. Although an order of magnitude smaller than the induced stretches, this means that the reversibility is not complete.

Table 1. Extracted Strain Parameters and Standard Deviation

	θ [deg]	ϵ_1 [%]	ϵ_2 [%]	R^2	ΔA_{rel} [%]
Value	-2.45	1.29	-1.18	0.71	0.17
St. Dev	2.94	0.25	0.05	0.05	0.02

3 Discussion and Outlook

Reversibility in anisotropic deformation of azopolymeric structures is a key issue for several potential applications and has received increasing attention in recent years. The approach presented above is a powerful tool to assess the quantitative degree of reversibility of such material systems. What's more, it will be useful for determining the effect of changing both material and experimental parameters such as azopolymer content and distribution, as well as irradiation time or intensity. We also note that the propagation of light-induced deformations outside of the irradiated area due to the propagation of strain in the SEBS matrix is an interesting feature.

In perspective, we aim to explore the use of similar substrates for biological applications, where inducing local surface stretches through either direct or indirect irradiation can be particularly advantageous for remote control of cell metabolism and behaviour. Furthermore, new types of actuators, which not only respond to radiation intensity, but also to its polarization, can be foreseen.

References

[1] A. Puliafito, S. Ricciardi, F. Pirani, V. Čermochová, L. Boarino, N. De Leo, L. Primo, E. Descrovi, Adv. Sci. **6**, 1801826 (2019)

[2] F. Pirani, A. Angelini, F. Frascella, R. Rizzo, S. Ricciardi, E. Descrovi, Sci. Rep. **6**, 31702 (2016)

[3] A. Ryabchun, A. Bobrovsky, Adv. Opt. Mater. **7**, 1901486 (2019)

[4] S. Loebner, N. Lomadze, A. Kopyshev, M. Koch, O. Guskova, M. Saphiannikova, S. Santer, J. Phys. Chem. B p. 9 (2018)

Selective ultrashort laser annealing of amorphous Ge/Si multilayer stacks

Nadezhda M. Bulgakova[1,*], *Vladimir A. Volodin*[2,3], *Yuzhu Cheng*[3], *Yoann Levy*[1], *Jiří Beránek*[1,4], *Siva S. Nagisetty*[1,5], *Martin Zukerstein*[1], *Alexander A. Popov*[6], *Alexander V. Bulgakov*[1]

[1]HiLASE Centre, Institute of Physics of the Czech Academy of Sciences, 25241 Dolní Břežany, Czech Republic
[2]Rzhanov Institute of Semiconductor Physics SB RAS, 630090 Novosibirsk, Russia
[3]Novosibirsk State University, 630090 Novosibirsk, Russia
[4]Czech Technical University, Faculty of Nuclear Sciences and Physical Engineering, 11519 Prague, Czech Republic
[5]Coherent LaserSystems GmbH & Co. KG, 37079 Göttingen, Germany
[6]Valiev Institute of Physics and Technology, Yaroslavl Branch, Russian Academy of Sciences, 150007 Yaroslavl, Russia

Abstract. We report on single-short laser crystallization of Ge/Si multilayer stacks consisting of alternating amorphous nanosized films of silicon and germanium using near- and mid-infrared femtosecond and picosecond laser pulses. The phase composition of the irradiated stacks was investigated by the Raman scattering technique. Several non-ablative regimes of crystallization were found, from partial crystallization of germanium without intermixing the Ge/Si layers to complete intermixing of the layers with formation of Ge_xSi_{1-x} solid alloys. The roles of one- and two-photon absorption, thermal and non-thermal (ultrafast) melting processes, and laser-induced stresses in selective pico- and femtosecond laser annealing are analysed based on theoretical estimations and comparison with experimental data. It is concluded that, due to a mismatch of the thermal expansion coefficients between the adjacent stack layers, efficient explosive solid-phase crystallization of the Ge layers is possible at relatively low temperatures, well below the melting point. The possibility of ultrafast non-thermal phase transition in germanium in the studied regimes is also discussed.

1 Introduction

Low-temperature crystallization of thin amorphous semiconductor films is in high demand for fabrication of "flexible electronics" devices. Recently, germanium thin films with a record-high hole mobility were obtained on flexible plastic using post-growth annealing at relatively low temperature of 500 °C [1]. However, for using inexpensive plastic flexible substrates, it is necessary to reduce their heating to temperatures below 120 °C. Pulsed laser annealing appears to be the only suitable technique for crystallization of amorphous semiconductor films without overheating substrates [2,3].

A number of attempts were made to selectively crystallize germanium in amorphous Ge/Si multilayer (ML) systems. However, usually laser annealing leads to intermixing of Si and Ge layers and to formation of their alloy. Recently, we achieved successful selective crystallization of germanium layers in an amorphous Ge/Si ML stack using picosecond 1030-nm laser annealing [4]. In the present study, we have extended the irradiation conditions to longer wavelength (1500 nm) and shorter pulse duration (70 fs) and compared ps and fs annealing regimes with the aim to find optimal conditions for selective crystallization of Ge layers in Ge/Si ML stacks without crystallization of silicon and intermixing Ge and Si layers.

2 Experimental

Multilayer a-Si:H/a-Ge:H stacks consisted of 4 Si (40-nm thick) and 3 Ge (15-nm thick) alternating amorphous layers were produced by plasma enhanced chemical vapor deposition on a Si(100) substrate. Monosilane (SiH_4) and monogermane (GeH_4) gases were used as precursors for the Si and Ge layers.

The produced Si/Ge stack were annealed by a ps laser (HiLASE PERLA-B, λ=1030 nm, pulse duration 1.4 ps) and by an fs laser (Astrella, Coherent) in combination with an optical parametric amplifier (TOPAS, Light Conversion) delivering a beam at λ=1500 nm with a pulse duration of 70 fs. The beams were focused at normal incidence on the sample surface onto circular spots of diameters of 2.1 mm for 1030 nm and 0.53 mm for 1500 nm. The laser fluence F_0 was varied in the range 20-200 mJ/cm^2. The phase composition of as-deposited and annealed ML stacks was studied by Raman spectroscopy using a T64000 spectrometer with micro-Raman setup. The excitation 514.5-nm laser beam was focused to a 10-µm-diam spot at the center of the annealed spots and thus the Raman spectra were collected from sites annealed uniformly at the peak laser fluence F_0.

* Corresponding author: bulgakova@fzu.cz

3 Results and discussion

Figure 1 shows Raman spectra obtained after irradiation of the Si/Ge MLs at low laser fluences for fs and ps pulses. The Raman spectrum of the as-deposited ML stack is also shown. The latter contains broad bands at ~480 cm^{-1} and ~275 cm^{-1} indicating that all the layers are completely amorphous. The spectrum of the as-deposited MLs has no features of the vibration frequencies of Ge–Si bonds (in range from ~390 cm^{-1} to ~440 cm^{-1}) pointing that the concentration of such bonds is too low to be detected.

The laser treatment of the Si/Ge stack at F_0 below 50 mJ/cm^2 for fs pulses and 40 mJ/cm^2 for ps pulses does not lead to changes in the Raman spectra, indicating no noticeable structural transformations in the ML samples. Ps irradiation at F_0 = 46 mJ/cm^2 results in appearance of a narrower feature at a frequency of ~ 296 cm^{-1} against the background of the amorphous component (see Fig. 1). The analysis shows that the observed Raman shift of ~5.5 cm^{-1} corresponds to the average size of Ge nanocrystals (NCs) of about 3 nm [5]. In the case of fs pulses at 64 mJ/cm^2, the NC-Ge peak is higher and narrower, and its position corresponds to the average NC size of ~4.5 nm. No peaks corresponding the Si crystalline phase have been observed.

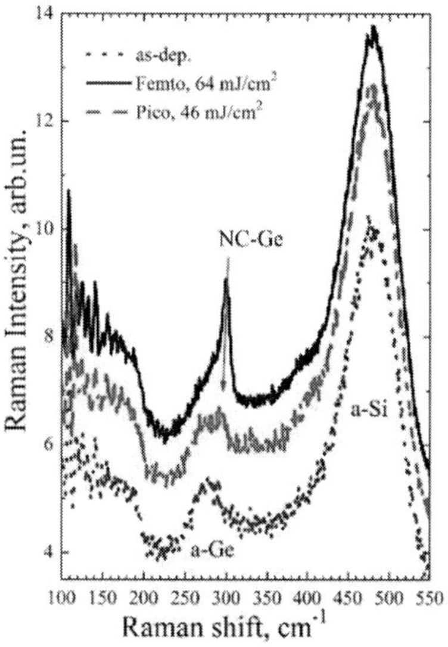

Fig. 1. Raman spectra of as-deposited and low-fluence laser annealed Si/Ge ML stack. The arrow shows a small peak of Ge nanocrystals produced by ps laser pulses.

At $F_0 \geq 70$ mJ/cm^2, the Ge layers are already almost completely crystallized but the Raman spectra show slight intermixing with the silicon layers. Starting from 100 mJ/cm^2, silicon crystallizations is observed while, at $F_0 \geq$ 150 mJ/cm^2, a strong peak of Ge-Si bonds appears in the Raman spectra indicating formation of a Ge-Si alloy.

We have analysed the major processes induced by ultrashort laser pulses in amorphous Ge and Si layers under the conditions of our experiments and considered their consequences for Ge/Si multilayer stacks. This includes laser ionization kinetics in Si and Ge under excitation by ultrashort near-infrared laser pulses, heating of a-Ge and a-Si layers, their melting and melt fraction, and laser-induced thermal stresses. The analysis is based on the ionization rate equations, energy balance, and the theory of thermoelasticity.

It has been found that the laser energy directly absorbed by Si layers can provide only a slight heating much below the melting point. At the same time, Ge layers experience strong ionization close to or higher than the critical plasma density. Interestingly, for ps laser pulses, Ge ionization is provided by one-photon absorption while, in fs ionization regime at the wavelength of 1500 nm, two-photon absorption plays a significant role in generation of a so-called solid plasma.

According to the energy balance analysis for ps laser pulses at F_0 below ~70 J/cm^2, partial melting of germanium occurs that should lead to formation of NC-Ge phase without affecting the Si layers. At slightly higher fluences, increasing the fraction of molten germanium may lead to the formation Ge-Si bonds at the interfaces between the layers. Irradiation with laser fluences ≥ 100 J/cm^2 provides complete melting of the Ge layers and their heating well above the melting point. As a result of the heat transfer from Ge to Si layers, the latter may receive enough heat for partial or even complete melting that depends on laser fluence and the adjacent molten layer should experience interdiffusion.

At fs irradiation regimes, the theory shows that a-Ge layers start to melt only at fluences of ~100 J/cm^2. However, estimations of thermal stresses yield the high values up to 0.97 GPa at 100 mJ/cm^2 and we note that such a level of stresses can cause solid-phase explosive crystallization of the Ge layers at low laser fluences [6]. Additionally, the evaluated levels of the density of electron-phonon pairs can induce destabilization of the amorphous lattice leading to ultrafast nonthermal phase transition that however calls for further studies.

References

1. T. Imajo, T. Ishiyama, N. Saitoh, N. Yoshizawa, T. Suemasu, K. Toko, ACS Appl. Electron. Mater. **4**, 269-275 (2022).
2. A.V. Kolchin, D.V. Shuleiko, A.V. Pavlikov, S.V. Zabotnov, L.A. Golovan, D.E. Presnov, V.A. Volodin, G.K. Krivyakin, A.A. Popov, P.K. Kashkarov, Tech. Phys. Lett. **46**, 560-563 (2020).
3. C. Korkut, K. Çınar, İ. Kabaçelik, R. Turan, M. Kulakcı, A. Bek, Cryst. Growth Des. **21**, 4632-4639 (2021).
4. V.A. Volodin, G.K. Krivyakin, A.V. Bulgakov, Y. Levy, J. Beránek, S. Nagisetty, Z. Bryknar, N.M. Bulgakova, P.V. Geydt, A.A. Popov, Proc. SPIE **12157**, 1215702 (2022).
5. V.A. Volodin, D.V. Marin, V.A. Sachkov, E.B. Gorokhov, H. Rinnert, M. Vergnat, J. Exp. Theor. Phys. **118**, 65-71 (2014).
6. N. Misra, L. Xu, M.S. Rogers, S.H. Ko, C.P. Grigoropoulos, M. Stutzmann, Phys. Status Solidi C **5**, 3264– 3270 (2008).

Thermal radiation in dipolar many-body systems

Svend-Age Biehs[1,*]

[1]Instut für Physik, Carl von Ossietzky Universität Oldenburg, Germany

Abstract. The framework of fluctuational electrodynamics for dipolar many-body systems is one of the working horse for theoretical studies of thermal radiation at the nanoscale which includes dissipation and retardation in a naturally way. Based on this framework I will discuss near-field thermal radiation in non-reciprocal and topological many-body systems. The appearance of the Hall and non-reciprocal diode effect for thermal radiation illustrates nicely the interesting physics in such systems as well as the edge mode dominated heat transfer in topological Su-Schrieffer-Heeger chains and a honeycomb lattices of plasmonic nanoparticles. In the latter, the theory allows for quantifying the efficiency of the edge-mode dominated heat transfer as function of the dissipation. Finally, I will present how the theoretical framework can be generalized to study far-field thermal emission of many-body systems close to an environment like a substrate, for instance. This theory might be particularly interesting for modelling thermal imaging microscopes.

1 Introduction

The theory for near-field thermal radation in dipolar many-body systems has been established more than ten years ago [1]. Since then it has been extended and due to its simplicity applied by many theoretical groups to study a great number of near-field effects as summarized in the review [2]. Here, I want to review some of our recent results in non-reciprocal and topological systems and finally show how the dipole approach can be applied for treating far-field thermal emissions of a nanoparticle assembly in a given environment.

2 Non-reciprocity

Magneto-optical or more generally non-reciprocal materials show interesting effects due to their non-reciprocal properties like the thermal radiative Hall effect [3], persistent currents [4], giant magneto-resistance [5, 6],and circular heat fluxes [7] as well as persistent angular momentum and spin of thermal radiation [7] have been highlighted. We review some of the recent developments in this direction [8] and discuss in some detail the possibility to rectify nanoscale radiative heat fluxes by means of non-reciprocal surface waves and propose a nanoscale heat flux rectifier or diode which can be controlled actively by means of externally applied fields [9]. We furthermore, reveal the spin coupling mechanism behind the observed heat flux rectification [10] as depicted in Fig. 1.

3 Topology

The fluctuational electrodynamics framework can also be applied to topological nanoparticle assemblies where near-field energy densities [11, 12] and heat fluxes [13] can be studied in the topological phase including retardation and dissipation. We discuss these properties for a Su-Schrieffer-Heeger chain as well as a honeycomb lattice of nanoparticles as depicted in Fig. 2. We show the impact of the edge mode dominated heat transfer with an emphasis on the impact of dissipation.

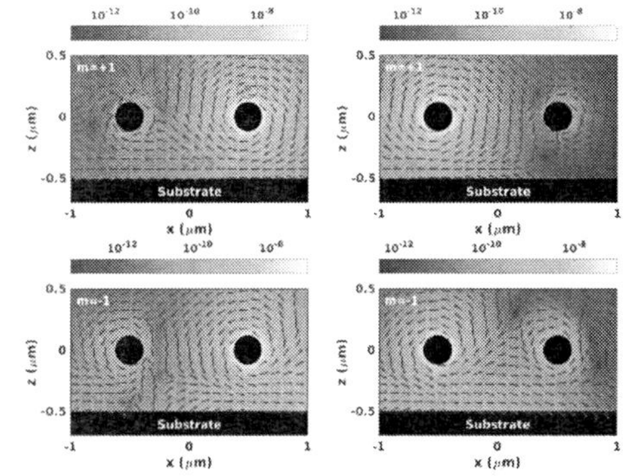

Figure 1. Directional spectral heat flux in forward and backward direction due to the coupling of the $m = \pm 1$ resonances with the spin-momentum locked non-reciprocal surface modes for InSb.

4 Far-field emission

Finally, a theory based on the fluctuational electrodynamics formalism is presented in order to treat thermal emission of dipolar objects close to an arbitrary environment which can be a substrate. This approach generalizes the single particle theory in Ref. [14] to the many-body case where the temperatures of the particles and environment

*e-mail: s.age.biehs@uol.de

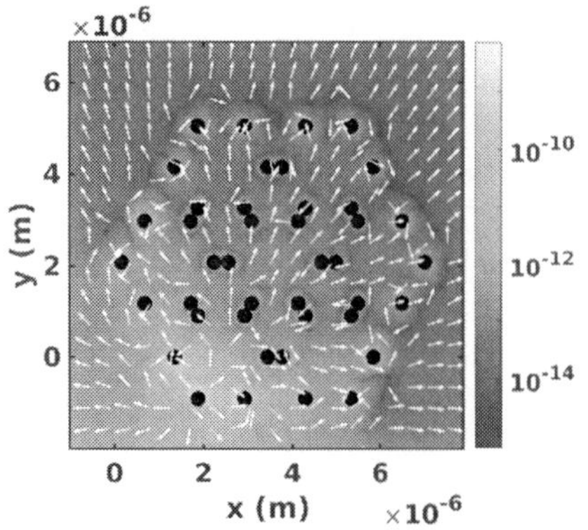

Figure 2. Spectral mean Poynting vector at the edge mode frequency in a honeycomb lattice of InSb nanoparticles where the particle at the lower left is heated with respect to the other particles.

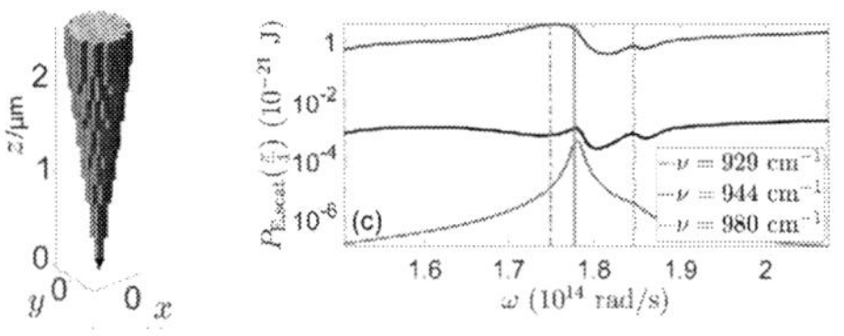

Figure 3. Far-field spectrum of a Si scattering tip (black curve) 10nm above a planar SiC substrate compared to the result of a single dipole model (small: red curve, large: black curve). The tip is treated within a discrete dipole approximation as in Ref. [16].

can be fixed separately. The versatility of the theory is demonstrated by discussing thermal emission spectra of nanoparticle assemblies and macroscopic objects within a discrete dipole approximation in close proximity to a planar substrate [15]. The latter can be used for modelling signals of scattering type near-field thermal microscopes as shown in Fig. 3.

References

[1] P. Ben-Abdallah, S.-A. Biehs, and K. Joulain, "Many-body radiative heat transfer theory," Phys. Rev. Lett. **107**, 114301 (2011).

[2] S.-A. Biehs, R. Messina, P. S. Venkataram, A. W. Rodriguez, J. C. Cuevas, and P. Ben-Abdallah, "Near-field radiative heat transfer in many-body systems," Rev. Mod. Phys. **93**, 025009 (2021).

[3] P. Ben-Abdallah, "Photon Thermal Hall Effect," Phys. Rev. Lett. **116**, 084301, (2016).

[4] L. Zhu and S. Fan, "Persistent Directional Current at Equilibrium in Nonreciprocal Many-Body Near Field Electromagnetic Heat Transfer," Phys. Rev. Lett. **117**, 134303 (2016).

[5] I. Latella and P. Ben-Abdallah, "Giant Thermal Magnetoresistance in Plasmonic Structures," Phys. Rev. Lett. **118**, 173902, (2017).

[6] R. M. Abraham Ekeroth, P. Ben-Abdallah, J.C. Cuevas, and A. Garcia Martin, "Anisotropic Thermal Magnetoresistance for an Active Control of Radiative Heat Transfer," ACS Photonics **5**, 705 (2017).

[7] A. Ott, P. Ben-Abdallah, and S.-A. Biehs, "Circular heat and momentum flux radiated by magneto-optical nanoparticles," Phys. Rev. B **97**, 205414 (2018).

[8] A. Ott, R. Messina, P. Ben-Abdallah, S.-A. Biehs, "Magnetothermoplasmonics: from theory to applications," J. Photon. Energy **9**, 032711 (2019).

[9] A. Ott, R. Messina, P. Ben-Abdallah, S.-A. Biehs, "Radiative thermal diode driven by nonreciprocal surface waves," Appl. Phys. Lett. **114**, 163105 (2019).

[10] A. Ott and S.-A. Biehs, "Thermal rectification and spin-spin coupling of non-reciprocal localized and surface modes," Phys. Rev. B **101**, 155428 (2020).

[11] A. Ott and S.-A. Biehs, "Radiative heat flux through a topological Su-Schrieffer-Heeger chain of plasmonic nanoparticles," Phys. Rev. B **102**, 115417 (2020).

[12] A. Ott, Z. An, A. Kittel, and S.-A. Biehs, "Thermal near-field energy density and local density of states in topological one-dimensional Su-Schrieffer-Heeger chains and two-dimensional Su-Schrieffer-Heeger lattices of plasmonic nanoparticles," Phys. Rev. B **104**, 165407(2021).

[13] A. Ott and S.-A. Biehs, "Topological near-field heat flow in a honeycomb lattice," Int. J. Heat Mass Transf. **190**, 122796(2022).

[14] K. Joulain, P. Ben-Abdallah, P.-O. Chapuis, Y. De Wilde, A. Babuty, and C. Henkel, "Strong tip sample coupling in thermal radiation scanning tunneling microscopy," J. Quant. Spectr. Rad. Transf. **136**, 1 (2014).

[15] F. Herz and S.-A. Biehs, "Generalized coupled dipole method for thermal far-field radiation," arXiv:2112.12016.

[16] S. Edalatpour, V. Hatamipour, and M. Francoeur, "Spectral redshift of the thermal near field scattered by a probe," Phys. Rev. B **99**, 165401 (2019).

https://doi.org/10.1051/epjconf/202226607002

Quantum levitation of photonic structures

Sol Carretero-Palacios[1,*]

[1]Universidad Autónoma de Madrid, Departamento de Física de Materiales, 28049 Madrid, Spain

Abstract. The Casimir-Lifshitz force originates from the quantum vacuum fluctuations of the electromagnetic field. This force is especially intense between interacting objects at nanoscale distances, and it can be attractive or repulsive depending on the optical properties of the materials (amongst other parameters). This fundamental phenomenon is at the heart of the malfunctioning of nano- and micro-electromechanical devices (NEMS and MEMS) that integrate many of the gadgets we use in our daily lives. Absolute control over these forces would make it possible to suppress adhesion and friction in these NEMs and MEMs. Here, we will show the possibility of controlling the Casimir-Lifshitz force by tuning the optical properties of the interacting objects. Specifically, we will present diverse examples of quantum levitation based on the Casimir-Lifshitz force of self-standing thin films comprising multilayer structures and films with spatial inhomogeneities (caused by imperfections, pores, inclusions, density variations, etc).

As it follows from the extended Lifshitz theory, the nature (attractive or repulsive) and intensity of the Casimir-Lifshitz force ($F_{(C-L)}$) depend, amongst others, on the optical properties of the interacting materials [1].

Here we show theoretically how a fine tuning of the materials optical properties allows modifying $F_{(C-L)}$ in a controlled manner, and in turn, the quantum trapping of planar strutures. The selected materials (polystyrene (PS), silicon dioxide (SiO_2), silicon (Si), and glycerol) display remarkable optical quality, experimentally manageable thicknesses, and unlike alcohols, stable dissipation properties. Based on previous results where we showed that thin films of SiO_2 and PS facing the same Si substrate in glycerol display $F_{(C-L)}$ of different sign and comparable intensity [2], we create photonic structures that combine such materials in either multilayer structures (ML) [3] or in composites [4], to modulate $F_{(C-L)}$ and the corresponding quantum trapping of such structures.

Results and Discussion

To calculate $F_{(C-L)}$, we employ the Lifshitz's theoretical formalism expressed for an arbitrary layered system [5]. In the case of multilayer structures, optical properties are computed by means of the Transfer Matrix Method formalism [6,7], while for nanocomposites (i.e, a homogeneous matrix embedding spherical nanoinclusions), our developed approach is based on the Monte Carlo method, which integrates Fresnel coefficients, and scattering Mie theory. The complex effective dielectric permittivity of the inhomogeneous material,

$$\varepsilon_{eff}(\omega) = \varepsilon'_{eff}(\omega) + \varepsilon''_{eff}(\omega) \quad (1)$$

is then extracted in a reverse process using an oscillatory model that fits the optical characteristics of the composite material. The resulting value of $\varepsilon_{eff}(\omega)$ is employed to calculate $F_{(C-L)}$ when the object approaches a substrate.

First, we spot a physical mechanism that enables the manipulation of $F_{(C-L)}$ in multilayer dielectric nanostructures due to optical interference effects (Fig. 1a). Optical interference effects yield high electric-field localization within the nanostructure, which in turn reshapes the partial absorptance of the constituent materials at the UV frequency range, modifying the $F_{(C-L)}$ experienced by the system. Figure 1b shows the spectral and spatial absorption per unit volume, δA, for a multilayer structure of 40 nm total thickness, comprising 50% of SiO_2 and 50% of PS arranged in 2 bilayers, immersed in glycerol over a Si substrate. We find that strong absorption modifications within the multilayer structure due to light trapping inside the system at UV frequencies allow to tune $F_{(C-L)}$ by keeping the material volume constant and stratifying it in different number of bilayers.

Second, we examine composite materials in which an otherwise homogeneous SiO_2 film contains PS nanospherical inclusions of different size and concentration. These nanoinclusions could account for the presence of imperfections, pores, inclusions, or density variations in real materials. In this case, we propose an effective medium approximation that accounts for the effect of inclusions and find that an unprecedented and counterintuitive intense repulsive Casimir−Lifshitz force arises as a result of the strong optical scattering and absorption size-dependent resonances caused by their

* Corresponding author: sol.carretero@uam.es

presence. Our results imply that the proper analysis of quantum trapping effects requires comprehensive knowledge and a detailed description of the potential inhomogeneity present in the materials. Figure 2 shows the effective dielectric function, in Matsubara frequencies, attained for a 1000 nm thick SiO_2 film containing 10% of nanospherical PS inclusions using our developed approach in which photon resonances are taken into account (in green). Contrary to effective medium approximations such as the Maxwell-Garnett model, which neglect photon resonances of nanoinclusions (in grey), here we demonstrate that the effects of scattering on optically dense inclusions lead to a decrease in the effective dielectric function when the radius of the inclusions increases (Fig. 2a). This counterintuitive effect, that does not agree with the predictions of effective medium approach, gives rise to a less attractive $F_{(C-L)}$ at short separation distances, bringing the system to larger quantum trapping distances (Fig. 2b).

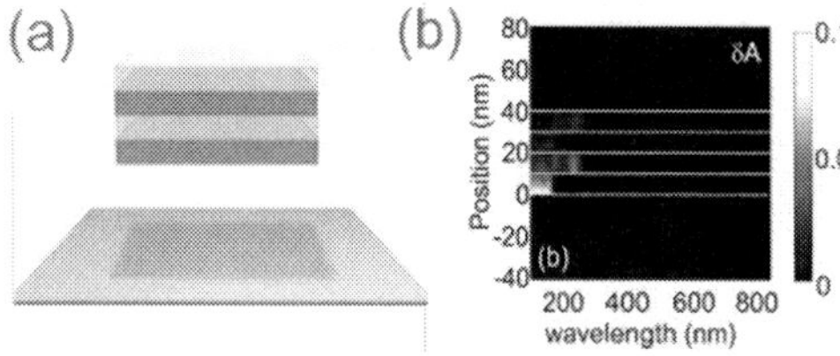

Fig. 1. (a) Schematics of a multilayer structure made of SiO_2 and PS slab bilayers, immersed in glycerol over a Si substrate. (b) Spectral and spatial distribution of absorption per unit volume, δA, of a representative example of a multilayer made of 2 bilayers. The specific geometrical parameters are: nanostructures comprising 50% SiO_2 and 50% PS, with a total thickness of 40 nm distributed in 2 bilayers (i.e., each layer thickness corresponds to 10 nm), immersed in glycerol with the SiO_2 layer facing a Si wall. As a guide for the eye, limits of each material layer of 10 nm thickness are shown with horizontal lines.

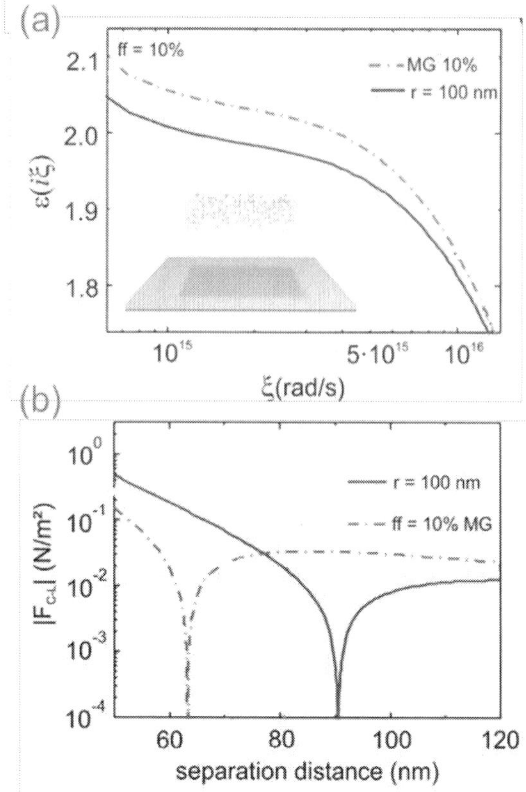

Fig. 2. (a) Effective dielectric function, evaluated at Matsubara frequencies, of a 1000 nm thick SiO2 film with PS nanospherical inclusions of radious r = 100 nm, in a 10% volume concentration (in green). For comparison, results attained using the Maxwell-Garnett effective medium approximation are shown in grey. (b) $F_{(C-L)}$ attained for the same system as in panel (a), when the composite material is immersed in glycerol and faces a Si substrate. Results obtained with our novel approach and using the Maxwell-Garnett method are depicted in green and grey colors, respectively.

Conclusions

We have shown that it is possible to control and manipulate $F_{(C-L)}$ through the optical properties of the interacting materials in the plane-parallel configuration. On the one hand, nanostructuring the same volume of dielectric materials in diverse multilayer configurations allows accessing $F_{(C-L)}$ of attractive or repulsive nature due to optical interference effects that alter the absorption losses of the multilayer. On the other hand, have shown that quantum trapping is strongly dependent on the characteristics of the potential and, likely, spatial inhomogeneity present in the system (porosity, impurities, nanoinclusions, etc), as the photon resonances they present, alter significantly the absorption properties of the composite material, and thus, the $F_{(C-L)}$. These results are of potential interest in the design and development of novel strategies to reduce malfunctioning of NEMs and MEMs.

References

1. E. M. Lifshitz. Sov. Phys. JETP, **2**:73-83 (1956).
2. V. Esteso et al. J. Phys. Chem. C., **119**, 5663–5670 (2015).
3. V. Esteso et al. Phys. Rev. A., **101**, 033815 (2020).
4. V. Esteso et al. J. Phys. Chem. Lett. 13, 4513-4519 (2022).
5. Dzyaloshinskii, I.E.; Lifshitz, E.M.; and Pitaevskii, L.P. Adv. Phys. **10 (38)**, 165-209 (1961).
6. Thompson, R. C. Optical Waves in Layered Media. J. Mod. Opt. (1990).
7. Hecht, E. Hecht Optics. Addison Wesley (1998)

Polycrystalline MoO_3 films fabricated by pulsed laser deposition for infrared polarization manipulation.

Daniele Ceneda[1,*], *Federico Vittorio* Lupo[2], *Maria Pia* Casaletto[3], *Roberto* Macaluso[2], Marco Centini[1], *Sina* Abedini Dereshgi[4], *Koray* Aydin[4], *Maria Cristina* Larciprete[1].

[1]Department of Basic and Applied Sciences for Engineering, SAPIENZA, University of Rome, Via A. Scarpa 16, 00161, Roma, Italy
[2]Department of Engineering, University of Palermo, Viale delle Scienze, Ed. 9, Palermo, 90128, Italy
[3]National Research Council (CNR), Institute of Nanostructured Materials (ISMN), Via Ugo La Malfa 153, Palermo, 90146, Italy
[4]Department of Electrical and Computer Engineering, Northwestern University, Evanston, Illinois 60208, United States

Abstract. We performed infrared optical characterization of polycrystalline MoO_3 films deposited by pulsed laser deposition on fused silica substrates. Several samples have been fabricated using different parameters such as temperature and oxygen pressure. Our analysis shows that under appropriate fabrication conditions it is possible to obtain a dominant α-phase film, with a well-defined, normal to surface (z-axis) orientation. These results are confirmed by reflection spectra performed at 45° incidence angle revealing a strong modulation of the sharp z-phonon Reststrahlen band as a function of the incident field linear polarization.

1 Introduction

The development of a photonic platform in the mid infrared range is motivated by applications, such as detection of traces of biomolecules and harmful/explosive substances, passive radiative cooling as well as devices for thermal imaging and for medical diagnostics. Recent promising solutions are based on the exploitation of polar materials [1] including ultra-thin van der Waals (vdW) materials such as MoO_3, MoS_2, Ga_2O_3, hBN [2-3]. Several works reported on the great potential of polar materials for IR sensing applications up to terahertz (THz) regime [4-7] and for the realization of compact IR photonic devices [8-10]. Among vdW materials, Molybdenum trioxide (α-MoO_3) is attracting great attention [11] as it supports surface phonon polaritons (SPhPs) in three different wavelength bands for the three orthogonal directions (range 10-20 μm) being a biaxial natural hyperbolic material [12-14].

Despite this huge natural potential of the material, the development of a new highly versatile and compact MoO_3 based platform for IR photonics is hampered by the lack of availability of good quality scalable films and/or multilayer stacks. According to literature, MoO_3 for IR photonics and polaritonics is mostly used in the form of crystalline flakes, coming from bulk MoO_3 crystals through a complex mechanical exfoliation process followed by a transfer to an appropriate substrate. The realization of single alpha phase, oriented, large area MoO_3 film is still an open technological challenge.

2 Discussion

In this work we report on fabrication and characterization of polycrystalline MoO_3 films deposited by pulsed laser deposition (PLD) on quartz substrates. Our results show that PLD can be properly adapted to obtain α-phase MoO_3 films at reasonably low temperatures (e.g. 500 °C) with a considerable strong off plane anisotropy related to a well oriented z-axis. All the samples have been analysed by X-ray diffraction (XRD) and atomic force microscopy (AFM) to confirm the orthorhombic α-phase of MoO_3 films, the polycrystalline nature and assess films morphology.
Optical characterization in the IR has been carried out by Fourier-transform infrared spectroscopy (FT-IR) in reflection mode with an incidence angle of 45° and a rotating linear polarizer. As an example we report in Fig. 1 and Fig. 2 reflection spectra for two samples obtained with different fabrication conditions exhibiting very different optical properties in the IR range.

In Fig. 1 we show the reflection spectrum of a 500 nm-thick MoO_3 film deposited on fused silica substrate 500° C and 10^{-1} mbar oxygen pressure. It can be noticed a strong tunable peak around 1000 cm^{-1} related to phonon modes along the z-axis. High reflectivity is obtained for p-polarized input field corresponding to a strong z-component of the e.m. signal. The reflection peak

* Corresponding author: daniele.ceneda@uniroma1.it

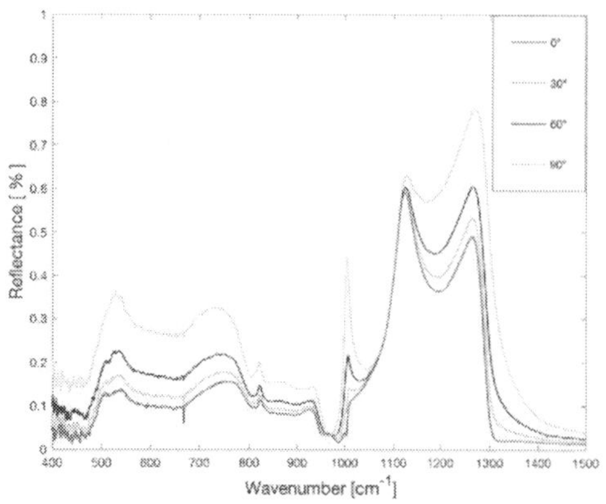

Fig. 1. FTIR reflection spectrum measured at an angle of incidence of 45° as a function of the different input polarization on MoO_3 film 500 nm-thick, grown by PLD on fused silica substrate, T=500°C, $P_{O2}=10^{-1}$ mbar.

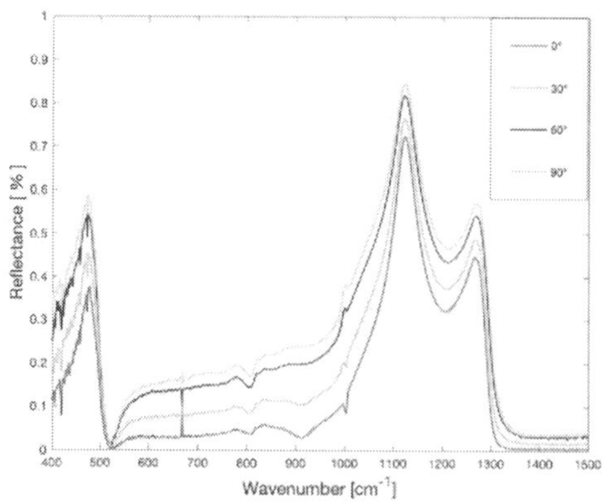

Fig. 2. FTIR reflection spectrum measured at an angle of incidence of 45° as a function of the different input polarization on MoO_3 film 500 nm-thick, grown by PLD on fused silica substrate, T=400°C, $P_{O2}=10^{-1}$ mbar.

is strongly suppressed for s-polarized incident field, corresponding to in-plane fields components. This is a signature of a well oriented z-axis of the sample. On the other hand, the spectrum depicted in Fig. 2 does not show modulation of the reflection peak as a function of the input polarization. In this case the 500 nm-thick MoO_3 film has been fabricated under the same oxygen pressure conditions as the previous sample but at lower temperature (400°C). XRD shows that both films are polycrystalline and exhibit only the orthorhombic α-phase. This means that, despite the presence of only the single alpha phase, the MoO_3 film grown at 400 °C does not present any peak modulation.

3 Conclusions

We have fabricated and characterized MoO_3 films deposited by PLD. Our analysis shows that under proper fabrication parameters a dominant α–phase is obtained, and a good alignment of the z-crystal axis is achieved. We also show that the samples fabricated with different deposition parameters do not display the same behaviour. Thus our results highlight a path to fabricate large scale and low cost MoO_3 films for infrared polarization management devices in the IR.

References

1. Y.Chen et al., ACS Photonics 1, 718724 (2014).
2. M. Tamagnone et al., Science Advances 4, eaat7189 (2018).
3. D. N. Basov et al., Science 354, 6309 (2016).
4. J.D.Caldwell et al., Nanophotonics 4, 44-68 (2015).
(2010).
5. I. Balinet al., Appl. Phys. Lett. 94, 111112 (2009).
6. B.Neuner et al., J.Phys. Chem. C, 114, 7489–7491
7. T.G. Folland et al., ACS Photonics, 7, 614621 (2020).
8. M.C. Larciprete et al., Journal of the Optical Society of America B, 34, 1459 (2017).
9. S. Abedini Dereshgi et al., Nature Communications 11, 5771 (2020)
10. G Álvarez-Pérez et al. Advanced Materials 32 (29), 2070220.
11. G.Álvarez-Pérez et al., Adv.Materials 32, 1908176 (2020).
12. Mingyuan Chen et al., Nature Materials 19, 1307–1311(2020).
13. G.Hu et al., Nature 582, 209-213 (2020).
14. Z. Zheng et al., Sci. Adv. 5, eaav8690 24 (2019).

Mid-infrared narrowband polarization management with Al doped ZnO-$ZnWO_4$ eutectic composites.

Marco Centini[1], *Maria Cristina* Larciprete[1], *Concita* Sibilia[1], *Dorota. A.* Pawlak[2]

[1]Department of Basic and Applied Sciences for Engineering, SAPIENZA, University of Rome, Via A. Scarpa 16, 00161, Roma, Italy
[2]ENSEMBLE3 sp. z o.o., Wolczynska 133, 01-919 Warsaw, Poland

Abstract. We report a narrowband polarization-dependent reflectivity from Al-doped ZnO/ZnWO4 self-assembled eutectic composites in the mid-infrared range. Our results show a reflectivity modulation from 0.05 to 0.75 for two orthogonal polarizations of the incident field with a 10% Al concentration. Acting as natural polarizing filters these eutectic composites could open the way to the future development of low-cost photonics components in the IR.

1 Introduction

Control and manipulation of mid-infrared (mid-IR) radiation by low cost and compact devices is a crucial task towards the development of a photonic platform for sensing of biomolecules and harmful/explosive substances as well as for thermal imaging and diagnostics. Recently proposed polarization sensitive applications in mid-IR relies on artificial metamaterial design and fabrication[1] or in the peculiar properties of van der Waals (vdW) materials exhibiting natural hyperbolicity[2]. Another interesting strategy might be provided by the exploitation of self-assembled ZnO/ZnWO4 eutectic composites [3].

We have recently shown that the optical linear properties of these kind of structures exhibits a natural narrowband strong polarization dependence of transmitted signal at $\lambda=(397\pm3)$ nm [4]. The effect is due to the refractive index matching of the two constituents so that the composite medium appears homogenous to light propagating at a specific wavelength and polarization while it produces strong scattering losses in every other conditions. Here we show that a similar polarization sensitive behaviour can be obtained in reflection in the IR range (around 11.6 μm).

2 Discussion

We characterized the polarization dependent reflectance in the 2-16 micron range from Al doped eutectic ZnO/ZnWO4 with different concentration of Al. Reflection spectra have been performed with a FTIR microscope (Bruker Hyperion). Our results show that it is possible to obtain a polarization sensitive reflection band by taking advantage of the optical phonon properties of ZnWO4 and excitation of surface phonon polaritons at Al-ZnO/ZnWO4 interfaces. Indeed, the composites under investigation have a lamellar geometry as shown in the inset of Fig.1 (image taken with an optical microscope). The dark regions are composed of Al-ZnO while the brighter regions correspond to ZnWO4. We show polarization dependent reflection spectra for two samples having different concentration of Al. The high reflectivity band around 850 cm^{-1} (11.6 μm) clearly visible in Fig 1a-b is due to a phonon Reststrahlen band of the $ZnWO_4$ and it reaches its maximum value when the incident field is polarized along the direction of the lamellas. On the other hand, when the input field is polarized along the orthogonal direction with respect to the lamellas a strong reduction of the reflected signal is observed. This effect is related to the excitation of surface phonon polaritons at the Al-ZnO/ZnWO4 interfaces. We note that the polarization sensitivity of the reflected signal in higher for Al concentration of 10% (Fig. 1a) and it is very weak for Al concentration of 1% (Fig. 1b). Higher concentration of Al seems to enhance surface waves coupling at the interfaces, however from the preliminary optical images we also note that higher Al concentration leads to thicker and more defined lamellas. This behaviour is under further investigation. Another interesting effect related to the presence of Al is evidenced at shorter wavelengths. We also report a modulation of the reflection spectra in the 3-5 micron wavelength ranges as a function of Al concentration when the input field is polarized in the orthogonal direction with respect to the lamellas. Our results suggest that this effect could be interpreted with the plasma frequency shift of the Al-ZnO [5] for different concentrations of Al and/or with the optical thickness of the lamellas.

* Corresponding author: marco.centini@uniroma1.it

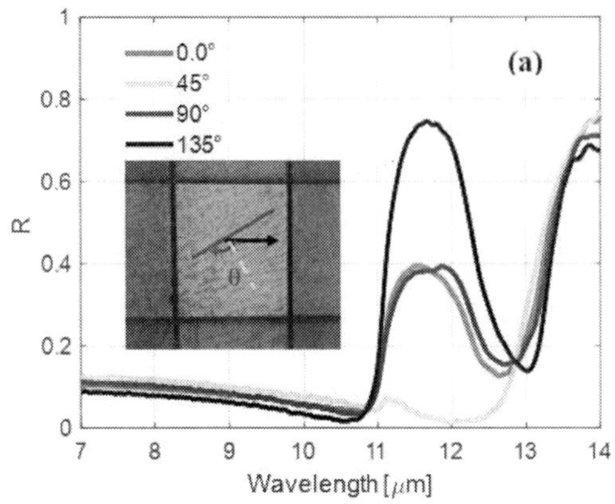

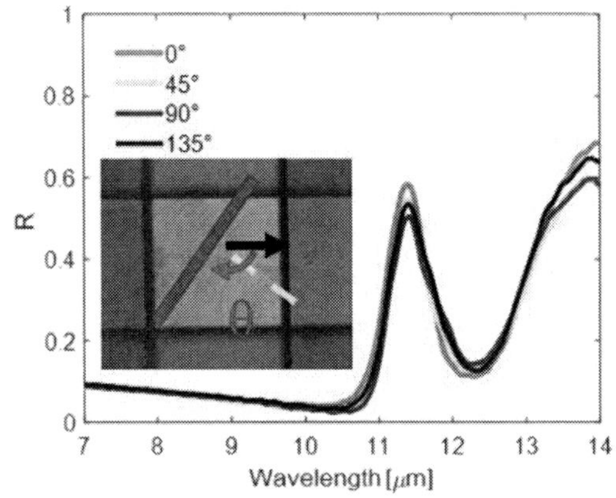

Fig. 1. **(A)** FTIR reflection spectra for the Al(0.1)-ZnO/ZnWO4 eutectic composite as a function of the input field polarization angle (45° corresponds to perpendicular to the lamellas, 135° is along the lamellas). **(B)** FTIR reflection spectra for the Al(0.01)-ZnO/ZnWO4 eutectic composite as a function of the input field polarization angle (30° corresponds to perpendicular to the lamellas, 120° is along the lamellas).

3 Conclusions

In conclusion the complex geometry of the eutectic composite reveals peculiar properties both in the visible and in the IR ranges, The investigated structures act as natural narrowband polarizing filters. The Al doping adds versatility opening the way to new interesting photonics applications in the IR. However the role of Al doping in these eutectic composites as well as its distribution is not clear and it will be the subject for future studies.

4 Acknowledgments

Authors would like to thank Ensamble3 UE project. Authors would also like to thank Mariangela Cestelli Guidi and Martina Romani at INFN-LNF Dafne-Light for their hospitality and support with the FTIR microscopy.

References

1. Y. Poddubny et al. Nat. Photon. 7, 948 (2013)
2. S. Abedini Dereshgi et al., Nat. Comm. 11, 5771 (2020)
3. M. Tomczyc et al, J Mater Sci, 56,11219 (2021)
4. P. Osewski et al., Adv. Opt. Mat. 8, 7, 1901617 (2020).

Enhancing the solar-to-thermal energy conversion in high vacuum flat plate solar collectors

Roberto Russo[1,*], *Davide* De Maio[1,2,#], *Carmine* D'Alessandro[1,2,#], *Daniela* De Luca[1,3], *Antonio* Caldarelli[1,2], *Eliana* Gaudino[1,2], *Marilena* Musto[2,1], *Emiliano* Di Gennaro[3,1]

[1]Consiglio Nazionale delle Ricerche, Istituto di Scienze Applicate e Sistemi Intelligenti, 80131 Napoli, Italy
[2]Deparment of Industrial Engineering, Università degli Studi di Napoli "Federico II", 80125 Napoli, Italy
[3]Department of Physics, Università degli Studi di Napoli "Federico II", 80125 Napoli, Italy
[#] Present Address: TVP Solar SA, 10 rue du Pré-de-la-Fontaine ZIMEYSA 1242 Satigny (GE), Switzerland

Abstract. In solar flat plate collectors, the high vacuum insulation suppresses the convective losses increasing the collector efficiency. The solar-to-thermal energy conversion efficiency in such solar thermal collectors is mainly defined by the optical and radiation losses of the selective solar absorber. We present the full process of design, optimization, fabrication, and characterization of multilayer coatings specifically thought for working in high vacuum flat solar thermal collectors at different operating temperatures, from 100 °C to 300 °C. We discuss the relative importance of absorptance and emittance in determining the collector thermal efficiency. The robustness of the performance of the coatings related to the unpreventable errors in layer thickness during the manufacturing stage is also considered through a genetic optimisation algorithm.

Exploiting renewable energies such as solar, wind, waves, hydropower, etc., could open a path towards a cleaner world, through the mitigation of the greenhouse gases in the atmosphere.
Solar energy is by far the most abundant source of renewable energy [1]: despite about half of the solar radiation being either absorbed or reflected by the clouds and the atmosphere, the Earth's surface still receives enough power to meet the demands of the whole world [2,3].
Among the different types of solar energy collectors, the high vacuum flat plates (HVFPs) present clear advantages. The high vacuum insulation reduces to zero the convective and conductive losses, guaranteeing the best performance at low and medium temperature operation. In such collectors, the maximum operating temperature and the efficiency are limited by the thermal radiation losses.
The core component of HVFP collectors is the Selective Solar Absorber (SSA): ideally, its spectral absorptivity $\alpha(\lambda)$/emissivity $\varepsilon(\lambda)$[a] can be defined as follows:

$$\alpha(\lambda) = \begin{cases} 1 & if\ \lambda \leq \lambda_c \\ 0 & if\ \lambda > \lambda_c \end{cases} \qquad (1)$$

where λ_c is the wavelength that maximizes the absorber thermal performance and it depends on the working temperature and the solar incoming power [4].

The solar absorptance α_S and thermal emittance $\varepsilon(T)$ can be evaluated from the reflectivity spectra $\rho(\lambda)$ of the multilayer solar absorbers using Eqs. (1) and (2), respectively [5]:

$$\alpha_S = \frac{\int_{0\,\mu m}^{\infty\,\mu m}[1-\rho(\lambda)]S(\lambda)d\lambda}{\int_{0\,\mu m}^{\infty\,\mu m}S(\lambda)d\lambda} \qquad (2)$$

$$\varepsilon(T) = \frac{\int_0^{\infty}[1-\rho(\lambda)]E_{bb}(\lambda,T)d\lambda}{\int_0^{\infty}E_{bb}(\lambda,T)d\lambda} \qquad (3)$$

where $S(\lambda)$ and $E_{bb}(\lambda, T)$ are the solar radiation spectrum and the blackbody radiation spectrum, respectively, depending on the radiation wavelength λ and temperature T.

Among all the possible designs investigated [6] (cermets, nano and metamaterials, photonic crystals, etc.) selective absorbers based on multilayer allow controlling the thermal emission while ensuring excellent thermal stability and high solar absorption[7]. Indeed, the low emissivity substrate is responsible for the low emittance and the absorption of the solar radiation is guaranteed from the multiple reflections at the interfaces between the layers composing the multilayer.
Multilayers also satisfy the requirement of an easy fabrication for large-scale production in roll-to-roll systems, however, the performance of multilayer-based

[a] According to the Kirchhoff's law of thermal radiation, for any object in thermal equilibrium $\alpha(\lambda) = \varepsilon(\lambda)$.

 https://doi.org/10.1051/epjconf/202226607005

absorbers can be strongly affected by the thickness variation of the various layers. Such variations can be difficult to avoid at the industrial level and one must take into consideration the robustness of the performance of the coating for unpredicted errors in the thickness of the various layers.

In this work, we present different multilayer selective absorbers based on SiO_2, Cr_2O_3, and Cr, sputter-deposited on a low emissive copper substrate. Those are optimized for different working temperatures, from 100 °C to 300 °C.

To assess which is the thickness package that ensures the highest performance and proper robustness for given errors in layer thickness, a fast and reliable genetic algorithm for the optimization of the coating, particularly for emerging HVFPC technology, has been developed in MATLAB. The code is based on the transfer matrix method and relay on the experimentally measured refractive index of the materials constituting the multilayers[8].

To optimize the absorber properties, the algorithm calculates the reflectance $\rho(\lambda)$ of the stack at each wavelength, allowing the evaluation of the solar absorptance α_S and thermal emittance $\varepsilon(T)$ according to Eqs. (1) and (2), respectively. The values of α_S and $\varepsilon(T)$ are then used to calculate the coating solar energy conversion efficiency according to the equation:

$$\eta_{coat} = \alpha_S - \frac{\varepsilon(T)\, \sigma_{SB}(T^4 - T_{amb}^4)}{H_{abs}}$$

and the coating efficiency η_{coat} is maximized at different operating temperatures (T_{amb}(K) is the environmental temperature, H_{abs} (Wm-2) the Sun irradiated power, and σ_{SB} ($Wm^{-2}K^{-4}$) the Stefan-Boltzmann constant). At a given operating temperature the maximum efficiency can be obtained using different combinations of α_S and $\varepsilon(T)$ resulting from different layer thicknesses arrangements. The numerical solutions are then tested against layer thickness variations and the most stable solution is retained.

Figure 1 shows the results of the simulated reflectivity and thermal emittance of the three selective solar absorbers.

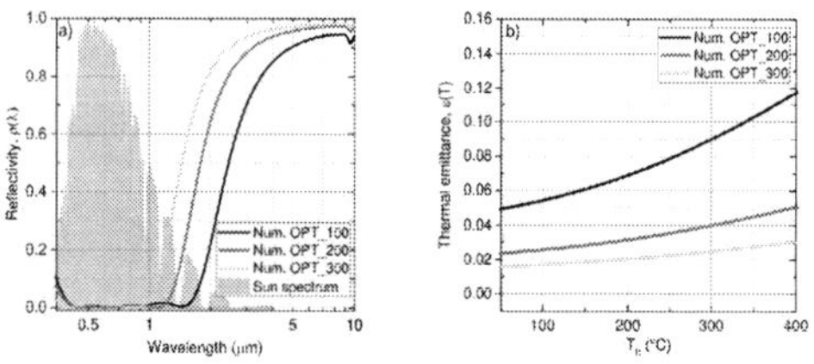

Fig. 1: Optical simulations. (a) Spectral reflectivity of three coatings optimized for a target temperature of 100 °C, 200 °C, and 300 °C (black, red, and green curve, respectively). Normalized Sun spectrum, grey filled area. (b) Temperature-dependent thermal emittance of three coatings optimized for a target temperature of 100 °C, 200 °C, and 300 °C (black, red, and green curves, respectively).

Multilayers have been deposited by a multi-target magnetron sputtering system and their optical and radiative properties were measured in a custom set-up [9]. The experimental reflectivity curves are in agreement with simulated ones and the thermal stability test indicates that the produced multilayer can safely work up to 300°C.

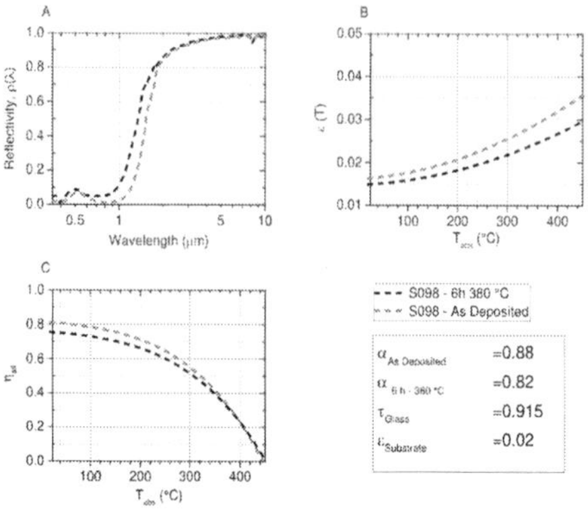

Fig. 2: Measured spectral reflectance of solar selective coating optimized to work at 300 °C before and after heating in vacuum at 380 °C for 6 hours. The

The obtained results indicate that multilayers are an excellent solution to achieving high efficiency with good thermal stability in HVFP collectors, contributing to the desired renewable energy transition.

References

1. R. Perez and M. Perez. "A fundamental look at energy reserves for the planet."
2. C. Lauterbach et al. Renewable and Sustainable Energy Reviews **16**, 5121–5130 (2012).
3. A. Shahsavari and M. Akbari. Renewable and Sustainable Energy Reviews **90**, 275–291 (2018).
4. Ning, Y. Et al. Sol. Energy Mater. Sol. Cells **206**, 110219 (2020).
5. Forouhi, A.R.; Bloomer, I. Phys. Rev. B **34**, 7018–7026 (1986).
6. P. Bermel et al. Annual Review of Heat Transfer **15**, 231–254 (2012)
7. D. De Maio et al. Energies **14**, 900 (2021)
8. D. De Maio et al. Solar Energy Materials and Solar Cells. **242**, 111749 (2022).
9. R. Russo R, M. Monti, F. di Giamberardino, V.G.Palmieri, Opt. Express **26**, A480 (2018).

Wide-field broadband CARS microscopy

Chiara Ceconello[1], *Federico* Vernuccio[1], *Alejandro* De la Cadena[1], *Arianna* Bresci[1], *Francesco* Manetti[1], *Subir* Das[1], *Renzo* Vanna[2], *Giulio* Cerullo[1,2], and *Dario* Polli[1,2,*]

[1]Department of Physics, Politecnico di Milano, P.zza Leonardo da Vinci 32, 20133 Milan, Italy
[2]CNR Institute for photonics and nanotechnologies (IFN), P.zza Leonardo da Vinci 32, 20133 Milan, Italy

Abstract. Coherent anti-Stokes Raman scattering is an extremely powerful non-linear optical (NLO) microscopy technique for label-free vibrational imaging allowing for a detailed study of biological samples in their native state. To overcome the long acquisition times associated with raster sample scanning required in NLO microscopy, which impair real-time investigation of fast biological dynamics, we employ here wide-field signal generation over a large field of view, covering tens of micrometers. To this aim, we exploit an innovative approach based on the use of an amplified femtosecond ytterbium laser source delivering high energy (≈μJ) pulses in the near infrared. This enables the generation of stable broadband Stokes pulses to measure the entire fingerprint region of the molecular vibrational spectrum, the richest in chemical information. Our results pave the way for future translational applications and clinical diagnostics with video-rate imaging capabilities. ©2022 The Authors

1 Introduction

Spontaneous Raman (SR) represents an extremely powerful investigation tool for life sciences allowing to characterize and map in 3D each component of a sample (like cells and tissues) in a label-free, remote and non-destructive manner. However, it suffers from a very low cross section, resulting in long acquisition times. The solution comes with coherent Raman scattering techniques that employ ultrashort pulses to promptly trigger the vibrational response [1]. Coherent anti-Stokes Raman Scattering (CARS) employs two narrowband synchronized picosecond pulses (pump and Stokes) to probe a single vibrational mode by detecting the signal at the blue-shifted anti-Stokes frequency. Broadband (B)-CARS increases the amount of extracted information by employing a narrowband pump pulse and a broadband Stokes that allows to probe an entire vibrational spectrum, thus combining high acquisition speed of CARS with the chemical specificity of SR. Traditionally B-CARS microscopy is performed in raster scanning fashion: tightly focused laser beams generate signal in one pixel at a time which is then acquired in a snake-like motion with a spectrometer. It can take up to minutes to generate a single image, even at the state of the art 3ms pixel dwell time [2], thus preventing real-time investigation of fast biological dynamics. Wide Field (WF) illumination on the other hand, allows signal generation over a large field of view, which is then acquired single shot with the help of a sCMOS camera. Non phase-matching WF illumination [3] represents a simple configuration for effective generation of WF-CARS. In this work, we present a novel configuration of WF-CARS microscopy with broadband and video-rate imaging capabilites. We employ an amplified ytterbium laser at 1035-nm central wavelength and 2-MHz repetition rate that enables white-light continuum (WLC) generation in bulk media: a compact, robust, simple and alignment-insensitive technique, that presents excellent long-term stability and low pulse-to-pulse fluctuations [4]. Unlike standard CARS setups, the red-shifted source (1035 nm pump, 1050-1300 nm Stokes) reduce multi-photon sample photo-damage, allowing the use of higher laser intensities on the sample. Furthermore, the reduced repetition rate allows for higher pulse energies, thus higher peak intensities that generate a stronger B-CARS signal thanks to the non-linear nature of the optical effect. As a test sample we used 8 μm Polystyrene (PS) beads in a solution of Dimethyl sulfoxide (DMSO). The spectral nature of the signal was assessed by performing single wavelenght WF-CARS imaging. Video-rate WF-CARS was demonstrated by collecting low acquisition time images at 2ms integration time.

2 Materials and methods

As shown in Figure 1, an amplified Ytterbium laser source delivers ≈ 270 femtosecond pulses at 1035 nm central wavelength, 2-MHz repetition rate and ≈ 5W average power. A polarizing beam splitter separates the fundamental in two beams. The first one goes through an etalon to generate narrowband pump pulses with ≈ 1.1 nm bandwith (10 cm^{-1}) guaranteeing sufficient spectral resolution for the vibrational Lorentian peaks. The second provides high enough pulse energy (≈ 1.2 μJ) to efficiently generate a stable white light supercontinuum in

*e-mail: dario.polli@polimi.it

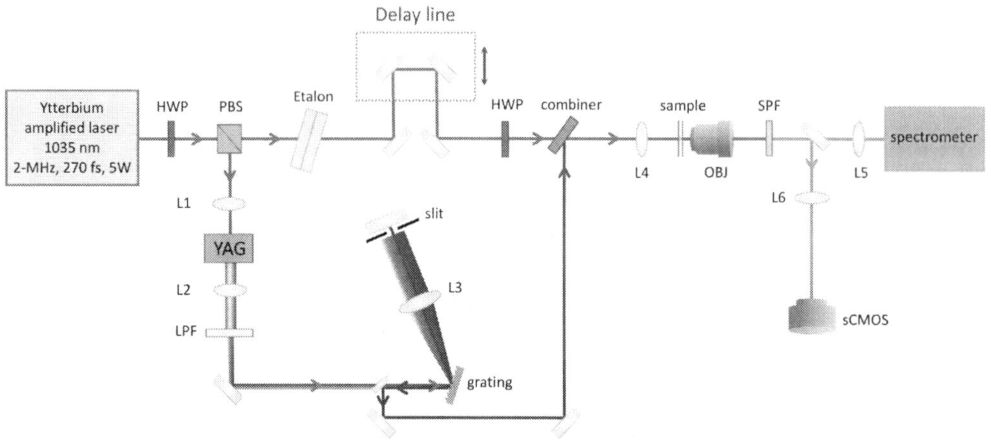

Figure 1. Scheme of the WF-CARS experimental setup. HWP: half-wave plate. PBS: Polarizing Beam Splitter. L1,L2,L3,L4,L5,L6: lenses. LPF: long-pass filter. OBJ: collection objective. SPF: short pass filter.

a 6mm thick YAG crystal. The fundamental is removed with a long pass filter that selects the red-shifted portion (1050-1300 nm) thus obtaining a broadband Stokes pulse that covers the whole fingerprint region (400-1800 cm $^{-1}$). In the Stokes path, a pulse shaper in 4f configuration with an adjustable mechanical slit mounted on a traslational stage allows for single-color selection. The two beams are then spatially and temporally recombined and loosely focused with a 100mm lens. The pump has a FWHM spot of 50-μm on the sample, with ≈210 mW average power, while the Stokes has a 60-μm spot and ≈10 mW average power. In this way it is possible to generate WF CARS signals from a 40-μm FWHM area, in a non-phase matched illumination configuration. The WF-CARS signal is collected with a 20x objective (NA=0.3), recorded with a 4.2 megapixels sCMOS placed after a short-pass filter to cut the Stokes and pump beams. WF-CARS imaging is performed single shot, acquiring the whole field of view at once.

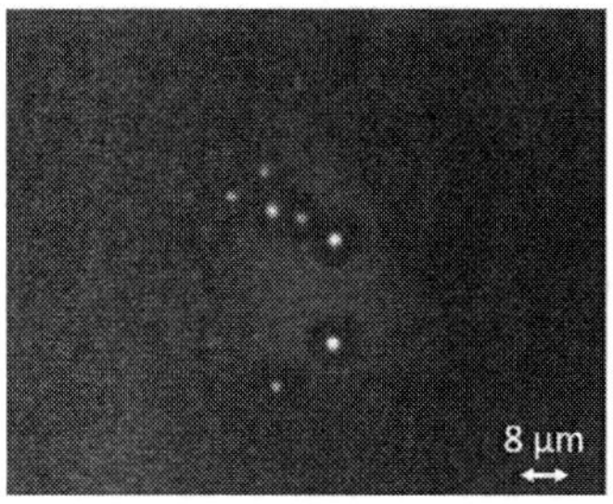

Figure 2. Wide-Field CARS image of 8-μm Polystyrene (PS) beads in a solution of Dimethyl sulfoxide (DMSO). 460x540 pixels, 2ms integration time.

For our experiment we imaged 8-μm Polystyrene (PS) beads in a solution of Dimethyl sulfoxide (DMSO) at various integration times. The video-rate capabilities of the setup were assessed by performing imaging with the mechanical slit of the 4f pulse shaper kept open, thus generating WF-CARS signal in the Ω 647-976 cm^{-1} spectral range. 460x540 pixels images of PS beads in DMSO were acquired with 2ms integration time (Figure 2). We then proved the spectral nature of the signal by performing hyperspectral WF-CARS measurements. We obtained quasi single wavelength Stokes beam by accurately closing the mechanical slit in the 4f pulse shaper varying its color by shifting the slit position thanks to a micrometric traslational stage. By acquiring different frames at the different Stokes wavelengths, we built three-dimensional hyperspectral WF-CARS data-cubes as a function of the vibrational frequency for the entire field of view. These results confirm the capability of generating stable WF-CARS signal, with forthcoming video-rate and broadband imaging possibilities.

References

[1] Zhang C., Cheng J.X., Perspective: Coherent Raman scattering microscopy, the future is bright, APL Photonics **3**, (2018)

[2] Camp C.H.J., Lee Y.J., Heddleston J.M., Hartshorn C.M., Hight Walker A.R., Rich J.N., Lathia J.D., Cicerone M.T., High-speed coherent Raman fingerprint imaging of biological tissues, Nature Photonics **8**, (2014)

[3] Lei M., Winterhalder M., Selm R., Zumbusch A., Video-rate wide-field coherent anti-Stokes Raman scattering microscopy with collinear nonphase-matching illumination, Journal of Biomedical Optics **16**, (2011)

[4] Dubietis, A., Tamošauskas, G., Šuminas, R., Jukna, V. Couairon, A., Ultrafast supercontinuum generation in bulk condensed media, Lith. J. Phys. **57**, (2017)

The comb waveguide: a new tool for strong interaction between atoms and light

Nikos Fayard[1,*], *Adrien* Bouscal[2], *Jeremy* Berroir[2], *Alban* Urvoy[2], *Tridib* Ray[2], *Sukanya* Mahapatra[3], *Malik* Kemiche[3], *Juan-Ariel* Levenson[3], *Jean-Jacques* Greffet[1], *Kamel* Bencheikh[3], *Julien* Laurat[2], and *Christophe* Sauvan[1,**]

[1]Université Paris-Saclay, Institut d'Optique Graduate School, CNRS, Laboratoire Charles Fabry, 91127 Palaiseau, France
[2]Laboratoire Kastler Brossel, Sorbonne Université, CNRS, ENS-Université PSL, Collège de France, 4 place Jussieu, 75005 Paris, France
[3]Centre de Nanosciences et de Nanotechnologies, CNRS, Université Paris-Saclay, 91120 Palaiseau, France

Abstract. Coupling quantum emitters and nanostructures, in particular cold atoms and waveguides, has recently raised a large interest due to unprecedented possibilities of engineering light-matter interactions. However, the implementation of these promising concepts has been hampered by various theoretical and experimental issues. In this work, we propose a new type of periodic dielectric waveguide that provides strong interactions between atoms and guided photons with an unusual dispersion. We design an asymmetric comb waveguide that supports a slow mode with a quartic (instead of quadratic) dispersion and an electric field that extends far into the air cladding for an optimal interaction with atoms. We compute the optical trapping potential formed with two guided modes at frequencies detuned from the atomic transition. We show that cold Rubidium atoms can be trapped as close as 100 nm from the structure in a 1.3-mK-deep potential well. For atoms trapped at this position, the emission into guided photons is largely favored, with a beta factor as high as 0.88 and a radiative decay rate into the slow mode 10 times larger than the free-space decay rate.

1 Introduction

Recently, the coupling of cold atoms with nanofibers or nanostructured waveguides has triggered the emergence of a new field of research known as waveguide QED. Photons travelling in the waveguide carry the information through long distances while atoms trapped in its vicinity can store it for long times.

These systems are a promising building block for quantum networks as shown recently by the experimental demonstrations of the heralded creation of a single collective excitation of atomic arrays [1], or a correlated photon transport. When the decay rate of the atoms inside the waveguide mode Γ_{1D} exceeds greatly the decay into every other mode Γ', we reach the strong coupling of waveguide QED where new phenomena in many-body physics appear such as the emergence of solitons dynamics or many-body localization.

A promising route to increase Γ_{1D} (or the beta factor $\beta = \frac{\Gamma_{1D}}{\Gamma_{1D}+\Gamma'}$) is to use periodic dielectric waveguides, i.e., waveguides with a periodic modulation of the refractive index along the propagation direction. Indeed, the radiative decay rate of a single atom in a guided mode is given by $\Gamma_{1D}/\Gamma_0 = n_g\sigma/(2A_{\text{eff}})$, where Γ_0 is the atomic decay rate in free space, σ is the absorption cross-section, n_g and A_{eff} are the group index of the mode and the effective area at the atom position [2, 3]. In periodic waveguides, the coupling between contrapropagating modes results in the opening of bandgaps in the dispersion relation and in the apparition of band edges where the group velocity $v_g = c/n_g$ goes to zero. Close to these peculiar points, periodic waveguides support slow guided modes with large n_g's, which, in turn, lead to increased values of Γ_{1D} and β.

The so called Alligator waveguide [3, 4], which is the most studied periodic waveguide up to now, was built upon this idea. However, it has two major flaws: a lack of accessibility of the trapping region (where the evanescent tail of the slow mode is large), and a very curved dispersion of the slow mode not robust enough against disorder in fabrication. Thus, only $N = 3$ atoms have been trapped close to this structure, with a moderate coupling to the slow mode $\Gamma_{1D} \sim \Gamma'$ ($\beta = 0.5$).

2 The comb waveguide

To improve those figures of merits, we designed an asymmetric comb waveguide [5] with the following assets:

- first, it can be suspended in air to provide easy access to the trapping region. As shown in Fig 1, the atoms can be trapped in the back of the structure
- second, the transverse asymmetry of the structure provides additional degrees of freedom to engineer the dispersion curve. We designed a completely new quartic dispersion $\omega \sim \Delta k^4$, that provides more resilience to our

*e-mail: fayardnikos@gmail.com
**e-mail: christophe.sauvan@institutoptique.fr

 https://doi.org/10.1051/epjconf/202226608002

structure against fabrication imperfections and enlarges its operation bandwidth.

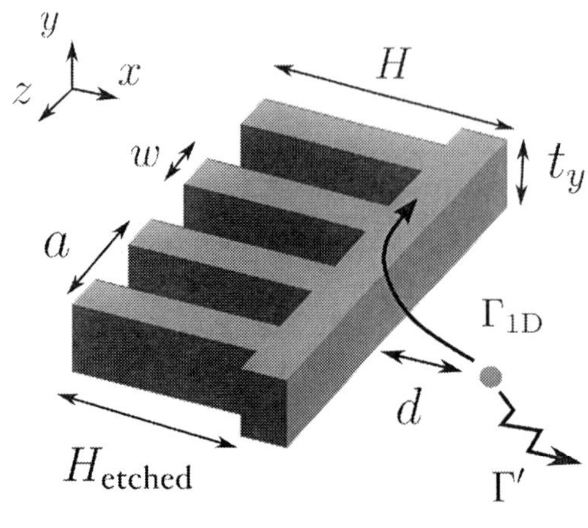

Figure 1. Scheme of the comb waveguide we designed to improve the experimental figures of merit of waveguide QED. We optimized the structure so that Rubidium atoms can be trapped in the back of the structure, in a deep potential well ($U_{trap} \sim 1$mK) and at a distance of $d = 100$nm from the waveguide.

We optimized the structure so that two additional fields, blue and red detuned with respect to the atomic transition frequency, can propagate through the waveguide. We computed rigorously the trapping potential produced by these fields using the recently-introduced open-source package Nanotrappy [6]. The calculation includes the Casimir Polder interaction that arises when an atoms is very close to a dielectric surface. We showed that Rubidium atoms can be trapped efficiently ($U_{trap} \sim 1$mK) as close as $d = 100$nm from the back of the structure, where the slow mode has a strong evanescent tail.

At this distance, the atoms emit preferentially into the slow mode. Using a rigorous modal method, we computed exactly the decay rate inside the slow mode and in all the other modes and showed that $\Gamma_{1D} = 10\Gamma_0 = 8.8\Gamma'$ for an atom inside the trap leading to a value of $\beta = 0.88$.

3 Conclusion and outlook

The fabrication of the comb waveguide is now in progress at c2n. We believe that the trapping of $N \sim 10 - 20$ atoms with a strong coupling to the comb waveguide can be seriously considered. Such high figures of merit would lead to the emergence of new physical phenomena.

First, the quartic dispersion of the slow mode can drastically modify the collective effets that emerges when many atoms are strongly coupled to the structure.

Second, thanks to the unprecedented value of $\beta = 0.88$ accessible with the comb waveguide, new quantum-non linear effects could be observed, such as soliton dynamics [7, 8] , or many-body localization [9] .

References

[1] N.V Corzo, J. Raskop, A. Chandra, A.S. Sheremet, B. Gouraud, and J. Laurat, Nature, **7744**, 359-362 (2019)

[2] G. Lecamp, , P. Lalanne and J-P. Hugonin, Physical Review Letters, **99**, 023902 (2007)

[3] A. Goban et al. Nature communications, **5**, 1 (2014)

[4] A. Goban, C.L. Hung, J.D. Hood, S.P. Yu, J.A Muniz, O. Painter and J.H Kimble, Physical Review Letters, **6**, 063601 (2015)

[5] N. Fayard, A. Bouscal, J. Berroir, A. Urvoy, T. Ray, S. Mahapatra, M. Kemiche, J.-A. Levenson, J.-J. Greffet, K. Bencheikh, J. Laurat and C. Sauvan, arXiv preprint, 2201.02507 (2022)

[6] J. Berroir, A. Bouscal, A. Urvoy, T. Ray and J. Laurat, Physical Review Research, **4**, 013079 (2022)

[7] B. Bakkensen, Y.X. Zhang, J. Bjerlin and A.S. Sørensen, arXiv preprint, 2110.06093 (2021)

[8] G. Calajó and D.E. Chang, Physical Review Research, **4**, 023026 (2022)

[9] N. Fayard, L. Henriet, A. Asenjo-Garcia and D.E. Chang, Physical Review Research, **3**, 033233 (2021)

Pulse self-compression down to the sub-cycle regime in hollow capillary fibers with decreasing pressure gradients

Marina Fernández Galán[1,*], *Enrique* Conejero Jarque[1], and *Julio* San Roman[1]

[1]Grupo de Investigación en Aplicaciones del Láser y Fotónica, Departamento de Física Aplicada, Universidad de Salamanca, E-37008, Salamanca, Spain

Abstract. We theoretically demonstrate an enhancement in the generation of clean, near-infrared sub-cycle laser pulses by soliton self-compression in gas-filled hollow capillary fibers using decreasing pressure gradients. Furthermore, we identify the optimal input parameters for high quality compression and the main advantages of this promising technique which paves the way towards ultrafast vacuum experiments.

1 Introduction

In the last years, pulse post-compression in gas-filled hollow capillary fibers (HCFs) has become a widely used method for generating energetic ultrashort laser pulses [1]. These pulses, with time durations of a few femtoseconds, are now enabling tools for ultrafast time-resolved spectroscopy and strong-field physics driving extreme-ultraviolet high-harmonic generation. Very recently, high-energy soliton dynamics in HCFs arising from the interplay between the negative group-velocity dispersion of the waveguide and self-phase modulation (SPM) have attracted a great interest as a direct route to extreme pulse self-compression down to the few- and sub-cycle regime in the optical spectral region [2, 3]. However, as this interaction can be very complex and dynamic, it is still necessary to identify the optimal experimental parameters for high-quality self-compression [4], and design guidelines are essential.

Some recent studies have suggested the use of decreasing pressure gradients in HCFs to avoid the distortions introduced by the transmission through the output windows and deliver the self-compressed pulses directly to vacuum experiments [5], and to minimize the nonlinear mode coupling and improve the purity of the output beam in the compression of the higher-order spatial modes of these fibers [6].

In this work, we explore further advantages of the negative pressure gradient technique, showing that it can be applied to the self-compression of standard near-infrared (NIR) pulses in the fundamental mode of typical HCFs to generate sub-cycle pulses with shorter durations, higher peak powers and a better temporal structure when compared to those obtained in the equivalent constant pressure situations [7], defined by matching the accumulated nonlinear phase shift, also known as B-integral, along the propagation [8].

*e-mail: marinafergal@usal.es

2 Results

To study the nonlinear propagation dynamics of ultrashort pulses in gas-filled HCFs we have numerically solved the one-dimensional generalized nonlinear Schrödinger equation for the pulse envelope in a regime of moderate intensity, including the linear losses of the fiber, the complete dispersion, SPM and self-steepening. To identify the optimal input parameters for obtaining high-quality sub-cycle pulses by soliton self-compression in typical experimental configurations, we have systematically simulated the propagation of a 30 fs pump pulse at 800 nm in the fundamen-

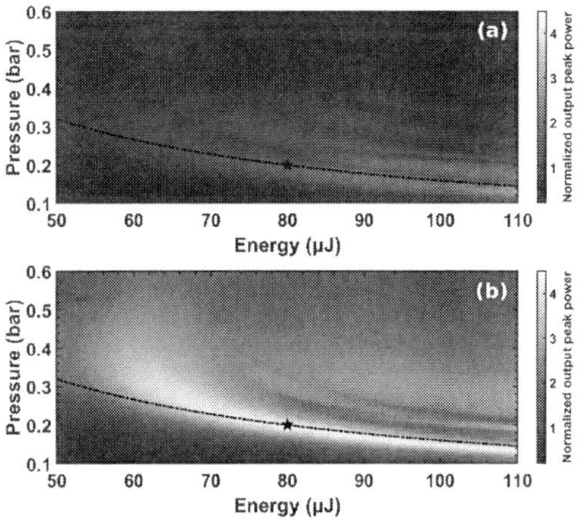

Figure 1. Peak power of the self-compressed pulses normalized to its initial value as a function of input pulse energy and equivalent gas pressure for (a) a statically argon-filled HCF and (b) its equivalent differentially pumped fiber with a decreasing pressure gradient. The dot-dashed line represents the curve of constant B-integral ($B = 8.5$ rad) which best fits the optimal region for high-quality self-compression and the star refers to the final situation shown in Fig. 2 (80 μJ, 200 mbar).

tal mode of a 3 m long, 100 μm core radius argon-filled HCF with both decreasing and constant pressure. In the parameter space ranging from input pulse energies of 50 to 110 μJ and gas pressures from 100 to 600 mbar, we have found a wide global optimal region for self-compression in the decreasing pressure gradient (Fig. 1). We have also verified that this optimal behavior is not affected by detrimental nonlinear effects such as self-focusing or ionization by comparing the results with a complete 2D propagation model [9].

As we can see in Fig. 1, the decreasing pressure gradient effectively prevents the onset of these higher-order nonlinear effects by shifting the optimal self-compression region towards lower energies when compared to the optimal parameters for constant pressure, which would lie further towards the lower pressure and higher energy corner of the grid. This allows for the generation of high-quality self-compressed pulses even in gases with low ionization potentials, such as argon. We have also demonstrated that with the differentially pumped fiber the optimal parameter region is wider that in the constant pressure case. This represents a clear experimental advantage as it would make the setup more robust against slight deviations from the optimal input parameters.

Furthermore, we have obtained a simple relation between the input pressure and the pulse energy based on the B-integral which allows to place the optimal region for self-compression in the 2D parameter space [7]. One of the output pulses along this curve is shown in Fig. 2, corresponding to the situation depicted with a star in Fig. 1. As we can clearly realize, in this regime the self-compressed

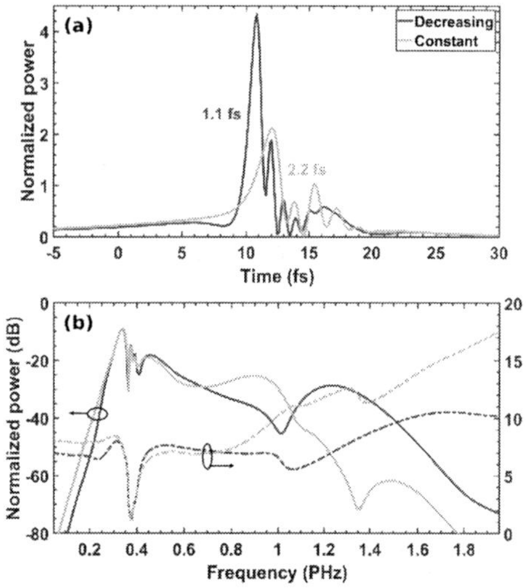

Figure 2. (a) Output pulses and (b) their corresponding spectra obtained after the propagation of a 80 μJ, 30 fs gaussian pump pulse at 800 nm through a 3 m long, 100 μm core radius HCF homogeneously filled with argon at 200 mbar (cyan) and with its equivalent decreasing pressure gradient ending in vacuum (purple). In both plots, the power is normalized to its initial peak value and the labels next to each pulse in (a) stand for the FWHM output durations.

pulses are much better in the decreasing pressure gradient case (purple) as they have shorter durations, higher peak powers and a cleaner temporal structure than those obtained with the equivalent constant pressure (cyan). This improvement is attributed to an effective suppression of higher-order dispersion and self-steepening in the last stages of the pulse compression [6], together with a continuous shift of the zero-dispersion frequency towards higher frequencies at the same time as the pulse spectrum broadens by SPM during its propagation in the fiber. The output pulse obtained from the differentially pumped fiber in Fig. 2(a) reached a sub-cycle full width at half maximum (FWHM) duration of 1.1 fs, corresponding to a peak power of 10.9 GW. Similar results were obtained at other energies and gas pressures lying in the optimal region [7].

3 Conclusions

In conclusion, we have identified the optimum regime to obtain single- to sub-cycle NIR pulses by soliton self-compression in HCFs filled with gas with a decreasing pressure gradient, showing that there is a relative broad region, defined by a simple relation between the input pulse energy and the average gas pressure, where high-quality self-compression can be achieved. In addition to the intrinsic experimental advantage of decreasing pressure gradients to couple the output pulses directly to vacuum experiments, we have also shown that they can considerably improve the robustness of the setup and the quality of the compression process when compared to the equivalent constant pressure situation, also preventing the onset of undesirable nonlinear effects. We believe that these findings will pave the way for a new generation of experiments in time-resolved spectroscopy and attosecond science.

This work was supported by grant PID2019-106910GB-I00, funded by the Spanish Ministry of Science and Innovation, MCIN/AEI/ 10.13039/501100011033.

References

[1] M. Nisoli, S. De Silvestri, O. Svelto, Appl. Phys. Lett. **68**, 2793 (1996).

[2] J.C. Travers, T.F. Grigorova, C. Brahms, F. Belli, Nat. Photonics **13**, 547 (2019).

[3] C. Brahms, F. Belli, J.C. Travers, Phys. Rev. Res. **2**, 043037 (2020).

[4] D. Schade, F. Köttig, J.R. Koehler, M.H. Frosz, P.St.J. Russell, F. Tani, Opt. Express **29**, 19147 (2021).

[5] C. Brahms, F. Belli, J.C. Travers, Opt. Lett. **45**, 4456 (2020).

[6] Y. Wan, W. Chang, Opt. Express **29**, 7070 (2021).

[7] M.F. Galán, E.C. Jarque, J. San Roman, Opt. Express **30**, 6755 (2022).

[8] M. Nurhuda, A. Suda, K. Midorikawa, M. Hatayama, K. Nagasaka, J. Opt. Soc. Am. B **20**, 2002 (2003).

[9] E.C. Jarque, J. San Roman, F. Silva, R. Romero, W. Holgado, M.A. Gonzalez-Galicia, B. Alonso, I.J. Sola, H. Crespo, Sci. Rep. **8**, 2256 (2018).

 https://doi.org/10.1051/epjconf/202226608004

Paraxial quantum fluids light in hot atomic vapors

Murad Abuzarli[1], *Tangui* Aladjidi[1], *Nicolas* Cherroret[1], and *Quentin* Glorieux[1,*]

[1]Laboratoire Kastler Brossel, Sorbonne University, CNRS, ENS-PSL University, Collège de France; 4 Place Jussieu, 75005 Paris, France

Abstract. Hot atomic vapors are widely used in non-linear and quantum optics due to their large Kerr non-linearity. This non-linearity induces effective photon-photon interactions allowing light to behave as a fluid displaying quantum properties such as superfluidity. In this presentation, I will show that we have full control over the Hamiltonian that drives the system and that we can engineer an analogue simulator with light.

Quantum fluids of light rely on the analogy between the non-linear Schrodinger equation (NLSE) describing the propagation of light in non-linear media and the Gross-Pitaevskii equation (GPE) describing a weakly interacting Bose gas [1]. :

$$i\hbar\frac{\partial}{\partial t}\psi = \Big(-\frac{\hbar^2}{2m}\nabla^2 + V + g|\psi|^2\Big)\psi \ \textit{(GPE)}$$

$$i\frac{\partial}{\partial z}E = \Big(\underbrace{-\frac{1}{2k_0}\nabla_\perp^2}_{\text{Kinetic}} - \underbrace{\delta n(\mathrm{r})\frac{k_0}{2}}_{\text{Potential}} + i\frac{\alpha}{2} - \underbrace{n_2 k_0 |E|^2}_{\text{Interaction}}\Big)E \ \textit{(NLSE)}$$

In this presentation, I will show that we have full control over the three terms and we can engineer an analogue simulator with light.

By studying the effect of interaction quenches in the fluid [2], we evidence the dynamical Casimir effect i.e the spontaneous emission of correlated pairs of phonons. Employing Bragg spectroscopy, we measure the static structure factor of our fluid of light [3], confirming the similarity between our photon-photon interactions and interatomic interactions in a BEC.

We merge novel characterization tools [4] and all optical potential in our latest experimental setup. Using beam shaping techniques, we realize arbitrary potential shapes and sign, allowing to explore new effects arising from confinement (see Fig. 1).

*e-mail: quentin.glorieux@lkb.upmc.fr

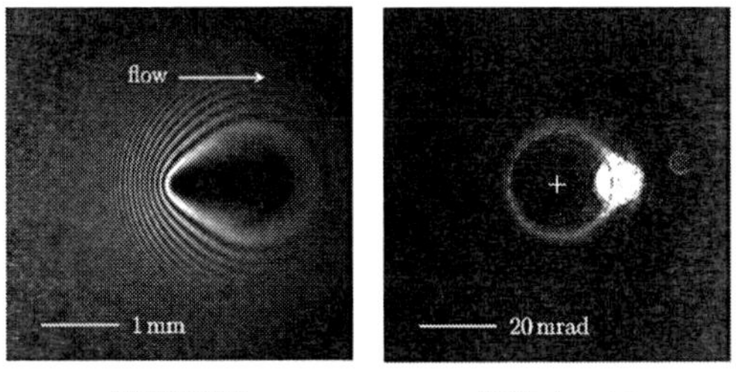

(a) Real space (b) Fourier space

Figure 1. Scattering of the fluid on an attractive or repulsive defect. a) Density distribution (real space imaging). b) Momentum space distribution (Fourier space imaging)

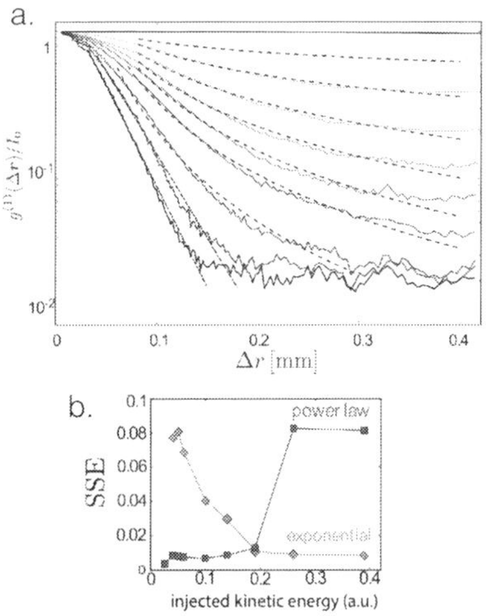

Figure 2. Emergence of an out-of-equilibrium BKT transition in optics. First order correlation function with increasing fluctuation strength ϵ, transition from an algebraic to an exponential order. quenches.

Finally, we control the initial kinetic energy distribution by sending intensity fluctuations on top of a gaussian background fluid, and measure the coherence of these fluctuations. We report on the observation of a pre-thermal state in a non-equilibrium, two-dimensional fluid of light. Direct measurements of the first order coherence function of the fluid reveal the dynamical emergence of algebraic correlations, a quasi steady-state with properties close to those of thermal superfluids. We observe the emergence of a prethermal state [5] with long range order, with a topological phase transition precursor. With increasing fluctuation intensity, the coherence function switches from an algebraic to an exponential decay, strongly reminiscent of a BKT transition (see Fig. 2).

References

[1] Q. Fontaine, T. Bienaimé, S. Pigeon, E. Giacobino, A. Bramati, and Q. Glorieux, Phys. Rev. Lett. **121**, 183604 (2018).

[2] J. Steinhauer, M. Abuzarli, T. Aladjidi, T. Bienaimé, C. Piekarski, W. Liu, E. Giacobino, A. Bramati, and Q. Glorieux, arXiv:2102.08279 (2021)

[3] C. Piekarski, W. Liu, J. Steinhauer, E. Giacobino, A. Bramati, and Q. Glorieux, Phys. Rev. Lett. 127 (2021).

[4] T. Aladjidi, M. Abuzarli, G. Brochier, T. Bienaimé, T. Picot, A. Bramati, and Q. Glorieux, arXiv:2202.05764 (2022).

[5] M. Abuzarli, N. Cherroret, T. Bienaimé, and Q. Glorieux, arXiv:2203.07399 (2022).

https://doi.org/10.1051/epjconf/202226608005

Quantum vacuum excitation of a quasi-normal mode in an analog model of black hole spacetime

Maxime J Jacquet[1,*]

[1]Laboratoire Kastler Brossel, Sorbonne Université, CNRS, ENS-Université PSL, Collège de France, Paris 75005, France

Abstract. Vacuum quantum fluctuations near horizons are known to yield correlated emission by the Hawking effect. In this talk, I will explain how a 1 dimensional flow of microcavity polaritons may be engineered to produce an effective curved spacetime with a black hole horizon. I will present numerical computations of correlated emission on this spacetime and show that, in addition to the Hawking effect at the sonic horizon, quantum fluctuations may result in a sizeable stationary excitation of a quasi-normal mode of the field theory. Observable signatures of the excitation of the quasi-normal mode are found in the spatial density fluctuations as well as in the spectrum of Hawking emission. I will explain how the driven-dissipative dynamics of the polariton fluid are key to observing the quantum excitation of the quasi-normal mode. Nonetheless, this observation suggests a general and intrinsic fluctuation-driven mechanism leading to the quantum excitation of quasi-normal modes on black hole spacetimes.

1 Introduction

Quantum fluctuations of fields in the vicinity of the event horizon of black holes (BHs) are predicted to cause the emission of correlated waves by the Hawking effect (HE): while Hawking radiation propagates away from the horizon to outer space, the partner radiation falls inside the horizon. Since signalling from inside the horizon is impossible, only the out-going Hawking radiation is observable and quantum correlations with the in-falling partner waves cannot be accessed.

The HE may also be observed in the laboratory thanks to analogue gravity setups [1], namely condensed matter or optical systems, engineered in such a way that their collective excitations propagate on effectively curved spacetimes [2]. This idea has been experimentally demonstrated in a variety of platforms: for example, a horizon for sound waves forms in a one-dimensional trans-sonic fluid where the flow velocity of the fluid exceeds the speed of sound. Crucially, observation on both sides of the horizon is possible with analogue setups and the HE has been detected via density correlations between Hawking radiation and its partner [3] in experiments based on both classical and quantum fluids.

2 Novel physics in a driven-dissipative fluid

In this talk, I use of a specific realisation of an effective spacetime in a driven-dissipative quantum fluid of exciton-polaritons in a semiconductor microcavity [4] to push forward the theoretical study of quantum fluctuations in the vicinity of horizons. I show how a one-dimensional trans-sonic configuration features a horizon and explain how the spatial shape of the quantum fluid and the external potential may be optimised together to maximize the strength of the HE [5], see Fig. 1. In addition to the typical signatures of the HE in the spatial correlation of sound waves, new features evidence the coupling of propagating waves with a localized mode living near the horizon, namely a quasi-normal mode (QNM) of the acoustic field.

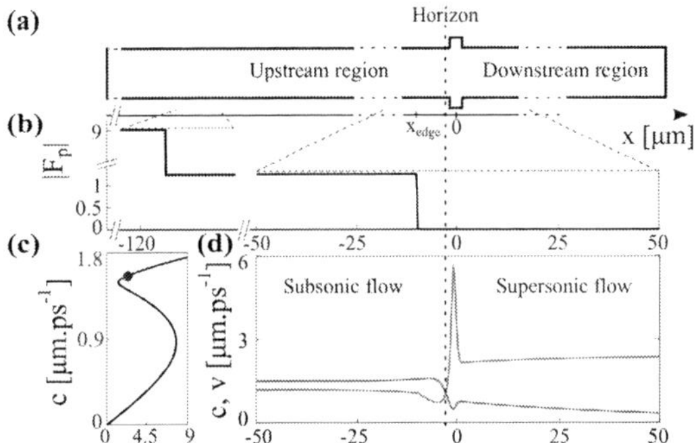

Figure 1. Acoustic horizon in a trans-sonic polariton fluid. A step-like laser field pumps polaritons, creating a trans-sonic fluid flow across an attractive obstacle. **(a)** Sketch of the polariton wire with a defect at $x_d = 0$. **(b)** Optical bistability of the spatially homogeneous polariton fluid. **(c)** Spatial profile of the pump near the defect. The pump intensity $|F_p|$ [ps^{-1}.μm$^{-1/2}$] drops abruptly to 0 at $x_{edge} = -10\,\mu$m. **(d)** Spatial properties of the fluid near the defect: orange, fluid velocity v; blue, speed of excitations c. Figure adapted from [4].

*e-mail: maxime.jacquet@lkb.upmc.fr

3 Quasi-normal modes in analogue gravity

Typically, localized modes have a finite lifetime due to radiative decay into waves propagating away from the horizon, for instance gravitational waves.

While classical ring-down oscillations of a scalar field have been observed experimentally at the surface of a rotating bathtub flow configuration [6] and the HE spectrum has been theoretically calculated in more complex flow geometries in conservative fluids [7], this work [4] establishes that QNMs of quantum fields get naturally excited by the same quantum fluctuations that are responsible for the HE. Translated back to the astrophysical context, these results suggest a general and intrinsic fluctuations-driven mechanism leading to quantum excitation of BH spacetimes.

References

[1] M.J. Jacquet, S. Weinfurtner, F. König, Philosophical Transactions of the Royal Society A: Mathematical, Physical and Engineering Sciences **378**, 20190239 (2020)
[2] W.G. Unruh, Physical Review Letters **46**, 1351 (1981)
[3] I. Carusotto, S. Fagnocchi, A. Recati, R. Balbinot, A. Fabbri, New Journal of Physics **10**, 103001 (2008)
[4] M.J. Jacquet, L. Giacomelli, Q. Valnais, F. Joly, M. Claude, E. Giacobino, Q. Glorieux, I. Carusotto, A. Bramati, arXiv:2110.14452 [gr-qc, physics:quant-ph] (2021)
[5] M.J. Jacquet, M. Joly, L. Giacomelli, F. Claude, E. Giacobino, Q. Glorieux, I. Carusotto, A. Bramati, arXiv:2201.02038 [gr-qc, physics:quant-ph] (2022)
[6] T. Torres, S. Patrick, M. Richartz, S. Weinfurtner, Physical Review Letters **125**, 011301 (2020)
[7] I. Zapata, M. Albert, R. Parentani, F. Sols, New Journal of Physics **13** (2011)

Fourier-limited attosecond pulse generation with magnetically pumped high-order harmonic generation

Rodrigo Martín-Hernández[1,*], *Luis* Plaja[1], and *Carlos* Hernández-García[1]

[1]Grupo de Investigación en Aplicaciones del Láser y Fotónica, Departamento de Física Aplicada, Universidad de Salamanca, E-37008, Salamanca, Spain.

Abstract. After more than two decades of attosecond physics, the generation and control of the shortest laser pulses available remains as a complex task. One of the main limitations of reducing the temporal duration of attosecond pulses emitted from high-order harmonic generation (HHG) is the *attochirp*. In this contribution, we demonstrate that HHG assisted by strong fast oscillating magnetic fields enables the generation of Fourier-limited attosecond pulses in the water window. In short, the magnetic field generates a nanowire-like structure, which transversally confines the electronic wavefunction in the HHG process. We demonstrate that the resulting HHG spectrum extends well beyond the semiclassical cutoff frequency, and most interestingly, it is emitted in the form of few-cycle, Fourier-limited, attosecond pulses.

1 Introduction

High-order harmonic generation (HHG) is an extreme non-linear process resulting from the non-perturbative interaction between matter and intense laser fields. In the standard picture, using a gas target and a linearly polarized laser, the harmonic generation stems from the recollision of previously ionized electrons with their parent ions. During its excursion along the continuum, the electron gains kinetic energy, that is subsequently radiated in the form of high-order harmonics of the incident field. In this simple approach, the detached electron behaves as a free particle following classical trajectories along the longitudinal direction, from the ionization time until the recollision [1]. The electron's excursion along the continuum is essential to build the main properties of HHG, and several approaches have been studied to control the electron trajectories in order to modify the HHG process. For example, the use of multicolor drivers, complex polarization schemes, or external magnetic fields [2]. In the latter case, however, only static magnetic fields have been considered, to the best of our knowledge, as in principle strong oscillating magnetic fields are only present in electromagnetic waves. In this latter case, the force exerted by an electric field is orders of magnitude over the magnetic field, therefore hindering any effect of this latter

On the one hand, one of the major interests of HHG is its ability to generate attosecond pulses. Attosecond pulse generation stands as one of the most robust tools to probe ultrafast processes at molecular and atomic scales. The extension of pump-probe experiments to these timescales allows for time-resolved observation of electronic dynamics, for example, in chemical reactions. Moreover, they are considered the main tool to understand, and ultimately control, charge migration in molecular systems, with huge implications for understanding the alteration of biological macromolecules, the development of more efficient photovoltaic systems, or more selective catalysts [3,4]. HHG stands as the standard tool for attosecond pulse generation thanks to the broad spectra generated towards the XUV or soft X-rays. Counterintuitively, the shortest duration of an HHG attosecond pulse is not limited by its bandwidth [5], but by its intrinsic chirp, also known as attochirp [6]. By using metallic foils, the attochirp can be partially compensated, reaching the record established at 43 as [7].

On the other hand, recently there has been an increasing interest and a huge development of structured light beams [8], opening new perspectives on light-matter interactions. For example, it has been shown how azimuthally polarized vector beams exhibit an isolated Tesla-scale femtosecond magnetic field on-axis oscillating at optical frequencies, paving the way to pure magnetic interaction with matter [9]. Isolated fast oscillating magnetic fields are of great interest in fields such as ultrafast magnetism or spintronics [10], but also in ultrafast and non-linear optics.

In this contribution, we perform theoretical simulations demonstrating that HHG assisted by a linearly-polarized magnetic field in coincidence with a circularly-polarized driver laser results in the generation of few-cycle, near Fourier limited, attosecond pulses

2 Model

We solve numerically the three-dimensional, time-dependent Schrödinger equation for the hydrogen atom. The Hamiltonian, using atomic units, for the electron under the action of both fields reads:

* Corresponding author: rodrigomh@usal.es

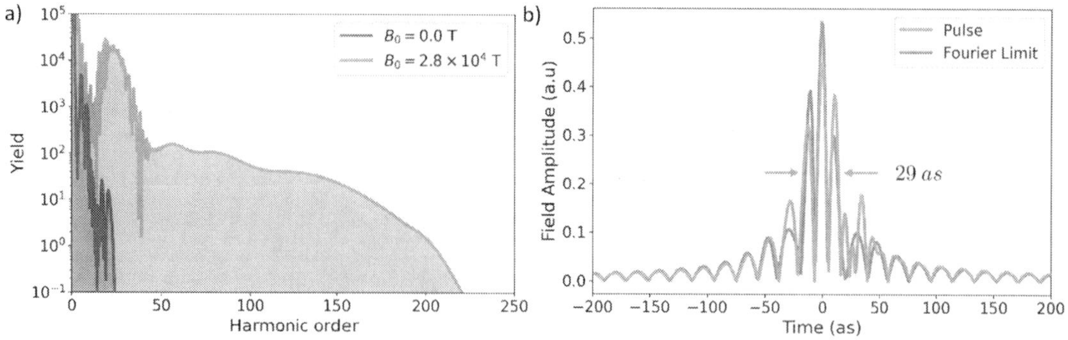

Fig. 1. (a) HHG spectra resulting from a circularly polarized driver laser without magnetic field (blue line) and with the magnetic field linearly-polarized. (b) Attosecond pulse arising from the spectra with the magnetic field in panel a), filtering above the 70th harmonic order (blue line). The measured FWHM pulse duration is approximately 29 as. The Fourier-limited pulse, plotted in green, shows a FWHM around 27 as.

$$\hat{H} = \frac{1}{2}\left[\boldsymbol{p} - \frac{1}{c}\boldsymbol{A}_E\right]^2 + \frac{1}{2c}B(t)\hat{L}_z + \frac{1}{8c^2}B^2(t)(x^2+y^2) - V(\boldsymbol{r})$$

The magnetic field B creates a time-dependent harmonic oscillator potential in the x-y plane, confining the detached photoelectron and its excursion along the continuum, inducing a restructuration of the continuum energy levels in the same fashion as quantum wire. Moreover, note that one of the components of the circularly-polarized electric field will exert a transverse force over the photoelectron inducing transitions to higher states inside the magnetic harmonic oscillator. These highly excited states will contribute to the HHG emission.

3 Results

We consider a four-period circularly polarized electric field at 800 nm with a peak intensity of 1.6×10^{14} W/cm2 and a linearly polarized oscillating magnetic field. While in the standard scheme, without a magnetic field, the harmonic yield drops immediately using drivers with circular polarization, we show the generation of broad supercontinuum by adding an external magnetic field with amplitudes around 2.8×10^4 T, oscillating at 1600 nm (Fig. 1 a). This emission, extending over 200 harmonic orders (110 eV), results from the excitation of the quantum wire transverse states by the circularly-polarized electric field. These highly excited states contribute to the harmonic emission explaining the broad spectrum observed. In fact, we have seen that the electron in the continuum is a semiclassical wave packet, following a well-defined trajectory under electric field driving. The emission yield is maximized when the collision with the parent ion occurs simultaneously in the longitudinal and the transverse directions, highly dependent on the magnetic field wavelength, phase, and amplitude.

By filtering the spectrum above the 70th harmonic order, we see how a 29 as FWHM pulse can be generated nearly Fourier-limited (Fig. 1 b). These attosecond pulses are centered at 6.7 nm (27 as period), meaning they are almost single cycle. To our knowledge, this is the first proposal to generate chirp-free HHG attosecond pulses. Thus, in this scenario, attosecond pulse generation is not limited by the use of negative GDD metallic foils, difficult to use at these wavelengths, and could be extended for broader spectra. More interestingly, by using longer wavelengths [5], we foresee that broader chirp-free spectra could be obtained, with the potential to reach few-cycle, few attosecond pulses.

We note that such isolated linearly-polarized magnetic fields could be obtained from structured Petawatt laser beams, with laser intensities around 10^{18} W/cm^2.This work opens the route to new and exotic HHG scenarios in which state-of-the-art Petawatt laser systems may play a determinant role, allowing the generation of nearly Fourier-limited attosecond pulses.

References

1. J. L. Krause, K. J. Schafer, and K. C. Kulander, Phys. Rev. Lett., **68**, 3535 (1992).
2. D. B. Milošević, A. F. Starace, Laser Phys., **10**, 278 (2000).
3. F. Calegari, et al., J. Phys. B: At. Mol. Opt. Phys., **49**, 142001 (2016).
4. H.J. Wörner, et al., Struct. Dyn., **4**, 061508 (2017).
5. T. Popmintchev, et al., Science, **336**, 1287 (2012)
6. K. Varjú, et al. J. Mod. Opt., **52**, 379 (2005).
7. T. Gaumnitz, et al., Opt. Express, **25**, 27506 (2017).
8. H. Rubinsztein-Dunlop et al., J. Opt., **19**, 013001 (2017).
9. M. Blanco, et al, ACS Photonics, **6**, 38 (2019).
10. J. Walowski, M. Münzenberg, J. Appl. Phys., **120**, 14091 (2016).

Acknowledgments

European Research Council (851201); Ministerio de Ciencia de Innovación y Universidades (PID2019-106910GB-I00, RYC-2017-22745); Junta de Castilla y León FEDER (SA287P18).

 https://doi.org/10.1051/epjconf/202226608007

Cooperative Spontaneous Four-wave Mixing in Single-channel and Dual-channel Sequences of Side-coupled Ring Resonators

Amideddin Mataji-Kojouri[1], *Massimo* Borghi[1], *Federico A.* Sabattoli[1,*], *Houssein* El Dirani[2], *Laurene* Youssef[3,**], *Camille* Petit-Etienne[3], *Erwine* Pargon[3], *John E.* Sipe[4], *Marco* Liscidini[1], *Corrado* Sciancalepore[2,***], *Matteo* Galli[1], and *Daniele* Bajoni[5]

[1]Dipartimento di Fisica, Università di Pavia, Via Agostino Bassi 6, 27100 Pavia, Italy
[2]Univ. Grenoble Alpes, CEA-Leti, 38054 Grenoble cedex, France
[3]Univ. Grenoble Alpes, CNRS, LTM, 38000 Grenoble, France
[4]Department of Physics, University of Toronto, 60 St. George Street, Toronto, ON, M5S 1A7, Canada
[5]Dipartimento di Ingegneria Industriale e dell'Informazione, Università di Pavia, Via Adolfo Ferrata 5, 27100 Pavia, Italy

Abstract. Cooperative photon pair generation by Spontaneous Four-Wave Mixing (SFWM) process in single-channel and dual-channel side-coupled ring resonator sequences is investigated. Our analysis shows that super-linear growth of generation rate with respect to the number of rings is possible even in presence of loss. Experimental evidence of super-SFWM is provided by comparing individual and collective generation rates obtained from a dual-channel ring resonator sequence. The results are in good agreement with theory and suggest that high photon pair generation rates can be achieved from integrated silicon ring resonator sequences without initiating nonlinear absorption processes.

1 Introduction

Despite its strong third order susceptibility, the rate of photon pair emission in silicon resonators is limited by nonlinear absorption at telecommunication wavelengths. Recently, it has been found that a process analogous to atomic super-radiance, *i.e.*, the enhancement of spontaneous emission from a collection of emitters [1], can occur in nonlinear parametric processes [2]. The authors considered photon pair generation by Spontaneous Four-Wave Mixing (SFWM) in a single-channel (SC) chain of N loss-less ring resonators, predicting that the rate would increase by a factor N^2 compared to that of the individual resonators. Hence, through super-radiant assisted SFWM, one can achieve a high pair generation rate without increasing the field amplitude in individual rings. However, linear absorption and scattering of the light in a real device limit the quadratic scaling of the generation rate. Here, we study the impact of linear loss mechanisms on the performance of ring resonators sequences with SC and dual-channel (DC) configurations. The DC configuration provides individual access to every emitter which is crucial for experimentally examining the cooperative emission. We provide experimental results to demonstrate the cooperative emission of ring resonators in a DC configuration.

*Current address: Advanced Fiber Resources Milan, via Fellini 4, 20097 San Donato Milanese (MI), Italy
**Current address: ENSIL-ENSCI, Centre Européen de la Céramique, 12 rue Atlantis, 87068 Limoges, France
***Current address: SOITEC SA, Parc technologique des Fontaines, Chemin des Franques, 38190 Bernin, France

Table 1: Transmission (T) and Field Enhancement (F)

Configuration	T	F
All-pass	$\frac{\sigma-e^{-\gamma l}}{1-\sigma e^{-\gamma l}}$	$\frac{j\kappa}{1-\sigma e^{-\gamma l}}$
Add-drop	$\frac{-e^{-\gamma l/2}\kappa^2}{1-\sigma^2 e^{-\gamma l}}$	$\frac{j\kappa}{1-\sigma^2 e^{-\gamma l}}$

2 Results

Figure 1(a) and Fig. 1(b) respectively show a SC and a DC ring resonator sequence. In the lossless case, the SC configuration has an all-pass frequency response with unity transmission. In presence of loss, transmission drops at the resonant frequencies and limits the efficiency of emission due to absorption of both the pump and the generated photon pairs, named the signal and the idler. The DC configuration behaves as a cascade of band-pass filters, hence the pump power decreases more slowly compared to the lossy SC sequence. But successive spectral filtering of the generated photon pairs can still limit the generation rate. To theoretically investigate the differences of the two configurations, we follow the approach described in [2, 3]. Field enhancement F and transmission functions T for these configurations are summarized in Table 1. Here, l is the circumference of the rings, $\gamma = \alpha + jk(\omega)$ is the complex propagation constant, and we assume an energy-conserving point coupler for each resonator with σ being its self-coupling coefficient and $j\kappa$ being its cross-coupling coefficient, satisfying $\sigma^2 + \kappa^2 = 1$. For a continuous-wave excitation, the generation rate is calculated from [2]

$$|\beta|^2 \propto \frac{\omega_p^2}{v^2(\omega_p)} \int d\omega \frac{\omega\omega'}{v(\omega)v(\omega')} \left|J(\omega, \omega', \omega_p, \omega_p)\right|^2, \quad (1)$$

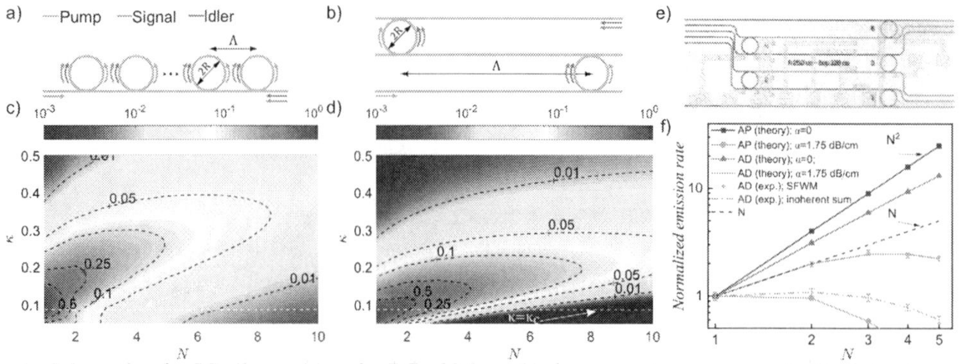

Figure 1: Schematic of a SC all-pass (a) and a DC add-drop (b) ring resonator sequence. Radii of the rings are R and the rings are separated by Λ. Asymptotic-in fields for pump, signal and idler waves are respectively shown by green, red and blue arrows. Photon pair generation rate $|\beta|^2$ for all-pass (c) and add-drop (d) configurations in presence of loss normalized to that of the sequence with the highest generation rate, calculated for different coupling coefficients (κ) and sequence lengths (N). (e) Optical microscope image of the fabricated DC ring resonator sequence. The heater layer (gold) is overlaid to that of the waveguides (black). (f) Theoretical and experimental normalized generation rate of the all-pass (AP) and the add-drop (AD) configurations for loss-less and lossy cases.

where ω and $\omega' = 2\omega_p - \omega$ span respectively over the frequency bandwidth of the signal and the idler waves, ω_p is the angular frequency of the pump and $v(\omega)$ is the group velocity. Here

$$J(\omega_1,\omega_2,\omega_3,\omega_4) = J_0F(\omega_1)F(\omega_2)F(\omega_3)F(\omega_4)\frac{\sinh(N\mu/2)}{\sinh(\mu/2)}$$
$$\times e^{-N(j\theta_T(\omega_1)+j\theta_T(\omega_2)+\alpha_T(\omega_1)+\alpha_T(\omega_2))-N(\gamma(\omega_1)+\gamma(\omega_2))\Lambda+(-N+1)\mu/2}$$

is the overlap integral of the asymptotic-in fields for the waves involved in the four-wave mixing process over all rings. The propagation direction of the asymptotic-in fields is schematically shown in Fig. 1(a,b). $T(\omega) = e^{-\alpha_T(\omega)-j\theta_T(\omega)}$, $\mu = j(\theta_T(\omega_3)+\theta_T(\omega_4)-\theta_T(\omega_1)-\theta_T(\omega_2)) + \alpha_T(\omega_3)+\alpha_T(\omega_4)-\alpha_T(\omega_1)-\alpha_T(\omega_2)$, and J_0 incorporates the remaining terms. Figure 1(c) and Fig. 1(d) respectively show the generation rate calculated for the lossy SC and DC configurations for different coupling coefficients (κ) and number of resonators (N). Here, we normalized $|\beta|^2$ to that of the sequence which shows the highest generation rate. Radii of the rings are $R = 29.3\ \mu m$, the spacing between resonators is $\Lambda = 500\ \mu m$, and the effective mode index and group index are 2.56 and 4.08, respectively. We assumed $\alpha = 1.75\ dB/cm$. Signal, pump, and idler resonant wavelengths are respectively around 1570 nm, 1560 nm, and 1550 nm. All these values correspond to the ones measured in the experiment discussed later. We see that for each value of κ, there is an optimum length N_{opt} that provides the highest generation rate. For under-coupled or critically-coupled resonators $N_{opt} = 1$. Figure 1(f) compares the scaling behavior of the generation rate of both configurations under loss-less and lossy scenarios. Here coupling coefficient is assumed to be $\kappa^2 = 0.03$, *i.e.*, its value in the experiment. For the SC configuration and in absence of loss (black), the scaling is quadratic with the number of resonators, as demonstrated earlier [2]. When losses are introduced, the generation rate quickly drops after two resonators (green). For the lossless DC configuration (blue), the normalized rate can be shown to grow as $N^{\frac{3}{2}}$. The different scaling with that of the all-pass configuration is set by the spectral filtering imparted to the photon pairs at each drop event in the sequence. When experimental value of loss is introduced in the calculations (red), the generation rate initially grows linearly with N, then it reaches a maximum and after that it decreases. We experimentally tested a DC sequence whose microscope image is shown in Fig. 1(e), and found a scaling law (red points) that agrees very well with the theoretical predictions. To prove the existence of cooperative emission, we independently recorded the coincidence rate of each individual ring in the sequence, and used them to predict the normalized rate that would be observed at the end of the sequence if their emission were incoherent. This is shown in Fig. 1(f) (magenta). The clear discrepancy with the case where all rings were coherently pumped demonstrates the presence of cooperative emission from the array, which enhances the pair generation probability.

3 Conclusions

In conclusion, we investigated SFWM in single- and dual-channel ring resonator sequences. We found that the super-radiant enhancement originally predicted in a loss-less chain of all-pass rings is severely affected by the presence of loss. On the other hand, the DC configuration is more loss-tolerant, and a super-linear scaling can be achieved even with moderate loss. We also provided experimental evidence of cooperative emission of photon pairs from an array of microresonators on a silicon photonic chip whose behaviour is in very good agreement with our theory.

References

[1] M. Gross, S. Haroche, Phys. Rep. **93**, 301 (1982)

[2] T. Onodera, M. Liscidini, J. Sipe, L. Helt, Phys. Rev. A **93**, 043837 (2016)

[3] M. Liscidini, L. Helt, J. Sipe, Phys. Rev. A **85**, 013833 (2012)

 https://doi.org/10.1051/epjconf/202226608008

Static Multi-Vortex Structures in Nonlinear Optical Media

Angel Paredes[1], *José Ramón* Salgueiro[1], and *Humberto* Michinel[1,*]

[1]Universidade de Vigo. Departamento de Física Aplicada. Edificio Manuel Martínez Risco. Campus de As Lagoas. 32004 Ourense, Spain.

Abstract. We demonstrate through numerical simulations the existence of a new type of nonlinear waves in optical media: structures of vortex solitons that remain static in certain configurations, which depend on their relative positions and topological charges. Several examples are presented to illustrate this surprising behavior.

1 Introduction

Optical vortices are singularities on the phase distribution of a light beam with an integer number l of windings around a point where the amplitude of the field vanishes. The integer l is usually called the *topological charge* of the vortex, although we must point out that this definition is only valid for boundless fields, where the charge is a conserved quantity, as it is the case of the present work.

The interactions amongst optical vortices contained in the same coherent field have been studied since long ago [1] and it is well known that different types of dynamics take place depending on the relative values of l. In this work we present for the first time numerical evidence of the existence of static configurations of optical vortices embedded in a constant background, that can be formed in nonlinear materials with a self-defocusing nonlinearity of the Kerr type.

Our results show that intricate phase interactions yield exotic distributions of the phase dislocations that remain stable for long propagations, without changes in the amplitude profile of the light distribution, which is therefore static. On the other hand, there is a complex phase structure that remains stationary as the beam propagates. This surprising behavior can be heuristically understood as a combination of topological optical "forces" that arrange themselves in such a way that a complex structure is formed that supports itself by the phase gradient profile generated by the sum of all the vortices of the structure. If any of the vortices were removed, the entire structure would rapidly unravel.

We must stress that the solutions we present here are very different from the vortex lattices or clusters obtained in similar systems like Bose-Einstein condensates in rotating traps[2]. First of all, no trap exists in our case. Moreover, the existence of the static structures described here depend on a delicate combination of vortices with different topological charges (in particular, vortices and antivortices must be present). Finally, these structures are certainly not the minimal energy configuration for a particular set of conserved quantities and thus, they are, at most, metastable. However, our numerical simulations show that they can survive for long propagations, in fact surviving in some cases for our whole simulation time.

*e-mail: hmichinel@uvigo.es

2 Theoretical model

Paraxial beam propagation in a nonlinear Kerr-type optical material is described by a nonlinear Schrödinger equation, which in its adimensional form reads as:

$$i\frac{\partial\psi}{\partial z} + \frac{1}{2}\nabla_{\perp}^{2}\psi + |\psi|^{2}\psi = 0 \qquad (1)$$

being $\nabla_{\perp}^{2}$ the nabla operator in the transverse plane x, y and z the propagation distance. All the spatial dimensions are normalized by the wavelength in the material and $|\psi|^2$ in average gives the beam irradiance divided by the second-order nonlinear refractive index.

Within the previous model we use as an input function for our simulations a continuous light background with N phase dislocations located in different positions with coordinates x_j, y_j ($j = 1 \dots N$), yielding an input beam of the form $\psi(x, y, 0) = \prod_{j=1}^{N} \psi_j(r_j)e^{il_j\theta_j}$ being $l_j = \pm1, \pm2, \dots$ the topological charge of each vortex, $r_j = \sqrt{(x-x_j)^2 + (y-y_j)^2}$ the distance to each phase dislocation and $\theta_j = \arctan((y-y_j)/(x-x_j))$ the polar coordinate in the transverse plane for each vortex. The $\psi_j(r_j)$ are the amplitude profiles for the vortices of topological order j.

It is well known [1, 3] that if a vortex dislocation is placed within a fluid flow, it simply moves along it. Thus, when several vortices coexist in the same background, each of them induces a particular motion on each other. Then, static configurations can be engineered in cases in which all the induced velocities induced on each vortex by all the other vortices sum up to zero, assuming that the separation between vortices is much larger that the core size of the vortices themselves. As we will show in detail elsewhere, this type of configurations can be achieved

https://doi.org/10.1051/epjconf/202226608008

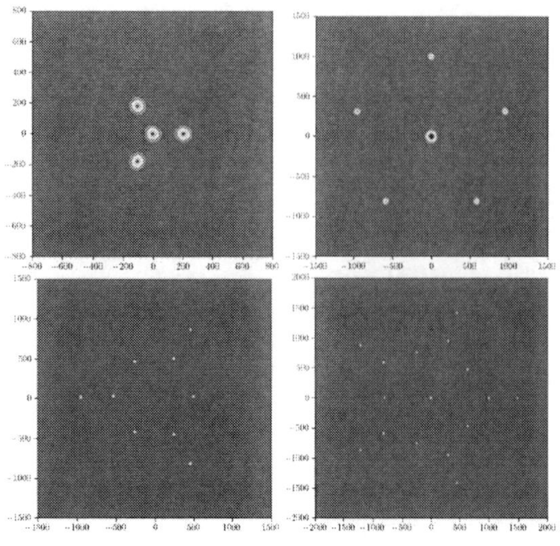

Figure 1. Transverse intensity distributions in the (x, y) plane of several vortex structures that remain unaltered during propagation. See the text for the description of each configuration.

by using ansatze with Z_n symmetry, where Z_n is a subgroup of the $U(1)$ rotation group around a particular center. Namely, we use regular n-polygons with the same center, each of which with vortices of a particular l on their vertices. Then, we look for relations between their sizes and the aforementioned topological charges to fulfil the staticity condition. It turns out that families of solutions exist, which, in particular, must satisfy the Diophantine equation:

$$\left(\sum_{i=1}^{N} l_i\right)^2 = \sum_{i=1}^{N} l_i^2 \qquad (2)$$

This shows that the l_i cannot all of them have the same sign. It is also worth remarking that Eq. (2) can be derived by using a Pohozaev identity[4].

3 Numerical results

The numerical simulations show that the previous calculations yield several possible structures in the form of vortex clusters with different configurations and topological charges that display a static intensity configuration as can be seen in fig. 1. In this figure we plot transverse intensity distributions in the (x, y) plane of several vortex clusters that remain unaltered during propagation. From top-left to right bottom is show: a central $l = -1$ vortex surrounded by three $l = 1$ phase dislocations forming an equilateral triangle (Z_3 symmetry), a central $l = 2$ vortex surrounded by five $l = -1$ singularities located in the vertices of a regular pentagon (Z_5 symmetry), an arrangement of nine single-charged vortices of which the six that form the inner core (which is almost a hexagon) are positive and the external three are negative (Z_3 symmetry) and finally an arrangement of 16 singly-charged vortices, where the central and the external vortices are positively charged, whereas the singularities forming the internal pentagons are negative (Z_5 symmetry). Notice that all of these structures of course satisfy Eq. (2).

It must be also underlined that the static configuration of the intensity distribution is obtained within a dynamical equilibrium in the phase space, which is evolving in time as can be seen in 2.

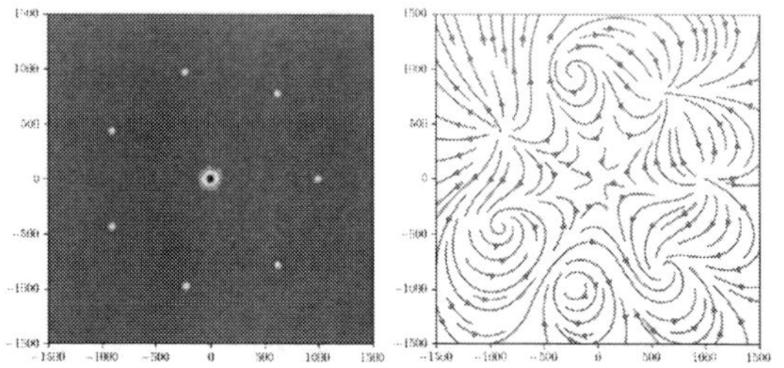

Figure 2. Transverse sections in the (x, y) plane showing the intensity (left) and phase gradient lines (right) of a vortex soliton structure with a central $l = 3$ singularity surrounded by an heptagonal regular distribution of negative single-charged vortices (Z_7 symmetry).

4 Conclusions

We have demonstrated through numerical simulations the existence of static structures of vortex solitons that remain unaltered during propagation for certain configurations, which depend on their relative positions and topological charges. Several examples have been presented illustrating that this surprising behavior can be achieved in very different configurations.

References

[1] D. Rozas, C. Law, G. Swartzlander, JOSA B **14**, 3054 (1997)

[2] J.R. Abo-Shaeer, C. Raman, J.M. Vogels, W. Ketterle, Science **292**, 476 (2001)

[3] Pismen LM, Pismen LM. Vortices in nonlinear fields: from liquid crystals to superfluids, from non-equilibrium patterns to cosmic strings. Oxford University Press; 1999.

[4] S. Pohožaev, Dokl. Akad. Nauk SSSR **165**, 36–39 (1965)

Dynamics of passive modelocking in class-B lasers with saturable absorber

Franco Prati[1,*], *Auro* Perego[2], and *Germán J* de Valcárcel[3]

[1]Università dell'Insubria, via Valleggio 11, 22100 Como, Italy
[2]Aston University, Birmingham B4 7ET, United Kingdom
[3]Universitat de València, Dr. Moliner 50, 46100 Burjassot, Spain

Abstract. We address the problem of passive modelocking in class-B lasers with saturable abosrber taking into account the fast dynamics of both gain and absorption. Our model, which is derived from a delay differential equation model, treats in a rigorous way the definition of the fast and slow times which are typically used in the master equation approach. In that way all the dynamical variables obey exact periodic boundary conditions and this makes the model suitable for analytic and numerical treatment. The model accounts for behaviours different from fundamental modelocking, such as Q-switching modelocking and harmonic modelocking.

1 Introduction

Passive modelocking (PML) is a widespread method used to generate coherent light pulses and optical frequency combs. The basic mechanism of PML has been understood since the fundamental work of Haus based on the master equation (ME) approach [1]. Yet it is known that Haus ME may fail in the common case of a class-B laser where the cavity photon lifetime is shorter than or comparable to the gain recovery time. Here we show how a sound ME for PML in class B lasers can be derived, which is valid for gain and absorption both slow and fast. We start from a well-known delay-differential equation model, transform it into a field map and pass to the continuous limit. Then we propose a rigorous definition of a two-time description of the problem, and finally split gain and absorption into slow and fast components. This procedure leads to exact periodic boundary conditions and it is essential to keep the problem simple, both analytically and numerically.

2 From the delay-differential equation model to the master equation

The use of the standard Haus ME for PML in class-B lasers is particularly problematic when the absorber is slow, as it happens for instance in dyes and semiconductors. In the recent years the problem has been addressed [2, 3] starting from a delay-differential equation (DDE) model for PML [4]. The DDE governs the dynamics of the slowly varying amplitude of the laser field at a reference plane as

$$\gamma^{-1}\dot{E}(t) + E(t) = \sqrt{K}e^{M(t-T_c)}E(t-T_c), \qquad (1)$$

where the derivative term (the overdot denotes d/dt) mimics the effect of a Lorentzian spectral filter of fixed bandwidth γ that could approximate the finite gain medium bandwidth or be due to another generic intracavity bandwidth limiting element, K is the power attenuation factor per cavity roudtrip due to nonresonant losses, $M = (1-i\alpha_g)G/2-(1-i\alpha_q)Q/2$ describes the net amplification per roundtrip due to the saturated gain G and absorption Q, with $\alpha_{g,q}$ the corresponding linewidth-enhancement factors of semiconductors, and T_c is the cold-cavity roundtrip time without the filter contribution.

The ME applies in the limit of small gain and losses in which one can assume that the system changes little over any cavity roundtrip [1]. In that limit it is possible to make the approximation $e^M \simeq 1+M$ and the DDE model can be written as [5]

$$\gamma^{-1}\dot{E} = -E + [1-\ell+M(t-T_c)]\,E(t-T_c), \qquad (2a)$$

$$\dot{G} = \gamma_g\,[G_0 - G(1+I)]\,, \qquad (2b)$$

$$\dot{Q} = \gamma_q\,[Q_0 - Q(1+sI)]\,, \qquad (2c)$$

with all (dimensionless) laser variables E, G, and Q evaluated at time t except when a different argument is explicitly written. $I(t) \equiv |E(t)|^2$ is the field intensity scaled to saturation intensity of the amplifier $I_{g,\mathrm{sat}}$, and the saturation parameter is defined as $s \equiv I_{g,\mathrm{sat}}/I_{q,\mathrm{sat}}$, where $I_{q,\mathrm{sat}}$ is the saturation intensity of the absorber. The decay rates $\gamma_{g,q}$ respectively are the inverses of the gain and absorber recovery times. Finally $\ell \equiv 1-\sqrt{K}$ represents linear losses, and G_0 and Q_0 are the unsaturated gain and absorption parameters. The initial assumption of small gain and losses is verified if $G, Q, \ell \ll 1$.

Our procedure for the derivation of the ME can be summarised as follows. First, we transform Eq. (2a) into a map relating the field at two consecutive roundtrips. Next, we introduce as it is customary a slow time T' which grows unbounded, and a fast time τ' restricted to the in-

*e-mail: franco.prati@uninsubria.it

 https://doi.org/10.1051/epjconf/202226608009

terval $[0, T_R]$, where T_R is the cavity roundtrip time normalised to take into account the effect of the spectral filter. However, as outlined in [6] our methodology differs from the usual one in that we introduce the new times $T = T' + \tau'$, $\tau = \tau'$ in such a way that all dynamical variables obey *exact* periodic boundary conditions of the form $X(T, \tau) = X(T, \tau + T_R)$. Another key point of our approach

Figure 1. Example of Q-switched modelocking (QSML). The pulse is modulated from zero to a maximum value with a period of about 150 roundtrip times.

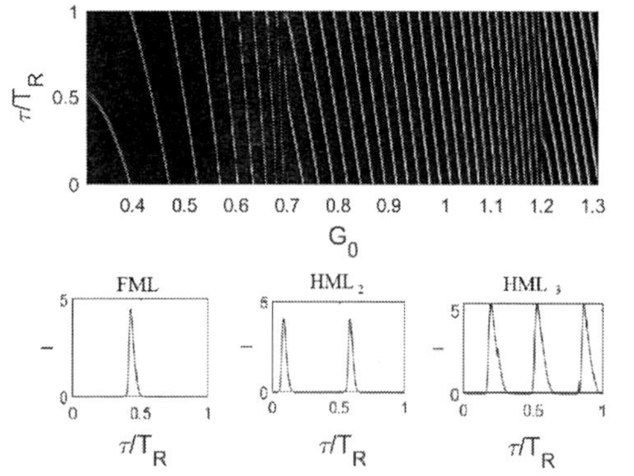

Figure 2. Example of transition from fundamental modelocking (FML) to harmonic modelocking with two pulses (HML_2) and harmonic modelocking with three pulses (HML_3).

consists in splitting the gain and absorption variables into slow and fast components [6]. The slow variables are defined as averages over one roundtrip time and are functions only of T, whereas the fast ones are defined as the difference between the total variables and the averages and can be regarded as functions only of τ.

3 Numerical results

Numerical simulations show different dynamical behaviours besides fundamental modelocking (FML). In the choice of the parameters we are guided by [2]. In all simulations we kept fixed the unsaturated absorption parameter $Q_0 = 0.3$ and the cavity losses $\ell = 0.005$, so that the (single mode) lasing threshold is $G_{0,th} = Q_0 + 2\ell = 0.31$. We also assumed that the amplifier is much slower than the absorber ($\gamma_g \ll \gamma_q$).

In Fig. 1 we considered a short cavity ($T_c = 2\gamma_q^{-1}$), relatively large gain bandwidth ($\gamma = 40\gamma_q$), linewidth enhancement factors different from zero ($\alpha_g = 1.5$, $\alpha_q = 1$), $\gamma_g = 0.01\gamma_q$, $s = 0.3$, and $G_0 = 0.5$. With these parameters both the stationary solution and FML are unstable. The long-term solution is a regime of Q-switched mode locking (QSML) with one pulse circulating in the cavity whose intensity is modulated from almost zero to a maximum value with a period of about 150 roundtrip times.

In a second simulation, shown in Fig. 2, we considered a longer cavity ($T_c = 30\gamma_q^{-1}$), relatively small gain bandwidth ($\gamma = 10\gamma_q$), null linewidth enhancement factors ($\alpha_g = \alpha_q = 0$), $\gamma_g = 0.05\gamma_q$, and $s = 0.15$. In this case we varied adiabatically the gain parameter G_0 from the lasing threshold 0.31 to 1.31. At threshold the laser already switches to the FML solution. At $G_0 \simeq 0.7$, FML is replaced by a modelocked solution with two pulses circulating in the cavity (HML_2) and finally at $G_0 \simeq 1.2$ we observe the transition to the modelocked solution with three pulses (HML_3). The transitions from one regime to the next are not abrupt, in between there are small intervals of G_0 for which no stable pulses exist. The different slope of the traces indicates that the pulses travel faster as their number increases.

References

[1] H. Haus, IEEE J. Sel. Top. Quantum Electron. **6**, 1173-1185 (2000)

[2] J. Hausen, K. Lüdge, S. V. Gurevich, and J. Javaloyes, Opt. Lett. **45**, 6210-6213 (2020).

[3] M. Nizette and A. G. Vladimirov, Phys. Rev. E **104**, 014215 (2021)

[4] A.G. Vladimirov and D. Turaev, Phys. Rev. A **72**, 033808 (2005)

[5] T. Kolokolnikov, M. Nizette, T. Erneux, N. Joly, and S. Bielawski, Physica D **219**, 13–21 (2006)

[6] A.M. Perego, B. Garbin, F. Gustave, S. Barland, F. Prati, and G.J. de Valcárcel, Nat. Commun. **11**, 311 (2020)

Spontaneous Parametric Down-Conversion from GaAs Nanowires at Telecom Wavelength

Grégoire Saerens[1,*], *Ngoc My Hanh* Duong[1], *Alexander S.* Solntsev[2], *Artemios* Karvounis[1], *Thomas* Dursap[3], *Philippe* Regreny[3], *Andrea* Morandi[1], *Robert J.* Chapman[1], *Andreas* Maeder[1], *Alexandre* Danescu[3], *José* Penuelas[3], *Nicolas* Chauvin[3], and *Rachel* Grange[1]

[1]ETH Zuerich, Optical Nanomaterial Group, Institute for Quantum Electronics, Department of Physics, 8093 Zuerich, Switzerland
[2]University of Technology Sydney, School of Mathematical and Physical Sciences, Ultimo NSW 2007, Australia
[3]Univ. Lyon, CNRS, ECL, INSA Lyon, UCBL, CPE Lyon, INL, UMR 5270, 69130 Ecully, France

Abstract. We report on the generation of photon pairs at 1550 nm from free-standing epitaxially grown self-assisted micrometre long GaAs nanowires. The efficiency of the spontaneous parametric down-conversion process has a rate of 320 GHz/Wm normalized to the transmission of the setup, the pump intensity, and the volume of the nanostructure. GaAs is a high index dielectric that can support electromagnetic Mie modes, therefore we model how shorter nanowires could improve the second-harmonic signal and we found that sub-micro long nanowires (600 nm length and 250 nm diameter) can support quality factors up to 15 at the pump wavelength (780 nm). We anticipate that the near field enhancement compared to micrometre long nanowires will boost the second-harmonic generation and, correspondingly, the biphoton rate efficiency.

1 Introduction

Photon pair sources are needed for applications in quantum cryptography [1], quantum imaging [2] or quantum metrology [3]. Spontaneous parametric down-conversion (SPDC), first demonstrated in 1970 by Burnham and Weinberg [4], is one method to generate photon pairs at room temperature with high spatial and temporal coherence. In the last years, increasingly smaller sources have been developed. Centimetre-sized nonlinear crystal [5] and periodically poled waveguides [6], requiring temperature and wavelength control, have given way to thin films [7], metasurfaces [8] and nanoantennas [9,10]. The lower photon pair rate that is inherent to miniaturizing the source is compensated by an increased generation efficiency (Hz/Wm). The overall photon pair rate can be improved when several sources, pumped by the same laser, are integrated next to each other. Furthermore, the typically stringent energy and momentum requirements for SPDC are relaxed for smaller structures, leading to a broad emission spectrum.

III–V materials such as GaAs have valuable properties for nonlinear applications such as a direct band gap and a non-centrosymmetric crystal structure. Additionally, compared to $LiNbO_3$ or beta-BaB_2O_4 (BBO) crystals, the bulk quadratic $\chi^{(2)}$ nonlinearities of GaAs are very high with a value for d_{36} of 370 pm/V [11].

2 III-V Nanowires for SPDC

We demonstrate photon pair generation from epitaxially grown III-V GaAs nanowires and present a pathway to increase the generation efficiency by slicing a nanowire to a pre-determined size using focused ion beam milling [12]. With this technique, we can shape high-quality nanoantennas, combining also lattice-mismatched materials, different refractive indices and crystal phases that are not available in bulk III–V materials [13]. For instance, a strong tuning of the SHG efficiency was recently demonstrated by simply tuning the GaAs NW crystal phase [14]. Mie-resonant nanostructures can be engineered to localize the electromagnetic field inside the nanostructure and therefore increase efficiencies of nonlinear processes [15,16].

A schematic of the SPDC process is shown in Figure 1a). A GaAs nanowire is pumped at 780 nm and the photon pairs (signal and idler) are collected in transmission. The goal is to compare the SPDC generation efficiencies of a long and a sliced nanowire (see Figure 1b) such that it shows a strong Mie-type resonance at the pump (780 nm) and conversion wavelengths (1560 nm). To determine this size, several spectra of second-harmonic generation (SHG), which is the reverse process of degenerate SPDC, were calculated for nanowires with different diameters and lengths. Figure 1c) shows a simulation using a finite element method (COMSOL Multiphysics) for a nanowire with a 250 nm diameter and sliced to a length of 600 nm. The resonances are located around 780 and 1560 nm as set for the experiment, with a dominant electric dipole emission profile. The quality factor Q at the pump wavelength of 780 nm is 15,

* Corresponding author: gsaerens@ethz.ch

indicating a field intensity enhancement inside the nanostructure, enabling an increase in the biphoton conversion efficiency.

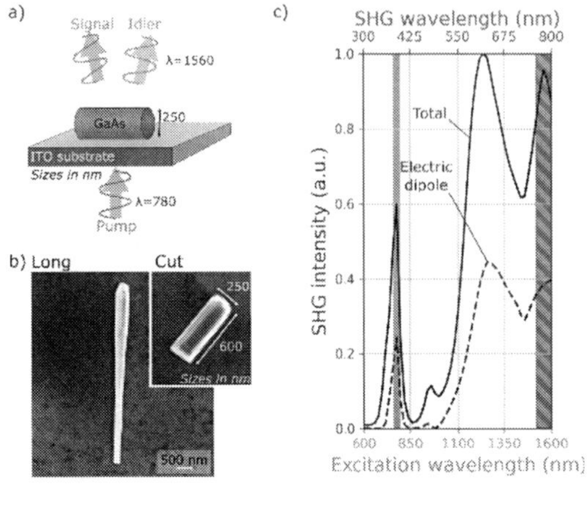

Figure 1. a) Schematic of the SPDC process for a GaAs nanowire pumped at 780 nm. b) Electron microscopy image of a nanowire before and after being sliced by focused ion beam milling. c) Simulation spectrum of second-harmonic generation (SHG), the reverse process of degenerate SPDC, for a sliced nanowire with a 600 nm length and a 250 nm diameter. The resonances are located around 780 and 1560 nm as set for the experiment, with a dominant electric dipole emission profile.

3 Experimental Results

Figure 2 shows the coincidence counts for a long nanowire measured in the Hanbury Brown–Twiss configuration [7]. This was achieved by integrating over 1 hour and pumping with a cw laser at 780 nm with 4 mW incident power. The signal to noise ratio (SNR) was above 2, indicating photon pair emission from the nanowire.

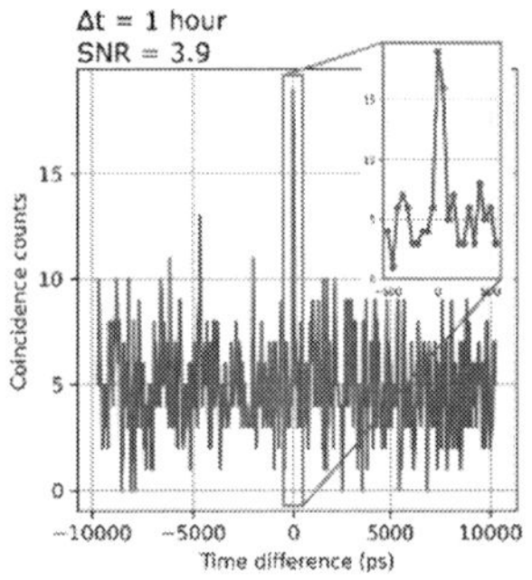

Figure 2. Coincidence counts of a long nanowire integrated over 1 hour and pumping with a cw pump laser at 780 nm with 4 mW incident power. The signal to noise ratio (SNR) is greater than 2 indicating photon pair emission from the nanowire.

After taking into account the 0.4% transmission of photon pairs in the setup and after normalizing to the pump intensity (0.05 mW/μm^2), we obtain an SPDC conversion efficiency of 320 GHz/Wm. This value is in comparison as high as in periodically poled waveguides (photon rate of 10^7 Hz with $L_0 \sim 50$ mm, $P_{in} \sim 1$ mW) [6,17] or two orders of magnitude higher than similar GaAs nanoantennas [9].

In conclusion, we have shown that GaAs nanowires are efficient compact sources of biphotons. The advantage of nanowires is that they can be easily deposited on any substrate, allowing the use of their full length to slice several nanoantennas, for example to form a cavity structure or a phase array where all emit in phase.

References

1 H. K. Lo, M. Curty, and K. Tamaki, *Nat. Photonics* **8**, 595–604 (2014).
2 G. B. Lemos, V. Borish, G. D. Cole, S. Ramelow, R. Lapkiewicz, and A. Zeilinger, *Nature* **512**, 409–412 (2014).
3 E. Polino, M. Valeri, N. Spagnolo, and F. Sciarrino, *AVS Quantum Sci.* **2**, (2020).
4 D. C. Burnham, and D. L. Weinberg, *Phys. Rev. Lett.* **25**, 84–87 (1970).
5 P. G. Kwiat, K. Mattle, H. Weinfurter, A. Zeilinger, A. V. Sergienko, and Y. Shih, *Phys. Rev. Lett.* **75**, 4337–4341 (1995).
6 E. Pomarico, B. Sangulnettl, N. Gisin, R. Thew, H. Zbinden, G. Schreiber, A. Thomas, and W. Sohler, *New J. Phys.* **11**, (2009).
7 C. Okoth, A. Cavanna, T. Santiago-Cruz, and M. V. Chekhova, *Phys. Rev. Lett.* **123**, 263602 (2019).
8 T. Santiago-Cruz, A. Fedotova, V. Sultanov, M. A. Weissflog, D. Arslan, M. Younesi, T. Pertsch, I. Staude, F. Setzpfandt, and M. Chekhova, *Nano Lett.* **21**, 4423–4429 (2021).
9 G. Marino, A. S. Solntsev, L. Xu, V. F. Gili, L. Carletti, A. N. Poddubny, M. Rahmani, D. A. Smirnova, H. Chen, A. Lemaître, G. Zhang, A. V. Zayats, C. De Angelis, G. Leo, A. A. Sukhorukov, and D. N. Neshev, *Optica* **6**, 1416 (2019).
10 N. M. H. Duong, G. Saerens, F. Timpu, M. Teresa B., Vincenzo Buscaglia, Andrea Morandi, Jolanda S. Mueller, A. Maeder, F. Kaufmann, A. Solntsev, and R. Grange, *ArXiv Quantum Phys.* 19 (2021).
11 R. W. Boyd, Nonlinear Optics (Elsevier Science Publishing Co Inc, 2008).
12 M. Timofeeva, L. Lang, F. Timpu, C. Renaut, A. Bouravleuv, I. Shtrom, G. Cirlin, and R. Grange, *Nano Lett.* **18**, 3695–3702 (2018).
13 K. Tomioka, Y. Kobayashi, J. Motohisa, S. Hara, and T. Fukui, *Nanotechnology* **20**, (2009).
14 B. Zhang, J. E. Stehr, P. P. Chen, X. Wang, F. Ishikawa, W. M. Chen, and I. A. Buyanova, *Adv. Funct. Mater.* **31**, 1–7 (2021).
15 R. Camacho-Morales, M. Rahmani, S. Kruk, L. Wang, L. Xu, D. A. Smirnova, A. S. Solntsev, A. Miroshnichenko, H. H. Tan, F. Karouta, S. Naureen, K. Vora, L. Carletti, C. De Angelis, C. Jagadish, Y. S. Kivshar, and D. N. Neshev, *Nano Lett.* **16**, 7191–7197 (2016).
16 G. Saerens, I. Tang, M. I. Petrov, K. Frizyuk, C. Renaut, F. Timpu, M. Reig Escalé, I. Shtrom, A. Bouravleuv, G. Cirlin, R. Grange, and M. Timofeeva, *Laser Photonics Rev.* **14**, 1–10 (2020).
17 J. Zhao, C. Ma, M. Rüsing, and S. Mookherjea, *Phys. Rev. Lett.* **124**, 163603 (2020).

Nonlinear pulse routing in plasmonic couplers

José R. Salgueiro[1,*] and *Albert* Ferrando[2,3,**]

[1]Universidade de Vigo, Departamento de Física Aplicada, E. E. Aeronáutica e do Ezpazo, As Lagoas, s/n, 32004 Ourense, Spain.
[2]Institut de Ciència dels Materials, Universitat de València, Paterna 46980 Valencia, Spain.
[3]Departament d'Òptica i Optometria i Ciències de la Visió, Universitat de València, Dr. Moliner, 50, Burjassot, 46100 Valencia, Spain.

Abstract. We present a system able to discriminate pulses according to their duration with potential applications in all-optical signal multiplexing. The device is based on a directional coupler with nonlinear cores and metallic claddings with dimensions in a nanometric scale. Simulations are carried out using the FDTD technique for ultrashort pulses of femtosecond order. It is shown that the device is able to separate such pulses respect to a time-width threshold which depends on the total energy of the pulse.

1 Introduction

Modern technology pursues the increase of communication capabilities requiring from fast processing systems to discriminate and route information and also from miniaturized components leading to a high encapsulation rates. In such a way, plasmonics has come into play to put optical devices into the nanoscale [1] even though important drawbacks, as the high losses associated to such devices, have to be overcome.

The implementation of all-optical operations is very important in order to achieve high processing speed and so exploiting nonlinear properties is one of the the most valuable resources available at the time. In such a way there is a great interest on the use of pulses to increase optical power density and induce the nonlinear response with a reasonable power. Up to now different nonlinear systems using metals have been studied mainly in the continuous wave regime (CW), particularly the nonlinear directional coupler for optical power switching [2]. Nevertheless, modeling such systems for their use with ultrashort pulses has proved to be difficult as it is revealed by the scant number of works available at the moment on the matter, which additionally deal with quite simple systems as metal-dielectric interfaces [3, 4] or single slot waveguides [5, 6].

In this work we present a device to perform all-optical switching based on the input pulse duration. In a previous work [7] we already explored the possibility of all-optical switching based on total pulse energy. The mechanism presented here is an alternative operation mode for the same device which in principle would not require from any physical modification to operate in one or another way, being consequently more versatile.

*e-mail: jrs@uvigo.es
**e-mail: Albert.Ferrando@uv.es

2 Device description

The directional coupler is assumed to lie in the plane XZ, being z the propagation direction, and is formed by two slot waveguides. The two dielectric cores show the nonlinear Kerr response and are embedded into metallic linear claddings. The field is assumed to be a TM mode with one magnetic and two electric components, H_y, E_x and E_z. A reference wavelength $\lambda_0 = 800\,\text{nm}$ is chosen and normalized units used, scaling the spatial coordinates by the wavenumber $k_0 = 2\pi/\lambda_0$ and the time by the reference frequency $\omega_0 = 2\pi c/\lambda_0$, being c the vacuum speed of light. This means that a space normalized unit equals to about 127 nm while a normalized time unit equals to 0.42 fs.

Cores are modeled using an instantaneous response. On the other hand metallic claddings are modeled by the *additional differential equation* (ADI) technique [8] using a classical Drude model dependent on two parameters, the plasma frequency and the electronic collision frequency (which accounts for losses). Those parameters for silver at the reference wavelength are $\omega_p = 1.30 \times 10^{16}$ Hz and $\Gamma = 2.93 \times 10^{13}$ Hz. The computation domain is surrounded by proper perfect matching layers (PML) and a source is implemented using the Total field/Scattered field technique, to generate pulses of Gaussian shape (spatial as well as temporal) at one of the cores (first core).

Energy of pulses is measured at a particular z-position in any of the cores by integrating in space (transversal coordinate, x) and time the power flux in the propagation direction, i.e. the z component of the Poynting vector $S_z = E_x H_y$.

3 Operation results

The key of the device operation is the ability of the coupler to pass energy to the second core from the first one (the excited one) after propagation along a distance named beat length (linear regime). For the simulated coupler this

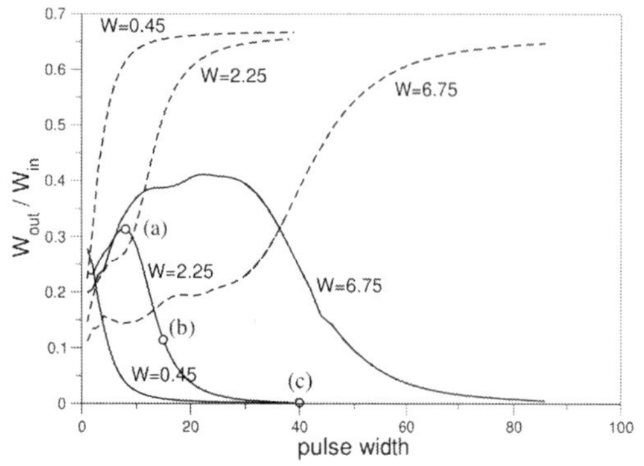

Figure 1. Normalised output energy at the excited core (solid lines) and second core (dashed lines) against input pulse width for three different energy pulses. Points (a), (b) and (c) correspond to cases with the same label in figure 2.

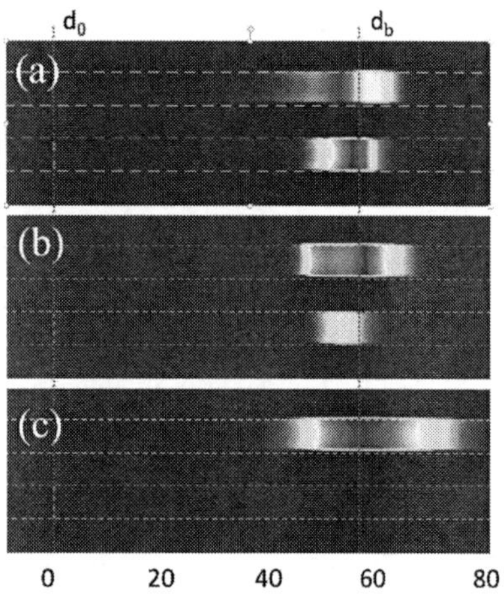

Figure 2. Simulation for three pulses of energy $W = 2.25$ and different time width: a) $\Delta t = 10$, b) $\Delta t = 15$, c) $\Delta t = 40$. Lines d_0 and d_b mark the source and output positions.

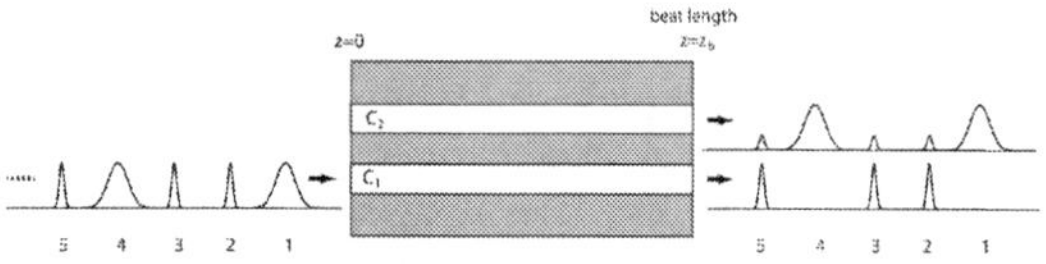

Figure 3. Sketch of the device operation to discriminate different time-width pulses.

length is about $z_b \sim 57$. When pulses have enough energy a nonlinear response is triggered, beat length changes and inter-core energy transmission is incomplete up to the point that it is fully suppressed if pulse energy is enough (energy switching) [7]. Pulse time-width also influences coupling as power density changes (if total energy is the same) and so it is also expected a change in the switching response when the pulse duration is modified. Actually, for a fixed energy, a longer pulse carries a lower density and so the system should work more as it were in a linear regime. This is clearly seen in figure 1 where the output energy after propagation for a beat length is plotted against the pulse duration for three different pulse energies. The output at the excited core (continuous lines) is large for low duration pulses as the system operates in a nonlinear regime and almost all the energy remains in the excited core. Transferred energy (dashed lines) is consequently low. Increasing the pulse duration makes the solid curves to decrease to zero as the nonlinear effect is weaker and a larger fraction of the energy passes to the second core. The effect takes place at larger pulse-width for high pulse energy as the nonlinear response depends on energy density. Optical losses are responsible for a total output energy fraction less than the unity (see figure 1, but due to the short length of the device (beat length around 7.5 μm) such losses are assumable (about 30% of the input energy). In figure 2 we show a particular simulation for three pulses of the same energy but different time width illustrating the three switching regimes. In subfigure (a) a short pulse triggers the nonlinearity and the coupler only passes a fraction of the energy to the second core. In (b) the longer pulse makes a big amount of the energy to switch to the second core, reducing the signal at the output of the excited core. In the third case an even longer pulse produces a full energy transfer to the second core since it is now in a linear regime.

According to this system performance the device is useful to separate shorter pulses from larger ones as illustrated in figure 3. The time-width threshold for discrimination is determined by the total energy of pulses and can be adjusted according to the requirements of every particular case. Also this threshold can be easily reconfigurable (changing pulse energy) resulting a very versatile device.

We thank the support of MCIN of Spain through the projects PID2020-118613GB-I00 and PID2020-120484RB-I00, as well as Xunta de Galicia, Spain (grant ED431B 2021/22) and Generalitat Valenciana for the grant PROMETEO/2021/082.

References

[1] N.C. Panoiu, W.E.I. Sha, D.Y. Lei, G.C. Li, Journal of Optics **20**, 083001 (2018)

[2] J.R. Salgueiro, Y.S. Kivshar, Appl. Phys. Lett. **97**, 081106 (2010)

[3] Z.L. Sámson, P. Horak, K.F. MacDonald, N.I. Zheludev, Opt. Lett. **36**, 250 (2011)

[4] A. Baron, T.B. Hoang, C. Fang, M.H. Mikkelsen, D.R. Smith, Phys. Rev. B **91**, 195412 (2015)

[5] A. Pusch, I.V. Shadrivov, O. Hess, Y.S. Kivshar, Opt. Express **21**, 1121 (2013)

[6] O. Lysenko, M. Bache, N. Olivier, A.V. Zayats, A. Lavrinenko, ACS Photonics **3**, 2324 (2020)

[7] J.R. Salgueiro, A. Ferrando, EPJ Web of Conferences (2020)

[8] A. Taflove, *Computational electrodynamics, the finite difference time domain* (Artech House Publishers, Norwood, Massachusetts, 2005)

Multi-beam vortex generation induced by the non-linear optical anisotropy of graphene.

Luis Plaja*, *Ana* García-Cabrera, *Roberto* Boyero-García and *Óscar* Zurrón *Julio* San Román and *Carlos* Hernández-García.
Grupo de Investigación en Aplicaciones del Láser y Fotónica, Departamento de Física Aplicada, Universidad de Salamanca, E-37008, Salamanca, Spain

Abstract. We analyse the high harmonic emission from single-layer graphene driven by infrared vector beams. We demonstrate that graphene's anisotropy offers a privileged scenario to explore non-trivial light spin-orbit couplings, which substantially extends the possibilities for the generation of high-harmonic structured beams currently studied in atomic and molecular targets. In our case, graphene's crystal symmetry introduces a spin-dependent diffraction pattern that, coupled with the fundamental conservation of the driver's topological phase, leads to the splitting of the harmonic field in a multi-beam structure, composed of spatially diverging vortices. Our work demonstrates that anisotropic targets are extraordinary tools to sculpt complex structured short-wavelength beams.

1 Introduction.

High-order harmonic generation (HHG) is well established as the main pathway to the production of high-frequency coherent radiation. The extreme non-linear, non-perturbative, response of atoms and molecules under intense driving leads to the emission of high harmonics. This process can be described using a strikingly simple mechanism [1]: electrons are tunnel ionized from the target elements, then evolve as quasi-free particles under the field excitation, following a trajectory redirected to the parent ion. Harmonics are radiated during the electron recollision with the ion, liberating its kinetic energy in the form of high-frequency radiation.

Recently the intense-field community has raised interest in the study of HHG in crystalline solids. This new playground offers evident advantages, as an increased harmonic conversion efficiency, together with promising scenarios, where crystal symmetries play a role during the harmonic conversion. Among them, for instance, HHG has been postulated as a realization of an ultrafast nanoscale temporal mater Talbot-Lau interferometer [2]. Driven in normal incidence, HHG in solids can be understood from a similar mechanism to that in atoms or molecules, now in terms of recolliding electron-hole trajectories in the reciprocal space [3].

Recent studies have demonstrated graphene as a very particular target for HHG. The presence of point degenerations in the energy dispersion, i.e. Dirac points, modifies the first step of HHG, as tunnel excitation is replaced by electron-hole pair creation during the non-adiabatic crossing of electron trajectories near the singular points [4]. Also, graphene confers HHG with a strong anisotropy [5,6]. Therefore, it is a natural playground to explore polarization-dependent modifications in HHG. In particular, the possibility of non-trivial structuring of harmonic vector beams, i.e. short-wavelength beams with an inhomogeneous distribution of polarization in the transverse plane.

2 Geometry.

We consider graphene single layer (SLG) driven at normal incidence by a 3-micron radial vector beam with waist 90 μm (see Fig. 1).

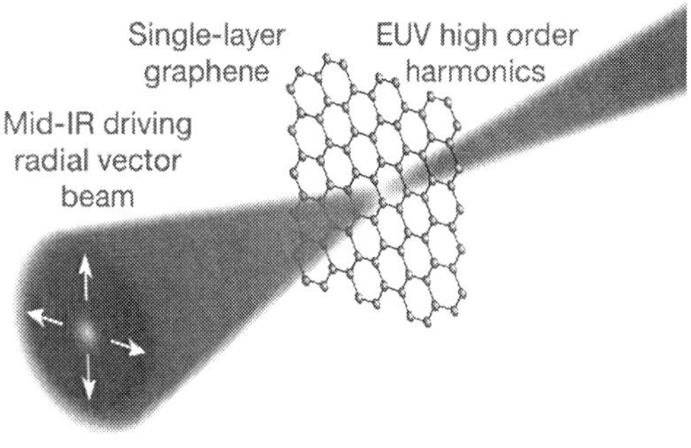

Fig. 1. Interaction geometry. A radial vector beam drives HHG in a graphene layer.

The vector beam is represented by a superposition of two Laguerre-Gauss vortices with topological charges $\ell_{1,2} = \pm 1$ and opposite polarization state $\sigma_{1,2} = \mp 1$, so that the net angular momentum of each beam component $k = 1,2$ is $j_k = \ell_k + \sigma_k = 0$. This condition is translated to the invariance of the field's geometric phase in each of the

* Corresponding author: lplaja@usal.es

two beam components when circling around the vortex ring, since the azimuthal phase introduced by the topological charge compensates the geometrical phase acquired by a circularly polarized beam when rotating around the axis [7]. It has been demonstrated [8] that the phase invariance is preserved in HHG therefore, in isotropic media, harmonics driven by vector beams are generated as vector beams. As we report here, graphene's response being strongly anisotropic introduces a polarization-dependency in HHG that, for vector beam drivers, turns into an azimuthal modification of the harmonic emission and, therefore, the introduction high OAM charges. This modifies substantially the isotropic scenario.

3 Harmonic near-field distribution.

To compute HHG from SLG we follow [9]. First, we discretize the target layer into elemental surface elements, small enough to consider the driver's profile along them as constant. HHG from each element is computed in the nearest-neighbour tight-binding approximation, and propagated to the far field using the integral solution of Maxwell's equations. The total field is then computed as the addition of the elemental contributions.

Figure 2a shows the harmonic intensity profile at the target, for the total beam and its left circular polarized component, demonstrating a rotationally shifted necklace pattern depending on the polarization. This shifted near field configuration translates to a shifted rotation in the far-field pattern, therefore demonstrating graphene's spin-dependent diffraction.

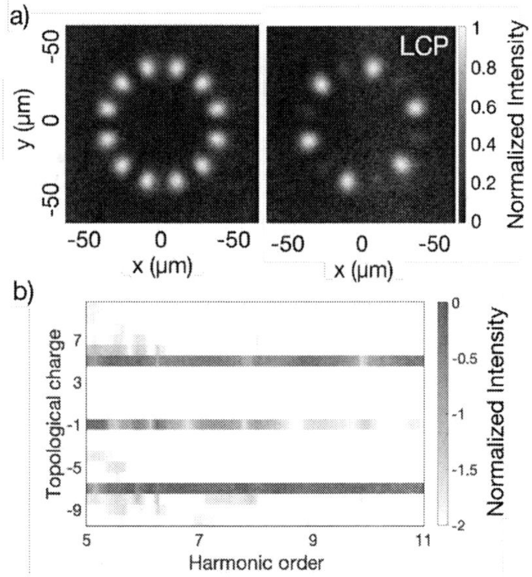

Fig. 2. (a) 9th harmonic near field intensity profile for (left) the total vector beam and (right) the left circular polarized component.

Figure 2b shows the angular momentum content of the near field harmonic. Most strikingly, harmonics are generated with high vortex charges, therefore apparently breaking the conservation rule for the total angular momentum $j_k = 0$ and, consequently, the preservation of the topological phase invariance that is found in HHG from isotropic charges.

4 Harmonic far-field distribution.

The analysis of the harmonic far-field distribution unveils the structural details of the harmonic field and resolves the apparent breaking of the topological phase invariance in HHG. We plot in Figure 3 the far-field intensity distributions of the 9th harmonic. Note that the harmonic field is composed of thirteen independent beams. The central beam is a ring structure corresponding to a radial vector beam, i.e. the composition of two vortex beams with opposite helicities. On the other hand, the twelve diverging beams correspond to vortex beams, with different azimuthal distributions depending on their polarization. Each of these field beams has a net angular momentum $j_k = 0$ and, therefore preserves the topological phase invariance of the driving beam.

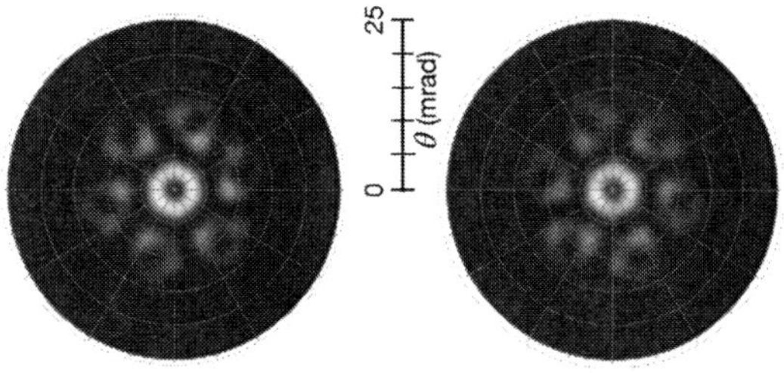

Fig. 3. Far field intensity distributions of the left (left) and right (right) polarized components of the 9th harmonic beam in terms of the azimuthal and divergence angles.

References

1. K. J. Schafer, *et al,* Phys. Rev. Lett. **70**, 1599 (1993).
2. A. García-Cabrera *et al,* New J. Phys. **23** 093011 (2021)
3. G. Vampa, *et al,* Phys. Rev. Lett. **113**, 073901 (2014).
4. Ó. Zurrón-Cifuentes *et al,* New J. Phys. **20**, 053033 (2018).
5. N. Yoshikawa, *et al,* Science **356,** 736 (2017)
6. Ó. Zurrón-Cifuentes *et al,* Opt. Express **27** 7776 (2019)
7. K. Y. Bliokh, *et al,* Nature Photonics **9**, 796 (2015).
8. C. Hernández-García *et al,* Optica **4**, 520-526 (2017)
9. R. Boyero-García, *et al,* Opt. Express, **29**, 2488 (2021).

Non-locality and single meta-atom spectroscopy in THz Landau polaritons

Giacomo Scalari[1,*], *Shima* Rajabali[1], *Elsa* Jöchl[1] , Sergej Markmann[1], *Simone* de Liberato[2], *Erika* Cortese[2], *Mattias* Beck[1], *Jérôme* Faist[1]

[1]Institute for Quantum Electronics, Department of Physics, ETH Zürich, Switzerland
[2]School of Physics and Astronomy, University of Southampton, UK

Abstract. We will discuss, theoretically and experimentally, the existence of a limit to the possibility of arbitrarily increasing electromagnetic confinement in polaritonic systems, where strongly sub-wavelength fields can excite a continuum of high-momenta propagative magnetoplasmons. This leads to peculiar nonlocal polaritonic effects, as certain polaritonic features disappear and the system enters in the regime of discrete-to-continuum strong coupling. We will as well present experiments reporting spectroscopy of a single, ultrastrongly coupled, highly subwavelength resonator operating at 300 GHz.

INTRODUCTION

Sub-wavelength electromagnetic field localization is a central theme in photonic research, as it allows sensing capabilities as well as increasing the light-matter coupling strength. Recently, the strong and ultrastrong light-matter coupling regime [1] in the THz range with split-ring resonators coupled to magnetoplasmons [2] has been widely investigated, achieving successive world-records for the largest light-matter coupling ever achieved. Ever-shrinking resonators have allowed to approach the regime of few electrons strong coupling, in which single-dipole properties can be modified by the vacuum field [3].

We will discuss, theoretically and experimentally, the existence of a limit to the possibility of arbitrarily increasing electromagnetic confinement in polaritonic systems. Strongly sub-wavelength fields can excite a continuum of high-momenta propagative magnetoplasmons [4]. This leads to peculiar nonlocal polaritonic effects, as certain polaritonic features disappear and the system enters in the regime of discrete-to-continuum strong coupling. We will as well discuss experiments reporting spectroscopy of a single, ultrastrongly coupled, highly subwavelength resonator operating at 300 GHz. By using a combination of immersion lenses we unravel the linewidth dependence of planar metamaterials as a function of the meta-atom number indicating quenching of the superradiance. On these grounds, we investigate ultrastrongly coupled Landau polaritons at the single resonator level [5], measuring a normalized coupling ratio $\frac{\Omega}{\omega} = 0.6$.

Polaritonic non-locality

Nanophotonic structures confine electromagnetic radiation below the Abbe diffraction limit by storing part of the electromagnetic energy into the kinetic energy of moving charges. Primarily relying on metals as charge reservoirs, plasmon- ics has mainly targeted the visible portion of the electromagnetic spectrum. The possibility of extending plasmonic excitation to low frequencies (mid-infrared and terahertz)by employing semiconductors and two-dimensional (2D) systems with plasma frequency tunable via the carrier concentration has allowed extreme electro-magnetic field confinements in nanostructures. An important question arises on the physical limitations to reducing the cavity volume and the subsequent increase of light–matter coupling. We show theoretically that the nonlocal effects are driven by the confinement of the resonator's electromagnetic field coupling to a continuum of propa- gating magnetoplasma excitations.

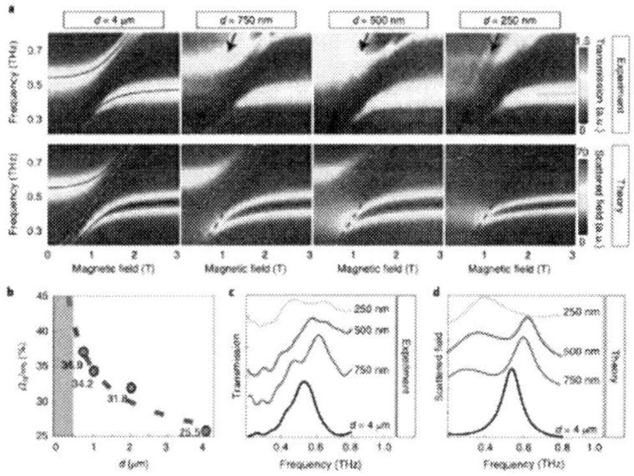

Fig. 1. a) Top row: transmission of cSrrs coupled to the Landau level transition in 2DEG versus magnetic field for four different gap sizes, *d*. . bottom row: the calculated scattered field for different gap feature sizes, *d* (same as the gap sizes of the experiment results). **b**, Normalized coupling versus *d*. The red-shaded region corresponds to where the large broadening of UP does not allow a measurement of the coupling. **c**,**d**, Sections (offset for clarity) of the colour plots in **a** at zero magnetic field for the experimental result (**c**) and the theoretical calculation (**d**) [4].

We experimentally observed such a physics using Landau polaritons at sub-terahertz frequencies by employing nanometre-sized resonators. First, we show that, when

* Corresponding author: scalari@phys.ethz.ch

nonlocal effects are not relevant (gap size d>750nm), the normalized coupling ratio between light and matter scales as √1d, increasing by ~50% from d = 4 μm to 750 nm, and reaching 37% (Fig.1 (b)). For smaller gap values, nonlocal effects become dominant and the sys- tem is no longer well described by a standard Hopfield model. As predicted by our multimode dissipative bosonic Hamiltonian, we observe a broadening of the UP branch and partial disappearance on the lower one (Fig.1a,c)

2 A single, ultrastrongly coupled THZ meta-atom

Free-space coupling to subwavelength individual optical elements is a central theme in quantum optics, as it allows the control over individual quantum systems. In this second experiment we show that, by combining an asymmetric immersion lens setup and a complementary resonating meta-surface we are able to perform terahertz time-domain spectroscopy of an individual, strongly subwavelength (50 um wide) meta-atom oscillating at 300 GHz, corresponding to a free space wavelength of 1 mm. We show that the linewidth of the metasurface scales as 1/N where N is the number of the meta- atoms, indicating quenching of the superradiant coupling. We propose an explanation of the observed signals based on a Fourier optics argument (Fig.2(a-c)).

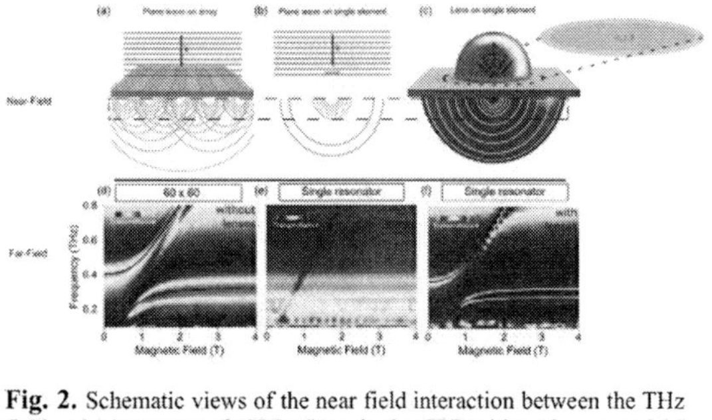

Fig. 2. Schematic views of the near field interaction between the THz field and (a) an array of cSRRs (b) a single cSRR without lenses, and (c) a single cSRR with lenses. (d)-(f) show the colormaps corresponding to the normalized transmittances of cSRRs on top of a GaAs quantum well measured under a varying B-field from 0-4T with the same lens configurations.

We successively investigate ultrastrongly coupled Landau polaritons at the single resonator level. We clearly show how the lens system allows the measurements of the polaritonic branches that are completely lost in the case of the single resonator excited by a plane wave, whose radiation is collected without an immersion lens. In this case, we do not measure signs of ultrastrong coupling since the correct resonant LC mode cannot be excited and detected. In Fig.2 f we report a single-resonator ultrastrong coupling measurement yielding a coupling ratio Ω/ω=0.33 with a single resonator cooperativity C = 94. Employing a more doped 2DEG we could measure a coupling ratio of Ω/ω=0.6 [5].

Our findings pave the way towards the control of ultrastrong light-matter interaction at the single electron/ resonator level. The proposed technique is way more general and can be useful to char- acterize the complex conductivity of micron-sized samples in the terahertz domain.

References

1. Forn-Diaz, P., et al., Rev. Mod. Phys.91, 025005 (2019)
2. Scalari, G. et al., Science, 335, 1323–1326 (2012)
3. R Chikkaraddy et al., Nature 535 (7610), 127-130 (2016)
4. S. Rajabali, E. Cortese, M. Beck, S. De Liberato, J. Faist and G Scalari, Nat. Phot, 15, 690 (2021)
5. S. Rajabali, S. Markmann, E. Jöchl, M.Beck, C. Lehner, W. Wegscheider, J. Faist and G. Scalari, Nat. Comm., **13 :2528** (2022)

 https://doi.org/10.1051/epjconf/202226608014

Supercontinuum generation in the enhanced frequency chirp regime in multipass cells

Víctor W. Segundo Staels[1,*], *Enrique* Conejero Jarque[1], *Daniel* Carlson[2], *Michaël* Hemmer[2], *Henry C.* Kapteyn[2], *Margaret M.* Murnane[2], and *Julio* San Roman[1]

[1]Grupo de Investigación en Aplicaciones del Láser y Fotónica, Departamento de Física Aplicada, Universidad de Salamanca, E-37008, Salamanca, Spain
[2]Department of Physics, JILA and STROBE NSF Science & Technology Center, University of Colorado and NIST, Boulder, Colorado, United States

Abstract. We identify, via numerical simulations, the regime of enhanced frequency chirp during nonlinear propagation in multipass cell. This regime - used before the dawn of chirped pulse amplification to generate ultrashort pulses - paves the way for the generation of temporally clean few-cycle pulses. Here, we demonstrate numerically that the spectra of pulses from an Yb-based laser system can be broadened into a flat supercontinuum with a smooth spectral phase compatible with a clean few-cycle pulse with temporal secondary structures with peak intensity below 0.5% that of the main peak.

1 Introduction

A limited number of laser systems - such as Ti:Sapphire or Cr:ZnSe - feature emission bandwidths that support the direct generation and amplification of ultrashort pulses. Other laser gain media - such as most rare-earth doped material - feature advantageous spectroscopic (enabling direct diode pumping) or thermo-optic properties (enabling high average power operation) but fail to provide emission spectra supporting ultrashort pulses. In order to derive ultrashort pulses from such systems, nonlinear post-compression schemes must be implemented. In the early 80s, the nonlinear propagation of laser pulses through optical fibres was proposed to compress optical pulses in the anomalous [1] and normal dispersion regimes [2]. This scheme was adapted in 1996 to compress ultra-intense laser pulses by using hollow-capillary fibres [3] or photonic crystal fibres [4], which are the most used post-compression schemes nowadays. However, these schemes have some intrinsic limitations related to propagation losses, damage threshold, fiber length and pulse energy that set an upper limit on practical energy scaling capabilities. For these reasons, new and complementary nonlinear compression methods have been investigated to enable operation in regimes of input pulse parameters where hollow-fibers become impractical. One of these schemes is nonlinear propagation of optical pulses in multipass cells (MPC) [5,6]. MPCs are cavities formed by two mirrors, typically filled with a noble gas or containing a thin dielectric plate in which a laser pulse propagates several roundtrips. During this propagation, the pulse experiences nonlinear propagation yielding spectral broadening while self-focusing is mitigated by the resonating nature of the propagation. In order to obtain clean short pulses, we propose to propagate the pulse in a gas-filled MPC in the same enhanced frequency chirp regime used with optical fibers in the 80s [2]. In this regime, the spectrum is broadened to a smooth spectral shape and phase - contrary to the typically strongly modulated spectra obtained via SPM.

In this work we use numerical simulations to investigate the enhanced frequency chirp regime in MPCs and identify the most promising parameter sets to achieve broad and clean spectra for experimentally relevant input pulse parameters.

2 Numerical Model

To simulate the propagation of a laser pulse in the gas-filled MPC we use a (3+1)D model [7], including the three spatial and temporal dimensions. We apply the Split-Step Fourier Method to propagate the envelope of the pulse accounting for diffraction, dispersion, SPM, self-focusing and self-steepening. We neglect the Raman effect contribution since our simulations are performed in monoatomic gases, as well as the ionization since we ensure the peak intensities involved remain below the ionization threshold. We also assume that the mirrors of the MPC have negligible dispersion. The spatio-temporal window has a minimum grid of $128 \times 128 \times 512$ points and is defined so that the beam perfectly fits during the entire propagation in the MPC.

* Corresponding author: vwsstaels@usal.es

3 Results and discussion

In order to delimit the region in which the enhanced frequency chirp regime in MPCs occurs, we impose the following three conditions: (i) the maximum peak power must remain below the critical peak power to prevent self-focusing; (ii) $L_{NL} < L < L_D$ and (iii) $1 < N = \sqrt{L_D/L_{NL}} < 20$, where L is the total propagation length in the MPC, L_{NL} is the nonlinear length and L_D is the dispersion length as defined in [8]. To find the region in which these conditions are fulfilled, we simulate 20 roundtrips of a Yb-based pulse with a duration of 177 fs at full width half maximum (FWHM) centred at 1030 nm in a 40 cm confocal cavity filled with argon. With these fixed parameters, we perform a gas pressure and pulse energy scan looking for the region in which the enhanced frequency chirp regime is optimum for the generation of a smooth supercontinuum, which we find around 10 bar and 100 μJ. In this region of parameters, the nonlinearity generates an important spectral broadening, mainly due to SPM, while the linear and nonlinear dispersion contribute to temporally stretch the pulse so that the spectrum losses its characteristic SPM modulations and acquire a quite smooth profile. Moreover, the beam presents a very good spatial homogeneity as it is well known for MPCs systems.

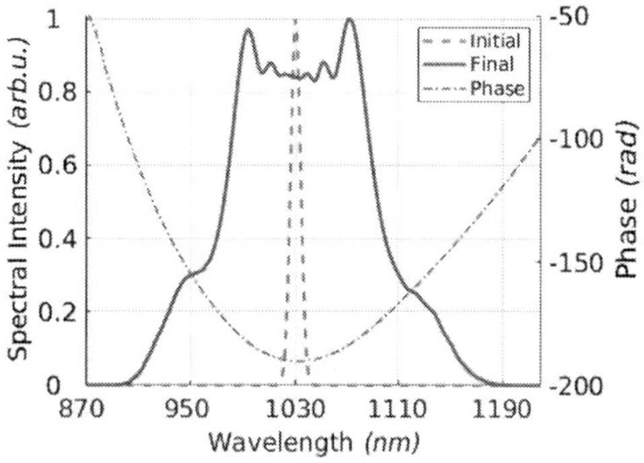

Fig. 1. Input (dashed red) and output (solid blue) on-axis spectral intensity profiles of a 100 μJ, 177 fs FWHM pulse in a confocal cavity filled with 10 bar of argon. The dashed-dotted line represents the output spectral phase.

Figure 1 shows the input and the output on-axis spectra, together with the spectral phase of the output spectrum for the case of 10 bar of argon and a pulse energy of 100 μJ. The broadened spectrum presents a relatively flat structure with a FWHM of 116 nm (~205 THz). The spectral phase presents a smooth, parabolic-like behaviour, as expected in this regime. The output spectral shape allows a temporal transform limited pulse of 14 fs FWHM (a compression factor >12) without visible sidelobes (less than 0.5% of its peak intensity). The input and output temporal pulses together with the transform limited output pulse are shown in figure 2. For better comparison, the transform limited output pulse shown in figure 2, which has a peak intensity of 2.8 TW/cm^2, has been normalized to the same peak intensity as the input pulse.

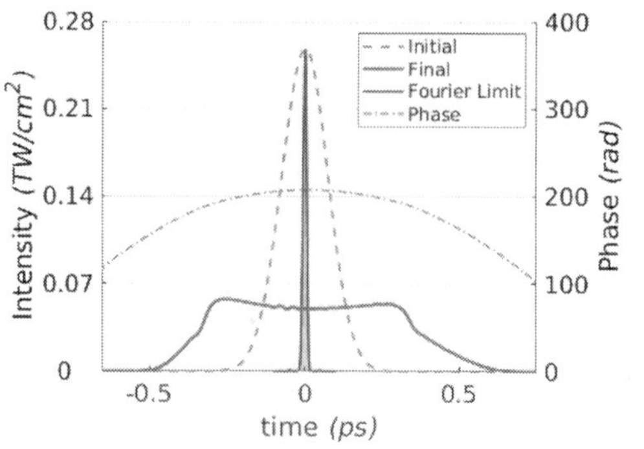

Fig. 2. Input (dashed red) and output (solid blue) on-axis temporal intensity profiles corresponding to the spectrum in Fig. 1. The transform limited pulse is plotted in purple and has been normalized to the peak intensity of the input pulse (it reaches a peak intensity of 2.8 TW/cm^2).

4 Conclusion

In this work, we show that it is possible to reproduce the enhanced chirp regime in multipass cells, yielding broad and smooth spectra compatible with clean and ultrashort transform limited pulses. For the case of 20 roundtrips of a Yb-based pulse with a duration of 177 fs at FWHM centred at 1030 nm in a 40 cm confocal cavity filled with argon, this regime appears around 100 μJ and 10 bar and opens a way to obtain clean ultrashort optical pulses.

This work was supported by: PID2019-106910GB-I00, funded by the Spanish Ministry of Science and Innovation, and PRE2020-092181, funded by the Ministry of Economy and Competitivity.

References

1. L. F. Mollenauer, R. H. Stolen, and J. P. Gordon, Phys. Rev. Lett. **45**, 1095 (1980).
2. H. Nakatsuka, D. Grischkowsky, and A. C. Balant, Phys. Rev. Lett. **47**, 910 (1981).
3. M. Nisoli, S. De Silvestri, and O. Svelto, Appl. Phys. Lett. **68**, 2793 (1996).
4. J. C. Knight, T. A. Birks, P. St. J. Russell, and D. M. Atkin, Opt. Lett. **21**, 1547 (1996).
5. M. Hanna, F. Guichard, N. Daher, Q. Bournet, X. Délen, and P. Georges, Laser & Photonics Reviews **15**, 2100220 (2021).
6. A.-L. Viotti, M. Seidel, E. Escoto, S. Rajhans, W. P. Leemans, I. Hartl, and C. M. Heyl, Optica **9**, 197 (2022).
7. M. Hanna, X. Délen, L. Lavenu, F. Guichard, Y. Zaouter, F. Druon, and P. Georges, J. Opt. Soc. Am. B, JOSAB **34**, 1340 (2017).
8. A. Couairon and A. Mysyrowicz, Physics Reports **441**, 47 (2007).

https://doi.org/10.1051/epjconf/202226608015

Light propagation in disordered aperiodic Mathieu photonic lattices

Jadranka M. Vasiljević[1,*], *Dejan V.* Timotijević[2], and *Dragana M.* Jović Savić[1]

[1]Institute of Physics, University of Belgrade, P.O. Box 68, 11001 Belgrade, Serbia
[2]Institute for Multidisciplinary Research, University of Belgrade, Kneza Višeslava 1, 11030, Belgrade, Serbia

Abstract. We present the numerical modeling of two different randomization methods of photonic lattices. We compare the results of light propagation in disordered aperiodic and disordered periodic lattices. In disordered aperiodic lattice disorder always enhances light transport for both methods, contrary to the disordered periodic lattice. For the highest disorder levels, we detect Anderson localization for both methods and both disordered lattices. More pronounced localization is observed for disordered aperiodic lattice.

1 Introduction

Anderson localization (AL), a well-known phenomenon in solid-state physics [1] is transferred to other fields like ultracold atoms, matter, light or sound waves [2], and demonstrated in various customized configurations [3–6]. The physics of periodic photonic systems has fundamental importance. Still, deviations from periodicity are significant as they may result in higher complexity, like the realization of photonic quasicrystals, the structures that are between periodic and disordered ones. Heretofore, light propagation properties have been studied in periodic photonic lattices [7, 8], as well as in disordered ones [3, 9, 10]. However, the quasiperiodic and aperiodic photonic lattices are merged as a further attractive research field for light propagation.

In our previous studies, we introduced aperiodic Mathieu structures with controllable complexity [11] and we studied light localization in them [12]. In such lattice, light expansion is hindered in comparison to periodic lattice and nonlinear light localization is demonstrated [12]. Randomization of periodic lattices can lead to AL [3, 9] or its suppression [13], while disordered quasiperiodic Penrose lattice can support AL and disorder-enhanced transport (DET) [14].

In this paper, we present two numerical methods for controllable randomization of photonic lattices and study disorder level (DL) influence on light propagation. For both methods, we numerically investigate the linear light propagation in disordered aperiodic Mathieu (DA) and disordered periodic (DP) lattices. For all DLs, we observe DET and AL is verified for higher DLs in DA lattices, for both methods. In contrast, in DP lattices disorder suppress diffraction and AL is observed for higher DLs. Localization length differs along different transverse directions owing to the crystal and lattice anisotropy. We confirm a more pronounced localization for DA lattices in both directions and both methods.

*e-mail: jadranka@ipb.ac.rs

2 Light propagation in DA and DP lattics

Two-dimensional disordered structures $\boldsymbol{DS}$, with an adjustable DL, are numerically realized by combining an original structure $\boldsymbol{S}$ with a disorder pattern $\boldsymbol{D}$ according to the relation $A_{DS} = (1-p)*A_S + p*A_D$, where A is the field amplitude, and parameter p is the relative contribution of the original structure and disorder pattern, i.e. DL. To ensure propagation invariant structures with the same propagation constant, we preset the Fourier spectrum of the disorder pattern, numerically calculated by interfering plane waves with constant amplitude and random phases, to be located on the same circle with radius k as the original structure [15]. As the original structure we use an aperiodic Mathieu structure created as in our paper [11], or square lattice with period d equal to the characteristic structure size $a = 25\,\mu$m of Mathieu beams used for the creation of the aperiodic structure. Disorder pattern's mean grain size $2\pi/k$ is equal to a of Mathieu beams. A case when the maximum lattice intensity I_{m} of the disordered lattice $I_{DL} = |A_{DS}|^2$ for each DL is unscaled, we will refer as M1, and M2 is the case when I_{DL} is scaled with I_{in} for each DL. For both methods, an increase of DL modifies the transverse intensity distribution of the original structure until completely substitutes it with the disorder pattern. For the same DL, the spatial intensity distributions of the disordered lattices are the same for both methods, but they differ in waveguides depths. For both methods, I_{m} dependencies of DL for DA and DP lattices are almost the same (Fig. 1 (A)). Opposite to the periodic lattice, our aperiodic lattice is not uniform in waveguide's distances, and their depths vary. We calculate averaged lattice intensity $I_{\mathrm{avg}} = \sum_{\mathbf{r}} I_{DL}(\mathbf{r})$ of DA and DP lattices for both methods (Fig. 1 (B)). For both methods, I_{avg}s are lower for DA than for DP lattices. For M1, I_{m} and I_{avg} are lower than for M2.

We study the difference in light propagation in DA and DP lattices realized with these methods. We use intensity distributions of disordered structures I_{DL} in numerical simulation of the light propagation along the z-axis in disordered lattices in a photorefractive crystal, numerically

https://doi.org/10.1051/epjconf/202226608015

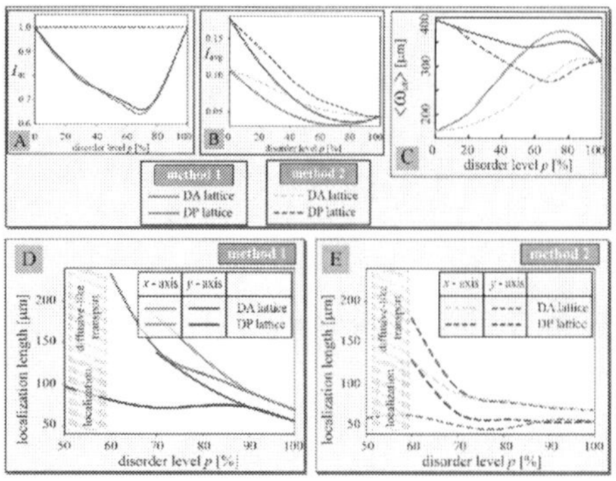

Figure 1. Methods differences of realization and light propagation in DA and DP lattices. (A) I_m, (B) I_{avg} of DA and DP lattices versus DL for different methods. (C) $\langle\omega_{eff}\rangle$ and (D)-(E) localization lengths along the x- and y-axis for M1 and M2, respectively versus DLs after 10 cm of propagation.

described by solving a system of equations as explained in Ref. [12]. We statistically analyze the propagation of a narrow Gaussian probe beam, of an FWHM of 8 µm, for different excitation positions selected to involve various local environments of the disordered lattice [14]. We realize such analysis for various DLs, using only one disorder pattern, averaging 64 different intensity distributions after 10 cm propagation distance.

According to relation $W_{eff} = \sqrt{IPR(z)}$, where IPR(z) is the inverse participation ratio [3], we calculate the effective beam width W_{eff} and determine a range of DL where DET and light localization are obtained. In Fig. 1 (C) we show scaled averaged effective beam widths $\langle\omega_{eff}\rangle = W_{eff}(z)/(W_{eff}(0)/FWHM)$. In the DA lattice with any percent of disorder, $\langle\omega_{eff}\rangle$ is greater than in the lattice without disorder, specifying DET for both methods. A maximum DET is indicated with $\langle\omega_{eff}\rangle$ highest value noticed at 80% for M1 and 90% for M2. Further increase of DL decreases $\langle\omega_{eff}\rangle$, indicating the possibility of AL occurrence. $\langle\omega_{eff}\rangle$ has a greater value for M1 than for M2, denoting a more pronounced DET for M1. Opposite in the DP lattice, $\langle\omega_{eff}\rangle$ decreases up to the minimum values which occur at 60% for M1 and 70% for M2. With the further increase of DL, $\langle\omega_{eff}\rangle$ increases for M2, while for M1 fluctuates. We examine the averaged transverse intensity distributions and corresponding log-plots of such averaged intensity distributions (not shown here). When the log-plots intensity profiles are linearly fitted around the center, we consider AL is confirmed [10]. In the region of DL where AL occurs, we obtain localization length by fitting intensity profiles with the exponential function [10], along the x and y-axis. For DA and DP lattices, in Figs. 1 (D)-(E) we show localization lengths for M1 and M2, respectively. For both methods, more pronounced localization is visible along the y-axis where localization lengths have lower values compared to the x-axis, due to the crystal and lattice anisotropy. AL occurs at different DLs along different directions, along the y-axis appears for lower DL than along the x-axis.

3 Conclusion

We presented two different methods for the creation of disordered photonic lattices with adjustable DL. We numerically studied light propagation in DA and DP lattices. For both methods, we observed enhanced light transport for all DLs but AL of light for higher DLs in DA lattice, contrary in DP lattice, we only observed AL for higher DLs. More pronounced localization is demonstrated in DA lattices for both methods and for M2 in both DA and DP lattices. Due to the crystal and lattice anisotropy, localization lengths differ in different directions.

Acknowledgements

This paper is supported by the Institute of Physics Belgrade and Institute for Multidisciplinary Research through the grants by the Ministry of Education, Science and Technological Development of the Republic of Serbia and by the Science Fund of the Republic of Serbia, Grant No 7714356, IDEAS - CompsLight.

References

[1] P. W. Anderson, Phys. Rev. **109**, 1492 (1958).
[2] A. Lagendijk, B. Tiggelen, D. S. Wiersma, Phys. Today **62**, 24 (2009).
[3] T. Schwartz, G. Bartal, S. Fishman, M. Segev, Nature **446**, 52 (2007).
[4] D. M. Jović, Y. S. Kivshar, C. Denz, M. R. Belić, Phys. Rev. A **83**, 033813 (2011).
[5] D. U. Naether, J.M. Meyer, S. Stützer, A. Tünnermann, S. Nolte, M. I. Molina, A. Szameit, Opt. Latt. **37**, 485 (2012).
[6] D. M. Jović, C. Denz, M. R. Belić, Opt. Lett. **37**, 4455 (2012).
[7] H. S. Eisenberg, Y. Silberberg, R. Morandotti, J. S. Aitchison, Phys. Rev. Lett. **85**, 1863 (2000).
[8] J. W. Fleischer, M. Segev, N. K. Efremidis, D. N. Christodoulides, Nature **422**, 147 (2003).
[9] Y. Lahini, A. Avidan, F. Pozzi, M. Sorel, R. Morandotti, D. N. Christodoulides, Y. Silberberg, Phys. Rev. Lett. **100**, 013906 (2008).
[10] M. Boguslawski, S. Brake, J. Armijo, F. Diabel, P. Rose, C. Denz, Opt. Express **21**, 31713 (2013).
[11] J. V. Vasiljević, A. Zannotti, D. V. Timotijević, C. Denz, D. M. Jović Savić, Phys. Rev. A **96**, 023840 (2017).
[12] J. V. Vasiljević, A. Zannotti, D. V. Timotijević, C. Denz, D. M. Jović Savić, Appl. Phys. Lett. **117**, 041102 (2020).
[13] S. Longhi, Opt. Lett. **46**, 2872 (2021).
[14] L. Levi, M. Rechtsman, B. Freedman, T. Schwartz, O. Manela, M. Segev, Science **332**, 1541 (2011).
[15] Z. Bouchal, Czech. J. Phys. **53**, 537 (2003).

EPJ Web of Conferences **266**, 08016 (2022)
EOSAM 2022
https://doi.org/10.1051/epjconf/202226608016

Single atom photon pair source

Jürgen Volz[1,*], *Xinxin* Hu[1], *Gabriele* Maron[1], *Luke* Masters[1], *Lucas* Pache[1], and *Arno* Rauschenbeutel[1]

[1]Department of Physics, Humboldt-Universität zu Berlin, 10099 Berlin, Germany

Abstract. Sources of entangled photon pairs are a crucial ingredient for many applications in quantum information and communication. Of particular interest are narrow-band sources with bandwidths that are compatible with solid state systems such as atomic media for storage and manipulation of the photons. Here, we experimentally realize a source of energy-time entangled photon pairs where the photons pairs are generated by scattering light from a single two-level atom and separated from the coherently scattered light via a narrow-band filter. We verify the performance of our pair-source by measuring the second order correlation function of the atomic fluorescence and we observe that one can continuously tune the photon statistics of the atomic fluorescence from perfect photon anti-bunching to strong photon bunching expected for a photon pair source. Our experiment demonstrates a novel way to realize a photon pair source for photons with spectral bandwidths and resonance frequencies that are inherently compatible with atomic media.

1 Introduction

For a single quantum emitter one typically assumes that it can only emit photons one by one. As a consequence, light scattered by such emitters exhibits strong non-classical properties that e.g. manifest themselves in perfect photon-antibunching in the second order correlation function. For this reason, quantum emitters are typically used in many experiments as a source of single photons. Here, we show that in general photon emission from a single quantum emitter can be described as an interference phenomena between two different two-photon amplitudes in the scattered light that are typically referred to as coherently and incoherently scattered component of the atomic fluorescence. In this picture, the perfect photon anti-bunching observed for single emitters arises from the complete destructive interference of the two components[1? , 2]. Due to their different spectral properties[4] it is possible to experimentally separate the coherently and incoherently scattered light. In this way, in our experiment we collect only the incoherently scattered part of the fluorescence and realize a single-atom-based source of ultra narrow-band energy-time entangled photon pairs.

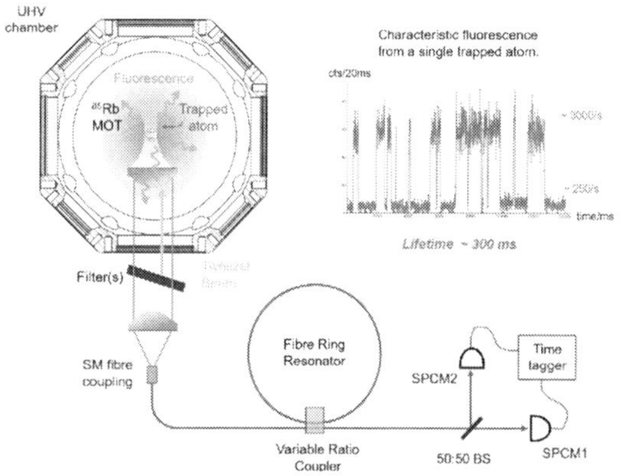

Figure 1. Schematic of out experimental setup. We trap a single ^{85}Rb atom inside an optical dipole trap and collect its fluorescence using a high NA microscope. The collected light is subject to a narrow-band band-block filter that is realized using a fiber-based ring resonator with a variable coupler. Depending on the incoupling rate, we can change the photon statistics of the collected atomic fluorescence from anti-bunching to strong photon bunching which we measure using a Hanbury-Brown Twiss setup. Inset: Time trace of the atomic fluorescence measured without resonator.

2 Experimental principle and setup

In general, the light scattered by a single quantum emitter can be decomposed into two parts that are coherent or incoherent with respect to the excitation light field. The coherent part has the same spectral properties as the excitation field, while the spectrum of the incoherent light is composed of energy-time entangled photon pairs that are contained within two frequency sidebands. These sidebands are separated from the excitation frequency and have a width given by the line-width of the atomic transitions. Using frequency selective components the amplitude of these two components can be individually controlled which we make use of to spectrally remove the coherently scattered part from the incoherently scattered light.

In our experiment, we use a single ^{85}Rb atom as photon-pair source which is trapped in an optical optical

*e-mail: juergen.volz@hu-berlin.de

dipole trap located in the center of a magneto-optical trap (MOT). The fluorescence of the atom in the MOT beams is collected with a high NA lens and coupled into a single mode optical fiber, see Fig 1. The coherently scattered light has the same spectral properties as our MOT laser while the entangled photon pairs have a spectral width of 6 MHz and are spectrally separated from the MOT laser by the effective Rabi frequency of the atom of about 50 MHz. To remove the coherent part, we introduce an ultra-narrow-band optical band block filter into the collected fluorescence. The filter is realized using a fiber-integrated ring resonator with an adjustable incoupling rate and a line-width of about 4 MHz. The resonator is set to resonance with the MOT light and removes the coherently scattered light from the atomic fluorescence while the photon pairs are not affected. In this way, the remaining light contains only the incoherently scattered light from the atom made of photon pairs.

3 Experimental characterization of our source

In order to study the performance of our photon pair source we record the photon statistics of the light after our collection and filtering optics. Using a Hanbury-Brwon Twiss setup, we record the second order correlation function of the fluorescence light. In our experiment, we can arbitrarily adjust the coupling rate to the filter-resonator, and thus the transmitted fraction of the coherently scattered light from the atom. This allows us to continuously change the second order correlation function of the collected light from perfect photon anti-bunching without frequency filter to strong photon bunching when setting the resonator to critical coupling such that it absorbs all coherently scattered light. In this way, we characterize the key features of our source such as photon pair rate and its ratio to single photon rate which we compare to the theoretical expectation for our system.

4 Outlook

Our results shed light on the physical mechanism that underlies atomic resonance fluorescence. At the same time it provides a new way to generate ultra narrow-band photon pairs that inherently match the spectral properties of the employed quantum emitter. This scheme is universal and will work with any type of quantum emitter. Thus, it can be employed to generate entangled photons with any spectral width for any part of the electromagnetic spectrum as long as a two-level quantum emitter with the desired spectral properties is available.

References

[1] S. Mahmoodian *et al.*, "Strongly Correlated Photon Transport in Waveguide Quantum Electrodynamics with Weakly Coupled Emitters." Phys. Rev. Lett., **121**, 143601 (2018).

[2] L. Hanschke *et al.* "Origin of Antibunching in Resonance Fluorescence." Phys. Rev. Lett. **125**, 170402 (2020).

[3] A. S. Prasad *et al.*,"Correlating photons using the collective nonlinear response of atoms weakly coupled to an optical mode." Nature Photonics **14**, 719 (2020).

[4] J. Hinney *et al.*, "Unraveling Two-Photon Entanglement via the Squeezing Spectrum of Light Traveling through Nanofiber-Coupled Atoms." Phys. Rev. Lett. **127**,123602 (2021).

https://doi.org/10.1051/epjconf/202226608017

Demonstration of propagation-invariant 3D space-time wave packets

Murat Yessenov[1,*], Justin Free[2], Zhaozhong Chen[3], Eric G. Johnson[2], Martin P. J. Lavery[3], Miguel A. Alonso[4,5], and Ayman F. Abouraddy[1]

[1]CREOL, The College of Optics & Photonics, University of Central Florida, Orlando, Florida 32816, USA
[2]Micro-Photonics Laboratory, the Holcombe Department of Electrical and Computer Engineering, Clemson University, Clemson, South Carolina 29634, USA
[3]James Watt School of Engineering, University of Glasgow, UK
[4]CNRS, Centrale Marseille, Institut Fresnel, Aix Marseille Univ., Marseille, France
[5]The Institute of Optics, University of Rochester, Rochester, NY, USA

Abstract. We present the first demonstration of propagation-invariant space-time (ST) wave packets localized in all dimensions - 2D space and 1D time. By introducing orbital-angular-momentum into the wave packets, we produce propagation-invariant ST-OAM wave packets traveling at arbitrary group velocities. Our results show 4× improvement in propagation distance of 3D ST wave packets compared with a separable Bessel beam having similar spatial bandwidth.

1 Introduction

Creating a localized bundle of energy overcoming diffraction and dispersion has been a long-standing challenge in optics [1]. Previous solutions include optical solitons that require a nonlinear medium, or waveguides that combat diffraction but introduce dispersion [1]. Hence, there has been a significant interest in so-called 'propagation-invariant' optical fields that propagate freely without suffering diffraction or dispersion [3]. In 1983 Brittingham proposed 'focus wave modes' (FWM) – a self-similarly propagating pulsed field in free space traveling ata group velocity equal to the speed of light c [2], which led to subsequent discoveries of other propagation-invariant wave packets. However, synthesizing such wave packets requires introducing spectrally precise non-differentiable angular dispersion (AD) in two transverse dimensions, a capability that has eluded optics to date. Consequently, with the exception of X-waves that are AD-free, no propagation-invariant optical wave packets that are localized in all dimensions have been observed in free space [3]. The challenge of introducing arbitrary AD into a generic pulsed beam along *one* transverse dimension has enabled the realization of space-time (ST) wave packets in the form of light sheets [4], which exhibit a broad host of sought-after effects [5]. Here, we demonstrate a spatio-temporal spectral modulation strategy that efficiently produces arbitrary yet precise AD in two transverse dimensions, and thus yields ST wave packets localized in all dimensions [6].

*e-mail: yessenov@knights.ucf.edu

2 Concept

Space-time (ST) wave packets are pulsed beams in which each spatial frequency is assigned to a single wavelength, thus resulting in propagation invariance at an arbitrary group velocity [4–6]. The spectral support for a propagation-invariant ST wave packet is a 1D conic section at the intersection of the light-cone $k_r^2 + k_z^2 = (\frac{\omega}{c})^2$ with

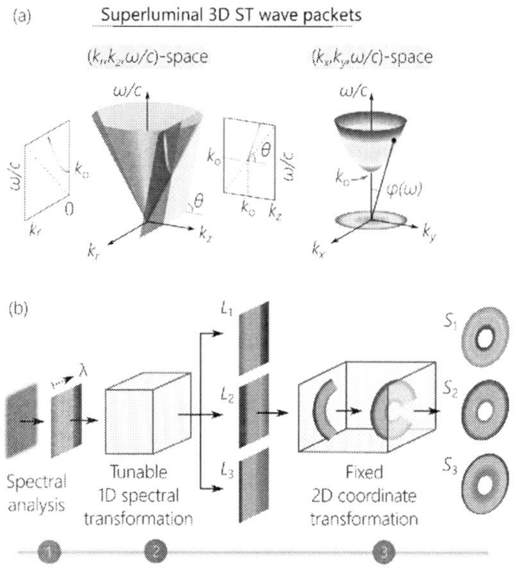

Figure 1. (a) Visualization of the spectral support domain for 3D ST wave packets on the surface of the free-space light-cone. (b) Three-stage synthesis scheme for 3D ST wave packets.

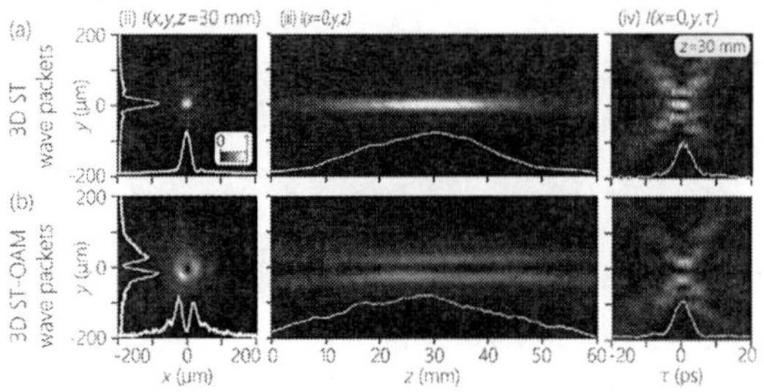

Figure 2. Measurement results for (a) superluminal 3D ST wave packets ($\widetilde{v}$ = 1.8c) and (b) for a superluminal ($\widetilde{v}$ = 1.3c) wave packet endowed with the OAM mode $\ell = 1$. In the first column, we plot the measured transverse intensity $I(x, y, z)$ at $z = 30$ mm, in addition to sections through $x = 0$ and $y = 0$ (white curves); and, in the second, we plot the measured intensity in a meridional plane $I(0, y, z)$. The white curve at the bottom of the panel in the third column is the on-axis intensity $I(0, 0, z)$ in (a) and $y = 30\,\mu$m in (b). In the third column we plot the measured $I(0, y, z; \tau)$ at $z = 30$ mm; For both cases, $\Delta\lambda = 0.3$ nm, the pulse width is ≈ 6 ps and the transverse beam size is $\approx 30\,\mu$m.

a spectral plane $\Omega = (k_z - k_o)c \tan\theta$, whose projection onto the $(k_z, \frac{\omega}{c})$-plane is a straight line making an angle θ (the spectral tilt angle) with the k_z-axis; here $k_r = \sqrt{k_x^2 + k_y^2}$ is the radial component of the transverse wave vector, k_z is the axial wave number, and ω is the angular frequency; $\Omega = \omega - \omega_o$, ω_o is the central frequency, and $k_o = \frac{\omega_o}{c}$ is the corresponding wave number [Fig.1(a)]. A wave-packet incorporating such a spatio-temporal spectral structure is localized in all dimensions and travels rigidly in free space.

3 Experimental setup

The synthesis scheme for 3D ST wave packets is implemented in three stages [Fig. 1(b)]. In the first stage, the spectrum of a generic plane-wave pulse from a Ti:Sapphire laser (100-fs pulse at $\lambda = 800$ nm) is spatially resolved along one dimension after a double-pass through a volume chirped Bragg grating (OptiGrate, L1-021) [7]. In the second stage, a spectral transformation 'reshuffles' the wavelengths into a prescribed sequence, which is performed by making use of a pair of spatial light modulators (SLM, Meadowlark 1920×1152 series). In the third stage, a log-polar-to-Cartesian conformal coordinate transformation converts the spatial locus of each wavelength from a line into a circle [8]. A lens finally converts the spectrally resolved wave front into a 3D ST wave packet localized in all three dimensions. By modulating the spatio-temporal spectral structure, we realize group velocities extending from the subluminal to the superluminal regimes over the range from 0.7c to 1.8c.

Moreover, by providing access to both transverse dimensions in a ST wave packet, degrees of freedom that cannot be accessed in only one transverse dimension, such as orbital angular momentum (OAM), can now be included. We characterize the synthesized field first in the Fourier domain (k_x, k_y) by scanning the fiber tip connected to spectrum analyzer along the radial direction. Then we record the axial evolution of the time-averaged intensity measurements by scanning a CCD camera along the propagation axis z. Moreover, the time-resolved wave packet envelope is reconstructed using linear interferometry with the initial pulses used as a reference [5]. Finally, the complex field profile is recorded by exploiting off-axis digital holography (see [6] for more details).

4 Results

In Fig. 2(a) we present the measurement results for 3D ST wave packets travelling at a superluminal group velocity $\widetilde{v} \approx 1.8c$. The measured transverse intensity profile $I(x, y, z = 30$ mm) shown in column (i) confirms the localization along both transverse spatial dimensions. The measured axial evolution of the time-averaged intensity $I(x = 0, y, z)$ depicted in column (ii) confirms the diffraction-free propagation over a distance of $L_{\text{max}} \approx 55$ mm, which corresponds to a 4× improvement compared with the separable Bessel beam and a 60× improvement compared with a Gaussian beam of the same spatial bandwidth. Lastly, in column (iii) we plot the time-resolved measurements $I(x = 0, y, \tau, z = 30$ mm), which reveal clearly the expected X-shaped profile that remains invariant over the propagation distance L_{max}. We observe a similar behavior for a superluminal ST-OAM wave packet ($\widetilde{v}$ = 1.3c) with $\ell = 1$ in Fig. 2(b).

5 Conclusion

In conclusion we have presented the first experimental demonstration of ST wave packets localized in all dimensions. In addition, we demonstrate propagation-invariant pulsed OAM wave packets with controllable group velocity in free space. Such 3D ST wave packets that are fully localized in all dimensions have potential uses in areas such as free-space optical communications, imaging, and nonlinear optics.

References

[1] B. E. A. Saleh, M. C. Teich, *Principles of Photonics*, Wiley (2007)
[2] J. N. Brittingham, J. Appl. Phys. **54**, 1179 (1983).
[3] J. Turunen and A. T. Friberg, Prog. Opt., **4**, 1–88 (2010).
[4] H. E. Kondakci and A. F. Abouraddy, Nat. Photon. **11**, 733 (2017).
[5] M. Yessenov, B. Bhaduri, H. E. Kondakci, and A. F. Abouraddy, Opt. Photon. News **30**, 34 (2019).
[6] M. Yessenov et al. arXiv preprint arXiv:2111.03095 (2021).
[7] Glebov et. al, Opt. Engineering 53(5), 051514 (2014).
[8] O. Bryngdahl, J. Opt. Soc. Am. A **64**, 1092–1099 (1974).

Photonic Maxwell's Demon: Feed-forward methods for photonic thermodynamic tasks

Maxime Jacquet[1,2,*], *Guilherme* Zanin[1,2], *Michael* Antesberger[1], *Paulo* Souto Ribeiro[3], *Lee* Rozema[1], and *Philip* Walther[1,4]

[1]*Work done at* University of Vienna, Faculty of Physics, Vienna Center for Quantum Science and Technology (VCQ), Vienna, Austria
[2]*Now at* Laboratoire Kastler Brossel, Sorbonne Université, CNRS, ENS-Université PSL, Collège de France, Paris 75005, France.
[3]Departamento de Física, Universidade Federal de Santa Catarina, Florianópolis, Santa Catarina 88040-900, Brazil
[4]Christian Doppler Laboratory for Photonic Quantum Computer, Faculty of Physics, University of Vienna, 1090 Vienna, Austria

Abstract. Maxwell's Demon is at the heart of the interrelation between quantum information processing and thermodynamics. In this thought experiment, a demon extracts work from two thermal baths at equilibrium by gaining information about them at the single-particle level and applying classical feed-forward operations. In this talk I will show how to implement a photonic version of Maxwell's Demon with active feed-forward in a fiber-based system using ultrafast optical switches. This is the first realisation of an active Demon. The experiment [1] shows that, if correlations exist between the two thermal baths, the Demon can extract over an order of magnitude more work than without correlations. This demonstrates the great potential of photonic experiments – which provide a unique degree of control on the system – to access new regimes in quantum thermodynamics.

1 Maxwell's demon as feed-forward

Thermodynamics was conceived as a phenomenological theory for the equilibrium properties of macroscopic systems ranging from gas (ensembles of 'many' small systems) to black holes, of which the theory provides a complete description in terms of quantities like temperature or work. On the other hand, the manipulation of thermal states (as well as their emergence from fluctuations at the smallest scale) is described in terms of (quantum) information processing. This interrelation is perhaps best manifested in the protocol known as 'Maxwell's demon' (MD): As illustrated in Fig. 1 **(a)**, the MD monitors the motion of gas particles inside a partitioned box. Based on its previous knowledge of the particle's temperature, the MD can sort hot particles from cold particles by opening an aperture in the partition such that all hot particles eventually end up on one side of the partition and all cold particles on the other. So the MD brings the system out of equilibrium, and can exploit the engineered temperature differential to extract work without violating the second principle of thermodynamics. In other words, the MD generates work by measuring individual particles to gain information, and depending on these measurement results, it applies different operations to the bath (i.e. opening the door, or leaving it closed). These measurement-dependent operations are called feed-forward operations.

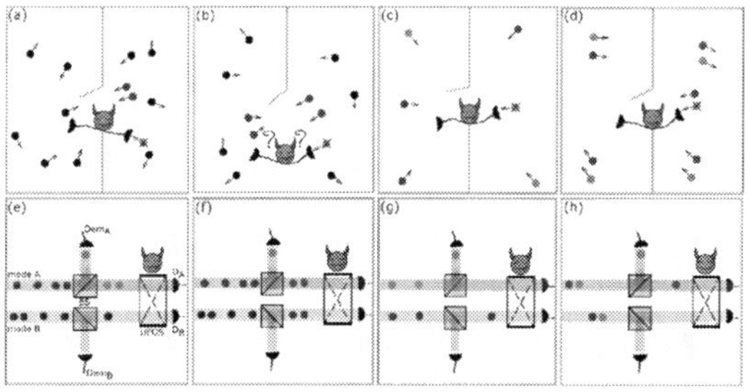

Figure 1. Principle of operation of the photonic Maxwell's Demon (MD): the MD controls the path of particles/photons in the system. The colours red and blue represent correlated photons; i.e. when the demon measures a red particle, it learns the location of the blue particle. Top row, particle picture of the photonic MD; bottom row, actual implementation of the photonic MD. **(a)** and **(e)**, uncorrelated thermal states; **(b)** and **(f)**, split thermal states; **(c)** and **(g)** correlated states; **(d)** and **(h)**, anti-correlated states. Figure from [3].

2 Photonic implementation

The 'power' of the MD, the amount of work it can extract per measurement, can be greatly enhanced by operating with correlated particles: Consider a situation where, for every particle moving in a certain direction in the left bath, there is a twin particle moving in the same direction in the right bath, as illustrated in Fig. 1 **(c)**. If the MD observes a particle in the left bath it also learns something about an (unobserved) particle in the right bath. Because of this ad-

*e-mail: maxime.jacquet@lkb.upmc.fr

https://doi.org/10.1051/epjconf/202226608018

ditional information, the MD can extract much more work per measurement [2] by simply modifying its feed-forward protocol to make use of the new information.

In a photonic MD, the baths are realised with optical modes [3], each containing a thermal state of photons whose average photon number directly sets the temperature of the bath. These two optical modes are then sent to individual detectors, D_A and D_B. When the two optical modes have an equal photon number (the baths have an equal temperature), the detectors register the same counts and no work can be extracted. When a photon is subtracted from this state at a beamsplitter and detected in the reflection port with detector Dem_A or Dem_B, bunching will occur, meaning that the photon number in the transmitted beam will be temporarily doubled (as depicted in red in Fig. 1 (**e**)). This photon extraction provides information about the average photon number in the beam, based on which a classical feed-forward operation may then be applied to create an imbalance in the photon number between the two beams. This energy imbalance can in turn be used to extract work.

3 First active Maxwell's Demon

We have implementd a photonic MD with active feed-forward (more details in [1]) to physically swap the paths of the beams dependent on detection events at Dem_A and Dem_B using an ultra-fast optical switch (methods adapted from [4]). Furthermore, we have used correlations as a resource to increase the power of the MD.

The choice of correlations is essential in enabling the MD to extract work. The experiment [1] evidences that particular correlations allow for an order of magnitude increase of the power of the MD, while others prevent the MD from extracting any work at all. In conclusion, information processing with correlations and anti-correlations vastly outperforms operation with the other states.

References

[1] G.L. Zanin, M. Antesberger, M.J. Jacquet, P.H.S. Ribeiro, L.A. Rozema, P. Walther, arxiv:2107.09686 (2021)

[2] A. Shu, J. Dai, V. Scarani, Physical Review A **95**, 022123 (2017)

[3] M.D. Vidrighin, O. Dahlsten, M. Barbieri, M.S. Kim, V. Vedral, I.A. Walmsley, Phys. Rev. Lett. **116**, 050401 (2016)

[4] G.L. Zanin, M.J. Jacquet, M. Spagnolo, P. Schiansky, I.A. Calafell, L.A. Rozema, P. Walther, Opt. Express **29**, 3425 (2021)

https://doi.org/10.1051/epjconf/202226609001

Polarization singularities and optical chirality in dielectric metasurfaces

Lucio Claudio Andreani[1,*], *Dario* Gerace[1], *Momchil* Minkov[2], *Luca* Zagaglia[1], and *Simone* Zanotti[1]

[1]Physics Department, University of Pavia, via Bassi 6, 27100 Pavia, Italy
[2]Flexcompute, Inc., 130 Trapelo Rd., Belmont, MA 02478, USA

Abstract. We theoretically study the relation between polarization singularities and optical properties in dielectric metasurfaces, or photonic crystal slabs. We focus on nondegenerate photonic bands leading to symmetry-protected Bound States in a Continuum (BICs). First, we discuss how BICs lead to polarization singularities in the far field, whose winding numbers – or topological charges – follows from the symmetry of the lattice. Then, we determine the polarization properties via the Stokes parameters, focusing on the conditions for the occurrence of a nonvanishing circular polarization. Finally, we calculate the optical response in reflection and the degree of circular dichroism. The results shed light on the role of polarization singularities and symmetry in determining the optical chirality.

1 Introduction

Photonic crystal slabs (PCS), or periodic dielectric metasurfaces, support Bound States in a Continuum (BICs) that lie in the spectral region of the quasi-guided modes, but have infinite quality factor and no coupling to the radiation field. Such BICs lead to vortex centers in the far-field polarization, or topological singularities, which are characterized by a winding number or topological charge [1]. Polarization singularities have attracted much interest in recent years within the area of topological photonics, as they lead to a number of processes that involve the topological charge [2], and they can be used to tailor the polarization properties in the far field.

Another research field of current interest in metasurfaces involves the chiral properties of the electromagnetic field, i.e., the differences in optical properties between states with right- and left circular polarization. Chiral effects like circular dichroism have been thoroughly studied [3], often with the goal of enhancing the sensitivity of chiroptical sensing, i.e., sensing of chiral molecules.

In this work, we present an extensive study of photonic crystal slabs that sheds light on the relation between polarization singularities and chiral optical properties in the far field.

2 Topological singularities and polarization

We calculate the photonic band dispersion and losses above the light cone in photonic crystal slabs by means of two complementary methods, namely Guided-Mode Dispersion (GME) [4,5] and finite-difference time domain (FDTD Lumerical-Ansys) simulations. Both methods lead to the k-space coefficients of the polarization d_x, d_y that allow to identify the topological singularities and the far-field behaviour of the polarization. Moreover, we develop simple models that implement the symmetry properties of the lattice, which are crucial in determining the topological invariants.

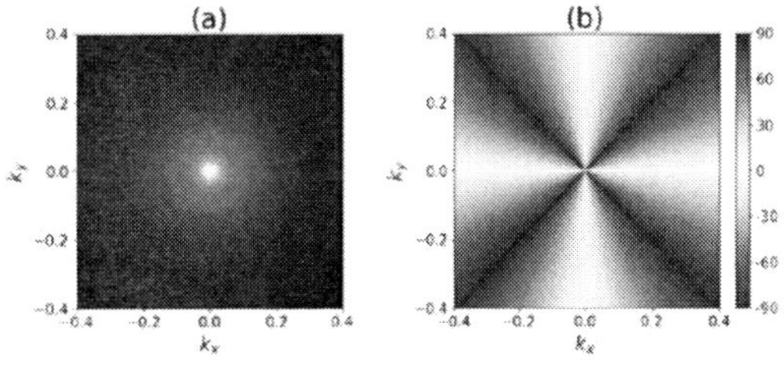

Fig. 1. (a) Q-factor and (b) phase of the polarization field in a PCS with hexagonal symmetry.

As an example, we show in Fig. 1 the quality factor and the phase of the polarization in a hexagonal PCS. The second photonic band has B-symmetry at the Γ point, leading to a BIC with infinite Q-factor. The polarization winds around the BIC with a topological charge $q = -2$.

For this kind of structures, we determine the degree of polarization in the far field by means of the Stokes parameters. The full Brillouin zone is mapped through GME and FDTD simulations. The normalized S_3 parameter is especially interesting as it determines the degree of circular polarization, which is nonvanishing only when the ratio d_y/d_x has a finite imaginary part. The conditions for the occurrence of a finite circular polarization, depending on the symmetry of the lattice and on the directions in k-space, are discussed.

https://doi.org/10.1051/epjconf/202226609001

3 Optical reflection and circular dichroism

Optical spectra like reflection and transmission are well suited to excite the photonic modes of the PCS and to probe the far-field properties of the polarization. Here, we study reflection spectra for linear and circular polarizations, and focus on the Circular Dichroism (CD) in reflection as a measure of chirality of the metasurface. This quantity is defined as

$$\mathrm{CD}=\frac{R_{lcp}-R_{rcp}}{R_{lcp}+R_{rcp}}, \quad (1)$$

where R_{lcp} (R_{rcp}) is the reflectance for left- (right-) circular polarization. The CD can be nonvanishing at normal incidence for a chiral lattice (intrinsic chirality), or at oblique incidence along non-symmetry directions of the lattice (extrinsic chirality).

We calculate reflectance and CD spectra by a generalized scattering matrix method that relies on Bloch-mode expansion in each layer of a stratified medium, implemented in the EMUstack code [6]. Considering extrinsic chirality in nonchiral lattices, the CD has a nontrivial dependence on the incidence angle and on the azimuthal orientation. Importantly, the CD is correlated to the degree of circular polarization of the PCS resonances via the S_3 Stokes parameter. This relation, which was introduced recently by means of temporal coupled-mode theory [7], is here determined in a systematic way.

4 Conclusions

Topological singularities in dielectric metasurfaces bear a subtle relation to chiral optical properties in PCS. Such relation is analysized in this work by means of various symmetry-based and numerical approaches.

Acknowledgements

This work is partially supported from the Italian Ministry of University and Research (MUR) through 2017 PRIN project 2017MP7F8F-004 "NOMEN".

References

1. B. Zhen et al., Phys. Rev. Lett. 113, 257401 (2014).
2. T. Yoda and M. Notomi, Phys. Rev. Lett. 125, 053902 (2020).
3. J.T. Collins et al., Adv. Opt. Mater. 5, 1700182 (2017).
4. L.C. Andreani and D. Gerace, Phys. Rev. B 73, 235114 (2006).
5. M. Minkov et al., ACS Photonics 7, 1729 (2020).
6. K.B. Dossou et al: J. Opt. Soc. Am. A 29, 817 (2012).
7. W. Chen et al., Phys. Rev. Lett. 126, 253901 (2021).

https://doi.org/10.1051/epjconf/202226609002

Demonstration of Cognitive Bias on a Noisy Intermediate-Scale Quantum Processor

Fabio Antonio Bovino [1,*], *Alessandro Comi* [2]

[1]Dept. . SBAI, University of Rome, via Antonio Scarpa 14/16, 00161Rome, ITALY
[2]IT & Digital Solutions, TIM S.p.A., via Oriolo Romano 240, 00189 Rome, ITALY

Abstract. The decision-making process of the individual is analogous to the collapse of a quantum system at the moment of measurement, whereby quantum system we mean the decision-maker, his mental processes and the environmental context that influences him in his choices. Classical probability depends on the lack of information, quantum probability is instead intrinsic to the system to be studied, meaning that the system, consisting of the decision-maker and the context, is in a superposition of states, each with its own amplitude of probability. Only the measurement process (the decision) selects in the decision maker a specific state among those in which he was overlapping. We demonstrated on a Noisy Intermediate -Scale Quantum Processor the model of the "order effect", one of the most important cognitive bias.

1 INTRODUCTION

Classical financial models have been developed with reference to several theories (e.g., Efficient Markets Theory [1] and Expected Utility Theory [2]) based on the assumption that market provides all the information necessary for decision-making processes and that investors, fully informed and therefore able to calculate the probability of all different possible alternatives and the consequent risks, make their decisions in a completely rational manner. Markets, however, frequently exhibit trends not consistent with the presumed rationality of investors, in clear violation of the classical theories. It does not always seem true that an individual rationally chooses the alternative that maximises expected utility, especially regarding situations of ambiguity, in which investors must decide between different alternatives to which it is not possible to associate an objective probabilistic evaluation [3]; well known in this context are the Ellsberg [4] and Allais [5] paradoxes. Many studies have confirmed that the rational choices are not fulfilled in practice because the way individuals think is different; some specific conceptual structures (optimism, pessimism, suspicion, etc.) act as a decision-making context and influence the choices of subjects in situations of ambiguity, driving them away from the rational behaviour predicted by the classical theories [6].

There is therefore a need to define a new mathematical model that can include also cognitive bias that are not captured by rational approaches to behavioural finance based on efficiency and rationality. Quantum formalism allows to model, at a statistical level, behavioural phenomena influenced by subjective contexts due to the specific situation in which the subject must make a choice, including many complex cognitive biases not consistent with classical models, such as the conjunction [7] and disjunction [8] effects. Just as Quantum Physics responds to the need to define a theoretical apparatus capable of explaining physical phenomena observed at the microscopic level that cannot be explained by classical physics (e.g. the black-body radiation, the stability of atoms, the photoelectric effect), the quantum approach to decisional cognitive processes originates from the need to find new theories that can justify the occurrence of important cognitive biases, resulting in numerous and systematic violations of the rational logic underlying classical behavioural processes [9]; "quantum-like" mathematical models allows to make more reliable forecasts regarding the "real" choices made by individuals. One of the most important cognitive bias is: the "ordering effect". In numerous experiments it has been observed that the order in which subjects are presented with decisions to be taken determines how decisions are taken. For example, in case of questions regarding the same context, each of which involves a number of possible alternative answers, the answers given by subjects will change if the same set of questions are asked in a different order. This behaviour has been observed in numerous statistical surveys. Quantum formalism provides a reason and simulates the influence of context on decisional processes. The projection operators do not commute and the answer to a first question in a survey changes the mental state of the interviewee, so he/she will be differently predisposed to give a specific answer to a second question.

We defined a quantum circuit allowing to reproduce the effect of the context (answer to the first question) on the subsequent decision process (answer to the second question), and we implemented the quantum circuit on a NISQ photonic processor.

https://doi.org/10.1051/epjconf/202226609002

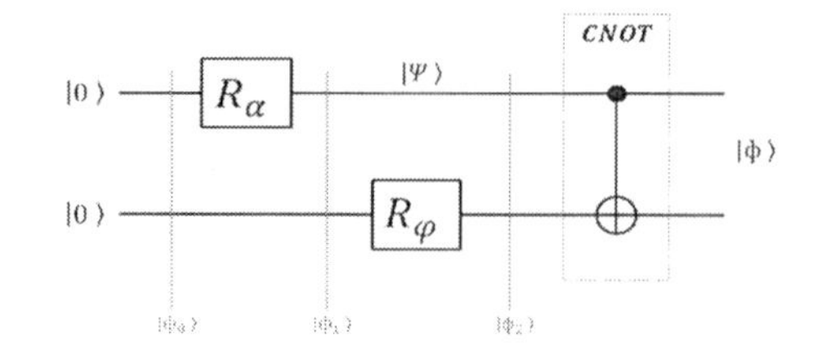

Fig. 1. Quantum circuit simulating the influence of context (operator $R\alpha$) on decision making process (operator $R\varphi$)

2 EXPERIMENTAL IMPLEMENTATION ON A NISQ PHOTONIC PROCESSOR

There are several architectures on which a quantum computer can be built, each of which is based on a different physical system: ion or atomic traps, magnetic resonance, Jopheson junctions, quantum dots, nuclear spin or photonic circuits. An interesting perspective is related to quantum computation that uses **linear photonic circuits**: it has the advantage that the smallest unit of quantum information, the qubit, is potentially free from decoherence, i.e. the information stored in a photon tends to remain unchanged. Light-based "on chip systems" deliver on the promise of scalability for quantum information applications through high-density component integration and parallelization. **The novel on chip photonic device [10]**, reduce the overhead due to multiple single photon sources with very high degree of reciprocal coherence and exploits the possibility to encode the whole state space in a complex optical circuit based on **MultiRail (MR) Architecture**: a particular state in input is encoded in the quantum superposition of a coherent light beam trough the different waveguides, so that it is possible to design quantum gate that show deterministic and not probabilistic behavior, as quantum entanglers, Bell measurements and teleportation schemes. Multirail Architecture provides a novel class of NISQ processors. that -although far from fault tolerant- support the execution of heuristic quantum algorithms, with a quantum advantage when applied to combinatorial optimization problems. In this case we implemented the circuital model shown in Fig.1 where the first qubit (at the top of the circuit) represents the context, i.e. the answer to the first question of a survay; and the second qubit represents the decision process, i.e. the answer to the second question of a survay. The results (see Fig.2) are in good agreement with the theory. More in general, within the Quantum Cognition the examined circuit can be used in order to simulate every scenario in which a decision B can be influenced from a context A (a previous decision, the result of a test, some meaningful information for the decision B).

References

1. Fama E., *Efficient Capital Markets: A Review of Theory and Empirical Work*, Wiley, 1970

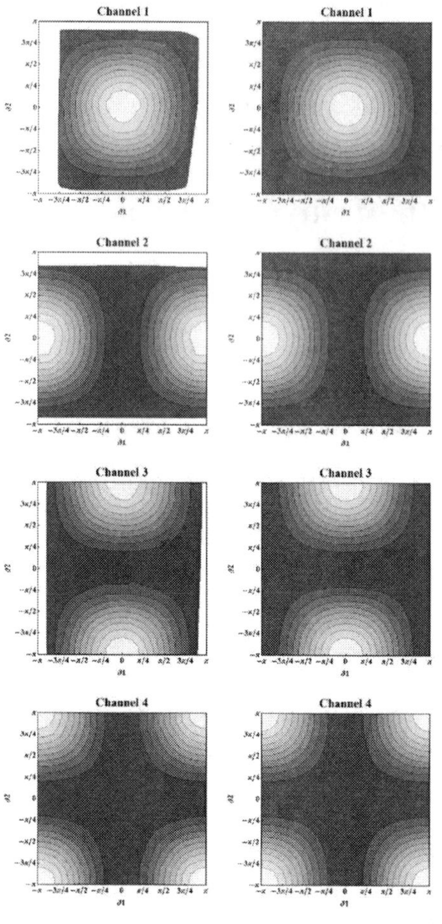

Fig. 2. Experimental Demonstration of Cognitive Bias (Left: Experimental results. Right: Theory)

2. Von Neumann J., Morgenstern O., *Theory of Games and Economic Behavior*, Princeton University Press, 1944
3. Knight F. H., *Risk, Uncertainty, and Profit*, Boston, MA: Hart, Schaffner & Marx; Houghton Mifflin Co., 1921
4. Ellsberg D., *Risk, Ambiguity, and the Savage Axioms*, Quarterly Journal of Economics vol. 75 n. 4, 1961
5. Allais M., *Le comportement de l'homme rationnel devant le risque: critique des postulats et axiomes de l'école Américaine*, Econometrica, 1952
6. Aerts D., D'Hooghe B., Sozzo S., *A quantum cognition analysis of the Ellsberg paradox*, in Proceedings of QI 2011-Fifth International Symposium on Quantum Interaction, D. Song, M. Melucci, and I. Frommholz, 2011
7. Tversky A., Kahneman D., *Extensional versus intuitive reasoning: the conjunction fallacy in probability judgment*, Psychological review, 1983
8. Tversky A., Shafir E., *The Disjunction Effect in Choice under Uncertainty*, Psychological Science, 1992
9. Sozzo S., *Applicazioni del Formalismo Quantistico alle Scienze Cognitive*, Ithaca, 2021
10. Fabio Antonio Bovino - Intrasystem Entanglement Generator and Unambiguos Bell StatesDiscriminator on Chip - ICASSP 2019 - 2019 IEEE International Conference on Acoustics,Speech and Signal Processing (ICASSP) - IEEE DOI: 10.1109/ICASSP.2019.8683820

https://doi.org/10.1051/epjconf/202226609003

Plasmonic and 2D-TMD nanoarrays for large-scale photon harvesting and enhanced molecular photo-bleaching

Giulio Ferrando[1], *Matteo* Gardella[1], *Matteo* Barelli[1], *Debasree* Chowdhury[1], *Pham Duy* Long[2], *Nguyen Si* Hieu[2], *Maria Caterina* Giordano[1] *and Francesco* Buatier de Mongeot[1,*]

[1]Dipartimento di Fisica, Università di Genova, Via Dodecaneso 33, I-16146 Genova, Italy

[2]Institute of Materials Science, Vietnam Academy of Science and Technology (VAST), 18 Hoang Quoc Viet, Cau Giay, Hanoi, Vietnam

Header The urgent environmental and energy challenges require novel solutions for efficient light harvesting and conversion in new-generation ultra-thin devices. Plasmonic nanoantennas and flat optics nanogratings can promote light matter interaction at the nanoscale being very attractive for ultra-thin photonics and sensing applications. In this work we developed two light trapping solutions based on large-scale nanomaterials. The first system is a large-scale (cm^2) plasmonic metasurface based on self-organized gold nanostripes. The second is based on the periodic re-shaping of ultra-thin semiconducting MoS_2 layers forming large-area flat-optics nanogratings. Under this condition Rayleigh Anomalies can be resonantly excited thus promoting in-plane light confinement and photon absorption into the few-layers material. To demonstrate the impact of these nanopatterned systems in photon harvesting we probed their efficiency into a prototypal photochemical reaction: the photo-bleaching of Methylene Blue (MB). We demonstrate the resonant enhancement of the photo-bleaching of these polluting dye molecules promoted either by the localized plasmon resonance in Au nanostripes or by the Rayleigh Anomaly in flat-optics MoS2 nanogratings. We investigate this effect through a quantitative analysis of the solution photodissociation induced by a monochromatic light. These results show the strong potential of flat-optics templates for light-harvesting and energy conversion in ultra-thin photonic devices.

ABSTRACT

Light harvesting platforms are crucial in the development of novel solution for energy and environmental applications. An emerging class of materials that is particularly interesting are the Transition Metal Dichalcogenides (TMDs). In fact, when confined in a 2D regime, they present a semiconducting behavior [1] characterized by tunable optoelectronic properties. Nevertheless, a big issue deals with the poor photon absorption of these ultra-thin layers, limited to about 10% in the few-layer regime. In parallel conventional light trapping solutions adopted in thick semiconducting devices are not viable. To address these issues nanophotonic systems such as plasmonic nanoantennas and/or flat-optics gratings are very attractive [2,3]. The latter can be interesting itself, because it enables to strongly confine the incident light in a subwavelength volume enhancing light scattering and absorption cross sections. These properties are gaining an increasing interest for their application in photochemical processes [4] with a promising implementation in waste water treatment and energy storage (exploiting a water splitting process).
However, for this kind of applications is crucial to develop fabrication techniques that are usable over large-areas. To overcome these constraints we developed a self-organized method [5] capable to fabricate (cm^2, scalable to m^2) glass nanorippled templates that can substain the growth of laterally separated gold nanowires. Thus by means of this type of substrates it was possible to obtain a system that shows a Localized Surface Plasmon Resonance (LSPR).
In order to demonstrate the effectiveness of the sample functionalization induced by his surface nanostructuration we employed that kind of sample in a photochemical reaction: the photobleaching of the Methylene Blue (MB). In particular we investigate [6] this effect through a quantitative analysis of the photodissociation of the solution performed using the illumination set-up reported in figure 1.

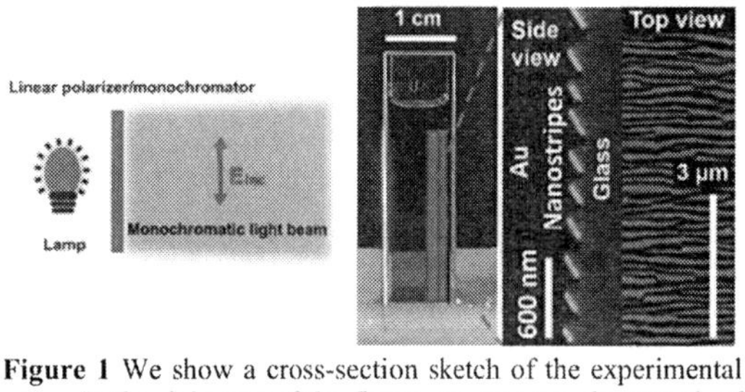

Figure 1 We show a cross-section sketch of the experimental setup. In the right part of the figure, we report a photograph of the sample inside a cuvette containing the MB solution and two SEM images. The two SEM images are the side view and the top view of the sample, with the proper dimensions.

Specifically, using a monochromatized and polarized illumination, we observed a higher MB's photodissociation rate at the wavelength at which the LSPR is tuned in resonance with the MB's maximum optical absorption. That kind of observation is crucial to

* Corresponding author: buatier@fisica.unige.it t

https://doi.org/10.1051/epjconf/202226609003

demonstrate the active role of the plasmonic resonance in the photodissociation process and to optimize the fabrication of the plasmonic substrate for the photodissociation of colored dye molecules. Therefore, these findings underline the potential of nanophotonics solutions in environmental applications such as waste water treatment.

Into a second experiment we develop a photon harvesting strategy that can be applied directly into 2D Transition Metal Dichalcogenides (TMDs) materials, such as MoS_2. For this type of materials, we explored a different approach based on a "flat optics" solution that is particularly interesting as it allows to make the material the active element for light trapping [7-9]. Specifically, we used a custom Laser Interference Lithography (LIL) method to nanostructure periodically a large-area surface. On his top we conformally deposited an ultra-thin film of MoS_2 enabling a film re-shaping.

Under this condition Rayleigh Anomalies can be observed and they promote in-plane light confinement and photon absorption into ultra-thin MoS_2 film. In order to demonstrate the potential of this solution we used that kind of samples in the photo-bleaching of the MB using the set-up reported in figure 1. We observed a strong enhancement of the MB photodegradation for the wavelength at which the Rayleigh Anomaly condition is satisfied, thus demonstrating the effectiveness of the light trapping solution.

Thus, both the light trapping solution shows a great potential in environmental application such as the waste water treatment. In addition, using the described illumination system it was possible to have wavelength resolved information of the photobleaching process paving the way for an improvement of the samples. Furthermore, the properties of the analysed samples are not limited to photochemical applications but can find a wide use in other fields such as solar harvesting and sensor application.

ACKNOWLEDGMENTS

F. Buatier de Mongeot and M. C. Giordano acknowledge financial support by Ministero dell'Università e della Ricerca, within the project 'Dipartimento di Eccellenza 2018-2022' art. 1, c. 314-337, Legge 232/2016.

M. C. Giordano acknowledges financial support by Ministero degli Affari Esteri e della Cooperazione Internazionale within the bilateral project Italy-Vietnam 2021-2023 'Large-area 2D/plasmonic heterostructures for photocatalysis and energy storage (H2D)'

References

1. Molina-Sánchez, A., Hummer, K. & Wirtz, L. *Surf. Sci. Rep.* **70**, 554–586 (2015).
2. Martella, C. *et al. Adv. Mater.* **30**, 1–9 (2018).
3. Martella, C. *et al. Adv. Mater.* **29**, 1605785 (2017).
4. Tagliabue, G. *et al. Nat. Commun.* **9**, (2018).
5. Giordano, M. C. & de Mongeot, F. B. *Adv. Mater.* **30**, 1–8 (2018).
6. Barelli, M., Ferrando, G., Giordano, M. C. & Buatier de Mongeot, F. *ACS Appl. Nano Mater.* **5**, 3470–3479 (2022).
7. Ferrando, G. *et al. Under submission.*
8. Bhatnagar, M. *et al. Nanoscale* **12**, 24385–24393 (2020).
9. Bhatnagar, M. *et al. ACS Appl. Mater. Interfaces* **13**, 13508–13516 (2021)

THz frequency combs form dispersion-compensated antenna-coupled ring quantum cascade lasers

Paolo Micheletti[1,*], *Urban* Senica[1], *Andres* Forrer[1], *Sara* Cibella[2], *Giudo* Torrioli[2], *Mattias* Beck[1], *Jerome* Faist[1], and *Giacomo* Scalari[1]

[1]Institute of Quantum Electronics, Physics Department, ETH Zurich, Zurich, Switzerland
[2]CNR-Istituto di Fotonica e Nanotecnologie, Rome, Italy

Abstract. We report comb operation of RF injected ring Quantum Cascade Lasers. A coupled waveguide approach is implemented for dispersion compensation while passive bullseye antenna improves the device power extraction and far field. Phase sensitive measurements are presented which hints at the presence of soliton state.

1 Introduction

Quantum cascade ring lasers are a promising alternative to the commonly used Fabry-Perot cavities to obtain soliton states. Whispering gallery modes exhibiting a distinct direction of rotation were observed in defect-free circular cavities in mid-IR[2]. In this case the absence of spatial hole burning, in combination with an anomalous dispersion, leaded to the demonstration of solitons. Here we report frequency combs formation in THz QCL ring cavities. The double metal waveguide laser is encapsulated in benzocyclobutene (BCB) on which the top contact is deposited. This allows to explore alternative designs such as ultrathin ring cavities and coupled double ring waveguides for dispersion compensation avoiding issues with the electrical connection of the device and the introduction of unwanted sources of back-scattering. Furthermore, passive metallic structures can be deposited onto the polymer for light extraction and far field engineering.

2 Experimental Results

We report the frequency comb operation in coupled waveguides ring lasers. The two waveguides are designed such that the propagation vectors of the modes in each waveguide are equal for a resonant frequency. These modes then couple forming a symmetric and anti-symmetric supermode where the global GVD is respectively enhanced or depressed [1] (Fig. 2 a). Thanks to the different overlap factor of the supermodes the laser naturally selects the anti-symmetric one which exhibits negative GVD (Fig. 2 b-c).

With this design spectra featuring sech2-shaped envelopes are obtained in free running and RF injected devices (Fig 1 d,e,f), hinting at the existence of soliton regimes in the QCLs. Nevertheless, the absence of any

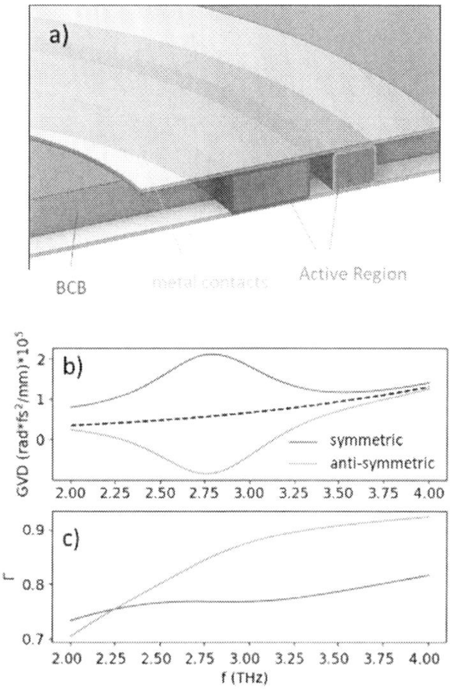

Figure 1. a) Schematic representation of a coupled waveguide ring laser. b) Group velocity dispersion of the symmetric (blue trace) and anti-symmetric (orange trace) super-modes of a coupled waveguide ring laser. The black dotted line represents the GVD of the bare GaAs. c) Overlap factor (Γ) of the symmetric (blue trace) and anti-symmetric (orange trace) supermode. Both GVD and Γ are computed with 2D axisymmetric simulations performed with Comsol Multiphysics.

*e-mail: pmicheletti@phys.ethz.ch

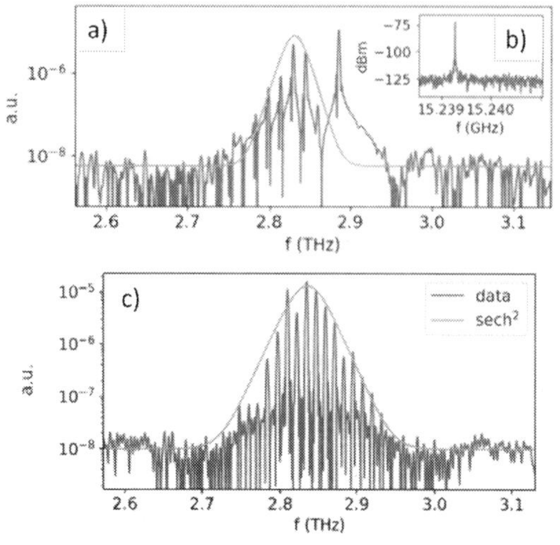

Figure 2. a) Free running Spectrum of a coupled waveguide ring laser with 800 µm radius driven in CW at 1.045A and kept at 37K with its electrically measured beatnote (b). c) Spectrum of a 1 mm radius double waveguide ring laser driven at 700mA in CW and kept at 35K. The device is strongly injected (~ 35 dBm) at the repetition frequency (12.45 GHz).

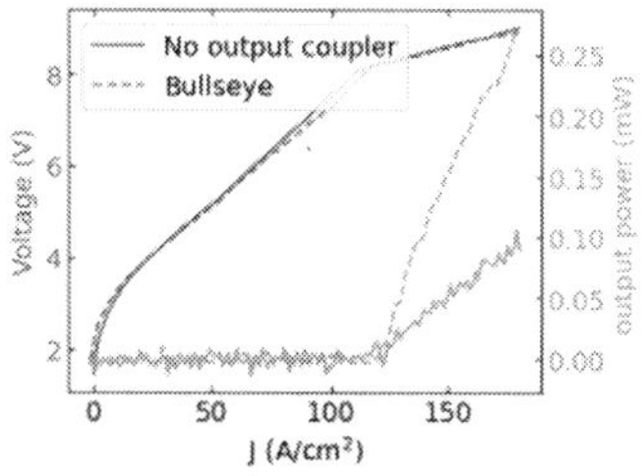

Figure 3. a) Voltage and output power as a function of current density of two devices with (solid lines) and without (dashed lines) bullseye antenna.

source of scattering inevitably, fundamental to obtain soliton states, also leads to low output power. It is therefore difficult to fully characterize the comb state performing phase sensitive measurements such as Shifted Wave Interference Fourier Transform Spectroscopy (SWIFTS)[3] as they rely on fast detectors which typically lack in responsivity in the THz domain. To face this challenge a bullseye passive antenna is integrated on top of the double ring laser. Thanks to the strictly axi-symmetric geometry no additional source of backscattering is introduced, while the output power and the far-field of the device are sensitively improved (Fig. 4 a). On these devices SWIFTS can be finally performed under moderate RF injection. Here we report a comb state featuring almost flat phases and a reconstructed waveform consisting of ~5 ps pulses (Fig. 4 b-c).

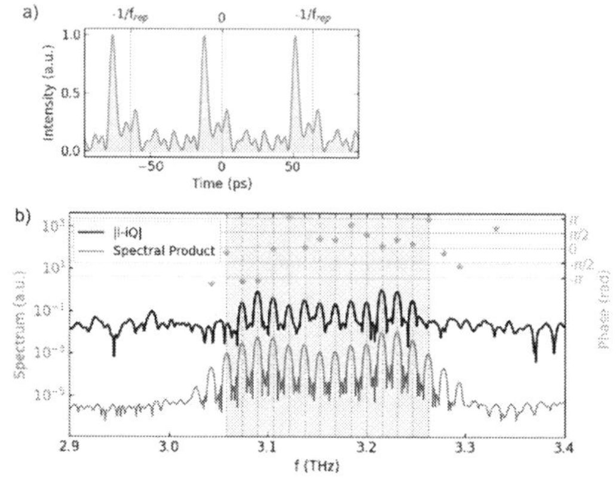

Figure 4. a) Reconstructed intensity profile of an 800 μm radius double ring laser with bullseye antenna. The relative Spectral product (i.e. $I(\omega)I(\omega + f_{rep})$ (light blue), SWIFT spectrum (dark blue) and the modes phases (orange stars) are reported in b).

References

[1] Y. Bidaux, F. Kapsalidis, P. Jouy, M. Beck, J. Faist, "Coupled-Waveguides for Dispersion Compensation in Semiconductor Lasers " , *Laser & Photonics Reviews*, 12, 5, 1700323, (2018), Wiley Online Librar

[2] B. Meng, M. Singleton, J. Hillbrand, M. Franckié, M. Beck, J. Faist, "Dissipative Kerr solitons in semiconductor ring lasers " , *Nature Photonics*, 16, 2, 142–147, (2022), Nature Publishing Grou

[3] D. Burghoff, Y. Yang, D. Hayton, J. Gao, J. Reno, Q. Hu, "Evaluating the coherence and time-domain profile of quantum cascade laser frequency combs " , *Optics express*, 23, 2, 1190–1202, (2015), Optical Society of Americ

https://doi.org/10.1051/epjconf/202226609005

Characterization of output circular polarization degree in low-cost asymmetric metasurfaces

Emilija Petronijevic[1,*], *Alessandro* Belardini[1], *Zakaria* El-ansary[2], *Bilal* Brioual[3], *Hari* Prasath Ram Kumar[1], *Tiziana* Cesca[4], *Carlo* Scian[4], *Giovanni* Mattei[4], and *Concita* Sibilia[1]

[1]Sapienza Università di Roma, SBAI Department, Via Antonio Scarpa 14-16, Rome, Italy
[2]Ibn Tofail University, 6CW7+CVP, Av. de L'Université, Kénitra, Morocco
[3]Abdelmalek Essaâdi University, P4P4+HWM, Tanger, Morocco
[4]University of Padova, Physics and Astronomy Department, via Marzolo 8, I-35131 Padova, Italy

Abstract. Addition of asymmetry in plasmonic nanostructures can lead to chiro-optical phenomena, usually monitored as different absorption of left and right polarization, i.e. circular dichroism. Moreover, interesting features arise when the nanostructure changes the polarization state of the input beam. In this work, we perform extrinsic chirality characterization in a widely tuneable near-infrared range, by monitoring both polarization of the input and of the transmitted beam. We characterize low-cost metasurfaces based on polystyrene nanospheres asymmetrically covered by Ag, by exciting them at different angle of incidence with left, right and linear polarization. We then resolve the circular polarization degree of the transmitted beam, demonstrating resonance-governed circular polarization degree in the output, showing the interplay of both intrinsic and extrinsic chirality.

1 Introduction

Metasurfaces with asymmetric features can present chiro-optical behaviour with consequences in both near- and far-field. Nowadays, symmetry of metasurfaces can be easily broken even with low-cost, self-assembling methods, e.g. by tilting the deposition of the plasmonic layer [1-3]. We previously proposed metasurfaces based on polystyrene nanospheres and covered by metal for various chiro-optical effects governed by extrinsic chirality. Namely, we measured different absorption [4], extinction [5] and diffraction [6] of impinging circular polarizations of opposite handedness. In these works, we excited the nanostructures with left or right circular polarizations (LCP and RCP, respectively), and monitored the total intensity of the transmitted or absorbed light. We used conventional (transmission) methods [5], as well as highly sensitive photothermal effects [4,6]. Another interesting degree of freedom is the control of the polarization state in the output light, after the beam interacts with the metasurface.

In this work, we present experimental approach where we study wavelength, spin, and angle-of-incidence dependent circular polarization degree in the transmitted field from asymmetric metasurfaces. We use a low-cost nanosphere lithography to fabricate dielectric metasurface based on self-assembled polystyrene nanospheres, and then asymmetrically cover them with 55 nm of Ag. Additionally, we introduce the in-plane tilt, which has been shown to give intrinsic planar chirality [7]. We investigate the influence of the incident polarization on the circular polarization degree map in the output. In particular, when we shine the metasurface with linear polarization, left and right circularly polarized output exhibit interesting, resonant features with the influence of both intrinsic and extrinsic chirality. Moreover, to characterize the circular polarization conversion, we repeat this mapping with LCP or RCP input. Previous works give us the confidence of numerical model which we use to gain insight into near-field confinement and behaviour of the nanoshell and nanohole arrays at the nanoscale. The implementation of this model will be a subject of future work.

2 Experimental set-up

The experimental set-up scheme is shown in Fig. 1. We use a near-infrared laser (Chameleon Ultra II by Coherent Inc.) to excite the nanostructures with a broad range of the incident field wavelengths, from 700 nm to 1000 nm. This wave impinges on a linear polarizer (LP) and a rotating quarter wave plate (QWP), which define the polarization state of the incident beam. In this work, we first investigate the excitation by linear polarization, and then LCP and RCP excitation. We can further rotate the sample in the xz plane to investigate chiro-optical behavior at oblique incidence (from -45° to +45°). After the interaction with the sample, the zeroth order transmission passes through another set of QWP followed by LP, in order to resolve the circular polarization degree in the output. Finally, the intensity is monitored by a Si photodiode (PD).

* Corresponding author: emilija.petronijevic@uniroma1.it

In the inset of Fig. 1 we show a Scanning Electron Microscope (SEM) image of the investigated sample. In the experimental set-up, the y-axis of the experimental coordinate system as aligned with the direction of the tilted deposition projection on the xy-plane. All the measurements are done at room temperature and the transmission data are normalized with respect to the transmission of light without the sample.

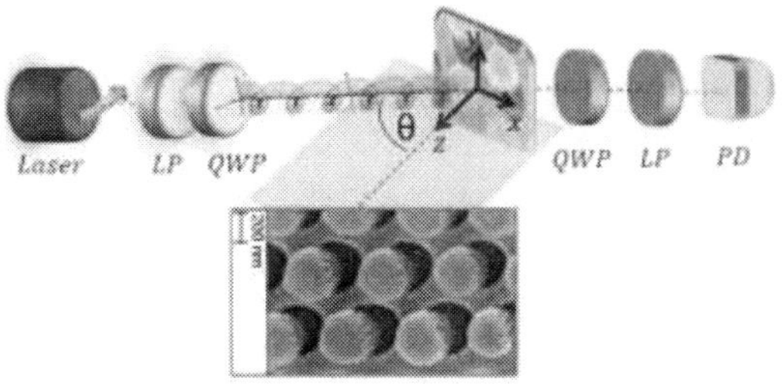

Fig. 1 Set-up of the chiro-optical characterization with both input and output beams resolved in polarization. Inset : SEM image of the investigated sample.

3 Results

In Fig. 2, we show the wavelength-incidence angle maps when the metasurface is excited with linearly polarized light; we plot the transmission maps at different output QWP angles, where T_0, T_{LCP} and T_{RCP} denote transmission into linear, LCP and RCP polarized light. Being in a highly reflecting region, transmission intensity is low, but the resonant features which appear could be due to lattice resonances [5]. Moreover, the sample positioned in this way shows intrinsic chiral features, with different LCP/RCP ratio at normal incidence across broad wavelength range. Next, by changing the incident polarization to LCP or RCP, we were able to almost completely extinct one polarization state in the output. Numerical analysis is next needed for detailed analysis of the nanoshell asymmetry on the polarization state in the transmitted beam.

4 Conclusions

In conclusion, we performed thorough chiro-optical characterization of low-cost asymmetric metasurfaces by measuring wavelength, incidence angle and polarization-dependent transmission, and resolving the circular polarization degree of the output beam. Many degrees of freedom of such characterization could pave the way to new interesting resonant chiro-optical effects in this low-cost nanoscale geometry.

A. B. acknowledges LASAFEM Sapienza Università di Roma Infrastructure Project prot. n. MA31715C8215A268.

Z.E, B.B., C.S. and E. P. acknowledge Erasmus+ programme between Sapienza Università di Roma and Universities in Morocco.

References

1. E.S.A. Goerlitzer et al., Adv. Optical Mater. 7, 1801770 (2019)
2. T. Hakkarainen et al., Sci. Rep. 9, 5040 (2019)
3. E.S.A. Goerlitzer et al., Nanoscale Adv. 2, 394–400 (2020)
4. E. Petronijevic et al., Appl. Phys. Lett. **114**, 053101 (2019)
5. E. Petronijevic et al., Phys. Rev. Appl. 16(1), 014003 (2021)
6. G. Leahu et al., Adv. Opt. Mater. 2100670 (2021)
7. E. Petronijevic et al., Opt. Quantum Electron. 52, 176 (2020)

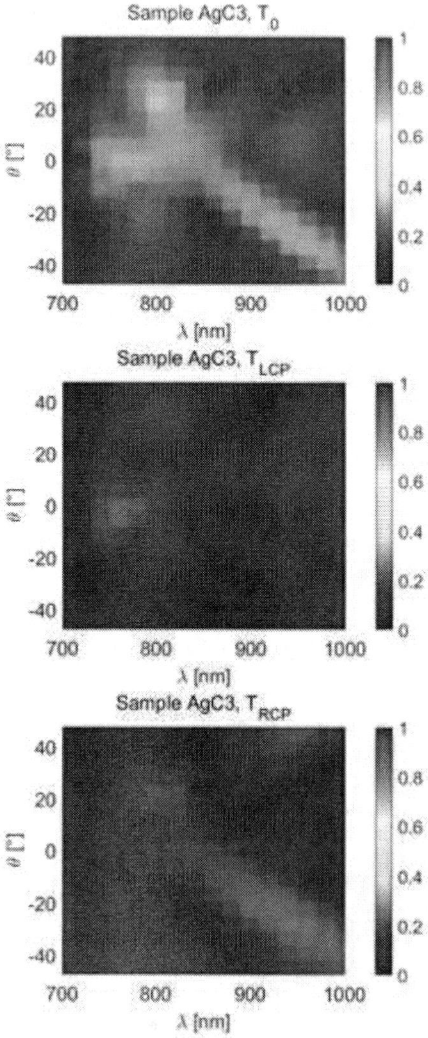

Fig. 2 Circular polarization resolved response of the sample when it is excited with linearly polarized laser beam, widely tunable in wavelength and angle of incidence. T_0, T_{LCP}, and T_{RCP} denote the output QWP angle of 0°, -45° and +45°, respectively.

Experiments and simulations of chiro-optical response in low-cost nanohole arrays in silver

Emilija Petronijevic[1,*], *Zakaria* El-ansary[2], *Bilal* Brioual[3], *Alessandro* Belardini[1], *Hari* Prasath Ram Kumar[1], *Tiziana* Cesca[4], *Carlo* Scian[4], *Giovanni* Mattei[4], and *Concita* Sibilia[1]

[1]Sapienza Università di Roma, SBAI Department, Via Antonio Scarpa 14-16, Rome, Italy
[2]Ibn Tofail University, 6CW7+CVP, Av. de L'Université, Kénitra, Morocco
[3]Abdelmalek Essaâdi University, P4P4+HWM, Tanger, Morocco
[4]University of Padova, Physics and Astronomy Department, via Marzolo 8, I-35131 Padova, Italy

Abstract. 2D metasurfaces based on periodic nanoholes in metal have been proposed in various plasmonic platforms. Specifically, their resonant features have led to applications spanning in biosensing. Here we investigate additional degree of freedom in elliptical nanohole arrays with hexagonal geometry: chiro-optical effects. Namely, the in-plane asymmetry and a slightly elliptical shape of nanoholes were previously shown to differently extinct light of opposite handedness, even at normal incidence. We now fully characterize nanoholes in Ag, fabricated by low-cost nanosphere lithography. We first measure the dependence of the transmitted intensity for opposite handedness, in a broad spectral and angle of incidence range. We then resolve the circular polarization degree of the transmitted light when the nanohole array is excited with linear polarization. Finally, we numerically investigate the origin of the chiro-optical effect at the nanoscale. We believe that circular polarization resolving of the transmitted degree could be further adapted as a highly sensitive tool in chiral sensing.

1 Introduction

Plasmonic nanohole arrays have been popular for decades due to their ability to couple light into plasmon oscillations at the metal-dielectric interface, thus enabling resonant electromagnetic field confinement and enhancement at the nanoscale [1]. The nanohole shape asymmetry and/or the in-plane tilt allow for additional degrees of freedom. Specifically, properly designed nanoholes can differently interact with left and right circularly polarized light (LCP and RCP, respectively), thus exhibiting chiro-optical effects and circular dichroism (CD). We have previously proposed the in-plane and shape breaking of symmetry in elliptical nanohole arrays (ENHA), and showed that such simple geometries lead to interesting chiro-optical effects in the near-infrared range [2-4]. These samples can be fabricated by means of low-cost, nanosphere lithography [2,3,5,6] and further optimized to giant CD in absorption and transmission [4].

In this work, we perform a detailed chiro-optical characterization of Ag-based ENHA, by means of two types of experiments, and two corresponding set-ups of numerical simulation. We first use a broadly tuneable near-infrared laser to excite ENHA with LCP and RCP at various degrees of incidence. As expected, the array shows rather low extinction CD due to the low ellipticity of the nanoholes. Next, we excite the ENHA with linear polarization and measure the polarization state of the transmitted beam. This nanostructure shows peculiar same sign circular polarization degree in the output, and we believe that this is due to the intrinsic chirality, even though it is rather low. This is the reason why believe that one needs to perform a complete characterization of the input and output state of the light interacting with the ENHA, and choose the one which is more dependent on the asymmetric features. To corroborate our assumptions, we perform numerical simulations which finally show the origin of the existence of circular polarization degree in the output.

2 Experiments

Our sample is made of ENHA in 55 nm of Ag, with hexagonal periodicity of 518 nm. It was previously investigated in Ref. [3], where we obtained low extinction CD at normal incidence. Here we apply the incidence-angle dependent set-up such as in Ref. [5]. The sample shows characteristic features of both extrinsic and intrinsic chirality, reaching only 5% at larger angles.

Next, we turn to the set-up presented in Fig. 1(a), which measures the circular polarization degree of the field transmitted through the sample We make use of a near-infrared laser (Chameleon Ultra II by Coherent Inc.), and we monitor the wavelengths range between 700 nm

* Corresponding author: emilija.petronijevic@uniroma1.it

to 1000 nm. We use laser light polarized in the x-direction, which impinges on a sample, and gets transmitted to the output rotating quarter-wave plate (QWP), then linear polarizer (LP). The measurements are done at different angle of incidence, by rotating the sample between -45° and 45°. To measure the circular polarization degree in the output, QWP can be rotated, too. Finally, a Si photodiode (PD) caught this transmission.

In Fig. 1(b) we show spectral results at normal incidence, when QWP is fully rotated from -90° to 90°, where -45° and +45° represent LCP and RCP in the output, respectively; results are normalized to the transmission with QWP at 0°. We note that across the whole spectrum, light is prevalently transmitted to RCP, suggesting intrinsic chirality. Remarkably, we do not see abrupt differences of this behaviour at oblique incidence.

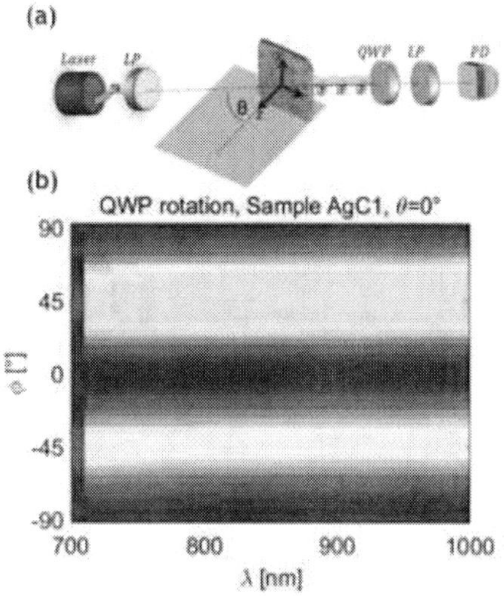

Fig. 1. (a) Set-up for the measurement of angle-dependent, spectrally tuneable circular polarization degree in the output. (b) Wavelength-QWP angle transmission map at normal incidence.

3 Simulations

Next, we numerically investigate the variation of polarization state of light passing through a plasmonic nanohole array with hexagonal unit cell, using the Lumerical (FDTD) software. The FDTD simulation region is defined as a rectangular unit cell (a and $a\sqrt{3}$, where $a = 518nm$) with periodic boundary conditions in xy plane. The light source is a linearly polarized plane-wave in x-direction, at 800 nm. The structure is a semi-infinite rectangle made of glass with a refractive index of 1,5, covered with 55 nm of Silver (Ag). Then an array of nanoholes is etched in the Ag layer. Finally, the detector is a polarztn_ellipse analysis group.

We analyze the polarization of linear incident light in output passed through normal and tilted elliptic nanohole array with diameters Dx=350nm, and Dy=400 nm. As expected, the tilting of the array changed the polarization from linear to elliptical. For the tilted array, we next simulate different nanoholes dimensions. Fig. 2 shows the evolution of the light in output for different diameters. As seen, at first the circular nanoholes do not change the incident polarization. Then by increasing D_y, the detected elliptical polarization keeps increasing.

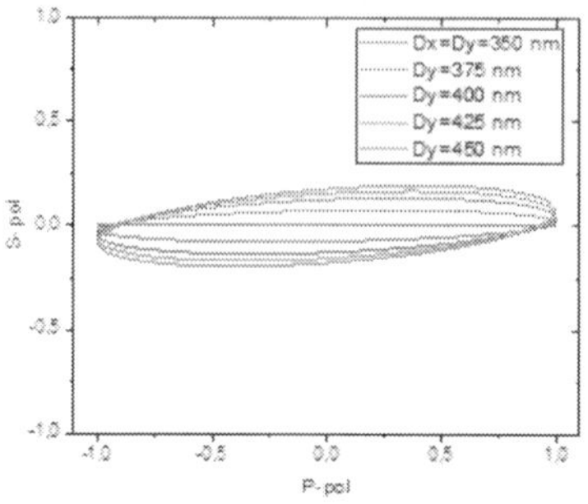

Fig. 2. (a) Set-up for the measurement of angle-dependent, spectrally tuneable circular polarization degree in the output. (b) Wavelength-QWP angle transmission map at normal incidence.

4 Conclusions

We investigate, experimentally and numerically, chiro-optical properties of hexagonal nanohole arrays in Ag. The slightly elliptical shape and the in-plane tilt leads to different circular polarization degree of the output beam. The samples are obtained by low-cost self-assembling approach. When excited by linear polarization, they give interesting circular polarization features in the transmitted beam. We believe such measurements are highly sensitive to asymmetric features of the nanostructured devices.

A. B. acknowledges LASAFEM Sapienza Università di Roma Infrastructure Project prot. n. MA31715C8215A268.

Z.E, B.B., C.S. and E. P. acknowledge Erasmus+ programme between Sapienza Università di Roma and Universities in Morocco.

References

1. F. J. García de Abajo, Rev. Mod. Phys. **79**, 1267 (2007)
2. E. Petronijevic et al., Appl. Phys. Lett. 114, 053101 (2019)
3. E. Petronijevic et al., Appl. Sci. 10, 1316 (2020)
4. E. Petronijevic et al., Opt. Quantum Electron. 52, 176 (2020)
5. E. Petronijevic et al., Phys. Rev. Appl. 16(1), 014003 (2021)
6. G. Leahu et al., Adv. Opt. Mater. 2100670 (2021)

Field-enhancing tapered planarized waveguides for THz quantum cascade laser frequency combs

Urban Senica[1*], *Andres Forrer*[1], *Tudor Olariu*[1], *Paolo Micheletti*[1], *Sara Cibella*[2], *Guido Torrioli*[2], *Mattias Beck*[1], *Jérôme Faist*[1], *Giacomo Scalari*[1*]

[1]Quantum Optoelectronics Group, Institute of Quantum Electronics, ETH Zürich, 8093 Zürich
[2]Istituto di Fotonica e Nanotecnologie, CNR, Via Cineto Romano 42, 00156 Rome, Italy

Abstract. We present a new planarized waveguide geometry for THz quantum cascade laser frequency combs with improved waveguide losses, RF and thermal dissipation properties. Ridge devices display broadband free-running comb states, and the THz emission can be further broadened by RF injection. Tapered waveguide devices feature a strong field-enhancement effect, which results in an improved comb performance. This includes free-running comb states with strong single beatnotes up to nearly -30 dBm at 90 K, almost three orders of magnitude stronger than for ridge devices. Improved comb operation is maintained also for high operating temperatures, up to 115 K.

1 Introduction

Terahertz (THz) quantum cascade lasers (QCLs) [1] are compact sources of broadband THz radiation, appealing for spectroscopy and sensing applications due to their operation as frequency combs [2] and dual combs [3]. The double metal waveguide is one of the most commonly used cavities for THz QCLs, as it features low losses and near-unity overlap of the fundamental mode with the active material, which is sandwiched between two metallic plates. It also acts as a microwave waveguide in the radio-frequency (RF) range, a property crucial for frequency combs, where an efficient guiding, readout and injection of RF signals is required.

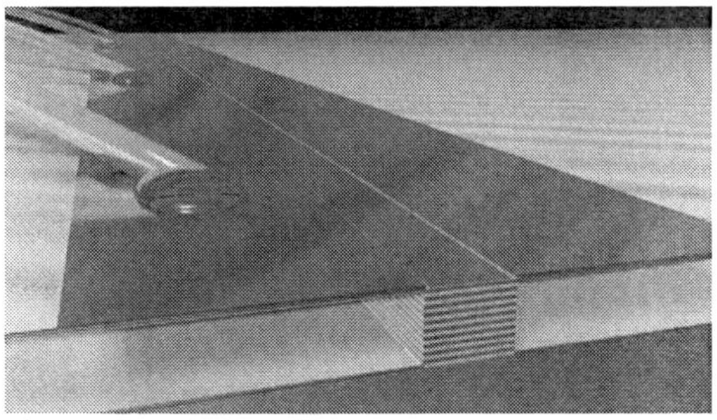

Fig. 1. Illustration of the planarized double metal waveguide geometry, where the active region waveguide (ridge with stripes) is embedded in a low-loss BCB polymer (on the sides). The bonding wires are placed on the extended top contact over the BCB-covered area, which improves the waveguide properties.

2 Results

2.1. Fabricated devices

We present an improved waveguide geometry: the planarized waveguide. After a standard double metal waveguide process with dry-etched active waveguides, the surrounding region is planarized with BCB, a low-loss polymer [4, 5]. This enables the fabrication of a wider top metallization, extending beyond the width of the active waveguide, and placing wire bonds above the BCB-covered area, as shown in Fig. 1. The latter prevents the formation of any defects or hotspots forming on top of the waveguide, which would degrade the device performance. Additionally, the active waveguide can now be narrower than the wire bond, which improves the heat dissipation properties and enhances high-temperature performance in continuous wave (CW).

The narrow waveguide dimensions can act also as a selection mechanism for lasing in the fundamental transversal mode for a stable comb operation. The extended top contact enhances the RF properties, as it forms an enclosed metallic cavity with the ground plane. Besides ridge waveguides, we also fabricated waveguides with a tapered profile, similar as demonstrated previously for mid-IR QCLs [6]. While the narrow waveguide sections act as a transversal mode filter, the wider sections have lower losses and provide more gain. Another crucial aspect is a strong field-enhancement effect in the narrower sections, which is roughly proportional to the width ratio. Consequently, four wave mixing, which is the fundamental frequency comb formation mechanism in QCLs, is greatly enhanced as it is a third order nonlinear

* Corresponding authors: usenica@phys.ethz.ch, scalari@phys.ethz.ch

process, proportional to the cube of the field intensity inside the active material.

2.2 Measurement results

We first present results on simple ridge devices with a cleaved facet, as shown in Fig. 2.(a). The waveguide is 40 μm wide, which is a good balance between low waveguide losses and fundamental transversal lasing mode selection. Due to the use of low-loss Cu-Cu planarized waveguides and a low-dissipation broadband active region [7], the LIV measurements in Fig. 2.(b) display a low threshold current density of 140 A/cm^2. The peak ouput power is around 3 mW in pulsed and CW operation at a heatsink temperature of 40 K. When measuring the output spectrum, free-running comb states with bandwidths above 600 GHz and single beatnotes up to nearly -60 dBm are found, as shown in Fig. 2.(c). With a strong RF injection (+35 dBm at the source), the emission can be broadened to over 1.4 THz, as shown in Fig. 2.(d) (these are typically not frequency comb states).

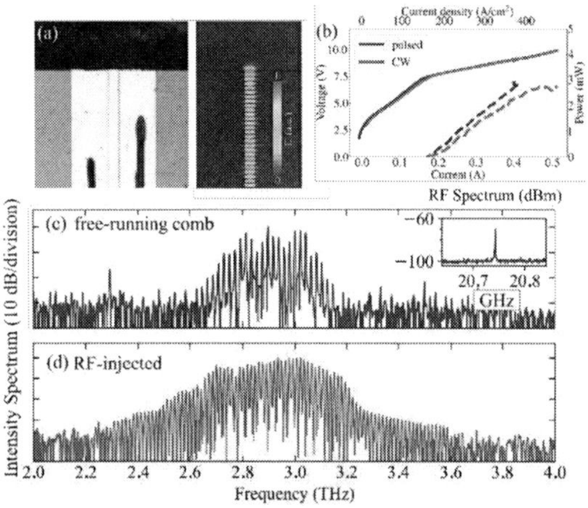

Fig. 2. (a) Microscope image of a ridge waveguide device. The cleaved facets act as mirrors, and standing waves form in the cavity, as shown on the right image with the E-field profile from a 3D numerical simulation of the propagating fundamental waveguide mode. (b) LIV curves of a 40 μm wide and 2.7 mm long device with a low threshold current density of around 140 A/cm^2 and peak powers around 3 mW at a heatsink temperature of 40 K. (c) Spectral measurement of a free-running device in a frequency comb state with a bandwidth over 600 GHz and a single RF beatnote (inset). (d) Injecting with a strong external RF signal (+35 dBm at source) close to the natural roundtrip freqeuncy broadens the THz emission to span over 1.4 THz.

Next, we present results on the tapered waveguides. Free-running devices produce very strong single beatnotes, measured up to nearly -30 dBm, as can be seen in Fig. 3.(a). This nearly three orders of magnitude stronger RF signal comes from the significant field enhancement in the narrow waveguide sections. The improved comb properties are observed also for high operating temperatures (Fig.3.(b)). While ridge devices are typically in a single mode regime close to the maximum operating temperature, for tapered devices we observe comb operation up to a heatsink temperature of 115 K, very close to the maximum CW lasing temperature of 118 K.

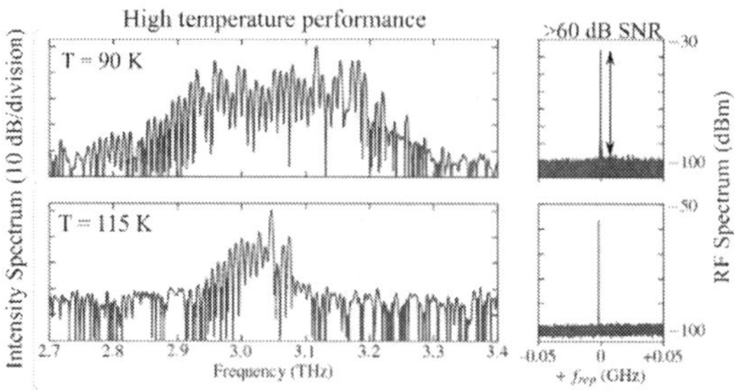

Fig. 3. (a) Free-running tapered waveguide devices produce very strong beatnotes, measured up to nearly -30 dBm at 90 K. (b) The improved frequency comb performance is maintained also for high temperatures, with a bandwidth of 200 GHz and a single beatnote measured above -60 dBm at a heatsink temperature of 115 K.

3 Conclusion

We have presented an improved planarized waveguide geometry for THz QCL frequency combs. The planarization with a low-loss BCB polymer and the extended top metallization result in improved waveguide losses, RF and heat dissipation properties. Ridge devices display broadband free-running comb states, with the THz emission broadened further with an external RF signal.

Tapered devices show three orders of magnitude stronger RF beatnotes and high temperature comb operation due to a strong field enhancement effect. The planarized waveguide platform will also enable a relatively straightforward co-integration of active and passive waveguide elements on the same chip.

References

1. Köhler, R., et al. *Terahertz semiconductor-heterostructure laser.* Nature, 417 (6885), pp.156-159 (2002).
2. D. Burghoff, et al. *Terahertz laser frequency combs*, Nature Photonics, vol. 8, pp. 426–467 (2014).
3. M. Roesch, et al. *On-chip, self-detected terahertz dual-comb source*, Appl. Phys. Lett., vol. 108 (17), pp. 171104 (2016).
4. Z. Diao, et al. *Continuous-wave vertically emitting photonics crystal terahertz laser*, Laser & Photonics Reviews, vol. 7 (5), pp. 45-50 (2013).
5. C. Bonzon, et al. *Integrated patch and slot array antenna for terahertz quantum cascade lasers at 4.7 THz*, Appl. Phys. Lett., vol. 104 (16), pp. 161102 (2014).
6. R. Wang, et al. *Mid-infrared quantum cascade laser frequency combs based on multi-section waveguides*, Opt. Lett. 45, 6462-6465 (2020).
7. A. Forrer, et al. *Photon-driven broadband emission and RF injection locking of THz quantum cascade lasers*, ACS Photonics, vol. 7 (3), pp. 748-791 (2020).

Light emission from color centers in phosphorus-doped diamond

Florian Sledz[1], *Assegid* M. Flatae[1], *Stefano* Lagomarsino[1,2], *Savino* Piccolomo[3], *Shannon S.* Nicley[4,5], *Ken* Haenen[4], *Robert* Rechenberg[6], *Michael F.* Becker[6], *Silvio* Sciortino[2,7], *Nicla* Gelli[2], *Lorenzo* Giuntini[2,7], *Giorgio* Speranza[3], and *Mario* Agio[1,8,*]

[1]University of Siegen, Laboratory of Nano-Optics, 57072 Siegen, Germany
[2]Istituto Nazionale di Fisica Nucleare, Sezione di Firenze, 50019 Sesto Fiorentino, Italy
[3]Fondazione Bruno Kessler, Centro Materiali e Microsistemi, 38122 Trento, Italy
[4]Hasselt University & IMEC, Institute for Materials Research (IMO) & IMOMEC, 3590 Diepenbeek, Belgium
[5]University of Oxford, Department of Materials, Oxford OX1 2PH, United Kingdom
[6]Fraunhofer USA, Center for Coatings and Diamond Technologies, East Lansing, MI 48824, USA
[7]University of Florence, Dipartimento di Fisica e Astronomia, 50019 Sesto Fiorentino, Italy
[8]National Research Council (CNR), National Institute of Optics (INO), 50125 Florence, Italy

Abstract. Light emission from color centers in diamond is being extensively investigated for developing, among other quantum devices, single-photon sources operating at room temperature. By doping diamond with phosphorus, one obtains an n-type semiconductor, which can be exploited for the electrical excitation of color centers. Here, we discuss the optical properties of color centers in phosphorus-doped diamond, especially the silicon-vacancy center, presenting the single-photon emission characteristics and the temperature dependence aiming for electroluminescent single-photon emitting devices.

1 Introduction

Single-photon emitters in the solid state represent an essential ingredient for developing quantum photonics technologies, such as quantum cryptography and linear optical quantum computing [1,2]. Color centers in diamond are promising in this regard because they exhibit stable emission characteristics, and they can be operated at room temperature. In particular, the silicon-vacancy (SiV) center provides most of its emission in a narrow zero-phonon line (ZPL) [3] and it withstands large temperatures [4].
For device miniaturization and energy efficiency, it is desirable to excite the color center electrically. This requires electron-hole recombination at the color center, hence charge transport in diamond. Early work demonstrated emission from nitrogen vacancy and SiV centers in p-i-n diodes [5,6]. However, growing p-i-n diamond structures is still quite challenging, and the devices are poorly reproducible. An alternative approach relies on Schottky diodes, where the color centers are placed in n-type diamond and holes are injected through the metal contact, if the Schottky barrier is reasonably small. This would only require an n-type diamond layer and the deposition of metal contacts, hence greatly simplifying the device structure.

2 Silicon-vacancy centers in n-type diamond

2.1 Sample fabrication

To obtain n-type diamond, the material can be doped with phosphorus (P) during microwave (MW) plasma enhanced (PE) chemical vapor deposition (CVD) growth. Specifically, one P-doped single-crystal diamond film (sample A) was grown on a diamond substrate utilizing an in-house built 2.45 GHz MWPECVD reactor. The hydrogen (H_2) rich plasma contained 0.09% of methane (CH_4) with a PH_3/CH_4 ratio of 4300 ppm. Two more P-doped homo-epitaxial samples were grown similarly by MWPECVD in a 2.45 GHz ASTeX PDS17 reactor with a water-cooled molybdenum substrate holder. The H_2 plasma contained 0.15% CH_4. One of the samples (sample B) was grown with a P-gradient, with the PH_3/CH_4 ratio adjusted in steps of 0, 250, 500, 1000, 2500, 5000, 10,000, and 20,000 ppm. The other sample (sample C) was grown with a constant 5000 ppm PH_3/CH_4. The H_2 and CH_4 gasses are filtered to <1 ppb (9 N) purity, and the PH_3 is used from a source diluted to 200 ppm in H_2.
The three samples underwent Si-ion implantation based on a 3 MV Tandetron accelerator (High Voltage Engineering Europe) equipped with a HVEE 860 Negative Sputter Ion Source. Aluminum (Al) metal foils were used to decrease the ion energy down to a few tens of keV for shallow implantation (within 200 nm from the surface). A custom designed furnace (1200 °C in high-vacuum conditions ~ 10^{-7} mbar) enables the activation of the SiV color centers in P-doped samples. The samples have been Si-implanted with five different fluences (~10^{14}

* Corresponding author: mario.agio@uni-siegen.de

cm^{-2}, ~10^{13} cm^{-2}, ~10^{12} cm^{-2}, ~10^{8} cm^{-2} and ~10^{7} cm^{-2}). The expected implantation depth is ≤ 200 nm. Details on the implantation process can be found in Ref. [7].

2.2 Optical properties

Since the inclusion of phosphorous modifies the host environment, the first important question is whether SiV centers exhibit optical properties comparable with intrinsic diamond and if it is possible to obtain single-photon emission [5].

We investigated the three diamond samples with different phosphorus doping profiles. Firstly, we clarified that co-doping with nitrogen causes a significant fluorescence background related to nitrogen vacancy (NV) centers that prohibits the observation of single SiV centers. Secondly, we observed that for samples implanted with low silicon ion fluences the emission background is significantly reduced. Figure 1 illustrates the emission spectrum under different ion fluences.

For low implantation fluences on samples with low nitrogen content (sample C) we finally observed single-photon emission from SiV centers [7]. While the lifetime is comparable to that of SiV centers in intrinsic single-crystal diamond, the ZPL is broader, probably because the host environment is more defective than intrinsic diamond. In this regard, low-temperature measurements would be needed to better understand the broadening effect.

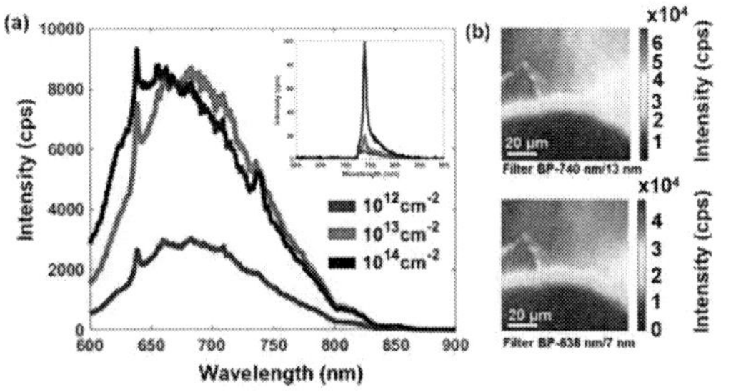

Fig. 1. Photoluminescence spectra in the regions of sample A implanted with a fluence of 10^{12} cm^{-2}, 10^{13} cm^{-2} and 10^{14} cm^{-2} using 532 nm CW laser show NV-related background. Inset: The background due to NV color centers is reduced by exciting the SiV color centers using 690 nm CW laser, as NV complexes have less absorption in this spectral range. (b) Spatially- and spectrally resolved confocal images indicate that the NV color centers (below) are mainly created at the location where the Si-ions are implanted (top) (Figure from Ref. [7]).

2.3 Temperature studies

Since free carriers in P-doped diamond suffer from the relatively high activation energy of donor levels (~ 0.6 eV), efficient electrical operation requires room or higher temperatures. We thus performed optical studies of SiV centers as a function of temperature, finding that the SiV center remains photostable up to the experimentally considered temperatures (100 °C in ambient conditions). Furthermore, the temperature dependence of the ZPL in intrinsic diamond and in P-doped diamond differ. This is likely to be attributed to the different crystalline environment between undoped and doped diamond. Details on the experiments can be found in Refs. [8].

3 Conclusions and outlook

We successfully created and characterized single SiV centers in P-doped (n-type) diamond in view of developing optoelectronics light emitting devices. While low-temperature studies would be necessary to more deeply understand the properties of SiV centers in n-type diamond for quantum applications, the temperature studies indicate that for classical light emission the system is promising. For instance, while conventional light-emitting devices (LEDs) degrade their performances when temperature increases, diamond-based LEDs would exhibit better performances under large temperatures [9] and operate in a much wider temperature range, thus offering a unexplored applications for LEDs.

Acknowledgments

The authors acknowledge funding from the University of Siegen, the German Research Association (DFG) (INST 221/118-1 FUGG, 410405168). The authors also acknowledge INFN-CHNet, the network of laboratories of the INFN for cultural heritage, for support and precious contributions in terms of instrumentation and personnel. FS, AMF and MA would like to thank P. Reuschel and N. Soltani for experimental support and D. Yu Fedyanin and I. Khramstov for helpful discussion.

References

1. S. Pirandola, *et al.*, Adv. Opt. Photonics **12**, 1012 (2020)
2. J. Wang, F. Sciarrino, A. Laing, M.G. Thompson, Nat. Photon. **14**, 273 (2020)
3. S. Lagomarsino, *et al.*, Diam. Relat. Mater. **84**, 196 (2018)
4. S. Lagomarsino, *et al.*, AIP Adv. **5**, 127117 (2015)
5. N. Mizuochi, *et al.*, Nat. Photon. **6**, 299 (2012)
6. A. M. Berhane, *et al.*, Appl. Phys. Lett., **106**, 171102 (2015)
7. A. M. Flatae, *et al.*, Diam. Relat. Mater. **105**, 107797 (2020)
8. F. Sledz, *et al.*, N. Cimento C **44**, 106 (2021)
9. D. Yu Fedyanin, M. Agio, New J. Phys. **18**, 073012 (2016)

Active tuning of the optical response of field-effect-gated transparent conductive oxides

Maria Sygletou[1,*], *Emilio* Bellingeri[2], *Maurizio* Canepa[1] and *Francesco* Bisio[2]

[1]OPTMATLAB Dipartimento di Fisica, Università di Genova, Via Dodecaneso 33, 16146 Genova, Italy
[2]CNR-SPIN, Corso F.M. Perrone 24, 16152 Genova, Italy

Abstract. Transparent Conductive Oxides (TCOs) are a class of materials with high optical transparency and electrical conductivity. This combination makes them extremely appealing for solar cells, optoelectronics and infrared-plasmonics applications. In this work, we report the active tuning of the optical response of aluminium-doped ZnO (AZO) films upon electrical gating within a parallel-plate capacitor configuration. We investigated the electrical-bias-dependent optical response of thin AZO films fabricated by pulsed laser deposition by means of spectroscopic ellipsometry (SE). Calculations based on an exponentially-decaying spatial distribution of injected/depleted charge density in the AZO film are in accordance with experimental observations, allowing the extraction of the Debye length.

1 Introduction

The realisation of active optoelectronic systems has drawn the attention of the scientific community over the last decade due to the possibility of merging optical and electrical control in a single device for next generation optoelectronics and photovoltaics. TCOs (such as ITO, ZnO etc.) are characterised by a combination of high optical transparency (~80%) and low electrical resistivity (~10^{-4} Ω·cm).

Zinc oxide (ZnO) is a n-type semiconductor with direct wide-energy band gap, and is a possible cheaper substitute to the commonly used indium-tin oxide. Aluminium doping increases its conductivity and makes it suitable for applications in optoelectronic devices, as a transparent conductive component and epsilon-near-zero (ENZ) material in the near infrared (NIR).

Actively modifying the optical response of TCOs via field effect is a potentially viable strategy to develop new functional optical materials for optoelectronics, photonics and photovoltaics ([1-2]). Under an applied electric field, the carrier concentration at a dielectric/conducting oxide interface induces charge accumulation/depletion in a thin layer whose width is usually less than 5 nm due to screening effects. Most of the studies about voltage-gated optically-active TCOs have focused on ITO-based systems, and have been limited to reflectance/transmission measurements within the visible and -NIR regimes.

In this study, we show the active tuning of the optical response of thin AZO films on $SrTiO_3(110)$ substrates (STO) via field effect, in a spectral range encompassing both the visible and near infrared (NIR) photon-energy range (0.735 to 3.22 eV). The TCO film acts as the top electrode in a parallel-plate-capacitor configuration.30 We successfully observed the active tuning of the optical response of AZO, and quantified the charge profile of the accumulation layer at the TCO/insulator interface.

2 Discussion

In order to realise a large charge modulation AZO films were grown on STO(110) substrates by Pulsed Laser Deposition (PLD). Spectroscopic Ellipsometry (SE) was performed by means of a J.A. Woollam V-VASE ellipsometer. For the SE measurements, the sample was mounted inside an optical cryostat (Oxford Instruments Optistat CF-V) and kept at either 4.2 K or 77 K temperature.

In order to investigate the variations of the optical response of the AZO films upon gating, we applied voltages of ±100 V across the STO substrate, and acquired SE spectra under bias (V=±100V) (Figure 1).

* Corresponding author: sygletou@fisica.unige.it

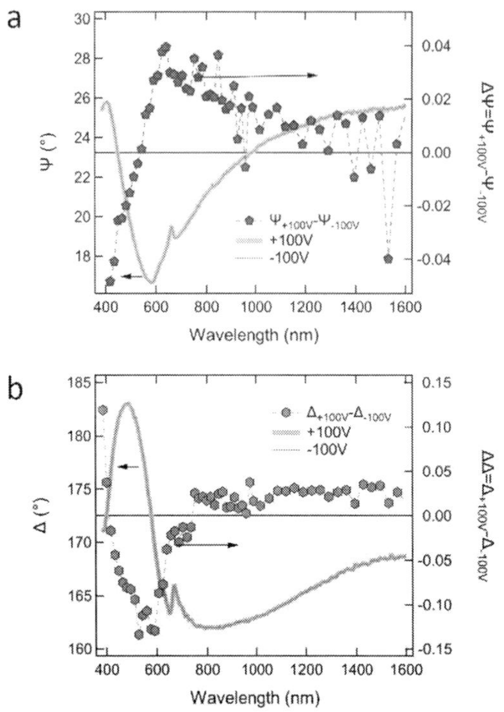

Fig. 1. a (b): Ellipsometry spectra Ψ(λ) (Δ(λ)) of 0.2 at.% AZO on top of STO (110) under bias (orange line: +100V, cyan line: -100V). Red (indigo) squares: Difference spectra δΨ(λ) (Δ(λ)), defined as Ψ(+100V)-Ψ(-100V) (Δ(+100V)-Δ(-100V)). All spectra were acquired at T=4.2 K with an incidence angle of 45°.

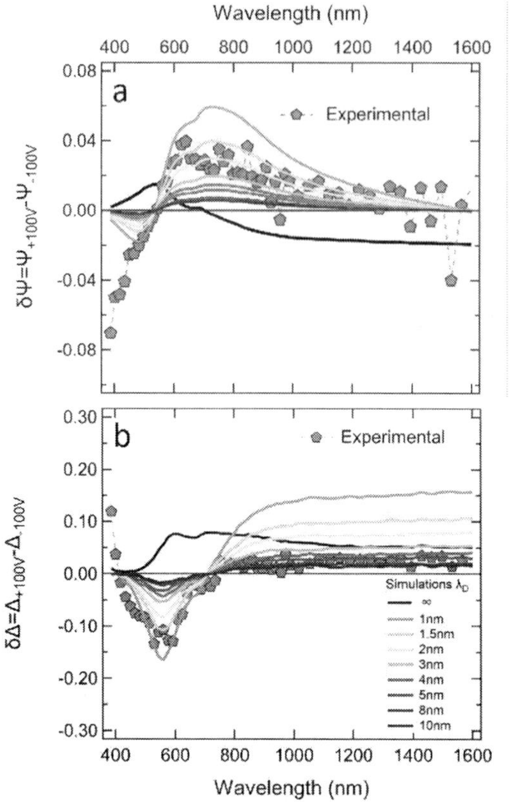

Fig. 2. a (b): Magenta (indigo) symbols: experimental difference spectra δΨ (δΔ)) as a function of wavelength, at T=4.2 K. Solid lines: calculated difference spectra obtained by modelling the charge injection/depletion in the AZO film. Different colours represent different values of the Debye length in the AZO film.

An estimate of the gating-induced charge variation in the AZO can be obtained modelling the system as a parallel-plate capacitor, where the STO acts as the dielectric, and the metallic contact at the back of the substrate and the AZO film are the two electrodes. Then , we assume that the injected/depleted charge density exhibit an exponential depth grading, that can be written as:

$$N_e(z) = N_0 \pm \left(\frac{\sigma_e}{\lambda_D}\right) e^{-\frac{z}{\lambda_D}} \qquad (1)$$

where σ_e is the surface charge density on the plates of the capacitor, N_0 is the charge density of unbiased AZO, λ_D is the Debye length, that accounts for the screening length inside the AZO while the ± case corresponds to the injection/depletion case, respectively. Knowing $N_e(z)$, it becomes possible to simulate the optical response of the system, just applying a corresponding variation to the Drude term (Figure 2).

3 Conclusions

We have performed an in-operando optical spectroscopy investigation of the AZO/STO(110) system under field effect. The optical data shown here, and the related modelling, represent a great enrichment with respect to electrical-transport measurements alone, contributing to a deeper understanding of the phenomena.

This project has received funding from the European Union's Horizon 2020 research and innovation programme under the Marie Skłodowska-Curie grant agreement no. 799126.

References

1. M. Sygletou, F. Bisio, S. Benedetti, P. Torelli, A. D. Bona, A. Petrov, M. Canepa, J. Vac. Sci. Technol. B **37**, 061209 (2019).
2. H. Yuan,H. Shimotani,A. Tsukazaki,A. Ohtomo,M. Kawasaki,Y. Iwasa, Adv. Funct. Mater. **19**, 1046 (2009)

EPJ Web of Conferences **266**, 09010 (2022)
EOSAM 2022
https://doi.org/10.1051/epjconf/202226609010

Tunable photoconductive devices based on graphene/WSe_2 heterostructures

Hongyu Tang[1] and *Giulia* Tagliabue[1*]

[1]Laboratory of Nanoscience for Energy Technologies (LNET), STI, École Polytechnique Fédérale de Lausanne, 1015 Lausanne, Switzerland

Abstract Optoelectronic tunability in van der Waals heterostructures is essential for their optoelectronic applications. In this work, tunable photoconductive properties were investigated in the heterostructures of WSe2 and monolayer graphene with different stacking orders on SiO_2/Si substrates. Here, we demonstrated the effect of the material thickness of WSe_2 and graphene on the interfacial charge transport, light absorption, and photoresponses. The results showed that the WSe_2/graphene heterostructure exhibited positive photoconductivity after photoexcitation, while negative photoconductivity was observed in the graphene/WSe_2 heterostructures. The tunable photoconductive behaviors provide promising potential applications of van der Waals heterostructures in optoelectronics. This work has guiding significance for the realization of stacking engineering in van der Waals heterostructures.

1 Introduction

Photo-excited electron-hole pairs usually increase the carrier density in the semiconductor, thereby increasing the channel conductivity and resulting in positive photoconductivity (PPC)[1]–[3]. However, in some low-dimensional materials, such as graphene, two-dimensional (2D) heterostructures, and nanowires, the phenomenon of negative photoconductivity (NPC), a decrease in electrical conductivity under light illumination, has been observed[4]–[6]. It mainly originates from the trapping of photocarriers by intrinsic or interfacial defects. Tunable photoconductivity could broaden the application in low-power, high-speed frequency-responsive photodetectors, and may also be useful for potential applications in optical communications. In recent years, the transition between NPC and PPC has been demonstrated by controlling the doping concentration, gate voltage, wavelength, and laser power[7]–[9]. However, there are few studies on the effects of the stacking order and the thickness of 2D on photoconductivity switching.

Van der Waals (vdW) heterostructures composed of graphene (Gr) and transition metal dichalcogenides (TMDCs) can achieve strong optoelectronic responses and wide photo response regions[7], [10]–[12]. Gr/TMDCs heterojunction devices based on photogating effect provide ultra-high gain for devices due to the trapping effect of TMDCs on photogenerated carriers[13], [14]. The gate-tunable phenomenon of NPC caused by photogating has been reported in Gr/TMDCs heterojunctions as well as 2D p–g–n heterojunction. Liu *et al.*[15]constructed a $MoSe_2$/Gr. vdW heterojunction which shows gate bias and laser power dependence NPC. Kim *et al.*[16] fabricated $MoSe_2$/Gr heterostructure device, where the switching from one state to the other depends on the light intensity as well as gate bias. The Gr/h-BN/MoS_2 vdW heterostructure device is time and gate voltage-dependence of the photoconductance[10]. Xing *et al.* [17] found that the WSe_2/Gr heterostructure exhibits ultrafast interfacial charge transfer, and the stacking order of Gr and WSe_2 can induce terahertz photoconductivity switching after photoexcitation. However, this phenomenon exists only in ultrafast photoelectric responses through pump–probe technique, there is no report of tunable photoconductivity in WSe_2/Gr heterostructure-based photodetectors.

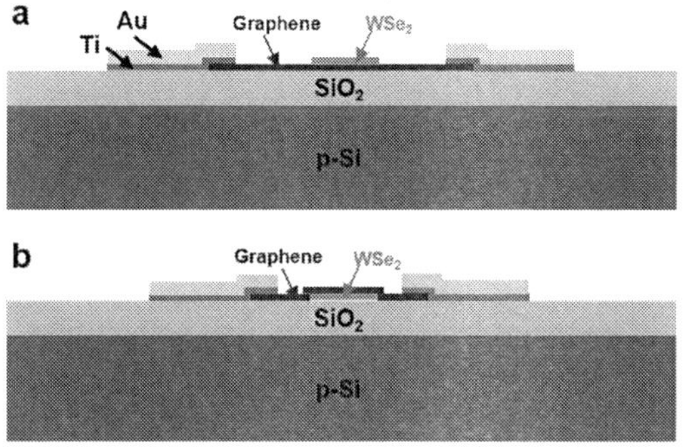

Fig. 1. Schematic diagram of the a) WSe_2/Gr and b) Gr/WSe_2 photoconductive devices.

In this work, we investigate the photoconductivity in 2D heterojunctions with monolayer (ML) graphene and few layer (FL) WSe_2 with different stacking orders and different thickness of WSe_2 on a SiO_2/Si substrate (**Fig. 1**). Firstly, we analyse the absorption of two kinds of heterojunctions comprising WSe_2/Gr and Gr/WSe_2 to find out the highest photoresponse wavelength. Next, we identify the neutral Fermi-Dirac point, output, and transfer curves at various gate bias (**Fig. 2**). Thus, the photo-doping level can be identified by analysing the responsivity of the devices under illuminations with different laser power.

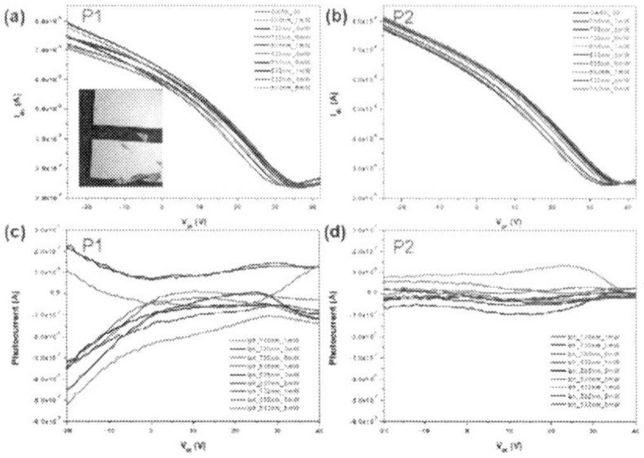

Fig. 2. The transfer curves of the a) graphene and b) Gr/WSe_2 heterojunction region in Gr/WSe_2 devices. The photocurrent of the a) graphene and b) Gr/WSe_2 heterojunction region in Gr/WSe_2 devices in different gate bias.

References

[1] H. Fang and W. Hu, "Photogating in Low Dimensional Photodetectors," *Adv. Sci.*, vol. 4, no. 12, p. 1700323, 2017, doi: https://doi.org/10.1002/advs.201700323.

[2] F. H. L. Koppens, T. Mueller, P. Avouris, A. C. Ferrari, M. S. Vitiello, and M. Polini, "Photodetectors based on graphene, other two-dimensional materials and hybrid systems," *Nat. Nanotechnol.*, vol. 9, no. 10, Art. no. 10, Oct. 2014, doi: 10.1038/nnano.2014.215.

[3] "Photodetectors of 2D Materials from Ultraviolet to Terahertz Waves - Qiu - 2021 - Advanced Materials - Wiley Online Library." https://onlinelibrary.wiley.com/doi/full/10.1002/adma.202008126 (accessed Jun. 02, 2021).

[4] Y. X. Boyao Cui(崔博垚) and Y. X. Boyao Cui(崔博垚), "Negative photoconductivity in low-dimensional materials," *Chin. Phys. B*, vol. 30, no. 2, pp. 28507–0, Jan. 2021, doi: 10.1088/1674-1056/abcf41.

[5] B. H. Kim, S. H. Kwon, H. H. Gu, and Y. J. Yoon, "Negative photoconductivity of WS2 nanosheets decorated with Au nanoparticles via electron-beam irradiation," *Phys. E Low-Dimens. Syst. Nanostructures*, vol. 106, pp. 45–49, Feb. 2019, doi: 10.1016/j.physe.2018.10.008.

[6] Y. Wang *et al.*, "Negative Photoconductance in van der Waals Heterostructure-Based Floating Gate Phototransistor," *ACS Nano*, vol. 12, no. 9, pp. 9513–9520, Sep. 2018, doi: 10.1021/acsnano.8b04885.

[7] K. Roy *et al.*, "Graphene–MoS2 hybrid structures for multifunctional photoresponsive memory devices," *Nat. Nanotechnol.*, vol. 8, no. 11, Art. no. 11, Nov. 2013, doi: 10.1038/nnano.2013.206.

[8] C. Biswas *et al.*, "Negative and Positive Persistent Photoconductance in Graphene," *Nano Lett.*, vol. 11, no. 11, pp. 4682–4687, Nov. 2011, doi: 10.1021/nl202266h.

[9] A. Grillo *et al.*, "Coexistence of Negative and Positive Photoconductivity in Few-Layer PtSe2 Field-Effect Transistors," *Adv. Funct. Mater.*, vol. 31, no. 43, p. 2105722, 2021, doi: 10.1002/adfm.202105722.

[10] T. Ahmed, K. Roy, S. Kakkar, A. Pradhan, and A. Ghosh, "Interplay of charge transfer and disorder in optoelectronic response in Graphene/hBN/MoS2 van der Waals heterostructures," *2D Mater.*, vol. 7, no. 2, p. 025043, Mar. 2020, doi: 10.1088/2053-1583/ab771f.

[11] Q. Bao and K. P. Loh, "Graphene Photonics, Plasmonics, and Broadband Optoelectronic Devices," *ACS Nano*, vol. 6, no. 5, pp. 3677–3694, May 2012, doi: 10.1021/nn300989g.

[12] M. Gmitra and J. Fabian, "Graphene on transition-metal dichalcogenides: A platform for proximity spin-orbit physics and optospintronics," *Phys. Rev. B*, vol. 92, no. 15, p. 155403, Oct. 2015, doi: 10.1103/PhysRevB.92.155403.

[13] A. Gao *et al.*, "Gate-tunable rectification inversion and photovoltaic detection in graphene/WSe2 heterostructures," *Appl. Phys. Lett.*, vol. 108, no. 22, p. 223501, May 2016, doi: 10.1063/1.4953152.

[14] H. Jiang *et al.*, "Enhanced Photogating Effect in Graphene Photodetectors via Potential Fluctuation Engineering," *ACS Nano*, Feb. 2022, doi: 10.1021/acsnano.1c10795.

[15] B. Liu *et al.*, "High responsivity and near-infrared photodetector based on graphene/$MoSe_2$ heterostructure," *Chin. Opt. Lett.*, vol. 17, no. 2, p. 020002, Feb. 2019, doi: 10.1364/COL.17.020002.

[16] H. J. Kim *et al.*, "Photoconductivity Switching in MoTe2/Graphene Heterostructure by Trap-Assisted Photogating," *ACS Appl. Mater. Interfaces*, vol. 12, no. 34, pp. 38563–38569, Aug. 2020, doi: 10.1021/acsami.0c09960.

[17] X. Xing *et al.*, "Influence of a substrate on ultrafast interfacial charge transfer and dynamical interlayer excitons in monolayer WSe2/graphene heterostructures," *Nanoscale*, vol. 12, no. 4, pp. 2498–2506, Jan. 2020, doi: 10.1039/C9NR09309E.

Nature engineered metasurfaces: spin-to-orbital angular momentum conversion in diatom frustules

Edoardo De Tommasi[1,*], *Maria Antonietta* Ferrara[1], *Giuseppe* Coppola[1], and *Gianluigi* Zito[1]

[1]National Research Council - Institute of Applied Sciences and Intelligent Systems "Eduardo Caianiello", Unit of Naples, Via P. Castellino 111, I-80131 Naples, Italy

Abstract. Evolution provided, through eras, several animal, vegetal, and protist species with sub-micrometric constituent structures able to manipulate light at the nanoscale in non trivial ways. In particular, diatoms are single-celled microalgae enclosed in a porous silica shell, the frustule, perforated by regular patterns of micro- and nano-pores and whose functionalities comprise mechanical stability, sorting of nutrients from harmful agents and optimization of sunlight harvesting. Photonic properties of frustules include focusing, photoluminescence, and optical activity, among others. In the present work we show preliminary results concerning the ability of single valves of *Arachnoidiscus ehrenbergii* diatom frustules to manipulate incoming, circularly polarized radiation in such a way to generate light beams provided with orbital angular momentum (optical vortices). The combination of cross-polarization imaging in different spectral ranges, polarization sensitive digital holographic imaging (PSDHI), and interferometry allowed characterizing the valves and detecting the presence of spin-orbit coupling induced by their ultrastructure.

1 Introduction

Several technological fields such as optical trapping, super-resolution imaging, multiplexed optical communication, and quantum computing can take great advantage by the use of optical vortices (OVs), i.e. light beams provided with a helical phase $\exp(-il\phi)$ and carrying an orbital angular momentum (OAM) equal to $l\hbar$ per photon, with l any integer [1]. OVs can be obtained exploiting phase discontinuities (e.g. by means of spiral phase plates [2], fork holograms [3] or spatial light modulators [4]), inducing spin-to-orbital angular momentum conversion by inhomogeneous birefringent elements (q-plates) [5] or by complex metallic or dielectric metasurfaces [6], or even by photonic crystal slabs supporting bound states in the continuum [7]. Surprisingly, we observed spin-orbit coupling induced by a nanostructured biomaterial. Several organisms and biological systems indeed (viruses, insects, bird plumage, flora, plants) are provided with structures at the micro- and nano-scale able to manipulate light in complex, non trivial ways. Some of the photonic nanostructures found in nature can cause coherent or incoherent scattering, others act as one-dimensional multilayer reflectors, polarization-selective reflectors, two-dimensional diffraction gratings, two- and three-dimensional photonic crystals or even hyperuniform photonic networks [8]. In particular, diatoms are ubiquitous microalgae enclosed in a porous hydrogenated silica shell, the frustule, characterized by regular patterns of pores whose dimensions range between some tens of nanometers up to about one micron, according to the species and the location within the frustule itself [9]. Several optical effects induced by the interaction of light with frustules have been observed, such as focusing, mode guiding, photoluminescence, and optical activity and applications in super-lensing, plasmonics, and optical sensing and biosensing have been successfully performed [10].

We studied the optical activity of single valves of *Arachnoidiscus ehrenbergii* diatom frustules, characterized by inner radiating *costae* connecting an inner flange to the mantle edge. Crossed-polarization imaging, polarization sensitive digital holography, and interferometry measurements revealed the ability of the valve to generate OVs due to spin-to-orbital angular momentum conversion of light at different wavelengths with different efficiencies.

2 Results and discussion

Single valves of *A. ehrenbergii* diatom frustules are characterized by a porous ultrastructure with pore dimensions ranging from some microns down to tens of nanometers, according to their position within the plate. The internal side of the valve hosts a system of ribs (*costae*) radiating from a flange around a central ring (see Fig.1a), giving rise to a birefringence characterized by an inhomogeneously oriented fast axis. Manipulation of polarization (optical activity) induced by the valve can be deduced from crossed-polarization imaging (see Fig.1b for incoherent, wide-spectrum incident radiation), where distinct intensity lobes associated to rotation of incident polarization and characteristic of a q-plate are clearly distinguishable. The lobes are visible also for coherent irradiation at single

*e-mail: edoardo.detommasi@na.isasi.cnr.it

wavelengths (λ = 406, 532 and 633 nm), the effect resulting more efficient for green incoming radiation.

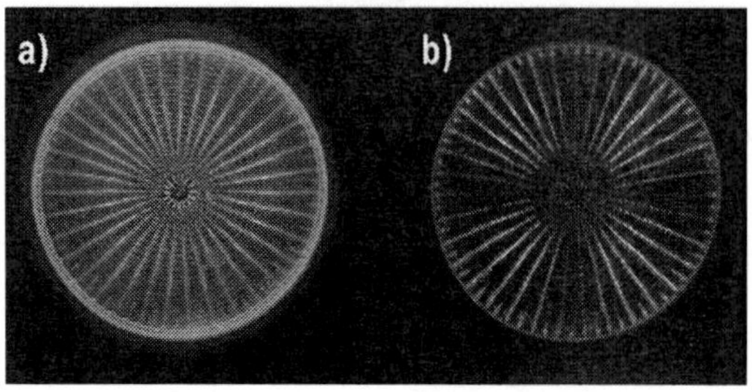

Figure 1. Dark field (**a**) and crossed-polarization dark field (**b**) images of a single valve of *A. ehrenbergii* diatom frustule. Diameter of the valve $\simeq$ 270 μm.

The state of polarization (SoP) of the valve when irradiated by a He-Ne laser beam has been fully characterized by a polarization sensitive digital holographic imaging (PSDHI) technique, which allows retrieving the degree of polarization of the outcoming wave in terms of the Stokes parameters [11]. The results confirm, for red irradiation, a partial depolarization of the incoming light induced by the valve ultrastructure, confirming the lower efficiency in optical activity in this spectral window.

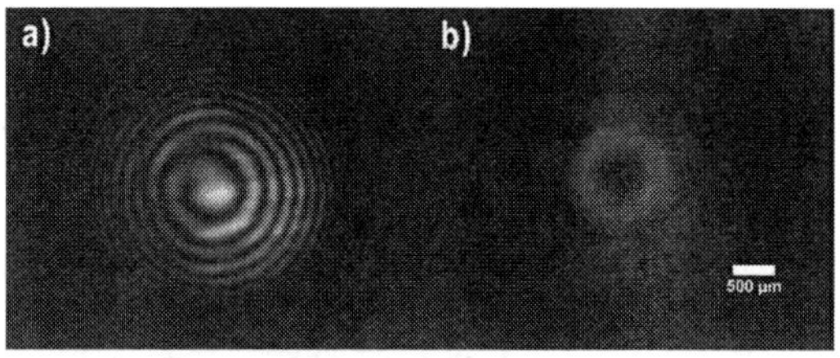

Figure 2. Interference pattern of the helical mode emerging from the diatom valve when illuminated by a LCP beam at λ = 532 nm, (**a**). Intensity distribution of the vortex obtained by blocking the reference beam (**b**).

In order to detect the presence of an OV generated by the interaction of light with the valve, the sample was first shined by a left-handed circularly polarized (LCP) laser beam at λ = 532 nm (where optical activity resulted more efficient), while the outcoming radiation (signal beam) was collected and superimposed to a reference beam after being properly filtered in order to allow the transmission of the sole light which undergoes spin-orbit coupling. The detected interference pattern shows a spiral shape which is associated to the presence of an OAM carried out by the signal beam (see Fig. 2a). Blocking the reference beam allows the detection of the intensity distribution of the vortex, with its peculiar doughnut-like shape (Fig. 2b).

3 Conclusions

Nanostructured, naturally occuring biomaterials can represent a low-cost, alternative way to obtain complex optical devices on a wide scale. We made recent observations on optical activity of single valves of *A. ehrenbergii*, which present an internal system of *costae* radiating from a central flange towards the mantle edge. As in the case of an artificially designed, inhomogenous birefringent waveplate with radial orientation of the fast axis, also *A. ehrenbergii* valves show the ability to induce spin-to-orbital angular momentum conversion of incoming, circularly polarized light with generation of an OV. The effect is more efficient around λ = 532 nm. While the biological significance of this phenomenon is far to be understood, being nevertheless probably related to light-plastid coupling, the ability of natural nanostructures to manipulate light in non trivial ways can stimulate the development of novel solutions for a new generation of bioinspired optical devices.

References

[1] Y. Shen, X. Wang, Z. Xie, C. Min, X. Fu, Q. Liu, M. Gong, X. Yuan, Light: Science & Applications **8**, 1 (2019)

[2] V.Y. Bazhenov, M. Vasnetsov, M. Soskin, Jetp Lett **52**, 429 (1990)

[3] M. Beijersbergen, R. Coerwinkel, M. Kristensen, J. Woerdman, Optics communications **112**, 321 (1994)

[4] A.S. Ostrovsky, C. Rickenstorff-Parrao, V. Arrizón, Optics letters **38**, 534 (2013)

[5] L. Marrucci, C. Manzo, D. Paparo, Physical review letters **96**, 163905 (2006)

[6] T. Liu, W. Li, Y. Meng, H. Ma, W. Tang, G. Li, H. Shi, S. Sui, J. Wang, S. Qu, Advanced Optical Materials **10**, 2102638 (2022)

[7] B. Wang, W. Liu, M. Zhao, J. Wang, Y. Zhang, A. Chen, F. Guan, X. Liu, L. Shi, J. Zi, Nature Photonics **14**, 623 (2020)

[8] E. De Tommasi, E. Esposito, S. Romano, A. Crescitelli, V. Di Meo, V. Mocella, G. Zito, I. Rendina, La Rivista del Nuovo Cimento **44**, 1 (2021)

[9] F.E. Round, R.M. Crawford, D.G. Mann, *Diatoms: biology and morphology of the genera* (Cambridge university press, 1990)

[10] E. De Tommasi, Journal of Spectroscopy **2016** (2016)

[11] G. Coppola, M.A. Ferrara, Applied Sciences **10**, 4520 (2020)

Hybrid optical measurement technique for detection of defects in epitaxially grown 4H-SiC layers

Elena Ermilova[1, *], *Matthias* Weise[1], and *Andreas* Hertwig[1]

[1]Bundesanstalt für Materialforschung und -prüfung (BAM), Division 6.1 Surface Analysis and Interfacial Chemistry, D-12203, Unter den Eichen 44 - 46, Berlin, Germany

Abstract. Recent developments in power electronics require the use of new wide bandgap compound semiconductor. We demonstrate the use of the ellipsometry and white light interference microscopy to detect defects in epitaxially grown SiC layers on SiC substrates. Such hybrid optical metrology methods can be used to better understand the mechanism of the development of the defects as well as their effects on the material's optoelectronic properties.

1 Introduction

Silicon carbide (SiC) is one of the most promising materials for power electronics due to its outstanding properties and commercially availability. However, defects in the SiC substrates or homoepitaxial SiC layers can affect the performance of electronic devices in a serious manner [1]. Some types of them, so called triangular "killer defects" in epilayers have been known to even make the operation of device impossible. To effectively exploit these materials, we need to understand mechanisms of defect formation. Moreover, the density of different types of defects is supposed to play a crucial role. In this contribution we demonstrate how optical methods such ellipsometry and white light interference microscopy can be applied for fast and non-contact investigation of epitaxially grown 4H-SiC layers on 4H-SiC substrates.

2 Ellipsometry measurements of 4H-SiC epitaxial layers

Ellipsometry can be used for the accurate characterization of thin layers in a broad spectral range. This is a powerful technique that measures changes in polarization during light reflection from the sample surface. The obtained ellipsometric transfer quantities Ψ and Δ represent material fingerprints. To determine the thickness of thin layers as well as optical/dielectric properties of layer materials a model-based fitting procedure is applied [2, 3]. Imaging ellipsometry combines the capabilities of ellipsometry and optical microscopy. Due to polarization sensitivity and very high contrast for surface structures, the different areas within a defect can be detected.

We have investigated 4H-SiC epilayers with a 12 µm thickness on 4H-SiC substrates using an EP4 imaging ellipsometer (Accurion).

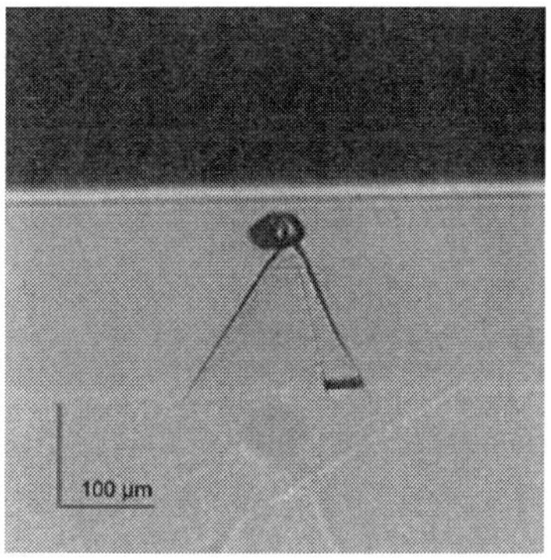

Fig. 1 Ellipsometric microscope image of a triangular defect on 12 µm 4H-SiC epilayer grown on 4H-SiC substrate

In the detected defect (Fig. 1) the spherical and triangle areas can be distinguished. The spherical area is reported to be caused by the landing of particles during the epitaxy process [4]. In the triangular area terrace-like structures are formed [4]. Using Raman spectroscopy, it was found previously that the triangular regions content 3C-SiC polytype inclusions. [4, 5] The high lateral resolution and the mapping of the physical quantities of interest over selected regions provide better opportunities for qualitative and quantitative investigations of defects using

* Corresponding author: elena.ermilova@bam.de

imaging ellipsometry. Line profiles of measured Ψ and Δ at certain wavelengths and angles of incidence enables first conclusions about the structure of different defect areas. Spatial resolution for these quantities were found to depend on the wavelength. In addition, spectroscopic measurements in wide spectral range were carried out for selected regions of interest within the defect area.

3 White light interferometry measurements of 4H-SiC epitaxial layers

Information about the surface structure contributes to understanding of the formation mechanisms of defects in 4H-SiC epilayers and help to reduce defect density already during production process and achieve the high quality of epilayers. Therefore, for investigation of the surface structure of the triangular defects white light interference microscopy (WLIM) was used.

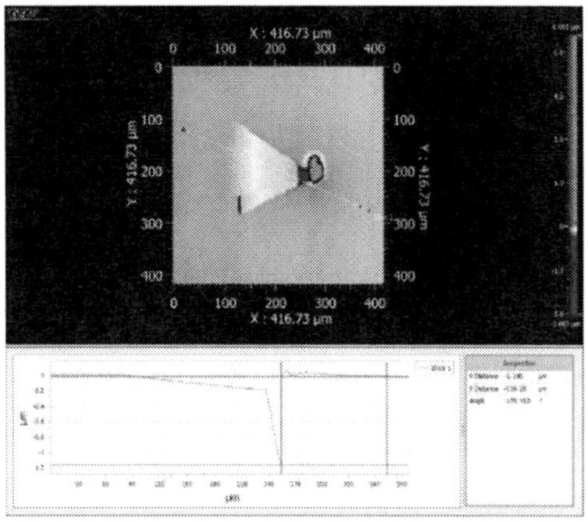

Fig.2 WLIM image and surface profile of a triangular defect on 12 µm 4H-SiC epilayer grown on 4H-SiC substrate

Measurements were carried up with the microscope Zygo NewView with a 20x objective (Fig. 2). The resolution was ~ 1 nm in vertical and ~ 0,7 µm in horizontal directions. These measurements brought into light some structures in the 4H-SiC epilayer, which were not visible by imaging ellipsometry.

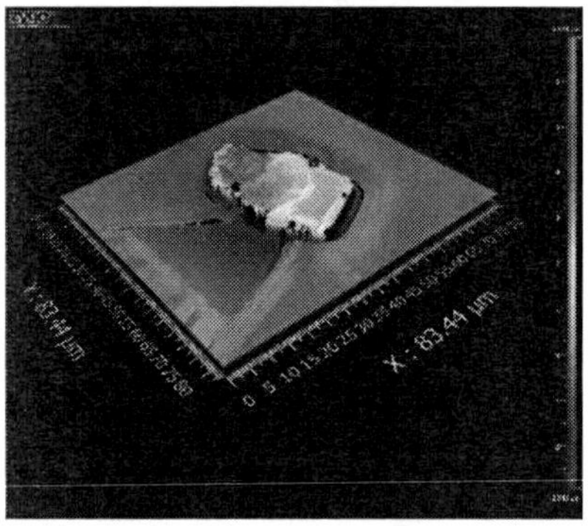

Fig. 3 WLIM 3D-image of the spherical area of defect on 12 µm 4H-SiC epilayer grown on 4H-SiC substrate

It was observed that the spherical area of the defects ("head") extends several µm above the layer surface (Fig. 3). It has a step-like structure, that may be expected for the case than a particle lands on the surface during the epitaxy process. The triangle extensions go up to 1 µm deep into the layer and become even steeper near the "head". They have different height profiles and slopes.

These non-contact 3D topography investigations can provide detailed information about morphology and propagation of defects in epitaxial 4H-SiC layers.

4 Conclusion

Hybrid measurement technique using spectroscopic and imaging ellipsometry as well as white light interference microscopy enables fast and non-destructive detection and traceable characterization of complex defects in epitaxially grown 4H-SiC layers. Ellipsometry and White Light Interference Microscopy are accredited measurement methods according to DIN EN ISO/IEC 17025 [6]. Therefore, both methods are very promising candidates for use already during production of SiC epilayers and for subsequent industrial quality control.

The PowerElec project has received funding from the EMPIR programme co-funded by the Participating States and from European Union's Horizon 2020 research and innovation programme.

References

1. M. A. Fraga, M. Bosi, and M. Negri, *Silicon Carbide in Microsystem Technology — Thin Film Versus Bulk Material*, (Advanced Silicon Carbide Devices and Processing. London, United Kingdom: IntechOpen, Online, 2015)
2. H. Fujiwara, *Spectroscopic Ellipsometry: Principles and Applications*, (John Wiley & Sons, 2007)
3. ISO 23131:2021 *Ellipsometry – Principles*
4. J. Guo, Y. Yang, B. Raghothamachar, T. Kim, M. Dudley, J. Kim, J. Cryst. Growth, **480**, (2017)
5. H. K. Kim, S. I. Kim, S. Kim, N. S. Lee, H. K. Shin, C. W. Lee, Nanoscale, **12**(15), (2020)
6. DIN EN ISO/IEC 17025: 2018 *General requirements for the competence of testing and calibration laboratories.*

https://doi.org/10.1051/epjconf/202226610002

Retardance measurement by spatially probing the sample with optical vortices

Tomáš Fordey[1,*] and *Petr* Schovánek[1]

[1]Department of Optics, Faculty of Science, Palacký University Olomouc, 771 46 Olomouc, Czech Republic

Abstract. Modern optical systems utilize various degrees of freedom, such as polarization, for shaping and controlling the light. Common representative of such a component is spatial light modulator (SLM), consisting of liquid crystal display, allowing for imposing predetermined retardation with given orientation of optical axis of anisotropy. Therefore, it is widely used for polarization coded phase shifting, polarization splitting of wavefront in digital holography etc. Narrowing tolerance in optical experiments puts higher demands on precise setting of the modulator and the parameters set. Consequently, measuring such devices and their parameters is essential for proper functionality. We present a single shot, common path method for measuring retardance map of the modulator, based on spatial probing the modulator with point images of spatially coherent light source and transforming them to optical vortices.

1 Introduction

Phase of light is important quantity providing optical metrologist with plenty of information about various properties of transparent samples. In phase measurements, different degrees of freedom, such as coherence and polarization of light are used to control the light and to achieve the measurement objectives with higher precision and accuracy. While the coherence of light is often at least partially predetermined by the nature of measurement taken and leaves only limited, nevertheless important space for being adjusted, polarization of light provides the metrologist with control of spatially varying phase difference between independent optical channels in common optical path. Recently, Jones matrix microscopy [1, 2], enabling access not only to the phase of the sample but also its polarization state, emerges in optics. Either for controlling and verifying the polarization manipulations as a measurement tool or for a measurement itself as in the case of Jones matrix microscopy, methods for measuring spatially distributed anisotropy are of increasing demands.

A well-known principle of double helix point spread function (DH PSF) [3] allows for measuring the phase difference between optical vortices. It has been already proved as a valuable tool for measuring topography of reflective as well as diffuse samples either by sampling the surface with point images created by microhole plate [4] or by marking the surface with fluorescent nanobeads [5]. We propose a different approach to spatially probing the sample via diffraction grating, providing us with equidistant point images, and apply it for the retardance measurement of a sample with a priori known orientation of the optical axis of anisotropy, the SLM.

2 Results and discussion

Experimental setup is as follows. Fibre coupled laser light is collimated and set diagonally polarized by means of a polarizer LP and half-wave plate HWP. The light then passes through diffraction grating, whose orders are focused on the SLM. Afterwards, the reflective SLM is reimaged by means of 4-f system with Q-plate [6] in its Fourier plane, preceded by quarter-wave plate QWP oriented by 45 ° relative to the axis of anisotropy . The quarter-wave plate converts the horizontally and vertically polarized light into left- and right-handed polarized light, which is subsequently converted into optical vortices with unit topological charge of opposite sign. Finally, the light is brought to interference by linear polarizer and detected on CMOS camera.

Fig. 1. Experimental setup

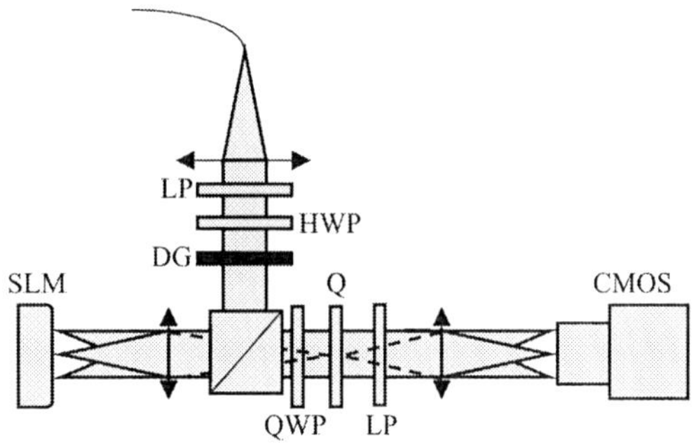

* Corresponding author: fordey@optics.upol.cz

Fig. 2. Central part of the image of the SLM. The SLM is addressed with a phase map containing a phase jump in the top left corner (left figure) and reconstructed retardance jump (right figure)

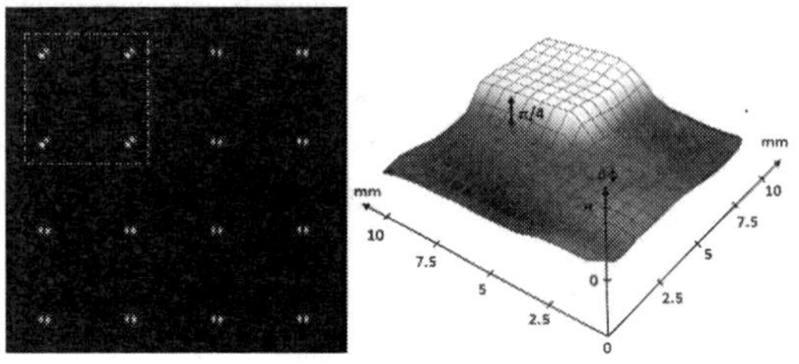

It is worth notice that the retardance can be, to a decent level, briefly evaluated even visually from the frame, of course for the precise evaluation the numerical processing is required. Diffraction efficiency of individual orders is not of a concern in principle, however, to maximize signal/noise ratio across the frame, it is better if there is same energy in individual PSFs. Advantage of the method is that optical aberrations of the imaging system used should not have impact on the accuracy of the measurement, nevertheless severe aberration can and will affect the precision of the method in various ways.

Fig. 3. Comparison of computer addressed map (left figure) and evaluated retardance (right figure).

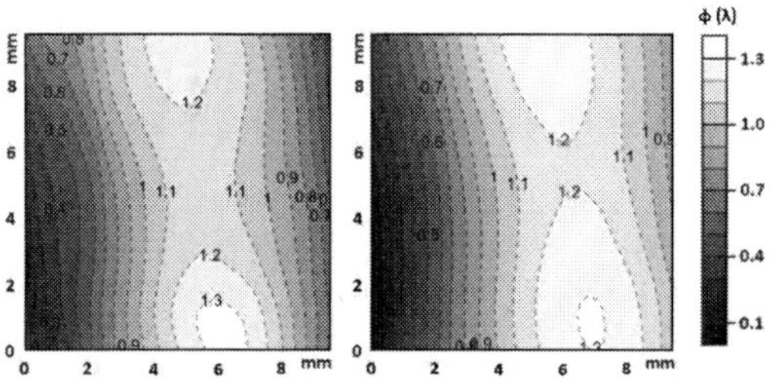

3 Conclusion

We presented a novel, single shot, common path method for measuring retardance map of spatial modulator. Albeit the principle of projecting the phase difference to the angular orientation of the double helix PSF is well known, we proposed and verified a new application of the phenomenon. Measuring spatial distribution of polarization properties of the sample is rapidly emerging area in optical microscopy and although our contribution provides only part of the information needed for determination of sample anisotropy and its application is therefore restricted, it proved as simple and utterly robust method.

We would like to thank Meopta-optika s.r.o. for providing us with their custom-designed diffraction grating.

References

1. Z. Wang, L. J. Millet, M. U. Gillette, G. Popescu, Opt. Lett., **33** (2008), p.1270-1272
2. Y. Kim, J. Jeong, J. Jang, M. W. Kim, Y. Park, Opt. Ex., **20** (2012), p. 9948-9955
3. A. Greengard, Y. Y. Schechner, R. Piestun, Opt. Lett., **31**, (2006), p.181-183
4. P. Bouchal, L. Štrbková, Z. Dostál, Z. Bouchal, Opt. Exp., **25**, (2017), p. 21428-21443
5. P. Schovánek, P. Bouchal, Z. Bouchal, Opt. Lett., **45**, (2020), p. 4468-4471
6. L. Marrucci, J.of Nanoph., **7**, (2013), p. 078598

Characterisation of nanowire structures with scatterometric and ellipsometric measurements

Jana Grundmann*, *Tim* Käseberg, and *Bernd* Bodermann

Physikalisch-Technische Bundesanstalt (PTB), Bundesallee 100, 38116 Braunschweig, Germany

Abstract. Nanowire structures arranged in a hexagonal lattice are to be characterized in terms of their diameter, height and pitch. A scatterometer and an imaging Mueller matrix ellipsometer, which is a combination of a commercial Mueller matrix ellipsometer and a microscope, have been used as measurement tools. These measurements are supported by numerical simulations using the finite element method to characterize the structure parameters.

1 Introduction

Within a European research project, the suitability of nanowire structures for energy harvesters of renewable sources, like solar, heat and movement, are investigated. These energy harvesters have great potential to address our global energy problem. Thus, nanowire based energy harvesters are particularly interesting due to their small physical size and high surface to volume ratio. For example, nanowire energy harvesters as solar cells have been shown to increase their efficiency from currently 17,8% up to 46,7% [1].

To achieve this, it is important to characterise dimensional parameters such as diameter, height and pitch of the nanowires to gain a better understanding of the performance of individual nanowires and an overall device, and to enable reliable and cost-effective device production. For this purpose, efficient and principally in-line capable measurement methods are required. Fast optical methods such as scatterometry and ellipsometry are very promising for this purpose. However, classical scatterometers and ellipsometers as integrating methods only provide global information about a nanowire array due to the spot size of typically 1 mm^2. An imaging ellipsometer in contrast is combining classical ellipsometry with microscopy. The data is no longer integrated over the entire illumination spot size, but evaluated for each pixel of the camera, which corresponds typically to about 1 μm^2 in the object space. Thus, areas much smaller than the illumination spot size and even non-periodic structures can be measured. In this way, data about the homogeneity of the nanowires over a large area as well as about the local polarization properties of individual nanowires can be obtained. Furthermore, full Mueller matrix ellipsometry providing the complete polarisation encoded information about the nanostructures has been shown to provide the best option to enable nanoscale dimensional metrology [2]. Thus, combining imaging and non-imaging methods add both fast integral as well as local information about the samples and therefore both are investigated here to develop the required nano-dimensional metrology. Both methods require numerical simulations to solve Maxwells equations and the inverse diffraction problem [3]. We use a finite-element-based Maxwell-solver for this purpose. In the following we present the experimental setups, the models used for the data analysis and first measurement examples for both investigated methods.

2 Experimental setup

In the first approach, measurements are performed with a goniometric DUV scatterometer. Thereby it can be checked if the distribution of the nanowires on the sample is stochastic or deterministic. In case of stochastically distributed nanowires, the surface is too inhomogeneous to perform measurements. For the second approach, an imaging Mueller matrix ellipsometer is used for the measurements to measure individual nanowires or local features in periodic nanowire gratings. In both cases, simulations are performed using JCMwave's finite element method to support the evaluation of the measured data with respect to the structure parameters.

2.1. Scatterometer

The used measurement device is a DUV scatterometer of own construction. The device has a high degree of automation, the detector is locating on a nearly 360° rotating arm, so that measurements in transmission and reflection are possible. The working wavelength is 266 nm and the illumination spot has a diameter of about 1 mm.

* Corresponding author: jana.grundmann@ptb.de

 https://doi.org/10.1051/epjconf/202226610003

2.2 Imaging Mueller matrix ellipsometer

For the Mueller matrix image measurements, an imaging Mueller matrix ellipsometer EP4 from Accurion is used (see figure 1). The Mueller matrix can be measured in a wavelength range from 400 nm to 1000 nm, the objectives have a magnification of 10, 20 and 50 and the angle of incidence can vary between 38° and 90°. The detector is a CCD camera with 1392 × 1040 pixel and a pixel size of 6,45 µm × 6,45 µm. A second compensator has been added to the commercial device in order to be able to measure the whole 4 × 4 Mueller matrix.

Fig. 1. Imaging Mueller matrix ellipsometer from Accurion.

2.3 Model structure

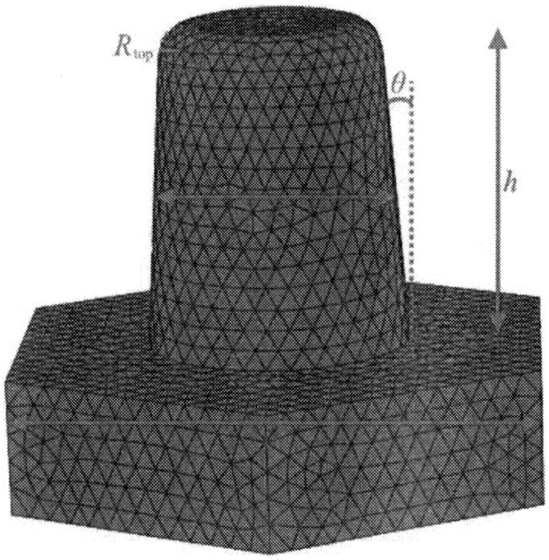

Fig. 2. Schematic layout of the used model with structure parameters: height h, pitch p, diameter at half maximum d, sidewall angle θ, top edge rounding R_{top}.

A sample from Eulitha AG in Switzerland is used for the measurements. The sample consists of a silicon substrate with silicon nanowires arranged as pillars on hexagonal lattice. For the model of the simulations in JCMwave the dimensions of the manufacturer are used:

- height h: 600 nm ± 15%
- pitch p: 1000 nm
- diameter d: 460 nm ± 10%
- sidewall angle θ: 2° - 6°

In addition, an edge rounding R_{top} with a radius of 50 nm is assumed. The final model with periodic boundaries can be seen in figure 2.

3 Measurements and simulations

Previous work [3, 4] has used 1D nanowire structures to demonstrate that measurements with a scatterometer provide reproducible results and that best fit results from simulations can be used to reconstruct the structure parameters. This procedure is extended for the scatterometric measurements on the sample presented here.

First simulations of the Mueller matrix images were performed (see figure 3). The images already provide meaningful results. When varying the structure parameters height, pitch, diameter, sidewall angle and top edge rounding, the individual matrix elements change, especially the off-diagonal elements. In this way, by comparing the measurements with the simulations in a non-linear optimisation the inverse problem can be solved to derive the searched structure parameters.

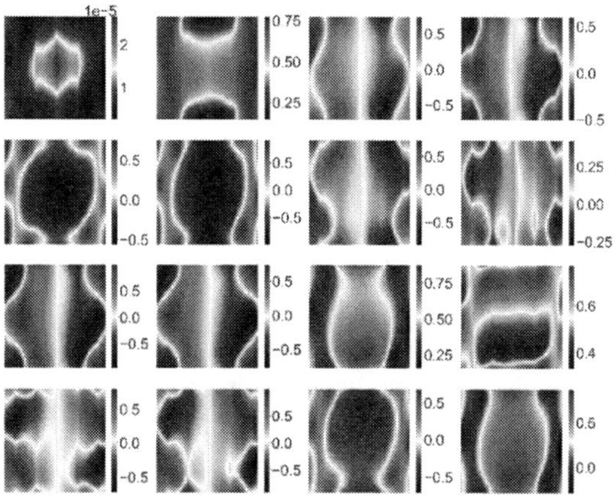

Fig. 3. Simulation of the Mueller matrix images at a wavelength of 500 nm with the structure parameters: h = 600 nm, p = 1000 nm, d = 460 nm, θ = 4°.

This project (19ENG05 NanoWires) has received funding from the EMPIR programme co-financed by the Participating States and from the Europeans Union's Horizon 2020 research and innovation programme.

References

1. J. E. M. Haverkort et al., Appl. Phys. Rev. **5**, 031106 (2018)
2. M. Wurm, et al., Opt. Express **28**, 8108-8131 (2020)
3. M. Henn, et al., Opt. Express **20**, 12771-12786 (2012)
4. J. Endres, et al., Proc. SPIE **9132**, 913208 (2014)

A virtual microscope for simulation of Nanostructures

Poul-Erik Hansen[1,*] *and Lauryna Siaudinyte*[2]

[1]Danish Fundamental Metrology A/S, DK-2970 Hørsholm, Denmark
[2]VSL – Dutch metrology institute, Thijsseweg 11, 2629 JA, The Netherlands

Abstract. Light-matter interplay is widely used for analyzing the topology of surfaces on small scales for use in areas such as nanotechnology, nanoelectronics, photonics, and advanced materials. Conventional optical microscope imaging methods are limited in resolution to a value comparable to the wavelength, the so-called Abbe limit, and cannot be used to measure nano-sized structures. Scatterometry is an optical method that can measure structures smaller than the wavelength. However, the relative uncertainties of the structure dimensions measured with scatterometry increase with decreasing structure size, and the industry is therefore looking for replacing simple intensity based scatterometry with a phase-sensitive measurement method such as coherent Mueller ellipsometry. In this work, we present a virtual microscope capable of simulating the coherent Mueller ellipsometry and scatterometry response from one-dimensional and two-dimensional periodic structures. Furthermore, we argue that coherent nonnormalized Mueller ellipsometry gives results with less uncertainties than standard normalized Mueller ellipsometry.

1 Introduction

Nanostructures have a wide array of applications in optics, diagnostics, food science, sensing, and process inspection monitoring. Some of these applications include enhancing waveguide coupling, improving linear encoders, making hyperspectral cameras and printing color images. Imaging technologies like Optical Microscopy (OM), Atomic Force Microscopy (AFM) and Scanning Electron Microscopy (SEM) are the dominating quality assessment technologies in low volume, high-cost nanoscale manufacturing, whereas scatterometry and Mueller ellipsometry are the preferred technologies for high volume manufacturing. However, the measurement accuracy for all of the above-mentioned technologies is decreasing with the ever-decreasing nanostructure sizes. OM cannot measure the shape of objects with lateral sizes less than one μm; AFM cannot accurately measure shape but can measure the nanostructure height if the separation width is longer than the tip width; lateral and vertical dimensions from SEM pictures are hard to obtain if the width of the borderline produced by the secondary electron becomes a significant part of the dimension to be measured. Scatterometry and Mueller ellipsometry can measure the shape of periodic nanostructures, however, the accuracy of the shape dimensions decreases with decreasing nanostructure sizes and increasing complexity. We proposed to use coherent Mueller ellipsometry for shape reconstruction of nanostructures.

Scatterometry and coherent scatterometry can be defined as the measurement and analysis of light diffracted by structures using fixed polarization settings. The scattered light is a signature which reflects the details of the structure itself. For a periodic device, the scattered light consists of distinct diffraction orders at angular locations specified by the grating equation. The fraction of the incident power diffracted into any order is sensitive to the shape and dimensional parameters of the structure and may therefore be used to characterize the structure itself. This is done using a mathematical model of the structure based on a priori information and a rigorous simulation of the light-structure interaction. The diffraction efficiency may be obtained by adding the diffraction orders as intensities (scatterometry) or as fields (coherent scatterometry). Mueller ellipsometry measures the polarization-dependent optical response from a sample.
Mueller ellipsometry may be further divided into two groups: Nonnormalized Mueller ellipsometers that measure all 16 Mueller matrix elements, and normalized Mueller ellipsometers in which the 16 Mueller matrix elements are normalized with the first Mueller matrix element m_{11}. For a microscope setup, Fig. 1, we may have multiple diffraction orders contributing to the field detected in one pixel. The sensitivity of coherent Mueller ellipsometry comes from the measurement of both the magnitude and phase of the Fresnel response from the sample.

* Corresponding author: peh@dfm.dk

 https://doi.org/10.1051/epjconf/202226610004

2 Virtual microscope

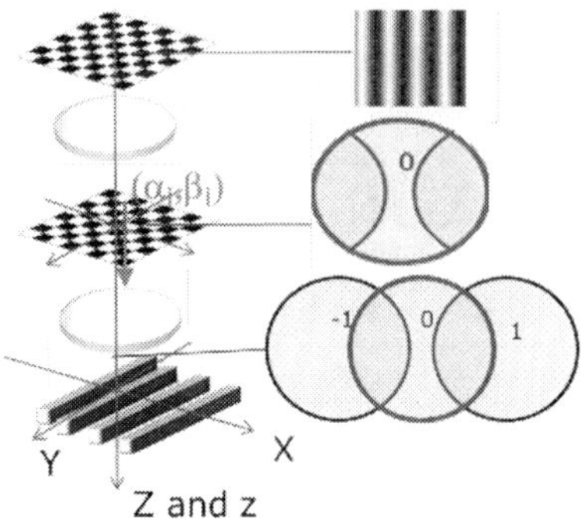

Fig. 1. Illustration of the basic principles of a microscope.

The working principle of a microscope may be explained in terms of how the scattered light from the surface is collected and processed by the microscope. The left column in Fig. 1 shows a schematic illustration of a grating (sample), lenses and planes (checkboard) in a microscope, while the right column shows the angular distribution of light before the first lens, the angular distribution after the first lens and the image after the second lens. The difference between the angular distribution before and after the first lens is due to the limited numerical aperture (NA) of the first lens, which sets the fundamental resolution limits for traditional optics. The virtual microscope works in the following way

1. The incident field is a normal incident wave with a normalized k-vector $(0,0,1)^T$ and the amplitude in the entrance pupil plane

2. The light beam coordinate system defined by the unit vectors for TE-amplitude, TM-amplitude and k-vector is used to describe the optical propagation

3. The response from the periodic structure is calculated as a Jones matrix using the RCWA method.

4. All the points that enter the same pixel in the exit pupil plane are added coherently if we consider coherent scatterometry or coherent Mueller ellipsometry and incoherently if we consider scatterometry.

5. The virtual microscope forms an image from the position depending angular distribution of light in the exit pupil plane. Each sample position gives one image point.

A silicon line grating with a pitch of 700 nm, height 190 nm, width 287 nm and sidewall angel of 90 degrees is investigated using incident light of 600 nm and an infinity corrected objective with a numerical aperture (NA) of 0.55.

3 Results and conclusion

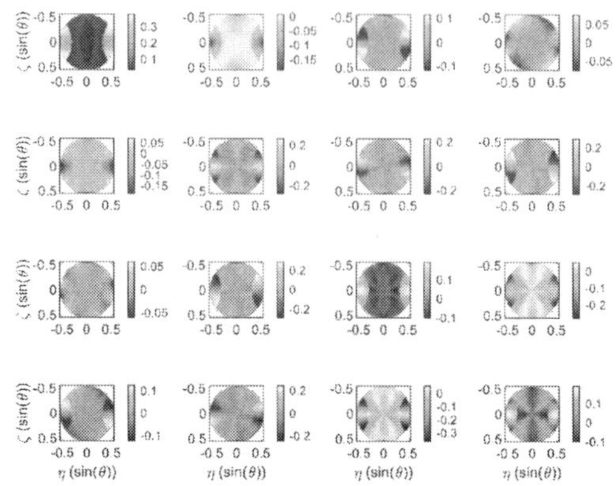

Fig. 2. Simulated nonnormalized coherent Mueller ellipsometry matrix elements m_{ij} i, j=1, … ,4. The m_{11} element is in the upper left corner and the m_{44} element is in the lower right corner. The η and ζ axis on the figure runs from -NA to NA.

Fig.2 shows the simulated nonnormalized coherent Mueller ellipsometry images in the exit pupil plane for the 16 Mueller matrix elements. The m_{11} element contains the coherent scatterometry signal and this element is frequently used to normalize all the Muller matrix elements. A recent study [1] shows that this is most unfortunately, since the normalization leads to higher correlation between the retrieved dimensional parameters than without the normalization. A microscope setup has the addition benefit that it only requires small metrology measurement areas on the wafer. We therefore suggest to use nonnormalized coherent Mueller ellipsometry for process inspection in high volume manufacturing.

Funding

This work has received funding from the EMPIR projects 20FUN02/ POLight and 20IND07 TracOptics. The EMPIR program is cofinanced by the Participating States and the European Union's Horizon 2020 research and innovation program. P-E. Hansen was supported by the Danish Agency for Institutions and Education

References

1. P. E. Hansen, S. R. Johansen, S.A. Jensen and J. S. M. Madsen Front. Phys. 9: 791459 (2022)

https://doi.org/10.1051/epjconf/202226610005

Holographic single-image depth reconstruction

Simon Hartlieb[1,*], *Christian* Schober[1], *Tobias* Haist[1], and *Stephan* Reichelt[1]

[1]Universität Stuttgart, Institut für Technische Optik, Pfaffenwaldring 9, 70569 Stuttgart, Germany

Abstract. In this article a camera-based single-image sensor is presented, that is able to measure the distance of multiple object points. The experimental results show an accuracy of 8,51 µm within a depth range of 20 mm. The sensor consists of a camera, whose lens is upgraded with a diffractive optical element (DOE). It fulfils two tasks: adding a vortex point spread function (PSF) and replication of the vortex PSFs to a predefined pattern of *K* spots. Both, shape and rotation of the vortex PSF is sensitive to defocus. By applying the depth reconstruction to each of the *K* replications and averaging the results, we experimentally show, that the reconstruction of the depth signal can be improved by a factor of up to 3.

1 Introduction

When an object point is imaged to a detector, its lateral position (x,y) can be detected rather easy by calculating the Center of Gravity (CoG) of the PSF, whereas the axial dimension (z), thus the distance of the object point, is lost. It can be reconstructed by evaluating the increasing diameter of the defocused PSF, but the achievable accuracy is bad. The accuracy of this reconstruction can be improved by increasing object-space numerical aperture (NA) of the imaging system. However, a large NA limits the depth of field and, therefore, the measurement range. Another possibility to improve the depth reconstruction of object points is to modify the PSF of the imaging system. A classic way of PSF modification is, for example, a superimposed astigmatism by the use of two orthogonal cylindrical lenses. In the last two decades mainly for the application in optical microscopy other ways of PSF manipulation based on diffractive optical elements (DOE) were developed. The purpose is, to modulate the phase of the light in a way, that the shape of the PSF corresponds to the changing distance z. Popular examples are the corkscrew PSF (CS-PSF) [1], self-bending PSF (SB-PSF) [2], tetrapod PSF (TP-PSF) [3] and double-helix PSF (DH-PSF) [4]. The ratio between measurement range and accuracy reached by those techniques ranges between 280 and 560.
This contribution addresses the question, to which extent it is possible to create a single-image depth reconstruction with both, good accuracy and large measurement range to use it as a single-camera 3D position sensor. Therefore, we combine a PSF modification method known from microscopy with a holographic replication technique called multipoint method (MPM) [5] and apply it to a low NA objective lens. The low NA lens ensures a large measurement range and the combination of the multipoint method and the PSF modification is used to increase the accuracy of depth detection.

2 Principle and results

For the PSF modification technique we use the DH-PSF introduced by Baránek et al. in [4]. The discrete spiral phase modulation (SPM) modifies the incoming light in a way, that the transversal component of the resulting intensity profile consists of two helixes rotating around each other. In the image plane, this forms two spots that rotate around a common axis. The angle of rotation is dependent on the defocus of the object point. The MPM is used to replicate this DH-PSF to a predefined pattern. Each copy consists of two rotating spots. So if the object point is moved in z, all DH-PSF replications show the same angle of rotation. By averaging all measured angles, errors caused by photon noise and discretisation are reduced and, therefore, the accuracy of the measured rotation angle can be improved theoretically by a factor equal to the square root of the number of replications. In Fig. 1 the principle of the multipoint double-helix PSF is shown for four replications.

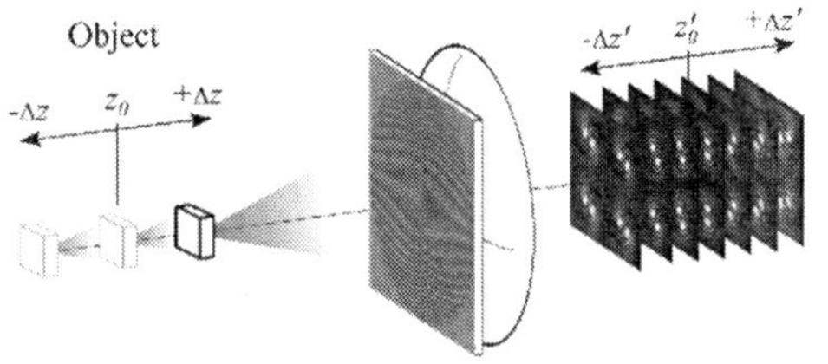

Fig. 1. Combination of the MPM and the DH-PSF. The two rotating spots created by the SPM are replicated to four copies [6].

The experimental setup consists of a linear stage (Walter Uhl GT6-BO01) that is used to move a point light source (fibre coupled laser, λ = 633 nm) in z. The camera system

* Corresponding author: hartlieb@ito.uni-stuttgart.de

has a low NA objective lens (f' = 50 mm, NA = 0,0595) with a DOE mounted in front, to perform the replication to $N = 25$ copies and induce the vortex phase modulation. The distance between light source and objective lens is 234 mm and the depth measurement range is 20 mm.

Simulations and experiments show, that the two rotating spots, created by the double-helix, form a tail, whose length is growing with the angle of rotation. More information regarding the simulations can be found in [6]. The tail formation is shown in Fig. 3 (d). There one DH-PSF replication is shown for a defocus of 8.9 mm. With growing defocus, the tail becomes a double spiral, as shown in Fig. 3 (e). This tail makes it difficult to evaluate the rotation angle using the CoGs of both spots. Therefore, we use the method of cross correlation with a reference image stack. The reference images are acquired at $K = 2000$ equidistant points in the measurement range of 20 mm. A measurement is performed by positioning the light source inside the measurement range, acquire an image and cross correlate it with the reference image stack. The peak of the resulting correlation energy curve marks the measurement result.

In Fig. 2 the measurement result is shown for a measurement consisting of 520 equidistant points acquired in the measurement range of 20 mm. The plot shows the error between measured distance change and the actual distance change of the linear stage, which is taken as reference. The standard deviation of the error signal is $\sigma_{520} = 7.86$ µm. In total three measurements with different number of equidistant measurement points are carried out. The mean standard deviation of all three measurements is $\bar{\sigma} = 8{,}51$ µm and the signals show good reproducibility [6].

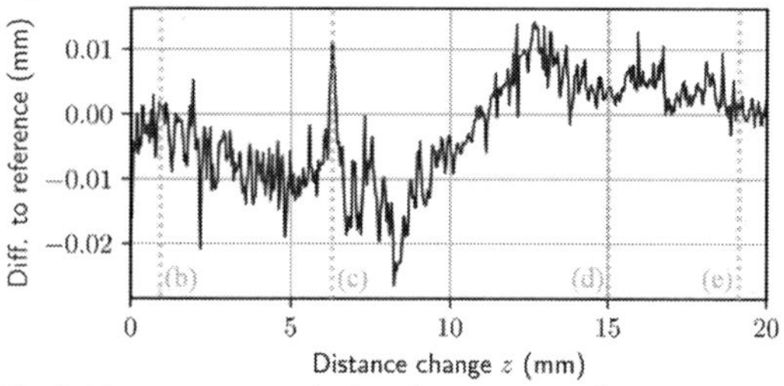

Fig. 2. Measurement result plotted as error over distance change. The error is calculated as the difference between measured distance and the linear stage position (reference). The blue dotted lines (b) to (e) refer to the images of Fig. 3 (b) to (e).

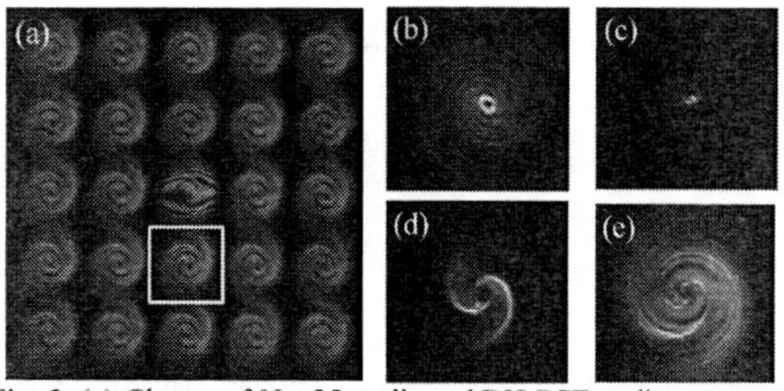

Fig. 3. (a) Cluster of $N = 25$ replicated DH-PSF at distance change $z = 20$ mm. The DH-PSF marked with a white square is shown for different distances and, therefore, defocus states in images (b) to (e) ; (b) $z = 1$ mm ; (c) $z = 6.1$ mm . Here the light source is in focus; (d) $z = 15$ mm ; (e) $z = 19$ mm .

This leads to the assumption, that the remaining error is of systematic nature and calibration could even improve the accuracy.
In Fig. 3 (a) the whole cluster of $K = 25$ replicated DH-PSF is shown. The spot marked with a white square is shown in the images (b) to (e) for different distances and defocus states. In (c) the light source is in focus and all information is stored in a few bright pixels, which makes it difficult to reach high accuracy with the method of image correlation. Therefore, the error signal in Fig. 2 shows a small peak at this position.
To evaluate the improvement achieved by the multipoint method, each of the 25 replicated DH-PSF is evaluated individually. This is done by cutting both, the measurement images and the reference image stack into 25 image sections containing only one DH-PSF each. The results show, that depending on which spot is evaluated, the reconstruction accuracy can be improved by a factor of up to 3.
The measurement range to accuracy ratio of the proposed position measurement system is 20 mm / 8.51 µm = 2350, which is very high compared to other single-image measurement systems. Nevertheless, there are still some points, such as 3D calibration, lateral shifted light source or lateral position measurement, that need to be addressed in future work to be able to use this method in a 3D position measurement sensor.

3 Summary

The presented single-image depth measurement system is based on the combination of a DH-PSF and a spatial replication method, which is both created by a phase modulating DOE placed in front of the imaging lens. The system is able to detect the distance of one or multiple point light sources with an accuracy of 8,51 µm within a measurement range of 20 mm. This offers the possibility to create a cost effective single-camera 3D position measurement sensor that can detect the 3D positions of multiple point light sources.

References

1. M. D. Lew, S. F. Lee, M. Badieirostami, and W. E. Moerner, *Optics Letters* **36**(2), 202–204 (2011).
2. S. Jia, J. Vaughan, and X. Zhuang, *Biophysical Journal* **104**(2), 668a (2013)
3. Y. Shechtman, S. J. Sahl, A. S. Backer, and W. E. Moerner, *Physical Review Letters* **113**(13), 133902 (2014)
4. M. Baránek and Z. Bouchal, *Proceedings of SPIE* **161**(9441), (2014)
5. T. Haist, S. Dong, T. Arnold, M. Gronle, and W. Osten, *Optics Express* **22**(12), 14450–14463 (2014)
6. S. Hartlieb, C. Schober, T. Haist, and S. Reichelt, *Optics Express* **30**, 23035-23049 (2022)

Polarization dependency of the 3D transfer behavior in microsphere enhanced interferometry

Lucie Hüser[1,*], *Tobias* Pahl[1], and *Peter* Lehmann[1]

[1]Measurement Technology Group, Faculty of Electrical Engineering and Computer Science, University of Kassel, 34121 Kassel, Germany

Abstract. Enhancing the lateral resolution limit in optical microscopy and interferometry is of great interest in recent research. In order to laterally resolve structures including feature dimensions below the resolution limit, microspheres applied in the optical near-field of the specimen are shown to locally improve the resolution of the imaging system. Experimental and simulated results following this approach obtained by a high NA Linnik interferometer are analyzed in this contribution. For further understanding of the transfer characteristics, measured interference data are compared with FEM (finite element method) based simulations with respect to the polarization dependency of the relevant image information.

1 Introduction

With the ongoing trend towards miniaturization overcoming Abbe's diffraction limit in optical metrology is of particular interest regarding research and industrial applications. Microsphere assistance as a near-field support is investigated as a technique to realize topographical interferometric [1–3] as well as microscopic [4–6] measurements of structures below the resolution limit. Photonic nanojets are mentioned frequently to be the decisive mechanism [7–9]. Also the role of evanescent waves and whispering gallery modes [10, 11] is discussed. It is also found that the polarization of the light source has an influence on the resolution capabilities using microspheres [6].

In this contribution the effects of the polarization of the illumination are examined with respect to interference microscopy. Thus, measured as well as rigorously simulated data sets are compared with respect to the polarization. Results are analyzed in the 3D spatial frequency domain in order to find out the transfer characteristics.

Experimental results are obtained by a high NA Linnik interferometer (100 x, NA = 0.9). The measurement process follows the principle of coherence scanning interferometry (CSI) and is described in further detail in [1]. Köhler illumination with an LED (royalblue, central wavelength: 440 nm) is used. A schematic representation of the experimental setup is shown in Fig. 1.

2 Analysis of the results in the 3D spatial frequency domain

The transfer characteristics of an interferometer can be obtained from the 3D spatial frequency domain representation [12, 13], which enables a profound understanding of the 3D measurement process. This method is applied to microsphere assisted interferometry to gain further insight into the mechanisms underlying the lateral resolution enhancement. Furthermore, Pahl et al. [14] developed a rigorous CSI simulation model based on FEM calculations of the electric field distribution close to the specimen and Fourier optics modeling of the imaging process. This model considers the conical 3D illumination in a microscope and has been extended by a microcylinder to approximate microsphere assisted interferometry.

For further analysis the data is transferred to the 3D spatial frequency domain also called $\mathbf{q}=\mathbf{k}_s-\mathbf{k}_{in}$ – space, with the incident wave vector $\mathbf{k}_{in}$ and the scattered wave vector $\mathbf{k}_s$. The resulting 3D spatial frequency distribution is limited by the Ewald-limiting-sphere which is shown

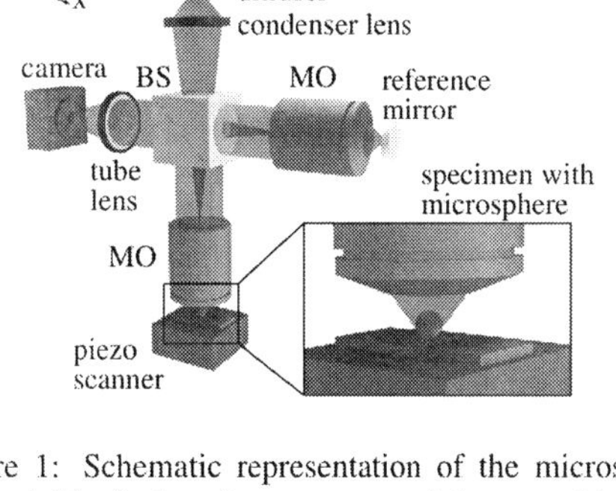

Figure 1: Schematic representation of the microsphere assisted Linnik interferometer containing two high NA microscope objectives (MO) and a beam splitter (BS). The illumination (blue) and imaging (red) beam paths are shown.

*e-mail: lucie.hueser@uni-kassel.de

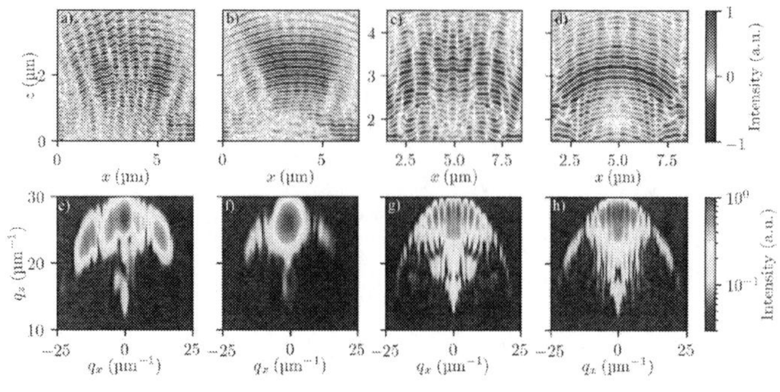

Figure 2: Interferometric measurement data obtained from a rectangular grating (SiMETRICS RS-N, Λ = 300 nm) using royalblue a) TM and b) TE polarized light through a microsphere (SiO_2, approx. 7 µm diam.), simulated data sets of a similar grating with c) TM and d) TE polarized light assuming a microcylinder of 5 µm diameter. The corresponding 3D spatial frequency representations are depicted in e) – h).

in further detail in [13]. Thus, the diffraction orders $q_{x,n} = \frac{2\pi n}{\Lambda}$ for a grating of period length Λ in x-direction appear as distinct vertical lines within the Ewald-limiting-sphere.

Figures 2a) – d) show measured and simulated interferometric data sets for a rectangular silicon grating (RS-N, SiMETRICS) with Λ = 300 nm and a nominal height of 140 nm. This phase grating evokes a phase modulation of the interference fringes along the x-axis, which is centered around the virtual image plane of the microsphere [1].

Comparing the different polarization configurations it can be observed that for TM polarized light the phase modulation induced by the grating is much stronger in both, measurement and simulation results. Besides, a qualitative correspondence of measured and simulated results can be seen. Therefore, the simulation approach in principle reproduces the transfer characteristics of microsphere assisted interferometry.

In Fig. 2e) – h) the 3D spatial frequency domain representations in the q_xq_z-plane of the different data sets are depicted. The diffraction orders for the measured data stacks are located at $q_x \approx \pm 15\,\mu m^{-1}$ (considering the additional magnification through the microsphere). Instead of distinct vertical lines the diffraction maxima appear blurred. This effect can be explained with the limited field of view while imaging through a microsphere. In Fourier optics the field of view corresponds to the area of integration for the Fourier integral leading to a convolution with an Airy-disc pattern in the Fourier domain leading to the blurring of the diffraction orders.

Comparing the frequency representations for the TM and TE case, the intensity maxima related to first order diffraction are lower for the TE case, which is in agreement with the weaker phase modulation visible in Fig. 2b). In comparison with the simulated results in the 3D spatial frequency domain, some effects occur differently. The magnification introduced by the microcylinder is higher (compare the extension in x-direction of the phase modulation in Fig. 2a) and c)), leading to slightly different positions of the diffraction orders. Furthermore, due to the use of a 2D microcylinder instead of a sphere in the simulation further effects appear [15]. Also slight differences in the diameter of the utilized microspheres influence the results.

3 Conclusion

Analyzing the polarization dependency of microsphere enhanced interferometry, TM orientation polarization should be preferred to obtain the relevant phase information of 2D topographies, which are shift invariant in y-direction. This is in agreement to previous results obtained for grating structures using an interference microscope without microsphere assistance and could be confirmed by rigorous simulations based on an FEM model [14].

References

[1] L. Hüser, P. Lehmann, Applied Optics **59**, 1695 (2020)

[2] P.C. Montgomery, S. Lecler, A. Leong-Hoï, S. Perrin, Physica Status Solidi (a) **1800761**, 1800761 (2019)

[3] S. Perrin, Y.J. Donie, P. Montgomery, G. Gomard, S. Lecler, Physical Review Applied **13**, 1 (2020)

[4] P.K. Upputuri, M. Pramanik, Optics Communications **404**, 32 (2017)

[5] Z. Wang, W. Guo, L. Li, B. Luk'Yanchuk, A. Khan, Z. Liu, Z. Chen, M. Hong, Nature Communications **2**, 1 (2011)

[6] A. Darafsheh, G.F. Walsh, L. Dal Negro, V.N. Astratov, Applied Physics Letters **101**, 141128 (2012)

[7] H. Yang, R. Trouillon, G. Huszka, M.A. Gijs, Nano Letters **16**, 4862 (2016)

[8] A. Darafsheh, Journal of Physics: Photonics **3**, 022001 (2021)

[9] P. Montgomery, S. Lecler, A. Leong-Hoi, P. Pfeiffer, Journal of Physics: Conf. Series **794**, 012006 (2017)

[10] Y. Ben-Aryeh, Journal of the Optical Society of America A **33**, 2284 (2016)

[11] R. Boudoukha, S. Perrin, A. Demagh, P. Montgomery, N.E. Demagh, S. Lecler, Photonics **8** (2021)

[12] J.C. Quartel, C.J. Sheppard, Journal of Modern Optics **43**, 591 (1996)

[13] P. Lehmann, S. Hagemeier, T. Pahl, Metrology **1**, 122 (2021)

[14] T. Pahl, S. Hagemeier, M. Künne, D. Yang, P. Lehmann, Optics Express **28**, 39807 (2020)

[15] L. Hüser, T. Pahl, M. Künne, P. Lehmann, *Mesophotonics: Physics and Systems at Mesoscale, Proc. SPIE* **12152**, 121528 (2022)

Nanoform evaluation approach using Mueller matrix microscopy and machine learning concepts

Tim Käseberg[1,*], *Jana* Grundmann[1], *Stefanie* Kroker[2], and *Bernd* Bodermann[1]

[1]Physikalisch-Technische Bundesanstalt, Bundesallee 100, 38116 Braunschweig, Germany
[2]Institut für Halbleitertechnik, Laboratory for Emerging Nanometrology, Technische Universität Braunschweig, Hans-Sommer-Str. 66, 38106 Braunschweig, Germany

Abstract. We realized an imaging Mueller matrix microscope for nanostructure characterization. For investigations on nanoform characterization via Mueller matrix images, we measured and simulated Mueller matrix images of specially designed nanostructures. As an approach towards machine learning evaluation in imaging ellipsometry, we calculated Haar-like features of the images and observed a higher sensitivity to subwavelength features in off-diagonal matrix elements compared to microscopy.

1 Introduction

Innovative devices that rely on shaping matter at the nanoscale are currently key for many technological advances in various fields, from photonics, nanotechnology and -electronics to metamaterial fabrication and semiconductor industries. Complementary to the technological progress in fabricating nanoscale structures, there is a need for precise metrology solutions for their characterization. For the examination of layer stack thicknesses or optical properties, Mueller matrix ellipsometry has been proven a useful tool [1]. By analysing the polarization properties of the sample and solving the inverse problem using numerical simulations, feature parameters at the nanoscale can be retrieved. As a purely optical method, ellipsometry benefits from being fast, non-invasive, and flexible to use. One disadvantage of conventional ellipsometry however is that measurement results are integrated over the whole illuminated area. For this reason, structural features that are smaller than the illumination area cannot be examined precisely, and their results are affected by signals from the surrounding media. For the local examination of polarization effects of nanostructures, imaging Mueller matrix ellipsometry is a reasonable alternative. It combines the conventional ellipsometric measurement principle with an imaging system, leading to a Mueller matrix for each pixel in an image of the structure under investigation and thus to so-called Mueller matrix images. These images allow for the observation of local changes of the samples' polarization properties, even in ranges much smaller than the illumination spot.

2 Measurement setup

For our investigations, we constructed and characterized an imaging Mueller matrix ellipsometry setup [2], which

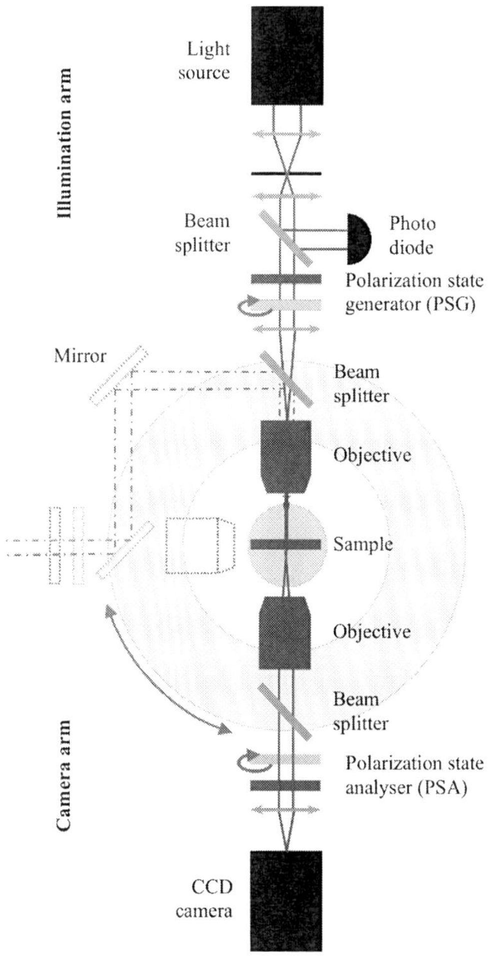

Fig. 1. Schematic representation of the Mueller matrix microscope setup.

[*] Corresponding author: tim.kaeseberg@ptb.de

is depicted in Fig. 1. It consists of a fixed illumination arm and a pivotable camera arm in a conventional dual-rotating retarder configuration. The illumination arm prepares the light from an LED via a polarization state generator (PSG) system. An objective combined with a Koehler illumination system is used to guide the prepared light onto the sample which is mounted on top of several translation and rotation stages. The camera arm is attached to a rotation stage with a large aperture of 30 cm in diameter around the sample. An infinity corrected objective with a long working distance of 25.5 mm, a magnification of 50x and a numerical aperture of 0.42 combined with a tube lens creates a microscope image of the sample at a CCD camera at the end of the camera arm. A polarization state analyser (PSA) identical to the PSG but in reverse order is positioned between the objective and the tube lens. This way, the full Mueller matrix can be determined for each pixel in the obtained image. The configuration of the two arms allows for a fluent transition from reflection to transmission measurements for arbitrary angles of incidence. Reflection measurements can be performed for angles from 37.5° to 90°. Additionally, the setup can be operated in a direct incidence mode as a Mueller matrix microscope. In this mode, the camera arm is moved to a position in a 90° angle to the illumination arm. The light that reflects from the sample back into the first objective is then guided via beam splitters and mirrors to the PSA of the camera arm. This way, reflection measurements at perpendicular incidences can also be performed.

3 Samples and measurement results

To investigate the influence of small geometrical features on different elements of the Mueller matrix, we have produced nanoform reference structures with sizes and shapes varying in the range between 50 nm and 5 µm, including square and circular structures as well as square structures with varying corner radii. Mueller matrix images of these structures were measured and compared to AFM measurements and numerical simulations. An example for the images of a square structure with 400 nm corner radius is shown in Fig. 2. For both, simulated and measured Mueller matrix images, a deconstruction of the images showed significant influences of the corner radius on off-diagonal Mueller matrix elements, even for changes in the subwavelength regime [3].

4 Haar-like feature examinations

The evaluation of Mueller matrix images is usually performed either by analysing the average signal of selected regions of interest in the image, or by modelling the structure and solving the inverse problem. The former is easier applied and faster to use, but as with conventional ellipsometry, it delivers no information about lateral features exceeding those from microscopy. The latter can retrieve complete structural parameters but requires a complex model and laborious numerical simulations which make the evaluation process difficult and time-consuming. A possible compromise would be the use of

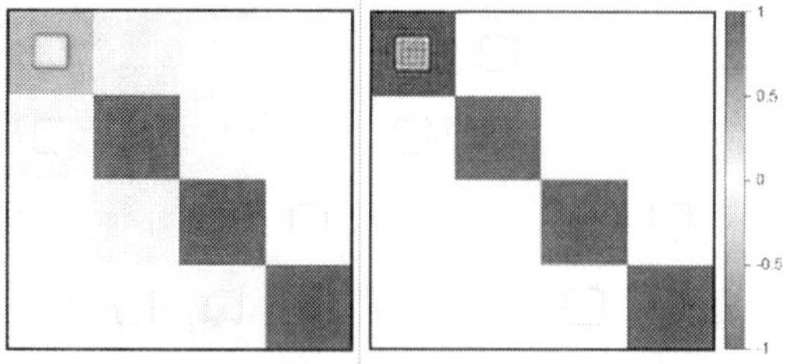

Fig. 2. Measurement (left) and simulation (right) of Mueller matrix images for a 5 µm sized square structure with 400 nm corner radius.

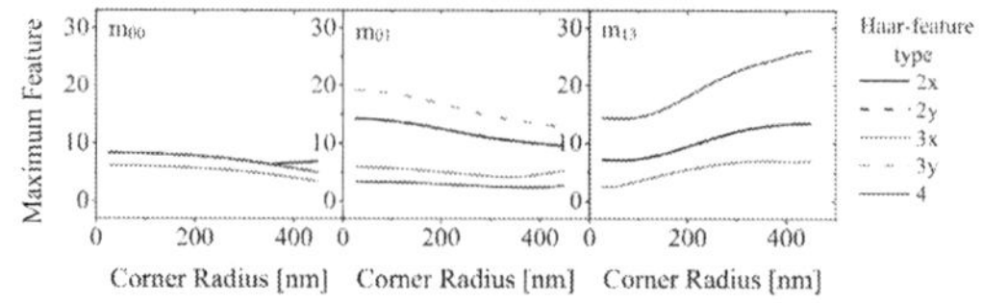

Fig. 3. Maximum Haar-like features extracted from simulated Mueller matrix images, depending on corner radius size.

machine learning, which was already successfully applied to the evaluation process in optical scatterometry [4]. Training a machine learning model for image evaluation initially also requires high computational costs, but once the training is finished, results are achieved fast and reliable. We approached this idea by calculating and evaluating Haar-like features [5] for the simulated Mueller matrix images of our nanoform structures. Haar-like features are certain patterns of intensity differences in images commonly used in object recognition techniques. An example for the features from simulated Muller matrix images is shown in Fig. 3. Especially in the images of off-diagonal matrix elements, a sensitivity of the Haar-like features to varying corner radii of the structures was observed that exceeded those from the microscope image alone. Thus, Haar-like features in Mueller matrix images present an information channel for nanostructure characterization, which hints towards a successful application of machine learning algorithms in Mueller matrix ellipsometry.

This project (20FUN02 "POLight") has received funding from the EMPIR programme co-financed by the Participating States and from the European Union's Horizon 2020 research and innovation programme.

References

1. M. Losurdo, K. Hingerl, *Ellipsometry at the Nanoscale* (Springer, Berlin, 2013)
2. T. Käseberg, J. Grundmann, J. Dickmann, S. Kroker, B. Bodermann, EPJ Web Conf. **238**, 06006 (2020)
3. T. Käseberg, J. Grundmann, T. Siefke, P. Klapetek, M. Valtr, S. Kroker, B. Bodermann, Front. Phys. **9**, 814559 (2022)
4. S. Liu, X. Chen, T. Yang, C. Guo, J. Zhang, J. Ma, C. Chen, C. Wang, C. Zhang, S. Liu, Measurement **191**, 110811 (2022)
5. C. P. Papageorgiou, M. Oren, T. Poggio, Proc. of ICCV, 555-562 (1998)

https://doi.org/10.1051/epjconf/202226610008

Particle detection enhancement by combining coherent Fourier scatterometry with synthetic optical holography

Haoyang Yin[1], *Dmytro* Kolenov[1], and *Silvania* Pereira[1,*]

[1]Optics Research Group, Imaging Physics Department, Faculty of Applied Sciences, Delft University of Technology Lorentzweg 1, 2628 CJ, Delft, The Netherlands

Abstract. By combining coherent Fourier scatterometry (CFS) with synthetic optical holography (SOH) we show that the sensitivity of detection of isolated nanoparticles on surfaces can be substantially increased. This improvement is a result of the boost in the signal at the detector due to the added reference beam, and the reduction of background noise caused by the electronics. We demonstrate an improvement of sensitivity of about 4 dB for the case of detection of a 60 nm polystyrene latex (PSL) particle on a silicon wafer at the wavelength of 633 nm ($\sim \lambda/10$).

1 Introduction

The detection of contamination due to isolated nanoparticles on surfaces that are used as substrates in nanofabrication has become one of the major challenges in the past decades, given the continuous decrease of the size of the features. Scatterometry is one of the most used techniques for this purpose since it is not limited by diffraction and it is based on far field detection. Recently, coherent Fourier scatterometry has been introduced to tackle this problem. The technique is similar to a confocal microscope operating in reflection where the surface to be inspected is scanned while being illuminated by a focused light beam. The difference with the latter is that instead of imaging the focal plane at the detector, in CFS the back focal plane is imaged into a split detector, where both halves of the detector are aligned perpendicular to the sample scan direction. The differential signal from both halves is recorded for each scan position. The sensitivity of the technique lies in the fact that although it operates in bright field mode (in the sense that both scattered and non-scattered fields are collected in reflection at the detector), the background is eliminated because when no particles are being illuminated by the focused beam, the far field is symmetric and the differential signal is zero. When a particle passes through the focused beam, the far field becomes asymmetric and position dependent, generating a differential signal that can be either positive or negative, depending on the relative position of the focused laser beam with respect to the particle. In CFS, the position of the particle can be determined with high accuracy and its size can be inferred by calibration [1].

In this work, we implement the synthetic optical holography (SOH) phase imaging technique [2] to enhance the sensitivity of CFS.

*s.f.pereira@tudelft.nl

2 Experimental setup

In order to combine SOH with CFS we simply add a piezo-actuated mirror at the open port of the beamsplitter of the existing CFS setup as in [1]. A collimated and linearly polarized He-Ne laser (λ = 633 nm) passes through a 50-50 non-polarizing beamsplitter which splits the beam into two halves: one beam goes to a microscope objective (NA = 0.9) and is focused on the sample of interest, and the other beam goes to a piezo actuated reference mirror. The sample is placed on a piezo stage that can be laterally scanned in a raster fashion. The reflected and scattered light from the sample arm is collected by the same objective and goes back into the beamsplitter, where it is combined with the reflected reference beam, similar to a Michelson interferometer. The Fourier plane of the objective is de-magnified by a telescopic setup consisting of two lenses and imaged on the split detector. The split detector has two halves, and here the intensity voltage from the left half is subtracted from the right half, resulting in a position dependent differential signal. While the sample is raster scanned, the reference mirror moves one step per scanned line creating a virtually tilted reference field. The reconstruction of the hologram is done as in the conventional off-axis digital holography [3, 4], which consists of Fourier transforming the scanned map containing the 0th and the $\pm$ first orders, isolating one of the first orders followed by shifting to zero and inverse Fourier transforming. The result of this operation is the retrieval of the complex field of the measured differential signal.

3 Results

We demonstrate the new combined technique by detecting a polystyrene latex (PSL) nanosphere with a diameter of 60 nm ($\sim \lambda/10$) on a silicon wafer. The laser power was set so low that the SNR of the detected signal from the nanoparticle in the conventional CFS setup was ≈ 3 dB. For the Fourier reconstruction procedure we used a band pass filter consisting of a 2D cosine window superimposed with a high pass filter. We also included the (noise filtered) map of the conventional CFS as reference.

After reconstruction, we observe that with the imaginary part of the first order signal one achieves a SNR of approximately 10 dB, despite the size of the particle and the wavelength of our laser. Compared to the reference measurement, our new technique has a higher signal amplitude and a SNR gain of 7 dB is achieved over the unfiltered signal or a gain of 4 dB over the filtered signal.

Beyond the detection of particles on wafers, our technique could also be implemented in other applications where the contrast between the nanoparticles and substrate is low due to the phase sensitive nature, such as PSL particles on glass or plastic substrates and pellicles. Furthermore, due to the modulation property, our technique is also suitable for low power applications, as in the case of biological samples.

References

[1] S. Roy et al., Review of Sci. Instrum. **86**, 123111 (2015).

[2] M. Schnell et al., Opt. Express 22, 15267–15276 (2014).

[3] T. Poon and J. Liu, *Introduction to modern digital holography: with MATLAB* (Cambridge University Press, 2014) 111.

[4] E. Cuche, P. Marquet, and C. Depeursinge, Applied Optics **39**, 4070 (2000).

Alignment autocollimator-based microscope adjustment and its quality assessment

Jan Krüger*, *Detlef* Bergmann, *Matthias* Sturm, *Wolfgang* Häßler-Grohne, *Rainer* Köning, and *Bernd* Bodermann

Physikalisch-Technische Bundesanstalt, Bundesallee 100, 38116 Braunschweig, Germany

Abstract. We report a custom microscope setup whose mechanical and optical components are adjusted by the means of an alignment autocollimator (AAC). Residual centring and angular misalignments of the components towards the microscope's optical axis are below 500 µm and 1 mrad, respectively. We further perform measurements of dot structures with diameters close to the diffraction limit (nominal diameter = 200 nm; chrome on glass mask) as suitable measures for the evaluation of the microscope's adjustment and to determine/ visualize the optical aberrations, which affect the image formation of microscopes.

1 Introduction

High-accuracy, dimensional, optical microscopy depends on model-based data evaluation. Only then it is possible to determine bidirectional measures such as diameters or widths, which refer to distances between two opposing edges, reliably. Rigorous simulations of the entire microscopical imaging process are the foundation of such model-based evaluations. The simulations require a variety of experimental parameter of the microscope setup to be known. For example, the optical aberrations are required, among other parameters, for a comprehensive model of the microscope's imaging system [1].

The optical aberrations like Coma, Astigmatism, and spherical aberration have a distinct impact on the microscope's imaging performance and are influenced by the state of the microscope's adjustment. So, misalignments of optical components can be responsible for additional optical aberrations.

At PTB, we started to employ AAC for the alignment of optical elements of high-accuracy dimensional microscopes. Thereby, the centering and angular misalignments of the microscope components are reduced to a minimum. In this contribution, we use measurements on a chrome dot with a nominal diameter of 200 nm to verify the quality of the alignment. If the acquired dot images are visibly affected by optical aberrations and the adjustment procedure should be repeated. Otherwise, the effect of the residual aberrations is not as prominent, and the microscope is operational. This contribution does not focus on the determination of the optical aberration based on the use of dot structures. These findings have been already reported in [2]. This work emphasizes on the alignment strategy applied as an example to a custom PTB microscope and the dot measurements as a fast and immediate measure for the adjustments' quality assessment. Therefore, the instructions and the sequence for the alignment strategy for this specific case is provided in section 2 before the dot measurements are described in section 3. Finally, we conclude in section 4.

2 Microscope alignment

The microscope under consideration is based on the optical components of the Vistek Semiconductor LMS IPRO4. However, the components are assembled in a custom setup. It is a UV-reflected light microscope with an illumination and imaging NA of 0.44 and 0.55, respectively. The magnification is 200x. The microscope is equipped with an overview microscope which has a magnification of 50x and uses visible light. The adjustment procedure for this microscope uses the same principles and is therefore omitted.

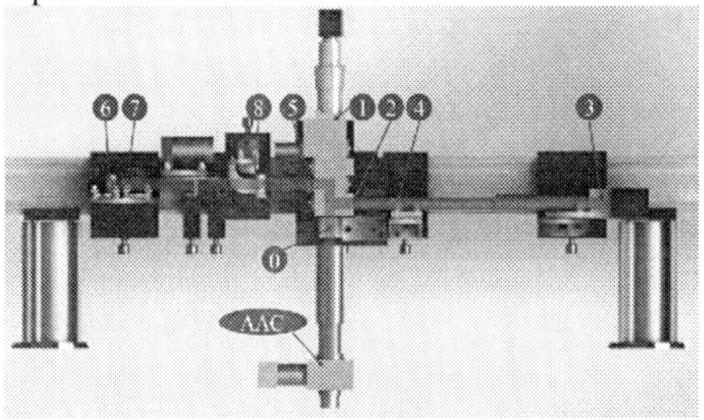

Fig. 1. Application of the AAC at the custom microscope of PTB

In figure 1, the AAC is attached to the custom microscope and the single alignments steps are indicated. Step 0 precedes the attachment of the AAC. Here the microscope's empty objective holder is aligned parallel to the surface of the supporting optical table. This is performed by employing a chromatic confocal distance sensor. Usually, a misalignment angle of the objective holder of 1 mrad is sufficient. Next, the AAC is attached to the empty objective holder. Subsequent steps are performed by the combined use of the AAC and different alignment targets at the position of the respective microscope component. The mounts of all components to be aligned can correct for lateral and angular

* Corresponding author: jan.krueger@ptb.de

https://doi.org/10.1051/epjconf/202226610009

misalignments. Through an iterative alternation between the two modes (imaging and autocollimator) of the AAC and the necessary adjustments at the mounts, the respective alignment target in the mount is aligned. Subsequently the target is replaced by the matching microscope component.

Concerning the sequence of the alignment steps, the main microscope body must be aligned first with an alignment target at position 1 of figure 1. The two beam splitters of the UV-microscope are positioned inside of that body. They separate the elevated illumination beam path from the lower imaging beam path. Their alignment steps are 2 and 5, assuming that the entire imaging path is configured before the illumination beam path. After the alignment of the beam splitter, the other components of the imaging beam path are aligned starting with the component which is the furthest away from the AAC. Thereby, lenses and other components can be already inserted without diffracting or blocking the AAC's emission. The measurement camera is aligned in position 3. Position 4 represents the location of a lens. The alignment of the imaging beam path follows the same order. The positions 5-8 correspond to the illumination path beam splitter, an optical fibre, and two lenses. The optical fibre feeds the light from an UV-LED into the microscope setup The supporting VIS-microscope is then adjusted likewise, before the AAC is removed and replaced by the objective. The resolution of the AAC for the centring alignment decreases with a larger distance between the AAC and the alignment target. While it is roughly 70 µm for the illumination beam splitter, it is just below 500 µm for the position of the fibre. The angular resolution is independent of the measuring distances at 1 mrad. The specimen table on top of a smooth (surface roughness < 1µm) granite block is placed below the objective. The last step of the microscopes adjustment requires a Bertrand lens (BL). With the BL inserted, the objective's back-focal plane is projected into the microscope's image plane on the camera. This is shown as the large aperture in the top left of figure 2. Now it is possible to insert and adjust the aperture diaphragm in a flexure stage of the illumination beam path, while monitoring the progress on the camera. When the aperture diaphragm is centred in respect to the objective aperture, it blocks most of the illumination beam path, too. Only a narrow well centred beam is allowed to pass perpendicular and then be focused by the objective onto the sample.

In a last step the field stop is centred and adjusted along the optical axis such that its edges reach maximum sharpness. The microscope is operational.

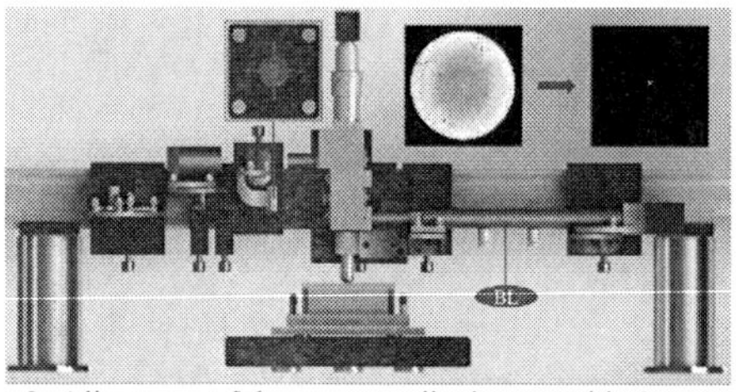

Fig. 2. Alignment of the aperture diaphragm with the BL

3 Dot measurements and evaluations

For the system evaluation a chrome dot on a high-quality photomask is measured. Prior structure widths measurements on the sample yield a fabrication offset of 20-30 nm for structures on that sample. Therefore, the nominal 200 nm wide dot has an effective diameter of 170-180 nm.

he dot measurements consist of repeated series of images in different height distances from both sides of the focal plane. The resulting stack of images is referred to as a focus series. Repeated focus series can be averaged due to the high measurement stability of the mechanical setup. In figure 3, the images a) and c) show the same dot after two different adjustments according to the above-described procedure. Asymmetries in the images are made visible by performing a least-squares fit of a Gaussian to the centre of the focal dot measurement. By subtracting the respective fit from a) yields to visible asymmetries in b) while the same procedure leads to a symmetric image in d) when c) is evaluated.

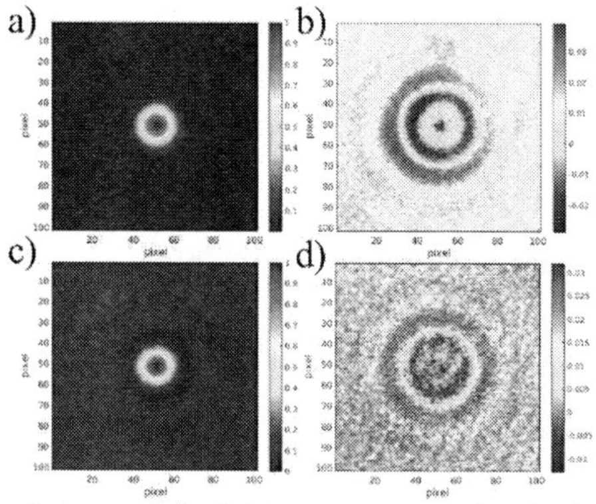

Fig. 3. Averaged, focal dot measurements and evaluation for an unsuccessful (a, b) and a successful (c, d) alignment of the microscope

4 Conclusion & Acknowledgement

This contribution provides an AAC-based alignment scheme for microscopist and a fast assessment technique, which is in-line compatible. Asymmetries in the dot measurement of the focal image uncover non ideal adjustments states. Since the asymmetries will be present in all subsequent measurements, a repetition of the entire alignment procedure might be the best decision.

The authors want to express their gratitude towards the German Federal Ministry of Education and Research (BMBF). Their funding of the project "SiM4diM" (project number: 01 IS 20 080) enabled this work.

References

[1] B. Bodermann et al., *Optical imaging and metrology: advanced technologies* (Wiley-VCH, Weinheim, 2012)
[2] P. Schneider, P. Manley, J. Krüger, et al., SPIE Photonics Europe, 12137-30 (2022)

EPJ Web of Conferences 266, 10010 (2022) https://doi.org/10.1051/epjconf/202226610010
EOSAM 2022

Elementary, my dear Zernike: model order reduction for accelerating optical dimensional microscopy

Phillip Manley[1,2,*], *Jan* Krüger[3], *Lin* Zschiedrich[1,2], *Martin* Hammerschmidt[1,2], *Bernd* Bodermann[3], *Rainer* Köning[3], and *Philipp-Immanuel* Schneider[1,2]

[1]JCMwave GmbH, Bolivarallee 22, 14050 Berlin, Germany
[2]Zuse Institute Berlin, Takustraße 7, 14195 Berlin, Germany
[3]Physikalisch-Technische Bundesanstalt, Bundesallee 100, 38116 Braunschweig, Germany

Abstract. Dimensional microscopy is an essential tool for non-destructive and fast inspection of manufacturing processes. Standard approaches process only the measured images. By modelling the imaged structure and solving an inverse problem, the uncertainty on dimensional estimates can be reduced by orders of magnitude. At the same time, the inverse problem needs to be solved in a timely manner. Here we present a method of accelerating the inverse problem by reducing images to their elementary features, thereby extracting the relevant information and distinguishing it from noise. The resulting reduction in complexity allows the inverse problem to be solved more efficiently by utilize cutting edge machine learning based optimization techniques. By employing the techniques presented here, we are able to perform for highly accurate and fast dimensional microscopy.

1 Introduction

Dimensional microscopy is an essential tool for non-destructive and fast inspection of manufacturing processes. One important application is the bidirectional measurement of line structures [1]. The current state of the art involves analysing the image data using threshold techniques. In many measurements, information about the sample and microscope is also available but remains unused. A more accurate estimate of the structure dimensions can be obtained by modelling the light interaction with the structure and comparing the simulated structure images to their experimental counterparts. By varying the model's structure dimension parameters (i.e. line width, particle diameter), the parameters which minimise the difference between model and experiment can be determined. This process is called solving the inverse problem.

In order for the inverse problem to be solvable, the model needs to be able to describe the measurements to a high degree of accuracy. To this end, we have developed a forward model using the finite element solver *JCMwave* [2], which takes into account an efficient sampling of the microscope's illumination pupil, nanooptical scattering at the structure and phase aberrations effects due to the imaging system [3]. Due to the sophistication of the model, we require a method of solving the potentially multi-dimensional inverse problem with relatively few model evaluations. Recently, the Bayesian target vector optimization (BTVO) method has been shown to both efficiently solve the inverse problem and provide information on the uncertainties of the predicted dimensional parameters such as line widths and particle diameters [4].

2 Method

Typical methods of solving the inverse problem attempt to globally minimise χ^2 (chi-squared) calculated from the simulated data and measured data and variances (denoted as σ^2). In the case of images the simulated and measured data consists of the intensity on each pixel (I_{sim} and I_{meas}, respectively) in multiple focal and afocal measurement planes. For high resolution images, the number of pixel data points can run in to the tens of thousands. Using these data points, χ^2 can be calculated by summing up the contribution of each ith pixel to the overall χ^2,

$$\chi^2(\vec{x}) = \sum_{i=1}^{N} \frac{(I^{(i)}_{\mathrm{sim}}(\vec{x}) - I^{(i)}_{\mathrm{meas}})^2}{\sigma_i^2}. \quad (1)$$

The inverse problem is then solved by finding the free model input parameters (represented by the vector $\vec{x}$), such as the line width or diameter, which minimize the total χ^2 for all N pixel data points. The method of Bayesian optimization (BO) performs this minimization by building a surrogate model (called Gaussian process regression) for χ^2, which can then be used to find the global minimum efficiently. However, in building our surrogate model for the sum of the χ^2 contributions, we lose information on the individual pieces of measurement data, and we approximate a multivariate Gaussian distribution for χ^2.

Alternatively, the BTVO method creates a surrogate model (i.e. Gaussian process regression) for each piece of

*e-mail: phillip.manley@jcmwave.com

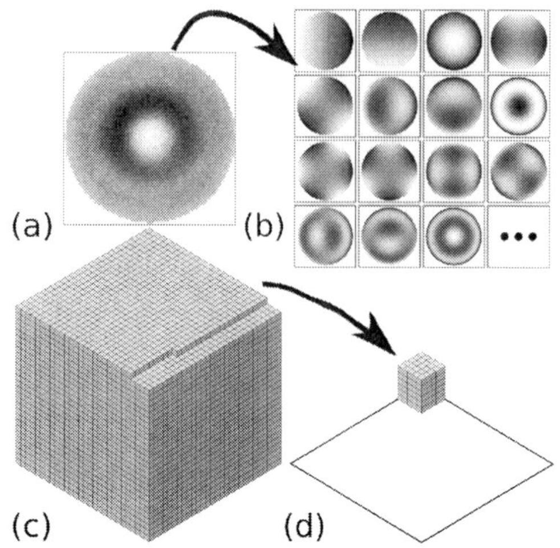

Figure 1. (a) Image of a Cr particle in a plane 2 μm from the focal plane, (b) first few Zernike basis functions used to expand the image in (a). (c-d) Schematic image displaying the number of Gaussian processes needed for different model approaches (one cube represents one Gaussian process). (c) If every pixel in the image stack is modelled with a Gaussian process, (d) if the coefficients of the image expansion are modelled with a Gaussian process.

measured data, which are then used to evaluate χ^2. This has the advantage of providing the correct distribution for χ^2 which results in accurate uncertainties on the reconstructed parameters.

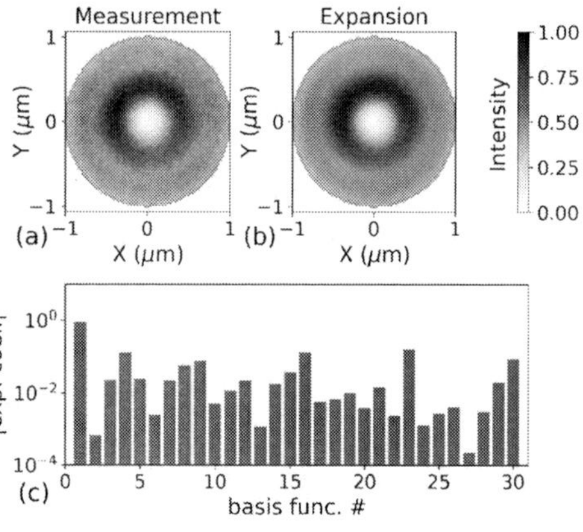

Figure 2. (a) Image of a Cr particle in a plane 2 μm from the focal plane. (b) The expansion of the image using Zernike polynomials. (c) The absolute value of the expansion coefficients used to in the expansion for (b).

The BTVO approach has a larger computational overhead, needing to train N Gaussian processes rather than a single one in the case of BO. In the example of our pixel data, training tens of thousands of Gaussian processes (one for each pixel) would be much too computationally demanding, as shown schematically in figure 1. Instead we seek to extract the essential information contained in the images, while removing the noise, by expanding the images in each plane in a set of orthogonal basis functions. For the particle Zernike polynomials are chosen due to the radial symmetry of the images. The coefficients of the basis function expansion then become the new measurement data. Expanding our simulated images in the same basis, we can solve the inverse problem by minimizing χ^2 for the basis coefficients instead of the pixel intensities.

3 Results

Figure 2 presents an example image expansion in a series of Zernike polynomials. Also shown are the absolute values of the basis coefficients determining the expansion. These values represent our dimensionally reduced measurement data which we use for solving the inverse problem in place of the pixels intensities of the image depicted in (a).

In this contribution, we will demonstrate the effectiveness of employing the Bayesian target vector optimization method for dimensional microscopy. We present results for model based estimates of line widths based on bidirectional optical measurements and compare with estimates based on other experimental techniques.

4 Acknowledgements

This project is funded by the German Federal Ministry of Education and Research (BMBF, project number 01IS20080A, SiM4diM). This project has received funding from the EMPIR programme co-financed by the Participating States and from the European Union's Horizon 2020 research and innovation programme (projects 20IND04 "ATMOC" and 20FUN02 "POLIGHT").

References

[1] B. Bodermann, R. Köning, D. Bergmann, E. Buhr, W. Hässler-Grohne, J. Flügge, H. Bosse, Proc. SPIE **8788**, 509 (2013)

[2] S. Burger, L. Zschiedrich, J. Pomplun, M. Blome, F. Schmidt, Proc. SPIE **8642**, 864205 (2013)

[3] B. Bodermann, E. Buhr, Z. Li, H. Bosse, Optical imaging and metrology: advanced technologies pp. 255–282 (2012)

[4] M. Plock, A. Andrle, S. Burger, P.I. Schneider, arXiv preprint p. 2202.11559 (2022)

https://doi.org/10.1051/epjconf/202226610011

Near-process indirect surface characterization of laser-chemically produced removal contours

Merlin Mikulewitsch[1,*], *Dirk* Stöbener[1,2], and *Andreas* Fischer[1,2]

[1]University of Bremen, Bremen Institute for Metrology, Automation & Quality Science, Linzer Str. 13, 28359 Bremen, Germany
[2]MAPEX Center for Materials and Processes, P. O. box 330 440, 28334 Bremen, Germany

Abstract. The manufacturing rate of laser chemical machining (LCM) is currently restricted to avoid disruptive boiling bubbles in the process fluid. An increase necessitates adjustments to the laser beam or fluid properties. However, the current understanding of the surface removal mechanisms is insufficient to achieve a consistent removal quality under these conditions. For an improved process modeling, in-process measurements of the surface geometry, the surface temperature and the boiling bubbles are required. Due to the complex process environment, no suitable in-process measurement technique for the geometry or surface temperature exists. This contribution presents an indirect geometry measurement approach based on confocal fluorescence microscopy that offers the potential for near-process application in the LCM process environment. As a result, the micro-geometry of different surfaces is shown to be indirectly measurable under LCM-equivalent process conditions such as thick fluid layers or gas bubbles in the beam path. Furthermore, a combined fluorescence-based measurement of geometry and temperature is proposed.

1 Introduction

Compared to conventional micro-manufacturing processes such as micro-milling, laser chemical machining (LCM) achieves higher dimensional accuracy at acute edge angles and small edge radii [1]. However, its manufacturing rate is significantly lower because the process energy applied to the component surface by means of a focused laser, and thus the removal rate, is severely limited in order to avoid disruptive boiling bubbles in the process fluid. The influence of boiling bubbles on the material removal rate can be reduced by different adjustments of the laser beam and fluid properties (e.g. beam shape or fluid viscosity). However, to achieve an increase in removal rate while maintaining removal quality, the current understanding of surface removal mechanisms must be fundamentally expanded. In this context, comprehensive LCM process modeling that considers the boiling bubble influence is only possible by means of in-process measurements of the surface geometry, the surface temperature and the boiling bubbles in the removal region. However, due to the complex fluid process environment, the gas bubbles occurring during the material removal, and the measurement requirements for the manufactured cavities, there is no suitable in-process measurement technique for the removal geometry or the process-relevant surface temperature [2]. Conventional optical geometry measurement methods are unsuitable for near-process applications due to various aspects. For example, refractive index variations in the chaotic fluid environment prevent the use of interferometric methods, while steep edge angles produce unavoidable artifacts due to unwanted reflections in measurements using confocal microscopy [3]. In contrast, an indirect measurement of the removal geometry using confocal fluorescence microscopy is not subject to these impairments. The method has already been successfully applied close to the process in manufacturing environments with fluid layers as thin as 120 µm [6] and in-situ in fluid layers several millimeters thick [7]. However, to date, no near-process application of the indirect measurement approach has been performed in the LCM process environment. Therefore, it has to be elucidated if the removal geometry can be measured indirectly under the LCM process conditions such as thick fluid layers and gas bubbles or particles in the beam path. In addition, it is important to investigate whether the near-surface temperature can also be measured by means of fluorescence, in order to be able to develop an extended understanding of the process.

2 Measurement approach

The indirect geometry measurement technique is based on confocal fluorescence microscopy with a model-based evaluation of the fluorescence signal to measure the microstructures in the mm-thick fluid layer present in laser chemical machining [7]. In contrast to conventional methods, which use the light scattered from the surface for geometry measurement, the indirect measurement principle determines the fluid boundary layer to the workpiece by detecting the fluorescence light emitted by the fluid, from which the geometry of the workpiece is inferred. Since light is detected even at angles > 75° to the surface normal [4], samples with steep edges can also be

* Corresponding author: m.mikulewitsch@bimaq.de

measured [5]. The confocal optical system used, consisting of pinhole, lens and objective (see Fig. 1a), limits the detection of the fluorescence signal to a confocal volume around the focal plane of the objective. When this volume is moved in z-direction through the fluid, a characteristic fluorescence signal $S(z)$ is generated, which can be modeled as follows:

$$S(z) = \left(\operatorname{erf}\left(\frac{z - z_0}{2\Xi} + \epsilon\Xi\right) - \operatorname{erf}\left(\frac{z - z_1}{2\Xi} + \epsilon\Xi\right)\right) e^{\epsilon(z - z_1)}. \quad (1)$$

Here, the normalized intensity is shown where Ξ represents a parameter describing the properties of the confocal volume, ϵ the absorption coefficient of the fluid, and z_0 and z_1 the respective positions of the measurand-fluid interface (measurand surface) and the fluid-air interface. From the pointwise measured fluorescence signal, this model function can be used to determine the surface position z_0 by using a least-squares approximation. If the measurement is repeated for several lateral surface positions (x, y), the surface geometry $z_0(x, y)$ is obtained. Additionally, the fluid temperature can be measured, since the fluorescence intensity and also the life-time are both temperature-dependent.

3 Measurement results

The results of the indirect in-situ geometry measurements under LCM-equivalent environmental conditions, such as gas bubbles present in the fluid generated during material removal, are shown in Fig. 1. It turns out that the model-based evaluation of the indirect measurement of the fluorescence signal enables indirect measurements even in the presence of interfering gas bubbles in the beam path, see Fig. 1b. Thus, even when gas bubbles are present in the fluid directly above the measured object, a surface position can be determined whose uncertainty correlates with the gas bubble density. Further measurements demonstrate the robustness of the measurement approach towards different practical geometries of both step geometries (Fig. 1c) and surfaces with different surface inclinations (Fig. 1d). As a result, the indirect geometry measurement technique allows the determination of relevant geometric parameters, such as step heights, of micro-objects immersed in fluid layers several millimeters deep. Even 85°-inclined surfaces, comparable to the flanks of laser-chemically manufactured cavities, can be measured indirectly. In contrast to conventional reflection-based confocal microscopy, no undesirably high measurement deviations occur in the regions of greatest slope. Thus, the indirect geometry measurement approach is shown to cope with realistic process conditions such as thick fluid layers, steep surface geometries and contaminated fluids by employing a model-based signal processing approach to determine the desired surface position.

Additionally, first investigations show that due to the temperature dependence of the fluorescence signal, spatially resolved measurements of a constant spatial surface temperature distribution are also possible, provided that the confocal volume is moved over the surface at a constant distance. This highlights the potential for simultaneous measurement of temperature and geometry when temperature is considered in a holistic modeling of the fluorescence signal.

Financial support was received from the Deutsche Forschungsgemeinschaft (DFG), project number 451385285.

References

1. H. Messaoudi, et al., Proc. of the 5th Int. Conf. on New Forming Techn. ICNFT (2018)
2. C. Gerhard, et. al., Prod. Eng. Res. Dev., **4** (2010)
3. J. Liu, C. Liu, J. Tan, B. Yang, T. Wilson, J. of Microsc., **261** (2016)
4. H. Hansen, et al., CIRP Annals, **55** (2006)
5. K. Maruno, M. Michihata, Y. Mizutani, Y. Takaya, Int. J. Autom. Technol., **10** (2016)
6. Y. Takaya, et. al., CIRP Annals, **65** (2016)
7. A. Fischer, M. Mikulewitsch, D. Stöbener, CIRP Annals, **69** (2020)

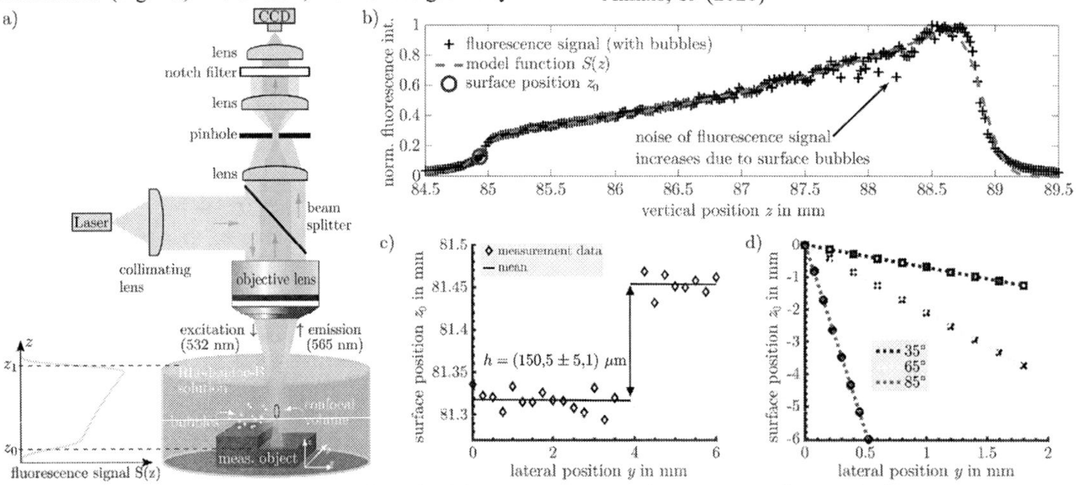

Fig. 1. Indirect geometry measurement : a) Schematic of the measuring system ; b) Fluorescence intensity in the presence of gas bubbles in the beam path ; c) Measurement results of a step and d) of differently inclined surfaces.

Reconstruction of coherence matrix in x-representation using nonclassical Hartmann sensor

Marek Vitek[1], Michal Peterek[1], Dominik Koutny[1], Martin Paur[1], Bohumil Stoklasa[1], Libor Motka[1,*], Zdenek Hradil[1], Jaroslav Rehacek[1] and L. L. Sanchez-Soto[2]

[1] Department of Optics, Palacky University, 17. listopadu 12, 771 46 Olomouc, Czech Republic
[2] Departamento de Optica, Facultad de Fisica, Universidad Complutense, 280 40 Madrid, Spain

Abstract: We show the coherence properties of a signal can be measured by a Hartmann wavefront sensor in a nonclassical regime. Recasting the detection theory of the classical Hartmann sensor in the sense of quantum tomography enables to measure the coherence function, which is an analogy to the density matrix of mixed quantum states. Two methods were tested for the reconstruction of the coherence matrix from the intensity scan in the nonclassical mode of the Hartmann sensor. The reconstruction was performed in a classic way using the POVM matrix and using data pattern tomography.

Introduction

Excimer lasers with an unstable resonator play a crucial role in DUV lithography applications. These laser sources provide a partial spatial coherent signal, and the wavefront description is not applicable in this case. The wrong signal description can lead to errors in production, where these lasers are used. Therefore, it is beneficial to have an experimental tool in hand able to characterize the coherence properties of the signal. Hartmann sensor is a standard technique allowing to test the quality of optical systems. Traditional wavefront sensors deal with fully spatially coherent signals. Going out of this assumption provided data are carefully reconstructed the coherence properties can be retrieved [1].

In cases a measurement apparatus is difficult to characterize and calibrate data pattern tomography can be preferably used. The method was proposed by J.Rehacek, D. Mogilevtsev, and Z. Hradil [2] as a tool for quantum state reconstruction without a priori knowledge of the measurement apparatus. The method was further developed by describing the efficient algorithm [3] and analyzed in different regimes together with the detector tomography [4,5].

Experiment

The key element of our experimental setup: Hartmann sensor is a standard technique allowing to test the quality of optical systems provided the measurement geometry separates the signal from the individual Hartmann mask apertures sufficiently. Hartmann mask can be realized via a digital micromirror device (DMD).

In the nonclassical approach the charge-coupled device (CCD) is placed behind the Hartmann mask in such a way the signal from the individual sub-apertures of the mask interferes mutually. In this case the information about coherence properties is encoded in the experimental data and can be reconstructed.

Fig. 1. DMD wavefront sensor based on Hartmann mask. The signal from the end of the optical fiber is collimated and measured.

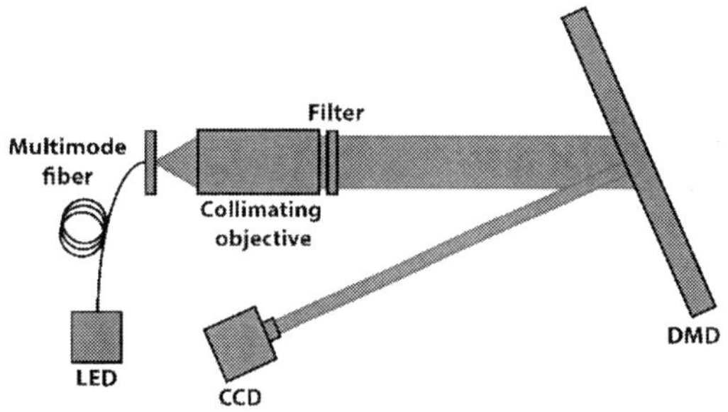

Tomography reconstruction

The state reconstruction can be performed via complete characterization of the measurement based on POVM matrices and careful detector calibration. Alternatively, data pattern tomography can be used provided the calibration states (probes) are measured in the same experimental apparatus.

In the case of DMD wavefront sensor based on Hartmann mask the calibration states can be generated directly using DMD chip by addressing the suitable set of the Hartmann masks within one exposition time.

In the experimental setup the end of the optical fiber was collimated and measured by the way described in the previous paragraph (see Fig. 1). The corresponding degree of coherence generally depends on the optical fiber core diameter, collimator focal length, wavelength and separation of the measured points. Theoretically derived values are depicted in the Fig. 2 together with the measured degree of coherence.

* Corresponding author: libor.motka@upol.cz

https://doi.org/10.1051/epjconf/202226610012

Fig. 2. True and measured coherence matrix of the collimated signal from the optical fiber (core diameter = 105μm).

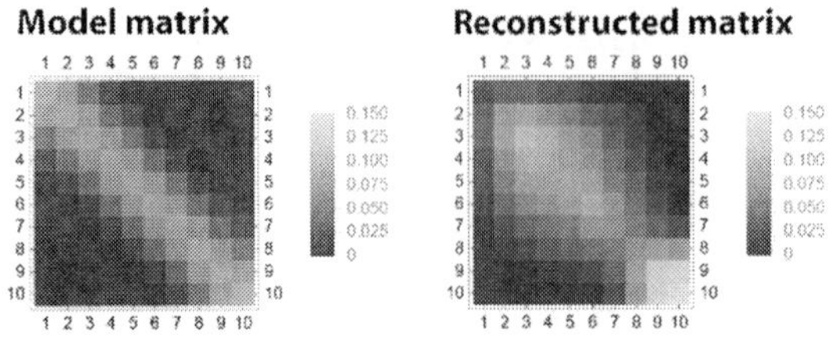

Conclusion

We have experimentally performed measurement of the coherence properties of the partly coherent signal. The key element is DMD wavefront sensor based on Hartmann mask operating in the nonclassical regime where the standard reconstruction approaches are not applicable. In case the interfering signal comes from spatial points we get the x-representation of the coherence matrix directly.

References

[1] B. Stoklasa, L. Moťka, J. Řeháček, Z. Hradil, and L. L. Sánchez-Soto, Wavefront sensing reveals optical coherence, Nature Communications **5**, 3275 (2014).

[2] J.Rehacek, D. Mogilevtsev and Z. Hradil, Operational Tomography: Fitting of Data Patterns, Phys. Rev. Lett 105, 010402 (2010).

[3] L. Moka, B. Stoklasa, J. Reháček, Z. Hradil, V. Karásek, D. Mogilevtsev, G. Harder, C. Silberhorn, and L. L. Sánchez-Soto, Efficient algorithm for optimizing data-pattern tomography, Phys. Rev. A **89**, 054102 (2014).

[4] L. Mota, M. Paur, J. Rehacek, Z Hradil and L. L. Sanchez-Soto, Efficint tomography with unknown detectors, Quantum Sci. Technol. **2**, 035003 (2017).

[5] L. Moťk, M. Paúr, J. Řeháček, Z. Hradil, and L.L. Sánchez-Soto, *Whe quantum state tomography benefits from willful ignorance*, New J. Phys. **23**, 073033 (2021).

Quasi-analytical and rigorous modeling of interference microscopy

Tobias Pahl[1,*], *Johannes* Breidenbach[1], and *Peter* Lehmann[1]

[1]Measurement Technology Group, Faculty of Electrical Engineering and Computer Science, University of Kassel, Wilhelmshöher Allee 71, Kassel 34121, Germany

Abstract. We present an extended vectorial Kirchhoff model of coherence scanning interferometry including several vector rotations occurring in the imagining and scattering process as well as polarization dependent reflection coefficients. For validation simulated results are compared to those of the conventional scalar Kirchhoff model and a rigorous finite element modeling.

1 Introduction

Optical surface topography measurement techniques such as coherence scanning interferometry (CSI) are widespread for fast and contactless measurement of surface features with lateral dimensions of several 100 nm and an axial resolution down to the subnanometer range. However, due to the wave characteristics of light and resulting diffraction effects, optical profilers suffer from systematic deviations occurring in context with surface characteristics [1–3]. In order to predict and analyze these deviations, numerical models, which can be subdevided into quasi-analytic and rigorous models depending on the assumptions made in the computation of the scattering process, are being developed. Usually, quasi-analytic models require less computation time and memory. They provide better insight in physical mechanisms of the imaging and scattering process. On the other hand, rigorous simulations show higher accuracy since no rough approximations are made. In previous studies quasi-analytic models are developed based on Fourier optics or scalar Kirchhoff theory to analyze systematic deviations and the transfer characteristics of CSI systems [3–6]. However, if imaging systems of high numerical aperture (NA) are used, a vectorial mathematical treatment is essential. In [7] a vectorial extension of the scalar Kirchhoff model is presented. However, the model shows inconsistencies since the electric field is not anymore perpendicular to the wave vector after the scattering process.

This contribution presents an extension of the vectorial model, where inconsistencies are fixed and local reflection coefficients depending on the local surface slope are considered. For validation, results obtained from the vectorial model are compared to those of the scalar Kirchhoff theory and a rigorous finite element method (FEM) based model described in [8].

*e-mail: tobias.pahl@uni-kassel.de

2 Model

The models presented in this study consider a Linnik interferometer of high NA and single color LED illumination. Hence, the influence of temporal coherence is neglected. Further, scattering geometries are considered to be invariant under translation in one direction (here y) and apodization as well as aberrations are neglected for simplicity. In general, the interference intensity I_{int} is computed by

$$I_{\mathrm{int}}(x,z) \sim \int_0^{2\pi} \mathrm{d}\varphi_{\mathrm{in}} \int_0^{\theta_{\max}} \mathrm{d}\theta_{\mathrm{in}} I_{\mathrm{int};\theta_{\mathrm{in}},\varphi_{\mathrm{in}}}(x,z)\sin(\theta_{\mathrm{in}})\cos(\theta_{\mathrm{in}}), \tag{1}$$

where θ_{in} is the incidence angle with respect to the optical (z-) axis, φ_{in} the azimuth angle and $I_{\mathrm{int};\theta_{\mathrm{in}},\varphi_{\mathrm{in}}}(x,z) \sim \mathrm{Re}\{\mathbf{E}_{\mathrm{o};\theta_{\mathrm{in}},\varphi_{\mathrm{in}}}(x,z)\cdot\mathbf{E}^*_{\mathrm{r};\theta_{\mathrm{in}},\varphi_{\mathrm{in}}}(x)\}$ the two-beam interference intensity with the electric fields $\mathbf{E}_{\mathrm{o/r}}$ of the object under investigation and the reference mirror, respectively [7]. The calculation of $\mathbf{E}_{\mathrm{o}}$ differs depending on the model.

2.1 Scalar Model

In the scalar model the electric field corresponds to a scalar field distribution $\psi_{\mathrm{o};\theta_{\mathrm{in}},\varphi_{\mathrm{in}}}$. Based on the Beckmann formulation of scattering [9], the field distribution scattered by a periodic phase object of height profile $h(x)$, amplitude ψ_0 and period length L_x is given by $\psi_{\mathrm{o};\theta_{\mathrm{in}},\varphi_{\mathrm{in}}}(x,z) = \psi_0 e^{iq_z(h(x)+z)}$, where $\mathbf{q} = (q_x, q_x, q_z)^T = \mathbf{k}_{\mathrm{s}} - \mathbf{k}_{\mathrm{in}}$ with the scattered and incident wave vectors $\mathbf{k}_{\mathrm{s/in}}$. Hence, the field in the image plane of a microscope obtained from the scattered field can be described by the Fourier series

$$\psi_{\mathrm{o};\theta_{\mathrm{in}},\varphi_{\mathrm{in}}}(x,z) = \sum_{n=n_{\min}}^{n_{\max}} \sqrt{\frac{k_{\mathrm{s};zn}}{k_{\mathrm{in};z}}} c_n e^{i2\pi nx/L_x - iq_{z,n}z}, \tag{2}$$

where $n_{\min}$ and $n_{\max}$ are limited by the NA of the objective lenses, c_n is the Fourier coefficient [3] and the root term considers energy conservation.

2.2 Vectorial extension

If objective lenses of high NA are used, a vectorial treatment of the scattering process increases the accuracy of modeling. Therefore, the electric field is calculated by

$$\mathbf{E}_{\mathrm{o};\theta_{\mathrm{in}},\varphi_{\mathrm{in}}}(x,z) = \sum_{n=n_{\mathrm{min}}}^{n_{\mathrm{max}}} \underline{R}^{\mathrm{T}}(\theta_{\mathrm{s};n},\varphi_{\mathrm{s};n})\underline{T}(\theta_{\mathrm{s};n},\varphi_{\mathrm{s};n}) \sqrt{\frac{k_{\mathrm{s};zn}}{k_{\mathrm{in};z}}}\mathbf{c}_n \mathrm{e}^{\mathrm{i}2\pi nx/L_x - \mathrm{i}q_{z;n}z}, \quad (3)$$

where $\underline{T}$ represents a rotation matrix in x, z plane ensuring that the electric field remains perpendicular to the corresponding wave vector, $\underline{R}^{\mathrm{T}}$ describes the rotation of the scattered electric field passing the objective lens and

$$\mathbf{c}_n = \frac{1}{L_x}\int_{-L_x/2}^{L_x/2} \mathrm{d}x \mathrm{e}^{-\mathrm{i}q_z h(x)}\mathrm{e}^{-\mathrm{i}2\pi nx/L_x} \left[R_{\perp;\theta_{\mathrm{in}},\varphi_{\mathrm{in}}}(x)\mathbf{E}_{\mathrm{in},\perp;\theta_{\mathrm{in}},\varphi_{\mathrm{in}}}(x) + R_{\parallel;\theta_{\mathrm{in}},\varphi_{\mathrm{in}}}(x)\mathbf{E}_{\mathrm{in},\parallel;\theta_{\mathrm{in}},\varphi_{\mathrm{in}}}(x)\right] \quad (4)$$

is the 3D Fourier coefficient. The reflection coefficients $R_{\perp;\theta_{\mathrm{in}},\varphi_{\mathrm{in}}}$, $R_{\parallel;\theta_{\mathrm{in}},\varphi_{\mathrm{in}}}$ as well as the components $\mathbf{E}_{\mathrm{in},\perp;\theta_{\mathrm{in}},\varphi_{\mathrm{in}}}$, $\mathbf{E}_{\mathrm{in},\parallel;\theta_{\mathrm{in}},\varphi_{\mathrm{in}}}$ of the incident electric field $\mathbf{E}_{\mathrm{in};\theta_{\mathrm{in}},\varphi_{\mathrm{in}}}$ perpendicular and parallel to the plane defined by the incident wave vector and the surface normal $\mathbf{n}(x)$ vary with respect to the x axis due to the dependency of $\mathbf{n}$ on the surface slope. The incident electric field $\mathbf{E}_{\mathrm{in};\theta_{\mathrm{in}},\varphi_{\mathrm{in}}} = \underline{R}(\theta_{\mathrm{in}},\varphi_{\mathrm{in}})\mathbf{E}_0$ is given by the rotation of the initial electric field vector $\mathbf{E}_0$ by the rotation matrix $\underline{R}$ considering the rotation of the field being focused. The normal vector is calculated numerically based on the derivative of $h(x)$.

3 Results

Figure 1 shows interference signals obtained from a sinusoidal nickel surface profile with maximum surface slope of 18° and $L_x = 6\,\mu\mathrm{m}$ simulated with monochromatic illumination of wavelength $\lambda = 440\,\mathrm{nm}$ using the scalar model (a), the vectorial model (c) and the FEM model presented in [8] (e). In all three cases, the sinusoidal profile can be seen clearly in the course of the maximum interference intensity. For better comparability Figs. 1(b, d, f) show the interference signals in hybrid $x\,q_z$-space, where a band pass filtering behavior according to the surface slope is observed as explained and experimentally confirmed by Künne et al. [10]. Since FEM calculates the light scattering process rigorously, FEM results are used for reference. Comparing Figs. 1(b) and 1(d), the vectorial modeling shows a slight improvement of the models accuracy. However, having regard to surface profiles including steep surface slopes, multiple scattering and shadowing effects become more important leading to still significant deviations between quasi-analytic and rigorous modeling. Nonetheless, in order to analyze transfer characteristics of microscopic imaging systems usually analytical models are preferred and hence, the vectorial extension may improve the accuracy of quasi-analytical models.

4 Conclusion

A scalar Kirchhoff model is extended by a vectorial treatment, where several rotations of the electric field during the focusing and scattering processes are considered. Comparing results obtained by scalar and vectorial Kirchhoff modeling to rigorous FEM results, the vectorial modeling provides slightly different and more realistic results compared to the scalar modeling. Therefore, considering objective lenses of high NA and surfaces of certain characteristics, a vectorial treatment of the imaging and scattering process may be preferred in order to analyze the transfer behavior of CSI systems in future studies.

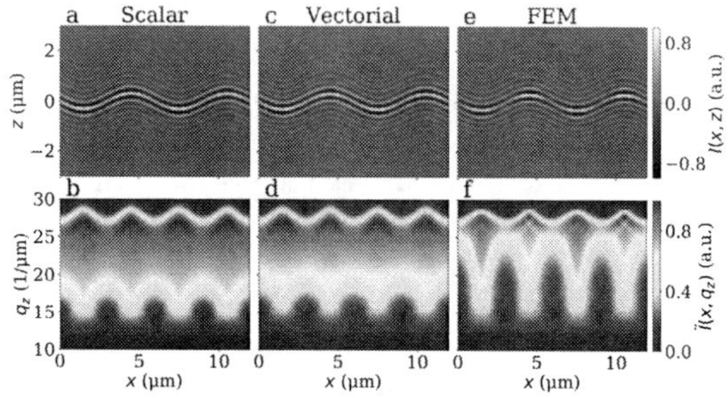

Figure 1: Offset reduced interference signals obtained from a sinusoidal surface profile with $L_x = 6\,\mu\mathrm{m}$ and maximum surface slope of 18° simulated with the scalar model (a), vectorial model (c) and FEM (e). The bottom row displays the Fourier transform of the interference intensities shown in the upper row with respect to the optical axis.

References

[1] A. Harasaki, J. Wyant, Applied Optics **39**, 2101 (2000)

[2] P. de Groot, X. Colonna de Lega, J. Kramer, M. Turzhitsky, Applied Optics **41**, 4571 (2002)

[3] W. Xie, P. Lehmann, J. Niehues, S. Terechenko, Optics Express **24**, 14283 (2016)

[4] P. de Groot, X. Colonna de Lega, JOSA A **37**, B1 (2020)

[5] R. Su, R. Leach, Light: Advanced Manufacturing **2**, 1 (2021)

[6] P. Lehmann, S. Hagemeier, T. Pahl, Metrology **1**, 122 (2021)

[7] T. Pahl, S. Hagemeier, M. Künne, C. Danzglock, N. Reinhold, R. Schulze, M. Siebert, P. Lehmann, Proc. SPIE **11783**, 117830G (2021)

[8] T. Pahl, S. Hagemeier, M. Künne, D. Yang, P. Lehmann, Optics Express **28**, 39807 (2020)

[9] P. Beckmann, A. Spizzichino, *The scattering of electromagnetic waves from rough surfaces* (Artech House, Inc., USA, 1987)

[10] M. Künne, T. Pahl, P. Lehmann, Proc. SPIE **11782**, 117820T (2021)

Coherent Fourier scatterometry for particle detection on structured surfaces

Anubhav Paul[1,*], *Dmytro* Kolenov[1], and *Silvania F.* Pereira[1]

[1]Optics Research Group, Imaging Physics Department, Faculty of Applied Sciences, Delft University of Technology Lorentzweg ,2628 CJ Delft, The Netherlands

Abstract. We demonstrate the detection of particles/contamination present on a structured surface using Coherent Fourier scatterometry (CFS) by applying Fourier filtering to the scanned maps, which eliminates background effects due to the electronic noise as well the structure itself. We show that by using filters in the Fourier space we can significantly improve the detection capabilities of the particles present on the structure.

1 Background

One of the most revolutionizing technological advancement in the past few decades have been observed in the semiconductor/nano-tech industry, which resulted in drastic modification in our society and in general helped improve the living standard of human beings. However, as we strive towards betterment the need for fabrication of more compact and accurate nano-structures is of paramount importance. For that accurate inspection of contamination on surfaces and gaseous environments is extremely crucial to guarantee the quality of the fabricated devices. Particle detection and particle metrology plays a role in the quantification of large number of particles on environments, but is also highly important in the detection of a very small number of particles (contamination) on ultra-clean surfaces and vacuum environments. Quantification of the latter is of great importance since it assures the quality and ultimately determines the yield of the fabricated structures.

In recent years, it has been shown that Coherent Fourier scatterometry (CFS) is a suitable technique for this purpose [1]. CFS is an optical metrology technique based on the light scattered from the object in the far field, which is very sensitive to detect small isolated particles or small changes in parameters of nanostructrues. In addition, the technique is not limited by diffraction, and the detection of deep subwavelength-size particles present on top of a plane substrate has been recently demonstrated [2]. However, as we move towards patterned surfaces, the detection of contamination becomes more complicated as we have to consider the effects of the background structure in addition to that of the contamination itself. In this work, we show a procedure on how to tackle this problem by filtering out the signal due to the background structure, creating a significant contrast between the scattering of the structure and the scattering of the defect/particle, making the problem analogous to particle detection on a plane substrate.

*Corresponding Author: A.Paul-1@tudelft.nl

2 Experimental setup

The experimental setup of CFS is similar to a confocal microscope where a collimated and linearly polarized He-Ne laser (λ = 633 nm) beam passes through a non-polarizing beamsplitter which latter goes to a microscope objective (NA = 0.9) and is focused on the sample of interest. The sample is placed on a piezo-stage that can be laterally scanned in a raster fashion (i.e., the stage scans consecutive lines at the time, separated by a certain distance between the lines). The reflected and scattered light from the sample is collected by the same objective and goes back through the beamsplitter. The back focal plane of the objective is de-magnified by a telescopic system consisting of two lenses and imaged at a split detector. The split detector has two halves that are aligned at direction perpendicular to the scan direction of each line. The intensity voltage from the left half is subtracted from the right half, resulting in a position dependent differential signal. With this detection system, the specular reflections for which the far field is symmetric is eliminated and we observe the differential signal to be zero. As the focused beam passes through a change in the sample structure, the far field becomes asymmetric and we observe a position dependent differential signal. In the case of a flat surface, the particle detection is straightforward since the the detector signal due to the surface only (without particles) is always close to zero. However, this is not the case if the substrate is not uniform but contains structures on it.

3 Results and discussions

We demonstrate that Fourier filtering technique along with the CFS can be used to improve the detection capability of particles on a (nano)structure. Here we intend to filter out all the spatial frequencies present in the scanned maps except that of the particles/defect present on the structure in order to isolate the frequencies due to the substrate from the frequencies due to the particle/defect. As we know, the frequencies of the light scattered by the particles are in the higher-frequency ranges. We study different filters on the Fourier space, like discrete high-pass filter, 2D cosine window superimposed with a high pass filter, and Gaussian notch filter.

We observe that, for each filter, there is a significant improvement in the detection capabilities of the CFS. However, we have to keep in mind that some frequencies of the background signal and the particles/defects may overlap and filtering those might decrease the SNR of the signal due to the particle slightly.

Furthermore, combination of weighted filters may further improve the SNR which was lowered due to the elimination of the overlapping signals.

Using this principle, we show the detection of subwavelength defects on grating structures that are used as quality control of lithographic systems.

References

[1] S. Roy et al., Rev. Sci. Instrum. **86**, 123111 (2015)
[2] D. Kolenov et al., Optics Express **29**, 16487 (2021)

https://doi.org/10.1051/epjconf/202226610015

Coherent Fourier Scatterometry for defect detection on SiC samples

Jila Rafighdoost[1,*], *Dmytro* Kolenov, [1], and *Silvania F. Pereira* [1]

[1]Optics Research Group, Imaging Physics Department, Faculty of Applied Sciences, Delft University of Technology, Lorentzweg 1, 2628 CJ Delft, The Netherlands

Abstract. Coherent Fourier Scatterometry (CFS) is a scatterometry technique that has been applied for grating and nanoparticle detection. Here, it has been challenged to verify the detectability of the so-called killer defects on SiC samples for power electronic applications. It has been shown that CFS is able to precisely recognize these defects regardless of their shape or size. CFS could be considered as a possible alternative for this purpose.

1 Introduction

Scatterometry is a well-known and successful metrology technique in the semiconductor industry, and it has been mainly for the optical characterization of periodic samples [1, 2]. It is non-imaging method that is not limited by diffraction and also can be fast and non-destructive.
Coherent Fourier Scatterometry (CFS) is one of those techniques that have been developed based on scatterometry in which a coherent light beam is focussed on the sample and the Fourier plane of the objective lens is imaged on a CCD camera. [3]

CFS has been applied for grating characterization [4] and for detection of isolated spherical nanoparticles [5,6]. The implementation of the latter is similar to the one used for grating reconstruction except that the CCD camera is replaced by a split detector. Here, we adapt the CFS particle detection scheme to detect isolated defects of various shapes. High quality large area samples with low concentration of typical defects known for SiC technology are provided. The defects have roughly been located with a polarization microscope with magnification of 20x, NA=0, 4, in cross-polarization mode. So far different methods have been used to analyse and detect these defects [7]. The main goal of this work is to verify whether CFS is able to detect, locate and classify these isolated defects.

2 Experimental Setup

In a conventional CFS setup, as shown in Figure 1, a linearly polarised He-Ne laser with a wavelength of λ=632 nm is coupled into a single-mode fibre and collimated by an off-axis parabolic mirror before it passes through the beam splitter (BS1) and focused on the desired sample by a high numerical aperture objective (NA=0.9). The sample is mounted on a 2D piezo-electric stage whose position can be controlled with sub-nm precision (P-629.2CD by Physik Instrumente). During the experiment, the sample is moved in the lateral direction (x-y) in raster scan mode (parallel lines along the x axis). The z-position can also be finely tuned to find the best focus position on the sample. The reflected and scattered field is captured by the same objective and directed back to BS1. Then a telescopic arrangement formed by L1, L2, L3, and (BS2) is used to image the back focal plane of the lens into the camera aperture and the split detector (ODD3W2 BiCell Silicon Photodiode). The camera is only used here for localizing features on the sample. The split detector is aligned with both sides perpendicular to the direction of the scan of the stage (x axis). Both sides of the split detector are integrated and subtracted from each other generating a photocurrent I. that is dependent on the x-y position of the stage.

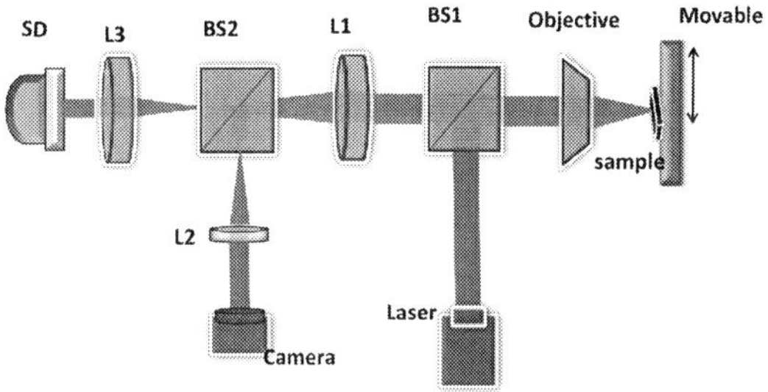

Fig. 1. The schematic of the CFS setup. The sample is placed on the stage with a piezo-electric XYZ translator, BS1 and BS2: 50-50 non-polarizing beam splitters, L1, L2, and L3: positive lenses; Camera: CCD camera used for localizing features on the sample; SD: split detector (bi-cell silicon photodiode). A He-Ne laser is used. The split detector is aligned perpendicular to the scan direction of one line of the raster scan (x axis).

2.1. Experimental results

* Corresponding author: j.rafighdoost@tudelft.nl

Samples containing a few types of special defects of interest provided by AIXTRON (ref. WF0053-40-5) have been analysed. The defects are called carrots and triangles. Two examples of the results of the detection of defects are shown in Figure 2. The raster scan takes place line by line. This is a method of construction of an image through the use of the signal on the split detector that is recorded in some fixed time intervals throughout each line scan (X axis). Then, the scan table is displaced in the y-axis by a Δy step. The values of Δy can be adapted. The smaller Δy is, more details can be obtained from the defect, but of course, it needs more lines and consequently, it takes a longer time to cover the same area than if it is scanned with a bigger Δy. Here we define the total scan T= $t_x \times$ N lines, where t_x is the scan time per line and N is the number of lines. In the present system, the optimum value for t_x is 150 ms, and it is presently limited by the response of the piezo table. In Figure 2 we show the results of the scattered maps of two types of defects, showing that one can clearly distinguish these two types of defects. Scatterometry is not an imaging technique and in the scan maps one sees the changes in the scattered patterns at the borders of the object, and in this way, the defect can be recognised. The scan time can be considerably reduced if bigger scan steps are taken at the cost of loss of resolution in recognizing a particular shape.

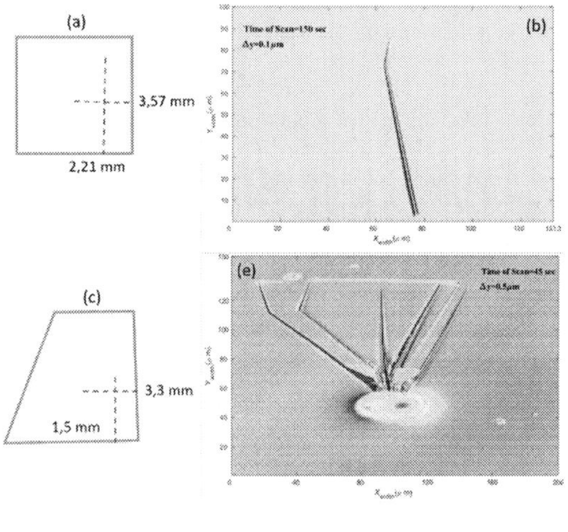

Fig. 2. (a) and (c) the rough location of the defects on SiC samples provided by AIXTRON. (b) and (e) represents the high-resolution scans of the defects, carrot and triangle, respectively. The scanned area and total time of scan T are indicated in each scan map.

3 Conclusion

We have shown how CFS can be applied for detection of non-symmetrical defects so-called killer defects in power electronics applications. The samples with such defects have been provided by AIXTRON together with a rough information about their locations and shapes. It has been shown that regardless the size or shape of the defects, they are all recognizable using our technique. CFS has the advantages of being a compact setup that needs very low power laser (60 μW in this case) and can be installed in line. However, in this stage, it is relatively slow compared to other known techniques especially when it comes to inspection of large areas. The speed could indeed be improved using different raster scan schemes and also by implementing parallel multiple laser spots.

We acknowledge AIXTRON for providing high quality state of the art samples that have been used for these measurements and the financial support from the Project 20IND09 "PowerElec" within the program EMPIR. The project is jointly supported by the European Commission and the participating countries within the European Association of National Metrology Institutes (EURAMET).

References

1. P. Boher, M. Luet, T. Leroux, J.Petit, P.Barritault, J.Hazart, P.Chaton, *Metrology, Inspection, and Process Control for Microlithography XVIII*, SPIE, 1302-1313, (2004).
2. C. Raymond, *AIP Conference Proceedings*, American Institute of Physics, 394-402, (2005)
3. O. El Gawhary, N. Kumar, S. F. Pereira, W. M. J. Coene, H. P. Urbach, Appl. Phys, B **105**, 775-781 (2011).
4. N. Kumar, P. Petrik, G. K. P. Ramanandan, O. El Gawhary, S. Roy, S. F. Pereira, W.M. J. Coene, H. P. Urbach, Opt. Express, **22** , 24678-24688, (2014).
5. S. Roy, Maryse Bouwens, Lei Wei, S. F. Pereira, H. P. Urbach, P. van der Walle, Rev. Sci. Instrum. **86**, 123111 (2015).
6. D. Kolenov, I. E. Zadeh, R. C. Horsten, S.F. Pereira, Opt. Expr. **29**, 16487-16505 (2021).
7. P.C. Chen, W.C. Miao, T. Ahmed, Y.Y. Pan, C.L. Lin, S.C. Chen, H.C. Kuo, B.Y. Tsui, D.H. Lien, Nanoscale Res. Lett., **17**(1), 1-17, (2022)

 https://doi.org/10.1051/epjconf/202226610016

Stokes CMOS polarimetry limits studied at non-classical polarisation states

Eva Roiková[1,*], and *Štěpán* Kunc[2]

[1] Technical University of Liberec, Faculty of Mechatronics, Informatics and Interdisciplinary Studies, Institute of New Technologies and Applied Informatics, 461 17 Liberec 1, The Czech Republic
[2] Technical University of Liberec, Faculty of Science, Humanities and Education, Department of Physics, 461 17 Liberec 1, The Czech Republic

Abstract. We present the study of the two polarisation state analysers. The first consists of a polarisation camera with a removable QWP, and the second consists of a non-polarisation camera with a rotating QWP and a stationary linear polariser. The theoretical analysis and experiment focus on studying the influence of polarimeter optical components accuracy and errors such as retardation errors, misalignments and extinction ratio on Stokes parameters precision. This research is a cornerstone to understanding polarisation state analysers limits. We examined laser beams with non-classical polarisation distribution, namely the Poincaré beam and the beam with radial polarisation.

1 Introduction

In recent years, laser beams with spatially variable polarisation distribution have been widely studied. A wide range of potential applications as various possible generation methods undoubtedly causes an increase in their popularity [1-4].

We aim to analyse the Bessel beam (BB) and the structured laser beam (SLB) with non-classical polarisation distribution. A longitudinal electric field could arise using an eligible input polarisation in focused electromagnetic fields like BB [5]. The knowledge of Stokes parameters (eq. 1) is required to detect the longitudinal component [6]. Therefore, our present research emphasises the influence of the polarimeter optical components accuracy and errors such as retardation errors, misalignments and extinction ratio on Stokes parameters precision.

$$\mathbf{S} = \begin{pmatrix} S_0 \\ S_1 \\ S_2 \\ S_3 \end{pmatrix} = \begin{pmatrix} I_0 + I_{90} \\ I_0 - I_{90} \\ I_{+45} - I_{-45} \\ I_R - I_L \end{pmatrix} \quad (1)$$

$$2\psi = \mathrm{atan}\frac{S_2}{S_1};\quad 2\chi = \mathrm{atan}\frac{S_3}{\sqrt{S_1^2 + S_2^2}} \quad (2)$$

$$Dolp = \frac{\sqrt{S_1^2 + S_2^2}}{S_0} \quad (3)$$

Our research focuses on studying two Stokes polarimetry approaches regarding spherical coordinates (eq. 2) and degree of linear polarisation (Dolp) with real optical parameters (eq. 3).We examined laser beams with non-classical polarisation for analytical and experimental analysis, namely the Poincaré beam and radial polarisation.

The most general case of the laser beam with arbitrary spatially variable polarisation is the full-Poincaré beam. The Poincaré sphere describes the polarisation vectors of the beam (see Figure 1) [7]. A particular case of the Poincaré beam is a cylindrical vector beam with spiral polarisation distribution. This polarisation is linear at radial directions and axially symmetric. A particular case of spiral polarisation is azimuthal and radial polarisation [8].

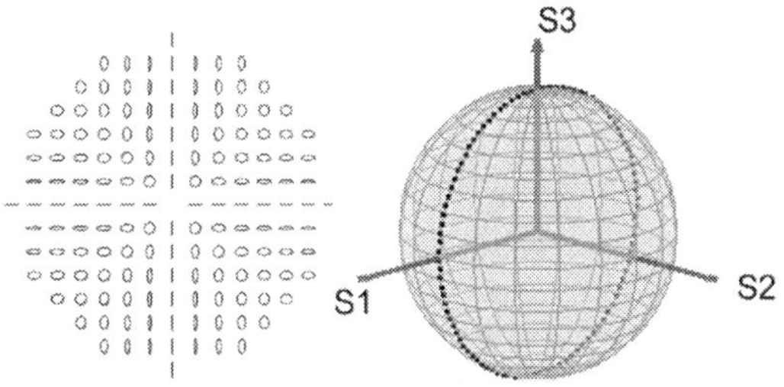

Figure 1: Visualisation of the ideal polarisation vector distribution in the cross-section of the Poincaré beam. The polarisation vectors/ellipses in the cross-section symbolise linear, right-handed and left-handed polarisation. Polarisation handedness is the same for diagonal quadrants. The polarisation vectors of the beam are depicted on the Poincaré sphere.

* Corresponding author: eva.roikova@tul.cz

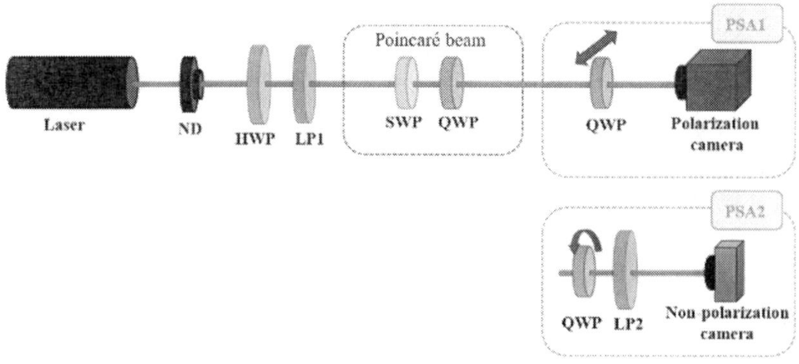

Figure 2: Schematic representation of polarisation distribution measurement in Poincaré beams. The laser beams were analysed with two different polarisation state analysers, PSA1 with a polarisation camera or PSA2 with a non-polarisation camera (Laser: He-Ne 632,8 nm, ND: neutral density filter 2.0, HWP: half-wave plate, LP1: linear polariser, LP2: linear polariser, SWP: structured half-wave plate, QWP: quarter-wave plate)

2 Methodology

The experimental optical setup (see Figure 2) consisted of a neutral density filter and a half-wave plate, which were used to set the input intensity of the laser beam. A linear polariser LP1 indicated a direction of input polarisation. To obtain a vector beam or Poincaré beam, a structured half-wave plate (SWP) (Altechna), SWP and a quarter-wave plate (QWP), respectively, were placed in the beam path.

The laser beams were analysed with two different polarisation state analysers (PSA). PSA1 consists of the polarisation camera (Thorlabs Kiralux®) with a removable QWP for Stokes parameters measurement. PSA2 is a classical approach using a CMOS camera with a rotating QWP and a stationary linear polariser. Stokes parameters evaluation was based on B. Schaefer et al. work [9].

A detailed understanding of the limits of CMOSs and PSAs will enable future imaging of BBs and SLBs with a fixed longitudinal electric field component.

3 Conclusions

The polarisation vector distribution in the cross-section of the real Poincaré beam and its depiction on the Poincaré sphere analysed by PSA1 is shown in Figure 3.

The experiment and the theoretical analysis of measurement uncertainty of PSA1 and PSA2 suggested that although the wire grid polariser of the polarisation camera has a relatively low extinction ratio (100:1) compared to LP2 used in PSA2 (10^5:1), it could provide better polarimetry approach for the beam with non-classical polarisation. The main benefit of PSA1 lies in the smaller number of required beam cross-section images. If we do not require knowledge of the handedness of the polarisation ellipse, only one image is needed, whereas PSA2 requires eight images.

Detailed information about the theoretical analysis of the real optical component influence (retardation errors, misalignments, and extinction ratio) to spherical coordinates evaluation and Dolp as the experiment specifics and results will be presented at the conference.

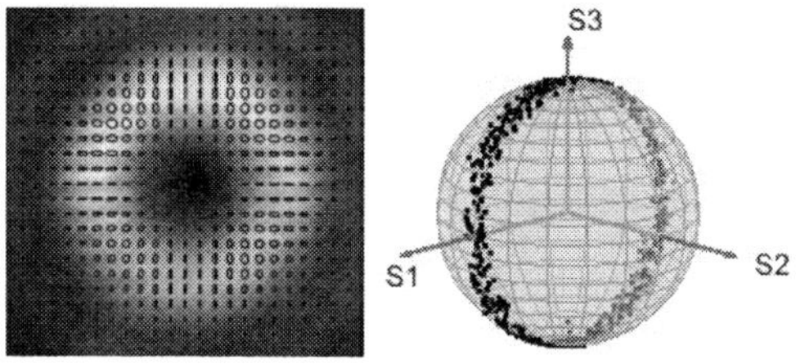

Figure 3: Polarisation vector distribution in the cross-section of the real Poincaré beam analysed by PSA1. The polarisation vectors/ellipses in the cross-section symbolise linear, right-handed and left-handed polarisation. Polarisation handedness is the same for diagonal quadrants. The polarisation vectors of the beam are depicted on the Poincaré sphere

This research was supported by the project Student Grant Contest from TUL (SGS-2021-3058).

References

[1] Q. Zhan, Opt. Express, **12**, 3377–3382, (2004)
[2] L. J. Wong, F. X. Kärtner, Opt. Express, **18**, 25035–25051, (2010)
[3] G. Machavariani, Y. Lumer, I. Moshe, A. Meir, S. Jackel, Opt. Commun., **281**, 732–738, (2008)
[4] F. Ding, Y. Chen, Y. Yang, S. I. Bozhevolnyi, Adv. Opt. Mater., **7**, 1900724, (2019)
[5] S. N. Khonina, S. V. Alferov, S. V. Karpeev, Opt. Lett., **38**, 3223-3226, (2013)
[6] D. Maluenda, M. Aviñoá, K. Ahmadi, R. Martínez-Herrero, A. Carnicer, Sci Rep, **11**, 17992 (2021)
[7] A. M. Beckley, T. G. Brown, M. A. Alonso, Opt. Express, **18**, 10777–10785, (2010)
[8] Q. Zhan, Adv. Opt. Photon., **1**, 1–57, (2009)
[9] B. Schaefer, E. Collett, R. Smyth, D. Barrett, B. Fraher, Am. J. Phys., **75**, 163 (2007)

Achieving the ultimate optical resolution

Luis L. Sánchez-Soto[1,2,*], *Zdeněk* Hradil[3], *Jaroslav* Řeháček[3], *Benjamin* Brecht[4], and *Christine* Silberhorn[4]

[1]Departamento de Óptica, Facultad de Física, Universidad Complutense, 28040 Madrid, Spain
[2]Max-Planck-Institut für die Physik des Lichts, 91058 Erlangen, Germany
[3]Department of Optics, Palacký University, 771 46 Olomouc, Czech Republic
[4]Integrated Quantum Optics Group, Paderborn University, 33098 Paderborn, Germany

Abstract. The accurate estimation of the separation between two signals is at the core of many modern technologies. We show new quantum-inspired schemes able to estimate that separation at the quantum limit. The method works in the spatial, temporal, and frequency domains. The question of whether the optical coherence brings any metrological advantage to mode projections is discussed.

In optics, when attempting to find the separations s between two incoherent emitters, the standard way of proceeding is to take a full image and extract the single parameter $s$from it. In such scenarios, the smallest precisely measurable separation is limited by the point-spread function (PSF), with a vanishingly small amount of information available for smaller and smaller separations, especially when photon shot noise is the dominant noise source (as in, for example, astronomical observations).

This limit is usually formalized in terms of the Fisher information, which quantifies the amount of information accessible per photon detection and is directly associated with the time-honored Cramér-Rao lower bound. For direct intensity imaging, the Fisher information drops to zero for separations smaller than the spread of the optical field, sometimes referred to as the Rayleigh curse, which limits the usefulness of photon counting for metrology.

Surprisingly, when one calculates the quantum Fisher information [1, 2] (i.e., optimised over all the possible quantum measurements), the associated quantum Cramér-Rao lower bound maintains a fairly constant value for any separation of the sources. This shows the potential for parameter estimation of distributions with precision unaffected by Rayleigh's curse. The key behind these techniques is phase-sensitive measurement in mode bases other than intensity [3].

In this work, we extend these techniques and show how mode-selective measurements can be utilised to estimate separations well below the spread of the source light [4]. These quantum-inspired measurements are capable of estimating the separation accurately in a regime where intensity-only measurements would be ineffective.

We implement these ideas in the temporal domain by performing spectrotemporal mode-selective measurements through the quantum pulse gate (QPG), a sum-frequency generation process in a group-velocity engineered waveguide. By shaping the strong QPG pump pulse and measuring the upconverted photon, the QPG implements a projective measurement onto the spectrotemporal mode defined by the QPG pump [5]. For Gaussian pulses, the optimal measurement is the Hermite-Gaussian basis. Measuring only the first two Hermite-Gauss modes allows us to estimate the sub-bandwidth spectral separation between two incoherently mixed pulses, well below the Cramé-Rao lower bound for conventional intensity detection.

Furthermore, we go beyond single-parameter and show how our technique can be used for simultaneous multiparameter estimation of the centroid, the separation, and the relative intensities of two incoherent sources [6, 7], in both frequency and time. This is possible by performing optimal detection, in this case, extending our projective measurements to higher-order Hermite-Gauss modes and their superpositions. By outperforming standard intensity-only detection, our quantum multiparameter estimation scheme can bring benefit to many practical scenarios.

The discussion thus far assumes incoherence between the signals. This conforms with the conditions underlying Rayleigh's criterion. However, a recent heated debate addressed the role of coherence in the resolution limits [8–11], with diverging conclusions.

We show here that coherence by itself does not provide a direct metrological advantage; incoherent superpositions set the ultimate limits in all cases [12]. However, we stress that coherence can be exploited as an information sorter, distributing information about different parameters into different channels.

References

[1] M. Tsang, E. Nair, and X. M. Lu, Phys. Rev. X **6**, 031033 (2016)
[2] M. Paúr, B. Stoklasa, Z. Hradil, L. L. Sánchez-Soto, J. Rehacek, Optica **3**, 1144 (2016)

*e-mail: lsanchez@fis.ucm.es

[3] J. Řeháček, M. Paúr, B. Stoklasa, Z. Hradil, and L. L. Sánchez-Soto, Opt. Lett. **42**, 231 (2017)
[4] J. M. Donohue,V. Ansari, J. Řeháček, Z. Hradil, B. Stoklasa, M. Paúr, L. L. Sánchez-Soto, and C. Silberhorn, Phys. Rev. Lett. **121**, 090501 (2018)
[5] V. Ansari, G. Harder, M. Allgaier, B. Brecht, and C. Silberhorn, Phys. Rev. A **96**, 063817 (2017)
[6] J. Řeháček, Z. Hradil, B. Stoklasa, M. Paúr, J. Grover, A. Krzic, and L. L. Sánchez-Soto, Phys. Rev. A **96**, 062107 (2017)
[7] V. Ansari, B. Brecht, J. Gil-Lopez, J. M. Donohue, J. Rehacek, Z. Hradil, L. L. Sánchez-Soto, C. Silberhorn, PRX Quantum **2**, 010301 (2021)
[8] W. Larson and B. E. A. Saleh, Optica **5**, 1382 (2018)
[9] M. Tsang and R. Nair, Optica **6**, 400 (2019)
[10] Z. Hradil, J. Řeháček, L. Sánchez-Soto, and B.-G. Englert, Optica **6**, 1437 (2019)
[11] K. Liang, S. A. Wadood, and A. N. Vamivakas, Optica **8**, 243 (2020)
[12] S. De, J. Gil-Lopez, B. Brecht, C. Silberhorn, L. L. Sánchez-Soto, Z. Hradil, J. Rehacek, Phys. Rev. Res. **3**, 033082 (2021)

Influence of camera temperature on MTF measurements with finite image distance

Markus Schake[1,*] and *Michael* Schulz[1,**]

[1]Physikalisch-Technische-Bundesanstalt, Bundesallee 100, 38116 Braunschweig

Abstract. Line Spread Function (LSF) based Modulation Transfer Function (MTF) measurements with finite image distance are sensitive to displacement errors in axial direction. Axial displacements between the sample and camera detector cause defocusing and thus, a MTF error proportional to the axial gradient of the sample's MTF. This article demonstrates the influence of the camera temperature on the focus position in the MTF reference setup at PTB.

1 Introduction

The MTF is a commonly employed key performance imaging parameter to asses the quality of optical lenses [1, 2]. It is a metric specifically defined for optical systems with linear transfer behaviour with respect to the electric field or intensity. The MTF is linked to the spatial resolution of the imaging system and describes the systems amplitude transfer characteristic for input signals of different spatial frequencies. The PTB employs a LSF based MTF measurement system with a slit object at infinity and finite image distance to the sample as proposed in ISO 9335 [3]. A sketch of the measurement setup is depicted in Fig. 1.

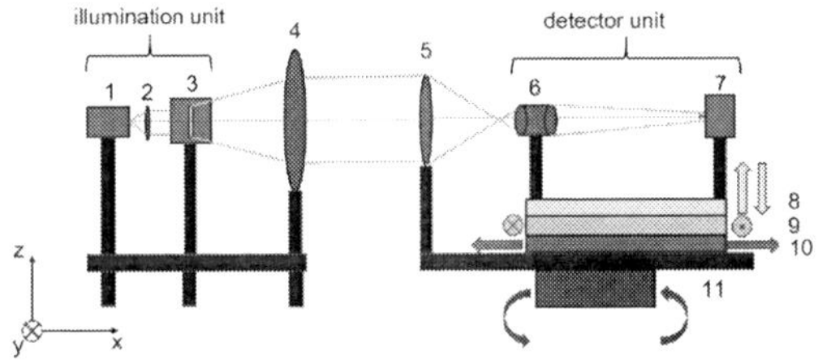

Figure 1. Schematic depiction of the MTF measuring setup [4]: 1) Fibre coupled led light source (Schott KL 2500), 2) optical system comprising a diffusor, condensor and wavelength filter ($\lambda \approx 546$ nm), 3) object slit, 4) optical collimator $D_c = 110$ mm, $f = 1100$ mm, 5) sample mount, 6) microscope objective for finite imaging, 7) camera detector, 8) z-positioner, 9) y-positioner, 10) x-positioner, 11) rotation axis to assign different field angles.

A thin slit object (3), which is unresolvable by the optical system, is imaged through the combination of a collimating lens (4) and the sample under test (5). The sample employed here is a calibration objective with infinite to finite imaging, a focus distance of $f_s = 50$ mm and a f-number $a_s = 4$. The image produced by the sample under test is magnified employing a relay microscope objective (6) and finally recorded by the camera detector (7). In the imaging plane the sharp rectangular slit object will appear as a blurred LSF. The form of the LSF encodes the characteristic transfer behaviour of the optical system, which is analysed in the frequency domain by its corresponding MTF. For more information about the setup at PTB the reader is referred to [4, 5]. MTF measurement systems are sensitive to defocus effects [2, 3]. Therefore, it is important to analyse the influence of thermal defocus as part of the measurement uncertainty contributions.

*e-mail: markus.schake@ptb.de
**e-mail: michael.schulz@ptb.de

2 Experiment and Results

A characteristic datum point of the sample under test is its on axis focus position P_f. To determine the focus position the focal criterion as described in [3] (5.1.5 Focusing a)) is employed. By this focal criterion, the test specimen is focused on-axis, if the MTF value at a given spatial frequency ν_0 reaches its maximum. The influence of camera temperature on MTF measurement with finite image distance is determined by measuring the shift in the focus datum point $\Delta P_f(t, \tau)$, while the camera is heating up. For this purpose the MTF measurement setup is shut down for multiple hours until a thermal equilibrium at room temperature is established. The setup is then turned on ($t_0 = 0$ min) and the required settings for a repeated focus scan measurement are applied. This takes approximately $t_1 = 3$ min, during this time the camera already heats up. The detector unit scans the focal region of the sample under test M times, measuring the MTF at discrete positions $x_n \forall n \in [1, \cdots, N]$. This results in M through focus curves of the MTF evaluated at the frequency ν_0, which describe the changes in the MTF when the sample gets defocused. The maximum of the through focus curve indicates the focus position according to the focal criterion [3] (5.1.5 Focusing a)) described above $P_f = x|\max_x(\mathrm{MTF}(x, \nu_0))$. Fig. 2 shows the through focus curves of a measurement at camera temperature $\tau_1 = 39.55\,°\mathrm{C}$ at time $t_1 \approx 3\,\mathrm{min}$ in comparison to the stationary camera temperature $\tau_M = 48.44\,°\mathrm{C}$ at time $t_M \approx 193\,\mathrm{min}$ at the end of the recording. It is visible that the focus position has changed by $\approx 3\,\mu\mathrm{m}$, which causes a high deviation $\max(|\Delta\mathrm{MTF}|) \approx 0.05$ between the recorded MTF curves. This is in good agreement with the defocusing tolerances for a lens with f-number $a_s = 4$ at $\lambda = 500\,\mathrm{nm}$ and $\nu_0 = 100\,\mathrm{lp/mm}$ given in table 1 of ISO 9335 [3]. Fig. 3 shows the shift of the focus position

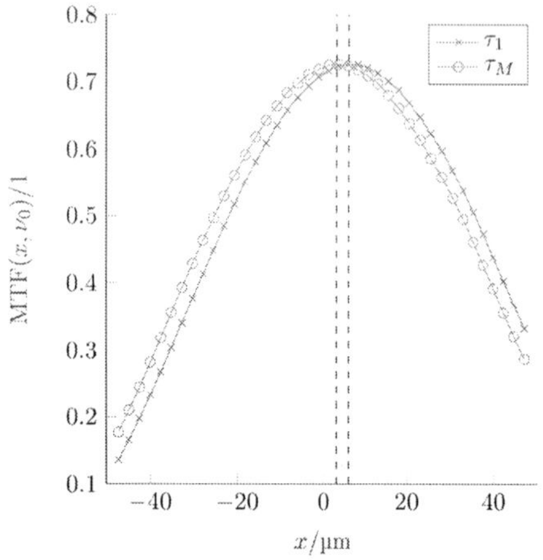

Figure 2. Comparison of the recorded MTF through focus curves evaluated at the spatial freuqency $\nu_0 = 100\,\mathrm{lp/mm}$ at temperature $\tau_1 = 39.55\,°\mathrm{C}$ at time $t_1 \approx 3\,\mathrm{min}$ (blue crosses) and the stationary camera temperature $\tau_M = 48.44\,°\mathrm{C}$ at time $t_M = 193\,\mathrm{min}$ (red circles). The black dashed lines indicate the position of the maximum of the MTF through focus curves, which coincide with the focus position P_{f1}, P_{fM} respectively.

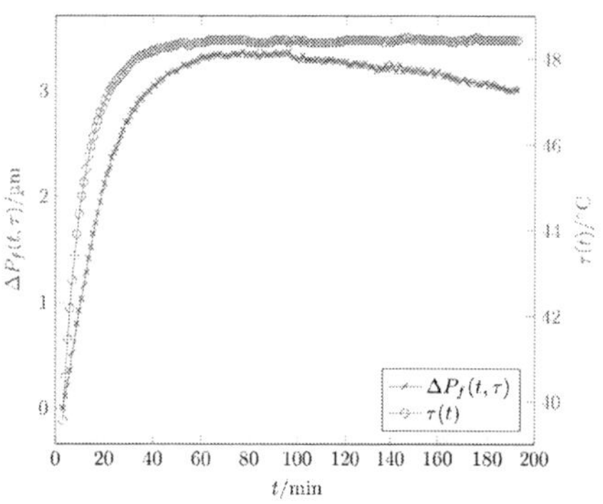

Figure 3. Shift of the focus position $\Delta P_f(t, \tau)$ with respect to the focus position observed in the first measurement (blue crosses). Camera temperature $\tau(t)$ over time t (orange circles).

$\Delta P_f(t, \tau)$, with respect to the focus position observed in the first measurement, in comparison to the camera temperature. It appears, that the shift of the focus position follows the temperature. It is likely, that the observed shift in the focus position is caused by the thermal expansion of the camera and thus, the thermal displacement of the camera chip. The camera temperature is measured at the camera core board inside the housing. The camera model is the a2A5320-23umBAS from Basler, the housing is a combination of zinc diecasting mount ($\alpha_{\mathrm{Zn}} \approx 30.2\cdot10^{-6}\,\mathrm{K}^{-1}$) and a steel sheet ($\alpha_{\mathrm{Steel}} \in [11, 13]\,10^{-6}\,\mathrm{K}^{-1}$) with dimensions of $42.8\,\mathrm{mm} \times 29\,\mathrm{mm} \times 29\,\mathrm{mm}$. Assuming that the housing temperature will converge to a stationary temperature that is lower than the core board temperature, the shift in focus position of $\Delta P_f \approx 3\,\mu\mathrm{m}$, which occurs for a core board temperature change of $\Delta\tau \approx 9\,°\mathrm{C}$ may reasonably be assigned to a shift of the image sensor caused by the thermal expansion of the housing. The ambience room temperature was recorded during the experiment and showed a peak to valley deviation of $\mathrm{PV}_{\tau_{\mathrm{amb}}} = 0.38\,°\mathrm{C}$, thus its influence on the measurement is considered to be negligible.

3 Conclusion

The presented results demonstrate, that the heating of the camera sensor may cause large deviations in the measured MTF. The MTF deviations are dependent on the specimen specific MTF gradient and the axial displacement of the camera chip. The measurement should not be started before the system has reached a stationary temperature.

References

[1] J.W. Goodman, *Introduction to Fourier Optics*, 2nd edn. (McGraw Hill, New York, 1996)

[2] G.D. Boreman, *Modulation Transfer Function in Optical and Electro-Optical Systems*, 1st edn. (SPIE, United States, 2001)

[3] International Organization for Standardization, *Optics and photonics — Optical transfer function — Principles and procedures of measurement* (1995), "https://www.iso.org/standard/17017.html"

[4] M. Schake, M. Schulz, *Identification of aliasing effects in measurements of unknown MTFs*, in *Optical Metrology and Inspection for Industrial Applications VIII*, International Society for Optics and Photonics (SPIE, 2021), Vol. 11899, pp. 67 – 80, https://doi.org/10.1117/12.2601176

[5] S. Fohrmann, P. Erichsen, M. Schake, M. Schulz, *Impact of detector position accuracy on the MTF measurement of optical lenses*, in *Optifab 2021*, International Society for Optics and Photonics (SPIE, 2021), Vol. 11889, pp. 15 – 30, https://doi.org/10.1117/12.2602744

Investigation of dynamic influences in tilted-wave interferometry

Gregor Scholz[1,2,*], *Michael* Schulz[1], *Ines* Fortmeier[1]

[1] Physikalisch-Technische Bundesanstalt (PTB), Bundesallee 100, 38116 Braunschweig, Germany
[2] Physikalisch-Technische Bundesanstalt (PTB), Abbestraße 2-12, 10587 Berlin, Germany

Abstract. Aspherical and freeform lenses allow for compact optical systems and have therefore gained high interest in optics. The interferometric measurement of these forms is a challenge, for which the tilted-wave interferometer (TWI) has been developed. To evaluate the measurement uncertainty of the TWI, both the static and the dynamic influence parameters have to be investigated. In this work, we focus on the dynamic influences on the measurement data of the interferometer. To this end, the individual influences as well as their point of insertion into the process chain are identified. Then the measurement of the interferogram data is modelled as a Monte Carlo simulation. The propagation of different influences through the data process chain to the optical path length differences (OPDs) is also simulated, and the resulting variation of the OPDs is estimated. Furthermore, the variation of the OPDs resulting from measured interferogram data is investigated for comparison. The analysis and quantification of variation of the OPDs along with its contributing influence sources are important steps on the way towards a full uncertainty estimation of optical form measurement with the TWI.

1 Introduction

Aspherical and freeform surfaces have gained high interest in optics due to their ability to form compact lens systems that can correct image aberrations with few optical elements. To enable their highly accurate production, accurate measurement systems capable of measuring these surfaces are in high demand. However, large deviations from flat or spherical shapes lead to high fringe densities in classical interferometry, making these surfaces challenging to measure. Therefore, the tilted-wave interferometer (TWI, [1]) offers a promising solution by utilizing a 2D microlens array and solving an inverse problem to reconstruct the surface form under test from the measured interferograms. To achieve high surface reconstruction accuracy, both the optical model of the TWI and the optical path length differences (OPDs) between the test arm and the reference arm of the interferometer, which are determined from the measured interferograms, have to be known accurately. The OPDs are especially affected by dynamic disturbances such as unstable illumination, imprecise phase measurement, image detector noise, etc. To investigate the measurement uncertainty of a TWI, the knowledge of the magnitude and effect of these parameters is of high importance.

In this investigation, we have used a digital twin of the TWI [2] together with a Monte Carlo approach to identify individual sources of uncertainty. We have also analysed them in terms of their impact on the variation of the OPDs. To this end, we have analysed the propagation of the sources of uncertainty through the data process chain of the TWI. Moreover, we have related the results to estimates of the variation of the OPDs gained by conducting experiments. The investigations presented here are based on the software and hardware realization of the TWI at the Physikalisch-Technische Bundesanstalt (PTB) [3].

2 Method

2.1. Investigation of individual parameters

The distribution and magnitude of individual uncertainty influence parameters of the OPDs can be investigated either by experiment, as, for example, in the case of the image sensor noise that is analysed from many consecutive measurements, or by prior knowledge. In addition to the magnitude, the way in which the OPDs are affected is also of importance. For example, different parameters can either directly influence the physical OPDs within the interferometer, the phase measurement, or the image acquisition of the interferograms.

2.2 Monte Carlo simulation

To analyse the influence of the different parameters on the OPDs calculated from the measured interferogram data, a Monte Carlo simulation of the measurement process and data process chain is performed. Undisturbed OPDs of a virtual specimen are therefore simulated by raytracing using a digital twin of the TWI. Subsequently, variations directly affecting the physical OPDs, such as wavelength instabilities, can be added. After this, the OPDs are converted to phase information, which is then converted

* Corresponding author: gregor.scholz@ptb.de

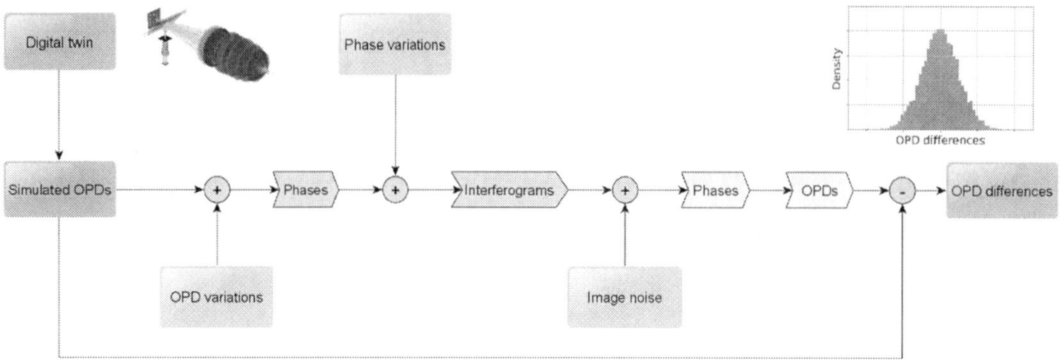

Fig. 1. Diagram of the Monte Carlo simulation process. The simulated OPDs are modified by the addition of OPD, phase, and image variations, and the OPD difference is calculated.

to a set of five different phase-shifted interferograms, which are used for the phase determination (see [4]). At the interface of this conversion, errors affecting the phase measurement, such as imprecise phase-shifting elements, are introduced. The resulting interferograms are converted to image data, and disturbances affecting the image data, for instance, image sensor noise or image clipping are added. The artificially generated interferograms are then processed in the same way as the measured interferograms in the TWI measurement process are usually processed: The five phase-shifted interferograms are used to retrieve the phase using Hariharan's phase retrieval [4] and Goldstein's phase unwrapping algorithm [5] to determine the OPDs. The difference of the disturbed OPDs and the original simulation can then be analysed by statistical methods. The process is depicted in Fig. 1. With this framework, the effects of different uncertainty parameters can either be analysed individually or in combination.

2.3 Experimental measurement

The experimental estimation of the variation of the OPDs can be performed by consecutive measurements of the same test specimen at the same measurement position within the TWI. With an appropriate number of measurements and precautions to keep the experimental setup stable for the required time, the mean of the OPDs and their variations can be determined. With this method, only the total variations can be determined, without knowledge of the error sources contributing to the variations. Nevertheless, the experimental result is important to compare the results to the virtual experiments and check the assumptions made.

3 Discussion

Quantifying the variation of the OPDs is crucial to the estimation of the measurement uncertainty of the TWI. Therefore, Monte Carlo methods are used to investigate the influence of individual uncertainty sources on the resulting OPDs. This not only helps to predict the variation of the OPDs from the uncertainty sources, but also helps future optimizations of the system. Furthermore, the investigation of the variation of the OPDs calculated from consecutive experimentally measured interferogram data helps to evaluate the quality of the simulation. The comparison between both approaches can moreover be used to evaluate the fraction an individual source contributes to the overall OPD variation.

The estimated OPD variation can be used for further analysis of the form measurement uncertainty of the TWI by applying the respective error influences to simulation experiments with the digital twin of the TWI. In combination with other uncertainty sources and appropriate data analysis methods, this can lead to an uncertainty budget of the TWI and therefore help to establish the TWI as a highly accurate measurement system for aspherical and freeform surfaces.

The authors thank Manuel Stavridis for the development of the SimOptDevice software tool and his support.

References

1. G. Baer, J. Schindler, C. Pruss, J. Siepmann, W. Osten, Opt. Express, **22**(25), 31200, (2014)
2. G. Scholz, I. Fortmeier, M. Marschall, M. Stavridis, M. Schulz, C. Elster, Metrology, **2**(1), pp. 84–97, (2022)
3. I. Fortmeier, M. Stavridis, M. Schulz, C. Elster, Meas. Sci. Technol., **33**(4), 045013, (2022)
4. P. Hariharan, B. F. Oreb, T. Eiju, Appl. Optics, **26**(13), pp. 2504–2506, (1987)
5. R. M. Goldstein, H. A. Zebker, C. L. Werner, Radio Science, **23**(4), pp. 713–720, (1988)

https://doi.org/10.1051/epjconf/202226610020

Influence of displacement gradients on laser speckle photography

Schweickhardt León[1,*], *Tausendfreund* Andreas[1], *Stöbener* Dirk[1,2], and *Fischer* Andreas[1,2]

[1]University of Bremen, Bremen Institute for Metrology, Automation and Quality Science (BIMAQ), Germany
[2]University of Bremen, MAPEX Center for Materials and Processes, Germany

Abstract. The influence of first and second order displacement gradients on laser speckle photography is investigated in a simulative study that is supported with experimental data. The systematic error is found to scale linearly with the second order gradient, while the random error scales with the first order gradient. The gradient-based error dominates the uncertainty budget of an in-process measurement during single tooth milling close to the machined surface.

1 Introduction

Laser speckle photography (LSP) is a fast and non-invasive measurement technique for displacement and strain fields. The optical measurement principle enables in-process applications with high requirements on both spatial and temporal resolution such as surface inspection during manufacturing. Laser speckles are used as surface markers in order to calculate local in-plane displacements with a two-dimensional digital image correlation (DIC) [1]. The spatial discretization along with under-matched shape functions for the subsets in the DIC algorithm lead to systematic measurement errors in the case of displacement fields with higher order displacement gradients [2]. High displacement gradients occur when either the load or the material properties are highly inhomogeneous, i.e. close to the tool during manufacturing or around cracks and notches [3].

For DIC with white-light speckles, theoretical and experimental studies were conducted on systematic errors caused by displacement gradients [2]. However, laser speckles are an interference pattern and do not deform like white-light speckles that are a part of the deforming surface. While a rigid body movement of the sample leads to a global displacement of the entire speckle pattern, displacement gradients alter the pattern of the speckles, which leads to a decorrelation of the two speckle images (before and after deformation) and, thus, affects the DIC evaluation [1]. It remains to be investigated to what degree the systematic errors are caused by displacement gradients in LSP. Of particular interest is the comparison of systematic to random errors, the latter of which are mainly dominated by speckle noise.

2 Methods

In order to statistically study the effect of different displacement gradients on speckle patterns and quantify the resulting systematic errors, a speckle simulation was employed. The simulation setup is schematically shown in Fig. 1. First, an isotropic, Gaussian surface topography with a desired roughness S_q and lateral correlation length L is simulated with the moving average method [4]. A displacement gradient is then applied to the surface topography. In order to more accurately represent subpixel displacements and avoid aliasing, the surface is upsampled by a factor of 30 via bicubic interpolation before the deformation. The speckle image is then simulated according to [1]. The aperture D of the optical 4f-system can be varied in order to adjust the speckle size. An augmented-Lagrangian digital image correlation [5] is used to calculate the resulting displacement fields, which can then be compared to the predefined displacement. The simulated displacements vary only along the abscissa. Thus, the deviation of the mean along the abscissa from the respective predefined displacement value yields the systematic error, while the standard deviation of the mean along the ordinate yields the random error.

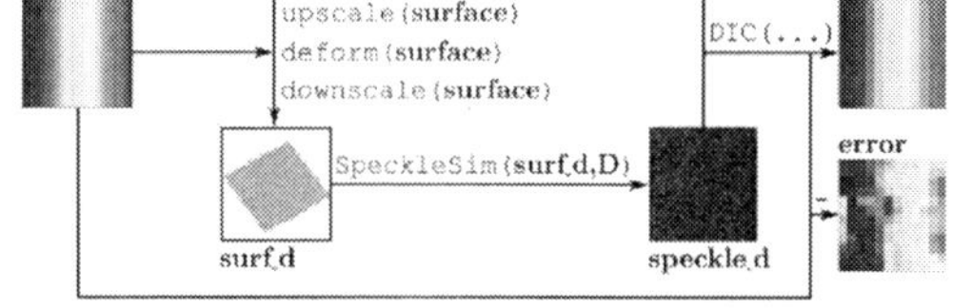

Fig. 1. Speckle simulation setup.

3 Results

We investigate systematic errors that are caused by first and second order displacement gradients by simulating a displacement field, which is composed of two parabolic segments (one with a positive and one with a negative

[*] Corresponding author: l.schweickhardt@bimaq.de

Laplacian), i.e. the displacements d_x scale quadratically along the x-axis. The specified displacement profile and the displacements calculated with DIC from the simulated speckle patterns are shown in Fig. 2a. Deviations of the DIC results from the true values occur where the displacement field has higher order gradients. In addition, the moving average of the predefined displacement is calculated in Fig. 2a with a subset size equal to that of the DIC. The deviations of the DIC result from the moving average indicate that the systematic errors are not only caused by the spatial averaging over an evaluation window. Fig. 2b shows the random error and the displacement gradient over x. Evidently, the displacement gradient not only causes systematic errors, but also increases the random error due to decorrelation. In Fig. 2c the displacement interval in x-direction is varied. Thereby the first and second order displacement gradients, i.e. the Laplacian, are adjusted. Identical to the findings from Xu [2] on white-light speckles, the systematic error of LSP depends linearly on the Laplacian of the displacement field. Decreasing the DIC evaluation window size W_{ev} reduces the systematic error, but in turn increases the random error.

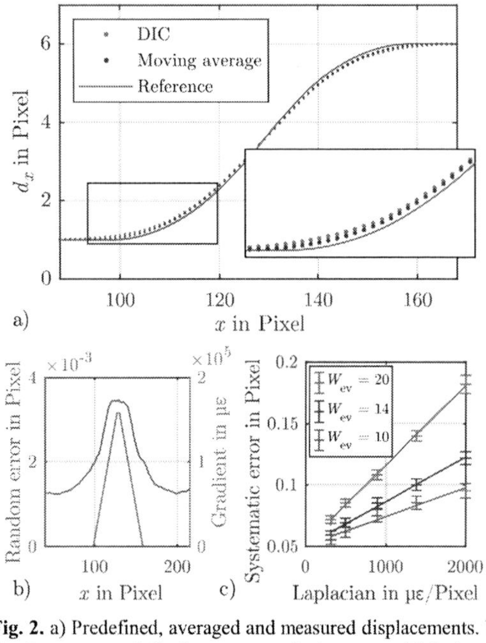

Fig. 2. a) Predefined, averaged and measured displacements. b) Random error and displacement gradient over x. c) Maximum systematic error over Laplacian of displacement for three evaluation window sizes W_{ev}.

In order to demonstrate the relevance of the systematic errors studied here, an in-process measurement during a single tooth milling process from [6] is considered. Fig. 3a shows the displacement field captured with LSP. The highly localized load causes displacement gradients at the machined surface. In Fig. 3b the measured displacements are plotted in y-direction, along with the modelled displacements in the simulation. From the simulation, we estimate a maximum systematic error of around 60 nm with $W_{ev} = 10$ Pixel. The maximum cumulative uncertainty of the measurement reported in [6] is 22 nm. Thus, systematic errors due to displacement gradients are a significant and previously not considered contribution to the measurement uncertainty budget, which deserve further attention.

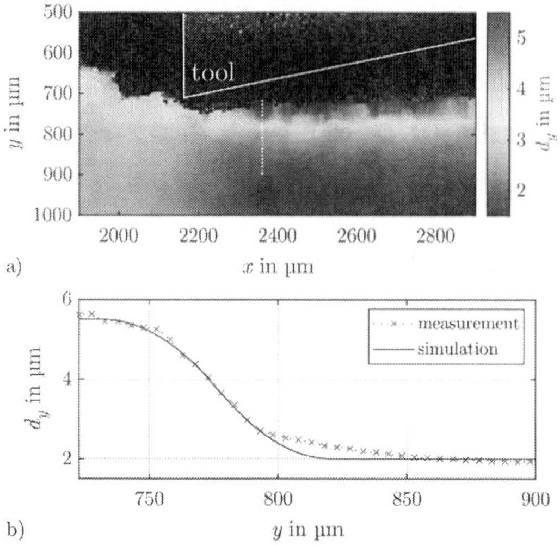

Fig. 3. a) Displacement field measured during single tooth milling. b) Measured displacements at $x = 2360$ µm (indicated as dotted white line in a) compared to simulated displacements.

4 Conclusion

Through a simulative study that is supported with experimental data it was shown that significant systematic errors in LSP measurements are caused by displacement gradients. The systematic errors scale linearly with the Laplacian of the displacement field and decrease with smaller DIC evaluation windows. Displacement gradients also cause the random errors to increase due to the decorrelation of the evaluated image pairs. For an in-process measurement of single tooth milling it is estimated that systematic errors due to displacement gradients are a dominant factor close to the machined surface, i.e. where a measurement is of particular interest due to the maximal load that the workpiece experiences. Further experiments with reference measurements are needed to validate the simulative findings.

The authors gratefully acknowledge the financial support by the DFG for subproject C06 'Surface optical measurement of mechanical working material loads' within the Transregional Cooperative Research Center SFB/TRR136.

References

1. J. W. Goodman, *Speckle phenomena in optics* (2007)
2. X. Xu, Y. Su, Q. Zhang, Opt. Lasers Eng., **88** (2017)
3. J. Tong, Fatigue Fract. Eng. Mater. Struct., **41**(9) (2018)
4. A. K. Fung, M. F. Cheng, Opt. Soc. Am. A, **2**(12) (1985)
5. J. Yang, K. Bhattacharya, Exp. Mech., **59**(2) (2019)
6. A. Tausendfreund, D. Stöbener, A. Fischer, J. Manuf. Mater. Process., **2**(1) (2018)

Fabrication influences on a miniaturised stokes polarimeter consisting of stacked nano-optical wire grid polarizer and retarders

Thomas Siefke*, Markus Walther, Carsten Stock and *Uwe D.* Zeitner

Friedrich Schiller University Jena, Institute of Applied Physics, Albert-Einstein-Str. 15, 07745 Jena, Germany

Abstract. The polarization properties of light can be fully controlled with nano-optical wire grid polarizers and artificial birefringent grating structures. We demonstrate an integrated polarimeter based on stacked layers of such elements. However, the optical performance of such elements is fundamentally limited and may be further altered by deviations arising from the fabrication processes. In this contribution we investigate the influences on the polarimetry performance for such a device.

1 Introduction

Nowadays optics technology is an important driving force in many aspects of our everyday life. Highly developed optical systems are enabling self-driving cars [1,2], high-resolution images with smartphone cameras and advanced medical services [3,4]. To achieve that, modern fabrication technology strives to ever smaller and more demanding structures. This is only achievable by closed loop process control. While many solutions are existing for that purpose, often elaborate equipment is required. As an example, ellipsometry can be utilized to retrieve information about thickness and refractive index of a deposited thin film [5]. While the achievable results are excellent with very small uncertainties, the required devices are large and expensive. By more sophisticated devices, it is even possible to deduce detailed information about the geometry of nanostructures [6]. Unfortunately, the size of such tools reduces their application range drastically. We propose a miniaturized polarimeter consisting of pixelated and stacked polarizer and retarder structures on an image sensor.

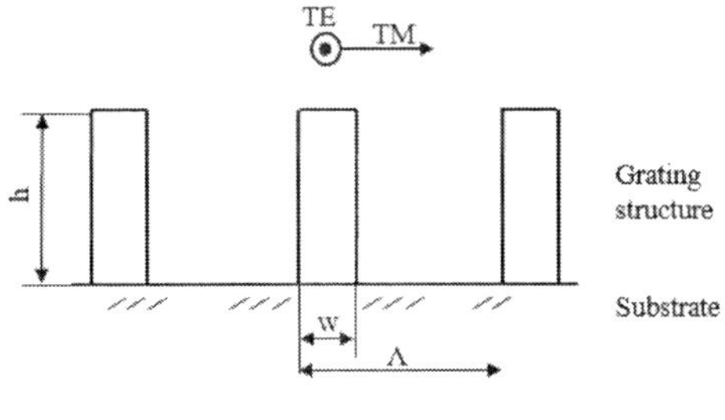

Fig. 1. Schematic grating structure for polarisation control. Direction for TE and TM polarized light is marked.

Zero-order grating structures (see Fig. 1) can be utilized to control the polarisation state of light. For this purpose, the grating period must be sufficiently small, such that only the zeroth diffraction order is transmitted. Depending on the utilized material either artificial birefringence or di-attenuation can be observed. For an absorbing grating material an exponential damping occurs for TE polarized light (TE-transversal electric, electric field vector parallel to ridges) while TM-polarized light is mostly transmitted [7]. Hence, the transmittance of such structures strongly depend on the polarisation direction, so that these elements can be used as linear polarizers (LP). If the grating material is dielectric the effective refractive index depends on the polarisation direction, hence such structures can be used as quarter wave plates (QWP) [8].

In order to measure the Stokes parameter of incident light a QWP and a LP are typically utilized. Classically, four (or six) subsequent measurements at different rotation angles of the linear polarizer, with and without the quarter wave plate, are performed to deduce the full set of stokes parameters[9]. In the here shown concept the LP and QWP are stacked in different layers of the optical element on a microscopic scale. Since the structures are fabricated by lithography, their direction can be locally changed (see Fig. 2). Thereby, multiple channels can be achieved on an image sensor. Combining the information of all channels, the stokes parameters can be evaluated.

Fig. 2. Schematic of an integrated Stokes polarimeter consisting of a (top to bottom) birefringent grating as QWP, wire grid polarizer as LP and an image sensor.

In Fig. 3, an SEM image of a cross section of one pixel of such a stacked nano-optical device is shown. The lower grating layer is the QWP structure consisting of titanium dioxide [8,10]. The upper layer is the LP structure consisting of iridium [11]. The intermediate layer is fabricated by filling the underlying grating with a sacrificial polymer and over coating of porous silicon oxide [12,13]. The sacrificial layer is removed by oxygen plasma ashing, creating a void in between the grating ridges. Thereby a high refractive index contrast is achieved. The wire grid polarizer is capped by over coating with ion beam deposition under an oblique angle.

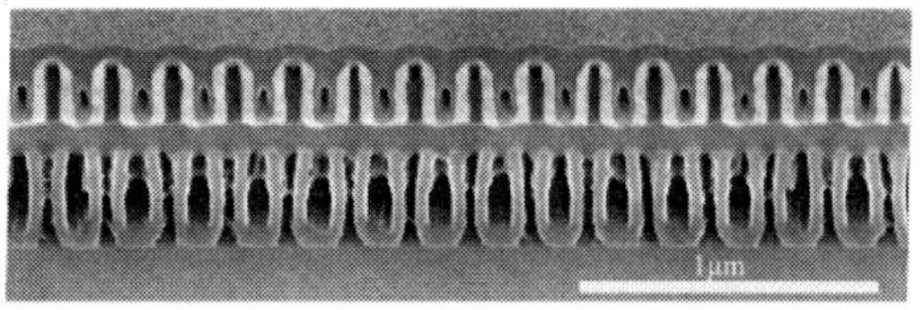

Fig. 3. Cross sectional SEM image of a layered grating structure. The upper layer consists of Ir and serves as linear polarizer, the lower layer consists of titanium dioxide and serves as retarder.

Typically the optical performance achievable with such nano-optical elements is limited compared to their classical counterparts [14,15]. That means the phase of the QWP may deviate from the ideal value and the extinction ratio (ratio between transmittance for TE and TM polarized light) of the LP is limited to values <1000 [16]. In this contribution we investigate the influence of these deviations on the optical performance of such a Stokes polarimetry by rigorous coupled wave analysis, Monte Carlo simulations and ray tracing. Furthermore, we propose advanced designs exploiting the unique possibilities available due to the lithographic fabrication techniques.

2 Conclusion

Nano-optical wire grid polarizers and retarders allow the full control of the polarization properties of light. Furthermore, since these elements are fabricated by lithography, they can be arranged on an image sensor in a pixelated manner. Thereby, multiple polarization channels are achieved enabling the single-shot evaluation of the Stokes parameter of incident light. However, the optical performance of such elements is limited. In this contribution, we investigate the influence on the performance to miniaturised polarimetry.

3 Acknowledgement

This work is funded through the project 20FUN02 POlight within the Programme EMPIR. The EMPIR initiative is co-founded by the European Unions Horizon 2020 research and innovation program and the EMPIR Participating Countries.

References

1. C. Häne, L. Heng, G. H. Lee, F. Fraundorfer, P. Furgale, T. Sattler, and M. Pollefeys, "3D visual perception for self-driving cars using a multi-camera system: Calibration, mapping, localization, and obstacle detection," Image Vis. Comput. **68**, 14–27 (2017).
2. C. Badue, R. Guidolini, R. V. Carneiro, P. Azevedo, V. B. Cardoso, A. Forechi, L. Jesus, R. Berriel, T. M. Paixão, F. Mutz, L. de Paula Veronese, T. Oliveira-Santos, and A. F. De Souza, "Self-driving cars: A survey," Expert Syst. Appl. **165**, 113816 (2021).
3. D. Qiao, F. Zulkernine, R. Masroor, R. Rasool, and N. Jaffar, "Measuring Heart Rate and Heart Rate Variability with Smartphone Camera," Proc. - IEEE Int. Conf. Mob. Data Manag. **2021-June**, 248–249 (2021).
4. K. Kaur, W. Chelangat, S. I. Druzhinin, N. W. Karuri, M. Müller, and H. Schönherr, "Quantitative E. coli Enzyme Detection in Reporter Hydrogel-Coated Paper Using a Smartphone Camera," Biosens. 2021, Vol. 11, Page 25 **11**(1), 25 (2021).
5. A. Gottwald, K. Wiese, T. Siefke, and M. Richter, "Validation of thin film TiO <inf>2</inf> optical constants by reflectometry and ellipsometry in the VUV spectral range," Meas. Sci. Technol. **30**(4), (2019).
6. T. Käseberg, J. DIckmann, T. Siefke, M. Wurm, S. Kroker, and B. Bodermann, "Mueller matrix ellipsometry for enhanced optical form metrology of sub-lambda structures," in *Proceedings of SPIE - The International Society for Optical Engineering* (2019), **11057**.
7. T. Siefke, E.-B. Kley, A. Tünnermann, and S. Kroker, "Design and fabrication of titanium dioxide wire grid polarizer for the far ultraviolet spectral range," in *Proceedings of SPIE - The International Society for Optical Engineering* (2016), **9927**.
8. C. Stock, T. Siefke, and U. Zeitner, "Metasurface-based patterned wave plates for VIS applications," J. Opt. Soc. Am. B **36**(5), D97 (2019).
9. A. Forbes, A. Dudley, K. Singh, and N. Tabebordbar, "Digital Stokes polarimetry and its application to structured light: tutorial," JOSA A, Vol. 37, Issue 11, pp. C33-C44 **37**(11), C33–C44 (2020).
10. T. Siefke, S. Kroker, K. Pfeiffer, O. Puffky, K. Dietrich, D. Franta, I. Ohlídal, A. Szeghalmi, E. Kley, and A. Tünnermann, "Materials Pushing the Application Limits of Wire Grid Polarizers further into the Deep Ultraviolet Spectral Range," Adv. Opt. Mater. **4**, 1780–1786 (2016).
11. T. Weber, T. Käsebier, A. Szeghalmi, M. Knez, E.-B. Kley, and A. Tünnermann, "Iridium wire grid polarizer fabricated using atomic layer deposition," Nanoscale Res. Lett. **6**(1), 558 (2011).
12. L. Ghazaryan, E.-B. Kley, A. Tünnermann, and A. Szeghalmi, "Nanoporous SiO $_2$ made by atomic layer deposition and atomic layer etching," Nanotechnology **27**(25), 96270P (2015).
13. "DE102016100907A1 - Verfahren zur Herstellung eines reflexionsmindernden Schichtsystems und reflexionsminderndes Schichtsystem - Google Patents," https://patents.google.com/patent/DE102016100907A1/fr.
14. T. Siefke, M. Heusinger, C. B. R. Hurtado, J. Dickmann, U. Zeitner, A. Tünnermann, and S. Kroker, "Line-edge roughness as a challenge for high-performance wire grid polarizers in the far ultraviolet and beyond," Opt. Express **26**(15), 19534 (2018).
15. T. Siefke, D. Lehr, T. Weber, D. Voigt, E.-B. Kley, and A. Tünnermann, "Fabrication influences on deep-ultraviolet tungsten wire grid polarizers manufactured by double patterning," Opt. Lett. **39**(22), (2014).
16. T. Siefke and S. Kroker, "Polarization Control by Deep Ultra Violet Wire Grid Polarizers," in *Optical Chaacterization of Thin Solid Films*, O. Stenzel and M. Ohlidal, eds. (Springer, 2018), pp. 359–374.

Practical limits and opportunities with speckle metrology

Mikael Sjödahl*

Luleå University of Technology, 97187 Luleå, Sweden

Abstract. In this presentation, the role of speckles as a carrier of information in phase-based optical metrology is re-visited. Starting with the fundamental mechanisms for speckle decorrelation it is shown that information about the state of an object is transferred through the modified mutual coherence function and can be detected either through the phase, speckle movement, speckle decorrelation or as a combination. The presentation is focusing on practical scale laws that set the limit for what is possible to achieve with present day technology and is demonstrated with a few examples incorporating measurements of microstructural changes, strain, shape, lenses and other refractive index objects.

1 Introduction

Applications in metrology were one of the original uses of the laser when introduced in the early 1960's. In particular the laser lay the foundation for the invention of holographic interferometry as a versatile tool for a great range of imaging measurement applications, nicely summarized by Vest [1]. Very quickly scientists noticed that holographic reconstructions always were accompanied by a grainy pattern called a speckle pattern and that this pattern limited the precision to which fringes could be located and resolved. However true, it was also realised that laser speckles carried unique information and that some properties of laser speckles could be used to enhance the performance of holographic interferometry as a metrology tool. This triggered scientists to investigate more deeply the origin and properties of laser speckles in all sorts of applications in optics, including metrology. The most complete summary is given by Goodman [2].

In this presentation, some basic properties that influence measurements utilizing laser speckles will be re-visited. It is shown that the precision of optical measurement systems in general can be understood from the correlation properties of laser speckles, both in a diffraction setting and in an imaging setting. It is further noted that image correlation in space and/or time gives information about the phase changes of the field that provides a route to perform interferometric measurements in disturbed environments. Examples will be shown how basic speckle properties can be used to calculate practical scaling laws for different types of measurements ranging from measurement of microstructural changes through measurements of strain, shape and refractive index to depth gated imaging. As such, the use of speckle theory in optical metrology stretches far beyond the original motivation to understand fringe formation in holographic interferometry. The rest of this paper is divided such that a few universal properties of laser speckles are sketched in section 2.1 followed by a discussion about limitations and possibilities with present day technology in section 2.2. The paper ends with a few examples from different areas of application.

2 Properties and tools

This section summarizes a few important speckle properties and tools used to extract useful information from time varying speckle fields.

2.1 Correlation properties

Consider the complex amplitudes $U_1(\mathbf{p_1};\mathbf{X_1})$ and $U_2(\mathbf{p_2};\mathbf{X_2})$, respectively, of a dynamic speckle field detected in a plane $\mathbf{X}$ and where the subscripts represent two different time instances. $\mathbf{p_1}$ and $\mathbf{p_2}$ are a list of parameters that specify a specific state of the measurement including object microstructure, illumination and detection directions, wavelength and geometry. The normalized correlation between the fields from these two instances may be expressed as [3],

$$\mu(\Delta\mathbf{X}) = \mu_\sigma(\Delta\mathbf{p};\boldsymbol{\xi})f(\Delta\mathbf{p};\mathbf{x})g(\Delta\mathbf{p};\mathbf{X}), \tag{1}$$

where $\Delta\mathbf{p}$ represents any change in the measurement state, $\boldsymbol{\xi}$ represents the speckle generating plane, $\mathbf{x}$ represents the aperture plane and $\mathbf{X}$ is the plane in which the measurement is performed. The correlation measured may hence be expressed as the product between three correlation functions associated with three distinct planes. The function $\mu_\sigma(\Delta\mathbf{p};\boldsymbol{\xi})$ describes the effect of microscopic phase changes associated with the scattering volume of the object. For a pure surface scatter this function may often be ignored as it will always be close to unity, but for a volume scattering material it may change quickly even for very small changes in the state. For the two remaining correlation functions one needs to consider the macroscopic phase change $\phi(\Delta\mathbf{p};\boldsymbol{\xi})$ associated with a change in state. The effect of this phase change is that the field will be rotated in space, which for moderate rotations may be approximated with a movement,

* Corresponding author: mikael.sjodahl@ltu.se

$$\mathbf{u} = Lk^{-1}\nabla\phi, \quad (2)$$

in the plane of interest; where L is a propagation length, k is the wavenumber and $\nabla\phi$ is the phase gradient. With this in mind, the function $f(\Delta\mathbf{p};\mathbf{x})$ may be expressed as $f(\mathbf{u}) = \int P(\mathbf{x}-\mathbf{u})P^*(\mathbf{x})d\mathbf{x}$, where $P(\mathbf{x})$ is the pupil function and * represents complex conjugation. In an imaging setting $P(\mathbf{x})$ is the entrance pupil and in the case of diffraction field detection it is in general the detector surface. The distance L is hence either the distance from the object surface to the entrance pupil plane or the distance between the object and the detector. The final correlation function, $g(\Delta\mathbf{p};\mathbf{X})$, represents the overlap between the speckles in the plane of evaluation. In an imaging setting, L becomes the distance between the object surface and the conjugate plane to the image. Hence, if the fields are evaluated in a plane in which the speckles remain stationary this correlation function can be ignored. This is however not true in general. The final correlation function, $\mu(\Delta\mathbf{X})$, has three parameters. The phase, ϕ, is the phase change associated with the change of state and is in general the parameter of primary interest. The displacement, $\mathbf{d} = \Delta\mathbf{X}_{max}$, given by the position of maximum correlation is related to the phase change through Eq. (2). Finally, the magnitude $|\mu|$ gives the precision with which ϕ and $\mathbf{d}$ may be determined. It may also be used to map areas with rapid changes in the microstructure.

2.2 Practical considerations

All measurements are limited by the hardware available and for measurements utilizing speckles often the limiting equipment is the detector. A general detector is characterized by its pixel pitch a, the number of pixels N (in both directions), and the sampling interval T. The importance of the pixel pitch is that it limits the numerical aperture at which a speckle field can be detected without aliasing. However, considering an object size S it may be shown that in practice all measurements utilizing speckles will be limited by the scaling law,

$$\nabla\phi < 2\pi\frac{N}{4S}, \quad (3)$$

regardless of whether the imaging is performed with traditional imaging or a holographic reconstruction. As determined from the correlation function $f(\Delta\mathbf{p};\mathbf{x})$, the acceptable phase gradient over the object surface before total decorrelation occur is hence only limited by the size of the object in relation to the number of pixels on the detector. In addition, the dynamic microstructural phase change reducing the correlation parameter μ_σ will add another time-scale, which can be determined from calibration. Because of this there are two different detection modes in which speckle fields are detected. If the sampling interval, T, is sufficiently small, the changes can be resolved within the correlation time of the event and the phase changes can be detected directly. If the event is too rapid, the event is detected in a time average mode, which results in a loss of speckle contrast. In this mode direct access to the phase is not possible and the outcome is evaluated from a model relating speckle contrast with the parameter of interest.

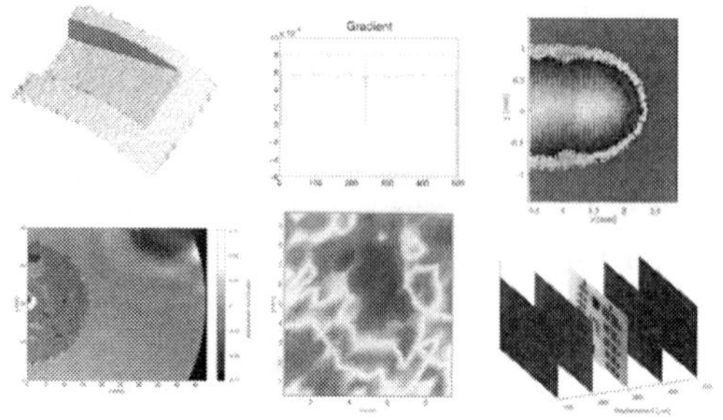

Fig. 1. Examples from various measurement results utilizing speckles.

3 Examples

A few examples from measurements utilizing speckles and the relations from section 2 are shown in Fig. 1. The top row shows examples in which Eq. (2) is used to robustify the calculation of the phase change. The top left image is an example from a dual-wavelength interferometry measurement of shape [4]. The middle image shows a measurement of uniaxial strain with symmetrical dual wavelength illumination [5], and the right image shows a measurement of an explosion using interferometric speckle deflectometry [6]. In the bottom row the left image shows the measurement of a transient wave on a rotating hard disc, in which the bulk rotation is withdrawn based on an image correlation calculation [7]. The middle image shows a measurement of microstructural activity in a sheet of copy paper and the right image shows an example of depth gated imaging utilizing the three-dimensional size of a rapidly changing speckle pattern as a spatial gate [8].

References

1. Charles M. Vest, *Holographic Interferometry* (John Wiley & Sons, USA, 1979)
2. Joseph W. Goodman, *Speckle phenomena in optics, theory and applications*, 2nd ed. (SPIE press, Bellingham, Washington, USA, 2020)
3. M. Sjödahl, *Opt. Eng.* **52**, 101908 (2013)
4. P. Bergström *et. al.*, *Appl. Opt.* **53**, 123-131 (2014)
5. M. Sjödahl and Y. Li, SPIE 9660, UNSP 966004 (2015)
6. M. Sjödahl *et. al.*, *Appl. Opt.* **48**, H31-H39 (2009)
7. P. Gren, *Appl. Opt.* **41**, 7237-7240 (2002)
8. M. Sjödahl *et. al.*, *Appl. Opt.* (to be published)

Multi-beam Coherent Fourier Scatterometry

Sarika Soman[1,*], *Silvania* Pereira[2]

[1]Optics Research Group, Imaging Physics Department, Faculty of Applied Sciences, Delft University of Technology, Lorentzweg 1, 2628 CJ Delft, The Netherlands

Abstract. Recent technological advancements in the past decades have been driven by the miniaturisation of devices using surfaces with nano-scale features. These advancements require fast, large area measurement techniques that can be used in process control to detect surface contaminations or to monitor fabrication quality. Here we present a modified version of the scanning coherent Fourier scatterometer with multiple beams that can be used to scan larger areas without increasing the scan time or decreasing the spatial resolution.

1 Introduction

The race to miniaturise devices in different industries such as photonics, electronics, biomedicine and security is driven by manufacturing surfaces with nano-scale structures [1-3]. The advent of fast, large area manufacturing requires inspection techniques capable of fast and large scan area measurements. Accurate inspection of surfaces for contamination is also critical to guarantee the quality of fabricated devices. Traditional metrology tools such as scanning electron microscopy (SEM) and atomic force microscopy (AFM) are often quite slow, destructive and require stringent measurement conditions. Optical techniques such as scatterometry provides a fast, low cost, precise, and non-destructive alternative, making it one of the most prevalent inline metrology techniques currently used in industries. In scatterometry, the change in optical properties of the light scattered from the target such as intensity and phase are measured to estimate the attributes of the target. The inverse scattering problem is generally ill-posed with respect to the existence, uniqueness, and stability of the solution. The ill-posedness of the problem is overcome by the use of prior knowledge of the system [4].

Coherent Fourier scatterometry is an advanced scatterometric technique that uses focused coherent light to probe the surface [5]. Using a focused spot, one can measure the scattering response due to several incident angles simultaneously by capturing the Fourier plane image of the target. The spot is scanned over the surface and the scattered light signal from each position is measured using a two-pixel detector. The point-by-point scanning provides high resolution in the localisation and high sensitivity in the particle detection but it comes at the cost of long scan times. Parallelisation is a good strategy to increase the scan area without compromising either the scan time or the spatial resolution. We propose a multi-beam coherent Fourier scatterometry setup for fast and efficient scans over large areas. The method can also be adapted to use with structured illumination and 3D scanning.

2 Experimental Setup

In a standard coherent Fourier scatterometry scheme, the target surface is scanned using a high numerical objective and the response in the Fourier plane is measured using a split detector. The source and the detector are static while the sample is raster scanned. The detector consists of two photo diodes. The scattered field is made to be incident exactly at the middle of the two diodes. The light intensity on each diode is then integrated separately and their difference is amplified to give the output signal. The calculation of the difference signal has been seen to reduce background noise and improve the signal to noise ratio [6]. A signal becomes non-zero when the scattering from a structure or a particle causes an asymmetry between the two halves of the light beam.

In the multi-beam setup, the parallel beams are first generated using a diffraction grating with suitable periodicity. The scattering response due to each beam is recorded separately by an array of photo diodes, two per beam. The periodicity of the diffraction grating is chosen by the wavelength of the incident light, the size of each photodiode and the separation between them such that each beam gets incident exactly in the middle of two photodiodes. The diffraction grating has a periodicity of 30 μm which when used with the He-Ne laser of 633 nm yields an angular separation of 1.2° between the multiple orders. The sample is scanned in the x and y-directions using two piezo motors. The objective has a numerical aperture of 0.4 with a magnification of 20x. After scattering from the surface, the light beam is de-magnified using a micro lens array with periodicity equal to that of the detector array.

* Corresponding author: s.soman@tudelft.nl

 https://doi.org/10.1051/epjconf/202226610023

3 Results and Discussions

The independent detection and measurement of structures by each beam was verified by scanning horizontally over a known structure and recording the output signal. The signal showed pulses at definite intervals corresponding to the measurement by each beam. The scattering response is dependent on the intensity of the diffraction order. Thus, for identical response signal, specialised gratings with equal intensity diffraction orders need to be used.

References

1. B. Javidi et al., J. Opt. **18,** 1-39 (2016)
2. S. Kim et al., ACS Appl. Mater. Interfaces **7**, 326-331 (2015)
3. Z. G. Chen, M. Segev, eLight **1**, 2 (2021)
4. S. S. H. Naqvi, et al., J. Opt. Soc. Am. A **11**, 2485-93 (1994)
5. S. Roy, O El Gawhary, N Kumar, SF Pereira, HP Urbach, J. Eur. Opt. Soc. Rap. Public. **7**, 12031 (2012)
6. S. Roy, M. Bouwens, L. Wei, S. F. Pereira, H. P. Urbach, P. van der Walle, Rev. Sci. Instrum. **86**, 123111 (2015)

Low Divergence Structured Beam In View Of Precise Long-Range Alignment

Miroslav Sulc[1,2*], *Jean-Christophe* Gayde[3]

[1]Technical University of Liberec, 46117 Liberec, Czech Republic
[2]Institute of Plasma Physics of the Czech Academy of Sciences, Toptec Department, 51101 Turnov, Czech Republic
[3]CERN – European Organization for Nuclear Research, 1211 Geneva 23, Switzerland

Abstract. A new method of generation of a Structured Laser Beam (SLB) with non-diverging central core was proposed and is promising for creating long distance multipoint alignment systems. This beam is generated by a set-up consisting of two convex lenses in Kepler telescope arrangement. The first one is a high refractive index ball lens, second one is a standard lens. The beam, in cross-section consisting of light and dark concentric circles, propagates over a large distance. The central core of the SLB has a very small divergence which can be tuned. A divergence of 10 µrad was proven experimentally. In this experiment, the small initial beam core diameter of 10 µm, and its diameter of 1.5 mm at a distance of 150 m, show its ability for use as a multipoint fiducial reference line. This small beam divergence seemingly lies beyond the diffraction limit for laser beams.

1 Introduction

The use of very well collimated laser beams to alignment is a common and precise part of many optical methods in metrology. Efforts to obtain very precise data on the position of objects have recently attracted interest in the use of so-called quasi-Bessel beams. These beams are often made up by different kinds of axicons, also used together with other optical elements. A characteristic feature of such beams is their structure, which is manifested in the cross-section of the beam by a symmetric distribution of light intensity, well approximated by the Bessel function. The central bright core of the beam is surrounded by a set of dark and light circles. Since the visibility of such circles is close to one and the diameter of the centre can be very small and comparable to the wavelength of light, it is possible to determine the position of the beam centre with significantly higher accuracy than is the case for laser beams with a Gaussian intensity distribution.
Such laser beams have been used as a fiducial line for high precision alignment system [1]–[3]. A limiting factor for the use of such beams in long-range meteorology is that their length does not exceed 20 m under standard conditions.
An almost nondivergent light beam that propagates across a long range with a narrow beam width was generated by a Galilean transmitting telescope with an eyepiece that has a spherical aberration [4]. The beam width was narrower than that of the diffraction limit of normal optics [5]. Additionally, the presented experimental results showed that such a type of beam can be much less influenced by atmospheric turbulence than laser Gaussian beams [6].
Discussions are underway at CERN, the European Organization for Nuclear Research, to build a new high-energy accelerator. During the first analysis of the requirements for a number of its properties, it became clear that the magnets and detectors must be prealigned with an accuracy at one sigma level which is typically of 10 - 100 µm along 200 m long sliding window section. Theoretical analysis and practical experiments showed that it is very difficult to create a reference line by conventional Gaussian beam on which multiple point positions can be measured at different distances with such an accuracy.
A new way of generating long-distance structured laser beams (SLB) with a transverse intensity profile, partly similar to quasi-Bessel beams, has been proposed [7]. It can be efficiently used for a development of long-distance alignment systems for accelerators and other applications. These types of SLB are under study and the beam propagation was already tested in the range from some centimetres up to a distance of 150 meters showing promising results. It would be a precise and elegant optical alternative to the currently used WPS and HLS alignment systems based on stretched wire and hydrostatic levels measured by capacitive sensors.

* Corresponding author: miroslav.sulc@tul.cz

2 Generation of Structured laser beam

In the search for the optimal method of generation of non-divergent optical beams for alignment, we have drawn on our experience with Bessel beams formed by axicon. These beams have a conical wavefront shape in the initial phase. The rays from the beam (perpendicular to the wavefront) form a Bessel beams by superposition of rays from opposite parts of the wavefront. If they intersect at points where the phase is the same, they form light rings, and in the opposite phase they form dark rings. The larger the angle between the rays, the narrower (but also shorter) the central part of the beam is.

The disadvantage of the limited range of the beam from the axicon can be overcome by using a different shape of the primary wavefront, where the beams from different opposite parts of the wavefront would meet at successively increasing distances. They could propagate to infinity for rays from parallel parts of the wavefront. The simplest form of such a wavefront is a wavefront with a spherical abberation and a defocus. This type of distorted concave spherical wave was generated by a Galilean transmitting telescope with an eyepiece concave lens with spherical aberration [4].

We designed and tested a different set-up. This new type of generator is similar to a Kepler telescope with both convex lenses. The input lens of the structured light generator is a 10 mm sphere with a high refractive index of 1.9938 for 632 nm. Second lens was standard lens with focal length 25 mm and 1 inch aperture. The system has strong spherical and defocus aberrations, see Fig.1.

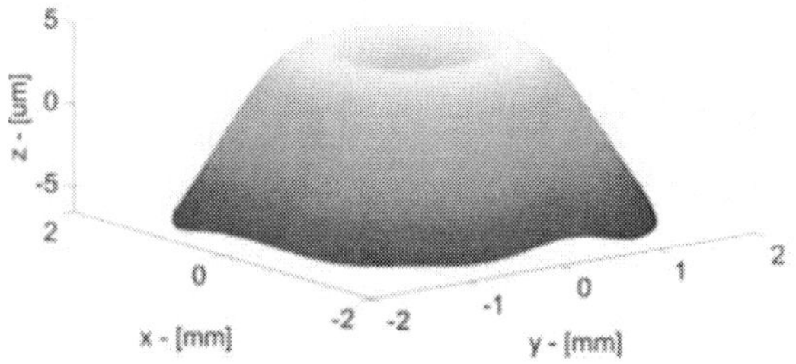

Fig. 1. Measured shape of the wavefront of the SLB, combined spherical and defocus aberrations

The beam generated in this arrangement, unlike the Bessel beam, has a distinct outer ring see Fig.2. Small amount of the light propagates outside this ring too. By changing the relative position of the lenses, the defocus aberration is changed and the outer beam can expand more or less. As it expands, the central core of the beam narrows to a certain limit. The central core of the beam remains very narrow and its diameter increases linearly with distance from approximately 10 μm at a distance of 2 m from the generator to 1.5 mm at a distance of 150 m. The core of the beam therefore has a significantly smaller divergence than that of a Gaussian beam. If we consider only the divergence of the beam core, we would be several orders of magnitude beyond the diffraction limit for light propagation.

The significant optical aberrations shown in Fig. 1 can be achieved with optical elements other than those described here as examples. Thus, beam parameters such as the divergence of the outer ring, the relative intensity of the light and dark rings, relative distances of both light and dark rings, and the diameter of the central core can be well varied. A number of simulations and experiments have been performed to optimize the beam for the specified requirements.

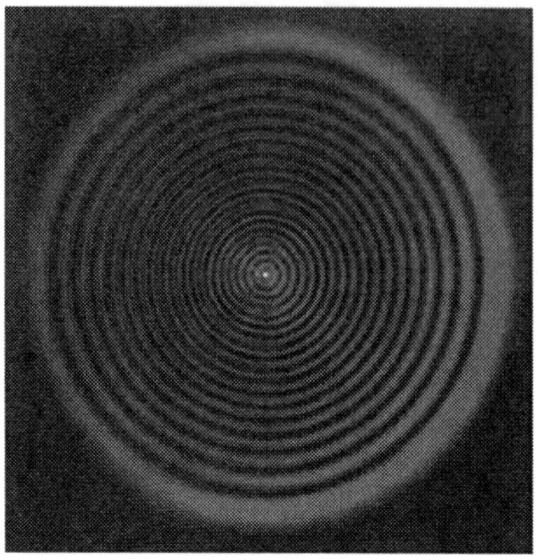

Fig. 2. Typical intensity distribution at transverse cross-section of SLB measured at 5 m from the SLB generator. Diameter of outer ring is 12.6 mm.

3 Conclusions

A new method of generation of pseudo-non-diverging structured laser beams is described in this paper. These beams can propagate with a very low divergence of the central core, which is an order of magnitude lower than the divergence of classical Gaussian laser beams. These beams are foreseen to be used as reference for very precise alignment and positioning of optical elements, or objects such as physics detectors, particle accelerator parts, etc. It would be a precise and elegant optical alternative to the currently used WPS and HLS alignment systems based on stretched wire and hydrostatic levels measured by capacitive sensors.

References

1. D. M. Gale, *Optical Measurement Systems for Industrial Inspection VII* **8082**, 80823D (2011)
2. R. E. Parks, *Optical System Alignment, Tolerancing, and Verification XIII* **11488**, 114880J-1 (2020)
3. R. E. Parks, *Optomechanics and Optical Alignment* **11816**, 1181603 (2021)
4. T. Aruga, Appl. Opt. **36**, 3762 (1997)
5. T. Aruga and S. W. Li, Appl. Opt. **38**, 2795 (1999)
6. T. Aruga, S. W. Li, S. Yoshikado, M. Takabe, and R. M. Li, Appl. Opt. **38**, 3152 (1999)
7. M. Sulc and J.-C. Gayde, (2019), An Optical System for Producing a Structured Beam. European patent application, 3564734 A1 (2019)

Study of laser induced temperature variation in silica nanofibers

Yanis Abdedou[1] *Mondher* Besbes[1], *Philippe* Delaye[1] and *Sylvie* Lebrun[1] *
[1]Laboratoire Charles Fabry, Institut d'Optique, CNRS, Université Paris-Saclay, 91127 Palaiseau, France
Jean-Charles Beugnot[2], *Jacques* Chrétien[2], *Maxime* Romanet[2], *Simon* Colombel[2], *Jérôme* Salvi[2], *Maxime* Zerbib[2], *Kien* Phan Huy[2]
[2]Institut FEMTO-ST, CNRS, Université Bourgogne Franche-Comté, 25030 Besançon, France

Abstract. After presenting a theoretical modelling based on the heat equation, we show two different experiments to measure the laser induced temperature variation in silica nanofibers in air, a direct one and an indirect one based on Brillouin scattering, leading to an estimated value of the convective parameter h.

1 Introduction

Optical tapered nanofibers are fabricated by heating and stretching standard silica fibers until reaching diameters comparable or smaller than the light wavelength which are homogeneous over several cm lengths. At such diameters, the optical mode is very strongly confined in the silica leading to the enhancement of optical non linearities. The mode can also exhibit an intense evanescent tail that can be used for optical sensing or traps in the vicinity of the nanofiber. One other strong advantage of tapered nanofibers compared to bulk components is their ability to be inserted in fibered networks with very low losses. All these properties make them significantly exploited for an increasing number of applications. One limitation of nanofibers that can lead to instabilities, degradation of the performances or even their destruction is their temperature increase due to the coupling of an intense laser light inside them. Only a few studies exist in the literature about these thermal effects and are limited to vacuum [1,2]. Controlling these thermal effects can also be used to propose self-cleaning of nanofibers [2]. Our aim in this paper is to study laser induced thermal effects in nanofibers, in different fluids such as air. Indeed, thermal exchanges with air at the micro and nanoscales are not well known, for example the convective coefficient (h) is still debated [3]. In our work, we propose a theoretical modelling based on the heat equation, that can be used for different gas and pressures. We take into account not only the nanofiber part but also its tapers shape. Preliminary experimental measurements of the temperature of nanofibers by two different methods (direct method and method based on Brillouin scattering) show good agreements with our simulations.

2 MODELLING

Our simulations are based on the differential heat equation valid for a cylindrical symmetry [1]:

$$c_p \rho \partial_t T \pi a^2 = -\partial_z H_{rad}(T) + \partial_z H_{rad}(T_0) + \lambda_c \partial_z^2 T \pi a^2 + \partial_z P_{heating} + dP_{gas}$$

∂_t and ∂_z are the derivative versus t (time) and z (direction of propagation), c_p and ρ are the specific heat and density of silica, λ_c is the heat conduction of silica. a is the radius depending on z. H_{rad} represents the radiative heat transfer between the nanofiber at temperature T and the environment at temperature T_0 and is calculated from the fluctuational electrodynamics [4].

$P_{heating}$ is the heat source, which in our case is due the laser propagation and is written:

$$dP_{heating} = kI(a)2\pi a dz$$

k is an absorption coefficient due to the presence of pollutants at the surface of the nanofiber [1], $I(a)$ is the intensity of the laser at the surface.

dP_{gas} represents the thermal exchanges with the surrounding gas. For pressures smaller than 1 μbar, dP_{gas} is proportional to the pressure [1]. For higher pressures, we use a linear approximation to compute this term:

$$dP_{gas} = h(T - T_0)2\pi a$$

h is the convective coefficient in W/m²K and is our single adjustable parameter.

Fig 1 shows the evolution of the temperature of the nanofiber and its tapers in the stationary regime and in air (nanofiber radius r_{NF} = 450 nm, nanofiber length L_{NF} = 2 cm, pump power = 53 mW, pump wavelength λ_p = 760 nm, h = 7.2 W/m²K). The maximum temperature is achieved in the nanofiber part, with an increase of 50 K with respect to the room temperature.

* Corresponding author: sylvie.lebrun@institutoptique.fr

https://doi.org/10.1051/epjconf/202226611001

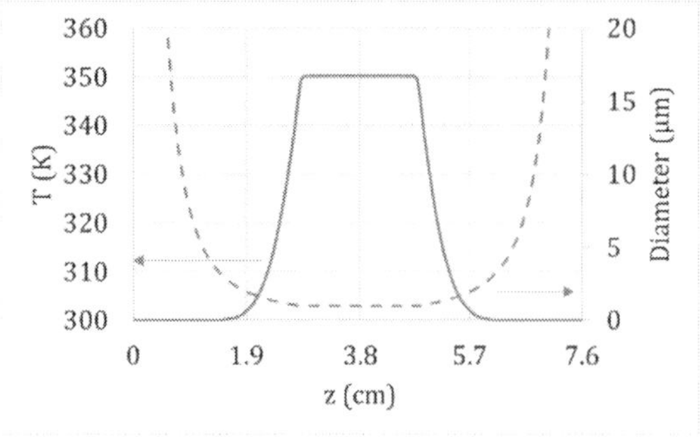

Fig. 1. Solid curve (left scale): temperature profile along the nanofiber and its tapers. Dotted curve (right scale): profile of the nanofiber and its tapers.

3 Direct method

In Fig. 2, we have represented the experimental setup for the direct measurement of the temperature of the nanofiber. The pump source is a cw laser emitting at 780 nm, which is injected in the nanofiber with a microscope objective. The nanofiber is drawn from a standard telecom fiber and preceded by a single mode filter to excite the fundamental mode only. We used a thermistor (diameter of 400 µm) that we put in contact with the nanofiber.

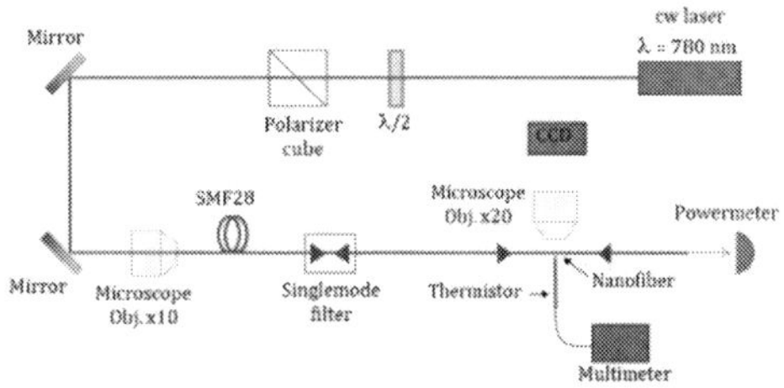

Fig. 2. Experimental setup for the direct measurement of the laser induced temperature of the nanofiber.

In Fig. 3, we have plotted the nanofiber temperature versus h which is our single fitting parameter and superposed experimental datas. The best fits give a value for h which is between 2.6 and 7.2 W/m²K.

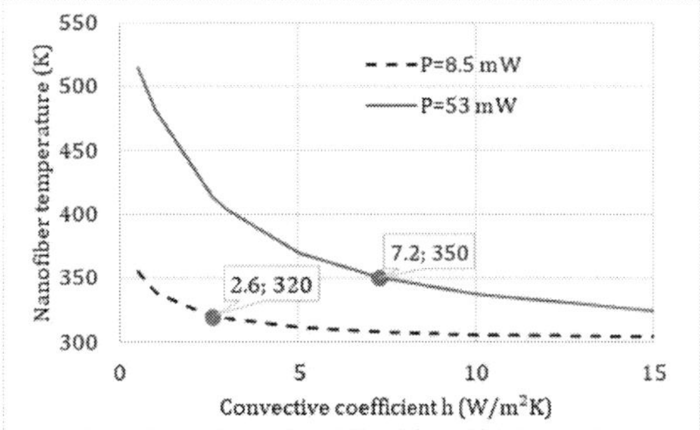

Fig. 3. Simulated nanofiber temperature versus the convective coefficient h for two different pump powers. Red points are experimental data.

4 Brillouin scattering

We have experimentally investigated the laser induced heating in nanofiber by monitoring the Brillouin scattering. In heterodyne detection, a small part of the laser pump source used to generate Brillouin scattering in nanofiber is absorbed to heat the nanofiber. We observe a linear dependence of Brillouin resonance from compression elastic wave as a function of input pump power for a nanofiber diameter of 940 nm (see Fig. 4). The slope is 0.85 MHz/°C, giving a temperature variation of 10°C between 20 mW and 40 mW, corresponding to a value of h around 5 W/m²K.

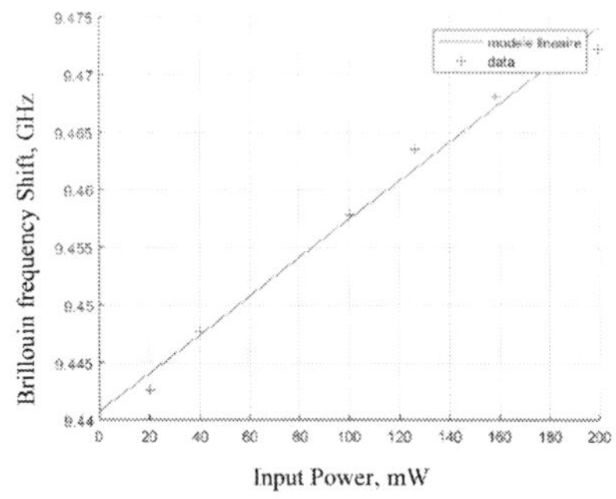

Fig. 4 Brillouin frequency shift measurement as a function of cw input power at 1550nm for a diameter of nanofiber of 940 nm and a length of 40 mm.

5 Conclusion and perspectives

We have presented a numerical model of the laser induced temperature variation of nanofiber and its tapers in air. Two different methods are used to measure the temperature. The results are consistent by fitting the experimental data with the model, which gives a range for the convective parameter h between 2.6 and 7.2 W/m²K. Further experiments will be conducted to confirm and precise these values.

References

1. P. Anderson, S. Jalnapurkar, E.S. Moiseev, D. Chang, P.E. Barclay, A. Lezama, and A.I. Lvovsky, AIP Advances 8, 055005 (2018).
2. C. Wuttke and A. Rauschenbeutel, Phys. Rev. Lett. 111, 024301 (2013).
3. X. J. Hu, A. Jain and K. E. Goodson, Intern. J. of Thermal Sciences, 47 820-824 (2008).
4. A. Vladyslav, M. Golyk, M. Kardar, Phys. Rev. E, 85, 046603 (2012).

Tapered multicore fibers for energy-scalable fiber laser systems

Christopher Aleshire[1,*], *Albrecht* Steinkopff[1], *Arno* Klenke[1,2], *Cesar* Jauregui[1], *Steffen* Böhme[3], *Tobias* Koch[3], *Stefan* Kuhn[3], *Johannes* Nold[3], *Nicoletta* Haarlammert[3], *Thomas* Schreiber[3], and *Jens* Limpert[1,2,3]

[1]Institute of Applied Physics, Friedrich-Schiller-University Jena, Albert-Einstein-Straße 15, 07745 Jena, Germany
[2]Helmholtz-Institute Jena, Fröbelstieg 3, 07743 Jena, Germany
[3]Fraunhofer Institute for Applied Optics and Precision Engineering, Albert-Einstein-Straße 7, 07745 Jena, Germany

Abstract. With active multicore fibers (MCFs), many parallel amplifying waveguides can be densely assembled into a common glass cladding. A tapered fiber geometry applied to MCFs enhances the power- and energy scalability of these systems by increasing the doped waveguide volume and reducing peak irradiance while maintaining low output mode order. Recent high-energy experiments have achieved 37 mJ ns-class pulse energies with Yb-doped tapered MCFs, with potential application to a new generation of compact MCF-based coherently-combined laser systems. In this submission, latest experimental results with tapered MCFs and flexible fabrication of taper profiles as a post-draw processing step will be discussed. Numerical analyses of MCF tapers will be presented, using beam propagation method (BPM) and mode-decomposition techniques to study mode coupling and inter-core crosstalk. These simulations are used to guide the tapering of existing fibers and aid the design of future "taper-ready" MCFs.

1 Introduction

Coherent beam combination (CBC) of parallel fiber amplifiers has enabled the development of high average power laser sources with near-ideal beam characteristics capable of driving new scientific and industrial applications [1]. The physical size of such a system is determined primarily by the unit size of its constituent lasers, limiting the practical scaling of CBC systems to 1-2 orders of magnitude. This scaling dependency is circumvented with the use of multicore fibers (MCFs) where many parallel amplifying waveguide cores are arranged in a single cladding. Recent ultrafast demonstrations of MCF CBC have achieved 500 W combined and compressed average power [2], and numerical simulations predict scalability to the ~ 10 kW level with several hundred mJ pulse energy [3].

As with single-core fibers, the capability of MCFs to generate and handle high powers and high pulse energies can be improved with mode area scaling techniques. A tapered fiber design can significantly increase the mode area and available doped volume while maintaining near-ideal output beam quality [4]. A tapered MCF amplifier was recently demonstrated, achieving 37 mJ total energy in ns-class pulses with improved beam quality in comparison to an untapered fiber (see Fig. 1), with a proposed application in CBC systems [5]. In this context, where the MCF cores are intended to be used as independent parallel amplifiers, the design and fabrication of a tapered MCF presents unique constraints in addition to those encountered with standard tapered gain fibers. Numerical simulations can analyze mode evolution and inter-core crosstalk in MCF tapers to produce optimized taper designs for a given MCF. These designs can be flexibly realized using a flame or CO_2 laser tapering system as a post-draw processing step.

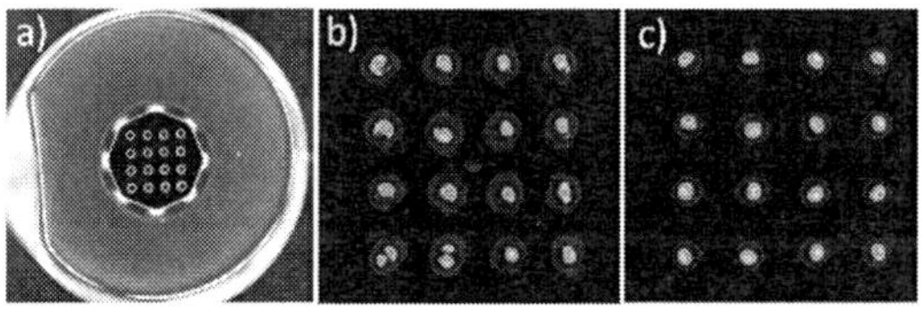

Fig. 1. a) Polished MCF facet. b) Beams of untapered MCF with 50-micron core diameter. c) Beams of tapered MCF with 19 micron input core diameter and 50-micron output diameter.

2 High energy tapered MCF experiment

A primary benefit of a tapered fiber geometry is the fiber's greater doped volume, and thus higher extractable energy, per unit fiber length. This and the lower peak intensity due to the larger mode field area lead to lower accumulated nonlinear effects and increased damage threshold. While fiber lasers are typically considered low pulse energy/high average power systems, tapered MCF systems can extend the working range of fiber systems into a higher energy regime.

A recent experiment to demonstrate the energy handling capability of a tapered MCF is depicted schematically in the upper portion of Fig. 2. A Q-switched MCF oscillator seeded a counter-pumped tapered 4x4 MCF amplifying stage. The tapered fiber had a total

* Corresponding author: christopher.aleshire@uni-jena.de

https://doi.org/10.1051/epjconf/202226611002

length of 98 cm with a ~20 cm taper from 19-micron core diameter to 50-micron core diameter. The amplified energy is plotted in the lower portion of Fig. 2, reaching a maximum of 37 mJ before damage due to a defect in the fiber. An untapered version of the fiber reached a pump-limited 49 mJ without damage. The output beams of both amplifying MCFs are shown in Fig 1 b), c), with the tapered fiber achieving an average single-beam $M^2 = 1.5$ vs. $M^2 = 2.1$ in the untapered fiber. With further taper refinement, combinable beams with pulse energies >10 mJ from a single fiber are likely feasible.

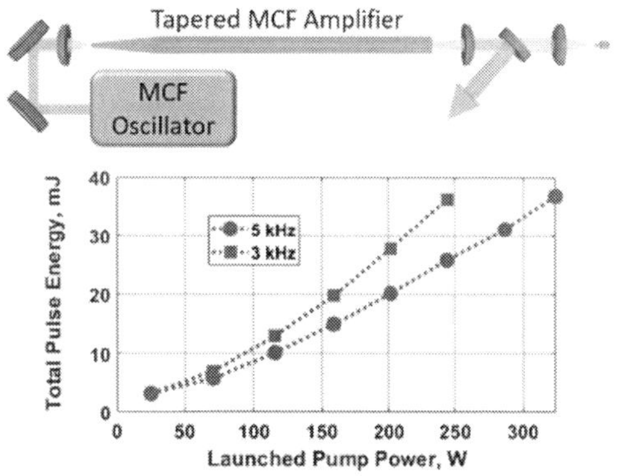

Fig. 2. (Top) schematic of MCF energy extraction experiment. (Bottom) extracted pulse energies at 3 kHz and 5 kHz repetition rate.

3 MCF taper analysis

Taper designs can be modelled and optimized for a given fiber design using numerical techniques such as beam propagation method (BPM). As an example, a facet image of a double-clad Yb-doped MCF produced by Fraunhofer IOF is shown in Fig. 1a). The core separation is 2.5x the core diameter and the doped cores have a low NA of 0.04. Cross-talk may occur between adjacent cores, depending on the drawn core diameter and chosen fiber length and is strongest in the strictly-single-mode regime in which tapers typically begin. This motivates a short taper length and a large taper angle to minimize crosstalk through the taper. A short taper also benefits laser operation by maximizing dopant volume and mode field diameter through the remainder of the fiber. In the case of single-core tapers, the minimum length is theoretically given by the adiabaticity criterion for axiosymmetric tapers [6]. With MCFs, additional losses and perturbations occur due to the off-axis arrangement of cores. Final intensity distributions of linear tapers simulated with BPM based on the fiber described above are shown in Fig. 3. The core diameters are tapered from 15 µm to 25 µm (single-mode to few-mode). Fig. 3 a) shows bend loss and mode distortion in a 5 mm taper, and Fig. 3 b) shows a 50 mm taper with noticeable power coupled from the two initially seeded cores into adjacent cores. Within this range of taper lengths, a working interval which minimizes crosstalk and mode distortion can be found.

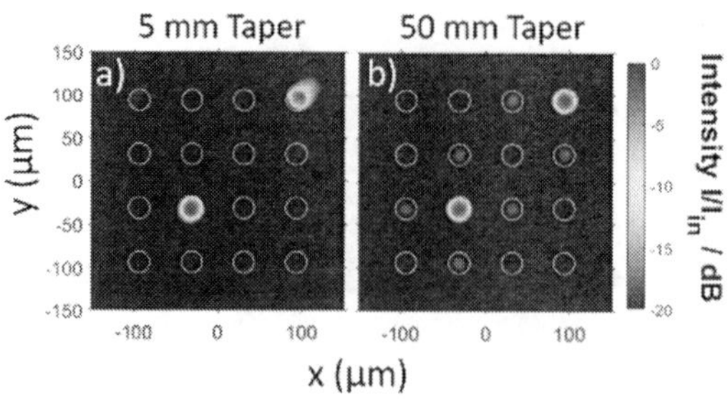

Fig. 3. Final intensity distributions of a) a 5 mm MCF taper and b) a 50 mm MCF taper.

Tapers with length scales of ~10-100 mm as in Fig. 3 can be fabricated using a flame-brush or CO2 laser tapering system after fiber drawing. Such systems are also capable of generating arbitrary longitudinal profiles [7]. This may be applicable to MCFs, for example, to accomplish a rapid initial taper for minimized core-to-core coupling, followed by a more gradual taper to minimize mode coupling in the multimode region.

4 Conclusion

Using numerical simulation techniques and post-draw processing, optimized designs for tapered multicore fibers can be realized which enable independent operation of large multimode waveguides with single-mode beam characteristics. While CBC with MCFs is already proving to be an attractive format for compact laser system scaling, applying a tapered fiber geometry extends and optimizes energy handling potential for a given MCF. Recent demonstrations have achieved ~50 mJ of output energy in ns-class pulses, and further refinement may enable Joule-class operation and fundamental mode output from highly multimode large core MCFs. These advances can enable CBC systems with power and energy scaled by multiple orders of magnitude.

References

1. M. Müller, C. Aleshire, A. Klenke, E. Haddad, F. Légaré, A Tünnermann, J. Limpert, Opt. Lett. **45**, 3083-3086 (2020).
2. A. Klenke, A. Steinkopff, C. Aleshire, C. Jauregui, S. Kuhn, J. Nold, C. Hupel, S. Hein, S. Schulze, N. Haarlammert, T. Schreiber, A. Tünnermann, J. Limpert, Opt. Lett. **47**, 345-348 (2022).
3. A. Steinkopff, C. Jauregui, C. Aleshire, A. Klenke, J. Limpert, Opt. Exp. **28**, 38093-38105 (2020).
4. V. Filippov, Y. Chamorovskii, J. Kerttula, K. Golant, M. Pessa, O. G. Okhnotikov, Opt. Exp. **16**, 1929-1944 (2008).
5. C. Aleshire, A. Steinkopff, A. Klenke, C. Jauregui, S. Kuhn, J. Nold, N. Haarlammert, T. Schreiber, J. Limpert, Opt. Lett. **47,** 1725-1728 (2022).
6. J. D. Love, W. M. Henry, W. J Stewart, R. J. Black, S. Lacroix, F. Gonthier, IEEE Proc.-J. **138**, 343-354 (1991).
7. J. Nold, M. Plötner, S. Böhme, B. Sattler, O. deVries, T. Schreiber, R. Eberhardt, A. Tünnermann, Proc. SPIE **10513**, 2018.

EPJ Web of Conferences 266, 11003 (2022) https://doi.org/10.1051/epjconf/202226611003
EOSAM 2022

Gas-pressure tuning of wavelength of photon pair emitted by Four-Wave-Mixing in Nanofibers

Agathe Bonifacio[1], *Sylvie Lebrun*[1], *Maxime Zerbib*[2], *Maxime Romanet*[2], *Jean-Charles Beugnot*[2] *and Philippe Delaye*[1*]

[1]Université Paris-Saclay, Institut d'Optique Graduate School, CNRS, Laboratoire Charles Fabry, 91127 Palaiseau, France
[2]Institut FEMTO-ST, CNRS, Université Bourgogne Franche-Comté, 25030 Besançon, France

Abstract. We present experimental results demonstrating the possibility to tune the wavelength of the photon pair emitted through four wave mixing in a nanofiber, using the pressure of a gas surrounding the nanofiber. Using Argon, a shift of idler wavelength of -1.1nm/bar is measured demonstrating fine adjustment possibility of emission wavelength, allowing to choose between different WDM channels.

1 Introduction

Nanofibers are a very interesting system to realize generation of pairs of photons, either using third order nonlinearity using Four Wave Mixing [1, 2], or using new second order nonlinear mechanisms allowed by surface nonlinearities brought by the small dimension of the fiber [3]. The nanofiber is pulled with tapers that gives good injection and extraction of pump beam and emitted photons. Moreover, the small diameter permits to realize phase matching with large wavelength shifts for which the detrimental influence of Raman scattering can be minimized. Thanks to these properties bright source operating either in the pulsed or in the Continuous wave regime with Coincidence to Accidental Ratio (CAR) higher than 20000 could be realized [4].
To improve the flexibility of the realized source, several methods were proposed to allow the tuning of the emitted wavelength of the given pulled nanofiber, using either temperature of the fiber or a longitudinal stress [2]. In the present paper we choose to use, as a tuning method, the pressure of a gas that surrounds the nanofiber. This method was already successfully used in Photonic Crystal Fibers [5,6] but, in that case, the time necessary to fill the fiber might limit the dynamics of the tuning like in the case of the thermal tuning.

2 Nanofiber design and experiment

The nanofiber characteristics were choosen in order to realize phase matching for a Pump beam issued from a Ti:Sapphire laser operating in the near IR, an Idler beam emitting in the telecom band (typically between 1520nm and 1620nm) and thus a Signal beam in the Red (typically around 600-650nm). This means that the nanofiber has to be in the 850nm diameter range. The phase matching characteristics can be calculated using the model of a silica rod surrounded by Argon [4], which wavelength dependences of the refractive index are known from the literature [7, 8], as well as pressure dependence of Argon. Other gases with higher refractive index such as Xenon may be used to obtain higher emitted wavelength shifts.
The 2cm nanofiber and its tapers is fixed on a 3D printed support and sealed in a ½" inox tube connected through a Swagelok's quickfit connector to the gas bottle. The pigtail outputs are used to easily insert the fiber on the Four Wave Mixing set-up. The emitted spectrum is characterized using a Stimulated Emission Tomography set-up (see [4] for details of the set-up), using a tunable telecom laser diode as a seed beam injected in the nanofiber together with the pump beam. At the output of the nanofiber, Pump, Idler and Signal beams are separated using dichroic mirrors and filters. Stimulated emission in the visible due to interaction at energy conservation, of Pump and Seed beams is detected using a Si-APD single photon detector, which signal is recorded as a function of the Idler seed beam wavelength. The measurement of the emitted spectrum as a function of Argon pressure shows a clear shift of the photon pairs emission (Fig.1).

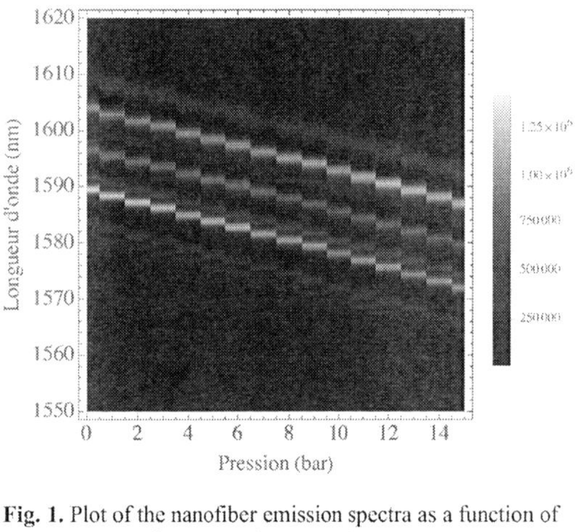

Fig. 1. Plot of the nanofiber emission spectra as a function of Argon pressure for a pump beam at 885.4nm.

* Corresponding author: Philippe.Delaye@institutoptique.fr

The emitted spectrum of that specific nanofiber presents multiple peaks due to non-uniformity of the fiber diameter (Fig.2). The three-main-peaks structure is compatible with a sinusoidal fluctuation of nanofiber diameter [4] with an amplitude 3.6nm around a mean value of 854.24nm and a period 4.1mm along the fiber of total length 20mm (See insert in Fig.2).

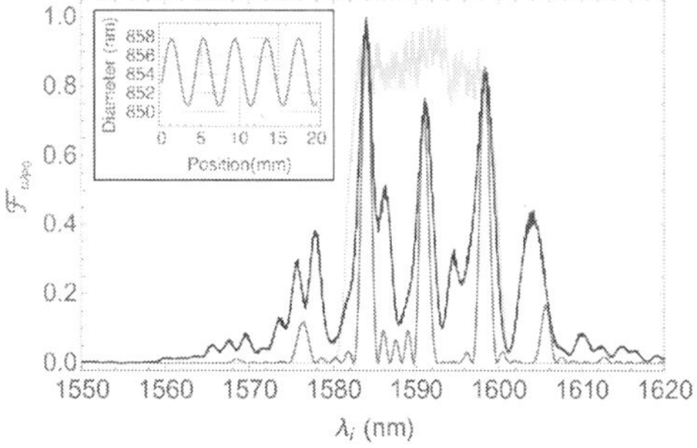

Fig. 2. Plot of the emitted spectrum at energy conservation at a pressure of 5 Bar (Blue curve), with the theoretical calculation (Red curve) for a nanofiber presenting a sinusoidal fluctuation of its diameter (shown in the insert). The pressure has been chosen so that all the main peaks, and thus most of the emitted idler photons, are inside the transmission band of the 1590nm channel of our Coarse WDM filter (Green curve).

The measured peak shift for two nanofibers with slightly different characteristics is similar within the current uncertainty of pressure measurement. When compared to the theoretical calculation (Fig.3), the accordance is very good despite the absence of free parameters. This indicates that the values of the used parameters, especially the refractive index of Argon, are correct.

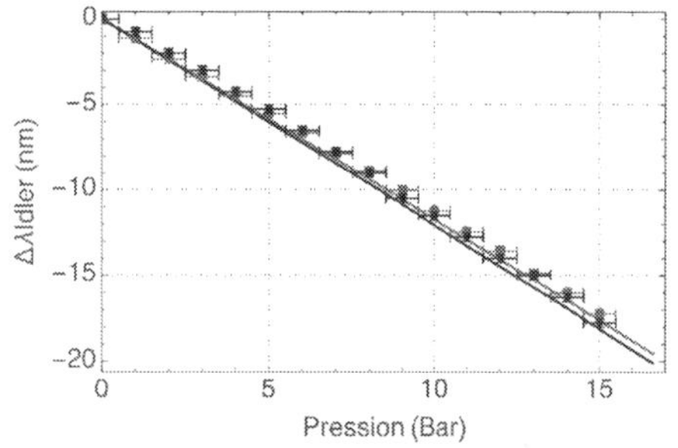

Fig. 3. Plot of peak shift as function of pressure for two nanofibers and the corresponding theoretical calculation with known parameters of the experiments. Red Symbols and line, nanofiber diameter 853nm, pump à 887nm. Blue Symbols and line, nanofiber diameter 867nm, pump at 916.5nm.

Playing with pump wavelength and pressure we can position the emitted spectrum either in the 1590nm or the 1570nm channel of the CWDM filter what allows to collect most of the emitted photons (Fig.2) and to limit the noise related to the broadband Spontaneous Raman Scattering of silica. We can thus reach very high Coincidence to accidental ratio around 70000 (Fig.4), which is the highest value measured in a fibered structure.

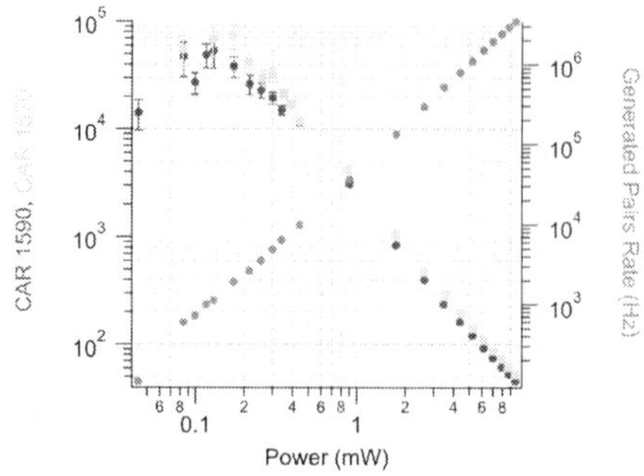

Fig. 4. Measured Coincidence to accidental ratio (Blue Symbols for pump at 887nm and pressure 4.75 Bar to work with the 1590nm output of the CWDM filter, and Green Symbols for pump at 895nm and pressure 9.25 Bar to work with the 1570nm output) and rate of pairs generated at fibre output (Red Symbols) as a function of the mean pump power.

3 Conclusion

We have shown that it is possible to finely tune the emission spectrum of a source of pairs of photons based on Four Wave Mixing in a nanofiber surrounded by a gas. Despite being preliminary, these results show potential of this technique. Using pressure tuning with an optimized fiber, both more uniform in order to have an emission peak closer to the ideal Sinc shape, and longer to have a smaller width of emission peak, we should be able to position the emission peaks in different output of a Dense WDM filter, and dynamically send the emitted idler photon in a specific WDM channel, together with addressing a specific wavelength on the signal side using pump wavelength tuning.

The authors want to thank Vincent Pécheur and Alexandre Matic for the conception and 3D impression of the nanofiber supports.

References

1. L. Cui, X. Li, C. Guo, Y.H. Li, Z.Y. Xu, L.J. Wang, W. Fang, Opt. Lett. **38**, 5063 (2013).
2. E. Ortiz-Ricardo, C. Bertoni-Ocampo, Z. Ibarra-Borja, R. Ramirez-Alarcon, D. Cruz-Delgado, H. Cruz-Ramirez, K. Garay-Palmett, A. B. U'ren, Quantum Science and Technology **2**, 034015 (2017).
3. A. Azzoune, Ph. Delaye, G. Pauliat, J. Opt. Soc. Am. B **38**, 1057 (2021)
4. Ph. Delaye, T. Liu, E. Mer, M. Bouhadida, S. Lebrun, Phys. Rev A **104**, 063715 (2021).
5. M. Cordier, Ph. Delaye, F. Gérôme, F. Benabid, I. Zaquine, Sci. Rep. **10**, 1650 (2020).
6. J. Hammer, M.V. Chekhova, D.R. Häupl, R.Pennetta, N.Y. Joly, Phys. Rev. Research **2**, 012079 (2020).
7. I. H. Malitson, J. Opt. Soc. Am. **55**, 1205 (1965).
8. A. Bideau-Mehu, Y. Guern, R. Abjean, A. Johannin-Gilles, J. Quant. Spectrosc. Radiat. Transfer **25**, 395, (1981)

Operating range of efficient Raman converters based on nanofibers immersed in different liquids

Maha Bouhadida, *Pierre-Enguerrand* Verdier, *Philippe* Delaye and *Sylvie* Lebrun*
Laboratoire Charles Fabry, Institut d'Optique, CNRS, Université Paris-Saclay, 91127 Palaiseau, France

Abstract. We present the operating range of Raman wavelength converters based on silica nanofibers immersed in liquids for the design of all fibered wavelength converters. This range is bounded on the lower limit by the pump energy necessary to reach the Raman threshold and on the upper limit by the laser induced breakdown of the nanofiber. These breakdown energies are measured in the ns regime for different liquids (water, ethanol, isopropanol) and for air. We finally define guidelines that open the way to a new family of low-cost compact and efficient all-fibered Raman converters that can be directly inserted in optical fibered networks with very low losses.

1 Introduction

Optical tapered nanofibers are fabricated by pulling fibers until reaching diameters comparable or smaller than the light wavelength. Silica nanofibers are significantly exploited for a wide range of potential applications. Indeed, at such diameters, nanofibers exhibit a strong confinement of light which enables the generation of nonlinear effects [1]. Nanofibers can also exhibit an intense evanescent field which has been exploited for optical sensing, optical traps or spectroscopy. However, the experimental investigation of the optical nonlinearities in the evanescent field of the nanofiber remains limited to a few studies [2,3]. In this work, we focus on the Raman scattering in the evanescent field of a nanofiber immersed in a liquid. The Stokes photons are generated in the immersing liquid by the Raman scattering of the pump photons present in the evanescent field, then couple to a guided mode of the nanofiber and can be directly collected at the output of the un-tapered fiber.

The Raman conversion operating range of these wavelength converters is bounded on the lower limit by the pump energy necessary to reach the Raman threshold and on the upper limit y the laser induced breakdown of the nanofiber. To optimize the extracted Stokes energy, it is then necessary to decrease the Raman threshold energy and to increase the nanofiber breakdown energy. The Raman threshold energy can be estimated with a quite good accuracy by using a simple modelling based on modal Raman gain [2]. The breakdown threshold energy is much more difficult to predict, and experimental data have to be used. To increase this breakdown energy, one way is to increase the diameter of the nanofiber. However, higher diameters would increase the Raman threshold because the intensity of the evanescent field would be decreased. A compromise must be found in order to determine the nanofiber diameter enabling the extraction of the maximal output energy at the Stokes wavelength. In this work, we use ethanol as the Raman liquid. Laser breakdown energies with different liquids are also tested. We finally define guidelines that open the way to a new family of low-cost compact and efficient all-fibered Raman converters that can be directly inserted in optical fibered networks with very low losses.

2 Experimental setup

The nanofiber, attached to two un-tapered fibers through two sections called the tapers, is fabricated by following the "pull and brush" technique [2]. After the fabrication, the component is totally immersed in the Raman liquid. The radius of the nanofiber is determined by the modal Raman gain and its length is determined by the pump energy necessary to reach the Raman threshold. The tapers are designed to be adiabatic when immersed in the liquid. We focus our study on three different nanofiber radii: 220 nm, 300 nm and 350 nm. The length of all nanofibers is 8 cm. The transmission of the nanofibers with their tapers is routinely higher than 90%.

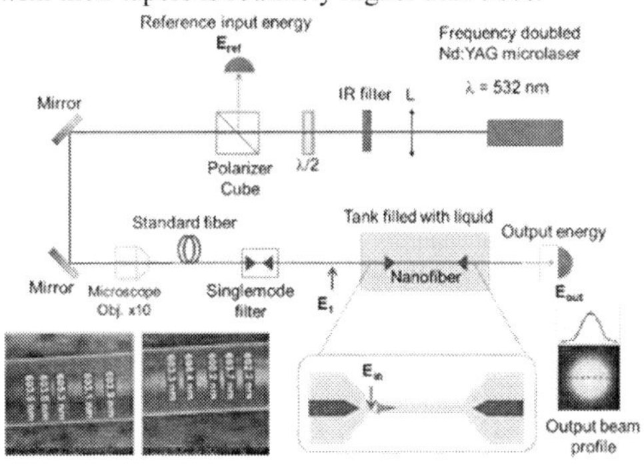

Fig. 1. Experimental setup. Bottom right: far-field picture of the output mode with cross section and Gaussian fit.

* Corresponding author: sylvie.lebrun@institutoptique.Ffr

Fig. 1 shows the experimental setup. The pump source is a frequency doubled pulsed pump laser emitting a beam at $\lambda p = 532$ nm with a pulse duration of 900 ps (FWHM), a frequency rate repetition of 4.7 kHz and a maximum pump energy of 6.3 µJ. The pump beam is injected in the un-tapered input of the nanofiber and forward beams (pump at 532 nm and first Stokes order of ethanol at 630 nm) are sent after the un-tapered output to an optical spectrum analyzer.

3 Results

We present in Fig. 2 the experimental Raman threshold energies, incident nanofiber breakdown energies and maximum extracted Stokes energies for 20 samples of nanofibers having a radius of 300 nm. The conversion efficiency from the pump to the first Stokes order of ethanol is higher than 60%.

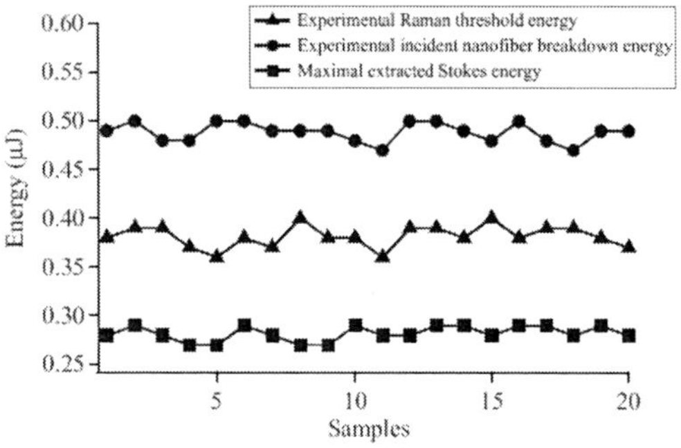

Fig. 2. Experimental Raman threshold and breakdown energies, extracted Stokes energy of 20 samples of nanofibers with a radius of 300 nm.

Fig. 3 shows the operating range, delimited by the Raman threshold and the laser breakdown energy of the nanofiber obtained with the three different radii studied. Each experimental point corresponds to the average of the results obtained with 20 nanofibers. These results illustrate the high reproducibility of the performances obtained.

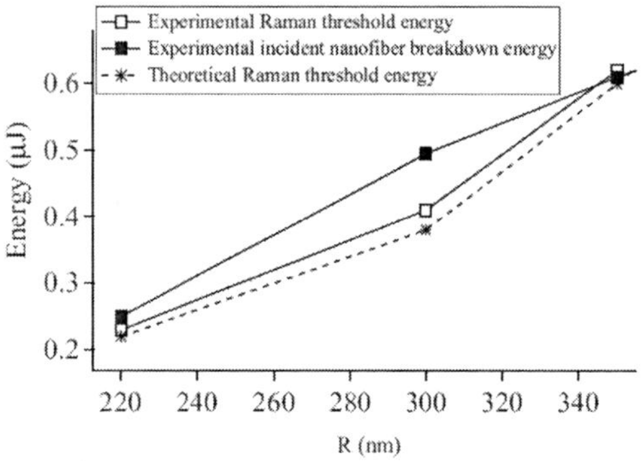

Fig. 3. Operating range of the Raman converters.

Since the nanofiber breakdown energy limits the performance of the Raman converters, we have experimentally studied this physical quantity for different liquids (water, ethanol, isopropanol) and different nanofiber radii [4]. Results are presented in Fig. 4. By knowing the pump energy at breakdown and by estimating the Raman threshold, it is possible to define the operating range of the converters for different immerging liquids and radii, by adapting, if necessary, the length of the nanofiber as we will show during the conference.

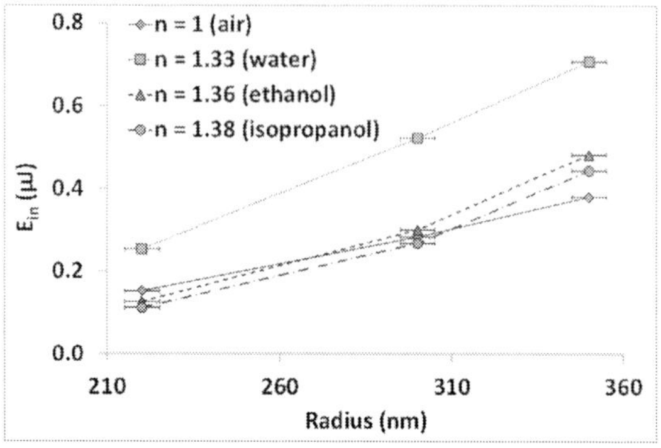

Fig. 4 Pump energy inside the nanofiber at breakdown for different radii and different liquids.

4 Conclusions and perspectives

We have demonstrated efficient and reproducible evanescent Raman converters in the nanosecond regime based on a silica nanofiber immersed in ethanol. We have defined an operating range which is a guideline for the conception of other evanescent Raman converters based on other liquids and/or fibers. Other perspectives are foreseen such as the deposition of nonlinear polymers on the nanofiber surface opening the way to a new family of robust all-fibered components that can be directly inserted in optical networks without additional losses.

References

1. M.A. Foster, A.C. Turner, M. Lipson, A.L. Gaeta, "Nonlinear optics in photonic nanowires", Opt. Express, 16(2), 1300 (2008).
2. M. Bouhadida, J. C. Beugnot, P. Delaye, K. Phan Huy and S. Lebrun, "Highly efficient and reproducible evanescent Raman converters based on a silica nanofiber immersed in a liquid", Applied Physics B, Volume 125, Issue 12, pp.1-7 (2019).
3. G. Fanjoux, J. Chrétien, A. Godet, K. Phan-Huy, J.-C. Bugnot and T. Sylvestre, "Demonstration of the evanescent Kerr effect in optical nanofibers ", Opt. Exp., vol. 27, n°20, pp. 29460-29470 (2019).
4. M. Bouhadida, P. -E. Verdier and S. Lebrun, "Laser-Induced Damage in Silica Nanofibers in Air and Immersed in Different Liquids in the Nanosecond Regime," IEEE Photonics Technology Letters, vol. 33, no. 17, pp. 967-970 (2021).

https://doi.org/10.1051/epjconf/202226611005

Tapered hollow-core photonic crystal fibers

Frédéric Gérôme[1,2*], *Jonas H.* Osório[1], *Foued* Amrani[1,2], *Benoit* Debord[1,2], and *Fetah* Benabid[1,2]

[1]GPPMM Group, XLIM Institute, CNRS UMR 7252, University of Limoges, Limoges, France

[2]GLOphotonics, 123 Avenue Albert Thomas, Limoges, France

Abstract. In this communication, we will first review the recent advances of hollow-core photonic crystal fibers. Then, the possibility offered to tailor their optical properties by making tapers will be discussed.

1. Introduction

The recent results on the technology of hollow-core photonic crystal fibers (HCPCF) consolidate them as building blocks for the new advances in photonics. Today, we distinguish two families of HCPCF, namely the photonic bandgap (PBG) [1] and inhibited coupling (IC) guiding ones [2]. Particularly, IC fibers have experienced tremendous progress in the latest years. The results concerned improvements in the fiber designs and fabrication procedures, which allowed attaining fibers with special properties and optimized transmission loss. Moreover, the unique properties of HCPCF can also be further optimized by using tapering process. Recently, several techniques have been developed to tailor specific parameters (*e.g.*, dispersion) and to efficiently integrate HCPCFs with the existing fiber devices and systems. Here, we present a review of such advances.

2. Recent advances in hollow-core fibers

The research endeavors on HCPCF have allowed identifying distinct limiting factors for loss decreasing depending on the spectral range of operation. For fibers working in the infrared (IR), the performances are limited by the fiber design. Conversely, for fibers working in the visible and ultraviolet ranges, the surface scattering loss (SSL) arising from the fiber core surface roughness hinders the loss decreasing. In the following sections, we discuss advances regarding IC fibers with ultralow loss in the short-wavelength range and a special fiber structure for low loss and effective single-mode operation.

2.1 Ultralow loss in the short-wavelength range

Since SSL limits the loss decreasing in fibers working at short wavelengths, improvements in the performances for the visible and ultraviolet (UV) ranges must rely on controlling the surface roughness of the fiber core. Considering this, we revisited the HCPCF fabrication methods and obtained a reduction of the core surface roughness levels. Fig. 1a presents comparative plots of the surface roughness profiles for fibers drawn via typical HCPCF fabrication methods and by using our innovative techniques. The reduction of the core surface roughness (peak-to-peak reduction by a factor ~3), allowed us to obtain new state-of-the-art HCPCF working at short wavelengths [3]. Fig. 1b shows, as an example, the loss spectrum of a single-ring tubular-lattice (SR-TL) HCPCF with record low loss at wavelengths <600nm. It is remarkable that the new record low values lie below the silica Rayleigh scattering limit, a fundamental restriction for solid-core fibers.

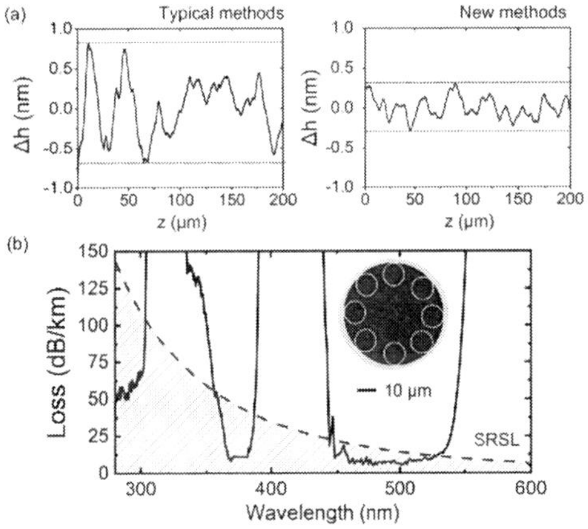

Fig. 1. Core surface profiles for HCPCF drawn by (a) typical and (b) innovative approaches. (c) Loss spectrum of a HCPCF with record low loss in the short-wavelength range. SRSL: silica Rayleigh scattering limit.

2.2 Robust single-mode operation

For fibers working in the IR, the design of the fiber cladding plays a crucial role in providing strong light confinement. Additionally, the judicious design of the fiber cladding allows tailoring the fiber modal properties for achieving robust single-mode operation. Remarkably, the six-tube SR-TL HCPCF design allows to efficiently filter the contaminating LP_{11}-like modes and provides effective single-mode operation. However, this design displays high loss. To circumvent this limitation, the confining power of the structure must be improved. An opportunity for having low loss and robust single-mode

* Corresponding author: frederic.gerome@xlim.fr

HCPCF is to use the so-called hybrid kagome-tubular (HKT) HCPCF [4]. Fig. 2a presents the cross-section of the HKT HCPCF, which exhibits an inner cladding formed by a ring of six tubes and an outer kagome one for confinement improvement. This structure allows having low loss (Fig 2b) and robust single-mode operation. Fig. 2c shows that the HKT HCPCF has a robust fundamental mode-like near-field output profile even if the light launching condition is shifted from its optimum (X=Y=0).

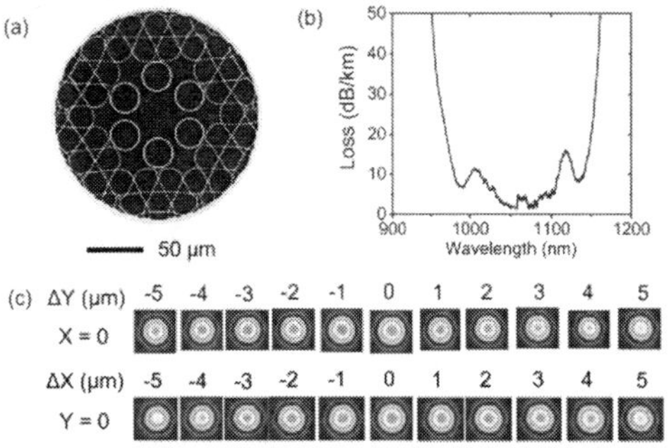

Fig. 2. HKT HCPCF (a) cross-section and (b) loss spectrum. (c) Robustness of the near-field output profile for varying input conditions.

3. Tapering process for tailoring the properties of hollow-core fibers

The optical performances of the HCPCFs can also be tailored by using tapering process. In fact, by varying the different geometrical parameters of the fiber along its length, it is possible to manage the characteristics of the guided mode. The tapered section can be managed over a few tens of mm to several meters combined with an evolution of the overall dimension of the fiber (*i.e.*, the outer diameter) by several factors, according to the need. To illustrate this, two examples are given.

The first one concerns the issue of HCPCF integration into systems. Indeed, due to the improved performances presented in the previous section, such fibers became today very attractive, and it is thus worthy to splice them to standard single mode fiber (SMF) for further application in gas photonics and telecommunications. However, achieving low loss and mechanically robust splicing between HCPCF to SMF remains technically challenging especially due to the large hollow-core size. Also, the delicate shape of the HCPCF (*i.e.*, negative curvature core contour) needs to be preserved during the post-processing to keep the waveguide performance. Thus, we proposed different tapering processes based on commercial splicers to answer this call. By tapering down a kagome-lattice HCPCF (core diameter: 61-75 μm, Fig. 3) from 300 μm to 125 μm outer diameter over an adiabatic length of 12 mm, a down-taper loss of 0.07 dB and a record SMF to kagome splice loss of 0.48 dB at 1550 nm was achieved [5]. However, the reverse splicing loss reached 3.5 dB due to the excitation of the HCPCF high-order modes in this scheme. More recently, by using a two-step reverse-tapering method, we could improve the control and the splicing loss in both directions. A record-low insertion loss of 0.88 dB for a SMF/HCPCF/SMF chain at 1310 nm was demonstrated [6].

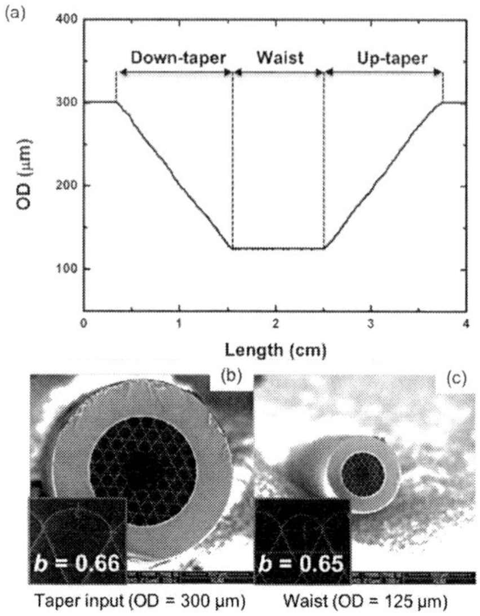

Fig. 3. (a) Experimental evolution of the taper diameter versus length; (b) Cross-section of untapered and (c) tapered kagome HCPCF fiber at the waist region. Inset: zoom in the core contour.

The second example concerns the use of long-tapered HCPCF specially designed to manage the dispersion profile of the fiber. Indeed, the position of the zero-dispersion wavelength and the value/slope of the dispersion curve are some of the key parameters in the nonlinear effects. Here, the tapering process is done directly during fiber fabrication by controlling the drawing speed. By using a homemade software integrated into the drawing tower, the variation of the fiber's outer diameter can be easily adapted in terms of shape and length. Such long-tapered HCPCFs were successfully fabricated. Adiabatic soliton pulse compression [7] and matched cascade frequency-conversion using stimulated Raman scattering have then been demonstrated [8].

To conclude, recent advances in HCPCF combined with tapering process open the way to a myriad of applications ranging from optical telecommunications, nonlinear and ultrafast optics, atom and quantum optics to plasma photonics.

The research was funded through PIA program 4F, Labex ΣLim and la région Nouvelle Aquitaine.

4. References

1. Birks *et al.*, *Electron. Lett.* **31**, 1941-1943 (1995).
2. Couny *et al.*, *Science* **318**, 1118-1121 (2007).
3. Osório *et al.*, CLEO Conference, SW4K.6 (2022).
4. Amrani *et al.*, *Light : Sci. Appl.* **10**, 7 (2021).
5. Zheng *et al.*, *Opt. Express* **2**, 14642-14647 (2016).
6. Wang *et al.*, *Opt. Express* **29**, 22470-22478 (2021).
7. Gérôme *et al.*, *Opt. Express* **15**, 7126-7131 (2007).
8. Beaudou *et al.*, *Opt. Express* **18**, 12381-12390 (2010).

Experimental investigation of light-matter interaction when transitioning from cavity QED to waveguide QED

Daniel Lechner[1,*], *Riccardo* Pennetta[1], *Martin* Blaha[1], *Philipp* Schneeweiss[1], *Jürgen* Volz[1], and *Arno* Rauschenbeutel[1]

[1]Institute of Physics, Humboldt University Berlin, 12489 Berlin, Germany

Abstract. Cavity quantum electrodynamics (cavity QED) is conventionally described by the Jaynes- or Tavis-Cummings model, where quantum emitters couple to a single-mode cavity. The opposite scenario, in which an ensemble of emitters couples to a single spatial mode of a propagating light field, is described by waveguide QED, where emitters interact with a continuum of frequency modes. Here we present an experiment where an ensemble of cold atoms strongly couples to a fiber-ring resonator with variable length containing an optical nanofiber. By changing the length of the resonator we can tailor the density of frequency modes and thus explore the transition from cavity QED to waveguide QED. We analyse the response of the ensemble–cavity system after the sudden switch-on of resonant laser light and find that for progressively longer resonators, the Rabi oscillations typical of cavity QED disappear and the single-pass dynamics of waveguide QED appear. Our measurements shed light on the interplay between the single-pass collective response of the atoms to the propagating cavity field and the ensemble–cavity dynamics.

In waveguide QED, emitters couple to a propagating light field via a continuum of frequency modes. Conceptually, the waveguide QED case can be interpreted as a cavity with mirrors at infinite distance. As the cavity is shortened progressively, the frequency spacing or free spectral range of the cavity modes increases until the emitters finally couple only to a single mode of the cavity. This regime is usually described in the framework of the Jaynes- or Tavis-Cummings model where a single or multiple emitters couple to a single mode of an optical cavity.

Here we experimentally and theoretically study light-matter interaction for emitters placed in a fiber-ring cavity with variable length where emitters interact with one, few or multiple resonator modes [1]. Systems in the transition region between cavity and waveguide QED, where the roundtrip time of the resonator is approximately as long as the lifetime of the excited state of the emitter, cannot correctly be described by the Jaynes- or Tavis-Cummings model, nor by the standard waveguide QED approach. This situation can, however, be described by a cascaded interaction model [2–4], which applies to cavities with arbitrary length and encompasses the cavity QED and waveguide QED limits. Based on this model, we study the temporal dynamics of the field transmitted through the resonator for different cavity lengths after the rapid switch-on of light that show typical features of cavity and waveguide QED, where in the intermediate cases features from both regimes are present simultaneously.

In Figure 1 (a), the experimental setup is sketched [1, 5]. A fiber-ring resonator with variable length is coupled to an input fiber via a fiber coupler. We prepare an ensemble of cold cesium atoms in a magneto-optical trap and interface it with the resonator field via the evanescent field of a tapered optical nanofiber (waist diameter $\approx$ 400 nm, waist length $\approx$ 1 cm) which is part of the fiber-ring.

In Figure 1 (b), we present the frequency response of our system for a resonator length of 45 m for an ensemble of cesium atoms corresponding to an optical density of 6. In this case, we are in the multimode strong coupling regime [1, 3], as the free spectral range of the cavity is smaller than the linewidth of cesium ($\nu_{\mathrm{FSR}} = 4.4\,\mathrm{MHz}$, $\Gamma/2\pi = 5.2\,\mathrm{MHz}$). As resonator modes are being pushed outwards compared to the reference spectrum, the atoms interact with multiple modes simultaneously. The cascaded interaction model fits the experimental data well and proves that it is well suited to describe the resonator response.

To measure the time domain dynamics, we record the response of the ensemble–cavity system after the sudden switch-on of light resonant with the D2 line of cesium ($6^2S_{1/2}$, $F = 4 \rightarrow 6^2P_{3/2}$, $F = 5$) using a single photon counting module (SPCM). To quickly turn on the laser, we use an electro-optic modulator with a rise time that is much shorter ($\approx$ 0.8 ns) than the lifetime of the exited state of cesium ($\approx$ 30.4 ns).

Figures 1 (c) - (e) show the predicted temporal response for three different resonator lengths using the cascaded interaction model and an optical density of 6. The resonator lengths we consider are (c) 5 mm, (d) 5 m and (e) 50 m which correspond to roundtrip times, t_{rt}, that are much smaller, comparable or much larger than the lifetime, τ_{at}, of the excited state of cesium. When $t_{\mathrm{rt}} \ll \tau_{\mathrm{at}}$ as in (c), the transmission shows the typical damped Rabi oscillation predicted by cavity QED. On the other hand,

*e-mail: daniel.lechner@physik.hu-berlin.de

 https://doi.org/10.1051/epjconf/202226611006

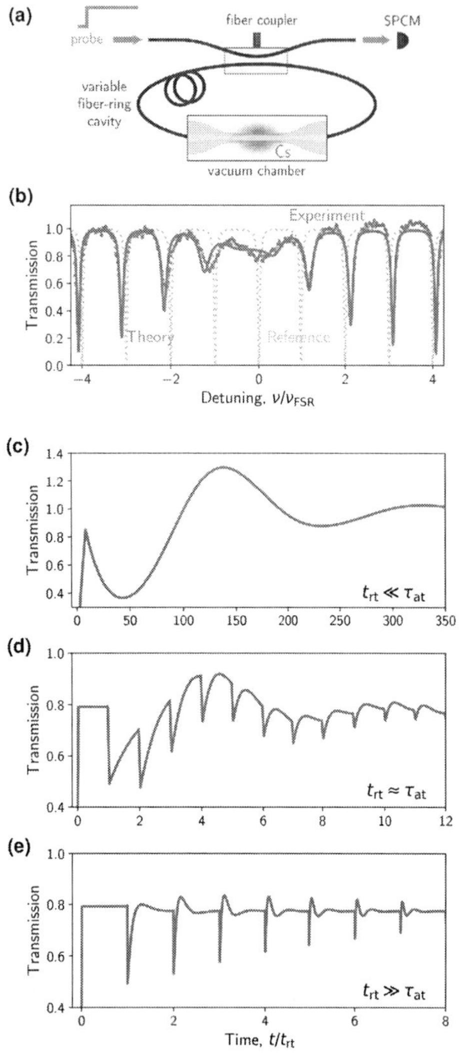

Figure 1: **(a)** Experimental setup of our fiber-ring cavity system. **(b)** Frequency scan of the cavity transmission for a 45-m-long cavity with frequency normalized to the free spectral range ν_{FSR} of the resonator. Experimental data is represented as blue dots, theory as a red solid line and reference cavity as a grey dotted line. In this case, the atomic ensemble interacts with multiple resonator modes at the same time. **(d)-(e)** The transmission through cavities with increasing lengths as predicted by the cascaded interaction model with time normalized to the roundtrip time of the resonator. The longer the resonator and the longer its roundtrip time, t_{rt}, is compared to the lifetime of the atoms, τ_{at}, the more the typical Rabi oscillations of cavity QED disappear and the more the repeating single-pass dynamics of waveguide QED dominate.

when $t_{rt} \gg \tau_{at}$ as in (e), the transmission is characterized by sharp features that repeat every roundtrip time, which result from the consecutive single-pass waveguide-QED-like response of the atomic ensemble to the light pulse propagating in the resonator. For $t_{rt} \approx \tau_{at}$ as in (d), the system is in a transition region between cavity QED and waveguide QED. In this case, the transmission displays a slowly oscillating envelope, which corresponds to a Rabi oscillation typical of cavity QED, while simultaneously showing the sharp, repeating features of the single-pass dynamics of waveguide QED. These predicted dynamics agree well with preliminary measurement results.

Our work is the first detailed experimental and theoretical study of the transition regime between cavity and wavguide QED. It improves the understanding of the response of emitters to a propagating cavity field for arbitrary resonator lengths and has the potential to enable new protocols and experiments in new regimes of light-matter interaction.

References

[1] A. Johnson, M. Blaha, A.E. Ulanov, A. Rauschenbeutel, P. Schneeweiss, J. Volz, Physical Review Letters **123**, 243602 (2019)

[2] J.T. Shen, S. Fan, Physical Review A **79**, 023837 (2009)

[3] M. Blaha, A. Johnson, A. Rauschenbeutel, J. Volz, Physical Review A **105**, 013719 (2022)

[4] R. Pennetta, D. Lechner, M. Blaha, A. Rauschenbeutel, P. Schneeweiss, J. Volz, arXiv:2112.10806 (2021)

[5] R. Pennetta, M. Blaha, A. Johnson, D. Lechner, P. Schneeweiss, J. Volz, A. Rauschenbeutel, Physical Review Letters **128**, 073601 (2022)

Trapping a single atom in the evanescent field of a WGM-resonator

Gabriele Maron[1,*], *Xinxin* Hu[1], *Luke* Masters[1], *Lucas* Pache[1], *Michael* Scheucher[2], *Elisa* Will[2], *Jürgen* Volz[1], and *Arno* Rauschenbeutel[1]

[1]Department of Physics, Humboldt-Universität zu Berlin, Germany
[2]Vienna Center for Quantum Science and Technology, TU Wien – Atominstitut, Austria

Abstract. We demonstrate the trapping of single atoms in the evanescent field of a whispering-gallery-mode (WGM) resonator with a standing wave optical dipole trap. We present our progress towards an improved trap loading scheme.

1 Introduction

WGM resonators are monolithic structures that guide light by total internal reflection. They exhibit ultra-high Q-factors combined with a small optical mode volume, and are conveniently tunable. They also provide lossless in- and out-coupling of light via tapered optical fibers, and allow one to reach the strong coupling regime of cavity-QED when coupling a single atom to their evanescent field. Furthermore, these resonators exhibit chiral (i.e. direction dependent) light-matter interaction [1], which makes them attractive atom-photon interfaces for novel quantum information processing devices [2]. However, so far, only free falling atoms have been coupled to WGM resonators, which severely limits the atom-resonator interaction time and results in a probabilistic operation. Here we describe a trapping scheme that led us, for the first time, to trapping of single ^{85}Rb atoms in the evanescent field of a WGM resonator [3]. With use of a two-colour magic wavelength trapping scheme, we are able to observe a vacuum Rabi splitting in the coupled atom-resonator spectrum. Furthermore, we present our ongoing implementation of a second generation trapping scheme, which will enable deterministic trap loading with improved trap lifetimes in addition to a well-defined coupling strength.

2 Experimental details

2.1 Dipole Trap Scheme

The trapping setup is sketched in Fig. 1. We use a bottle microresonator (BMR), i.e. a fiber-based WGM resonator with a quasi-cylindrical geometry, stabilized to the unperturbed D2 transition of ^{85}Rb. A tapered fiber is interfaced with the BMR to couple light into and out of it. A cloud of cold ^{85}Rb atoms is delivered from a magneto-optical trap (MOT) to the BMR's evanescent field by an atomic fountain. The system is set to critical coupling such that, when an atom enters the evanescent field, an increase in the coupling fiber's transmission is measured. This signal triggers an FPGA-based real time detection scheme, which switches on our dipole trap in ~250 ns, i.e. much less than the transit time of the atom in the evanescent field (~1-2 μs). The trap consists of a red-detuned light field tightly focused on the resonator surface [4]. A standing wave pattern results from the interference between the incident and back-reflected beams, such that an atom can probabilistically be trapped at the first potential minimum, at a distance of ~200 nm from the BMR's surface.

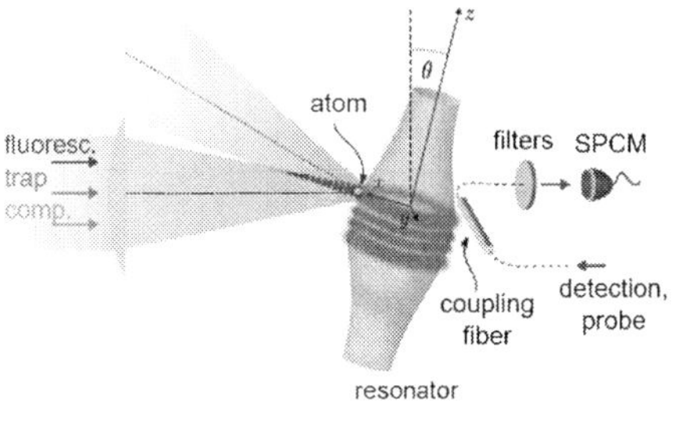

Figure 1. Simplified scheme of the setup for trapping single atoms close to the WGM resonator ssurface.

2.2 Light Shift Compensation and Atom Detection

The trapping field induces a position-dependent AC-Stark shift of the atomic transition, which impairs the atom-resonator coupling. In order to compensate for this effect, a dual-colour magic wavelength trapping scheme is employed: a second light beam, with proper power and detuning with respect to a higher atomic transition, is complementarily focused on the trapping position, with the ultimate effect of shifting the excited level back into resonance to remove the atom-resonator detuning. Furthermore, a fluorescence detection scheme is used to inspect

*e-mail: gabriele.maron@physik.hu-berlin.de

the trap loading: short resonant pulses are sent through the trap optics, a fraction of the scattered light is coupled to the resonator and collected through the coupling fiber, in order to detect the presence of a trapped atom at variable time delays. In this way, we can infer a trap loading efficiency of $\eta_0 \approx 0.7$ % and a trapping lifetime of $\tau \approx 2$ ms. This represents a 1000-fold increase in the atom-resonator interaction time. In Fig. 2, transmission spectra of the atom-resonator system with and without the Light-Shift Compensation (LSC) field are shown in comparison with the spectrum of the empty resonator. The spectrum with LSC shows a symmetric vacuum Rabi splitting, which agrees well with the theoretical prediction curve based on our setup parameters. From the latter, we estimate the coupling strength of the atom-resonator system to be $g = 2\pi \times 9.3$ MHz.

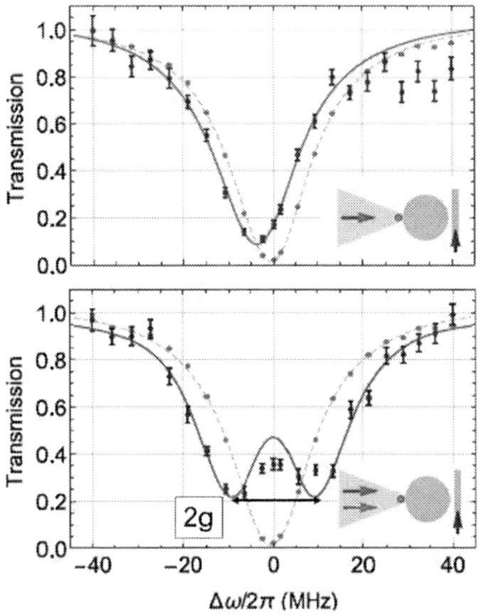

Figure 2. Transmission spectra of the coupled atom-resonator system. In violet (above): without LSC. In blue (below): with LSC. In grey: transmission spectrum of the empty resonator. The solid lines represent the theoretical predictions.

2.3 Outlook

We are currently working on the implementation of a second generation trapping setup. In this novel scheme, the dipole trap is directly focused into the MOT region, which is located at a short distance from the BMR (~1 mm). Once the trap is loaded, the atom is delivered to the resonator surface with a moveable optical tweezer, generated by a Focus Tuneable Lens (FTL), whose focal length is adjustable by an externally applied electric field. However, this poses a further complication: while moving to the resonator, the interference of the incoming and reflected beam builds up a high contrast standing wave potential, which inhibits transport to the evanescent field of the WGM. The envisaged solution is based on the use of a Spatial Light Modulator (SLM) to generate a superposition of higher-order Laguerre-Gaussian modes, to enable the suppression of the interference fringes of the standing wave potential [5]. With the prospected upgrades, our setup will allow trapping of colder atoms, which will result in a better control of the atomic position and, thus, on an even longer trapping lifetime with a better definition of the coupling strength g.

3 Summary

We have successfully implemented an experimental protocol that led, for the first time, to trapping of a single atom in the evanescent field of a WGM resonator. The trap-induced light shift of the atomic transition was overcome by employing a dual-colour magic wavelength trapping scheme, based on the addition of a second beam to restore the atomic resonance. This resulted on a 1000-fold increase of the atom-field interaction time in the strong coupling regime. The ongoing upgrades of the setup will enable us to reach a longer trapping lifetime and a better-defined coupling strength g.

References

[1] P. Lodahl et al., Nature **541**, 473 (2017)
[2] M. Scheucher et al., Science **354**, 1577 (2016)
[3] J. D. Thompson et al., Science **340**, 1202 (2013)
[4] E. Will et al., Phys. Rev. Lett. **126**, 233602 (2021)
[5] J.-B. Beguin et al., PNAS **117/42**, 26109 (2020)

EPJ Web of Conferences 266, 11008 (2022)
EOSAM 2022 https://doi.org/10.1051/epjconf/202226611008

Optical nanofibers for signal delaying

Alexandre Matic*, *Adrien* Godet, *Jacques* Chrétien, *Kien* Phan-Huy, and *Jean-Charles* Beugnot**

FEMTO-ST Institute, Université Bourgogne Franche-Comté, CNRS UMR 6174, 25030 Besançon, France

Abstract. In this abstract, we purpose an optical delay line based on optical nanofibers. Silica high elasticity and the low pulling force required to stretch a nanofiber allow to get optical delays up to 20 picoseconds with a 10 centimeter-long optical nanofiber at telecommunications wavelength.

1 Introduction

Optical nanofibers with sub-wavelength diameter allows high confinement of the light and strong interaction with environment due to evanescent fields, which enables many applications in non-linear optics [1–3]. In addition to these optical properties, optical nanofibers also offer excellent mechanical properties in terms of elasticity [4, 5].

In this abstract, we present an optical nanofiber stretcher which can produce an optical delay up to 20 ps. Firstly, we briefly describe the theoretical background behind this device. Then, we purpose to match the theory using Rayleigh Optical Frequency Domain Reflectometry (OFDR) measurements.

2 Theory and basic concepts

Application of a strain force on an optical fiber will result on increase of the optical path and produce an optical delay (figure 1) :

$$\Delta t = \Delta l * n_{eff}/c \quad (1)$$

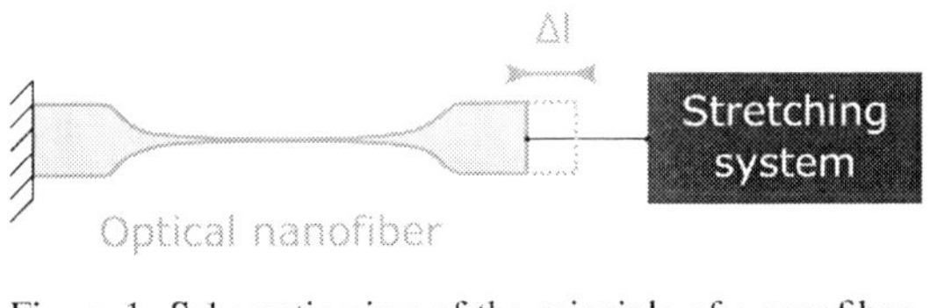

Figure 1: Schematic view of the principle of a nanofiber-based delay line

The force required to strain the fiber by a ratio $\Delta l/l$ depends on the diameter d of the optical fiber and the silica Young modulus E :

$$F = E * (\Delta l/l) * \pi(d/2)^2 \quad (2)$$

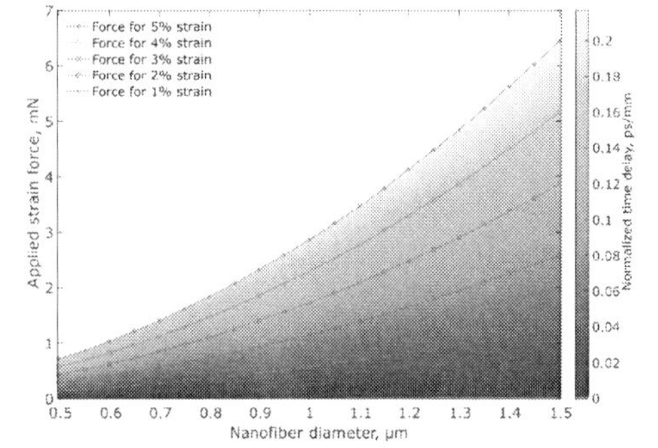

Figure 2: 3D-map linking required strain force F for different stretch ratio $\Delta l/l$ depending on nanofiber diameter d. Color map translates fiber stretching in normalized time delay in ps/mm

Nanofibers requires only a fraction of the force required to stretch classical 125 µm-diameter optical fibers. Moreover, higher elasticity of the silica at sub-micron scale and absence of polymer coating allow to reach higher $\Delta l/l$ ratios.

By combining Eq.1 and Eq.2, strain force can be linked to optical time delay :

$$\frac{\Delta t}{l} = \frac{4Fn_{eff}}{Ec\pi d^2} \quad (3)$$

For nanofibers, the effective optical index varies with the diameter of the nanofibers and the considered optical mode [1]. These results only consider the fundamental mode. The relation given in Eq.3 can be represented as a 3D-map given in figure 2.

For a given diameter, this chart links the stretch ratio $\Delta l/l$, the strain force F and the normalized optical delay $\Delta t/L$. As an example, for a 100 mm-long nanofiber of 1,0 µm diameter, a 3 mN strain force is sufficient to get an optical delay of 20 ps.

*e-mail: alexandre.matic@femto-st.fr
**e-mail: jc.beugnot@femto-st.fr

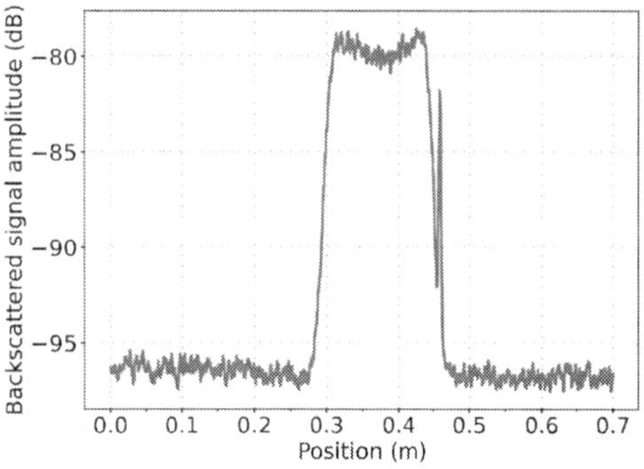

(a)

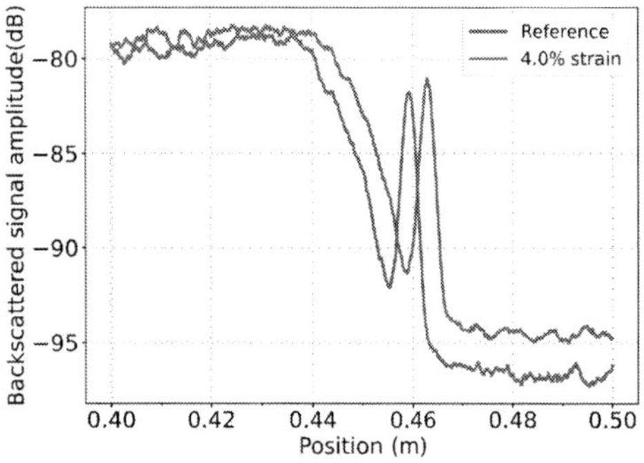

(b)

Figure 3: Rayleigh backscattered signal coming from an optical nanofiber (a) and zoom of the trace on the ending side of the nanofiber before and after stretching (b).

3 Time delay measurements

To verify our theoretical model given in figure 2, optical delay induced by stretching a nanofiber is measured with a high-resolution reflectometer Luna OBR 4600. This tool uses Rayleigh Reflectometry in Frequency Domain (OFDR) to achieve up to 10 μm resolution in spatial domain (i.e. 0.05 ps in time domain).

A 120 mm-long nanofiber with 1.0 μm diameter has been characterized with this tool. Nanofibers can be easily identified on reflectometry traces, as backscattered signal rises by 10-15 dB on the nanofiber part (figure 3a). This nanofiber has been fixed on two Newport XML210 translation stages. Different stretch ratios $\Delta l/l$, between 0 and 4%, are applied to the nanofiber by spreading both stages. Effective stretching of the nanofiber is then recorded with the OBR 4600 (figure 3b), which is converted as optical time delay according to Eq.(1).

Results given on figure 4 are in agreement with our theoretical model. We were able to produce a 18.8 ps optical delay by stretching up to 4% a 120 mm-long nanofiber. Stretch ratio can be push up to 6% on silica nanofibers and thus allow to get up to 28 ps time delay on such device.

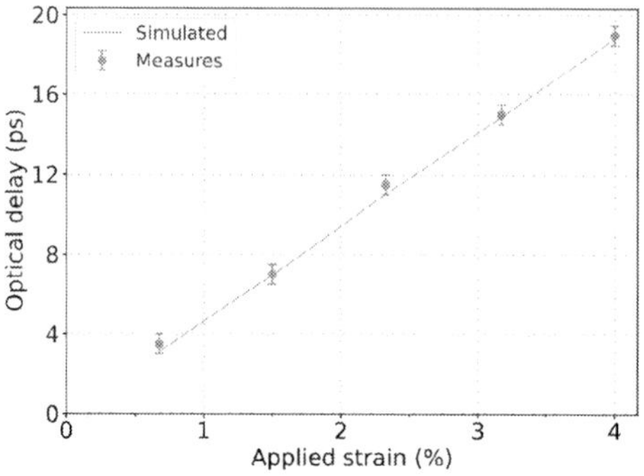

Figure 4: Measured and theoretical optical delays for different applied strains on a nanofiber.

Conclusion

We have demonstrated a novel fiber stretcher using nanofibers to produce optical time delays up to 20 ps. These devices only requires few millinewtons of strain force and 100 mm-long nanofiber to achieve time delays greater than 10 ps. This technology could lead to more compact and more stable fibered optical delay lines [6].

Acknowledgments

Authors thank the Regional Council of Bourgogne-Franche-Comté, the French National Research Agency (ANR-17-EURE-0002, ANR-21-ESRE-0040) and the Tech Transfer acceleration company Sayens.

References

[1] L. Tong, J. Lou, E. Mazur, Optics Express **12**, 1025 (2004)

[2] G. Brambilla, Journal of Optics A: Pure and Applied Optics **12**, 043001 (2010)

[3] J.C. Beugnot, S. Lebrun, G. Pauliat, H. Maillotte, V. Laude, T. Sylvestre, Nature Communications **5**, 1 (2014)

[4] A. Godet, A. Ndao, T. Sylvestre, V. Pecheur, S. Lebrun, G. Pauliat, J.C. Beugnot, K. Phan Huy, Optica **4**, 1232 (2017)

[5] A. Godet, J. Chretien, J.C. Beugnot, K.P. Huy, Optics Express **30**, 815 (2022)

[6] J.C. Beugnot, J. Chrétien, A. Godet, *Process for delaying an optical signal* (2019)

https://doi.org/10.1051/epjconf/202226611009

Using optical nanofibres to mediate cold atom interactions

Síle Nic Chormaic[1,*]

[1]Okinawa Institute of Science and Technology Graduate University, Onna, Okinawa 904-0495, Japan

Abstract. We discuss several recent advances related to optical nanofibres in cold atom systems, including two-colour dipole trap optimisation using an in-loop stochastic artificial neural network machine learner, upper bound limitations on Rydberg atom excitation due to localised ion formation, spectral lineshapes arising from the high intensity evanescent fields, and two-photon processes mediated via the evanescent field of the optical nanofibre.

1 Introduction

Optical nanofibres – very thin, tapered optical fibres where the waist diameter is less than the propagating light wavelength – have been shown to be very useful tools for atom-light interactions [1,2]. Their small size and relative ease of integration into optical fibre-based experimental setups, in addition to their minimal perturbation on magneto-optically trapped cold atoms, have ensured their adoption into cold atom physics. Here, we will discuss some recent applications of optical nanofibres to manipulate, trap, and control cold ^{87}Rb atoms in ground or Rydberg states. We will present some recent experimental and theoretical results related to the interactions between the atoms and the optical nanofibre field and introduce some of the limitations observed.

2 Details

We have performed several experimental and theoretical studies related to ground state and excited state alkali atom interactions with the evanescent light fields of optical nanofibres. These allow us to explore some hard-to-access transitions in the alkali atoms, such as quadrupole excitations [3,4], and to determine the influence of the dielectric nanofibre on spectral emissions and lineshapes [5-7]. Two specific examples of studies that we have completed are detailed in the following.

2.1. Machine learner optimisation of fibre-based dipole traps

One example of an application of optical nanofibres is the confinement of cold alkali atoms in two-color fibre-based traps [8]. The efficiency of these traps depends on several factors, such as the wavelength and intensity of light in the fibre's evanescent field, and the preloading laser-cooling process. Typically, only one or no atom is confined per trapping site. Increasing the number of filled trapping sites, thence the optical depth of the system, is desirable and relies on improving the trapping efficiency. We have used a machine learner that optimizes the absorption signal of a probe beam, hence the optical depth, by exploring a parameter space related to initial laser-cooling parameters. Through this process, we increased the number of atoms trapped in the fibre traps by 50% and the optical depth by 70% [9].

2.2 Rydberg atom generation near optical nanofibres

A second application is the creation of Rydberg atoms, which are highly excited neutral atoms, via the evanescent field [10]. As a first step, we have explored the upper limitations on the Rydberg level, n, that can be achieved for S and D state atoms due to dielectric interactions and ion formation. We envision that through judicious selection of the Rydberg level and by improving the trapping efficiency of ground state atoms in fibre-based traps via machine learning as mentioned above, we can explore novel schemes for generating 1D chains of Rydberg atoms near optical nanofibres.

3 Conclusion

In conclusion, we have discussed several recent theoretical and experimental advances in atom and optical nanofibre integrated systems. In future work, a combination of several techniques that we have developed will be used to create 1D chains of Rydberg atoms next to optical nanofibres to study many-body physics and to further explore nonlinear processes such as four-wave mixing.

This work was supported by OIST Graduate University and JSPS Grant-in-Aid for Scientific Research (C) Grant Number 19K05316. The author acknowledges contributions by E. Brion, D. Brown, T. Busch, J.L. Everett, V. Gokhroo, R.K. Gupta, R. Henke, K. Karlsson, D.F. Kornovan, P.K. Lam, F. Le Kien, T. Nieddu, K.S. Rajasree, A. Raj, T. Ray, Z. Shahrabifarahani, E.

* Corresponding author: sile.nicchormaic@oist.jp

Stourm, G. Tkachenko, A.D. Tranter, A. Vylegzhanin to the reported work.

References

1. M.J. Morrissey, K. Deasy, M. Frawley, R. Kumar, E. Prel, L. Russell, V.G. Truong, S. Nic Chormaic, Sensors **13**, 10449 (2013)
2. P. Lodahl, S. Mahmodian, S. Stobbe, A. Rauschenbeutel, P. Schneeweiss, J. Volz, H. Pichler, P. Zoller, Nature **541**, 473 (2017)
3. F. Le Kien, T. Ray, T. Nieddu, T. Busch, S. Nic Chormaic, Phys. Rev. A **97**, 013821 (2018)
4. T. Ray, R.K. Gupta, V. Gokhroo, J.L. Everett, T. Nieddu, K.S. Rajasree, S. Nic Chormaic, New J. Phys. **22**, 062001 (2020)
5. E. Stourm, M. Lepers, J. Robert, S. Nic Chormaic, K. Mølmer, E. Brion, Phys. Rev. A **101**, 052508 (2020)
6. F. Le Kien, D.F. Kornovan, S. Nic Chormaic, T. Busch, Phys. Rev. A **105**, 042817 (2022)
7. V. Gokhroo, F. Le Kien, S. Nic Chormaic, J. Phys. B : At. Mol. Opt. Phys. (accepted 2022)
8. F. Le Kien, V.I. Balykin, K. Hakuta, Phys. Rev. A **70**, 063403 (2004)
9. R.K. Gupta, J.L. Everett, A.D. Tranter, R. Henke, V. Gokhroo, P.K. Lam, S. Nic Chormaic, AVS Quant. Sci. (accepted 2022)
10. K.S. Rajasree, T. Ray, K. Karlsson, J.L. Everett, S. Nic Chormaic, Phys. Rev. Res. **2**, 012038(R) (2020)

Nonlinear Optical Fiber Couplers Made of Chalcogenide Glass

Mohsen Rezaei[1,*], *Md Hosne Mobarak* Shamim[1], *Mohammed* El Amraoui[2], *Younes* Messaddeq[2], and *Martin* Rochette[1]

[1]Department of Electrical and Computer Engineering, 3480 University Street, McGill University, Montréal, Canada, H3A 0E9
[2]Centre d'Optique, Photonique et Laser, 2375 Rue la Terrasse, Université Laval, Québec, Canada, G1V 0A6

Abstract. We demonstrate chalcogenide optical fiber couplers designed with a transmission spectrum response that varies with input power. The measured critical power is as low as 126 *W* at a wavelength of 1938 *nm*.

1 Introduction

Optical fiber couplers (OFCs) are typically four-port devices routinely used to divide or combine light in fiber optics devices and systems [1]. They generally operate in a linear regime, where the coupling ratio in between two output ports is independent of the power delivered to an input port. However, some OFCs can enter a nonlinear regime where the coupling ratio is a function of input power. Nonlinear OFCs are of particular interest because of their applications in all-optical switching and mode-locked fiber lasers. Initially proposed by Jensen [2], nonlinear OFCs have since been demonstrated experimentally in dual-core silica fibers, with a minimum reported critical power $P_c = 32\ kW$ at a wavelength of 620 *nm* in a 0.5 *cm* long fused-quartz dual-core fiber directional coupler [3]. Reported nonlinear OFCs are limited to critical powers in the order of tens of kW in response to the relatively low nonlinear refractive index of silica. In contrast, chalcogenide glasses are excellent candidates to make nonlinear OFCs with low P_c, thanks to their exceptionally high nonlinear refractive index that is ~3 orders of magnitude beyond that of silica. So far, chalcogenide OFCs have been demonstrated in the linear regime [5–9] including the first single-mode chalcogenide OFC [7] as well as chalcogenide OFCs such as power dividers, wavelength division multiplexers, and polarization beamsplitters [8]. To ensure a nonlinear response, the OFC design must ensure that the coupling length and nonlinear length are roughly equal [2].

Here, we present the first chalcogenide OFCs with a nonlinear coupling response. Two nonlinear OFCs made of As_2Se_3 are presented, showing a critical power as low as $P_c = 126\ W$, the lowest ever reported for any OFC. These OFCs will find applications for all-optical switching as well as act as saturable absorbers in mode-locked and Q-switch fiber lasers with low power threshold.

2 Experiment and Results

The output power of a nonlinear OFC is given by [2]: $P_{Through}(L) = \frac{P_0}{2}\left[1 + cn\left(\pi L/L_c | (P_0/P_c)^2\right)\right]$, and $P_{Cross}(L) = P_0 - P_1(L)$, where $P_{Through}$ is the output power on the same fiber where the input power is applied, P_{Cross} is the output power crossing to the coupled fiber, P_0 is the input power, L is the length of the coupling region, L_c is the length required for a complete power transfer from one fiber to the next, $cn(x|m)$ is a Jacobi elliptic function with argument x and modulus m. $P_c = \frac{2\pi}{\gamma L_c}$ is the critical power, where for $P_0 = P_c$ the power of through and cross output ports reach the same level, and γ is the waveguide nonlinearity. A practical nonlinear OFC requires a P_c that has access to low values and thus, its geometry needs to be designed in a way that provides both large γ and large L_c, while L_c should remain in the order of the nonlinear length. Nonlinear length, $L_{nl} = 1/P_0\gamma$, is the effective propagation distance at which the nonlinear phase shift is 1 rad.

The designed OFCs are prepared from two identical pieces of As_2Se_3 fiber with a core/cladding diameter of 20/240 μm and a numerical aperture of 0.2 pretapered down to a cladding diameter of 144 μm. They are cleaved, polished, and coupled to SMF-28 using UV-cured epoxy. The fibers are subsequently set side by side in close contact on a tapering setup. Both fibers are heated to their softening temperature and stretched into tapered OFCs [10]. Figure 1 shows a schematic of the tapering setup. The degree of fusion between both fibers is chosen in a way that leads to a polarization independent OFC by controlling the temperature and speed of stretching [8]. The tapering process is interrupted when the fiber cladding diameter at the waist of the OFC reaches the desired final diameter.

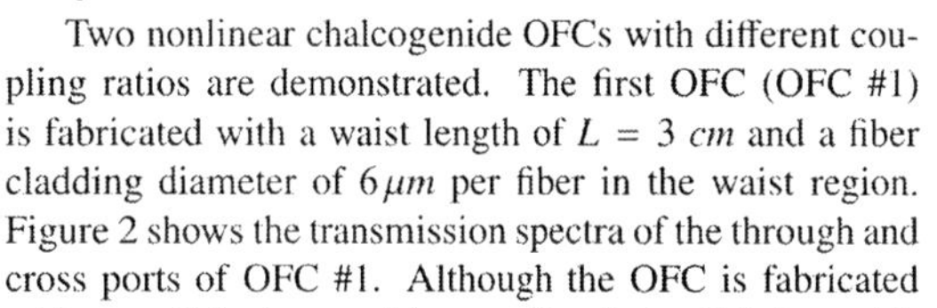

Figure 1. Schematic of the tapering setup. PS: pulling stage; HE: heating element; TS-HE: translation stage of the heating element; PM: power meter.

Two nonlinear chalcogenide OFCs with different coupling ratios are demonstrated. The first OFC (OFC #1) is fabricated with a waist length of $L = 3\ cm$ and a fiber cladding diameter of 6 μm per fiber in the waist region. Figure 2 shows the transmission spectra of the through and cross ports of OFC #1. Although the OFC is fabricated with an adiabatic transition profile, it is slightly multimode because fibers are multi-mode, leading to transmission spectra that are wavelength dependent. To investigate the nonlinear behaviour of the OFC, pulses with different peak powers are sent into the OFC while both through- and cross-port outputs are monitored. Figure 3 shows a schematic of the characterization setup. Pulses injected to the OFC have a full width at half maximum of 1.4 *ps*

*e-mail: mohsen.rezaei@mail.mcgill.ca

and a repetition rate of 30 *MHz* at a central wavelength of 1938 *nm*. The overall length of SMF-28 from the amplifier output to the OFC input is 1.6 *m*. With a maximum peak power of 200 *W* in silica, the maximum nonlinear phase-shift prior to entering the OFC is thus negligible with 0.25 *rad*.

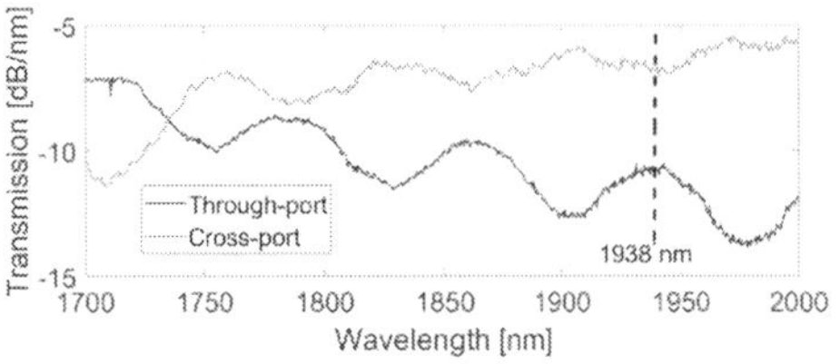

Figure 2. Transmission spectra of OFC #1.

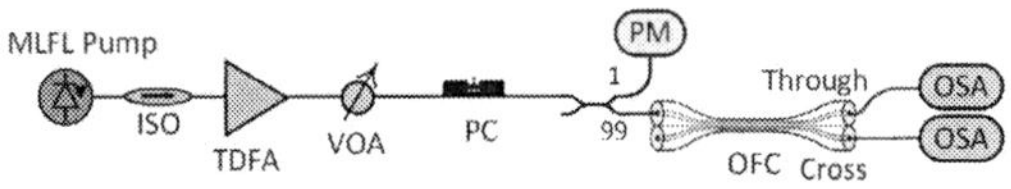

Figure 3. Schematic of the characterization setup.

Figure 4a shows the output power of OFC #1 as a function of input peak power. In a linear regime, the coupling length of this OFC is $L_c \approx 1.6\,L$. While at low input power most of the power comes out of the cross-port, an increase of input power leads to an increasing fraction of power at the through-port. No evidence of saturation or nonlinear absorption is observed from the total power emerging from both ports. The linear behaviour is identified by straight asymptotes in fig. 4a. Figure 4b shows the relative output power as a function of input peak power. At low input power, more than 70% of the output power is in cross-port. This value drops down as the input power increases. At an input peak power of ~ 140 *W*, through and cross ports provide the same output power. Further increase of the input power makes through-port the dominant output port. At an input peak power of ~200 *W*, again the coupling ratio is 70:30 but this time 70% of the output power appears in through-port. Full lines are theoretical fits following with $P_c = 126\,W$ and $L_c = 1.6\,L$.

Besides the change in the coupling ratio, pulses propagating in the OFC experience spectral broadening due to self-phase modulation and temporal changes in the pulse shape. Over a length of L_c, the tails of the input pulses mostly couple to the cross-port while the middle part with a higher power level mostly stay in the through-port. Figure 5 shows the cross-port optical spectra for input pulses with different peak powers along with the numerical fit derived from the coupled-mode equation for a symmetric coupler.

A second OFC with a waist length of $L = 3\ cm$ and a fiber waist diameter of 9 μm is also fabricated. With a larger waist diameter and thus a lower nonlinearity, a larger critical power is expected in comparison to OFC #1. Figure 6 shows the relative output power of OFC #2. At low input power, the coupling ratio is 80:20 with 80% of the power emerging from the cross-port. Increasing the input power changes the coupling ratio toward equalizing power of the two output ports. A theoretical fit matches the experimental data with $P_c = 239\,W$ and $L_c = 1.4\,L$.

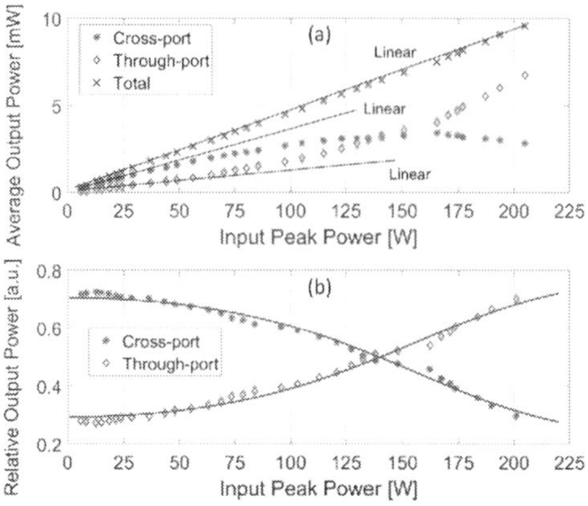

Figure 4. (a) Output power of OFC # 1 as a function of the input peak power. (b) Relative output power of OFC # 1.

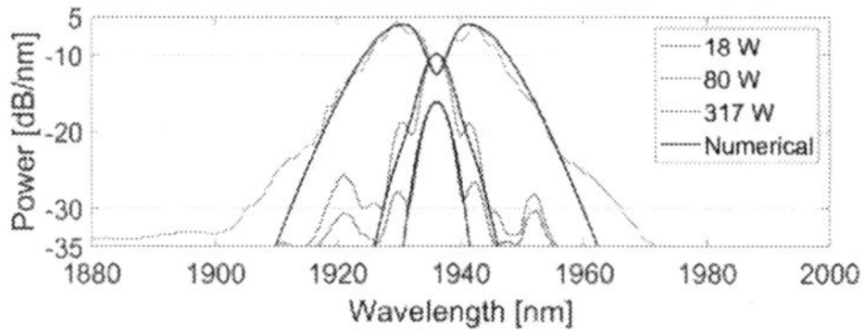

Figure 5. Cross-port optical spectra of OFC #1 for input pulses with different peak powers.

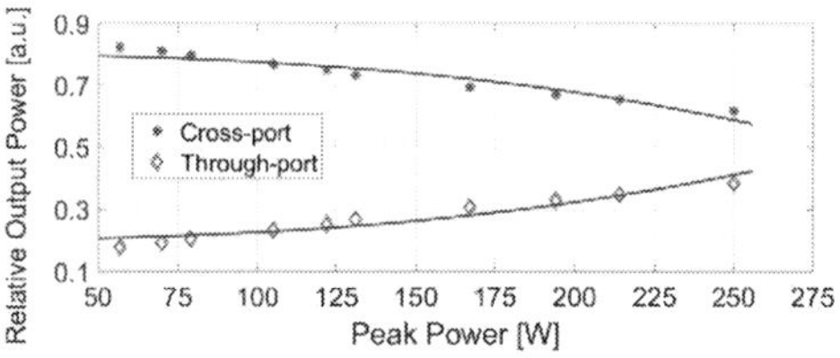

Figure 6. Relative output power of OFC # 2.

In conclusion, we have experimentally demonstrated the nonlinear behaviour of chalcogenide OFCs. These nonlinear OFCs show the excellent potential of nonlinear chalcogenide OFCs for applications such as all-optical switching and passively mode-locked lasers with low power threshold.

References

[1] M. Rezaei et al., Opt. Fiber Technol. **71**, (2022).
[2] S. Jensen, IEEE J. Quantum Electron. **18**, 15801583 (1982).
[3] S. Friberg et al., Opt. Lett. **13**, 904906 (1988).
[4] A. Weiner et al., IEEE J. Quantum Electron **25**, 26482655 (1989).
[5] D. Schaafsma et al., J. Light. Technol. **15**, 22422245 (1997).
[6] F. Tavakoli et al., IEEE Photon. Technol. Lett. **29**, 735738 (2016).
[7] M. Rezaei et al., IEEE Photonics Conference (IPC) (2018).
[8] M. Rezaei et al., Opt. Lett. **44**, 52665269 (2019).
[9] O. Benderov et al., Appl. Opt. **58**, 72227226 (2019).
[10] C. Baker et al., Opt. Mater. Express **1**, 10651076 (2011).

Single-Mode ZBLAN Fiber Couplers

Mohsen Rezaei*, *Gebrehiwot Tesfay* Zeweldi, *Md Hosne Mobarak* Shamim, and *Martin* Rochette

Department of Electrical and Computer Engineering, 3480 University Street, McGill University, Montréal, Canada, H3A 0E9

Abstract. We demonstrate single-mode ZBLAN optical fiber couplers. A controlled tapering procedure leads to coupling ratios of 5%/95% and 14%/86% at a wavelength of 2200 nm, with insertion losses of 1.6 dB and 1.8 dB, respectively.

1 Introduction

Optical fiber couplers (OFCs) are components used for power dividing/combining, wavelength multiplexing/demultiplexing, filtering, and polarization beam splitting, broadly used in optical fiber devices and systems. Just like optical fibers, the vast majority of OFCs are made of silica glass. However, intrinsic losses prohibit the use of silica-based OFCs at wavelengths greater than ~ 2 μm.

In contrast to silica, fluoride and chalcogenide fibers are transparent to mid-infrared light, up to 5 μm for ZBLAN and 12 μm for As_2Se_3 fibers. Chalcogenide OFCs have already been demonstrated in the form of single-mode OFCs [1–3], nonlinear OFCs [4], and also served in the fabrication of laser cavities [5, 6], but remain an emerging technology that needs refinement before commercialization. In contrast, fluoride glass OFCs have yet to be developed, mainly due to crystallization that spontaneously occurs close to the softening temperature [7]. Despite this major limitation and the competing development of chalcogenide-based OFCs, the realization of fluoride glass OFCs would lead to a complementary technology of strong interest because of their larger optical damage threshold and lower optical nonlinearity, compared to chalcogenide glass [8]. In 2016, G. Stevens et al. demonstrated the only fluoride glass OFC reported thus far [9]. The multi-mode ZBLAN OFC had a coupling ratio of up to 50%/50% at a wavelength of 2000 nm. The multimodedness of OFCs is however an impairment to the functionality of single-mode based optical fiber systems and devices. The fabrication of a single-mode OFC requires the use of single-mode fibers and the realization of appropriate adiabatic transition profiles on both sides of the OFC coupling section. The single-modedness of an OFC is validated from a smooth transmitted power versus OFC extension (during fabrication), as well as a smooth transmission spectrum, free from modal interference fluctuations [3].

Here, we demonstrate the successful fabrication of single-mode ZBLAN OFCs. The OFCs are fabricated using a multiple-sweep tapering technique that allows precise and repeatable control of the geometry. This results in single-mode OFCs with repeatable optical properties, while limiting crystallization. Complete transmission spectra of the OFCs over a 1500 – 2200 nm band are provided, showing evidence of single-modedness. Coupling ratios following the OFC fabrication recipe can be varied anywhere from 0.1%/99.9% to 14%/86% at the through/cross ports, with a maximum insertion loss of 1.8 dB. This research is an important step towards the extension of optical fiber technologies in the mid-infrared.

2 Experiment and Results

To fabricate the OFCs, two pieces of ZBLAN fiber (Le Verre Fluoré) with core/cladding diameters of 6.5/125 μm and a numerical aperture of 0.23 are cleaved, polished, and coupled to SMF-28 using UV-cured epoxy. The two fibers are then placed together on a tapering setup with a twist of half a turn to ensure close contact of the fibers. Figure 1 shows a schematic of the tapering setup. Using an electrical heater, the fibers are heated at their softening temperature (290 °C) and stretched following an adiabatic tapering profile to avoid higher order modes excitation. This choice of temperature is a compromise between a colder temperature where viscosity becomes too large to enable stretching without breaking the fibers, and a higher temperature where crystallization grows too fast, leading to propagation losses before coupling can occur. During the tapering process, the heater brushes the waist of the OFC ~100 times with a velocity of 254 μm/s by moving back and forth in between the two pulling stages, resulting in a total tapering time of ~200 min. This allows the carving of a smooth transition region that adiabatically links the fiber mode to a tapered fiber mode, a uniform OFC waist region, and OFCs reproducibility. The tapering setup is enclosed in a box continuously purged with Argon gas to limit the crystallization process since ambient air acts as a catalyst.

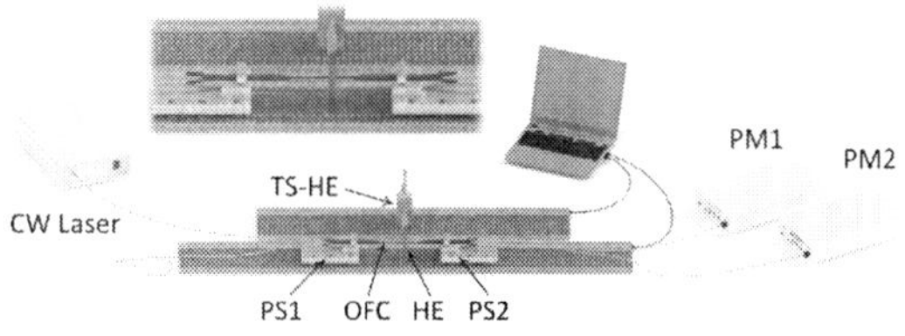

Figure 1. Schematic of the tapering setup. PS: pulling stage; HE: heating element; TS-HE: translation stage of the heating element; PM: power meter.

During tapering, continuous wave laser light is sent into one of the OFC's inputs and both though-port and cross-port outputs are monitored. Figure 2 shows the OFC

*e-mail: mohsen.rezaei@mail.mcgill.ca

transmission as a function of extension at a wavelength of 1550 nm, repeated for 3 OFCs with a waist length of 3 cm. Extension is the distance where the two pulling stages have moved away from each other since the beginning of the tapering process. At low extension, all the power remains in the through-port. However, as the tapering goes on with increasing extension, the fundamental modes guided by the two fiber cores increasingly overlap, leading to an increase in coupling. The tapering process is manually interrupted when the desired coupling ratio is achieved. The 3 OFCs demonstrate repeatable transmission patterns during the fabrication process. However, some OFCs break during the stretching. The rate of success in completing the extension of an OFC without breaking is of the order of 80%. Despite the actions taken to limit the crystallization, some crystallization is still triggered at the fiber surface and eventually gets deeper into the bulk material if tapering continues after a fiber diameter of ~30 μm. There is therefore a trade off in between tapering more to increase the coupling, and crystallization that eventually overlaps with the propagating light and increases losses.

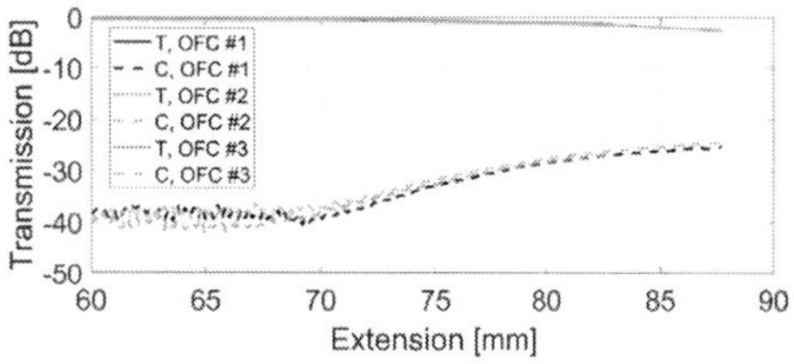

Figure 2. In situ transmission as a function of extension at the wavelength of 1550 nm, for 3 OFCs. T: through-port; C: cross-port.

Once the tapering process is completed, a broadband light source is launched into one input port of the OFC and the output power spectra from both arms are measured using an optical spectrum analyzer. Figure 3a shows the transmission spectra of OFC #3 made with a final waist fiber diameter of 27 μm and a waist length of 3 cm. The OFC has a larger coupling ratio at longer wavelengths due to a smaller V-number and weaker light confinement in both tapered fibers. The same reason is responsible for the larger insertion loss at longer wavelengths. More light interaction with the fiber surface leads to a larger scattering loss. OFC #3 provides a coupling ratio of 5%/95% with insertion losses of 1.6 dB at a wavelength of 2200 nm. To increase the coupling, the OFC waist length needs to be increased, and/or the fiber waist diameter needs to be decreased. Increasing the OFC waist length provides a longer interaction length for the light to couple to the second core. Decreasing the fiber waist diameter brings the two cores closer to each other, resulting in an increased modes overlap and larger coupling, but also more light interaction with the fiber-air interface where crystallization sets up and scattering loss occurs.

A fourth OFC is fabricated with the same fiber waist diameter but a longer waist length. Figure 3b shows the transmission spectra of OFC #4 with a waist fiber diameter of 27 μm and a waist length of 6 cm. Compared to OFC #3, OFC #4 has a larger coupling ratio. It shows a coupling ratio of 14%/86% with insertion losses of 1.8 dB at a wavelength of 2200 nm. The smooth transmission spectra of the OFCs and smooth transmission as a function of extension support the claim of single-modedness of the OFCs. Finally, ZBLAN OFCs are actually being kept at room temperature to observe their transmission as a function of time. For the time being, after two weeks of constant probing, no noticeable change could be detected on the OFC transmission. This measurement is still ongoing. In conclusion, we have demonstrated the first realization of single-mode couplers made of ZBLAN. This is an important step towards the extension of optical fiber technologies in the mid-infrared.

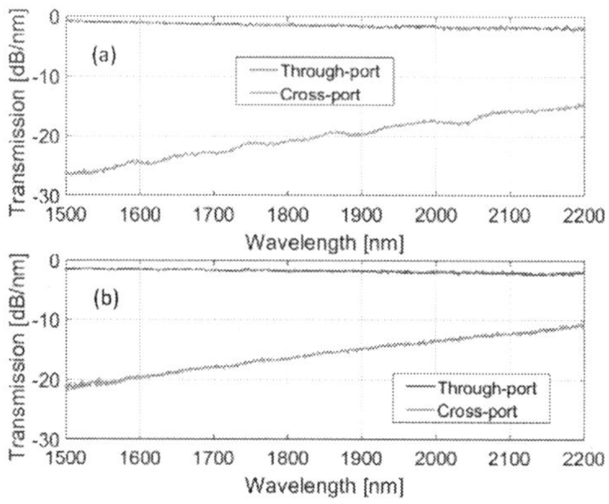

Figure 3. Transmission spectra of (a) OFC #3 with a waist length of 3 cm and (b) OFC #4 with a waist length of 6 cm.

The authors are thankful to Le Verre Fluoré for providing the ZBLAN fiber used in this experiment.

References

[1] F. Tavakoli et al., "Broadband and wavelength-dependent chalcogenide optical fiber couplers," IEEE Photon. Technol. Lett. **29**, 735-738 (2016).

[2] M. Rezaei et al., "All-chalcogenide single-mode optical fiber couplers," Opt. Lett. **44**, 52665269 (2019).

[3] V. Fortin et al., "Post-processing soft glass optical fibers," in Mid-Infrared Fiber Photonics, (Elsevier, 2022), pp. 233–302.

[4] M. Rezaei et al., "Nonlinear chalcogenide optical fiber couplers," accepted for publication in Opt. Express, (2022).

[5] I. Alamgir et al., "Optical parametric oscillation from soft-glass fiber cavity," accepted for publication at the Conference on Lasers and Electro-Optics (CLEO), (2022).

[6] M. Rezaei et al., "All-Chalcogenide Ring Fiber Laser," Opt. Fiber Technol. **71**, (2022).

[7] A. Torres et al., "A quantitative characterization of micro-gravity and unit-gravity processed zblan glass," Materials Characterization, **107**, 386–397 (2015).

[8] C. Xia et al., "10.5 W time-averaged power Mid-IR supercontinuum generation extending beyond 4 μm with direct pulse pattern modulation," IEEE J. Sel. Top. Quantum Electron., **15**, 422–434 (2009).

[9] G. Stevens et al., "Mid-ir fused fiber couplers," in Components and Packaging for Laser Systems II, **9730**, 973007 (2016).

 https://doi.org/10.1051/epjconf/202226611012

Non-Reciprocal Amplification of Light Using Cold Atoms Coupled to an Optical Nanofiber

Sebastian Pucher[1], *Christian* Liedl[1], *Shuwei* Jin[1], *Arno* Rauschenbeutel[1], and *Philipp* Schneeweiss[1,*]

[1]Department of Physics, Humboldt-Universität zu Berlin, 10099 Berlin, Germany

Abstract. Optical nanofibers realized as the waist of tapered silica fibers can be used to trap and optically interface laser-cooled atoms. Building on this system, we experimentally show a novel scheme for the non-reciprocal Raman amplification of light. While typically either the magneto-optical effect, a temporal modulation or an optical nonlinearity is employed to break reciprocity, in our approach, this results from the spin of the atoms forming the gain medium. By taking advantage of the inherent spin-momentum locking present in optical nanofibers, we perform an experiment in which we set the amplification direction by a suitable preparation of the atomic spin state. Our approach is general and, suitable quantum emitters provided, could also be implemented beyond the optical domain of the electromagnetic spectrum.

1 Introduction

Typically, non-reciprocal optical devices are realized by employing either the magneto-optical effect, time modulation or an optical nonlinearity in order to break Lorentz reciprocity [1]. In recent years, a new class of such devices based on spin-polarized quantum emitters have attracted increasing interest, and respective isolators and circulators have been demonstrated experimentally [2]. Recently, we have demonstrated non-reciprocal Raman amplification of light using spin-polarized atoms with directional coupling to a tapered optical fiber with a nanofiber waist [3]. We show that we can switch the operation direction of the amplifier by flipping the atomic spin.

2 Experimental Setup

Figure 1 a) shows a schematic of the experimental setup. We couple a 1D array of laser-cooled, optically trapped cesium atoms to the evanescent field surrounding an optical nanofiber of 500 nm diameter [4]. We then launch a probe laser field into either port 1 or port 2 of the nanofiber. We choose the polarization such that, due to the strong transverse confinement of the light, the local polarization at the position of the atoms is either σ^-- or σ^+-polarized, depending on the propagation direction [5]. This inherent link between polarization and propagation direction is also known as spin-momentum locking. In addition to the probe light, we illuminate the atoms with a free-space π-polarized coupling laser field that propagates in the x-direction. In Fig. 1 b), we show the energy levels relevant to our experiment. Initially, we prepare the atoms in state $|6S_{1/2}, F = 4, m_F = -4\rangle$. In order to stabilize this spin-polarization, we apply a magnetic offset field of $\sim$ 7 G along z. The coupling field is detuned by Δ=82 MHz from the excited state, and the probe laser is tuned to the respective light shifted two-photon resonance.

*e-mail: philipp.schneeweiss@hu-berlin.de

3 Experimental demonstration of non-reciprocal Raman amplification

First, we switch on the probe laser field, which does not couple to the atoms in their initial state, and measure a transmission of 9 pW. We then switch on the coupling laser field at time $t = 0\ \mu$s and measure the probe transmission dynamics, see Fig. 2 a). We observe an increase of the probe transmission by up to $\sim$ 130 % when launching the probe field into port 1 of the waveguide (red diamonds). However, if the probe laser is launched into port 2, the transmission is almost unchanged, since the atoms do not couple to σ^+-polarized probe light (blue circles). The data is in good agreement with the predictions of our model, which consists of solving the Master equation for a three-level system for each atom and consecutively calculating the transmission coefficient along an array of N = 1400 atoms, see solid lines in Fig. 2. By preparing the atoms in $|F = 4, m_F = +4\rangle$, we are able to switch the operation direction of the amplifier, and the probe laser field is amplified only when propagating from port 2 to port 1, see Fig. 2 b). These results demonstrate non-reciprocal amplification of the probe light field and that we are able to set the direction of amplification via the atomic spin state.

4 Outlook

Our results provide a new element to the novel class of spin-controlled, non-reciprocal optical devices. Our scheme could be extended to continuous-wave operation by the addition of a repump laser field and thus enable

 https://doi.org/10.1051/epjconf/202226611012

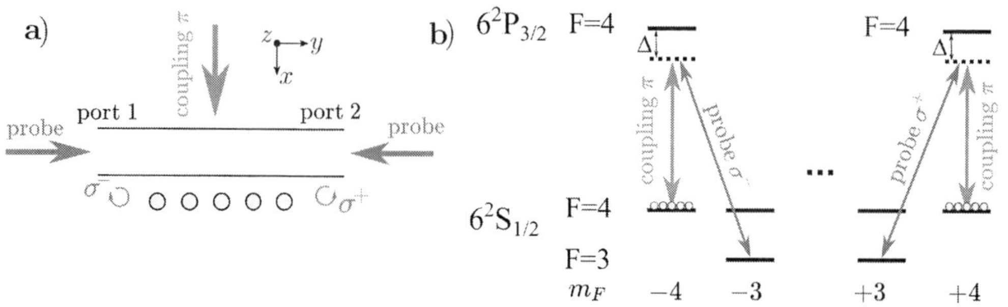

Figure 1. a) We trap cesium atoms (yellow circles) in a 1D array of trapping sites, close to an optical nanofiber. A probe laser field is launched either into port 1 or port 2 of the waveguide and interacts with the atoms via the evanescent part of the nanofiber-guided mode. The probe field is either predominantly σ^-- or σ^+- polarized at the position of the atoms. A π-polarized coupling laser field illuminates the atoms from the side. b) Energy levels relevant to our scheme. The atoms are initially prepared in $|6^2S_{1/2}, F = 4, m_F = -4\rangle$. The probe and coupling laser fields couple to different Λ-system, depending on the probe polarization and initial atomic state.

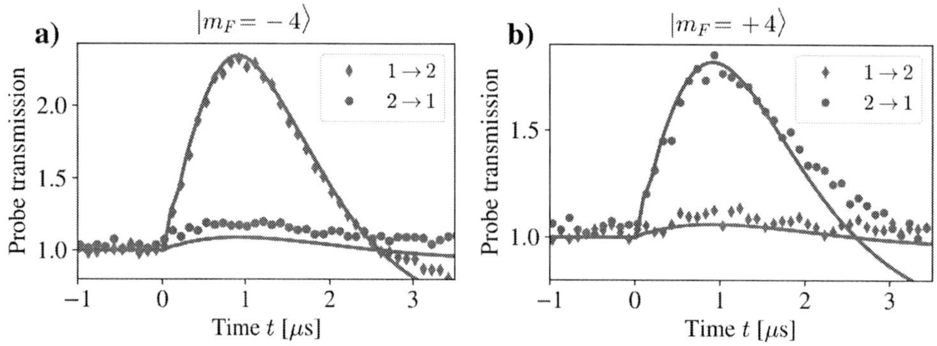

Figure 2. Non-reciprocal gain. Measured probe transmissions when launching the probe laser into port 1 or port 2 of the waveguide, see red diamonds and blue circles, respectively. a) We prepare the atoms in $|F = 4, m_F = -4\rangle$. At $t = 0\,\mu$s, we switch on the coupling field and the probe transmission increases by ~ 130 % in the 1 → 2 direction, while it is almost unchanged in the 2 → 1 direction. b) Here, we prepare the atoms in $|F = 4, m_F = +4\rangle$ and observe that the direction in which there is gain is now switched, showcasing non-reciprocal amplification and that the gain directionality can be controlled by suitably preparing the atomic spin state. The data is in good agreement with the theory prediction, see solid lines.

fundamental studies of lasing with spin-controlled gain directionality. Moreover, our scheme could be implemented using other quantum emitters and nanophotonic structures, allowing the extension of our method to other domains of the electromagnetic spectrum.

References

[1] V. S. Asadchy *et al.*, "Tutorial on electromagnetic nonreciprocity and its origins," Proc. IEEE **108**, 1684 (2020).

[2] P. Lodahl *et al.*, "Chiral quantum optics," Nature **541**, 473 (2017).

[3] S. Pucher *et al.*, "Atomic spin-controlled non-reciprocal Raman amplification of fibre-guided light," Nature Photonics **16**, 380 (2022).

[4] E. Vetsch *et al.*, "Optical interface created by laser-cooled atoms trapped in the evanescent field surrounding an optical nanofiber," Phys. Rev. Lett. **104**, 203603 (2010).

[5] R. Mitsch *et al.*, "Quantum state-controlled directional spontaneous emission of photons into a nanophotonic waveguide," Nature Communications **5**, 1 (2014).

https://doi.org/10.1051/epjconf/202226611013

Rydberg atoms in the vicinity of an optical nanofiber

Erwan Stourm[1], *Maxence* Lepers[2], *Jacques* Robert[1], *Síle* Nic Chormaic[3], *Klaus* Mølmer[4], and *Étienne* Brion[5,*]

[1]Université Paris-Saclay, CNRS, Laboratoire de physique des gaz et des plasmas, 91405, Orsay, France
[2]Laboratoire Interdisciplinaire Carnot de Bourgogne, CNRS, Université de Bourgogne Franche-Comté, 21078 Dijon, France
[3]Light-Matter Interactions for Quantum Technologies Unit, Okinawa Institute of Science and Technology Graduate University, Onna, Okinawa, 904-0495, Japan
[4]Department of Physics and Astronomy, Aarhus University, Ny Munkegade 120, DK-8000 Aarhus C, Denmark
[5]Laboratoire Collisions Agrégats Réactivité, IRSAMC and UMR5589 du CNRS, Université de Toulouse III Paul Sabatier, F-31062 Toulouse Cedex 09, France

Abstract. Highly excited (so-called Rydberg) atoms are the key ingredient of many quantum information schemes. In this presentation, we shall theoretically investigate how spontaneous emission properties and van der Waals interactions of such atoms are modified in the neighbourhood of an optical nanofiber with respect to the free-space (vacuum) case. This work constitutes a very preliminary step towards the realization of a quantum network based on atomic ensembles linked via optical nanofibers.

1 Introduction

Within the past two decades, the strong dipole-dipole interaction experienced by two neighboring Rydberg-excited atoms has become the main ingredient for many atom-based quantum information protocol proposals [1]. This interaction can indeed be so large as to forbid the simultaneous resonant excitation of two atoms if their separation is less than a specific distance, called the blockade radius, which typically depends on the intensity of the laser excitation and the interaction between the Rydberg atoms. The discovery of this "Rydberg blockade" phenomenon [2] paved the way to numerous proposals for atomic quantum registers [3] and repeaters [4].

Scalability is one of the crucial requirements for quantum devices : interfacing atomic ensembles into a quantum network is a possible way to reach this goal. Photons naturally appear as ideal information carriers. Photon-based protocols considered so far include free space setups which are relatively easy to implement but present the drawback of strong losses. An alternative option would be to use optical fibers and in particular nanofibers which permit strong coupling and transverse mode selection.

Such fibers have recently received much attention [5, 6] because the coupling to the evanescent guided modes of a nanofiber allows for easy-to-implement atom trapping [7] and detection [8]. This coupling increases in strength as the fiber diameter reduces and the atoms approach the fiber surface. It has also been shown that energy could be exchanged between two distant atoms via the guided modes of the fiber [9]. This suggests that optical nanofibers could play the role of a communication channel between the nodes of an atomic quantum network consisting of Rydberg-excited atomic ensembles.

Preliminary steps have been taken towards building a quantum network based on Rydberg-blockaded atomic ensembles linked via an optical nanofiber. On the experimental side, the excitation of cold ^{87}Rb atoms towards Rydberg 29D state was demonstrated at submicron distances from an optical nanofiber surface [10], in a two-photon ladder-type excitation scheme. On the theory side, the spontaneous emission of a highly excited (Rydberg) sodium atom in the neighborhood of a silica optical nanofiber was investigated [11]. In particular, the dependence of the emission rates into the guided and radiative modes on the radius of the fiber, the distance of the atom to the fiber, and the symmetry of the Rydberg state was studied. Since it used the so-called mode function approach, this work could unfortunately not account for the fiber's absorption and dispersion. This point is critical with Rydberg atoms since they can de-excite along many transitions of different frequencies for which the fiber index is different and potentially complex.

By contrast, here, we resort to the framework of Macroscopic Quantum Electrodynamics (MQE) based on the dyadic Green's function [12]. This formalism enables us to take the exact refractive index of silica into account and relaxes all constraints on the transitions we can address. This framework also offers a natural way to compute not only spontaneous emission rates and Lamb shifts of an atom close to the nanofiber (Sec. 2), but also van der Waals interaction potential between two atoms in the vicinity of the fiber (Sec. 3).

*e-mail: etienne.brion@irsamc.ups-tlse.fr

2 Spontaneous emission and energy shifts of a Rydberg rubidium atom close to an optical nanofiber

In this presentation, we shall present the numerical results we obtained in the MQE approach for a rubidium atom prepared in a Rydberg-excited state $|n = 30; L = S, P, D; JFM_F\rangle$ in the vicinity of a multimode silica optical nanofiber [13]. We chose ^{87}Rb as it is commonly used in Rydberg atom experiments [10]. In particular, we show that a non-negligible fraction of spontaneously emitted light is guided along the fiber and study how it depends on the principal quantum number, n, the radius of the nanofiber, a, the distance of the atom to the nanofiber axis, R, and the direction of angular momentum polarization. Interestingly, when the quantum and fiber axes do not coincide, spontaneous emission becomes directional, as already noticed for low-excited atoms [14] due to the peculiar polarization structure of the field in the neighborhood of the fiber. This effect is particularly strong for photons emitted into the fiber-guided modes and persists even for high principal quantum numbers, n. This is promising in view of potential applications in chiral quantum information protocols [15] based on a Rydberg atom-nanofiber interface. We also address Lamb shifts and associated dispersion forces that arise. In particular, we show that, as n increases, the contribution of quadrupolar transitions becomes more important. This contrasts with spontaneous emission rates for which quadrupolar transitions have negligible influence.

3 Van der Waals Interaction of two Rydberg rubidium atoms close to an optical nanofiber

Giant van der Waals interactions are among Rydberg atoms' most striking features. For two atoms prepared in levels of principal quantum numbers $n > 50$ and a few μm apart, such interactions can indeed induce energy shifts of the order of tens of GHz. In this area, the interatomic interaction between two atoms (A, B) separated by the distance r_{AB} follows the law identified by London [16]

$$U_{AB}^{(0)} = -\frac{C_6\,(A, B)}{r_{AB}^6}$$

The C_6 coefficient depends on the states in which atoms (A, B) are prepared as well as their geometric arrangement. It scales with the principal quantum number as n^{11}. For a pair of rubidium atoms in the state $|60S_{1/2}\rangle$ in vacuum it is of the order of $100\text{GHz}.(\mu\text{m})^{-6}$. In this presentation we shall investigate how the presence of the fiber modifies this interaction with respect to the vacuum case [17]. This study follows other works in plane geometries involving Rydberg atoms in front of a conducting half-space [18].

We shall first briefly recall the form of the interaction Hamiltonian between two Rydberg atoms in the presence of a dieletric medium. Then we shall study the effect of this Hamiltonian on two atoms prepared in the same state $|nS_{1/2}\rangle$, with $n \geq 30$ and compare with the case of dielectric half-space. On the example of two atoms prepared in the state $|nP_{3/2}, M_j = \frac{3}{2}\rangle$, with $n \geq 30$, we shall also demonstrate a strong enhancement of the interaction potential related to the existence of a Förster quasi-resonance. Furthermore, we shall show that it is possible to change the nature of the Van der Waals force for $n > 38$. Finally, we shall consider the influence of the fiber on the anisotropy of the C_6 coefficient, which also exists in vacuum.

References

[1] M. Saffman, T. G. Walker, and K. Mølmer, Rev. Mod. Phys. **82**, 2313 (2010).

[2] M. D. Lukin, M. Fleischhauer, R. Côté, L. M. Duan, D. Jaksch, J. I. Cirac, and P. Zoller, Phys. Rev. Lett. **87**, 037901 (2001).

[3] E. Brion, K. Mølmer, and M. Saffman, Phys. Rev. Lett. **99**, 260501 (2007).

[4] E. Brion, F. Carlier, V. M. Akulin, and K. Mølmer, Phys. Rev. A **85**, 042324 (2012).

[5] T. Nieddu, V. Gokhroo, and S. Nic Chormaic, J. Opt. **18**, 053001 (2016).

[6] P. Solano, J. A. Grover, J. E. Hoffman, S. Ravets, F. K. Fatemi, L. A. Orozco, and S. L. Rolston, Adv. At. Mol. Opt. Phys. **66**, 439 (2017).

[7] V. I. Balykin, K. Hakuta, F. L. Kien, J. Q. Liang, and M. Morinaga, Phys. Rev. A **70**, 011401(R) (2004).

[8] K. P. Nayak, P. N. Melentiev, M. Morinaga, F. Le Kien, V. I. Balykin, and K. Hakuta, Opt. Express **15**, 5431 (2007).

[9] F. Le Kien, S. Dutta Gupta, K. P. Nayak, and K. Hakuta, Phys. Rev. A **72**, 063815 (2005).

[10] K. S. Rajasree, T. Ray, K. Karlsson, J. L. Everett, and S. Nic Chormaic, Phys. Rev. Res. **2**, 012038(R) (2020).

[11] E. Stourm, Y. Zhang, M. Lepers, R. Guérout, J. Robert, S. Nic Chormaic, K. Mølmer, and E. Brion, J. Phys. B: At. Mol. Opt. Phys. **52**, 045503 (2019).

[12] S. Y. Buhmann, *Dispersion Forces I and II*, Springer-Verlag, Berlin, 2012.

[13] E. Stourm, M. Lepers, J. Robert, S. Nic Chormaic, K. Mølmer, and E. Brion, Phys. Rev. A **101**, 052508 (2020).

[14] F. Le Kien and A. Rauschenbeutel, Phys. Rev. A **90**, 023805 (2014).

[15] P. Lodahl, S. Mahmoodian, S. Stobbe, A. Rauschenbeutel, P. Schneeweiss, J. Volz, H. Pichler, and P. Zoller, Nature (London) **541**, 473 (2017).

[16] T.F. Gallagher, *Rydberg Atoms*, Cambridge University Press, Cambridge (1994).

[17] E. Stourm, M. Lepers, J. Robert, S. Nic Chormaic, K. Mølmer, E. Brion, Van der Waals Interaction of two Rydberg rubidium atoms close to an optical nanofiber, *in preparation*.

[18] J. Block and S. Scheel, Phys. Rev. A **96**, 062509 (2017).

High peak/average power picosecond pulsed MOPA system with tapered large mode area double-clad Yb-doped fiber

V. Ustimchik, E. Motorin and V. Filippov[*]
Ampliconyx Ltd, Lautakatonkatu 18, 33580 Tampere, Finland

Abstract. In this paper, a high peak/average power picosecond pulsed master-oscillator power-amplifier (MOPA) system utilizing tapered large mode area double-clad fiber (T-DCF) is presented. In a high average power regime, we report a pulsed MOPA laser system with average power up to 613 W at 20 MHz repetition rate with 50 ps pulse duration (613 kW of peak power and 30.6 μJ pulse energy). In high peak power regime, over 3MW of peak power pulses (at seed source bandwidth) at 1 MHz repetition rate (310 μJ pulse energy and 310 W average) was demonstrated. The outstanding properties of T-DCF for amplification of laser signals to high power/energy levels with maintaining the excellent beam quality open a new horizon for advanced laser material processing.

1. Introduction

High power/energy pulsed fiber laser systems have been a topic of intense research owing to their outstanding characteristics, demanding for various fields of laser-induced applications, ranging from science to industry. However, owing to some nonlinear effects, amplification of high power/energy pulses through the common double-clad fiber faced strong limitations.

In 2008, a new concept of double-clad fiber, called taper double-clad fiber (T-DCF) was introduced, providing a new horizon for amplification of pulsed laser to high power/energy levels [1]. In T-DCFs, both core and clad diameters vary along the fiber length. The T-DCFs provide a varying large core diameter of up to 200 μm, resulting in the minimum non-linear phase shift and the high pump absorption capability. Since the size of the core gradually increases along the fiber length, the mode content does not change through the amplification process, leading to excellent beam quality. There are some representatives, [2,3], which indicate the tremendous potential of T-DCFs as a gain medium for high power/energy amplification of pulsed laser systems. As a case in point, T-DCF can be utilized as gain mediums in coherent beam combining of ultrafast laser systems [4].

In this work, a polarization-maintained high peak/average power picosecond pulsed MOPA system utilizing a large mode area Yb-doped T-DCF is presented. Amplification of high average power and high peak power picosecond pulses are demonstrated in sections 2 and 3, respectively.

2. Experimental setup

To investigate the amplification of high power/energy pulses, a MOPA system based on Yb-doped T-DCF was fabricated. Fig.1 illustrates the schematic configuration of the experimental high-power MOPA system, consisting of a seed source, two low power optical pre-amplification stages and a high-power optical amplifier based on T-DCF. A commercially available fibre-coupled gain-switched (GS) diode is used as a seed source at 1039.5 nm, with a 50 ps pulse duration and variable repetition rate in the range of 1 to 20 MHz. The pre-amplification stages can provide up to 40 and 100 mW average power in 1 and 20 MHz, respectively. The amplified radiation of the seed source is coupled through a high-power optical isolator into the narrow end of an active polarization-maintaining tapered double-clad fiber (PM-T-DCF).

The Yb-doped PM-T-DCF has a length of approximately 7.7 m, core NA = 0.1, cladding NA = 0.26, and core/cladding ratio of 1/10 (input/output core diameter ≈10/ 98 μm). For effective thermal management, the entire T-DCF was coiled on a water-cooling Aluminium plate with custom made grooves. A 2-degree polished angle silica glass endcap (diameter of 9 mm and length of 20 mm) covered by antireflection coating at 976-1040 nm was spliced to the wide end of the T-DCF. The pump unit consists of seven commercially available wavelength-stabilized laser diodes with central emission at 976 nm, combined via a 7 × 1 combiner with an output fiber of 200/220 μm. Using a dichroic miror (to separate the pump and signal light) and a lense, up to 902 W pump light was launched into the T-DCF in free space.

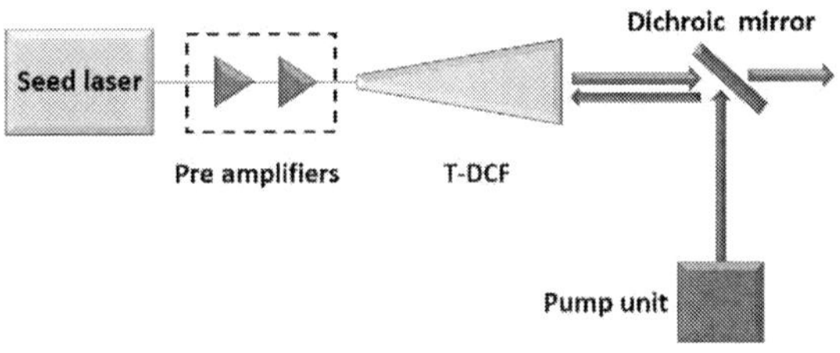

Fig.1. Scheme of the MOPA experimental setup with T-DCF.

[*] Corresponding author: valery.filippov@ampliconyx.com

3. High average power picosecond pulsed MOPA system with T-DCF

To investigate the T-DCF for amplification in high average powers, a GS seed laser was operated at 20 MHz. By utilizing 7 diodes at full power, a pulsed MOPA laser system with an average power of 613 W (613 kW of peak power and 30.6 μJ pulse energy) with 68.5% slope efficiency was achieved. At this power, the signal-to-noise ratio (SNR) of the Stimulated Raman Scattering (SRS) peak was 20 dB. The spectrum of the output laser at maximum power and the output power versus the total pump power is shown in Figures 2 and 3, respectively.

The profile of the beam, coming out of the fiber facet at the far field is shown in figure 4. The beam quality measurement (M^2 factor) at 100 W output average power is depicted in Figure 4, with the profile of the beam presented in the inset.

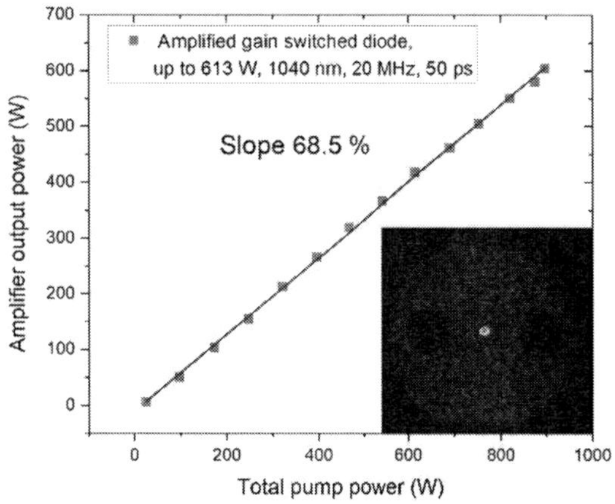

Fig.2. Output spectra of high power MOPA system at the maximum achieved power.Inset: the far-field beam profile and the fiber facet image.

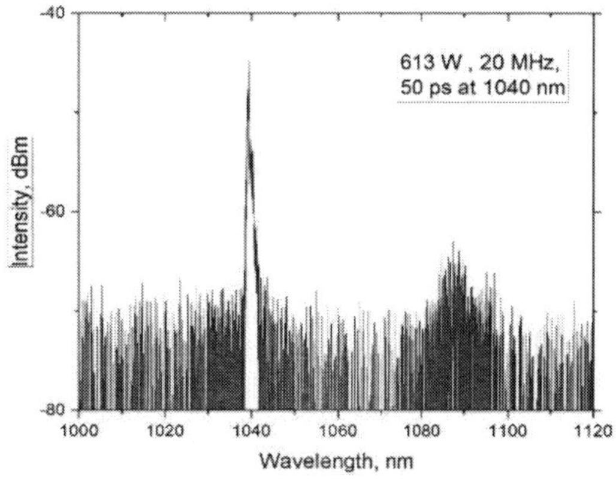

Fig.3. Output power of MOPA system vs total pump power.

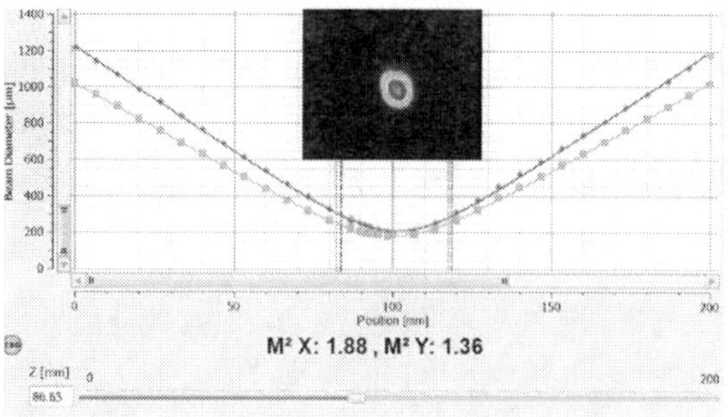

Fig.4. The result of the beam quality measurement at 100 W output average power at 20 MHz 1, with the beam profile inset

4. High average power picosecond pulsed MOPA system with T-DCF

To investigate the T-DCF in amplification to high peak powers /energy levels, we utilized a GS seed laser operating at 1 MHz. By injecting the seed laser and increasing the pump, a pulsed MOPA laser system delivered 310 W of average power and 6.2 MW peak power (310 μJ pulse energy), with an optical efficiency of 56 %. The spectrum of the output laser at the maximum power is depicted in Figure 5. As can be seen, from the perspective of inline amplification, around 13 dBm was the SNR of the SRS.

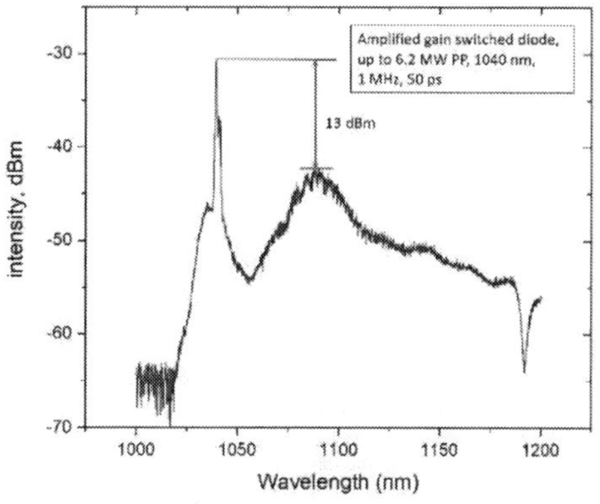

Fig.5. Output spectra of high power MOPA system at 1 MHz repetition at the maximum achieved output power.

5. References

1. V. Filippov, Y. Chamorovskii, J. Kerttula, K. Golant, M. Pessa, and O.G. Okhotnikov, Opt. Express, 16, 1929-1944 (2008).
2. V. Filippov, J. Kerttula, Y. Chamorovskii, K. Golant, and O.G. Okhotnikov, Opt. Express 18, 12499-12512 (2010).
3. A. Petrov, M. Odnoblyudov, R. Gumenyuk, L. Minyonok1, A. Chumachenko, and V. Filippov, Sci.Reports **10**, 1778 (2020).
4. H. Fathi, M. Närhi, and R. Gumenyuk, *Photonics*, **8**, no. 12: 566 (2021).

Bragg gratings and BIO-Bragg-gratings in tapered optical fibers

Martina Delgado-Pinar[1,*], *Antonio* Díez[1], *Jose Luis* Cruz[1] and *Miguel V.* Andrés[1]

[1]Laboratory of Fiber Optics, Departamento de Física Aplicada y Electromagnetismo – ICMUV, C/ Dr. Moliner, 50, 46100, Burjassot, Spain

Abstract. Tapered optical fibres are well-established devices for different applications, in order to exploit nonlinear effects, manage dispersion, excite azimuthal resonances in microresonators (so-called Whispering Gallery Modes). Also, the micro- or nanofibres guide optical-fields with large evanescent fields. In this talk, we will focus on the combination of tapers and Bragg gratings to perform novel optical devices. We will present two applications: the first, the fabrication of single-mode Bragg gratings in commercial multimode or few-mode tapered fibres by means of UV-photoinscription. The result is a grating that exhibits a single-mode reflection band and low insertion loss. The second application is the fabrication of Bio Bragg Gratings in micrometric tapers of single-mode fibres. In this case, the Bragg Grating is formed by a periodic pattern of biomolecules microstamped on the surface of the taper waist. As the molecules binds to its specific antibody, the reflectivity of this Bio Bragg Grating will increase, allowing quantification of the antibody concentration.

1 Introduction

A tapered waveguide is that which dimensions are tailored in a gradual way along its length, usually diminishing its diameter. This leads to a continuous variation of different guiding parameters, such as effective modal index, dispersion or modal area. The transitions from the pristine fibre section to the waist of the taper can be designed to be adiabatic even for subwavelength diameters, which ensure low insertion losses for these devices.

In the context of optical fibres, one of the first uses of tapered ones was the study of nonlinear phenomena. León Saval *et al.* fabricated submicrometric tapers (below 300 mm for the diameter waist) from conventional telecom fibre, for the generation of supercontinuum[1]. Also, a fine tuning of the core size of microstructured fibres can be performed by means of tapering techniques to match the pump wavelength to the zero dispersion wavelength of the fibre and, hence, optimize the supercontinuum spectrum[2]. Other phenomena, such as third- or second-harmonic generation via intermodal coupling has been reported using submicrometric fibers[3]. Micrometric tapers are also used for exciting Whispering Gallery Modes (WGM) thanks to their large evanescent optical field, with coupling efficiencies close to 100%[4]. Tapered capillaries with submicrometric thickness of the walls can be used as microfluidic channels, using WGM for their optical interrogation[4]. The combination of tapers and Fibre Bragg Gratings (FBG) have been exploited to fabricate tunable delay lines[6] and chirped gratings using a uniform phase mask for the photoinscription process[7]. In this talk we are going to present two examples of devices that combine gratings and tapered fibres to produce novel components with unique properties: (1) singlemode gratings (SMG) inscribed in multimode fibers (MMF)[8], and (2) Bragg gratings composed of molecules microstamped in a periodic fashion over the surface of a microtaper, BIO-Bragg-gratings (BBG)[9].

2 Singlemode gratings in multimode fibres

High power fibre lasers require of large mode area, hence multimode fibres, to prevent nonlinear effects in the laser cavity. This can lead to a poor quality of the beam and multiwavelength operation. The use of SMG can avoid these two drawbacks, thus there is an interest to offer methods that allow fabricating such gratings in MMF.

In this work, we present the fabrication of SMG in two commercial fibres:
- a four-modes step-index fibre from OFS (FMF), V-number ~ 5 @ 1550 nm, and
- the MM 50/125 graded index multimode fibre (GMMF) from Spectran Corporation, V-number ~ 14 @ 1550 nm.
According to the calculations, the FMF showed SM operation for tapered diameters below 55 μm, while GMMF was SM below 25 μm of clad-diameter.
The gratings were inscribed using a uniform phase mask (pitch: 1076.25 nm) and a doubled-argon UV laser. The different fibres were hydrogenated in a 30-bar atmosphere for 15 days to improve their photosensitivity. Figure 1 shows an example of a FBG written in a GMMF that was previously tapered down to 20 μm in diameter. The technique used for tapering was the fuse-and-pull technique; waist length: 4 cm, taper transition length: 5 cm with an exponential profile. The loss introduced ng the tapering was less than 0.1 dB. Several reflection bands are observed in the red curve, correspondent to the spectrum measured with the fibre in air. When a thin film

* Corresponding author: martina.delgado@uv.es

of oil-index-matching was applied to the fibre, the two reflection bands at shorter wavelengths disappeared, while that correspondent to the Bragg wavelength for LP_{01} mode remained intact, see green curve. This indicates that these reflection bands are cladding modes which are excited at the taper transition, interact with the grating in the waist region, and are coupled back to the counter-propagating cladding modes. Nevertheless, the energy carried by these modes is below 5%, value estimated from the measurement of the transmission spectrum. Thus, we demonstrated that SM gratings can be inscribed in a MMF with a large V-number.

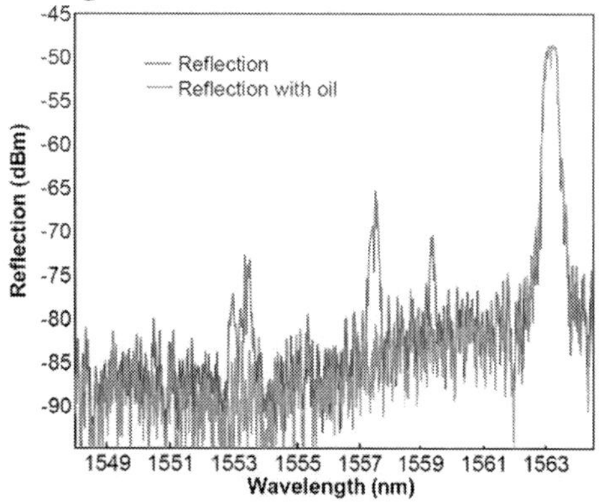

Fig. 1. Reflection spectra of a grating in a GMMF.

3 BIO-Bragg gratings in tapered fibres

In this section we present a label-free biosensor based on the combination of two techniques: first, the microstamping of molecules on glass surfaces[10] and, second, the use of micrometric tapered fibres that is, guiding modes with large evanescent fields.

A grooved PDMS stamp with a pitch of 555 nm was used for the microstamping of molecules over the taper surface (waist diameter: 3 μm, waist length: 2 cm; parameters were optimized for the best results). The stamp was incubated with BSA molecules; after two hours they were microstamped in the taper waist by applying mechanical pressure between the stamp and the microfibre. At this point, a periodic network of BSA molecules covers a section of the taper surface and interacts with the evanescent field of its fundamental mode. The reflectivity of such grating will be determined by the modulation depth, that is, the size of the molecules forming the pattern. Thus, an incipient, weak reflection band is observed at the correspondent Bragg wavelength. The device is rinsed in deionized water and then immersed in a solution containing a given concentration of the antibody, IgGs in this case. When BSA and IgGs bind each other, the modulation depth of the BBG increases and so does its reflectivity. By measuring the variation of the BBG the reflectivity, it is possible to quantify the concentration of the IgG in the solution.

Figure 2 shows an experimental immunoassay (dots), and their fit to a sigmoidal regression. Each concentration was measured at least three times in air, to ensure reproducibility. From the date, experimental detection and quantification limits were determined: 0.1 μg/mL and 0.4 μg/mL, respectively.

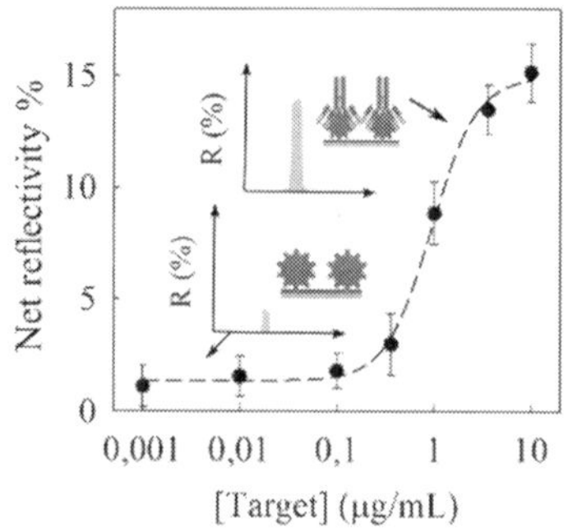

Fig. 2. Experimental immunoassay using BSA-IgG.

4 Conclusions

Two examples of the combination between two well developed devices, fibre tapers and fibre gratings, have been demonstrated. In the first case, it is shown how tapering a MMF fibre allow inscribing single mode gratings in them. The taper transitions make the role of impedance adapters for symmetric modes of the MMF, thus the insertion losses of these devices is below 0.1 dB. Some cladding modes can lead to additional reflection bands apart from that of the fundamental mode that can be easily eliminated with a thin film of index-matching oil. The second example consists on a label-free biosensor based on the microstamping of molecules (BSA) over the surface of a fibre taper. The measurement of the reflectivitiy of the BBG allows quantifying the concentration of the antibody. Detection and quantification limits comparable or better to other reported in the literature are obtained.

References

1. S. G. Leon-Saval *et al.*, Opt. Express **12**, 2864-2869 (2004)
2. M. Delgado-Pinar *et al.*, Nonlinear Photonics, Proc. paper NWD3, Karlsruhe – Germany (2010)
3. M. Delgado-Pinar *et al.*, CLEO/QELS 2010, Proc. pp. 1-2, San Jose - USA (2010)
4. X. Roselló-Mechó *et al.,* Opt. Lett. **41**, 2934-2937 (2016)
5. V. Zamora *et al.*, Photonics. and Nanostructures, **9**, pp. 149-158 (2011)
6. J. Mora *et al.*, Opt. Comm. **210**, 51-55 (2002)
7. J. Mora *et al.*, IEEE Photon. Technol. Lett. **16**, 2631-2633 (2004)
8. L. A. Herrera-Piad *et al.*, Opt. Lett., **44**, 4024-4027 (2019)
9. A. Juste-Dolz *et al.,* Biosens. & Bioelectron. **176**, 112916 (2021)
10. Miquel Avella-Oliver *et al.*, Anal. Chem., **89**, 9002-9008 (2017)

https://doi.org/10.1051/epjconf/202226612001

Imprinting characteristics of droplet lenses on liquid-repelling surfaces into light

Valeriia Bobkova[1,*], *Eileen* Otte[2,3], *Sarah* Trinschek[4], and *Cornelia* Denz[1]

[1]Institute of Applied Physics, University of Muenster, Corrensstr. 2/4, 48149 Muenster, Germany
[2]Geballe Laboratory for Advance Materials, Stanford University, 476 Lomita Mall, Stanford, CA 94305, USA
[3]Center for Soft Nanoscience, University of Muenster, Busso-Peus-Str. 10, 48149 Muenster, Germany
[4]Department of Engineering Physics, Muenster University of Applied Sciences, Stegerwaldstraße 39, 48565 Steinfurt, Germany

Abstract. We propose an experimental method that allows the investigation of droplets on liquid-repelling surfaces. The described technique goes beyond the standard imaging approaches and reveals a plethora of spatial droplet information, which is usually unavailable. Liquid droplet lenses shape the transmitted light field of a Gaussian laser beam passing though them, thereby forming refracted three-dimensional (3D) light landscapes. We investigate numerically and experimentally these 3D landscapes which are customized depending on the droplet shape as well as its refractive index, and demonstrate the encoding of droplet information. This approach can also be applied for analyzing droplets showing high-speed dynamics, in order to reveal even minimal shape deviations. The developed technique complements and therefor extend the existing conventional tools for the investigation of the droplets formed on liquid-repelling surfaces.

1 Introduction

The investigation of natural liquid-repelling surfaces inspires the invention of novel artificial substrates, which allow frictionless transport of liquids in different systems. A broad field of relevant applications includes self-cleaning, anti-icing phenomena, as well as liquid transport in lab-on-a-chip devices [2]. Thereby, the liquid-repelling effect can have various initial origins, among them are micro- and nano- structuring [3], chemical surface treatment [4], or thermal effects as the Leidenfrost mechanism [5]. For further development of advanced applications as tuneable optoelectronic devices, which enable control over adaptive liquid lenses [6], a deeper understanding of nonwetting behavior is essential.

Standard approaches for the analysis of droplets are based on the two-dimensional (2D) imaging of a volumetric droplet, usually revealing only a fraction of the information on the droplet such as, for instance, the contact angle [7]. When the droplet is illuminated by an expanded coherent laser beam [8], interferometric approach allows for temperature measurements close to the droplet surface. However, this approach does not reveal further properties of the droplet itself.

In this contribution, we propose a method for investigation of liquid droplets on nonwetting surfaces, employing these droplets as thick biconvex liquid lenses[1]. Thereby, the standard 2D imaging is advanced by encoding a plethora of droplet information into the Gaussian laser beam, passing through the liquid lens. The complete 3D shape of the droplet, its tilt with respect to the beam propagation axis, the refractive index, which strongly depends on the liquid and on the temperature, are simultaneously encoded into the transmitted light field.

We experimentally demonstrate the measurement of such a 3D light field, refracted by a water droplet on a natural liquid-repelling surface, a Taro *(Colocasia esculenta)* leaf. Numerical validation of this experiment is realized using a vectorial ray-based diffraction integral (VRBDI) method [9] and is comprehensively described in [1].

2 Experimental investigation of light-shaping features of liquid lenses

For analyzing the 3D light-shaping features of droplet lenses, we design the experimental setup, as shown in Figure 1a. A horizontally polarized Gaussian beam (wavelength $\lambda = 532$ nm) is focused to have the beam waist of $\omega = 450\,\mu$m at the first surface of a droplet resting on a Taro leaf substrate (shown symbolically as a grey plate). The beam waist is chosen to be approximately two times smaller than the droplet dimensions in order to avoid diffraction on the liquid-air surface. The volume of the droplet is controlled by a microfluidic pump. The 3D light field formed by propagation through the droplet is measured in the following way: the refracted beam is imaged by a 20× long working distance microscope objective (MO) and a $f = 200$mm tube lens onto a camera. The MO is mounted on a high-precision translation stage which is controlled by a computer. Within one minute the stage shifts for 1000 μm in 500 steps in z-direction (propagation direction of beam), capturing the transverse intensity distributions for each step. The subsequent longitudinal and

*e-mail: v.bobkova@uni-muenster.de

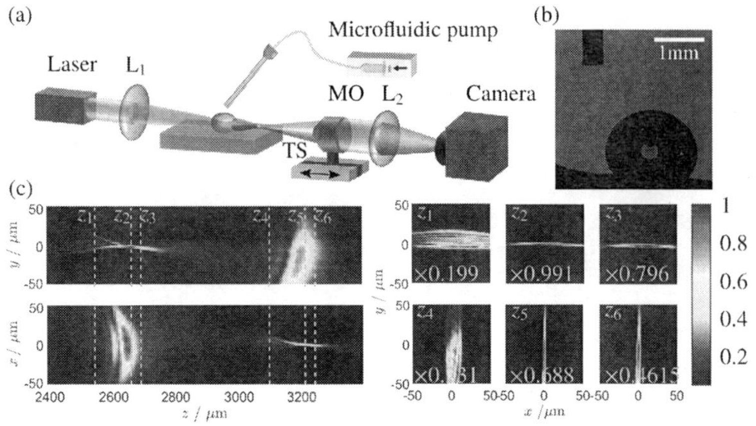

Figure 1. (a) Sketch of the experimental setup used to detect the 3D light landscape refracted by a droplet on a liquid-repelling surface (Taro leaf). A linearly polarized Gaussian beam is focused by the first lens (L_1) into the droplet. The refracted light is imaged by a microscope objective (MO) mounted on a translation stage (TS) and a lens L_2 onto a camera. (b) Droplet on a leaf (camera image, blue LED illumination). A needle (0.4 mm diameter) is shown in the left upper corner. (c) Experimentally measured intensity landscape behind the droplet. $z = 0$ is set to the middle of the droplet. Shown transverse distributions are normalized to the maximum per plane for z-positions $z_i = \{2550, 2660, 2690, 3100, 3210, 3240\}$ μm, j = [1, 6]. Adapted from [1], licensed under a Creative Commons Attribution 4.0 License.

transversal cuts of the captured light intensity are shown in Fig.1c.

The longitudinal (x, z) and (y, z) cuts demonstrate the presence of strong astigmatism in the optical system, which results in two separated vertical (tangential) and horizontal (sagittal) focal planes. The origin for the astigmatic focusing is the mismatch of the sagittal and tangential curvatures of the light-shaping object, namely, the curvatures of the approximately ellipsoidal droplet in (x,y) and (x,z) planes. This results in different focal distances in orthogonal planes. The corresponding intensity distributions in the transverse planes show the focal structures for different z-positions z_1 to z_6. Intensity fringes before the focal planes are caused by the presence of spherical aberration. Overall, the astigmatic distance is fully defined by the droplet's refractive index and its geometry, allowing conclusions on the liquid-repelling properties of the underlying surface.

3 Conclusion & Discussion

We have introduced a method for the investigation of droplets on liquid-repelling surfaces and present an exemplary experimental measurement of a light landscape sculpted by a water droplet on a Taro leaf. The described approach can also be applied for the imaging of high-speed dynamic movement of droplets, such as a bouncing Leidenfrost droplet on a hot surface [1]. In this case, a minimal displacement of the droplet can hardly be detected by standard imaging techniques, but leads to a significant restructuring of the light landscape. The described method can be extended by using structured laser beams [10] carrying non-zero orbital angular momentum. In this case, it has the potential to also detect further droplet properties, such as internal fluid flows. In total, standard imaging techniques can be complemented by our approach to extract additional information about the droplets and, thereby, the liquid-repelling surfaces they are formed on.

References

[1] V. Bobkova, S. Trinschek, E. Otte, C. Denz, Optics Express **30**, 5937 (2022)
[2] P. Zhang, F.Y. Lv, Energy **82**, 1068 (2015)
[3] M. Kumar, R. Bhardwaj, Sci. Rep. **10**, 1 (2020)
[4] S. Munekata, Progress in Organic Coatings **16**, 113 (1988)
[5] A. Bouillant, T. Mouterde, P. Bourrianne, A. Lagarde, C. Clanet, D. Quéré, Nature Physics **14**, 1188 (2018)
[6] C.C. Cheng, C.A. Chang, J.A. Yeh, Opt. Express **14**, 4101 (2006)
[7] T. Huhtamäki, X. Tian, J.T. Korhonen, R.H. Ras, Nature protocols **13**, 1521 (2018)
[8] M.A. Van Limbeek, M.H.K. Schaarsberg, B. Sobac, A. Rednikov, C. Sun, P. Colinet, D. Lohse, Journal of fluid mechanics **827**, 614 (2017)
[9] B. Andreas, G. Mana, C. Palmisano, JOSA A **32**, 1403 (2015)
[10] A. Forbes, M. de Oliveira, M.R. Dennis, Nature Photonics **15**, 253 (2021)

Deformation and shapping of optically trapped microdroplets: an ab-initio numerical study.

Hugo Chesneau[1,2,*], *Hamza* Chraibi[2], and *Jean-Pierre* Delville[2]

[1]Commissariat à l'Energie Atomique et aux Energies Alternatives, Centre d'Etudes Scientifiques et Techniques d'Aquitaine, F-33116 Le Barp, France.
[2]Université de Bordeaux, CNRS, LOMA, UMR 5798, F-33405 Talence, France

Abstract. We numerically study the deformation of optically trapped microdroplets with the optical radiation pressure using a house-made code based on the boundary elements method. Particular attention is paid to coupling between the electromagnetic waves propagation within the droplets and the resulting droplets morphologies.

1 Introduction

Recently, an interest has grown for the optical trapping and deformation of micro-droplets. To our knowledge, a numerical model allowing to access the whole dynamics of optically trapped droplet deformation has never been reported. We present here an ab-initio numerical model based on the boundary elements method. At each time step, we simultaneously solve Helmholtz and Stokes equations in order to compute the optical stress on the droplet interface and the resulting deformation. In what follows, we describe in broad terms the numerical method and present some results about droplet deformations with a low refractive index contrast in an optical tweezers; which is typical of an oil emulsion droplet in water.

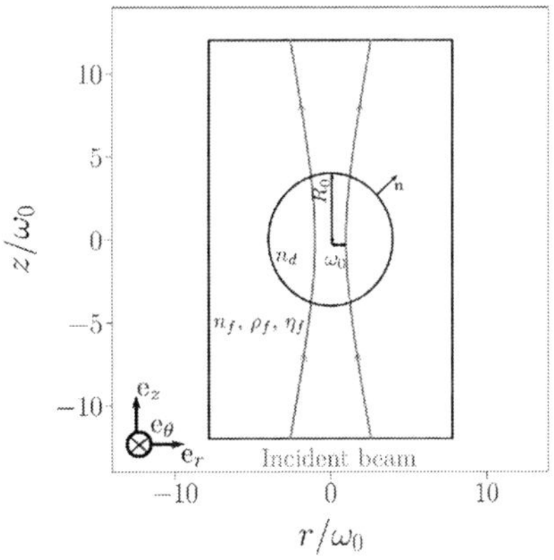

Figure 1. Numerical system scheme.

2 Physical and numerical systems

We consider an initially spherical droplet of radius R_0 which is immersed in a fluid at rest. Both fluid phases are characterised by their viscosity η_i, density ρ_i and refractive index n_i; with $n_d > n_f$. They are supposed to be immiscible and incompressible. The index i = d, f denotes the properties of the droplet and exterior fluid. At time $t = 0$ an initially gaussian beam of wavelength λ_0 and waist ω_0 is focused in the droplet center. When crossing the interfaces between the droplet and the exterior fluid, the laser beam eventually induces a normal stress called the radiation pressure. In a corpuscular point of view, this stress results from the optical momentum discontinuity at the interfaces. One should note that it always deforms the interface towards the less refractive phase. We consider an axisymmetrical system (droplet + gaussian laser beam) and describe it on a cylindrical orthonormal basis $(\mathbf{e}_r, \mathbf{e}_\theta, \mathbf{e}_z)$. A scheme of the system is depicted in Figure 1. Using a boundary elements method, the equilibrium of our numerical system can be solely expressed by the stress balance at the interface:

$$\left(\mathcal{T}_f - \mathcal{T}_d\right) \cdot \mathbf{n} = (\kappa(r)\sigma - \Delta\rho g z(r) + \Pi_{rad}(r))\,\mathbf{n} \qquad (1)$$

With $\mathcal{T}_i = p_i\mathcal{I} + \eta_i\left(\nabla\mathbf{v} + \nabla\mathbf{v}^T\right)$ the hydrodynamic stress tensor, $\mathcal{I}$ being the identity matrix and $p_i = p_{0i} + \rho_i g z$ a pressure term corrected by the gravity with $\mathbf{g} = g\mathbf{e}_z$, and p_{0i} the pressure in the fluid phase i. $\kappa(r) = (1/r)\left(rz''(r) + z'(r)^3 + z'(r)\right)/\left(1 + z'(r)^2\right)^{3/2}$ is the curvature in cylindrical coordinates, where the $'$ denotes the spatial derivative with respect to r and σ is the interfacial tension. Finally, $\Pi_{rad}(r)$ is the radiation pressure and $\mathbf{n}$ the normal to the interface pointing towards the exterior fluid. Solution to the Stokes equation can be written as the fol-

*e-mail: chesneau.hugo@gmail.com

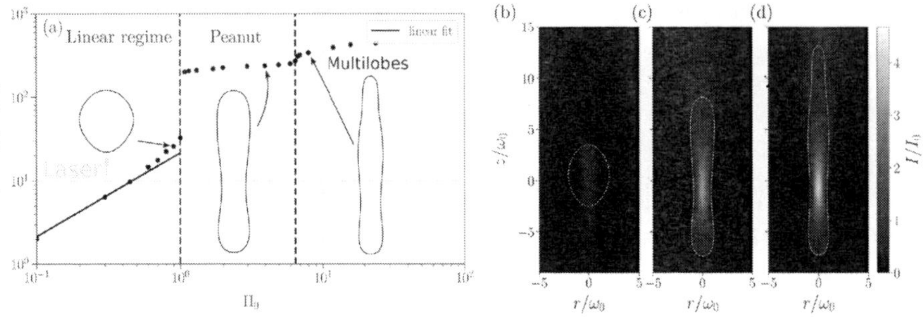

Figure 2. Droplet deformation and the electromagnetic field propagating in it. (a) Elongation with the radiation pressure amplitude. (b-d) Intensity of the incident laser beam propagating into the droplet at $\Pi_0 = 0.9, 4, 8$.

lowing integral form [1]:

$$
\begin{aligned}
&-\frac{1}{\eta_f}\int_{S_i} \mathcal{J}(x|y)\cdot \mathbf{n}\left(\kappa(r_y)\sigma - \Delta\rho g z(r_y) + \Pi_{rad}(r_y)\right) dS_y \\
&-(1-\lambda)\int_{S_i} \mathbf{n}\cdot\mathcal{K}(x|y)\cdot \mathbf{v} dS_y \qquad (2)\\
&-\frac{1}{\eta_f}\int_{S_W}\left(\mathbf{n}\cdot\mathcal{T}_f\right)\cdot\mathcal{J}(x|y) dS_y = \frac{1+\lambda}{2}\mathbf{v}(\mathbf{x})
\end{aligned}
$$

With $\mathcal{J}(x|y) = 1/(8\pi)(\mathcal{I}/d)(\mathbf{d}\otimes\mathbf{d})/d^3$ and $\mathcal{K}(x|y) = -3/(4\pi)(\mathbf{d}\otimes\mathbf{d}\otimes\mathbf{d})/d^5$ the Green's kernels representing respectively the velocity and stress at the point **x** induced by a point force localised in **y**. $\mathbf{d} = \mathbf{x} - \mathbf{y}$, $d = |\mathbf{d}|$ and the symbol $\otimes$ denotes the tensor product. $< \lambda > \eta_d/\eta_f$ is the viscosity ratio.

Finally, at each time step, we compute the different terms of equation (2) to get the interface velocity and then we advect its position with a Lagrangian scheme:

$$\frac{\partial \mathbf{r}(t)}{\partial t} = \mathbf{v}(\mathbf{t}) \qquad (3)$$

With $\mathbf{r}(t) = (r\mathbf{e}_r, z\mathbf{e}_z)$.

3 Droplets deformation

We report on Figure 2 the elongation of a droplet as a function of the ratio between the optical stress amplitude and Laplace pressure $\Pi_0 = \Pi_{rad}/(\sigma/R_0)$ for $n_r = 1.03$. Its initial radius is $R_0/\omega_0 = 2.3$ and the beam waist is set to $\omega_0/\lambda_f = 0.8$ with $\lambda_f = \lambda_0/n_f$. For each point in Figure 2(a), we start from a spherical droplet at rest and we set an incident optical stress amplitude Π_0.

First we can observe a linear regime of deformation with the radiation pressure amplitude for $\Pi_0 \in]0, 1]$. The droplet goes from a spherical shape to a spheroidal one.

When increasing Π_0 above unity, we reach an instability and the droplet takes spontaneously a peanut like shape. If this kind of shape has already been observed experimentally [2] and with a stationary numerical model [3], this dynamic transition has never been observed numerically. Finally, when further increasing the radiation pressure, we reach a third regime which can be called multilobes regime. Our method allows us to compute the electromagnetic field at each time step or for different stationary state in the whole numerical domain. Figure 2(b-d) depicts the field intensity at stationary state for $\Pi_0 = 0.9, 4, 8$. Below threshold, (b) the initally gaussian beam is focused in the droplet. As the droplet elongates with Π_0, the intensity increases at the droplet center which leads to the peanut shape (c). Finally, when further increasing Π_0 (d), the droplet become thinner and the incident beam seems guided towards the upper part of it.

4 Conclusion

This brief demonstration of numerical droplet trapping and deformation by the radiation pressure shows the intercoupling between a liquid object deformation and its waveguiding properties. However, other original behaviours could be highlighted. Indeed, we can show that when a droplet reaches a peanut-like shape, it presents an hysteretic behaviour as observed for droplets deformed with an electric field [4]; meaning that it keeps this shape even when we decrease the incident radiation pressure amplitude. When increasing the refractive index ratio, the droplet can take extremely original shapes such as step-like interfaces [5] on one side and have conical tip on the other side [6].

References

[1] J. Tanzosh, M. Manga, H.A. Stone, *Boundary element technology VII* (Computational Mechanics Publications, Southampton Boston, 1992) 19-39

[2] Ward, A. D., Berry, M. G., Mellor, C. D. and Bain, C. D., Chem. Comm., 4515-4517 (2006)

[3] Tapp, D., Taylor, J. M., Lubansky, A. S., Bain, C. D. and Chakrabarti, B., Opt. Express, **22**, (2014)

[4] Sherwood, J., J. Fluid. Mech., **188**, 133-146 (1988)

[5] Bertin, N., Chraibi, H., Wunenburger, R., Delville, J.-P. et Brasselet, E., Phys. Rev. Lett., **109**, 244304, (2012)

[6] Zhang, J.-Z. et Chang, R. K., Opt. Lett., **13**, 916–918 (1988)

https://doi.org/10.1051/epjconf/202226612003

Microdroplet lasers and their applications

Matjaž Humar[1,2,3,*]

[1]Condensed Matter Department, J. Stefan Institute, Jamova 39, SI-1000 Ljubljana, Slovenia
[2]Faculty of Mathematics and Physics, University of Ljubljana, Jadranska 19, SI-1000, Ljubljana, Slovenia
[3]CENN Nanocenter, Jamova 39, SI-1000 Ljubljana, Slovenia

Abstract. Bio-integrated lasers, that are lasers made of biological and biocompatible materials and implanted into cells and tissues, are gaining interest from the research community. Here we show how whispering gallery mode microlasers and microcavities made of solid beads or droplets can be used for sensing different processes in biological materials including inside cells. By making microcavities of a predefined size they can also be used to encode some information and for cell tracking. Sensing and tracking can be applied to highly scattering tissues.

Currently the most frequently used techniques to study complex processes in live cells employ fluorescent probes. However, fluorescent probes have several disadvantages including photobleaching, sensitivity to environmental factors and broad emission spectrum, which limits their sensitivity, multiplexing ability and imaging capabilities in biological tissues. The transition from detecting laser emission from bio-integrated lasers instead of fluorescence represents a paradigm shift. Due to narrow emission linewidth, high coherence, large intensity and highly nonlinear output from lasers, they open huge opportunities in ultrasensitive sensing, spectral multiplexing and microscopy.

Micro-sized lasers completely embedded within single live cells [1] and biological tissues [2] have been demonstrated. The lasers were made out of solid beads and droplets including biocompatible and biodegradable materials. We have observed that both macrophages and non-macrophage cells engulf beads up to 20 μm in diameter (Fig. 1a). The lasers inside cells can act as very sensitive sensors, enabling us to better understand cellular processes. By using a micro pipette droplets of high refractive index oil containing fluorescent dye were injected into a cell. By analyzing the light emitted by a droplet laser, we can measure that deformation and calculate the forces acting within a cell (Fig. 1b).

Further, lasers were used for cell tagging. Each laser within a cell emits light with a slightly different fingerprint that can be easily detected and used as a barcode to tag the cell [3], providing the ability the study cell migration. By using a micro pipette droplets of high refractive index oil containing fluorescent dye were injected into a cell. By injection the diameter of the generated oil droplets and polymerized microbeads could be reproduced with a standard deviation of 1 nm and 20 nm, respectively. Encoding of short words and numbers has been demonstrated by producing three beads with predefined sizes (Fig. 1c) [4]. The stored information has been read from the emitted spectrum.

Droplet lasers serve as well to precisely measure interfacial tension between to immiscible liquids [5]. The interfacial tension is calculated from the Young–Laplace equation where the radius and the internal pressure needs to be measured. The size of the droplet is monitored and pressure is applied to the microcapillary to measure the equilibrium pressure (Fig. 1d). The droplet shown in Fig. 1d has an equilibrium pressure of 1.75 kPa. With this method it is possible to measure interfacial tension of volumes in the picoliter range.

We have also demonstrated that small lasers can be used for novel nonlinear microscopy, including super resolution imaging [6] and imaging in highly scattering tissues [7]. When the lasers are embedded into deep into the tissue, because of scattering their image is blurred (Fig. 1e and f). Since the spectral positions of laser lines do not change with propagation through scattering and absorbing media, we are still able to measure their spectrum (Fig. 1g). From the spectrum we can determine the number of microcavities embedded in the tissue, their individual sizes and the surrounding refractive index. The size can be used for previously mentioned barcoding. For sensing the beads are functionalized with a coating which changes the refractive index based on the change in the environment, such as temperature and pH. Simultaneously we can also measure the position of each laser in all three dimensions. Even though the image of closely separated lasers is highly blurred (Fig. 1f), we can extract the contribution of each laser by capturing their unique spectra (Fig. 1g). Therefore, the contribution of each laser to the final image can be isolated and position can be determined. This is similar to PALM or STORM, but the signal of each emitter is not separated in time but in spectrum.

 https://doi.org/10.1051/epjconf/202226612003

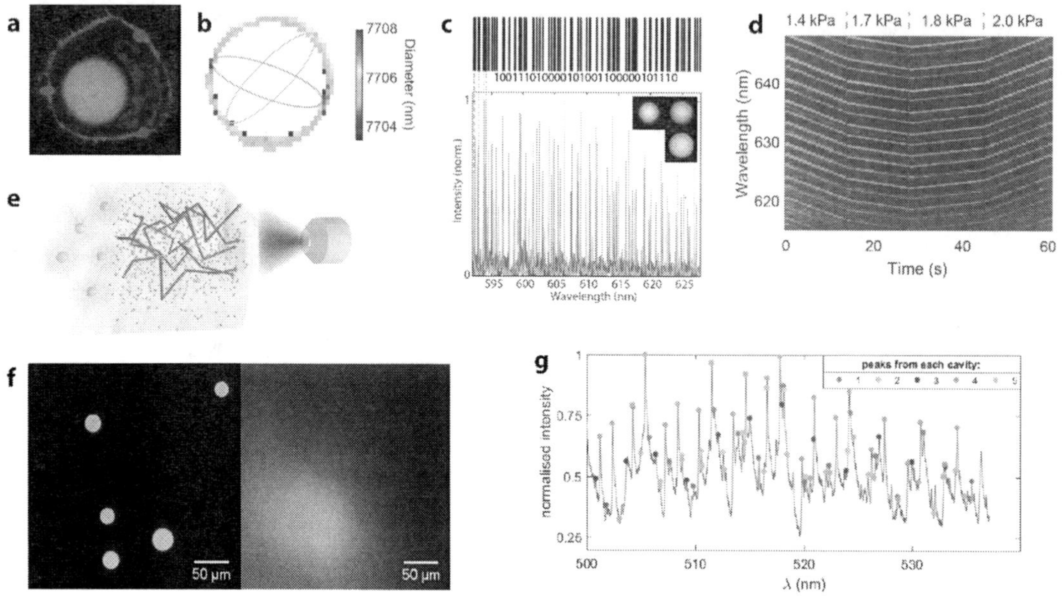

Fig. 1. a) Laser inside a cell. b) Diameter map shows deformation of the cavity. c) Spectra collected from three beads with predefined sizes. The sizes of the three generated beads calculated from their spectrum encode the word "STEFAN" which is written as a binary number. d) Change of the spectrum as the pressure to the droplet is changed. e) Scattering of light from lasers embedded in a scattering tissue. f) Microcavities without and with scattering layer on top. g) From the spectrum the size, surrounding refractive index and position of each microcavity can be determined.

Lasers made from liquid crystals which emit vector beams were also studied. Liquid crystals form an extremely rich range of self-assembled topological structures with artificially or naturally created topological defects. Liquid crystals have been used before inside laser cavities, however, until now only relatively simple liquid-crystal structures have been employed. Our study provides experimental and simulation insights into coupling of light with the complex liquid-crystal topological superstructures inside a laser cavity [8]. This results in non-trivial intensity and polarization of the generated structured light. The proposed soft matter microlaser approach opens new direction in soft matter photonics research.

Lasers made of soft and biological materials including liquds and integrated into the living biological systems have potential to enable novel applications including sensing and imaging [9,10].

References

1. M. Humar and S. H. Yun. Intracellular microlasers, Nature Photonics 9, 572 (2015).
2. M. Humar, A. Dobravec, X. Zhao, S. H. Yun. Biomaterial microlasers implantable in the cornea, skin, and blood. Optica 4, 1080-1085 (2017).
3. M. Humar, A. Upadhya, S.H. Yun. Spectral reading of optical resonance-encoded cells in microfluidics. Lab on a Chip 17, 2777-2784 (2017).
4. D. Richter, M. Marinčič, M. Humar, Optical-resonance-assisted generation of super monodisperse microdroplets and microbeads with nanometer precision, Lab on a Chip 20, 734-740 (2020).
5. G. Pirnat, M. Humar. Whispering gallery-mode microdroplet tensiometry, Advanced Photonics Research 2, 2100129 (2021).
6. S. Cho, M. Humar, N. Martino, S. H. Yun, Physical Review Letters 117, 193902 (2016).
7. A. Kavčič, M. Garvas, M. Marinčič, K. Unger, A. M. Coclite, B. Majaron, M. Humar. Deep tissue localization and sensing using optical microcavity probes, Nature Communications 13, 1269 (2022).
8. M. Papič, U. Mur, K. P. Zuhail, M. Ravnik, I. Muševič, M. Humar, Topological liquid crystal superstructures as structured light lasers, PNAS 118, e2110839118 (2021).
9. M. Humar, S.J.J. Kwok, S.Y. Cho, M. Choi, A.K. Yetisen, S.H. Yun. Toward biomaterial-based implantable photonic devices. Nanophotonics 5, 60–80 (2016).
10. D. Yu, M. Humar, K. Meserve, R. C. Bailey, S. N. Chormaic, F. Vollmer, Whispering-gallery-mode sensors for biological and physical sensing, Nature Reviews Methods Primers 1, 83 (2021).

Analysis of size and concentration of microplastics in water using static light scattering combined with PCA and LDA

Mehrdad Lotfi Choobbari[1*], *Leonardo Ciaccheri*[2], *Tatevik Chalyan*[3], *Barbara Adinolfi*[2], *Wendy Meulebroeck*[3], *Heidi Ottevaere*[3]

[1]Vrije Universiteit Brussel, Department of Applied Physics and Photonics, Brussel Photonics, Pleinlaan 2, 1050 Brussels, Belgium
[2]CNR-Istituto di Fisica Applicata "Nello Carrara", Via Madonna del Piano 10 - 50019, Sesto Fiorentino (FI) -Italy
[3]Vrije Universiteit Brussel and Flanders Make, Department of Applied Physics and Photonics, Brussel Photonics, Pleinlaan 2, 1050 Brussels, Belgium

Abstract. Quantitative analysis of size and concentration of microplastics is a crucial step for having a better understanding of plastic pollution in the environment. Such information is typically obtained in a single particle mode that significantly increases the analysis time and can be a cumbersome task. Therefore, we demonstrate here a measurement technique based on Static Light Scattering (SLS) combined with chemometric methods such as Principal Component Analysis (PCA) and Linear Discriminant Analysis (LDA) for resolving the size and concentration of multiple microplastic particles in water. Two sets of samples with uniform and non-uniform size distribution of microplastics, called "monodisperse" and "polydisperse", respectively, are fully investigated. It is shown that a relationship exists between the scattering signals of mono- and polydisperse samples on the PCA space. Hence, a PCA-LDA model that is constructed on the PCA space of monodisperse samples is used to discriminate the size of the microplastics in polydisperse samples. By specifying the size of the particles, their concentration is determined using a simple linear fit.

1 Introduction

Without any doubt, plastics are one of the key materials for today industries with applications in different areas [1]. Notwithstanding, the improper management of plastic waste and the low percentage of them being recycled has led to a new environmental challenge called "Microplastic (MP) identification". These plastic particles can be found everywhere, even into human placentas [2].

One of the necessary actions for tackling the problem of microplastics is to have a better understanding of their presence in nature through standardized regulatory protocols that are currently lacking in the field. Therefore, different analytical techniques such as Raman spectroscopy (RS), Fourier-transform infrared spectroscopy (FTIR) and static light scattering (SLS) have been proposed for the study of MPs. RS and FTIR have received great attention due to their valuable advantages. Nevertheless, these methods may not be suitable for sizing and enumeration of MPs. It was recently demonstrated that light scattering can be used to study the surface characteristics of MPs. Light scattering is a well-stablished technique that proved to be very useful for the characterization of small particles [3]. Nevertheless, less attention has been paid to this valuable technique for the analysis of microplastics in recent years. Accordingly, a practical approach based on the combination of SLS with chemometric methods is proposed here for the batch analysis of size and concentration of microplastics in water. PCA is applied on the SLS signals of polystyrene (PS) samples to classify the size of the particles. Subsequently, an LDA model is trained using the PCA results of monodisperse samples. Next, the PCA-LDA model is used to classify the size of the particles in unknown mono- and polydisperse samples. Finally, the concentration of particles is determined using a simple linear fit.

2 Materials and methods

Monodisperse PS microspheres with different diameters were used as the target MPs in this work. Light scattering measurements were performed using a REFLET-180S goniophotometer (Light Tec) which can work either in transmittance or reflectance mode. As shown in Fig. 1, the optical design consists of two symmetric arms being the illumination and the detection parts which are focused at the same point on the sample. The collimated light from a 100 W Tungsten Halogen lamp with a beam diameter of 0.2 millimetre passes through the sample container at a fixed position and the photodetector detects the scattered light by scanning from $\theta = -90°$ to $+90°$ with an angular

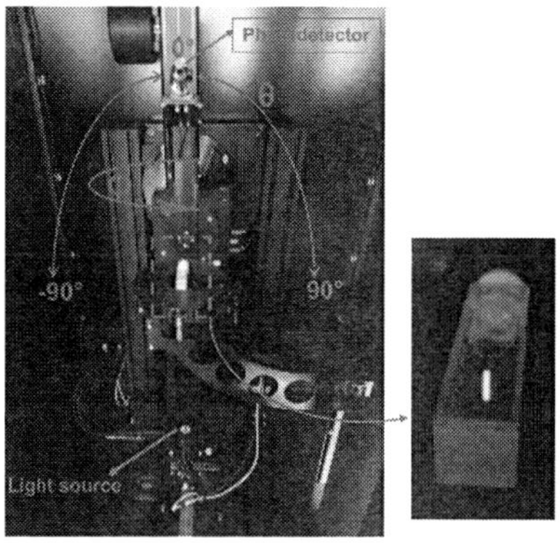

Fig. 1. Multi-angle static light scattering measurement setup.

resolution of 0.1°. The total duration of one measurement was 30 s. The format of the output file was a two columns matrix consisting of the Bidirectional Transmittance

* Corresponding author: Mehrdad.Lotfi.Choobbari@vub.be

Distribution Function (BTDF) versus the scattering angles. Further processing and analysis on the output data were performed using MATLAB and The Unscrambler (Aspen Technology).

3 Results and discussion

After measuring the BTDF signal of the MPs, the first derivative of the scattering intensity patterns was calculated using a second order, nine-point Savitsky-Golay filter as shown in Fig. 2. As a result, most of the information about the concentration of particles was removed, however, the difference between the shape of the scattering intensity patterns was significantly enhanced with respect to the size of the scatterers. Accordingly, this data was used as the input for the PCA analysis to discriminate the size of the particles.

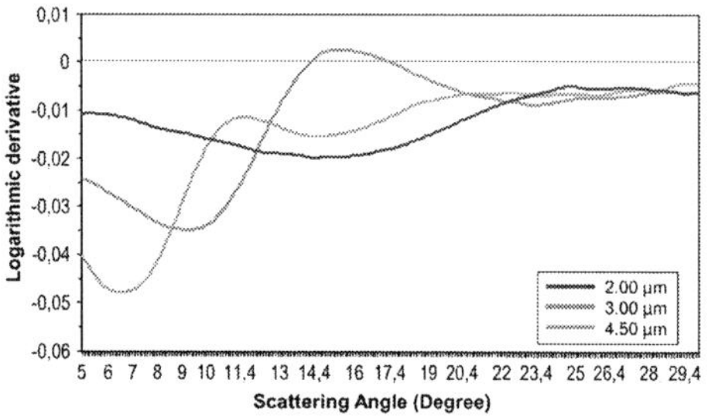

Fig. 2. Logarithmic derivative of the scattering intensity plots for different microplastic diameters.

PCA is known as an unsupervised chemometric method that can split between the groups of a dataset by identifying the existing patterns among them with no prior knowledge. Therefore, PCA was applied on the first derivative of the logarithmic scattering intensity patterns of the monodisperse samples to discriminate the size of the particles. As it is clear from Fig. 3, the first two PCs accounting for more than 80% of the total variance, were enough to discriminate between the size of the particles. Fig. 3 also demonstrates the projection of the PCA map of polydisperse samples (pink-diamond symbols). As seen, the PCA map of a polydisperse sample is located between the PCA maps of its constituent particles, i.e., the monodisperse samples. Thanks to this relationship, a PCA-LDA model was trained to classify the size of the particles in unknown mono- and polydisperse samples. The classification/prediction rate was 100% in both scenarios.

After specifying the size of the particles in a sample, their concentration can be determined using the scattering intensity at an arbitrary angle from the selected angular range. Fig. 4 represents the decimal logarithms of particle concentrations against the scattering intensity which was measured at 30° for all particle sizes. As seen, almost a linear relationship was obtained. Therefore, a simple linear fit was enough for measuring the concentration of MPs in a sample, provided that the size of the particles was known.

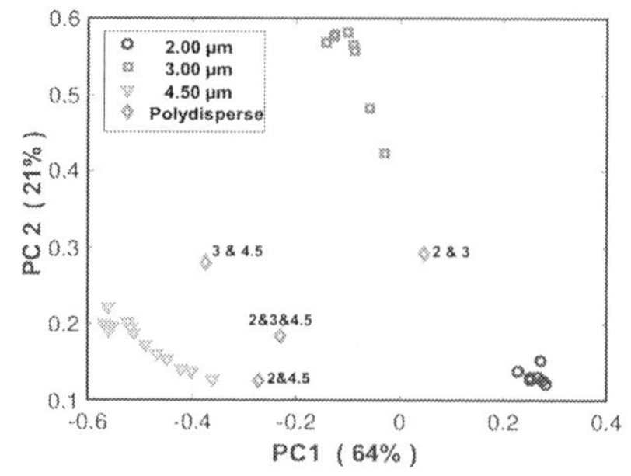

Fig. 3. PCA maps of mono- and polydisperse samples.

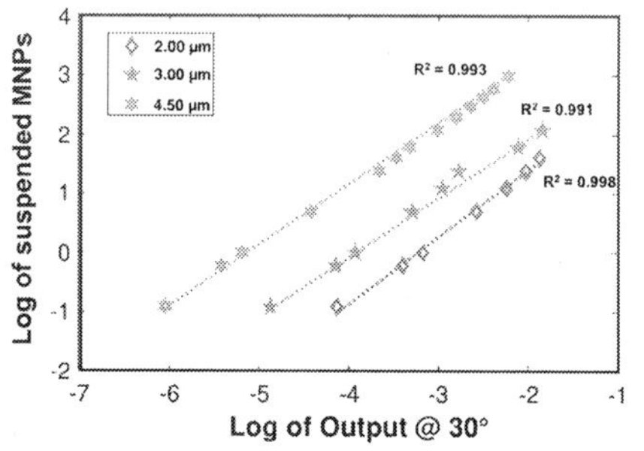

Fig. 4. The logarithm of MPs concentration in water versus the logarithm of scattering intensity at 30°.

4 Conclusion

Despite being a well-stablished technique for the characterization of small particles, light scattering received the past years less attention in comparison to other common optical techniques such as Raman spectroscopy and FTIR, for the analysis of microplastics. Here, it was demonstrated that the combination of light scattering with chemometric pattern recognition algorithms, such as PCA and LDA, can bring valuable advantages for the batch analysis of the size and concentration of microplastics in water. In a next step SLS will be combined with an optofluidic chip for in-situ monitoring of microplastics in water.

5 Acknowledgment

This work was supported by the "MONPLAS" European Union's Horizon 2020 research and innovation programme under the Marie Skłodowska-Curie grant agreement No 860775.

References

[1] D. M. Mitrano and W. Wohlleben, *Nat. Commun.*, vol. 11, no. 1, pp. 1–12, 2020, doi: 10.1038/s41467-020-19069-1.

[2] A. Ragusa *et al.*, *Environ. Int.*, vol. 146, p. 106274, 2021, doi: 10.1016/j.envint.2020.106274.

[3] G. Balakrishnan, F. Lagarde, C. Chassenieux, and T. Nicolai, *Preprints*, vol. 2020030189, no. March, pp. 1–27, 2020, doi: 10.20944/preprints202003.0189.v1.

 https://doi.org/10.1051/epjconf/202226612005

Trapping, characterization and reactions of biocolloids in a salinity gradient

Martin K. Rasmussen, *Jonas* N. Pedersen, and *Rodolphe* Marie*

Department of Health Technology, Technical University of Denmark, Oersteds Plads building 345C, 2800 Kongens Lyngby, Denmark

Abstract. The properties of soft matter nanoparticles like exosomes are interesting for drug delivery and diagnostics applications. However, the simultaneous characterization of multiple properties, e.g., size and zeta potential, can only be done serially and is highly sensitive to the purification prior to characterization. Here we show how a salt gradient established in a nanofluidic channel induces opposing transport of particles and liquid that trap the particles. Particles are thus accumulated in the trap. We show how optical microscopy images of the particle positions in the salinity gradient provide a measurement of the size and surface charge. We demonstrate the method on a sample of exosomes and on individual particles. Finally, we show how biomolecular reactions at the surface of the nanoparticle can be detected from the optical microscopy analysis of the particles' trapping position.

Nanometer-sized biocolloids play an important role in many research fields, e.g., in drug delivery and diagnostics, and their size and surface charge are important parameters for their performance. The zeta potential, which depends on the surface charge, is an indicator of the biocolloids' stability in dispersions and is crucial for their initial adsorption on the cell membrane. Hence the zeta potential is both a key parameter when designing liposomes for drug delivery and for understanding the body's cell-to-cell communication with exosomes [1]. The particle size influences the endocytotic uptake rate after adsorption on the cell membrane, so the size is also important for their efficiency as drug delivery vehicles [2]. Current nanoparticle characterization techniques all have challenges. Measurements of size and zeta potential are routinely performed by dynamic light scattering (DLS) and laser Doppler electrophoresis (LDE), but they require multiple experiments and are challenged by polydisperse samples [3]. Alternatively, the size of individual particles can be measured by nano tracking analysis (NTA). Combined with refractive index measurements by interferometric imaging, this provides the two parameters simultaneously [4].

Here we present a nanofluidic-based method for determining the size and zeta potential from a single measurement with data acquisition times less than a minute [5]. In our method, we establish a salinity gradient across a funnel-shaped nanochannel. The nanochannel has a height of 240 nm, and the widths at the ends are 5 μm and 20 μm, see Figure 1a. The salinity gradient induces both a diffusiophoretic particle migration towards higher salt concentration and an oppositely directed difussioosmotic fluid flow in the nanochannel [5]. Particles in the nanochannel are trapped where the diffusiophoretic velocity $v_{ph}(x)$ and the diffusioosmotic fluid velocity $v_{os}(x)$ balance each other. The trapping position depends on the particles' diameter and zeta potential.

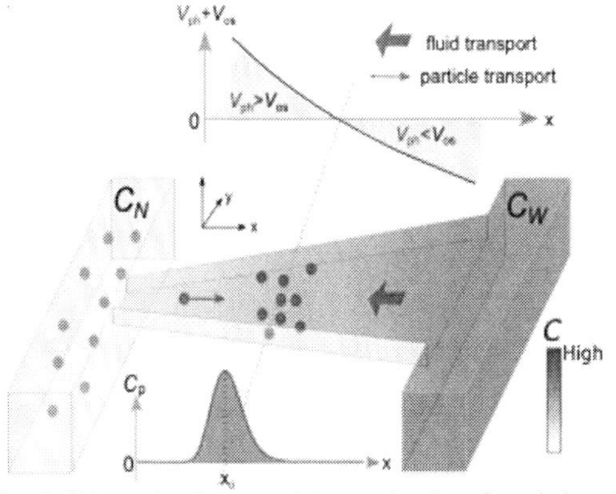

Fig. 1. Schematic of nanoparticle trapping in a funnel-shaped nanochannel by diffusioosmosis and diffusiophoresis. The salinity gradient is maintained by a continuous flow of solutions with PBS concentrations C_L and C_R (C_L<C_R) in the microchannels.

Figure 2a,b shows fluorescence microscopy images of trapped exosomes extracted from human blood serum and from a sample with two subpopulations of liposomes, respectively. The liposome subpopulations have different lipid compositions, and hence different zeta potentials, which causes a separation into two bands (Fig. 2b,c). The sizes and zeta potentials of both exosomes and liposomes are obtained from a fit of the fluorescence intensity along the nanochannel to a model that includes both the diffusion and the diffusio-phoretic and -osmotic velocities (Fig. 2c). The fitted values for the particle sizes and zeta

* Corresponding author: rcwm@dtu.dk

potentials are consistent with results from standard methods (DLS and LDE).

The method is also able to characterize the diameter and zeta potential of individual liposomes by tracking their positions in the nanochannel over time (Fig. 3). So we can explore sample heterogeneity.

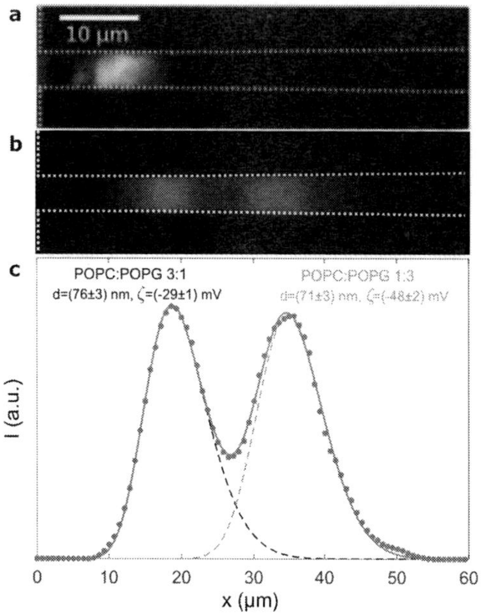

Fig 2. (a) Fluorescence image of a band of exosomes trapped in a nanochannel by a salinity gradient. From the analysis we obtain an exosome diameter of (78±7) nm and a zeta potential of (-18±1) mV. (b) Fluorescence image of trapped liposomes from a mixture. Two bands appear due to the different zeta potential of the two subpopulations. (c) Fluorescence intensity of the liposomes shown in panel c and a corresponding fit. The values of the fitted parameters are shown on the figure.

Finally, we use the chip as a label-free sensor for real-time sensing of ligands binding to liposomes [6]. In this experiment, the nanofluidic trap is populated with liposomes that have surfaces decorated with DNA probes. The DNA probes are partially double-stranded and carry a double cholesterol anchor at one end. The cholesterol anchors are inserted in the lipid bilayer. The other end is single-stranded and 15 bases long. The trapped liposomes are then exposed to a solution containing 15 bases long single-stranded targets complementary to the single-stranded probes on the liposomes. The hybridization of the targets on the liposomes' surfaces yields a change of their zeta potentials and results in a detectable shift of the trapping position. This is readily seen on the fluorescence images of the trap.

The concentration of nanoparticles by diffusiophoretic trapping has potential applications within exosome characterization, where biorecognition of, e.g., surface DNA, proteins, and antibodies, is a promising candidate for early-stage cancer diagnostics.

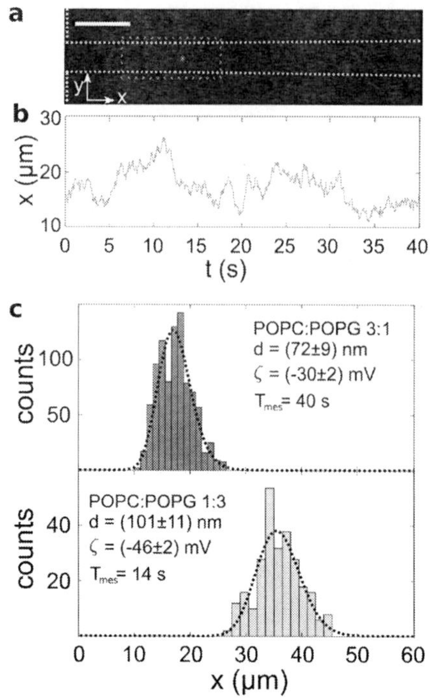

Fig. 3 (a) Fluorescence microscopy image of a single liposome trapped by a salinity gradient in a nanochannel. (b) The x-coordinates of a single, trapped liposome in a nanochannel. (c) Histogram of the x-coordinates shown in panel b and corresponding fit (top), and a similar histogram and corresponding fit for a liposome with a different size and zeta potential (bottom).

References

1. S Honary, & F Zahir, Tropical Journal of Pharmaceutical Research, **12**(2), 255–264 (2013).
2. E. Blanco, H. Shen, M. Ferrari, Nature Biotechnology **33**, 941 (2015).
3. Tomaszewska, E., Soliwoda, K., Kadziola, K., Tkacz-Szczesna, B., Celichowski, G., Cichomski, M., Szmaja, W., Grobelny, J, Journal of Nanomaterials, 313081 (2013).
4. Kashkanova, A. D., Blessing, M., Gemeinhardt, A., Soulat, D., & Sandoghdar, V., Nature Methods, **19**(5), 586–593 (2022).
5. M. K. Rasmussen, J. N. Pedersen, R. Marie, Nature communications, **11**, 1-8 (2020).
6. M. K. Rasmussen, J. N. Pedersen, R. Marie, ACS Sens. **5**, 12, 4057–4063 (2020)

Design of an all-liquid anamorphic imaging device

Daniel Sauter[1,*], *Pengpeng* Zhao, and *Hans* Zappe[1]

[1]Gisela and Erwin Sick Chair of Micro-Optics, Department of Microsystems Engineering-IMTEK, University of Freiburg, Germany

Abstract. The design of a novel anamorphic optofluidic imaging system based on a pair of liquid lenses whose toroidal surfaces create different optical powers in the symmetry-axes is presented. Using electrowetting-on-dieletrics for actuation, a cylindrical fluidic system is actuated by 32 azimuthally-distributed electrodes allowing the definition of non-rotationally-symmetric surface shapes. We present the design and simulation of this optical system and show that an anamorphic ratio of 1.43 at a maximum field of view of 6.82° is attainable.

1 Introduction

Anamorphic optics have different optical powers in two orthogonal planes, allowing realization of optical systems with different magnifications and fields of view in these planes. Anamorphic lenses were most prominently used in cinematography where objectives composed of normal and cylindrical glass lenses enabled recording of widescreen scenes on standard film [1]. In addition to their historical purpose, anamorphic optics are also used in applications such as optical processors [2] and imaging spectrometers [3].

Anamorphic systems made of conventional glass lenses require many elements, making them bulky and their assembly complex. However, other ways of realizing them have emerged, using tunable devices such as liquid crystal displays [4] or fluidic membrane lenses [5].

Recent work on electrowetting-on-dielectrics (EWOD) tunable lenses with increased electrode numbers have shown the ability to produce surfaces with great shape variability, including double-plane symmetric surfaces [6, 7]. Based on these concepts, we demonstrate that a completely fluidic anamorphic system may be realized and propose a design for an anamorphic imaging device using two EWOD tunable liquid lenses within a single device.

2 Concept

The concept is based on a previously demonstrated technology that allows the integration of multiple liquid lenses inside a single cavity as well as a high degree of surface control through sectioned electrodes [7, 8]. By deliberately shaping the lenses to have different curvatures in two orthogonal planes, the resulting optical system has a different field of view in these directions.

Shaping of the liquid interface is enabled by a sectioned electrode area with 32 sections per lens. These electrodes are embedded in a polymer foil that is located on the inner surface of a glass cylinder as shown in Figure 1 (a). Two immiscible liquids of different refractive index are then stacked inside this cylinder to form two refractive surfaces which act as the lenses. A set of glass substrates with coated aperture structures seals the cavity and provides the required electrical connection of the lenses.

For an anamorphic lens, the desired surface shapes are toroidal and their control and voltage calculation is carried out as described in [7]. Figure 1 (b) shows a *Surface Evolver* [9] simulation of one of the final lens configurations that was done to verify their feasibility. A sketch of both active sections is given in Fig. 1 (c)(d), showing the intended difference in magnification. The XZ-plane is chosen to have a larger, the YZ-plane to have a smaller FOV. The ratio of both is called anamorphic ratio (AR).

The general design approach was to start with a paraxial description of the system to find starting values for a ray tracing simulation. An anamorphic imaging system can, in the paraxial case, be treated as two separate rotationally-symmetric systems bound by a common set of constraints [1]. These constraints are that the image and object planes have to coincide and that the lens positions must be the same for both systems. Additionally, some constraints arise from the liquid nature of the device such as the dependence of lens position on curvature due to volume conservation, or the limits on lens curvature due to the achie-

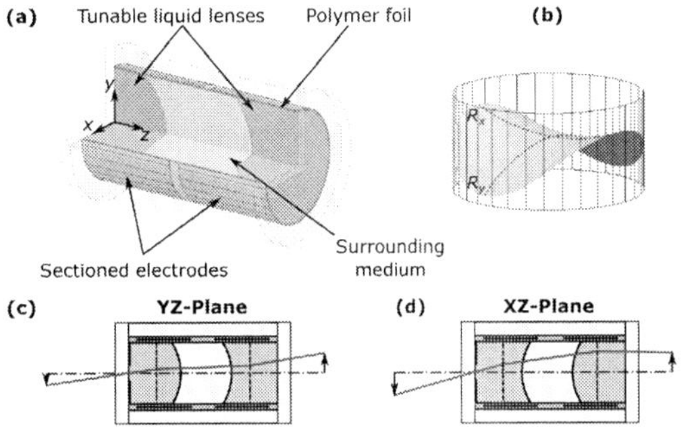

Figure 1. (a) Structure of the device, (b) double-plane symmetric interface simulated using *Surface Evolver* and (c)(d) simplified ray diagram of the two active planes (not to scale).

*e-mail: Daniel.Sauter@imtek.uni-freiburg.de

vable contact angles. The first-order solution combined with fixed technological parameters such as the cylinder diameter of 5 mm was then used to do further optimization through simulation.

3 Simulation

For an anamorphic imaging system, there are three parameters that are of interest: anamorphic ratio, field of view and resolution. They are all relevant for optimization but there is a trade-off between them, such that a compromise solution must be chosen.

For optimization, raytracing simulations were carried out in *OpticStudio* where a model of all surfaces including substrates, liquids and the sensor, was implemented. To account for the position change when varying the lens curvature, a macro was incorporated that calculates the position as a function of curvature, linking the two variables. The lens surfaces are modelled as biconic surfaces with their conic constants set to zero.

Using this model, parameters such as lens volume, cavity length, curvatures, distances were optimized with regard to the previously mentioned design objective. Initially, an AR of 2.0 was targeted but had to be lowered to maintain image quality. As a consequence, the chosen values are a compromise that does not sacrifice too much of either AR, FOV or image quality (here RMS wavefront was used for optimization). The final parameters are a device length of 11 mm, object distance of 200 mm, an aperture diameter of 1 mm, a horizontal FOV of 6.82° and a vertical FOV of 4.77° leading to an AR of 1.43.

The imaging quality is evaluated using the simulated MTF, as shown in Fig. 2. Since the focal length and thus the NA are different for the active planes, the MTF has to be considered separately for tangential and sagittal planes. Due to the dependence in variables and the spherical nature of liquid interfaces, it was not possible to achieve diffraction-limited performance in both planes at the same time. Evaluating the MTF at the Rayleigh criterion yields an optical resolution in the image plane of 30.8 and 22.7 cy/mm for the x and y directions, respectively.

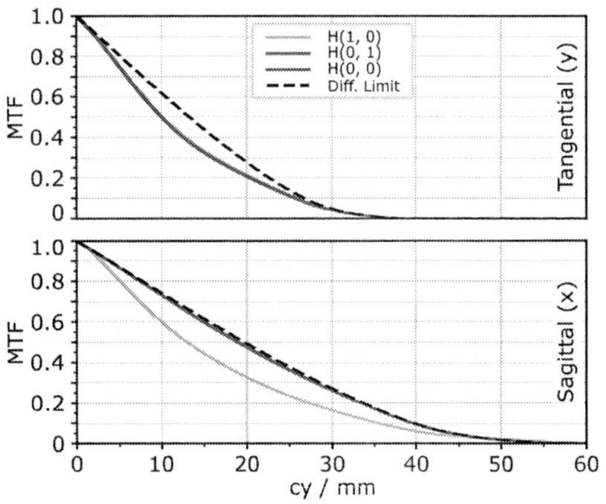

Figure 2. Simulated image-side MTF of the system with on-axis field and maximal field each in x and y. Fields are given in normalized field coordinates $H(x, y)$.

Finally, an image simulation was carried out to qualitatively judge the quality and demonstrate the anamorphic effect. Figure 3(a) shows the object and (b) the simulated image – the characteristic anamorphic compression along the x axis is clearly visible. Moreover, the difference in resolution along the axes can be identified by comparing the vertical and horizontal lines.

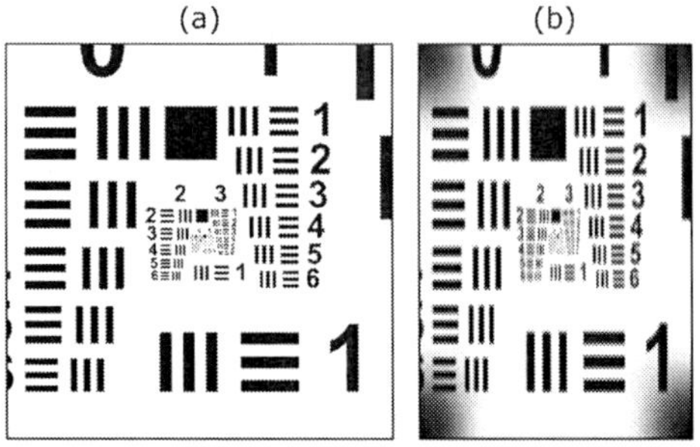

Figure 3. Image simulation of the system using *1951 USAF resolution test chart*: (a) imaged field of view for the given distances used as the source image and (b) the result of the image simulation exhibiting compression along the x-axis. Anamorphic data as given in the text.

In conclusion, we showed through fluidic and optical simulations the feasibility of creating an all-liquid anamorphic imaging device which has an AR of 1.43 at a maximum FOV of 6.82°.

References

[1] A. Dodoc, *Anamorphic Prime and Zoom Lenses*, in *Zoom Lenses VI*, edited by T. Yamanashi, A. Dodoc (SPIE, San Diego, United States, 2019), p. 2

[2] G. Unnikrishnan, J. Joseph, K. Singh, Applied Optics **40**, 299 (2001)

[3] R. Swanson, W.S. Kirk, G.C. Dodge, M. Kehoe, C. Smith, *Anamorphic Imaging Spectrometers*, in *Image Sensing Technologies: Materials, Devices, Systems, and Applications VI* (SPIE, 2019), Vol. 10980, pp. 14–28

[4] C. Iemmi, J. Campos, Journal of the European Optical Society: Rapid Publications **4**, 09029 (2009)

[5] E. Ozgur, D. Reetz, F. Akhoundi, N. O'Brien, J. Wycoff, R. Voorakaranam, P.A. Blanche, L. LaComb, C. Liang, G. Peyman et al., Applied Optics **60**, 4755 (2021)

[6] P. Zhao, D. Sauter, H. Zappe, Applied Optics **60**, 5302 (2021)

[7] D. Sauter, M. Sieben, P. Zhao, H. Zappe, Journal of Optical Microsystems **1**, 044002 (2021)

[8] D. Kopp, T. Brender, H. Zappe, Applied Optics **56**, 3758 (2017)

[9] K.A. Brakke, Experimental Mathematics **1**, 141 (1992)

https://doi.org/10.1051/epjconf/202226612007

Size-based chromosome separation in a microfluidic particle separation device using viscoelastic fluids

Therese R. Wassberg[1], *Mathilde L.* Witt[1], *Murat* Serhatlioglu[1], *Christian F.* Nielsen[2], *Ian D.* Hickson[2], and *Anders* Kristensen[1*]

[1]Department of Health Technology, Technical University of Denmark, Ørsteds Plads, Building 345C, 2800 Kgs. Lyngby, Denmark
[2]Center for Chromosome Stability, Department of Cellular and Molecular Medicine, Faculty of Health and Medical Sciences, University of Copenhagen, 2200 København N., Denmark

Abstract. Viscoelastic flow-based particle manipulation techniques enable bio-particle focusing, separation, and enrichment by precisely tuning the rheological parameters, flow conditions, and microchannel geometry. In this study, we fabricated a PDMS-based single inlet/outlet microchannel to separate bio-particles by their size ranging from 1-10 µm. Flow conditions and rheological properties are optimized using 2 µm and 4 µm Polystyrene beads to reach the best particle separation condition. We demonstrated the size-based separation of human chromosomes by separating 1-2 µm size small chromosomes from 8-10 µm size large chromosomes. Thanks to its miniaturized size and simplicity, the isolation chip and unique viscoelastic separation method have great potential to be used as a future pioneering tool for genetic applications to study chromosome abnormalities such as fragile-X and trisomy.

1 Introduction

Miniaturized optofluidic platforms are essential for in-vitro bio-analysis and diagnostic applications [1]. Low sample requirements and rapid analysis capabilities are advantageous for such systems.

Optofluidic sensory tools such as optical tweezers directly manipulate bio-particles, such as chromosomes; however, current systems have limitations due to the low purification level of the prepared chromosome samples. Metaphase chromosome samples are prepared using cell lysates, with many impurities such as cell debris and nuclei with chromosomes [2]. Such impurities generate misinterpretation during device operation and data analysis.

This study aims to improve sample preparation for the chromosome analysis platforms by developing a new microfluidic device to separate chromosomes and extend the purity of chromosome sample solutions. The proposed device is a planar PDMS channel consisting of contraction/expansion cascades (Fig. 1). The device uses the viscoelastic particle separation technique to generate size-based multiple particle trajectories at the outlet using viscoelastic force and the pinched flow fractionation method [3, 4]. Metaphase cell lysates (a mixture of cell debris, chromosomes, and nuclei) have particles spanning the 1-10 µm size range. The elutes will be mixed with biocompatible viscoelastic buffer solutions and pumped through the microchannel inlet. It will be possible to reach size-based multiple particle trajectories at the channel outlet by carefully tuning the flow parameters (flow rate and viscosity). As the last step, multiple trajectories of different size chromosomes and cell debris will be collected (not demonstrated here) from multiple chip outlets.

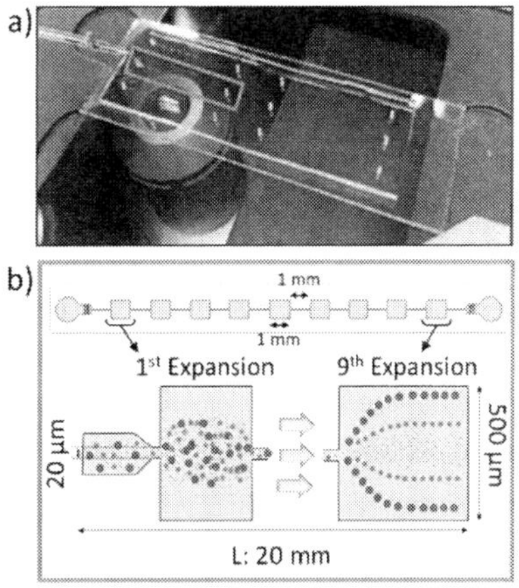

Fig. 1. a) The fabricated PDMS device and experimental setup, and b) schematic illustration of channel design and particle separation.

The demonstrated chromosome isolation chip and unique viscoelastic isolation method have the potential as a pioneering tool in chromosome research. It also has further potential applications for size-based isolation/enrichment of microorganisms in environmental science applications.

* Corresponding author: akri@dtu.dk
T. R. Wassberg, and M. L. Witt are contributed equally to this work as first authors.

2 Material and Methods

Two different molecular weights (600 kDa and 2 MDa) Polyethylene Oxide (PEO) viscoelastic polymers are used to prepare buffer solutions at various concentrations in 1X-PBS. PBS/PEO_{600kDa} buffer solutions were prepared at 2000, 5000, and 9000 ppm, and PBS/PEO_{2MDa} buffers were prepared at 1000, 2000, and 5000 ppm concentrations. Spherical polystyrene calibration beads of 2 μm and 4 μm diameter (Duke Scientific Inc.) were suspended in prepared viscoelastic buffer solutions at a concentration of 1×10^6 particles/ml. Flow experiments are demonstrated with calibration beads and optimized.

Human chromosomes are isolated from their cell lysates and suspended in viscoelastic buffer solution to a concentration of 2×10^6 chromosomes/ml. Chromosome separation experiments are performed at the best-optimized flow conditions.

The microfluidic PDMS device is fabricated using conventional soft lithography techniques in a cleanroom. A 10 μm thick SU-8 mold was prepared on a silicon substrate. Then 3-mm thick PDMS layer was cast and cured on Si mold. Later cured PDMS peeled and bonded to a glass slide to form the microchannel. The PDMS device is placed on an inverted microscope, and high-speed camera videos are recorded at 5000 fps. Recorded videos are image-stacked to show the formation of particle separation before the outlet, the 9th expansion.

3 Experimental Results

Initially, the performance of the microfluidic device was characterized by using various-sized model polystyrene microparticles. Later, in-vitro analysis of metaphase chromosomes was performed.

Calibration experiments were performed using 2 μm and 4 μm PS beads with mixture suspensions to determine the best particle separation condition. Results were analyzed for flow rates varying from 480 μl/h to 720 μl/h using six different viscoelastic buffer solutions. Multiple particle trajectories with clear size bands for the2 μm, 4 μm, and larger aggregates (>4 um) were achieved at 720 μl/h using 9000 ppm PBS/PEO_{600kDa} (Fig. 2).

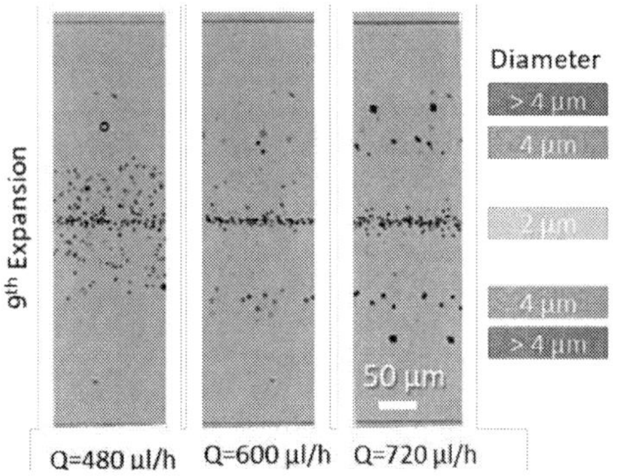

Fig. 2. Image stacks of 2 and 4μm PS bead mixture suspended in 9000 ppm PBS/PEO_{600kDa} at 480, 600, and 720 μL/h flow rates.

Size-based chromosome separation experiments were conducted at a 720 μl/h flow rate using 9000 ppm PBS/PEO_{600kDa} buffer solution (Fig. 3). Particle trajectories of individual chromosomes are generated for different size chromosomes. Single frame snapshots were inserted in Fig. 3 to show the size-based particle trajectories for chromosomes observed.

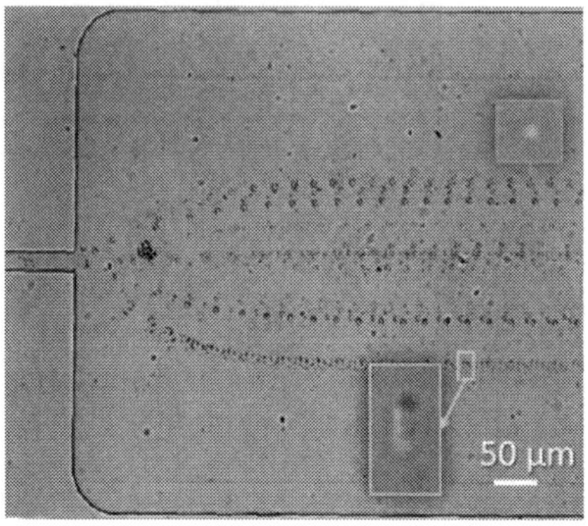

Fig. 3. Image stacked picture of chromosomes, suspended in 9000 ppm PBS/PEO_{600kDa} at 720 μL/h flow rate. The inserted images are the single-frame snapshots (7 times enlarged) from high-speed camera video recordings showing the location of 2.5 μm (blue outline) and 8.1 μm (yellow outline) size chromosomes.

4 Conclusion

The presented study investigated size-based human chromosome separation using PEO-based viscoelastic buffer solutions and viscoelastic pinched flow fractionation method in a cascaded configuration microchannel. Viscoelastic particle separation performance was optimized by tuning the fluid rheology and flow conditions using PS particles with diameters of 2 and 4 μm. Efficient size separation of chromosomes based on their size was achieved. The proposed device configuration is promising for the future of clinical studies in genetics, specifically aiming for the analysis of size-related chromosomal abnormalities.

This work was funded by Novo Nordisk Foundation Interdisciplinary Synergy Programme, under grant agreement No. 34948 (ChromoCapture).

References

1. M. Asghari, M Serhatlioglu, B Ortaç, M E Solmaz, C Elbuken, *Sci Rep* **7**, (2017)
2. H. Feng, M. Hockin, M. Capecchi, B. Gale, H. Sant, *Biomicrofluidics* **14** (2020)
3. D. Li, X. Lu, X. Xuan, *Anal Chem* **88**, (2016)
4. B. Ismail, *J Chromatogr A*, **30**, (2021)

Heliconical Cholesterics: new opportunities for optofluidics?

Francesco Simoni*

Institute of Applied Sciences and Intelligent System – CNR – Pozzuoli, Italy

Abstract. In this presentation the novelty represented by the heliconical cholesteric liquid crystals (Ch-OH) for easy electric and optical control of optical properties are highlighted. After a quick summary of their electro-optical properties, an account of the recent experimental and theoretical achievements about the nonlinear optical response of Ch-OH will be given. The peculiar conical structure allows an easy control of the spectral location of the Bragg resonance making possible effects never observed previously in pure liquid crystals, making these materials attractive for development of several optical devices.

1 Introduction

It is well known that cholesteric liquid crystals (CLC) are characterized by a continuous twist in space of the director **n** that describes a helix, with **n** always orthogonal to the helix axis. Of course, such a self-assembled periodic structure gives rise to selective Bragg reflection centered at a wavelength $\lambda_B = n_{av}P$, being n_{av} the average refractive index of the material and P the pitch of the helix. Meyer[1] and De Gennes[2] in 1968 have theoretically investigated the effects of a low frequency electric field (in the following mentioned as "static field") on this structure for materials with positive dielectric anisotropy ($\Delta\epsilon = \epsilon_{\parallel} - \epsilon_{\perp} > 0$). The unwinding of the helix due to a field applied perpendicular to the helix axis was expected and early demonstrated by Khan[3]. In case of field applied along the helix direction bend and twist deformations are involved therefore the ratio of the corresponding elastic constants K_3 and K_2, $k = K_3/K_2$, strongly affects the resulting phenomenon. Usually in CLC $k > 1$ and the main effect is a rotation by $\pi/2$ of the helical axis above a threshold for the applied static field. More interesting is what happens when $k < 1$: a conical deformation takes place over a threshold field, in which the director twists and bends around the heliconical axis. The main difference with respect to the conventional cholesteric state is that the director is not perpendicular to the heliconical axis but makes with it an angle $0 < \theta < 30$. The main consequence is the continuous change of the pitch (and of the Bragg wavelength λ_B) induced by an electric field applied parallel to the heliconical axis. A few years ago, materials with $k < 1$ have been synthesized thus allowing the realization of the conical deformation of cholesterics defined as "oblique heliconical cholesterics" (Ch-OH), in this way the possibility of tuning the Bragg resonance over the whole visible spectrum has been demonstrated by the group of Lavrentovich[4-6] by varying the applied static field.

2 Light-induced reorientation and nonlinear optical properties

After the discovery of the Giant Optical Nonlinearity (GON) in nematic liquid crystals (NLC) the possibility of using light-induced reorientation of the molecular director in CLC to get all-optical tuning of the Bragg resonance was investigated. However, it was demonstrated that GON cannot be achieved in CLC[7,8] therefore no significant light-induced modulation of the helical structure has never been observed in pure cholesterics.
On the contrary Ch-OH provide a new opportunity to observe this effect due to the bend deformation of the director that leads to the conical structure. In fact, this deformation allows an efficient coupling between the optical field and the director **n**. In this case we get an optical torque competing with the one due to the static field since it pushes the director towards a direction perpendicular to the helix axis while the static field torque pushes the director towards the direction of the helix axis. The equilibrium condition is given by a conical structure where the effective static field that stabilizes the structure is intensity dependent and is lower than the applied one, thus corresponding to a different pitch of the helix[9]:

$$E_{eff}(I) = \sqrt{E^2 - CI}$$

where:

$$C = \frac{\Delta\epsilon_{OPT}}{\Delta\epsilon\ \epsilon_0 c n_{av}}$$

being $\Delta\epsilon_{OPT}$ the dielectric anisotropy at optical frequencies and c the speed of light in vacuum. As consequence, we get light-induced tuning of the Bragg resonance because the pitch of the structure depends on the intensity I according to the expression:

* e-mail: f.simoni@photomat.it

$$P(I) = k\frac{E_{NC}P_0}{E_{eff}(I)}$$

being E_{NC} the threshold field over which the structure unwinds to a uniaxial nematic and P_0 the pitch of the corresponding planar cholesteric.

This effect has been experimentally demonstrated using a pump-probe technique. The pump beam (a CW frequency doubled Nd:YAG laser at λ = 532 nm), was focused on the Ch-OH sample 20 mm thick and probed by a white light (SLED) source, focused by the same optics, while a static field was applied to stabilize the conical structure. The inverted microscope set up allowed monitoring both the reflection spectra and the sample appearance in reflection mode. At a fixed value of the static electric field a light-induced red-shift of the Bragg wavelength higher than 100 *nm* has been measured[10]. Additional proof of the optical reorientation effect has been provided by measuring the light-induced change of the polarization state of the transmitted light that corresponds to the behavior observed without pump beam when varying the applied static field. For wavelength close to the Bragg resonance polarization changes from linear to elliptical to circular and back were detected by increasing the pump beam intensity, thus obtaining an all-optical control of the probe light polarization at the exit of the Ch-OH sample.

Furthermore, the nonlinear effects on the light propagation due to optical reorientation have been investigated[9]. We have found that for wavelengths close to λ_B the induced red shift of the Bragg resonance increases or decreases the reflectivity depending on the light intensity and polarization leading to a stop band in the light transmittivity for a definite range of intensity dependent on the applied static field.

Recently a more accurate calculation of the nonlinear propagation based on Ambartsumian's layer addition modified method[11] confirmed such behavior for moderate intensities while showing the onset of optical instabilities leading to self-oscillations for increasing value of the light intensity.

These peculiar optical properties have been shown to be attractive for developing tuneable optical devices such as optical notch and bandpass filters[12] and envisage the possibility of being exploited in advanced sensing devices in optofluidic configurations[13] or coupled with piezoelectric photo-mobile supports[14].

I acknowledge support from European Union's Horizon 2020 research and innovation programme H2020-FET-OPEN grant no. 863227 – Project PULSE-COM, "Photo-Piezo-ActUators based on Light SEnsitive COMposite"

References

1. P.G. De Gennes, Solid State Comm. **6**, 163 (1968)
2. R.B. Meier, Appl. Phys. Lett. **12**, 281 (1968)
3. F.J. Khan, Phys. Rev. Lett. 24, 209 (1970)
4. J. Xiang, S. V. Shiyanovskii, C. Imrie and O. D. Lavrentovich, Phys. Rev. Lett., **112**, 217801 (2014)
5. J. Xiang, Y. Li, Q. Li, D. A. Paterson, J. M. D. Storey, C. T. Imrie, O. D. Lavrentovich, Adv. Mat., **27**, 3014 (2015)
6. M. Mrukiewicz, O.S. Iadlovska, G. Babakhanova, S. Siemianowski, S.V. Shiyanovskii and O. Lavrentovich, Liquid Crystals doi: 10.1080/02678292.2019.1587528
7. B.Ya Zel'dovich, N.V. Tabiryan, Sov.Phys. JETP **55**, 99 (1982)
8. E. Santamato, G. Abbate, P. Maddalena, Y.R. Shen, Phy. Rev. A, 36, 2389 (1987)
9. F. Simoni, Opt. Lett. **45**, 6510 (2020)
10. G. Nava, F. Ciciulla, O. S. Iadlovska, O. D. Lavrentovich, F. Simoni, L. Lucchetti, Phys. Rev. Res. **1**, 033215 (2019)
11. A. H. Gevorgyan, F. Simoni, Photonics **9**, 139 (2022)
12. G. Nava, F. Ciciulla, F. Simoni, O. S. Iadlovska, O. D. Lavrentovich, L. Lucchetti, Liq. Cryst. doi.org/10.1080/02678292.2021.1884911
13. S. Bonfadini, F. Ciciulla, L. Criante, A. Zaltron, F. Simoni, V. Reshetnyak, L. Lucchetti, Sci. Rep. **9**, 1062 (2019)
14. PULSE-COM www.pulsecom-h2020.eu – EU H2020-FET-OPEN grant no. 863227

Characterisation of the stability and long-term evolution of the properties of a 45TW laser operating at 10Hz

Aarón Alejo[1,*], *Adrián* Bembibre[1], *Juan* Peñas[1], *José* Benlliure[1], *Lucía* Martín[2], and *María Teresa* Flores-Arias[2,3]

[1]Instituto Galego de Física de Altas Enerxías (IGFAE). Universidade de Santiago de Compostela (Spain)
[2]Laboratorio Láser de Aceleración y Aplicaciones (L2A2). Universidade de Santiago de Compostela (Spain)
[3]Instituto de Materials (iMATUS). Universidade de Santiago de Compostela (Spain)

Abstract. Industrial quality applications of high-power lasers working at high repetition rates, including laser-driven particle acceleration, will require laser systems capable of operating in a stable and prolonged manner. Several factors can affect this stability, including the environment conditions, such as temperature or humidity, and the progressive heating of the optical components involved. Here we report on the evolution of the main laser parameters for a 45 TW system operating at 10 Hz, showing that a significant change in the wavefront and direction of propagation, even when the laser energy remains sufficiently constant. These results highlight the importance for future laser systems to integrate closed-loop beam tracking diagnostics that can correct the temporal evolution.

1 Introduction

The progress in laser technologies over the past couple of decades has led to an increasing interest in laser-driven particle acceleration. These accelerators benefit from the huge acceleration gradients that can be sustained by the plasma, more than three orders of magnitude greater than the fields achieved in conventional accelerators. Impressive results recently reported include the acceleration of protons to energies up to 100 MeV from solid targets [1], or the acceleration of electron beams to energies up to 7.8 GeV from a single stage a few centimetres long [2].

Beyond the ongoing efforts to further improve the acceleration mechanisms, laser-plasma accelerators are reaching a level of maturity that enables their widespread use as an alternative to conventional accelerators. An example in this direction is the EuPRAXIA project [3], an European-funded development aiming to build a plasma-based electron accelerator of industrial quality capable of producing narrow-band and high-current electron beams with energies up to 5 GeV, in a stable and consistent manner. Additionally, such stable operation and consistent acceleration must be maintained while operating continuously at high repetition rates for extended periods of time. Significant work is being carried out to optimise the acceleration process, as well as on the production of targets compatible with repetitive operation at such rates, including studies up to the kHz range [4]. Beyond the production of improved target systems, a stable laser-based accelerator will also require the laser system to remain stable for extended periods of time.

There are several elements that can affect the evolution of a laser system, including changes in humidity or temperature of the room, or the long-term evolution of the laser pumps, which have a direct impact on the energy and direction of propagation of the laser beam. In addition to these, the effects of the heating of optical elements along the beamline in the case of systems operating at high repetition rates has recently attracted significant attention. In particular, given the high intensities reached on the compressor of systems based on chirped pulsed amplification (CPA), the effects of heating of the gratings responsible of the final pulse compression producing the ultra-short pulses on the beam properties are being actively analysed, including studies on the spatial and temporal evolution of systems operating at 5 Hz [5]. These effects are expected to become more relevant as the repetition rate is further increased.

Here we analyse the long-term evolution of a high-power (45 TW), commercial laser system operating at 10 Hz through a 6-hour period, particularly the evolution of the energy, pointing, and wavefront of the laser beam.

2 Experimental Setup

The experiment was carried out at the Laser Laboratory for Acceleration and other Applications facility (L2A2, Universidade de Santiago de Compostela), by employing the high-energy beamline of the STELA (Santiago TErawatt LAser) system, a Ti:Sapphire laser providing ultra-short (τ = 30 fs, Full-Width-at-Half-Maximum) pulses with an energy per pulse of ~ 1.2 J and a peak power of 45 TW, central wavelength λ = 800 nm, high contrast ratio (>10^{10} at 5 ps), and operating at a repetition rate of up to 10 Hz.

In order to fully characterise the stability and long-term evolution of the different laser parameters, a number of diagnostics is distributed throughout the system, including the laser chain, compressor, and target area.

*e-mail: aaron.alejo@usc.es

https://doi.org/10.1051/epjconf/202226613001

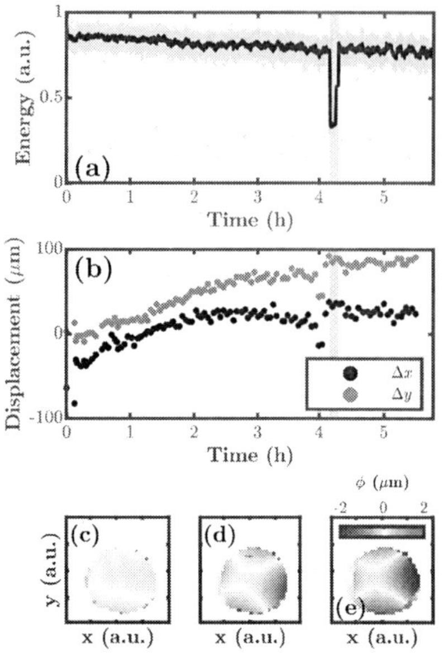

Figure 1. Evolution of the laser parameters. (a) Evolution of the laser energy. **(b)** Drift of the far field. **(c-e)** Wavefront of the laser beam after operating the laser for a period of (c) 45 min, (d) 1.5 h, and (e) 5.5 h.

The compressed beam is characterised using the leak through one of the mirrors inside the compressor vacuum chamber. The leaked beam is then propagated passing through a *BK7* window, and de-magnified ($M \sim 1/5$) using a pair of metallic spherical mirrors ($f_1' \simeq -200$ mm, $f_2' \simeq 800$ mm). The resized beam is then subsequently divided using a range of beam-splitters and propagated to a full suite of diagnostics, namely a calorimeter for energy characterisation, SPIDER system for temporal characterisation, a HASO Shack-Hartmann camera to characterise the beam wavefront, a spectrometer for spectral characterisation, and a far-field and near-field diagnostics to characterise the pointing and spatial profile of the laser beam.

Following compression, the high-energy beam is propagated into target area through an 8 m vacuum pipe. The beam is attenuated by a factor of $\sim 10^{-4}$ using a pair of uncoated dielectric wedges, and subsequently focussed using an $f/3$ off-axis parabolic mirror. The focal spot is characterised using a long working-distance microscope objective, imaging the focal plane onto a CMOS camera with a magnification of $M \simeq 220$. Additionally, the beam pointing and energy reaching the target area are measured using the leak passing through the one of the attenuating wedges.

3 Results

3.1 Energy

The evolution of the laser energy is shown in Fig. 1(a). Overall, the energy was found to remain stable throughout the 6-hour measurement, albeit exhibiting a slowly-decaying trend. In this case, the total energy drop was $\sim$ 20 mJ, equivalent to an energy degradation rate of 0.33% of the total laser energy per hour of irradiation. It should be noted that the the brief drop in energy observable in Fig. 1(a) for a time $\sim$ 4.15 h corresponds to a failure and recovery of one of the laser pumps.

In addition to the steady loss, the energy is found to exhibit an oscillatory behaviour with a characteristic periodicity of 9 minutes. Such temporal evolution is speculated to correspond to temperature oscillations in the laser area.

3.2 Pointing

The pointing of the laser beam was characterised at different stages throughout the laser system. The evolution of the pointing in the case of the compressor leak is shown in Fig. 1(b). A significant drift in the direction of the beam is observed, particularly along the vertical direction. It should be noted that this drift is not observed when the pump energy in the amplifiers is reduced, lowering the total energy contained in the high-power pulse, which indicates that such a drift would be associated with the heating of optics along the beamline, particularly the heating amplifier crystal or the compressor gratings.

3.3 Wavefront

The phase shifts of the beam wavefront at different times of the run are shown in Fig. 1(c-e). Such shifts are observed to be continuously increasing, with a tendency to introduce a phase that closely matches the phase for a vertical astigmatism. The phase shift is only observed to appear when the laser operates at full power, which indicates that it would be associated to the heating of optical elements, as previously reported.

4 Conclusions

The long-term evolution of a high-power laser system operating at 10 Hz through a 6-hour period has been studied, particularly the evolution of the energy, pointing, and wavefront of the beam. The energy is found to remain sufficiently stable throughout the run. The beam pointing evolves significantly, potentially associated to the heating of the compressor gratings. Also related to the compressor heating, the wavefront is found to evolve significantly, showing the appearance of a phase related to vertical astigmatism. These measurements highlight the need for closed-loop correction systems to ensure the stability of high-power laser systems, particularly in the context of demanding applications such as laser-plasma accelerators.

References

[1] A. Higginson, *et al.*, Nat. Comms. **9**, 1 (2018)
[2] A. Gonsalves, *et al.*, Phys. Rev. Lett. **122**, 084801 (2019)
[3] R. Assmann, *et al.*, The European Physical Journal Special Topics **229**, 3675 (2020)
[4] A. Alejo, *et al.*, Phys. Rev. Acc. Beam **25**, 011301 (2022)
[5] V. Leroux, *et al.*, Optics express **28**, 8257 (2020)

EPJ Web of Conferences 266, 13002 (2022)
EOSAM 2022
https://doi.org/10.1051/epjconf/202226613002

Fluorescence for non-contact detection of salmon lice in fish farms

Kari Anne Hestnes Bakke[1*], *Jon* Tschudi[1], and *Trine* Kirkhus[1]

[1]SINTEF Digital, Smart Sensors and Microsystems, Forskningsveien 1, 0373 Oslo, Norway

Abstract. This work presents a promising method for automatic non-contact detection and counting of salmon lice infested on salmon in an aquacultural farm setting. The method uses fluorescence in the visual part of spectrum to enhance the contrast between fish skin and lice. The wavelengths used are compatible with an underwater measurement system.

1 Introduction and background

The densities of salmon (*Salmo salmar*) in fish farms makes them ideal for the parasite salmon lice (*Lepeophtheirus salmonis*) to infect and reproduce [1]. This is the number one threat to Norwegian Aquaculture. There are governmental requirements to count and register lice in fish farms to ensure control with the parasite. Continuous surveillance helps in deciding on taking measures to reduce the lice infestation on salmon.

It is difficult to get good images of salmon skin due to its semi specular nature, and varying colouring. The lice are semi-transparent and changing in appearance and size (see Fig. 1), and it is desirable to detect and count lice in early stages (1-2 mm size) as well as late life stages (5-8 mm size). Several solutions based on underwater camera are granted dispensations from the manual counting regime by the Norwegian Food Safety Authority – Aquabyte, Stingray and Createview – none of these have published peer-review papers on their systems or their performance.

The lice have an exoskeleton made by chitin protecting the lice against its surroundings [2]. Using optical methods that enhance the appearance of the chemical components of chitin is preferable. Chemical components can typically be detected using spectroscopy or fluorescence.

We have characterized the lice's optical properties using several methods (polarisation, spectroscopy, fluorescence) in the lab. The results are evaluated with the purpose of making a measurement setup possible to industrialize and implement in the aquaculture industry. In this abstract we present our findings on fluorescence properties.

2 Methods and experiments

We had access to newly slaughtered salmons with living lice which was delivered in seawater. Properties of lice, fish skin and seawater were investigated.

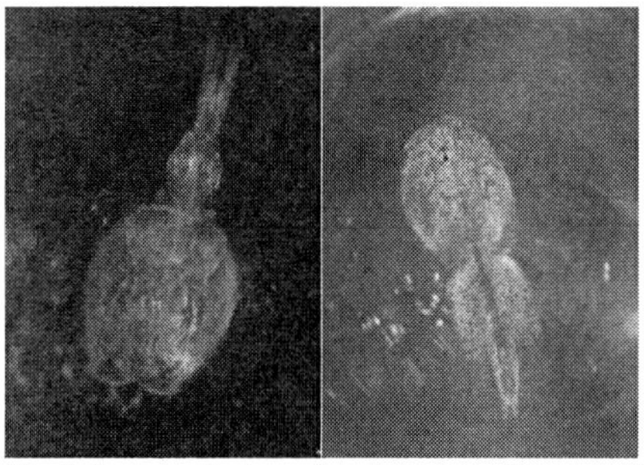

Fig. 1. Images of salmon lice, late stages (5mm and 8mm).

2.1 Fluorescence properties of salmon and lice

To assess the fluorescence properties, narrowband light sources were used for excitation. Both a spectrometer and a camera with excitation blocking filter, were used for emission detection. We tested UV and 532nm excitation. Four lice were measured in the 532nm fluorescence setup. The resulting fluorescence was recorded by an Ocean Optics VIS spectrometer, see Fig. 2. All the observed lice give fluorescence from major parts of the body (red curve). The tail of the largest lice (blue curve) did however not give a signal that could be separated from fish skin (black curve).

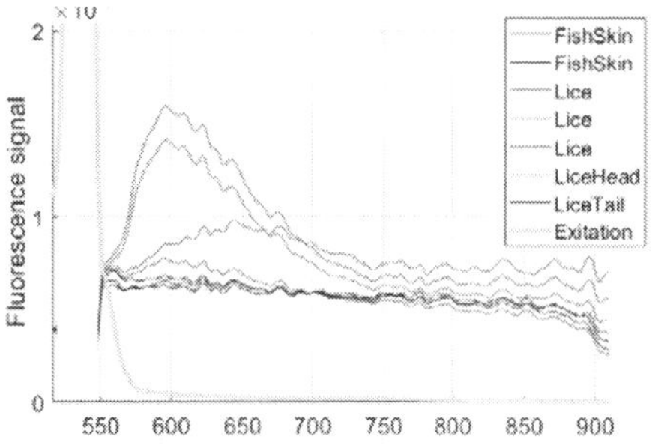

Fig. 2. Fluorescence signal using 532nm excitation.

 https://doi.org/10.1051/epjconf/202226613002

Images of lice illuminated with 532nm and imaged through filters are shown in Fig. 3.

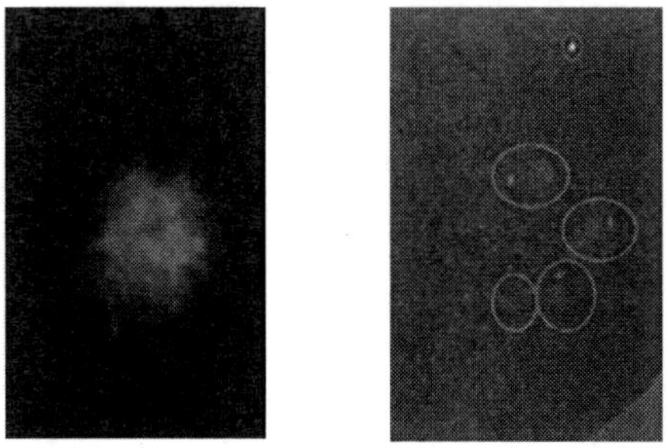

a) b)

Fig. 3. Fluorescence from a) lice on plastic, illuminated with 532nm, imaged through a long pass filter (lower cut-off at 600nm) and b) Four lice on salmon from a 2-meter distance, imaged using a mobile camera though laser goggles (black and white image).

2.2 Water properties

For continuous lice monitoring in a salmon farm relevant distance to the fish is 1-2 m.

The seawater 's absorption and scattering coefficients as a function of wavelengths are varying in Norwegian fjords during an annual cycle. They depend on e.g., the type of coastal water, the mixing ratio between freshwater and seawater, the concentration of chlorophyll and yellow substance [3]. In Fig. 4 we show the measured transmission spectrum taken through the seawater that came with the salmons and lice used in our experiments.

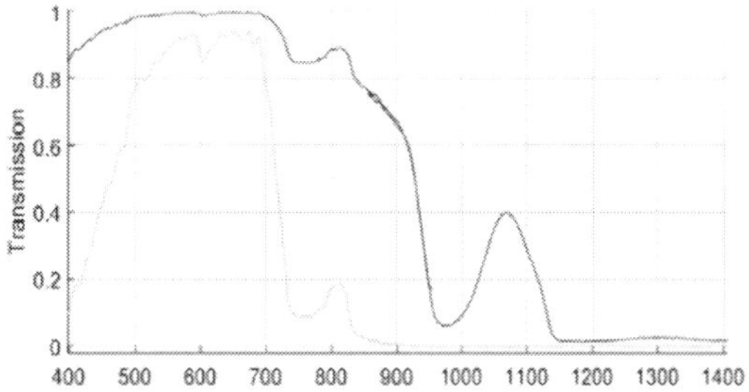

Fig. 4. Transmission measured through 69mm (blue curve) and calculated for 1m (green curve) of seawater in a salmon farm.

The measured transmission is a snapshot of seawater transmission from this specific sample – it is included to illustrate that selection of wavelengths in underwater measurement system needs considerations. For longer wavelength the measured water transmission is in good agreement with the findings in literature [3], which show that above 700nm the mean free optical path is <1m. If this wavelength range is used under water, the distance between sensor and object must be kept short and controlled, to enable correcting the optical effects of the water itself.

We observe a steep gradient in transmission for wavelengths below 500nm, and very low transmission below 450nm.

3 Results and discussion

The shell of the lice is chitin [2]. It produces a fluorescence signal when excited with green light (532nm). This is in accordance with our observations of fluorescence signal from the lice body. The peak wavelength of the fluorescence emission varies over the group of lice from 600nm to 650nm. The reason for these differences is unknown at this stage. The fish skin is not fluorescent when excited at 532nm.

Measuring fluorescence emission in the 550-700nm range to separate lice and fish skin is promising. Both the illumination/excitation wavelength of 532nm and the fluorescence emission wavelengths are within the range were seawater have low absorbance.

4 Conclusions

Methods for salmon lice detection have been investigated, and a main prerequisite have been to search for a method which can be applicable in the aquacultural industry.

Based on our experiments, we propose a system using illumination at 532nm and detection using an underwater camera with a filter blocking the illumination source. This system will have a more complicated light source than existing underwater camera systems, and it enhances the difference between lice and fish skin. Lice can then better be separated regardless of salmon skin colour, lice colour and lice transparency – since it is the chemistry of the lice's exoskeleton that is imaged. This will reduce the computational complexity needed for online automatic lice counting.

Next step in realizing this measurement system will be more experimental work in a larger scale, including measurements in water tank with seawater, living lice and living salmon.

References

1. A. Coates, B. L. Phillips, S. Bui, F. Oppedal, N. A. Robinson, T. Dempster, Rev. in Aquaculture **13**, 1397 (2021)
2. H. M. Hardardottir, *PhD thesis, Univ, in Bergen,* (2021)
3. Ø. Frette, S.R. Erga, B. Hamre, J. Aure, J.J. Stamnes, Sarsia, **89**, 276 (2004)

https://doi.org/10.1051/epjconf/202226613003

A ϕ-Shaped Bending-Optical Fiber Sensor for the Measurement of Radial variation in Cylindrical Structures

Victor Cardoso[1,4,6], *Paulo* Caldas[4,5], *M. Thereza* R. Giraldi[2], *Orlando* Frazão[3,4], *João* Costa[1], and *José* L. Santos[3,4]

[1]Federal University of Pará, Applied Electromagnetism Laboratory, Rua Augusto Corrêa, 01, 66075-110, Belém, Pará, Brazil
[2]Military Institute of Engineering, Laboratory of Photonics, Praça Gen. Tibúrcio, 80,22290-270, Rio de Janeiro, Brazil
[3]Department of Physics and Astronomy, Faculty of Sciences of University of Porto, Rua do Campo Alegre, 687, 4169-007 Porto, Portugal
[4]Institute for Systems and Computer Engineering, Technology and Science, Rua do Campo Alegre, 687, 4169-007 Porto, Portugal
[5]Polytechnic Institute of Viana do Castelo, Rua Escola Industrial e Comercial de Nun'Álvares, 4900-347, Viana do Castelo, Portugal
[6]victorcard@ufpa.br

Abstract. This work presents preliminary results of the ϕ -shaped sensor mounted on support designed by additive manufacturing (AM). This sensor is proposed and experimentally demonstrated to measure the radial variation of cylindrical structures. The sensor presents an easy fabrication. The support was developed to work using the principle of leverage. The sensing head is curled between two points so that the dimension associated with the macro bend is changed when there is a radial variation. The results indicate that the proposed sensor structure can monitor radial variation in applications such as pipelines and trees.

1 Introduction

Compared to traditional methods of monitoring diameter variation, fiber-based sensors have several advantages: high sensitivity, immunity to electromagnetic interference, compact size and low weight, multiplexed measurement possibility, fast response, and ability to communicate over long distances [1, 2]. On the other hand, there is an increasing demand for developing sensors and instrumentation to reduce costs, optimize processes, and increase production through technological advances in Industry 4.0. In this scope additive manufacturing(AM), or 3D printing, is gaining applicability in industries and is used to manufacture structures using digital models [3].

The search for sensors with the above mentioned characteristics has been the research object in several works. Among them, we can mention Sienkiewicz *et al*[1] that uses a sensor based on an optical fiber in ϕ-shape for displacement measurement over large ranges associated with optical power loss as a function of the macro bend. The sensor has an extremely simple design. This same configuration was used years later in respiratory monitoring [2].

This work presents preliminary results where we investigate a designed setup for radial variation monitoring using a simple sensor based on ϕ-shaped with 3D printer support. It was possible to analyze the radial variation of cylindrical structures, typical of some kind of trees and pipelines.

2 Sensor Operation and Preliminary Experimental Setup

The proposed sensor is formed by a piece of a singlemode fiber (SMF-28, Corning Inc.) in ϕ-shaped, as depicted in Fig. 1. When submitted to strain or displacement, this device acts as an intensity optical fiber sensor. The fiber is glued to the support at two points, one on each side of the loops [1]. Due to it, the radius of curvature of the sensor loops changes when subjected to an extension. This results in the stretching of the sensor loops and optical power loss in this setup. Contrariwise, a compressive strain would shorten the loops and increase intensity at the detector. These deformations take place in the plane of the loops. It is a simple optical sensor. These deformations take place in the plane of the loops. It is a simple optical sensor.

This structure was evaluated to use the lever principle, observed in several examples of everyday life as in pliers and scissors. The material used is Polyethylene Terephthalate Glycol (PET-G). This is a thermoplastic copolyester material, highly resistant and durable, chemical resistant, which is used commercially in desktop 3D printers [4]. It is an excellent material for various applications. Its use was intensified in 2020 in face shields production for the pandemic crisis caused by the SARS-CoV-2 virus, responsible for Covid-19 [5]. PET-G is a tough material, between 2.01 GPa and 2.11 GPa for Young's modulus and between 60 and 66 MPa for tensile strength [6, 7]. These char-

*This study was financed in part by the Coordenação de Aperfeiçoamento de Pessoal de Nível Superior – (CAPES) – Finance code 001, the Conselho Nacional de Desenvolvimento Cientifico e Tecnologico (CNPq) from Brazil, Institute for Systems and Computer Engineering, Technology and Science (INESC TEC) and the Department of Physics and Astronomy, Faculty of Sciences of University of Porto from Portugal)

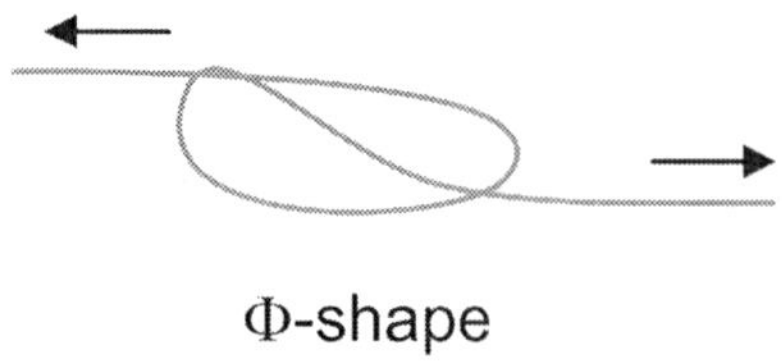

Figure 1. Schematic of the proposed sensor.

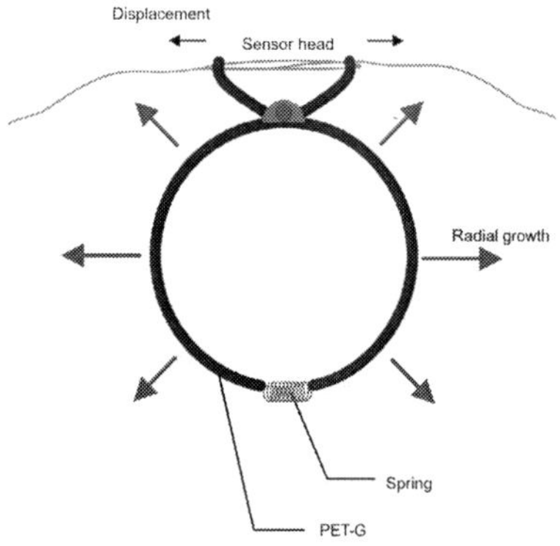

Figure 2. Schematic of the experimental piece.

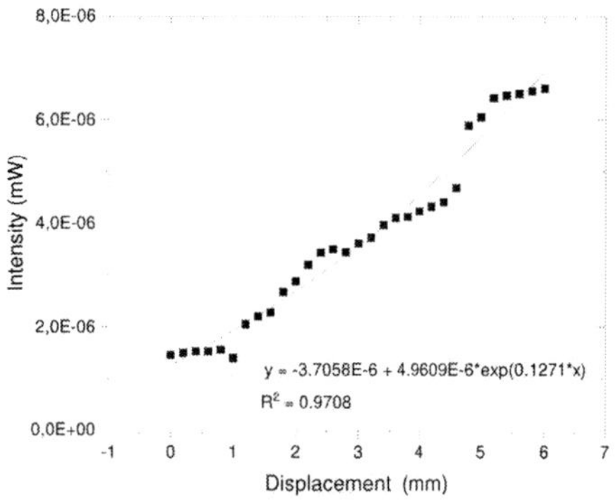

Figure 3. Gain intensity caused by the diameter variation.

acteristics make it an excellent material for the proposed application.

The schematic of the experimental piece is illustrated in Fig. 2. The support diameter is 90 mm, and it can be placed around the plant specimen or pipeline. The structure expands from the radial growth, changing a macro bend in the sensor, thus producing a variation in the optical power. The advantage of this design is the gap between the sensor head and the cylindrical structure because the sensor is free from possible transverse deformations. A spring was placed in the piece to be adjusted along with the radial variation.

3 Results and discussion

The preliminary results obtained from the experiments are depicted in Fig. 3. It is possible to observe that the results presented good behavior using the ϕ-shaped sensor when increasing the radius piece. This sensor allows greater control of the optical power for the dynamic range from 0 to 6 mm with a gain of the $\approx$ 11.8 dB. For this setup, the sensor has given good responses with R^2= 0.9708, and the sensitivity obtained was 1E-6 mW/mm.

4 Conclusion

In conclusion, we presented an experimental investigation to measure the cylindrical structures' radial variation using a ϕ-shaped sensor mounted on a 3D printer piece designed to support it. The setup is very simple and repeatable. The results showed that it is possible with the designed sensor to monitor the radial variation from 0 to 6 mm with good sensitivity and R^2=0.9708. This dynamic range can be used in many applications, such as obtaining the irrigation levels in agriculture by measuring variation of the stem of trees. The next step in this investigation will be to perform measurements using other kinds of optical sensors and optimized 3D printer structures.

References

[1] F. Sienkiewicz, A. Shukla, Optics and Lasers in Engineering **28**, 293–304 (1997)

[2] A.T. Augousti, F.X. Maletras, J. Mason, Physiological Measurement **26**, 585 (2005)

[3] H. Lasi, P. Fettke, H.e.a. Kemper, Bus Inf Syst Eng **6**, 239 (2014)

[4] J. Kechagias, N. Fountas, K. Ninikas, M. Petousis, N. Vidakis, N. Vaxevanidis, Materials and Manufacturing Processes (2021)

[5] K. Grzelak, J. laszcz, P.M. Jakub Polkowsk and, J. Kluczynski, J. Luszczek, J. Torzewski, I. Szachogluchowicz, R. Szymaniuk, Materials **14**, 4823 (2021)

[6] S. Guessasma, S. Belhabib, H. Nouri, Polymers **11**, 1220 (2019)

[7] C. Cianci, G. Pappalettera, G. Renna, C. Casavola, M. Laurenziello, G. Battista, C. Pappalettere, D. Ciavarella, Front. Mater. **7**, 104 (2020)

Celebrating a face-to-face congress of young researchers in Optics after the pandemic years: the I NW MYRO

Bastián Carnero[1,*], *Alba* de las Heras[2], *Alejandro* Doval[1], *Álex* Martín-Rodríguez[2], *Alicia* Muñoz-Ramos[1], *Ana* García-Cabrera[2], *Carlos Damián* Rodríguez-Fernández[1], *Damián* Insua-Costa[1], *Irene* Romo-Díez[1], *Isabel* Rodríguez-Pérez[2], *Javier* Prada-Rodrigo[2], *Javier* Varela-Carballo[1], *José* Paz-Martín[1], *María* Sánchez-Hernández[2], *María Jesús* Martínez-Morillo[2], *Mario* Guerras-Rodríguez[2], *Millán* Pérez[2], *Sabela* Fernández-Rodicio[1], *Verónica* Villa-Ortega[1], *Víctor* W. Segundo-Staels[2] and *María Teresa* Flores-Arias[1]

[1]Dissemination Group and Students Association LUZADA, USC-OPTICA Student Chapter and Santiago USC Young Minds Section, Universidade de Santiago de Compostela, E-15782 Santiago de Compostela, Spain.
[2]OSAL Student Chapter, Universidad de Salamanca, E-37008 Salamanca, Spain.

Abstract. This contribution reports the organization and celebration after the Covid-19 pandemic of a singular scientific conference focused on early-career researchers from the Spanish universities of Santiago de Compostela and Salamanca: the "I Northwest Meeting of Young Researchers in Optics (I NW MYRO)".

1 Introduction

In recent months, with the appearance of vaccines and the improvement of the situation resulting from the Covid-19 pandemic, new face-to-face events are starting to take place again. Scientific congresses have been the main affected by this scenario and despite the value of the online alternatives [1,2], it is difficult to match the enriching experience of on-site scientific meetings. It is our responsibility as young members of the research community to encourage the recovery of these events, as they will provide synergies of great interest for our future careers. This idea led to the birth of the "I Northwest Meeting of Young Researchers in Optics (I NW MYRO)", an exciting three-day congress focused on the research and outreach of optics and photonics within the northwest of Spain and north of Portugal. The event was organized by young researchers and students from the following outreach groups: LUZADA, USC-OPTICA Student Chapter and Santiago Young Minds Section, from Universidade de Santiago de Compostela (USC); and OSAL Student Chapter from Universidad de Salamanca (USAL).

The meeting was held in Santiago de Compostela, on 5th-7th May to commemorate the International Day of Light 2022 and it was devised as a unique opportunity to support young physics and optics and photonics students and researchers in their career development, as well as to strengthen scientific and industrial connections in the northwest of the Iberian Peninsula and even internationally. The event received the support of the main scientific societies such as Optica, European Physical Society (EPS), and Real Sociedad Española de Física (RSEF), and gathered more than 50 experts in optics from industry and academia. The event concluded with an outreach event that involved two local museums.

2 Congress programme

The event consisted in a three-day meeting focused on research exchanges and networking opportunities for the young scientific community in optics and photonics of our area. We highlight below the following central activities of the I NW MYRO, distinguishing between academic and outreach sections.

2.1 Academic section

2.1.1 *Plenary talks*

One of the core objectives of the congress was to bring together the main research institutions in the northwest area of the Iberian Peninsula. The I NW MYRO was able to assemble relevant experts and researchers from all universities and research centres in the area: USC, USAL, University of Vigo (UVIGO), Universidade da Coruña (UDC), Laser Laboratory for Acceleration and Applications (L2A2) and Iberian Nanotechnology Laboratory (INL). In this session, during the plenary talks, the invited experts discussed diverse optics and photonic topics such as: laser material processing, photonic devices in crystalline mediums, natural photonics, etc.

2.1.2 *Industrial talk*

Industry tends to be the great unknown at universities. Researchers themselves are often unaware of how valuable their training is for companies, and the great opportunities that industry offers after academia. With the intention of improving this scenario, an industrial talk was organised during the I NW MYRO, with the presence of

*Corresponding author: bastian.carnero.groba@usc.es

two companies from the northwest region (BFlow SL and AIMEN) and an international one (Menlo Systems). In all these companies, optics and photonics play a decisive role in their production processes, and their representatives had the opportunity to present their main R&D lines. Afterwards, a roundtable was held in which the representatives were able to discuss university spin-offs, patents, and R&D trends in photonics with the audience.

2.1.3 Women in Optics

Highlighting the work of female scientists and supporting their careers is a key part of our labour not only as members of scientific groups, but also as members of society. We can proudly state that all the sessions in the I NW MYRO were scheduled with a gender perspective, ensuring that both male and female scientists were provided with the same visibility. Furthermore, the programme included a "Women in Optics" session where different women scientists at different stages of their careers (Fig. 1) (PhD student, post-doc researcher and professor) could introduce their research and areas of expertise and, afterwards, integrate a roundtable to discuss the challenges that women are facing in their scientific careers. This session addressed interesting and thought-provoking topics such as deterring physicist stereotypes, facing gender bias, motherhood during a research career, or the importance of female references.

Fig. 1. Women in Optics roundtable, featuring female researchers at different stages of their careers.

2.1.4 Young researchers talks

Knowing how to present their investigation constitutes a key skill for young researchers. Therefore, during the I NW MYRO young researchers talks, all participants were able to face the podium (Fig. 2) (some of them for the first time in their careers).

Fig. 2. Young researchers a) moderating the session and b) presenting their research lines.

2.2 Outreach section

Apart from the academic section of the program, two outreach events were carried out for the general public in collaboration with two important local museums: the House of Sciences of A Coruña (Fig. 3a) and the Natural History Museum of the USC. These events were not only intended to commemorate the International Day of Light 2022, but also to bring science closer to our society. Spreading a scientific culture is a key action for inspiring new generations of scientists. In our case, we aim to foster the love for optics and physics with experiments covering diverse light phenomena like reflection, refraction, holography, fluorescence, or phosphorescence, as well as funny light experiments involving polarization glasses, long-exposure photography (Fig. 3b) or 3D cinema. Holding the events in parallel (they were celebrated on the same day in two different cities) was a logistical challenge but it allowed us to reach a very large audience (Fig. 3c) of approximately 1000 people of all ages.

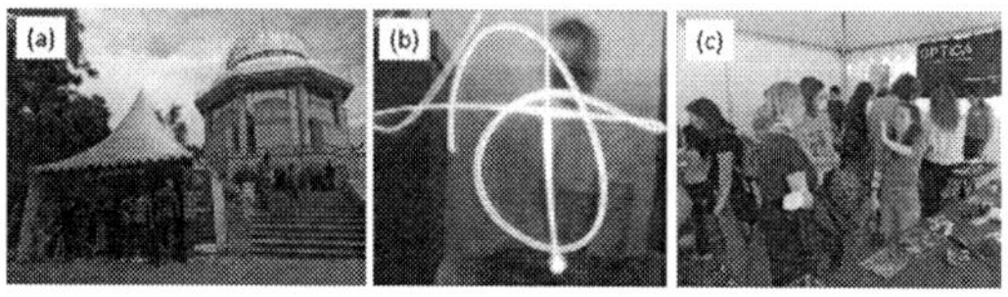

Fig. 3. A look at our outreach activities. a) Our stand next to the House of Sciences of A Coruña. b) Long-exposure photography experiment. c) Disseminating optics to a varied audience.

3 Conclusion

In conclusion, the I NW MYRO was successfully celebrated from 5th to 7th May 2022 in Santiago de Compostela, featuring plenary talks given by a large and egalitarian list of excellent researchers in optics and photonics, oral contributions from early-career researchers, industrial and women in science roundtables, and outreach activities organized in cooperation with two science museums. This was possible thanks to the hard work of student groups supported by scientific societies and universities. The face-to-face format was crucial in the creation of strong bonds among the young research community at I NW MYRO, which may lead to synergies between the future researchers in optics and photonics in the Iberian Peninsula.

We thank Optica, the European Physics Society (EPS), the Real Sociedad Española de Física (RSEF) and Universidade de Santiago de Compostela (USC) for their economical support to the I NW MYRO. We also thank Universidad de Salamanca (USAL) and our partners SEDOPTICA, BFlow SL, AIMEN and Menlo Systems GmbH.

References

1. C. D. Rodríguez-Fernández, B. Carnero, et al. *EPJ Web Conf.* **255,** 12003 (2021)
2. A. García-Cabrera, V. W. Segundo-Staels, et al. *XIII Reunión Nacional de Óptica.* Contribution #201 (2021)

Multi-well platform manufacturing combining stereolithography and pulsed laser ablation for cellular studies

Bastián Carnero[1,2], *Carmen* Bao-Varela[1], *Ana Isabel* Gómez-Varela[1], *Ezequiel* Alvarez[3,4,5] and *María Teresa* Flores-Arias[1,*]

[1]Photonics4Life research group, Departmento de Física Aplicada, Facultade de Física and Instituto de Materiais (iMATUS), Universidade de Santiago de Compostela, Campus Vida, E-15782 Santiago de Compostela, Spain
[2]BFlow S.L., Edificio Emprendia, Santiago de Compostela, Spain
[3]Departamento de Farmacoloxia, Farmacia e Tecnoloxia Farmacéutica. Universidade de Santiago de Compostela, 15782, Santiago de Compostela, Spain.
[4]Instituto de Investigación Sanitaria de Santiago de Compostela (IDIS); Fundación IDIS, SERGAS; 15706 Santiago de Compostela, Spain
[5]CIBERCV, Madrid, Spain

Abstract. Novel cell culture platforms, with more physiological surface roughness, require different technologies capable of precisely micropattern substrates. 3D printing offers a considerable accuracy and user-friendly procedures. For its part, pulsed laser ablation proves to be a versatile technology to perform detailed surface micropatterning. In this work, both technologies were combined to easily fabricate a versatile PDMS multi-well platform for performing cellular studies on a micropatterned biocompatible surface.

1 Introduction

Surface micropatterning of culture platforms is of great interest to obtain novel properties. Recent studies have found out how different micropatterns can offer antibacterial properties, improve hydrophobicity for self-cleaning, and even control cell attachment [1]. In the latter case, superficial microstructures help to obtain more physiological platforms than the normally used in cellular studies (with polished surfaces), capable of mimic the roughness of tissues or bones [2] to perform cell cultures.

Optical technologies have been traditionally used to micropattern different materials and until now, the most used optical technology has been photolithography, characterised by complex time-consuming protocols. This technology allows the precise micropattern of photoresists in 2D but produces highly polluting chemical waste that can be harmful for the cell cultures. Therefore, in recent years, the rapid development in laser applications has favoured the appearance of different technologies that can be used to microstructure diverse substrates in a more user-friendly way. This is the case of Pulsed Laser Ablation (PLA), that allows to achieve outstanding resolutions (in the micron range) when micropatterning 2D surfaces. On the other hand, 3D printers based on the selective Stereolithography (SLA) of liquid resins have facilitated the production of 3D objects with great structural complexity (in the millimetre range).

In this work, a multi-well platform is manufactured combining both technologies in each of their most suitable dimensional range. This platform can be easily used to perform cell attachment studies mimicking different surface roughness and patterns.

2 Materials and Methods

A Formlabs Form 3B SLA printer was used in this work to manufacture the multi-well master. SLA printers have a tank that contains the liquid printing resin coupled with a laser that performs a selectively layer-by-layer photopolymerization. Commercial Model resin from Formlabs was selected as printing resin given the high precision it offers, its good response to laser ablation, and its good performance when replicating polymers.

PLA of the 3D printed substrate was performed using a Rofin Nd:YVO_4 laser. This laser operates in Q-switch regime with 20 ns pulse duration and 1064 nm fundamental wavelength. The laser setup was combined with a galvanometer system and a flat field lens (f=160 mm) that focus the laser with an homogeneous energy distribution in a 120x120 mm^2 area.

Polydimethylsiloxane (PDMS) was selected as soft lithography polymer for replication of the substrate given its biocompatibility and optical transparency to visible and UV light. The polymer was prepared from Sylgard 184 elastomer by mixing the monomer and the curing agent in a ratio 10:1. Once the mixture covered the substrate, it was cured at 60 ºC during 12h in an oven.

3 Results and Discussion

3.1 Achievable micropatterns

By PLA, it is possible to create a series of micropatterns in the resin, to be later replicated by PDMS soft lithography. Several ablation studies were performed

* Corresponding author: maite.flores@usc.es

on Model resin surface to find the proper laser parameters and the achievable geometries, using different arrays of straight lines that were designed using a 2D Computer-Aided Design (CAD) software and micropatterned on the surface of the resin. Then, 700x700 μm^2 confocal images were taken to perform surface inspection (Fig. 1).

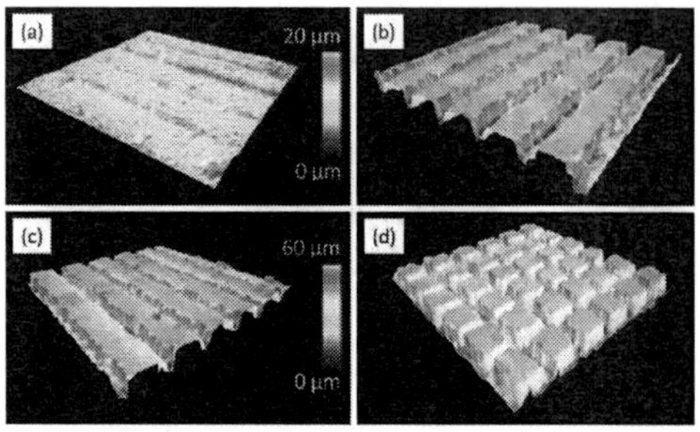

Fig. 1. Confocal images of a) unpatterned surface (control), b) and c) straight channels and d) matrix of square pillars micropatterned on the resin surface by PLA. The scale bar is the same in b), c) and d).

In the first place, a surface was left unpatterned (Fig. 1a) to serve as a control, as is usual in cellular assays. A smooth waviness of the surface was observed, with an amplitude of approximately 5 µm. This periodic behaviour has its origin in the layer-by-layer photo-polymerization working principle of stereolithography, and will be present on all printed surfaces, with the same orientation and period.

In the second place, regular straight channels of 70 µm width, 45 µm of height and 120 µm of period can be manufactured using an average power of 4.5 W, a repetition rate of 10 kHz, a velocity scan of 50 mm/s and 3 repetitions. Two orientations were selected: perpendicular (Fig. 1b) and parallel (Fig. 1c) to the background waviness of the resin, with the intention of analysing its influence on the surface finish. No obvious differences were observed, which makes sense given the large difference in magnitude between the two.

Furthermore, a matrix of rectangular pillars (Fig. 1d) can be obtained by crossing the previous lines and using the same parameters. The same dimensions were obtained.

3.2 Multi-well platform manufacturing

Once the proper laser parameters are found, manufacturing process starts with the SLA 3D printing of the multi-well master. This master consists of 35 mm diameter disk that featured four cylindrical elevations of 8 mm in diameter and 2 mm in height (Fig. 2a). A circular wall was added around the master to avoid replication polymer losses. In the next step, PLA of the upper part of the elevations was performed, using the previous introduced parameters and geometries (keeping a unpatterned well as control). The PDMS was then deposited over the master structures forming a sheet of 3 mm (Fig 2b) and introduced in an oven. Next, the PDMS multi-well platform was peeled off from the master, replicating the four different patterns made by PLA (Fig. 2c). Finally, the platform was placed at the bottom of a 6 well plate (Fig. 2d) commonly used to perform cell attachment studies.

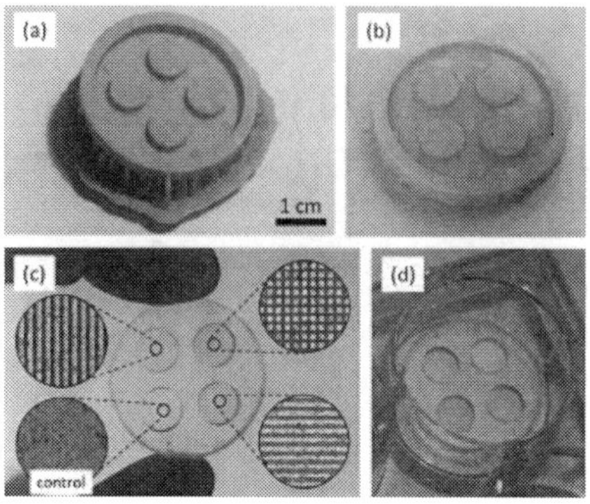

Fig. 2. a) Multi-well master 3D printed in Model resin. b) Soft lithography process in which the master is filled with liquid PDMS. c) PDMS multi-well platform with different replicated micropatterns in the bottom of each well. d) Multi-well platform coupled with a common 6 well plate ready to perform cellular studies.

4 Conclusions

Novel cell culture platforms, with more physiological surface roughness, require different technologies capable of manufacturing very precise structures reliably and accurately. 3D printing has the potential to revolutionize these areas, abridging manufacturing processes through simple procedures and very low waste production. For its part, laser ablation is a versatile and much easier technique than photolithography for making detailed surface micropatterns in the micron range.

In this work, both technologies were combined (taking advantage of each in the range of dimensions in which they show most suitable results) to easily fabricate a versatile PDMS multi-well platform for performing cellular studies on a micropatterned biocompatible surface. Some geometries were presented, but the proposed manufacturing process allows to obtain a wide range of different configurations in a simple way, depending only on the needs of the user.

This work has been supported by contracts: AEI RTI2018-097063-B-100 and Consellería de Educación Xunta de Galicia/FEDER ED431B 2020/29. B. Carnero acknowledges GAIN/Xunta de Galicia for contract 11-IN606D-2021-2604925. Ana I. Gómez-Varela acknowledges Consellería de Cultura, Educación e Ordenación Universitaria for ED481D-2021-019 postdoctoral fellowship.

References

1. S. Yu, D. Liu, T. Wang, Y. Z. Lee, J. C. N. Wong, X. Song, J Biomed Mater Res B Appl Biomater, **109.10**, 1525-1533 (2021)
2. A. Daskalova, B. Ostrowska, A. Zhelyazkova, W. Święszkowski, A. Trifonov, H. Declercq, C. Nathala, K. Szlazak, M. Lojkowski, W. Husinsky, I. Buchvarov, Applied Physics A, **124.6**, 1-15 (2018)

Free convection along pumped active mirror amplifying medium and its impact on laser wave propagation.

Hugo Chesneau[1,*], *and Sébastien* Montant[1]

[1] Commissariat à l'Energie Atomique et aux Energies Alternatives, Centre d'Etudes Scientifiques et Techniques d'Aquitaine, F-33116 Le Barp, France.

Abstract. Using COMSOL Multiphysics, we simulate free convection along a pumped active mirror amplifying medium. We study the induced boundary layers and how it affects the wave front propagation. To comfort our simulations, we set up a Mach-Zehnder interferometer to characterize and measure the local variation of air refractive index and temperature.

1 Introduction

Active mirrors amplification scheme is a solution to enhance laser beam mean power with a lower number of passes compared to thin disk amplifier [1]. The rise of mean power increases the thermal load. Hence, it induces thermal effects issues like free thermal convection in air.

To quantify these effects, we perform numerical and experimental investigations. Numerically, we simulate the heat transfer within the pumped active mirror (which can be referred to as the disk in the text) and in the air and we observe the induced thermal boundary layer along it. In order to validate our numerical model, we set up a Mach-Zehnder interferometer (MZI) to observe and measure local variation of both air refractive index and temperature. From the resulting interferograms we compare numerical and experimental results. Finally, we use an Algebraic Reconstruction Technique (ART) [2] to compute the 3D refractive index field and temperature.

In what follows, we describe the physical and numerical system, the experimental system and the chosen method for the 3D reconstruction.

2 Physical and numerical system

We describe the numerical system on an orthonormal Cartesian basis with the unit vectors $(\mathbf{e}_x, \mathbf{e}_y, \mathbf{e}_z)$. The fluid flow is considered as incompressible, and we make the assumption of the Boussinesq approximation; only the density of the fluid changes with temperature. To have a universal numerical model, we used only dimensionless variables. We take the following characteristic variables. D the active mirror diameter, $u_T = (g\beta q_w D^2/k)^{1/2}$ and $p_T = \rho_0 u_T^2$ respectively the thermal velocity and pressure. $\Theta = (T - T_a)k/(q_w D)$ a relative temperature to the ambient one, $Q_c = q_w/D$ a source term, and the dimensionless Grashof and Prandtl numbers which are denoted $\mathrm{Gr} = g\beta D^4 q_w/(k\nu^2)$ and $\mathrm{Pr} = \rho_0 \nu C_p/k$. With $\mathbf{g} = g\mathbf{e}_x$ the gravity, β the thermal expansion coefficient, ν the kinematic viscosity, k the thermal conductivity, ρ_0 the density, C_p the isobaric heat capacity and q_w a surface heat flux. All these quantities are related to the air at room temperature (20°C). The dimensionless governing system of equations is:

$$\nabla \cdot \mathbf{u} = 0 \quad (1)$$

$$(\mathbf{u} \cdot \nabla)\mathbf{u} = -\nabla p + \Theta \mathbf{e}_x + \mathrm{Gr}^{-\frac{1}{2}} \nabla^2 \mathbf{u} \quad (2)$$

$$(\mathbf{u} \cdot \nabla)\Theta = \mathrm{Gr}^{-\frac{1}{2}} \mathrm{Pr}^{-1} \nabla^2 \Theta + \mathrm{Q_c} \quad (3)$$

These equations are respectively, the mass, the momentum and the energy conservation. In our experimental conditions $\mathrm{Gr} \approx 10^6$ and $\mathrm{Pr} = 0.7$.

3 Experimental system

We set up a MZI to probe the heated air in front of the disk as shown in Figure 1.

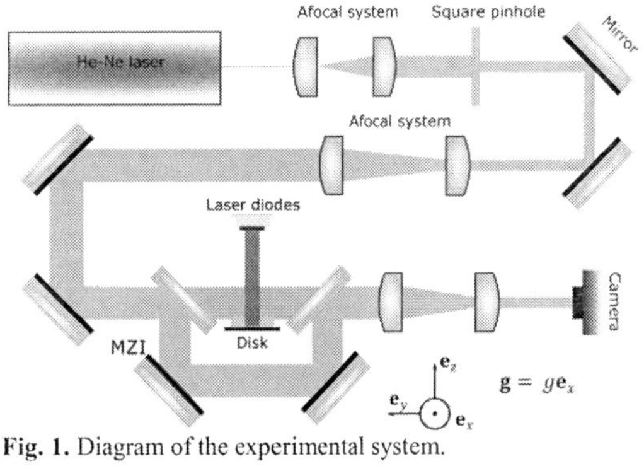

Fig. 1. Diagram of the experimental system.

To capture the whole behaviour of the boundary layers, we use a square-shaped He-Ne laser beam. The incident gaussian beam is magnified sixteen times

---

* Corresponding author: hugo.chesneau@cea.fr

through a first afocal system with two plano-convex (PC) lenses. Then, it is shaped with a square pinhole, expanded three times with another set of PC lenses and imaged in front of the active mirror in one of the interferometer arm. Finally, we use a last pair of PC lenses to decrease the size of the beam by a factor three and we image it on a camera. The active mirror is pumped with laser diodes at $\lambda_d \sim 800\text{nm}$, delivering $E_d = 1.6\text{J}$ at a frequency of $f_d = 4\text{Hz}$ with $\tau_d = 400\mu\text{s}$ pulses. We depict on Figure 2 the typical results we can get at room temperature and with the pumped active mirror. Figure 2 (b) clearly highlights the fringe shift resulting from the thermal boundary layer in front of disk and the plume above it.

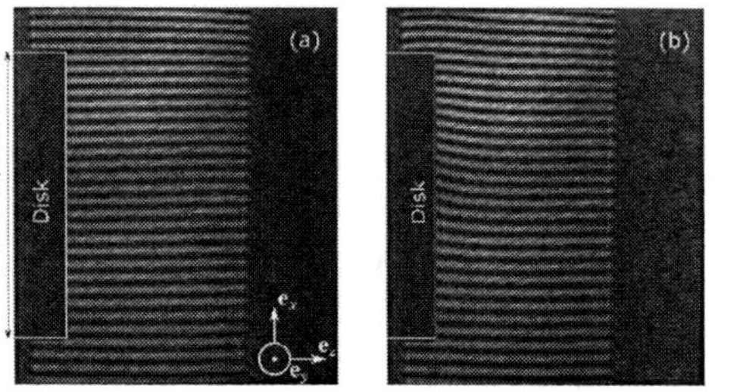

Fig. 2. Snapshots of the interference patterns obtained with the MZI (a) at room temperature and (b) with the pumped disk.

Once we have both interference patterns, we are able through Fourier analysis to compute the phase shift induced by the refractive index gradient [3]. As a thumbnail sketch, it consists in (i) Fourier transform the images, (ii) selecting and filtering the spatial frequency of the interferences and (iii) invert Fourier transform the filtered spectrum to get the wrapped phase of the interference pattern. Finally, (iv) we unwrapped the latter [4] to compute the phase shift as displayed in Figure 3 (b).

Thereafter, we perform the same experiment for different angles with the $\mathbf{e}_y$ axis to reconstruct the refractive index field with a Simultaneous Iterative Reconstruction Technique (SIRT)[5], which is a kind of ART. The method can be sum up in a few steps to. First, we write the phase shift induced by the integration of the spatially varying refractive index as a discrete ray-sum equation; each pixel of the camera is considered as the result of a single beam. Meaning that, the continuous equation becomes:

$$\varphi_{ij} = \sum_{ij} \frac{2\pi}{\lambda}\left(n_{ij} - n_0\right) w_{ijnm} = \sum_{ij} f_{ij} w_{ijnm} \quad (5)$$

With λ the laser beam wavelength, L the propagation distance and n_0 the unperturbed refractive index. The indexes (i,j) represent the coordinates of the grid on which we reconstruct the field, n is for a given angle of projection and the index m denotes each ray (pixel). f_{ij} is called the perturbation function and w_{ijnm} is a weight matrix which is the distance covered by the mth beam for the nth projection in each (i,j) cell.

$$f_{ij}^{(q+1)} = f_{ij}^{(q)} + \alpha \frac{\sum_{mn} w_{ijmn} \times \frac{\varphi_{mn} - \sum_{ij} w_{ijmn} f_{ij}^{(q)}}{\sum_{ij} w_{ijmn}}}{\sum_{mn} w_{ijmn}} \quad (6)$$

The letter (q) denotes the qth iteration. α is a relaxation coefficient for which the iterative algorithm is stable when $0 < \alpha \leq 2$.

4 Preliminary experimental results and conclusion

We show on Figure 3 a first comparison between numerical (a) and experimental (b) phase shift. We retrieve similar boundary layer shape and the same amplitude of phase shift in both cases. One could also note that we likely did not pump the active mirror at his center regarding the position of the minimum experimental phase shift.

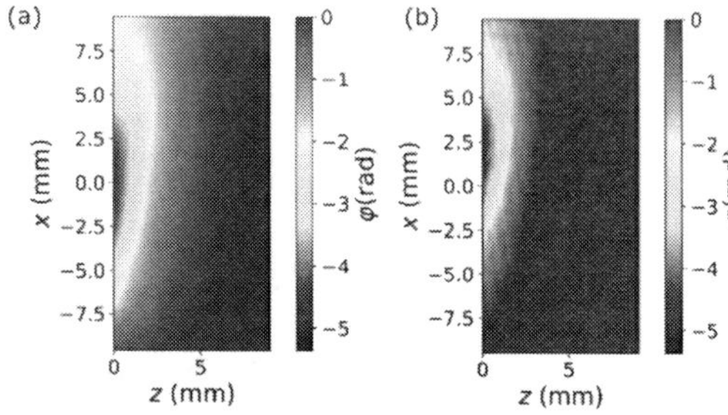

Fig. 3. Numerical (a) and experimental (b) calculated phase shift for a probe beam propagating along $\mathbf{e}_y$ axis.

5 Conclusion

We fully characterised numerically the boundary layers and the refractive index gradient induced by the active mirror pumping. We successfully compared the resulting phase shift computed from numerical and experimental results. Finally, we plan to perform more experiments with different viewing angles, to be able to reconstruct the 3D refractive index field.

This work was partly supported by the Conseil Régional de Nouvelle Aquitaine and the European Union through the project LEAP - Axes B et C (contracts CPER #16004205 and #DEE21-04-2019-5131820 and ERDF #2663710 and #3404410).

References

1. D. C. Brown, J. H. Kelly, J. A. Abate, IEEE J. Quantum Elect. **QE-17**, 9, 1981.
2. A. H. Andersen, A. C. Kak, Ultrasonic Imaging **6**, pp. 81-94, (1984)
3. D. J. Bone, H.-A. Bachor, R. J. Sanderman, Appl. Opt. **25**, 1653, (1986)
4. M. A. Herraez, D. R. Burton, M. J. Lalor, M. A. Gdeisat, Appl. Opt., **41**, 35, 2002.
5. G. Golovin, S. Benerjee, J. Zhang, S. Chen, C. Liu, B. Zhao, J. Mills, K. Brown, C. Ptersen, D. Umstadter, Appl. Opt. **84**, 11, (2015)

Convolutional neural network optimisation to enhance ESPI fringe visibility

J.M. Crespo[1,*], and V. Moreno[1]

[1] QMatterPhotonics Research Group. Optics Area. Department of Applied Physics. Faculty of Physics / Faculty of Optics and Optometry. University of Santiago de Compostela. E-15782 Santiago de Compostela. Galicia. Spain

Abstract. The use of convolutional neuronal networks (CNN) for the treatment of interferometric fringes has been introduced in recent years. In this paper, we optimize and build a CNN model, based U-NET architecture, to maximize its performance processing electronic speckle interferometry fringes (ESPI).

1 Introduction

Over the last few years, continuous advances in artificial intelligence techniques and specialized hardware for their use have eased their application to different areas of research, and more specifically to the application of these techniques to improvements in image visualization.
Its applications to denoise interferometric images to clean the interference fringes has been broadly studied, being these techniques a disruptive tool to use and enhance the qualitative results of interference data.

Specifically, the convolutional neural networks using U-NET architectures with the appropriate computational framework, are powerful and easy to use tools to enhance visibility of interference fringes. However, some initial decisions on the neural network design must be taken and the final network architecture must be designed to fit the problem to solve. Namely the design of the training dataset and the hyperparameters of the neural network: the layers and the kernels to be used.

1.1 Speckle Pattern Interferometry

Electronic speckle pattern interferometry (ESPI) is a technique used to measure from sub-micron to tenth of microns displacements in optically rough surfaces assessing the overlap of two speckle patterns. Since its initial uses in 1971, noise, and low visibility are the major drawbacks for this technique and the artificial intelligence has been proved as an effective technique to enhance and mitigate those major drawbacks.

The ESPI fringe pattern is obtained from a pair of images, I and I', taken from a specimen under study before and after deformation. The fringe pattern is obtained subtracting those images and the resulting intensity pattern is according to the formula [1]

$$|I - I'| = \left|4\sqrt{I_o I_r}\sin\left(\phi_o - \phi_r + \frac{\varphi}{2}\right)\sin\left(\frac{\varphi}{2}\right)\right| \quad (1)$$

with I_o, I_r, ϕ_o, and ϕ_r are the amplitude and phase of the object and reference speckle fields, and φ is the optical phase difference induced by the deformation in the surface of the specimen under study. The minimum value for the resulting intensity occurs due to the term $\left|\sin\left(\frac{\varphi}{2}\right)\right|$ where $\varphi = n2\pi$ and the maximum value for this term at $\varphi = (n+1)2\pi$, revealing the ESPI fringe patterns which can be used to measure sub-micron displacements in the structure.

1.2 U-Net Architecture

Our proposed design to denoise ESPI fringe fields, is the U-NET architecture [2], a powerful and easy to use type of convolutional neural networks, broadly used in interferometric denoise [3], wrapped phase denoise [3] and SAR image denoise

The U-NET is a convolutional encoder – decoder with internal connections between the encoding and decoding paths, where the noisy image to be cleaned is introduced in the encoder path where 2D-Convolutional (Conv2d) and maxpooling layers are combined to reduce the spatial resolution of the input image capturing the image detail (features). Once the image has been fully encoded, there is a decoder path reversing the encoding operations to get a cleaned image.

2 Network optimisation

The optimisation of the network design is reduced to the selection of the number of layers in the network and the size of the kernel to be used in the Conv2D operations accordingly to the problem to solve. The selection of the hyperparameters will be based in a small and quick training using a small training dataset maximising the

https://doi.org/10.1051/epjconf/202226613007

indicator used to measure the network performance. In our case, the Structural Similarity Index (SSMI) [4], commonly used indicator for the quantitative assessment of the perceived increase of quality across the reconstructed image.

2.1 Training dataset

To build the training dataset to be used for the hyperparameter selection, we use the formula (1) where for the estimation of φ we use random Zernike polynomials to build a random phase, $\varphi = \sum_0^n c_i Z_i$, where c_i and Z_i are a random coefficient and the i-Zernike polynomial respectively and n is the order of the Zernike expansion.

Using this approach, any displacement field can be simulated, setting the proper interval for c_i and moreover, selecting the order of the Zernike expansion will control the complexity of the simulated deformation.

To simulate the interfering true speckle fields, we use the method described by J. Goodman [1]. Following this procedure, we produced 5000 pair of images $[|I - I'|, \varphi]$ to be used in the selection procedure.

2.2 Hyperparameter selection

To select the depth and kernel size, we run several trainings with a 5000 pair of images in the training dataset, each one with different set of hyperparameters, levels in the encoding path varying from 3 to 6 and kernel size from [3x3] to [7x7] checking 9 potential combinations.

After the training procedure we run another test dataset composed of 1000 random pair of images to select the set of hyperparameters where the average SSMI index, comparing the output of the trained network and the ground true (our simulated φ), has its maximum value.

Table 1: SMMI value for the checked combinations of levels in the encoder path and kernel size

		KERNEL SIZE		
		3x3	5x5	7x7
Levels	4	0,896	**0,900**	0,801
	5	0,798	0,760	0,728
	6	0,880	0,859	0,760

For our specific case the selected outputs to define the final U-NET hyperparameters are a depth of 4 levels in the encoder path and a kernel size of 5x5.

3 Model Results

With the selected hyperparameters (kernel size = 5x5 and layers = 4), we build and train a U-NET using a training dataset composed of 15.000 pair of images $[|I - I'|, \varphi]$. The final training was implemented using ADAM optimisation [5] and MEAN_SQUARED_ERROR as the loss function. The training set was divided into training and validation (80% and 20%). The training process took 23 min using a GOOGLE COLLAB instance with GPU support, and it automatically stopped due to the early stop condition (3 epochs without improvement in the loss) after 115 epochs, reaching a loss value of 0.0084 on the validation set. The average SSMI index between the ground true (expected output) and the reconstructed image using the network was 0.899.

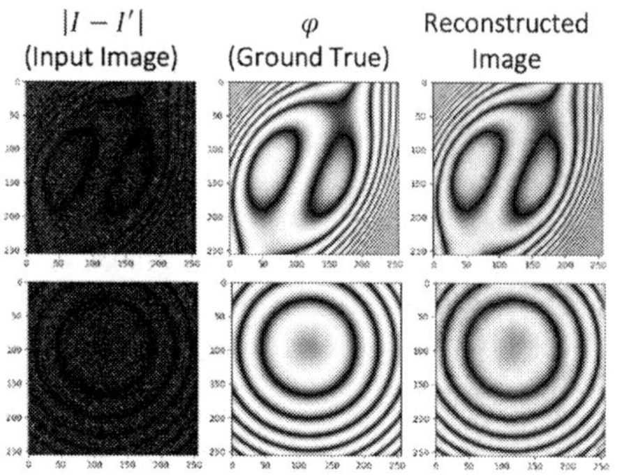

Fig. 1. Results of the Neural Network on a sample of Test Dataset

4 Conclusion

We built a specialised U-NET to process ESPI fringe images, optimising its architecture using an easy method to select the network hyperparameters.

The resulting network improves the perceived quality of the ESPI fringe field and U-NET networks are reliable for their use in speckle interferometry, and the use of synthetic generated training datasets eases the training process removing the need for specialised lab equipment for the design phase.

References

1. Joseph W. Goodman, *Speckle phenomena in optics*, SPIE (2020)
2. Ronneberger, O., Fischer, P., & Brox, T. *U-net: Convolutional networks for biomedical image segmentation*. International Conference on Medical image computing and computer-assisted intervention. 2015.
3. Zuo *et al.* Deep learning in optical metrology: a review. Light: Science & Applications (2022) 11:39
4. Zhou Wang, A. C. Bovik, H. R. Sheikh and E. P. Simoncelli. *Image quality assessment: from error visibility to structural similarity*. IEEE Transactions on Image Processing. 2004
5. Diederik P. Kigma, Jimmy BA. *Adam: A method for stochastic optimization*. 2014.

10 PW Peak Power Laser at the Extreme Light Infrastructure Nuclear Physics – status updates

Ioan Dancus[1*], *Gabriel V.* Cojocaru[1], *Robert* Schmelz[1], *Dan* Matei[1], *Lidia* Vasescu[1], *Dmitrii* Nistor[1], *Anda-Maria* Talposi[1,2], *Vicentiu* Iancu[1,2], *Gabriel P.* Bleotu[1,2,3], *Andrei* Naziru[1,2], *Alexandru* Lazar[1], *Alice* Dumitru[1], *Antonia* Toma[1], *Marius* Neagoe[1], *Stefan* Popa[1,2], *Saidbek* Norbaev[1], *Cristian* Alexe[1], *Andreea* Calin[1], *Costin* Enc[1], *Adrian* Toader[1], *Nicolae* Stan[1], *Mihai* Caragea[1], *Sorin* Moldoveanu[1], *Olivier* Chalus[4], *Christophe* Derycke[4], *Christophe* Radier[4], *Sandrine* Ricaud[4], *Vincent* Leroux[4], *Caroline* Richard[4], *François* Lureau[4], *Andrei* Baleanu[5], *Romeo* Banici[5], *Alexandru* Ailincutei[5], *Iulian* Moroianu[5], *Andrei* Gradinariu[5], *Constantin* Caldararu[5], *Cristian* Capiteanu[5], *Daniel* Ursescu[1,2], *Domenico* Doria[1], *Ovidiu* Tesileanu[1], *Takahisa* Jitsuno[1], *Razvan* Dabu[1], *Kazuo A* Tanaka[1], *Sydney* Gales[1], *Calin Alexandru* Ur[1]

[1] Extreme Light Infrastructure - Nuclear Physics, IFIN-HH, Street Reactorului 30, 077125 Bucharest Magurele, Romania
[2] Physics Doctoral School, Bucharest University, Magurele, Ilfov, 077125, Romania
[3] LULI-CNRS, CEA, Université Sorbonne, École Polytechnique, Institut Polytechnique de Paris, F-91128 Palaiseau Cedex, France
[4] Thales LAS France, 2 Avenue Gay Lussac, 78990 Élancourt, France
[5] Thales Systems Romania, Soseaua Orhideelor 15a, 060071 Bucuresti, Romania

Abstract. We have shown, for the first time in the world, the production of 10 PW peak power laser pulses and their propagation to an experimental area at the Extreme Light Infrastructure - Nuclear Physics (ELI-NP). We are also steadily running the laser system for experimental campaigns, increasing the output power levels delivered for experiments and fine-tuning the parameters of the laser pulses, the operational procedures, and the operational teams. During our presentation, we will show the laser developments at ELI-NP emphasizing the 10 PW peak power demonstrations and the latest results for the HPLS beam delivery.

1. Introduction

Extreme Light Infrastructure – Nuclear Physics (ELI-NP) is a new European Centre for Scientific Research built in Bucharest-Magurele, Romania [1, 2, 3]. One of the scientific tools of ELI-NP is the High-Power Laser System (HPLS) capable of delivering 10 PW laser pulses. The ELI-NP HPLS is a Ti: Sapphire, hybrid CPA, system with six outputs: two 0.1 PW outputs at 10 Hz repetition rate, two 1PW outputs at 1 Hz and two 10 PW outputs at 1 shot per minute. A detailed description of the HPLS is presented in the reference [4].

2. 10 PW peak power laser pulses demonstration at ELI-NP

Laser pulses with peak power above 10 PW were generated for the first time using the Extreme Light Infrastructure – Nuclear Physics HPLS during two demonstration campaigns. The generated 10 PW pulses were transported in vacuum using the Laser Beam Transport System (LBTS) to the 10 PW experimental areas.

The pulses were routed towards different parts of the 10 PW experimental areas during each of the two experimental campaigns.

To qualify the spatial parameters of the laser pulses at the end of the propagation line a dedicated diagnostic bench was deployed.

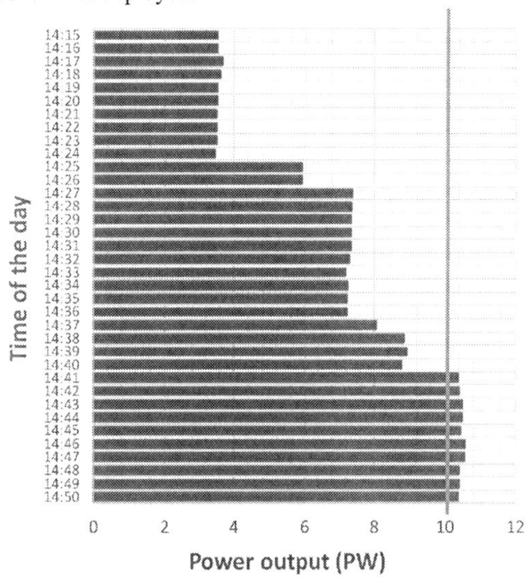

Figure 1. 10 PW shot sequence during one of the demonstrations at ELI-NP, showing the 1 shot per minute repetition rate.

The full aperture pulses are demagnified on this diagnostic bench, and the near field and wavefront are

* Corresponding author: ioan.dancus@eli-np.ro

https://doi.org/10.1051/epjconf/202226613008

measured. The measurement is performed at full HPLS amplification with the energy properly attenuated before the optical compressor. This measurement ensured us that the beam profile and wavefront are proper for the propagation of the high-power pulses. The HPLS diagnostic bench, described in [4], was used for pulse duration measurement and optimisation.

For full-power beam propagation (non-attenuated pulses), a beam dump, produced by Gentec EO, was installed at the end of the LBTS during the firing of the peak power pulses.

Sequences of shots were fired, gradually increasing the pump energy of the last amplifier until reaching the required level for the 10 PW output. Such a sequence of shots can be observed in Figure 1.

After the 10 PW pulse propagation, all the optics were inspected. No damage was observed using visual and computer-aided inspection.

A beam with low wavefront error (74 nm RMS) was propagated to the experimental areas (Figure 2). Based on propagation simulations we estimate that, when tightly focused with the parabolic mirrors that are under implementation, having a focal distance of 1500 mm, this peak power can produce irradiances in the range of 10^{22} – 10^{23} W/cm^2 [5].

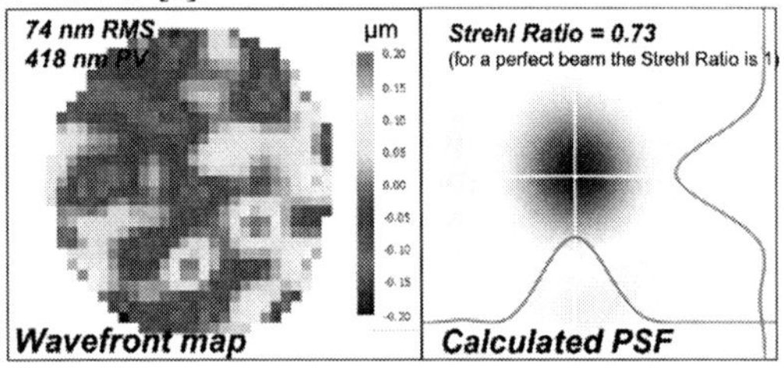

Figure 2. Wavefront map and calculated PSF at the output of the LBTS, in the experimental area, showing a good focusability of the laser pulses, with a calculated Strehl Ratio of 0.73.

3. Beam delivery

The HPLS is used for the Experimental Commissioning Program since January 2020. During this period the laser system and related operational procedure are tuned to deliver the proper parameters on targets and the equipment is prepared to be used for the coming user campaigns.

During the second half of 2021, we delivered 1192 pulses from the 1 PW output on target from which more than 700 had an energy of more than 20 J (Figure 3). A good day on a gas target in the 1 PW experimental area were delivered as shown in Figure 4.

In our presentation, we will include updated statistical data related to the HPLS beam delivery.

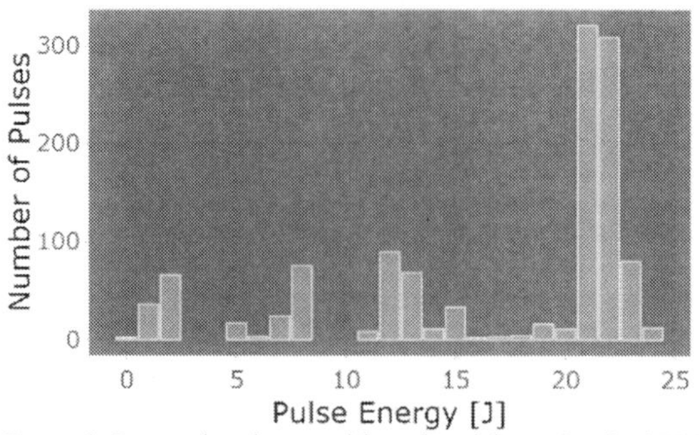

Figure 3. Energy distribution of the pulses delivered in the 1 PW experimental area (reproduction from [6]).

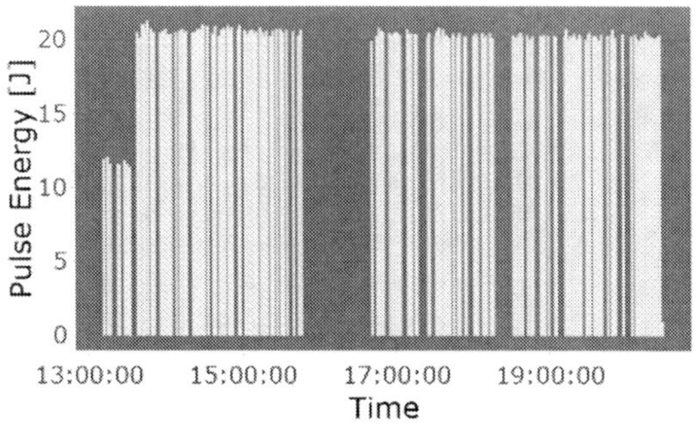

Figure 4. HPLS pulses delivered on a gas target in the 1 PW experimental area (reproduction from [6]).

Extreme Light Infrastructure Nuclear Physics (ELI-NP) Phase II, is a project co-financed by the Romanian Government and the European Union through the European Regional Development Fund and the Competitiveness Operational Programme (1/07.07.2016, COP, ID 1334). This work was supported by the contract sponsored by the Romanian Ministry of Research and Innovation: PN 19 06 01 05.

References

1. N.V. Zamfir, Rom. Rep. Phys., **68**(Supplement), S3–S4 (2016)
2. S. Gales, et al., Rep. Prog. Phys., **81**, 094301 (2018)
3. K. A. Tanaka, et al., MRE, **5**, 024402 (2020)
4. F. Lureau, et al., High Power Laser Sci. Eng., **8**, E43 (2020)
5. I. Dancus, et al., Frontiers in Optics / Laser Science, (Optical Society of America, 2020), p. LW5G.3
6. I. Dancus, et al., ELI-NP 2020-2021 Annual report, in press

https://doi.org/10.1051/epjconf/202226613009

Secure key distribution using an ultra-long fiber laser with bi-directional EDFA

Beatriz Soares[1*], *Ariel* Guerreiro[1,2], and *Orlando* Frazão[1]

[1]INESC TEC, Institute for Systems and Computer Engineering, Technology and Science, Rua do Campo Alegre, 687, 4150-179 Porto, Portugal
[2]Faculdade de Ciências da Universidade do Porto, Rua do Campo Alegre, 687, 4150-179 Porto, Portugal

Abstract. In this paper we describe the implementation of a secure key distribution system based on an ultra-long fiber laser with a bi-directional erbium doped fiber amplifier. The resilience of the system was tested against passive attacks from an eavesdropper, having been observed a similarity in spectrum for both secure configurations of the system.

1 Introduction

Many encryption protocols require the transmission of a secret key between two parties before communication between them can take place. The distribution of this key constitutes one of the weakest links in this type of communication system and is the driving force for the development of unconditionally secure key distribution schemes based on fundamental properties of quantum mechanics. With this in mind, a system based on a ultra-long fiber laser (UFL) that utilizes standard fiber optic components was proposed by Scheuer *et al.* [1]. This scheme, unlike the ideal implementation of QKD, is not unconditionally secure, relying instead on the technological difficulty of an eavesdropper to get access to the shared key. Such unconditional security has not, however, been a necessary pre-requisite for many encryption schemes, such as public key encoding schemes, which rely on the computational difficulty on part of the eavesdropper, rather than an absolute security proof.

In this paper we first discuss the principle of operation of the UFL key distribution system, next we present and analyse the results obtained for our own experimental implementation of the system, and finally, we discuss possible improvements and vulnerabilities of the protocol, both in general and to our setup, in particular.

2 Principle of operation

The setup used for the UFL key distribution system (KDS) is shown in figure 1. The scheme consists of a long erbium doped fiber laser that connects two users, Alice and Bob, positioned at opposing sides of the laser. Each user possesses an identical mirror which can be selected to have its peak reflectivity at two different frequencies, which can be assigned as f_0 and f_1. For the exchange of a single key bit, Alice and Bob choose, randomly and independently, one of these mirror states to reflect at. If both of them make the same choice of mirror state, there will be enough gain in the cavity to surpass the lasing threshold and a clear signal at either f_0 or f_1 will form, thus giving a potential eavesdropper (Eve) an easy access to the arrangement of both mirrors. However, if Alice and Bob choose different mirrors states, only a small signal at $f_c = 1/2\,(f_0 + f_1)$ will develop, and Eve will not be able to easily determine the exact configuration of mirror states used, only that Alice's and Bob's differ from one another. Thus, it can be assigned, for example, a bit value of '0' to the configuration of (Alice: f_0; Bob: f_1) and a bit value of '1' to the configuration of (Alice: f_1; Bob: f_0), which will allow Alice and Bob, knowing their own mirror state, to deduce the others choice, while preventing Eve getting access to the exchanged key bit. This protocol for key distribution using UFL is summed up in table 1.

Table 1. Protocol for a key distribution system using an UFL. Only when Alice and Bob choose different mirror states are the bits kept in order to obtain the sift-key.

Alice's mirror	0	0	1	1
Bob's mirror	0	1	1	0
UFL signal				
Sift-key	-	0	-	1

3 Experimental Implementation

3.1 Experimental Setup

The experimental setup is shown in figure 1 (a).

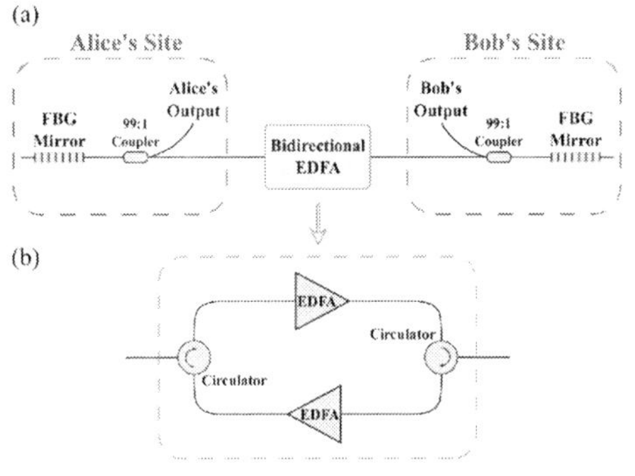

Fig. 1. (a) Lab setup and (b) internal schematic of the bidirectional EDFA utilized.

* Corresponding author: beatrizgsoares97@gmail.com

 https://doi.org/10.1051/epjconf/202226613009

It consists of two separate users, Alice and Bob, each with a fiber Bragg grating (FBG) mounted on a translation stage, allowing for the manual tuning of their peak reflectivity wavelengths by application of mechanical tension. Each user was further equipped with a 99:1 coupler in order to perform measurements on the cavity's signal. The gain of the system was provided by a customized bidirectional EDFA, whose internal schematic is depicted in figure 1 (b).

It consists of two circulators and two conventional EDFAs, i.e. an erbium-doped fiber and a co-directional pump, with each pump being able to be independently controlled by separate voltage sources. To perform the measurements required for the experiment, an optical spectrum analyser (OSA) model "AQ6370D" with wavelength range of 600 nm to 1700 nm from Yokogawa was used.

3.2 Experimental Results

The obtained spectrum of the four possible states of the UFL (read in Alice's output) are depicted in figure 3. The pump powers to the EDFA were chosen so the non-secure states were obtained near the lasing threshold. While the spectra for the non-secure states presents distinct well defined lasing peaks ($\lambda_{0,0}$=1561.098 nm at -17.703 dBm and($\lambda_{1,1}$=1561.484 nm at -18.280 dBm), the spectra for the secure states, (1,0) and (0,1), resemble optical noise and are thus difficult to distinguish. As expected, Eve would not find it technologically easy to access Alice's and Bob's mirror choice.

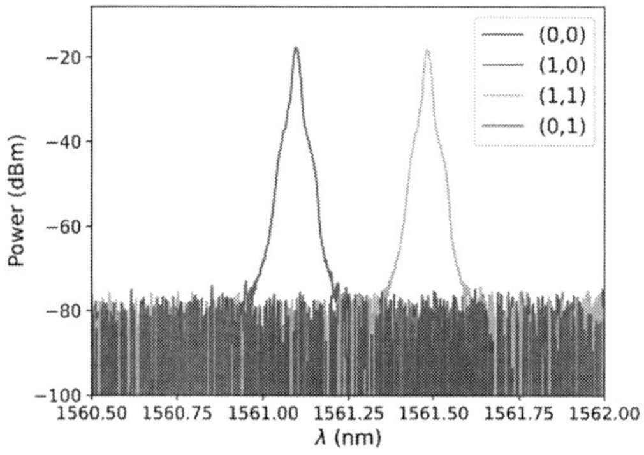

Fig. 2. Optical spectrum of the four states of the UFL-KDS.

4 Discussion

4.1 Improvements

The process of increasing the bit-rate of the setup, both to generally speed up the process and to give Eve less time to measure the cavity signal, is limited by the time required by Alice and Bob to identify the UFL state. That means the signal has to have enough time to pass through both mirrors and build up to a level capable of distinguishing a lasing from a non-lasing state. This translates to approximately 1.5 round trips [2], which results in a maximum frequency of $f = c/3dn$ bps, where d is the length of the link and n the refractive index of the fiber. This gives a maximum frequency in the order of 10^2 bps for a 100 km long link. Furthermore, the efficiency of a basic UFL protocol is only half of the total rate, as on average the number of secure bits sent will only be half of the total number. To address this relatively low effective bit-rate, modifications to the basic protocol have been proposed and numerically tested by Bar-Lev et al. [3], including using wavelength division multiplexing (WDM), or improving the 50% efficiency of the secure bit transmission.

4.2 Vulnerabilities

As previously discussed, even in its theoretical ideal case, the UFL protocol is still susceptible to attacks, relying instead on the technological difficulty of an eavesdropper to distinguish between different secure states. Some types of vulnerabilities not explored in this paper are those posed by active (as opposed to passive) attacks. Although they have been dismissed by most studies on the subject [1], [2], the paper by Garcia-Escartin et al. [4] suggests they need to be considered as important threats. The proposed attack is reliant on the ability of Eve to introduce a probing signal on the cavity, below the noise floor. She achieves this by spectrally broadening the signal using modulation, so that the total power is stretched out over the bandwidth of interest. While possible countermeasures were also proposed, to our knowledge, they have yet to be tested, and correspond to important future work in the study of the reliability of the UFL protocol.

5 Conclusions

Practical limitations of QKD have motivated searches for alternative, classically based solutions to key distribution. Accordingly, a system based on an ultra-long fiber laser has been proposed in previous studies. In this paper we discussed the basic operation principle of the UFL protocol, implemented our own UFL setup with a bidirectional EDFA and verified its feasibility as a secure key distribution system. Lastly, we discussed some possible improvements and vulnerabilities of the UFL protocol.

As the future of the UFL protocol goes and seeing most studies have been focused on passive attacks, it is important that the susceptibility to active attacks be fully tested to evaluate the viability of the protocol.

References

1. J. Scheuer, A. Yariv, Phys. Rev. Lett **97**, 1-4 (2006)
2. O. Kotlicki, J. Scheuer, Quantum Inf. Process **13**, 2293–2311 (2014)
3. D. Bar-Lev, J. Scheuer, Phys. Lett. Sect. A Gen. At. Solid State Phys. **373**, 4287–4296 (2009)
4. J. C. Garcia-Escartin, P. Chamorro-Posada, IEEE J. Sel. Top. Quantum Electron. **24**, 1–9 (2018)

Characterization of a lobster-eye type X-ray telescope

Thorsten Döhring[1,*], *Veronika* Stieglitz[2,3], *Rene* Hudec[3,4], *Iryna* Sapsai[1,5], *Peter* Friedrich[2], *Vadim* Burwitz[2]

[1] TH Aschaffenburg – University of Applied Science, Würzburger Str. 45, D-63743 Aschaffenburg, Germany
[2] Max-Planck-Institute for extraterrestrial Physics, Giessenbachstraße 1, D-85748 Garching, Germany
[3] Czech Technical University in Prague, Technicka 2, 166 27, Prague 6, Czech Republic
[4] Astronomical Institute, Czech Academy of Sciences, Fričova 298, 251 65 Ondřejov, Czech Republic
[5] Borys Grinchenko Kyiv University, 18/2 Bulvarno-Kudriavska Str., Kyiv, 04053, Ukraine

Abstract. Lobster-eye type X-ray telescopes use reflecting plano mirrors under grazing incidence and can observe a large field of view. As part of a Bavarian-Czech cooperation, two telescopes were build, equipped with mirrors coated with gold and iridium. Their X-ray characterization was carried out at the PANTER test facility, which simulates parallel starlight incident on the telescopes. The telescopes have an angular resolution of about 4 arc minutes in X-rays and a focal length of about 2 meters. The used X-ray mirrors reflect and focus visible light as well; their functionality in the optical regime was checked in laboratory tests. Now another test campaign will be carried out to examine the telescope resolution for real objects of the visible night sky and the imaging properties for star constellations.

1 Introduction

Astronomical X-ray optics were under development in the Aschaffenburg Competence Center ACCASI ("Aschaffenburg Competence Center for Astronomical and Space Instrumentation") since several years [1]. As part of a Bavarian-Czech cooperation between the Technical University of Aschaffenburg and the Technical University of Prague [2], two geometrically identical telescopes, called HORUS, were built (Fig.1).

Fig. 1. The two HORUS telescopes, installed next to each other in the vacuum test chamber of the PANTER test facility, are waiting for the "First Light Event". (Photo: MPE/PANTER)

One of these telescopes was equipped with conventional gold-coated mirrors, manufactured by the Czech project partners. The 34 X-ray mirrors of the second telescope use an innovative tri-layer coating system made of chromium and iridium, which was applied in the Aschaffenburg coating laboratory [3]. Both telescopes work according to the bionic principle of a reflecting lobster eye [4]. The optics works with two consecutive reflections on mutually perpendicular mirror surfaces. This enables a large field of view with many square degrees in diameter, which, however, comes at the price of reduced angular resolution.

2 The X-ray test facility PANTER

The extensive characterization of the two HORUS telescopes was carried out at the X-ray test facility PANTER of the Max Planck Institute for Extraterrestrial Physics. The measurements took place in a 12 meter long vacuum chamber of 3.5 meters in diameter, with the X-ray source being 120 meters away. An almost parallel X-ray beam reaches the test chamber through a tube, which is also evacuated and has a diameter of one meter. This experimental set-up simulates incident parallel radiation from cosmic X-ray sources. On Earth, the atmosphere absorbs such radiation, while it spreads unhindered through the vacuum of space or in the evacuated test facility. Usually, X-ray optics of large international satellite projects (e.g. in cooperation with ESA, NASA, IKI) are characterized in the PANTER test facility - but this time the two telescopes from Aschaffenburg and Prague are tested.

3 Successful start: "First light event"

On March 8th, 2021, the mirrors of the Aschaffenburg HORUS telescope saw the X-ray light for the first time, an event known as "First Light". In times of the Corona pandemic, unfortunately, the Aschaffenburg scientists could not be present on site; they were connected via video conference directly to the PANTER control room.

During the test campaign of 195.441 seconds duration (approx. 54 hours) 536 individual measurements were carried out at the two telescopes (Fig. 2), thereby photon energies and angles of incidence were varied. Hereby a

* Corresponding author: thorsten.doehring@th-ab.de

total number of 11,153,574 X-ray photons were individually detected, analysed and digitally processed by the sensitive CCD camera of the PANTER facility, using a prototype the eROSITA satellite camera [4].

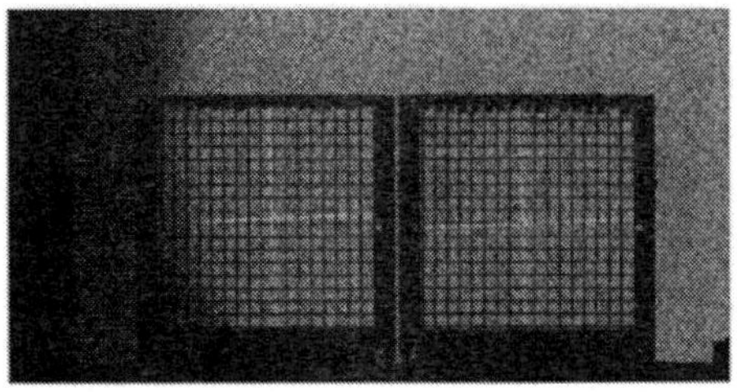

Fig. 2. The shadow of the two HORUS telescopes in X-ray illumination. (Photo: MPE/PANTER)

4 Mirror coatings are more effective

After evaluation of the measurement data, the two wide-angle X-ray telescopes have an angular resolution of around four arc minutes (Fig. 3) and a focal length of around two meters. In direct comparison with the gold-coated telescope, the new chromium-iridium tri-layer mirror coatings bring in some energy ranges the X-ray light up to 30% more effective on the focus point [3, 4]. The rich measurement datasets have been evaluated in detail, especially with regard to the imaging quality (on the optical axis and away from it), based on the analysis of individual focus images (Fig. 3).

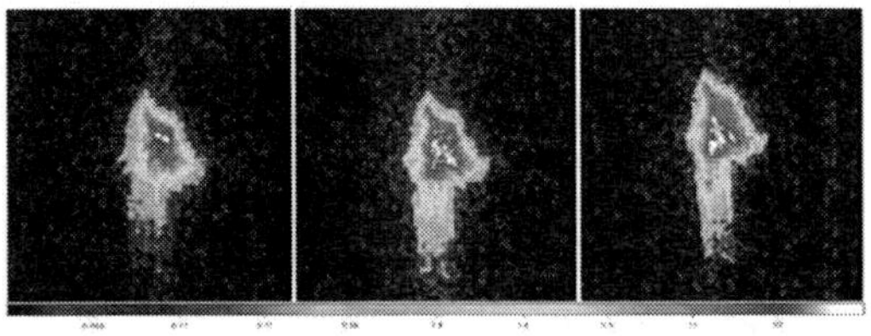

Fig. 3. The double reflection spots of the iridium coated telescope using a low energy continuum band. From left to right: Measurements at 1°, 1.6° and 2.2°. (Photo: MPE/PANTER)

5 Outlook

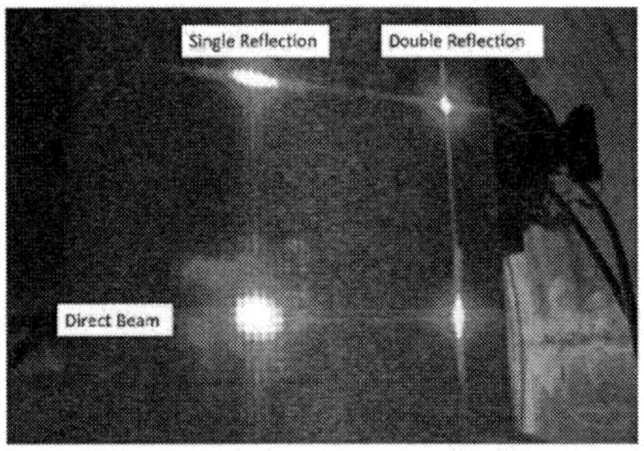

Fig. 4. Focal plane image with the laser alignment showing the location of the direct beam through the optic, two single reflection on the top and to the right side of the image, and the double reflection on the top right corner. Here the telescope optic was rotated by 2.3° in pitch and yaw. (Photo: MPE/PANTER)

As mentioned before, these Lobster eye X-ray mirrors reflect visible light in applicable quality as well. Therefore, visible light was used for a pre-alignment in standard atmosphere. An image of the laser pre-alignment inside the PANTER vacuum chamber is shown in Fig. 4. Now another test campaign is planned to examine the resolution for real objects of the visible night sky and the imaging properties of star constellations (Fig. 5). The measurements will take place at the MPE telescope in Garching and will be compared with parallel measurements using a classical astronomical telescope. The analysis of these data will provide valuable insights for the use of Lobster-Eye telescopes for intended satellite-based X-ray observations.

Fig. 5. Set-up for the night-sky measurements; here tested in the laboratory with a laser beam array. (Photo: TH Aschaffenburg)

Acknowledgement

This work was supported by an internal grant from the Czech Technical University under the number SGS21/120/OHK3/2T/13 and by several collaborative projects funded by the Bavarian-Czech Academic Agency (BTHA). We also thank the EU project H2020 AHEAD2020 at the Grant agreement 871158. The Cr-Ir-Cr mirror layers of the HORUS telescope were applied in the coating laboratory of the Technical University of Aschaffenburg.

References

1. T. Döhring et al.: „Development of iridium coated X-ray mirrors for astronomical applications", Proc. SPIE 10399,103991C (2017)
2. T. Döhring and R. Hudec: "JEUMICO and TRILAMICO: Dimensions of successful Bavarian-Czech cooperation", Contrib. Astron. Obs. Skalnaté Pleso 48, 430 – 436 (2018)
3. T. Döhring et al.: "Characterisation of X-ray mirrors based on chromium-iridium tri-layer coatings", Proc. SPIE 11776, 1177607 (2021)
4. V. Stehlikova et al.: "Lobster eye type X-ray telescope with chromium-iridium coated tri-layer mirrors", Proc. SPIE 11776, 1177605 (2021)

Gap in pagination due to unavailable paper.

Pages 339-340

Micro-reactors fabricated by Subaquatic indirect Laser-Induced Plasma-Assisted Ablation on soda-lime glass substrates.

M. Teresa Flores-Arias[1,*], *Ana I.* Gómez-Varela[1], *Alberto* P. Munuzuri[2] , *Alejandro* Carballosa[2] and *Carmen* Bao-Varela[1]

[1] Photonics4Life research group, Department of Applied Physics, Facultade de Física and Institute of Materials (iMATUS), Universidade de Santiago de Compostela, Campus Vida, E-15782 Santiago de Compostela, Spain
[2] Group of Nonlinear Physics, Department of Physics and Galician Center for Mathematical Research and Technology (CITMAga),, University of Santiago de Compostela, E-15782 Santiago de Compostela, Spain.

Abstract. Synchronization control of complex systems is a field that emerged with huge interest and aims to study new possible routes to synchronization in networks of non-locally coupled chemical oscillators. Light can be used to stimulate these systems and to be able to synchronize the different micro-reactors involved in the complex system. To this end, transparent reactors with good optical qualities are needed. Glass is the most appropriated material to be used for fabricating the micro-reactors. Subaquatic indirect Laser-Induced Plasma-Assisted Ablation is presented as a laser technique that combines underwater ablation with shock waves as a potential technique for fabricating these micro-reactors by using a Nd:YVO_4 laser.

1 Introduction

The phenomenon of synchronization among coupled oscillators is quite common in natural and man-made systems. It occurs in cooperative crowd effects [1,2] while in non-living physical systems synchronization is seen, for example, in arrays of Josephson junctions and semiconductor lasers [3, 4]. Last, systems of coupled chemical reactions provide representative examples in chemistry [5]. The coupling of these oscillators using light pulses at different intensities, increasing, or decreasing their frequency so the oscillators can adapt to each other and synchronize their signal, offers a huge potential in the field. However, to use light for this purpose, the microreactor needs to exhibit optical transparency in the excitation at the visible wavelength range. Thus, glass is a good candidate to be used for this kind of microreactors due to its robustness and chemical resistance.
During the last decades laser technologies have received high attention due to its versatility to perform different kind of structures for a wide variety of applications. Moreover, it is a non-contact method that implies no thermal or mechanical deformation of the sample if properly used. The speed of the process and its low cost, once the equipment is installed, as well as the non-toxicity of the process, make them a good option to fabricate many devices [6 -7]. Soda-lime glass substrates are used in this work to manufacture with a nanosecond IR laser these microreactors. Subaquatic indirect Laser-Induced Plasma-Assisted Ablation is used to achieve the dimension needed for the micro-reactor as well to obtain straight walls, that cannot be obtained with other laser-based techniques. Micro-reactors with dimensions of 3 mm radius and 2 mm depth are presented. An initial study on the synchronization of complex system is showed.

1.1 Fabrication of microreactors on sode-lime glass with a nanosecond IR laser.

The technique used in this work for fabricating micro-reactors over soda-lime glass substrates corresponds with the setup presented in the Fig.1. An Infrared Nd:YVO_4 laser pulsed in the nanosecond regime, with a pulse duration around 20 ns, is focused using a flat field lens on the surface of a metallic target. The glass substrate is located over this metallic target, but not in direct contact with it. To avoid the shift of the glass substrate during the laser processing due to the shock waves, two weights are used to fix it. This setup in introduced in water, in such a way the debris can escape from the inner of the structures that are being performed in the glass. The water also helps to reduce the increasing of the temperature due to the absorption of the energy of the laser in the nanosecond regime. With this technique micro-reactors with diameter of 3 mm and depth of 2 mm, with straight walls (Fig. 2).

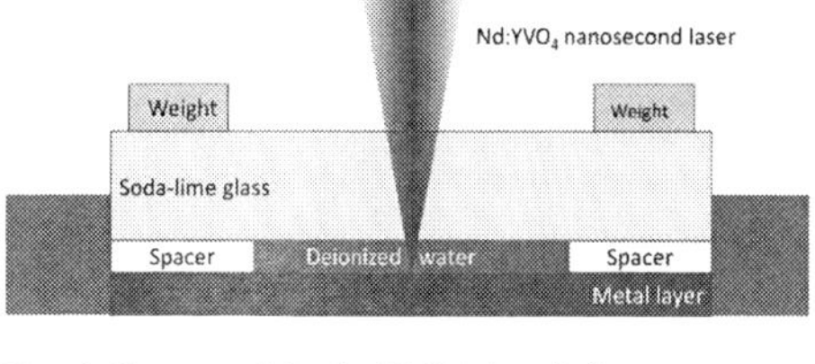

Fig. 1. Setup used for the SLIPAA technique.

* Corresponding author: maite.flores@usc.es

https://doi.org/10.1051/epjconf/202226613012

2 Preliminary studies on micro-reactor

Each "micro-reactor" contains a photo-sensitive chemical oscillator, composed of ruthenium gel and the Belousov-Zhabotinsky reaction. In the surface of the gel, the chemicals react, and the gel changes its color from orange to transparent (and viceversa) via a redox mechanism in the form of a wave, giving raise to periodic oscillations of colour. By varying the light intensity that we use to illuminate the reactor and due to the photosensitivity of the chemical reaction, we can modify the frequency of the oscillations. The idea is to couple the different oscillators using light pulses at different intensities, increasing, or decreasing their frequency so the oscillators can adapt to each other and synchronize their signal.

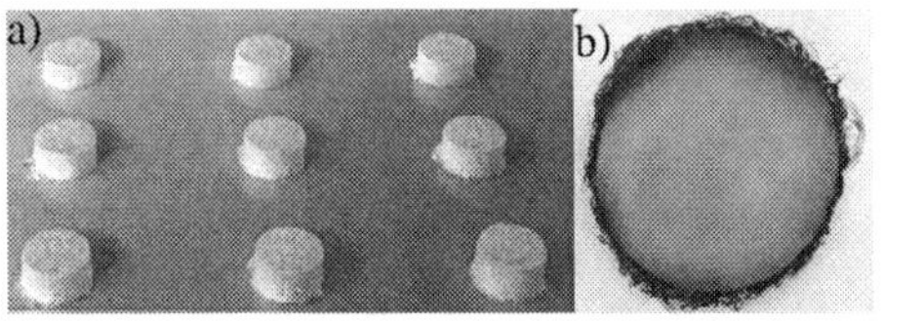

Fig. 2. a) photography of an array of micro-reactors and b) microscopic image of the up-side view of a micro-reactor, fabricated by SLIPAA

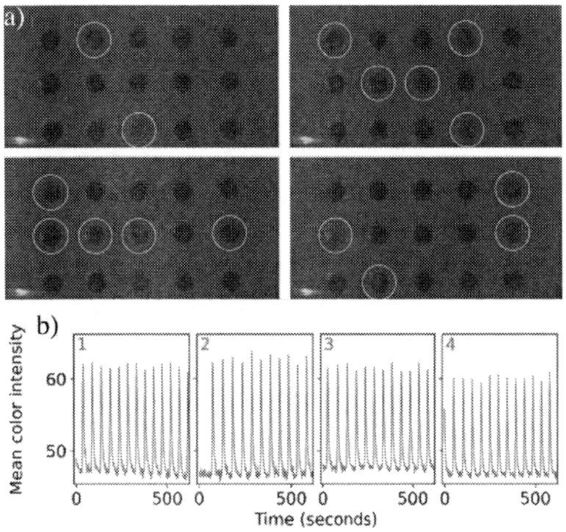

Fig. 3. From left to right and from top to bottom, consecutive images of the emerging oscillations in the micro-reactors. a) yellowed circle is drawn on the reactors where an oscillation is taking place. b) Measured time signals from the marked "micro-reactors" 1 to 4.

Figure 3 presents a picture of the reactor with a set of chemical oscillators. Yellow circles in Figure 3a) mark those oscillators that are experiencing an excitation peak. Different snapshots correspond with different moments of the experiment. Figure 3b) shows the evolution of the excitation state of several oscillators (those marked with red numbers).

In Figure 4a, a synchronization experiment is presented with some preliminary results. The leading oscillator is located at the center of the field of vision (Figure 4a) and periodically triggers a luminous pulse over the three other chemical oscillators in the picture. The excitation pulses, as recorded directly from the light intensity in the images, are plotted in Figure 4b. Note that after some transient time the three slave oscillators fully synchronize with the leading oscillator (marked with dashed black lines).

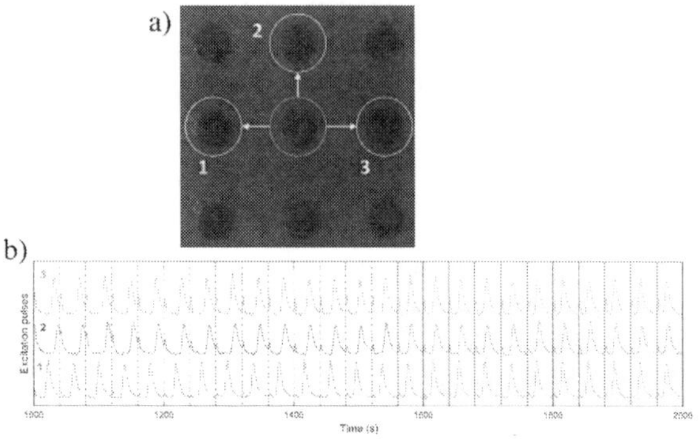

Fig. 4. Network of coupled chemical oscillators. When the central oscillator is activated, a shadowed pulse is projected (panel a) onto the oscillators of its network (marked in white in panel a). b) Excitation pulses in the slave oscillators synchronized after some transient.

3. Conclusions.

Micro-reactors for analysing the synchronization of complex systems with light have been fabricated by a subaquatic laser ablation, using a metal foil to generate the plasma and shock waves responsible of the ablation on soda-lime glass substrates. Results of the experiments carried out with these micro-reactors for synchronization of complex systems induced by light, prove that the micro-reactor arrays fabricated by SLIPAA are good enough.

References

1. Z. N´eda, E. Ravasz, Y. Brechet, and A.-L. Barab´asi, Nature **403**, 849–850 (2000).
2. H. Rheingold, Smart Mobs: The Next Social Revolution (Basic Books, 2003)
3. E. Goldobin, D. Koelle, R. Kleiner, and R. G. Mints, Physical Review Letters **107**, 227001 (2011).
4. D. V. Ramana Reddy, A. Sen, and G. L. Johnston, Physical Review Letters **80**, 5109–5112 (1998).
5. G. Ertl, Science **254**, 1750–1755 (1991).
6. J. R. Vázquez De Aldana, P. Moreno, and L. Roso, Opt. Mater., vol. 34, no. 3, 572–578, (2012).
7. M. Malinauskas, A. Zukauskas S. Hasegawa, Y. Hayasaki, V.Mizeikis, R. Buividas, and S. Juodkazis, Light Sci. Appl., **5**, e16133-e16147, (2016).

This work has been d supported by the Agencia Estatal de Investigación, under contract RTI2018-097063-B-100 and the Consellería de Educación Xunta de Galicia/FEDER under contracts ED431B 2020/29 and ED481D-2021-019, Spain.

https://doi.org/10.1051/epjconf/202226613013

Micro-diffractive optical element arrays for beam shaping

Shima Gharbi Ghebjagh[1,*], *Thomas* Handte, and *Stefan* Sinzinger

[1]Fachgebiet Technische Optik, Institut für Mikro- und Nanotechnologien, Technische Universität Ilmenau, Ilmenau, 98693, Germany

Abstract. We describe the design of Fourier type array generators and beam shapers as periodic configurations of refractive-diffractive optical elements in microscale to provide specific beam shaping and imaging functionalities. We investigate how the addition of micro-nanostructures to regular microstructure arrays enables new degrees of freedom for the design of micro-optical systems, in combination with adapted fabrication techniques yields a better optical performance and leads to enhancement of the array concept, uniformity and efficiency.

1 Introduction

Microlens arrays (MLA) as one of the most well-known microstructured arrays, are arrangements of microlenses on a single substrate. One of the most prominent applications is in beam shaping and generation of arrays of equidistant and uniform intensity peaks referred to array illuminators, which is based on aperture division [1] . The growth of the fields of applications of the microstructure arrays, is connected to the complexity of the corresponding optical structure and the advancement of the fabrication techniques.
New fabrication technologies pave the way for combining different complex diffractive and refractive structures on a single micro-optical element, which enables new degrees of freedom. In this contribution we investigate a method for enhancement of the array generators and the evolution of application areas by addition of diffractive microstructures to regular MLAs. We use analytic and numerical design and optimization methods to design micro structured micro-lens arrays and study their achievable functionalities. we propose a new fabrication sequence for the complex structure of the optimized array with standard regular MLAs as starting point for the fabrication. In this concept, we suggest to fabricate the diffractive microstructures with reduced requirements on the spatial complexity, on regular MLAs. We will further investigate the system advantages, tolerances and limitations as a result of lithographic fabrication of microstructures on the curved surfaces of refractive MLAs.

2 Design of array illuminators

Uniform laser-beam irradiation is needed e.g. for laser microfabrication, laser welding and multi-location, parallel and simultaneous processing but is increasingly also interesting in imaging applications such as light sheet microscopy [2] . An efficient method for designing the phase distribution for beam shaping applications is based on the iterative Fourier transform algorithms (IFTA), suggested by Gerchberg and Saxton. This optimization algorithm, allows the design of a DOE generating a specific diffraction pattern for a given illumination pattern. The algorithm can be started with a random phase function or an appropriate starting phase distribution.Based on this principle, modifications have been suggested e.g. by Fienup with the aim to increase efficiency and uniformity [3].
Here, we investigate the design and fabrication of array illuminators, based on Fienup algorithm and MLAs. Due to the periodicity of the MLA it is possible to regard it as diffraction gratings. Therefore, based on the theory of diffraction, the intensity pattern of the modulated beam is defined by an amplitude modulation due to the interference of the beams from individual microlenses. The beam shaping ability in array generators is achieved by replacing a conventional MLA by a micro-structured diffractive optical element array, in which individual diffractive optical elements is optimized to generate any given diffraction pattern. One advantage of the method is that a poor uniformity in the input beam does not affect the generated spot array uniformity. However, there is a limitation due to the complexity of the diffractive structure. The larger the number of output spots or the desired output intensity pattern, the larger the complexity of the diffractive structure. The uniformity in the spots is determined mainly by the fabrication precision. It is affected more severely if the complexity of the diffractive micro-structure is very high.
As the MLA itself generates a signal spot array with poor uniformity, we employ it as the initial guess for the subsequent optimization procedure. In our case, the basic period of the transmission function of MLA as the array generator, defines the pattern and the intensity distribution of the desired signal spots [4]. The fundamental period of the 2D pattern generator, which is a single square-framed microlens is given by:

$$t(x, y) = exp[i\phi(x, y)]rect(\frac{x}{d})rect(\frac{y}{d}) \quad (1)$$

*e-mail: shima.gharbi-gebjagh@tu-ilmenau.de

Where d is the spatial period and the microlens phase expression, $\phi(x, y)$ is defined by:

$$\phi(x, y) = \frac{-N\pi(x^2 + y^2)}{d^2} = -\frac{\pi(x^2 + y^2)}{f\lambda} \qquad (2)$$

The 2D pattern generator concentrates most of the diffracted light into the desired signal pattern of $N \times N$ spots. However, it does not provide satisfactory uniformity among the generated spots. The modulation profile generated by the MLA is the far field diffraction pattern, determined by the discrete Fourier transform (DFT) of the phase modulation.

In order to design a beam shaper array generator we replace the MLA by a micro-diffractive optical element (mDOE) array in which the individual diffractive structure is optimized to generate a given diffraction pattern with uniform intensity distribution. The phase profile is designed and optimized using the iterative error reduction Fienup algorithm with a starting phase distribution of a MLA to generate $N \times N$ spots. We defined the ideal phase modulation to provide a signal spot array with perfect uniformity and optimum efficiency, as a signal constraint during the optimization process. Fig. 1. Shows the starting MLA phase and the optimized mDOE array with the corresponding 2D signal spot array of $N = 17$.

Fig. 2 shows the single period of the optimized mDOE

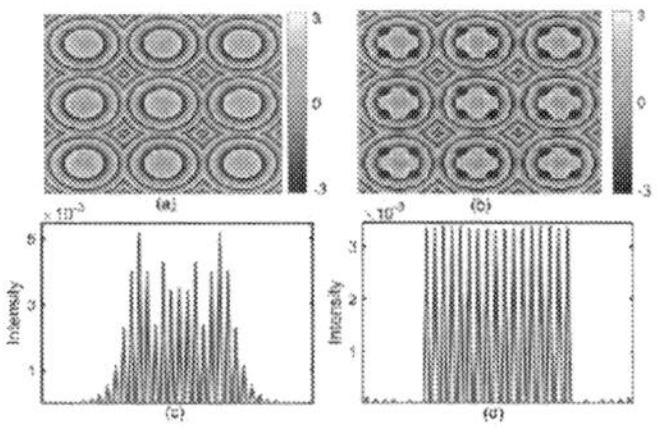

Figure 1. Starting MLA (a), optimized mDOE array phase structure of a 17×17 spot array generator (b), intensity distribution of the spot arrays generated by MLA (c) and mDOE array (d).

and the unwrapped subtracted structure of the mDOE from the single microlens and the corresponding unwrapped phase profiles of the elements. By subtracting the diffractive microstructures of the mDOE from the starting MLA phase, it is possible to fabricate the additional microstructure (STR), as the result of optimization process, on the curved surface of the conventional MLA, as the starting phase. Combining different complex diffractive and refractive structures on a single micro-optical element allows us to fabricate more complex structures, which in turn provides more complex functionality.

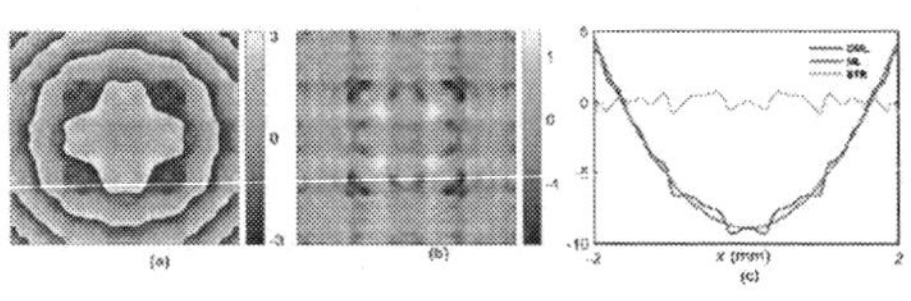

Figure 2. (a) Single period of the optimized mDOE array. (b) Single period of the subtracted structure of the mDOE from the microlens, (c) profiles of the phase structures at single periods of

We simulated the performance of the STR structure quantized into 8 phase levels, etched onto the curved surface of the MLA and the effects of possible alignment errors in the lateral directions. The simulation results show a good uniformity among the array spots.

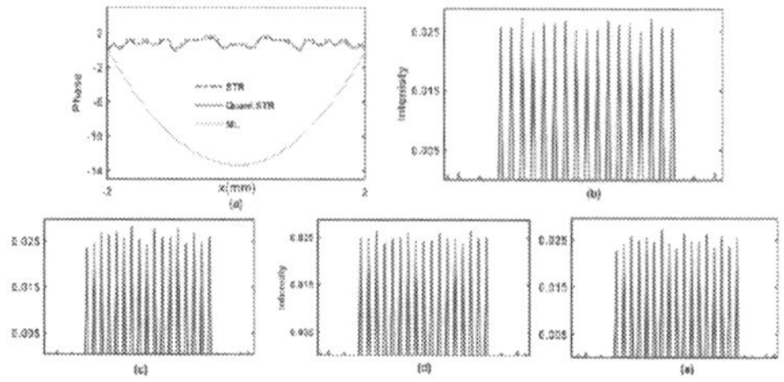

Figure 3. (a) Single periods of the quantized STR and the microlens. (b) Intensity distribution of the signal spots, (c-e) Intensity distribution of the signal spots for alignment error of 1μm in lateral x, y and both x and y directions .

3 Summary

In summary, we investigate a method for enhancement of the array generators and beam shapers by addition of diffractive micro-nanostructures to regular MLAs as the classical array generators. As a result, we propose a new fabrication sequence for the complex structure of the optimized mDOE which uses standard regular MLAs as starting point for the fabrication. In this concept, we suggest to fabricate the diffractive microstructures with reduced requirements on the spatial complexity on regular MLAs, e.g. by imprint lithography [5].

References

[1] Nörbert Streibl, Uwe Nölscher, Jürgen Jahns, Susan Walker, Appl. Opt. **30**, 2739 (1991)

[2] M. Hofmann, S.G. Ghebjagh, K. Lemke, S. Sinzinger, *Multi-sheet excitation and imaging of flow driven samples in an LSFM with a modified multi-focal diffractive lens*, in *OSA Imaging and Applied Optics Congress 2021* (Optica Publishing Group, 2021), p. CM1A.7

[3] J. R. Fienup, Optical Engineering **19**, 297 (1980)

[4] Victor Arrizón, Markus Testorf, Stefan Sinzinger, Jürgen Jahns, J. Opt. Soc. Am. A **17**, 2157 (2000)

[5] T. Handte, N. Scheller, L. Dittrich, M.W. Thesen, M. Messerschmidt, S. Sinzinger, Micro and Nano Engineering **14**, 100106 (2022)

https://doi.org/10.1051/epjconf/202226613014

Advancing fluorescence microscopy and spectroscopy for the benefit of cell science and drug delivery

Ana I. Gómez-Varela[1,2*], Adelaide Miranda[2], Dimitar Stamov[3], Ricardo Gaspar[2], Bruno Silva[2], and Pieter De Beule[2]

[1]Photonics4Life research group, Departmento de Física Aplicada, Facultade de Física and Instituto de Materiais (iMATUS), Universidade de Santiago de Compostela, Campus Vida, E-15782 Santiago de Compostela, Spain
[2]International Iberian Nanotechnology Laboratory, Avenida Mestre José Veiga s/n, Braga, Portugal
[3]JPK BioAFM, Bruker Nano GmbH, Am Studio 2D, 12489 Berlin, Germany

Abstract. Here we report on two recent advances we made applying advanced fluorescence microscopy and spectroscopy to further the study of cell biology and drug delivery. At first, we detail a new instrumental set-up combining atomic force microscopy in liquid and super-resolution fluorescence microscopy in novel configuration such that long-term simultaneous and co-localized observation with both techniques becomes feasible. We believe this will contribute to the study of fast membrane activated cell-signalling processes in the years to come. Furthermore, we report on a novel application of fluorescence cross-correlation spectroscopy (FCCS) for the characterization of lipid-nucleic acid complexes. We could determine the number of nucleic acid particles incorporated in each liposome, a parameter not readily accessible on ensemble basis with other methods. This parameter is crucial for co-delivery applications, and we believe FCCS can play an important role in the future development of new drug delivery systems.

1 Introduction

In this work we present the first demonstration of a hardware set-up capable of performing simultaneously super-resolution fluorescence microscopy and atomic force microscopy (AFM). We combine AFM with super-resolved structured illumination microscopy (SR-SIM), a super-resolution technique that allows to surpass the diffraction barrier by a factor of two in every spatial direction. In this set-up it is possible to acquire AFM data under constant SIM illumination without inducing distortions on the AFM cantilever. AFM features high-resolution imaging and quantitative nanomechanical analysis of biological samples, requiring minimum to no specimen modification, further enabling measurements at near physiological/native conditions. Due to its surface characterization nature, AFM is further combined with other microscopy modalities, thus unlocking a truly correlative potential for simultaneous acquisition of physical, chemical and biological information in complex systems.

We also present the successful application of dual-color fluorescence cross-correlation spectroscopy (FCCS) as a direct method to characterize and optimize the assembly of nonviral cationic liposome (CL)-DNA complexes based on a CL formulation consisting of the cationic lipid DOTAP and zwitterionic lipid DOPC [2]. We demonstrate that FCCS is able to quantitatively determine the extent of the association between DNA and the liposomes and assess its loading capacity.

2 Combining Super-Resolution Optical Microscopy and Atomic Force Microscopy

The schematic representation of the experimental set-up of the combined SR-SIM-AFM combined platform is depicted in Fig. 1. It is based on an atomic force microscope (JPK NanoWizard 3, Bruker Nano GmbH,

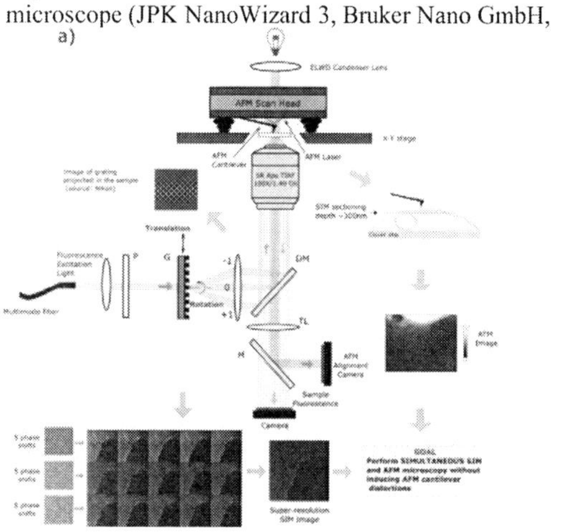

Fig. 1. Experimental set-up for the SR-SIM and AFM combined platform. Adapted from [1].

* Corresponding author: anaisabel.gomez@usc.es

https://doi.org/10.1051/epjconf/202226613014

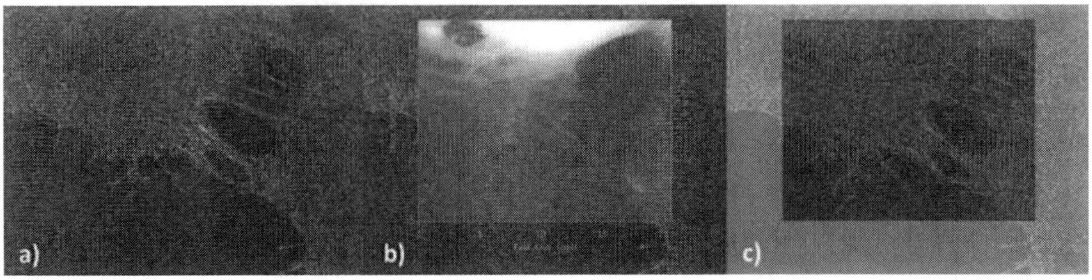

Fig. 2. SR-SIM images of the human bone osteosarcoma epithelial cells (a) superimposed with the height/topography channel of the AFM data (b). The correlated SR-SIM/AFM channels on cells in a semi-transparent mode (XY-scales of the AFM image in (b) (c), include a mean-curvature enhanced AFM-topographic information [1].

Berlin, Germany) and a structured illumination microscope (N-SIM E, Nikon Instruments Europe B.V., Amsterdam, The Netherlands). The AFM is mounted on an inverted microscope (Eclipse Ti2-E, Nikon Instruments Europe B.V., Amsterdam, The Netherlands) and a standard monochrome CCD camera (ProgRes MFCool, Jenoptik, Jena, Germany) is coupled for AFM laser spot cantilever alignment.

System performance is demonstrated by conducting simultaneous operation of SR-SIM/AFM using CRISPR/Cas9 genome-edited human bone osteosarcoma epithelial cells, with plasma membrane transporter 1 (MCT1) tagged with an enhanced green fluorescent protein (EGFP) at the N-terminal (Fig. 2).

3 Fluorescence cross-correlation spectroscopy for lipoplex characterization

Among nonviral vehicles for delivery applications, cationic liposome-DNA (CLDNA or lipoplexes) nanoparticles are among the most promising since they present low immunogenicity, potential for large-scale production and high flexibility of functionalization. FCSS is an optical technique that by labelling DNA and liposomes using green and red dies, respectively, is able to provide information the time traces generated by the two fluorescently labelled species simultaneously, enabling the study of specific bimolecular binding.

In the experiments, DNA was incubated for at least 30 minutes with green fluorescent dye YOYO-1 prior to use. For lipoplexes, Texas red DHPE was used as labelling agent. FCCS measurements were conducted using a Zeiss LSM 780-NLO confocal microscope (Carl Zeiss AG, Jena, Ger- many) with a Zeiss C-Apochromat 40×/1.2 W objective. YOYO-1 fluorescence was excited using the 488 nm line of the Argon (Ar) laser and Texas Red was excited using the 561 nm diode-pumped solid state (DPSS) laser.

4 Conclusions

In this work we present the first demonstration of a hardware set-up capable of performing simultaneously super-resolution fluorescence microscopy and AFM. We combine AFM with SR-SIM, a super-resolution technique that allows to surpass the diffraction barrier by a factor of two in every spatial direction. In this set-up it is possible to acquire AFM data under constant SIM illumination without distorting the cantilever operation. We believe that this scheme combination of microscopy techniques presents one of the most promising schemes enabling simultaneous co-localized imaging, allowing the recording of nanomechanical data and cellular dynamics at the same time.

Moreover, we report for the first time on the potential of fluorescence cross-correlation spectroscopy for the characterization of previously unavailable properties of cationic lipid-DNA complexes. This methodology promises to become a powerful guide to the development of similar formulations and accelerate research and innovation in mRNA vaccine and cancer vaccine development. As an example, in a recent development as a follow-up of this research, we report on the discovery of a new stability criterion for the preparation of stable hybrid lipid-polymer-nucleic acid nanoparticles [3].

Ana I. Gómez-Varela acknowledges Consellería de Cultura, Educación e Ordenación Universitaria for ED481D-2021-019 postdoctoral fellowship. Adelaide Miranda and Pieter De Beule acknowledge financial support from Norte's Regional Operational Programme 2014-2020–Norte2020 (NORTE-01-0145-FEDER-000019). This work was also supported by the Fundação para a Ciência e a Tecnologia through the project SAM – [PTDC/NAN-OPT/31596/2017].

References

1. A.I. Gómez-Varela, D.R. Stamov, A. Miranda, et al. Sci. Rep. **10**, 1122 (2020).
2. A.I. Gómez-Varela, R. Gaspar, A. Miranda, et al. J. Biophotonics: e202000200-e202000200 (2020).
3. J. L. Paris, R. Gaspar, F. Coelho, P. A.A. De Beule, B. F. B. Silva bioRxiv 2022.02.06.479316.

* Corresponding author: anaisabel.gomez@usc.es

 https://doi.org/10.1051/epjconf/202226613015

Comparison of simulation and experimental characterization of a 4-pass diode-pumped 4-active-mirrors 1053 nm laser amplifier

Thomas Hamoudi[1], Hugo Chesneau[1], Thomas Dubé[1], Emma Dubois[1], Manon Lamy[1], Cédric Maunier[1], Alain Braud[2], Cesare Meroni[2], Patrice Camy[2], Sébastien Montant[1]*

[1] CEA,CESTA, F-33116 Le Barp, France

[2] CIMAP-ENSICAEN, 6 Boulevard du Maréchal Juin, 14050 Caen Cedex, France

Abstract. In this paper, we will discuss the performances of a 4-pass diode-pumped 4-active-mirrors laser amplifier. Numerical simulations along with experimental characterization paves the way to reach 1 J at 10 Hz pulse rate at 1053 nm. Both Nd :glass and Nd :Lu :CaF_2 amplifier medium performances will be compared in this amplifier.

1 Context

The Laser MegaJoule (LMJ) is a French laser facility designed to perform extreme laser-matter interaction Physics. Each of the 22 laser chains is injected by a front-end composed of a master oscillator fiber amplifier and a bulk Pre Amplifier Module (PAM). The nJ pulse yielded by the fiber seeder is consecutively amplified by the regenerative and the 4-pass amplifiers of the PAM up to 1 J. In this paper, we will draw comparisons between simulations and experiments of the 4-pass diode-pumped 4 active-mirrors laser amplifier to overcome thermal effects in order to achieve 1 J at 10 Hz.

2 Amplifier architecture

2.1 Geometry of laser medium

To reach energies at the joule level, two types of laser materials geometry can be used: rod and thin disk. Some architectures of rod amplifiers have shown the possibility of obtaining energies close to the joule [1, 2]. However, surface/volume ratio of this rod geometry lead to important harmful thermal effects. For example, in the Laser MegaJoule preamplifier, thermal effects occurring in the flashlamp-pumped Nd-doped phosphate glass rod of the 4-pass amplifier limits the 1 Hz repetition rate of the regenerative amplifier to 1 shot every 5 min. Generally, thermal effects may appear as wavefront deformation [3]. To counteract thermal effects in the amplifier rods in high power laser, thin-disk amplifiers have been developed. With an interesting surface/volume ratio, extraction of heat capacity is optimal, hence thermal effects are limited. However, sub-millimeter thin-disk thickness triggers a lower gain per pass. A trade-off was made with the development of active-mirrors amplifiers (AMA), having a millimeter thickness. Thus, AMA have been demonstrated to be sustainable amplifiers for use in high energy laser [4, 5, 6].

2.2 Gain medium

In addition to the architecture modification to increase the heat extraction, a new gain medium is used. For equivalent laser parameters, CaF_2 codoped with Nd^{3+} et Lu^{3+} exhibits a 4.3 W/m/K thermal conductivity which is seven time higher compared to Nd-doped phosphate glass [7]. Longitudinal pumping of the AMA enforces high homogeneity of the pump beam profile to obtain a good homogeneity of the spatial gain. We design an optimized hollow duct (HD) consisting of a funnel-shaped part to shape the pump beam and a waveguide to optimize its homogeneity. Several configurations with collimated or non-collimated pump diodes have been compared along with experiments and simulations in terms of efficiency, homogeneity and gain.

3 One active-mirror characterization

3.1 Wavefront distortion

Firstly, we have to characterize the wavefront distortion of the new lasers materials. For this, one AMA is pumped by a laser diode with a top-hat square profile. A probe beam at 1053 nm is seeded into the AMA during pumping. Then, the probe laser is analyzed with a Shack-Hartmann wavefront sensor (SHWS). 5 active mirrors have been characterized: Nd: Phosphate with a thickness of 3 and 5 mm and 0.5%Nd:x%Lu:CaF_2 active-mirrors with x equals to 5%, 6.5% and 8% with a thickness of 4 mm.

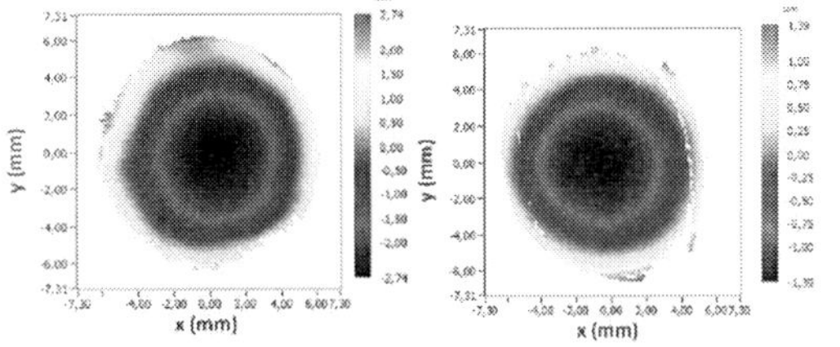

Figure 1 : Wavefront distortion measurment of 0.5%Nd: 5%Lu : CaF_2 at 10 Hz repetition rate (left) and Nd :Phosphate at 4 Hz repetition rate (right) active-mirror.

Figure 1 shows the wavefront distortion measured by SHWS of Nd: 5%Lu: CaF_2 and Nd: Phosphate active-mirror of 4 and 5 mm thickness respectively, with 1.6 J of pump energy at 4 Hz and 10 Hz respectively. The wavefront displays almost spherical thermal lens in

* Corresponding author: sebastien.montant@cea.fr

contrast with transversely pumped rod amplifier [3]. However, the peak-to-valley curvature value is higher for Nd: Phosphate than Nd: Lu: CaF_2 (5.48 µm vs 2.78 µm).

3.2 Hollow-Duct development

To optimize efficiency and homogeneity of the pump beam, a spatial diode shaping system has been developed. This hollow duct is composed of two distinct parts: a funnel to shape and a waveguide to homogenize the pump beam. To limit reflection losses, a gold coating is applied on the hollow duct. At the output of this HD, the pump beam is top-hat with a 8x8mm² square shape. Finally, different sizes of the funnel and waveguide have been compared. An influence has been shown on the gain and its spatial homogeneity with the different sizes. The gain homogeneity have been measured with a spatially resolved gain measurement. A trade-off was made between these two parameters to obtain a gain of 1.4 per pass with 4 J of pump energy with a good spatial homogeneity.

3.3 4-pass 1-active-mirror demonstration

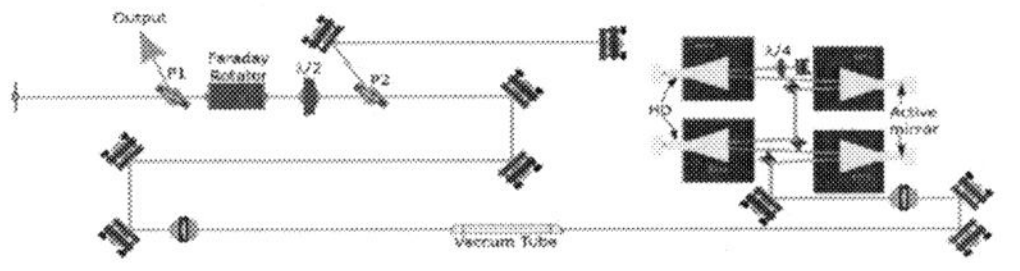

Figure 2. 4-pass amplifier scheme with P: polarizer, HD : hollow duct, λ/2 : half wave plate.

Figure 2 outlines the general scheme of the 4-pass amplifier, with 4 AMA each longitudinally pumped with 2 collimated diode stacks.

This 4-pass amplifier is seeded with 10 mJ, 1053 nm central wavelength, circular gaussian profile regenerative amplifier. A serrated aperture picks an almost square top-hat profile. The square pump beam and laser dimension are 8x8mm² and 6x6mm² respectively. All active-mirrors have a diameter of 15 mm and a thickness of 5 mm. They are each glued to copper heatsinks water-cooled at 20°C. One AMA is placed on a 4 axes precision positioning gimbal mount, with 2 rotation and 2 translation stages. The formers are used to optimize the position of the AMA center on the laser beam during pumping. Indeed, the absorption of pump beam induced an almost curvature wavefront distortion on AMA. If the laser beam is off-centre with respect to the centre of the curvature of the AMA, the beam may become tilted [8]. In our scheme, we measure the tilt of the laser beam with a SHWS. Figure 3a shows the influence of the vertical offset of the AMA with respect to the laser beam on its horizontal and vertical tilts. Theses measurements were conducted with a pump energy of 1.8 J on the AMA and a 300 s thermalization time.

To confirm experimental results, a propagation code with angular spectrum method has been implemented. Results of the simulation are presented in Figure 3a (lines). Figure 3b presented measurement of the evolution of the temperature of AMA (blue), radius of curvature (black), horizontal tilt (wine) and vertical tilt (green) versus time during pumping with AMA alignment optimization (cf Figure 3a).

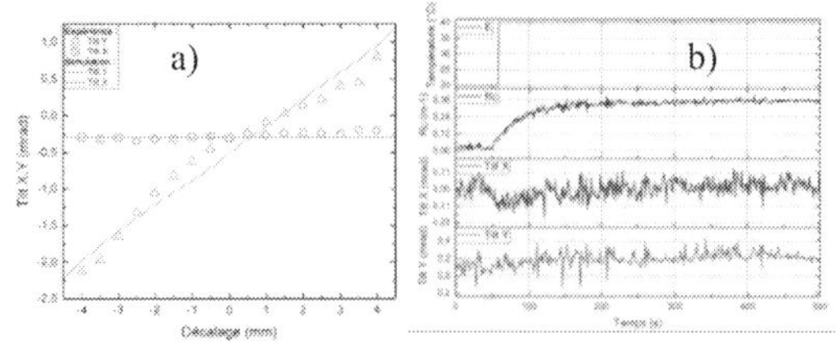

Figure 3. (a) Horizontal (green) and vertical (blue) tilt versus offset vertical offset of the AMA. Points: experimental ; Line: simulations. (b) Evolution of the temperature of AMA (blue), radius of curvature of active-mirror (black), horizontal tilt (wine) and vertical tilt (green) versus time during pumping.

These measurements show the possibility of compensating the induced vertical tilt by the rise in temperature during pumping

4 Conclusion

In this presentation, we will outline experimental performances of a 4-pass 4-active-mirrors amplifier with Nd: phosphate or Nd :Lu :CaF_2 compared to thermo-optical simulations using COMSOL [3] and propagation simulation using Miro [9].

This work was partly supported by the Conseil Régional de Nouvelle Aquitaine and the European Union through the project LEAP - Axes B et C (contracts CPER #16004205 and #DEE21-04-2019-5131820 and ERDF #2663710 and #3404410).

References

1. Pan L, Ji S, Huang W, Guo J, Lu X, Wang J, Fan W, Li X, Zhu J. *Photonics*; 8(4):96. (2021).
2. J. Guo, J. Wang, H. Wei, W. Huang, T. Huang, G. Xia, W. Fan, and Z. Lin. *High Power Laser Science and Engineering*, vol. **7**, p. e8, (2019).
3. C. Bernerd, B. Da Costa Fernandes, E. Boursier, J. Luce, C. Maunier, A. Braud, C. Meroni, P. Camy, and S. Montant. *Appl. Opt.* 59, 9905-9911 (2020).
4. D. Brown, J. Kelly, J. Abate, *IEEE Journal of Quantum Electronics*, vol. **17**, no. 9, pp. 1755-1765 (1981).
5. J. A. Abate, L. Lund, D. Brown, S. Jacobs, S. Refermat, J. Kelly, M. Gavin, J. Waldbillig, and O. Lewis. *Appl. Opt.* 20, 351-361 (1981).
6. X. Fu, Q. Liu, P. Li, Z. Sui, T. Liu, and M. Gong. *Appl. Physics*. Express, vol. **8** (2015).
7. S. Normani, A. Braud, R. Soulard, J.-L. Doualan, A. Benayad, G. Brasse, R. Moncorgé et J.-P. Goosens and P. Camy. *Cryst. Eng. Comm.*, pp. 9016-9025 (2016).
8. Z. Xiahui, and C. Xiyao. *Opt. Express,* vol **27**, pp. 1179-1198 (2019)
9. O. Morice, *Optical Enfineering*, n° %16, pp. 1530-1541 (2003).

 https://doi.org/10.1051/epjconf/202226613016

Determination of a source's temporal coherence function via depolarization in liquid crystal

Cristian Eduardo Hernández Cely[1,*] and *Rafael Ángel* Torres Amaris[1,**]

[1]Universidad Industrial de Santander

Abstract. Based on the fact that the degree of polarization of a light source is given in terms of the absolute value of its complex degree of coherence, we design a depolarization experiment using a variable retarder in which we measured the degree of polarization as a function of the retarder's birefringence. We show that if the incident light is previously linearly polarized we can perform a direct measurement of the light source's coherence using a Stokes meter.

1 Introduction

The temporal coherence function of a light source contains all of its first-order statistical information which specifies its spectral content and its polarization information [1]. Getting to know the coherence function of a given light source becomes useful in many optical situations. As an example we may consider when an interference experiment is performed in which the degree of coherence determines the visibility of the interference pattern. On other hand, a second example corresponds to when a depolarization experiment is performed, in this case, according to Wolf's polarization theory, the change in the degree of polarization depends on the degree of coherence, for instance, when a light beam propagates through a birefringent medium as liquid crystal or wave-plate retarders it not only changes its state of polarization but also its degree of polarization, thus in order to determine the depolarization effect, the source's degree of coherence have to be known [2–4]. Thus we can see that in many experimental setups which imply the phenomena of interference or polarization the measurements depends fundamentally on the light coherence [5].

Those experiments mentioned above are really useful nowadays in optical metrology and microscopy techniques, everything that involves ellipsometry and depolarization measurements, and of course, all the optical systems working with interferometers such as FT-IR spectrometers. In all these techniques it is vital to know, previous to starting any kind of measurement, the coherence function associated with the light source you are working with in order to keep your technique calibrated. For that reason, it is necessary to be able to measure the coherence of a light source. The common methods used to determine the temporal coherence of a light source consist of the measurement of its power spectral density which is the Fourier transform of the coherence function. The other common method is via Michelson interferometer, where the visibility of the interference pattern as a function of the phase difference is also a direct measurement of the degree of coherence [6]. Though both methods are the standard and nowadays can be performed with great precision sometimes it requires highly sophisticated devices such as spectrometers and high sensitive CCD cameras. Thus, in order to perform a much more simple determination of the temporal degree of coherence of a light source, we propose a method to measure it using depolarization. The method set-up consists of a linearly polarized light beam that propagates through a variable retarder, then the degree of polarization as a function of the birefringence induced is measured. We will show that the measurement of the degree of polarization under a given experimental condition is a direct and much more simple measurement of a light source's temporal degree of coherence.

2 Depolarization in birefringent media

In general when a polarized light beam propagates through a birefringent medium such as a wave-plate, characterized by a birefringence Δn and thickness l, an optical path difference Δnl is introduced between the x and y components of the incident light. As a consequence of this, a phase shift is introduced producing a change in the polarization state. For partially coherent light sources, i.e. sources for which δnl is comparable with its coherence length, the polarization change is also accompanied by a depolarization, where now the ellipticity and orientation are not the only polarization parameters that change but also the degree of polarization (DOP). The DOP changes as a function of the degree of cross-coherence γ_{xy} and the amplitudes E_i of each electric field component

$$DOP = \sqrt{1 - \frac{4E_xE_y}{(E_x^2 + E_y^2)^2}\left(1 - |\gamma_{xy}(\Delta nl)|^2\right)}, \quad (1)$$

*e-mail: cristian.hernandez17@correo.uis.edu.co
**e-mail: Rafael.Torres@saber.uis.edu.co

https://doi.org/10.1051/epjconf/202226613016

in that sense, the more $|\gamma_{xy}(\Delta nl)|$ approaches zero the more is the effect of depolarization introduced by the propagation in the wave plate because the DOP will also approach zero. In fact, for a given value of Δnl the maximum depolarization occurs in the particular case when $E_x = E_y$, which can be accomplished with the incident light beam being circularly polarized or linearly polarized. In such a case the DOP will reduce to the following expression

$$DOP = |\gamma_{xy}(\Delta nl)|, \tag{2}$$

which means that the degree of polarization is exactly equal to the absolute value of the degree of cross-coherence, which is basically the same light source's degree of coherence. Then, we can use this theoretical result to develop a new way to measure the coherence of a light source by measuring the change of the DOP as a function of the birefringence of a retarder wave-plate, and that is what we will explain in detail in the following section, the way in how we can accomplish that measurement with the information explained here. For other cases in which $E_x \neq E_y$ and the incident light beam is found to be in a totally polarized state or even a partially polarized state the expression of the DOP can be derived as is shown in the following papers [3, 4, 7].

3 Measurement technique

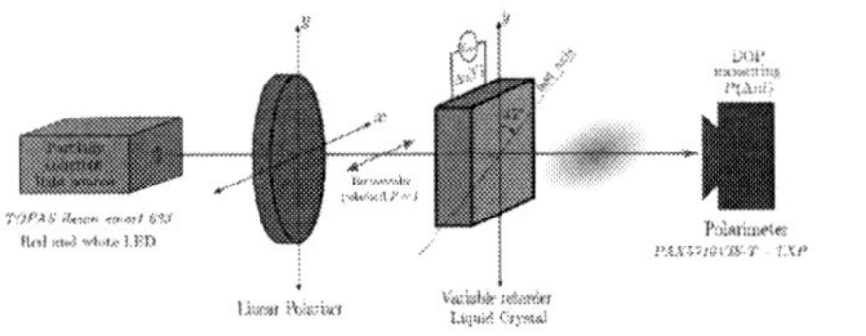

Figure 1. Experimental set up of the depolarization experiment

The experimental setup used to determine the coherence function through the depolarization of a light source is shown in Figure 1. We use a variable retarder consisting of a liquid crystal (*LCC2415-VIS/M* produced by *Thorlabs inc.*) whose birefringence can be varied through and applied voltage. This birefringence can be varied in a range of Δnl : $0 - 4\,[nm]$ (being l the thickness of the liquid crystal). The incident light beam of the partially coherent light source we want to characterize is polarized linearly along the x axis such that the variable retarder fast axis is rotated to an angle of $\theta = 45^o$ with respect the vertical axis, whit this condition the retarder will introduce the maximum phase shift in the incident horizontally polarized light beam. Thus, the DOP will be given by the equation 2. Now in order to determine the degree of coherence of a given source we just need to measure the DOP for different values of Δnl using a Stokes meter (*PAX5710VIS-T - TXP (Thorlabs, inc)*, and the data obtained is fitted with a linear combination of Gaussian functions such that the function that is adjusted with the experimental data is exactly equal to the light source's degree of coherence $|\gamma(\tau)|$.

In this experiment we use two kinds of light sources, a red LED and a white LED passed through a color glass filter, both being partially coherent light sources. In Figure 2 it is shown how the experimental data were fitted in order to obtain an explicit expression of $DOP = |\gamma|$ as a function of the variable Δnl and so, the explicit function for the degree of coherence for each source. We can see that the red LED source is more coherent than the white LED. If we want to change from distance variables δnl to time variables , we just need to replace the distance variable by $c\tau$ in the expression we found for $|\gamma|$ by the fitting. In that way, we can estimate and determine some statistical information about each source like the coherence length or coherence time associated with each source and so its spectral bandwidth.

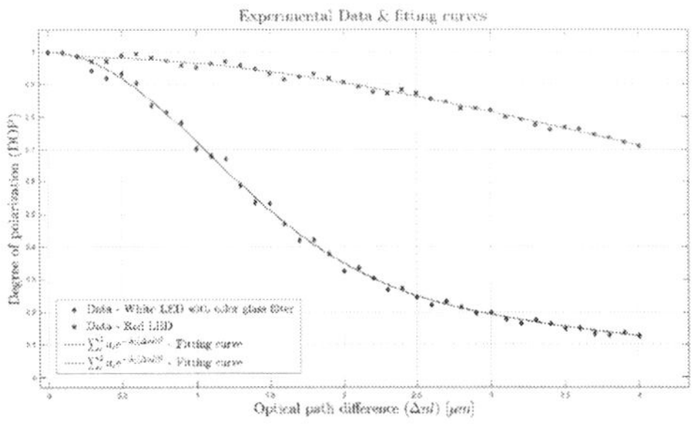

Figure 2. Graphic of the experimental data.

4 Conclusions

A method to determine the degree of coherence by a depolarization experiment was proposed and tested. For the sources used we could determine an explicit function of its degree of coherence which gives us the value of its correlation for optical path length differences in the range of $0 - 4[\mu m]$, if a more coherent source like a laser diode needs to be characterize we may use multiple liquid crystals in the experiment. Additionally we can estimate other statistical parameters as the spectral bandwidth since coherence and power spectral density are connected through Wiener-Khintchine theorem.

References

[1] E. Wolf, *Introduction to the Theory of Coherence and Polarization of Light* (Cambridge university press, 2007)
[2] A. Domański, Opto-Electronics Review **13**, 171 (2005)
[3] A.W. Domański, M. Redek, D. Budaszewski, Photonics Letters of Poland **1**, 67 (2009)
[4] C.E.H. Cely, R.Á.T. Amaris, pp. JTh5A–77 (2021)
[5] E. Wolf, Physics letters A **312**, 263 (2003)
[6] M. Born, E. Wolf, *Principles of optics: electromagnetic theory of propagation, interference and diffraction of light* (Elsevier, 2013)
[7] P.L. Makowski, A.W. Domański, **7745**, 77450J (2010)

Novel ultrafast structured EUV/x-ray sources from nonlinear optics

Carlos Hernández-García[1,*]

[1]Grupo de Investigación en Aplicaciones del Láser y Fotónica, Departamento de Física Aplicada, Universidad de Salamanca, E-37008, Salamanca, Spain.

Abstract. Coherent extreme-ultraviolet (EUV)/x-ray laser sources, structured in their temporal/spectral, spatial and angular momentum properties are emerging as unique tools to probe the nanoworld. One of the key ingredients for the emergence of such sources is the extraordinary coherence in the up-conversion of infrared laser sources through the highly nonlinear process of high-order harmonic generation. In this contribution we will review the advances during the last decade that led to the generation of structured EUV/x-ray sources, such as circularly polarized attosecond pulses, harmonic vortices with time-varying orbital angular momentum, ultrafast vector and vector/vortex beams, tunable high-order harmonic combs or attosecond pulse trains with time-dependent polarization states. The use of such sources is being already applied to the investigation of chiral matter or magnetic materials. In the latter case, structured ultrafast sources are very promising to achieve a complete understanding of the electronic and spin interactions that govern sub-femtosecond magnetization dynamics.

The development of structured ultrafast laser sources is a key ingredient to advance our knowledge about the fundamental dynamics of electronic and spin processes in matter. In particular, it has been widely recognized the relevance of ultrafast sources structured in their spin angular momentum (SAM, associated to the polarization of light) and orbital angular momentum (OAM, associated with the transverse phase profile, or vorticity of a light beam) to study chiral systems and magnetic materials in their fundamental temporal and spatial scales. In that scenario, structured coherent extreme-ultraviolet (EUV)/soft x-ray pulses are emerging thanks to the highly nonlinear process of high harmonic generation (HHG).

HHG is among the most extreme nonlinear up-conversion processes to date, and it can be semiclassically understood through the so-called three-step model. An intense, infrared laser field is focused into an atomic or molecular target, and an electronic wavepacket is detached from its parent ion through tunnel ionization. Afterwards, the electronic wavepacket acquires kinetic energy from the laser field, and due to the oscillatory behavior of the later one, it recollides with its parent atom, releasing the kinetic energy as high frequency radiation. The emitted radiation spectrum is composed of harmonics of the infrared driving field, that can extend towards the soft x-rays [1], being temporally emitted as attosecond pulses [2]. In the last decade, HHG in solid systems is being also investigated, where an analogy with the three-step model can be found. In HHG, some of the driving laser beam properties are imprinted on the dynamics of the radiating electron, and, in turn, on the emitted EUV/soft X-ray light, making it a very versatile process to engineer the temporal, spectral and spatial properties of high-harmonic and attosecond pulses.

In this contribution we review our recent work in the generation of structured laser pulses at the attosecond scale. Several works that revealed the mapping properties of SAM and OAM in the HHG process were crucial. First, circularly polarized attosecond pulses were obtained by using non-trivial geometries of the driving field [3-5]. Second, high harmonic vortices with high topological charge can be obtained through OAM conservation [6-8]. These two achievements over the SAM and OAM of high-order harmonic pulses, that resulted from several works from of many research groups, paved the way towards more sophisticated structured high harmonic and attosecond pulses: high-harmonic pulses with both OAM and SAM [9], high-harmonic vortices with time-dependent OAM or self-torque [10], attosecond pulses with time-dependent polarization states [11], or EUV vector and vector/vortex beams with spatially varying polarization states [12,13], among others.

Very recently, the proper use of OAM driving fields has allowed to achieve precise coherent control over the frequency content of high-harmonic pulses—a key property to perform high-harmonic spectroscopy. In particular, by harnessing OAM conservation in HHG, we can generate a transverse necklace-shaped spatial phased array of harmonic emitters that allow us to tune the line spacing and divergence of the emitted harmonic combs [14] (see Fig. 1). Our theoretical and experimental results

* Corresponding author: carloshergar@usal.es

show that the resulting on-axis HHG emission is composed of harmonics whose frequency separation can be controlled through the OAM of the driving field, extending towards the soft X-rays. In addition, such harmonic radiation presents extremely low divergence, well below that obtained when using Gaussian driving beams, which further decreases with the harmonic order. This work provides a new degree of freedom for the design of harmonic combs—particularly in the soft X-ray regime, where very limited options are available.

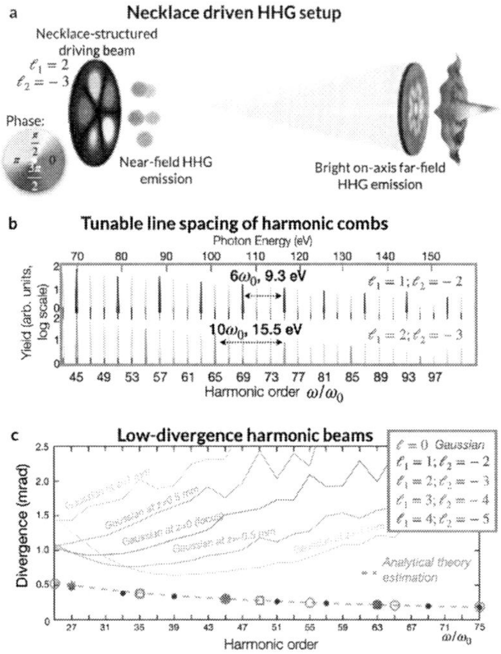

Fig. 1. a) Generation scheme of phased-necklace HHG, where the driving field is composed of two collinear linearly polarized vortex beams with opposite—and different—OAM content (ℓ_1, ℓ_2) and same frequency (ω_0). b) Tunability of the on-axis HHG spectra in He driven by a 800 nm field. The line spacing can be tuned through the OAM content of the driving fields. Panel c) shows the extremely low divergence of the on-axis harmonic beams. Remarkably, the divergence decreases with frequency [14].

Another way to structure ultrafast light pulses is to use a structured material. For example, HHG in an anisotropic material [15], such as graphene, possess the potential to structure the intensity and polarization state of light beams. Another recent proposal consists in the interaction of ultrafast azimuthally polarized beams with metals, where a large current oscillates at the laser frequency, generating ultrafast, intense longitudinal magnetic fields isolated from the electric fields [16] (see Fig. 2). Such structured sources offer an appealing alternative to study sub-femtosecond magnetization dynamics, where a complete understanding of the electronic and spin interactions remains unexplored.

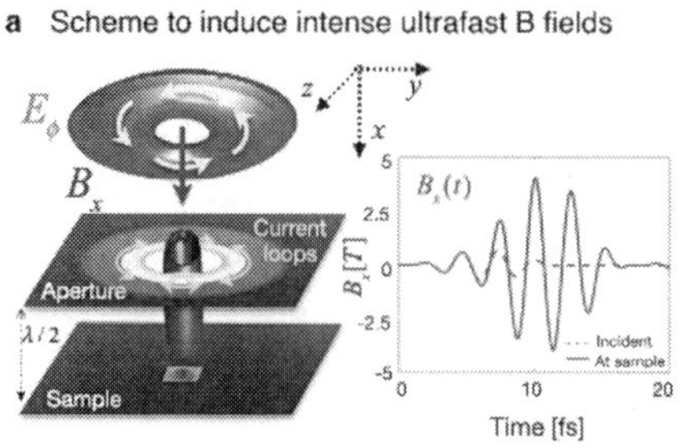

Fig. 2. Enhanced ultrafast magnetic field (blue) at a metal sample irradiated by an infrared femtosecond azimuthally polarized beam [16].

This project has received funding from the European Research Council (ERC) under the European Union's Horizon 2020 research and innovation program (Grant Agreement No. 851201, ATTOSTRUCTURA); Ministerio de Ciencia de Innovación y Universidades (PID2019-106910GB-I00, RYC-2017-22745); Junta de Castilla y León FEDER (SA287P18).

References

1. T. Popmintchev, et al. Science **336**, 1287 (2012).
2. X. Shi, et al. J. Phys. B **53**, 184008 (2020).
3. T. Fan, et al, Proceedings of the National Academy of Sciences **112** (46) 14206-14211 , (2015).
4. P. -C. Huang et al Nature Photonics **12**, 349-354 (2018).
5. K.-Y. Chang et al Optica, **8**, 484-492 (2021).
6. C. Hernández-García, A. Picón, J. San Román, and L. Plaja, Phys. Rev. Lett. **111**, *083602* (2013).
7. C. Hernández-García, Nature Physics *13*, 327-329 (2017).
8. A. Pandey, et al., ACS Photonics **9**, 944–951 (2022).
9. K. M. Dorney, et al. Nat. Photon. **13**, 123-130 (2019).
10. L. Rego, K. Dorney, N. Brooks, Q. Nguyen, C.-T. Liao, J. San Román, D. Couch, A. Liu, E. Pisanty, M. Lewenstein, L. Plaja, H. Kapteyn, M. Murnane, C. Hernández-García, Science **364**, eaaw9486 (2019).
11. L. Rego, J. San Román, L. Plaja and C. Hernández-García et al., Opt. Lett. **45**, 5636 (2020).
12. C. Hernández-García, A. Turpin, J. S. Román, A. Picón, R. Drevinskas, A. Cerkauskaite, P. G. Kazansky, C. Durfee, I. J. Sola, Optica 4, 520 (2017).
13. A. de las Heras, A. K. Pandey, J. S. Román, J. Serrano, E. Baynard, G. Dovillaire, M. Pittman, C. G. Durfee, L. Plaja, S. Kazamias, O. Guilbaud, and C. Hernández-García, Optica **9**, 71 (2022).
14. L. Rego, N. J. Brooks, Q. L. Nguyen, J. San Román, I. Binnie, L. Plaja, H. C. Kapteyn, M. M. Murnane, C. Hernández-García, Sci. Adv. **8**, eabj7380 (2022).
15. O. Zurrón-Cifuentes, et al, Optics Express **27**, 7776–7786 (2019).
16. M. Blanco, F. Cambronero, M. T. Flores-Arias, E. Conejero Jarque, L. Plaja, and C. Hernández-García, ACS Photonics **6**, 38 (2019).

https://doi.org/10.1051/epjconf/202226613018

Quadrics for Structuring Invariant Space-Time Wave Packets

Pierre Béjot[1] and *Bertrand* Kibler[1,*]

[1]Laboratoire Interdisciplinaire Carnot de Bourgogne, UMR6303 CNRS-UBFC, 21000 Dijon, France

Abstract. We provide a general approach for structuring invariant 3D+1 optical wave packets in both bulk and structured dispersive media, through a simple engineering of phase-matched space-time frequencies on quadric surfaces.

1 Introduction

The field of light structuring has made significant progress in the design, implementation and applications [1-3], in particular, propagation-invariant or non-diffracting beams. However, the pulsed nature of light waves and the structured property of optical media like waveguides are often overlooked. We here present a full spatiotemporal approach that takes advantage of tight correlations between the spatial modes, the topological charges, and the frequencies embedded in an optical field, in order to reveal propagation-invariant (dispersion- and diffraction-free) space-time wave packets carrying orbital angular momentum (OAM) in both bulk media or multimode fibers [4]. We demonstrate that such wave packets can evolve on spiraling trajectories in both time and space (i.e., helicon wave packets), and their phase-matched space-time frequencies generally lie on quadric surfaces [5]. Besides their intrinsic linear nature, we show that such wave structures can spontaneously emerge when a rather intense ultrashort pulse propagates nonlinearly in OAM modes [4].

2 Results

We investigate space-time helicon wave packets in the framework of cylindrically symmetric waveguides. The bulk case can then be seen as the limiting case of waveguide of infinite dimension. For simplicity, we restrict our study to a scalar approach by considering the weak guidance approximation. The partial differential equation driving the linear propagation of an electric field E is then given in cylindrical coordinates by:

$$\left[\partial_z^2 + \partial_r^2 + \frac{1}{r}\partial_r + \frac{1}{r^2}\partial_\theta^2 + \frac{n^2(r,\omega)\omega^2}{c^2}\right]\tilde{E}(r,\theta,z,\omega) = 0 \quad (1)$$

where $n(r,\omega)$ is the radial-dependent refractive index. By separating transverse and longitudinal variables, the linear evolution of the field along z-axis is given in the modal basis by: $\partial_z\bar{E} = iK_z(l,p,\omega)\bar{E}$, where $K_z(l,p,\omega)$ is the propagation constant of the OAM mode (l,p) at frequency ω. Recall that l and p refer here to azimuthal and radial indices, where $l(0,\pm1,\pm2,...)$ is the topological charge, related to the phase front of OAM modes.

Next, one has to look for invariant fields under uniform motion but also invariant at a given rotation around the z axis. Such fields belong to the kernel of the space-time screw axis symmetry differential operator $\Pi = \partial_z + K_1\partial_t + K_l L_z - iK_0$, where K_l is an arbitrarily chosen constant and $L_z = x\partial_y - y\partial_x = \partial_\theta$ is the z-component of the angular momentum operator. This implies that the decomposition of such wave packets only embeds a family of eigenvectors whose eigenvalues respect:

$$K_z(l,p,\omega_{lp}) = K_0 + K_1\omega_{lp} + K_l l \quad (2)$$

Accordingly, any electric field built from a given family is a diffraction- and dispersion-free space-time wave packet propagating at the group velocity $1/K_1$ whose intensity continuously rotates around the propagation axis with a spatial period $2\pi/K_l$. By considering only second-order dispersion around a given arbitrarily chosen frequency ω_0, we show that the propagation constant of a waveguide can be well approximated as:

$$K_z \simeq k_0 + k_1\Omega + \frac{k_2}{2}\Omega^2 + \Gamma(l,p,\Omega) \quad (3)$$

with Γ is a three-dimensional second-order polynomial. It then follows that spatiotemporal helicon wave packets are contained in surfaces whose algebraic expression is a quadric, whose surface depends on the considered dispersion regime (normal or anomalous).

Below, a simple and analytical case at the frontier between bulk media and fibers is studied, namely a dispersive medium of finite transversal dimension (radius $R = 100$ μm). In the paraxial approximation, the propagation constants can be approximated by:

$$K_z(l,p,\omega) \simeq k_0 + k_1\Omega + \frac{k_2}{2}\Omega^2 - \frac{\alpha_{lp}^2}{2k_0R^2} \quad (4)$$

where α_{lp} is the pth root of lth Bessel function of first kind J_l. Using the formalism described above, with a second-order polynomial fit of α_{lp}^2, one can find that the

* Corresponding author: bertrand.kibler@u-bourgogne.fr

quadric surface associated to spatiotemporal helicon wave packets in finite bulk media has a rank of 3 [5]. Corresponding typical examples are shown in Fig. 1 in both dispersion regimes of a fused silica rod.

Figure 1(a) shows the three-dimensional pattern (i.e., quadrics) of phase-matching obtained in the normal dispersion regime, when considering a central wavelength $\lambda_0 = 2\pi c/\omega_0 = 800$ nm, and the arbitrary choice of the constants $K_1 = k_1$ and $K_0 = k_0$ (i.e., propagation constants of the wave packet are those of ω_0). The rotation period was fixed to $2\pi/|K_l| = 1$ cm. The linear construction of a helicon wave packet then results from the superposition of the calculated frequencies Ω_{lp}. Figure 1(b) presents an example corresponding to the particular case of selected modes with various radial and angular indices ($0 \leq l \leq 8$, and $1 \leq p \leq 8$), and with equal spectral amplitudes and phases (see circles in Fig. 1a). The successive subplots display the iso-surface of the spatiotemporal intensity at half-maximum at distinct propagation distances. The helicon wave packet looks like a corkscrew in space-time coordinates, the points of high intensity being extremely localized in both space and time since all frequencies are in phase. This ultrashort structure does not disperse and simply rotates during its propagation around the z- and t-axis because of the inherent invariant nature of the constructed wave packet. Another example of quadric obtained in the anomalous dispersion regime is shown in Fig. 1(c), when considering the central wavelength at 1800 nm, and the constants $K_1 = k_1$ and $K_0 = k_0$. The rotation period is now fixed equal to 1.25 cm. Figure 1(d) presents an example of constructed helicon wave packet obtained from the superposition of modes: $0 \leq l \leq 22$, and $1 \leq p \leq 2$, and with equal spectral amplitudes and phases (see circles in Fig. 1c). Again we retrieve the typical invariant spiraling trajectory of the intensity pattern, with a more complex spatiotemporal pattern made of three consecutive high-intensity sub-pulses.

Finally, we will demonstrate numerically that spiraling wave packets can naturally emerge from the nonlinear propagation of rather intense ultrashort pulses [4]. As for conical waves emerging during filamentation in bulks, the formalism of helicon wave packets brings a proper understanding of the phenomena taking place when intense pulses embedding different OAM beams propagate in bulks and waveguides. Given the current strong interest of both OAM-carrying beams and fiber modes, we expect that this work will stimulate further research in the field.

References

1. H. Rubinsztein-Dunlop *et al.*, J. Opt. **19**, 013001 (2017)
2. A. Forbes, M. de Oliveira, M.R. Dennis, Nat. Photon. **15**, 253 (2021)
3. H.E. Kondakci, A.F. Abouraddy, Nat. Photon. **11**, 733 (2017)
4. P. Béjot, B. Kibler, ACS Photon. **8**, 2345 (2021)
5. P. Béjot, B. Kibler, ACS Photon. (in press, 2022)

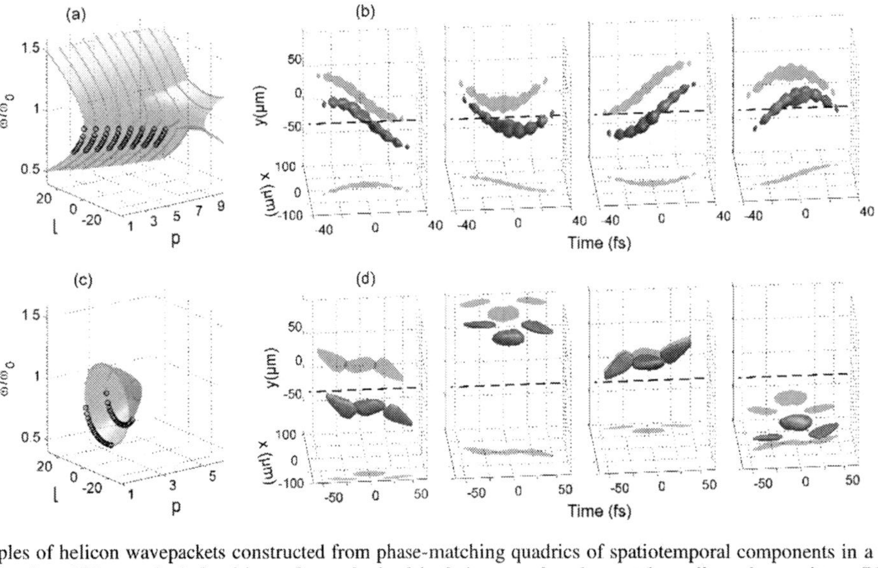

Fig. 1. Examples of helicon wavepackets constructed from phase-matching quadrics of spatiotemporal components in a fused silica rod with radius $R = 100$ µm. (a,c) Quadric surfaces obtained both in normal and anomalous dispersion regimes. Black circles correspond to the selected modes used for the linear construction of helicon wavepackets depicted in (b,d) through iso-surfaces of their spatiotemporal intensity pattern at half-maximum. The successive subplots obtained every $\pi/(2|K_l|)$ confirm the invariant nature of helicon wavepackets over propagation. The dashed line indicates the origin ($x = 0, y = 0$). Projections on planes (shadow plots) are provided for a clear observation of rotation.

https://doi.org/10.1051/epjconf/202226613019

Extracting complex refractive indices from THz-TDS data with artificial neural networks

Nicholas T. Klokkou[1,*], *Jon* Gorecki[2], *James S.* Wilkinson[3], and *Vasilis* Apostolopoulos[1]

[1]School of Physics and Astronomy, University of Southampton, Southampton, SO17 1BJ, UK
[2]School of Engineering and Materials Science, Queen Mary University of London, Mile End Road, London, E1 4NS, UK
[3]Optoelectronics Research Centre, University of Southampton, Southampton, SO17 1BJ, UK

Abstract. Terahertz time-domain spectroscopy (THz-TDS) benefits from high signal-to-noise ratios (SNR), however extraction of material parameters involves a number of steps which can introduce errors into the final result. We present the use of artificial neural networks (ANN) as the first step to achieve a comprehensive approach for the extraction of the complex refractive index from THz-TDS data. The ANN shows performance superior to approximation methods and has a more straightforward implementation than root finding methods. Deep and convolutional neural networks are demonstrated to accept an entire frequency range at once, providing a tool for fitting where SNR is low, producing a more stable result.

1 Introduction

By measuring the amplitude of the electric field of a broadband pulse, a terahertz time-domain spectrometer (THz-TDS) is able to retain the phase information of the radiation propagating through a sample when a Fourier transform is performed and gives us the possibility to extract the complex refractive index of a material. However, obtaining a material's frequency dependent complex refractive index in THz-TDS requires multiple analysis steps, each of which can introduce errors into the calculations. Along with difference in experimental procedures, the data analysis contributes to a concerning amount of variability in results between research groups around the world [1]. An artificial neural networks (ANN) of even a modest architecture has enough flexibility to be able to approximate any function [2], but unlike an analytical description of a function, these networks are trained to converge on the correct and accurate mapping as long as enough training data is provided and an appropriate training algorithm is used. It follows that such a network could analyse THz-TDS data and output the associated material's refractive index, with the accuracy being limited only by the size of the data set and number of neurons in the hidden layer. Here we show how ANNs of varying complexity can extract material parameters as a step to reduce the variability of data analysis in THz-TDS and help its adoption in industrial settings.

2 Frequency point wise extraction

The theoretical transfer function $\tilde{H}_{theo}(\omega)$ describes the mapping of a reference pulse propagating through air onto a pulse that is transmitted through a material, is given by

$$\tilde{H}_{theo}(\omega) = \frac{4\tilde{n}(\omega)}{(\tilde{n}(\omega)+1)^2} e^{-i\tilde{n}(\omega)\omega d/c}, \quad (1)$$

where the refractive index of the material is $\tilde{n}(\omega) = n(\omega) + ik(\omega)$, d is its thickness, c is the speed of light and ω is the angular frequency. The relation is only solvable analytically if the imaginary component of the refractive index is much smaller than the real part, otherwise, a root-finding method must be used.

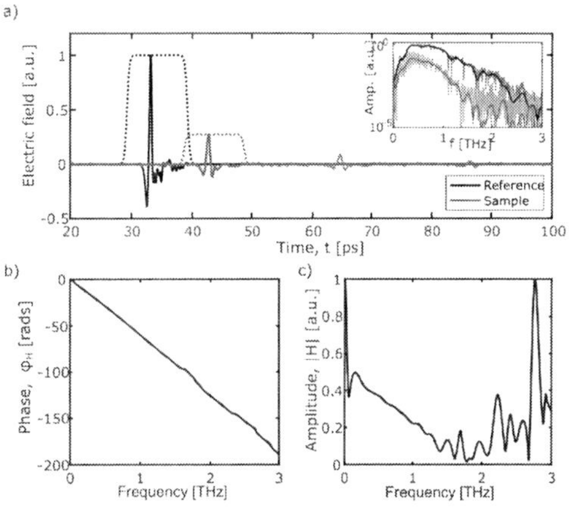

Figure 1. THz-TDS measurements of an air reference and a lithium niobate sample. The time-domain traces are windowed (a) with their spectra shown in the inset. The division of the sample and reference spectra give an experimental transfer function with both phase (b) and magnitude (c).

Figure 1(a) shows a time-domain measurement reference (air) and a 500 micron thick sample of lithium nio-

*e-mail: n.t.klokkou@soton.ac.uk

bate with a windowing function applied to remove internal reflections for equation 1 to hold. The Fast Fourier Transform (FFT) of the signals are shown in the inset. The phase of the FFT is wrapped between 0 and 2π so reconstruction is required before division including offsetting in the y-axis, the result of which is shown in Figure 1(b).

Training sets were generated by simulating the propagation of a THz pulse through a slab material with a random complex refractive index and random thickness. The range of these were selected to encompass the values for most material samples measured with THz spectroscopy ($2<n<8$, $0<k<0.3$) while the thickness was kept between $100\,\mu$m and 1 mm. In total, the data set used for training contains 10^5 random materials each contributing one training entry selected at a random frequency. Each training entry has: (a) an input that contains the frequency, the corresponding value of the complex transfer function and the thickness of the sample, and (b) an output containing the real and imaginary refractive indices at this frequency.

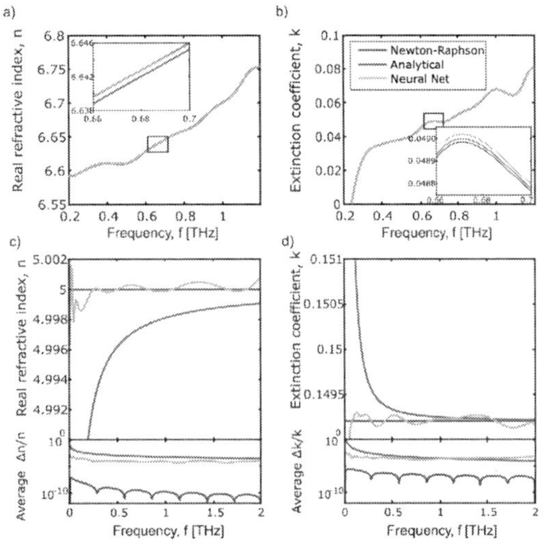

Figure 2. Examples showing the performance of different extraction methods for the real and imaginary refractive index of a measured sample of lithium niobate (a,b) and a simulated example of n = 5 (c), and k = 0.15, (d) published in [3]. The fractional difference across the entire parameter space of is shown underneath for each method.

3 Deep and Convolutional Neural Nets

The training of neural network architectures containing multiple hidden layers with different connections allows the recognition of features and trends that involve multiple inference steps. Two architectures were trialled, a fully connected deep neural network (DNN) of 512 and 128 neurons as hidden layers as well as a convolutional neural network (CNN) including a 200 neuron hidden layer. The networks accept a complete frequency range of the complex transfer function at 32 equally spaced points forming a 2x32 matrix. The thickness is fixed 500 microns and a noise floor is also simulated.

Figure 3 shows the extracted refractive index of the lithium niobate sample with the amplitude of the sample's FFT included as an indication of signal strength. The results of the CNN show a more stable result at low SNR in the low frequency range. Often, values at these frequencies are omitted, where their inaccuracies can results from the phase extraction. We aim to use CNNs to expand the usable frequency range further into this domain.

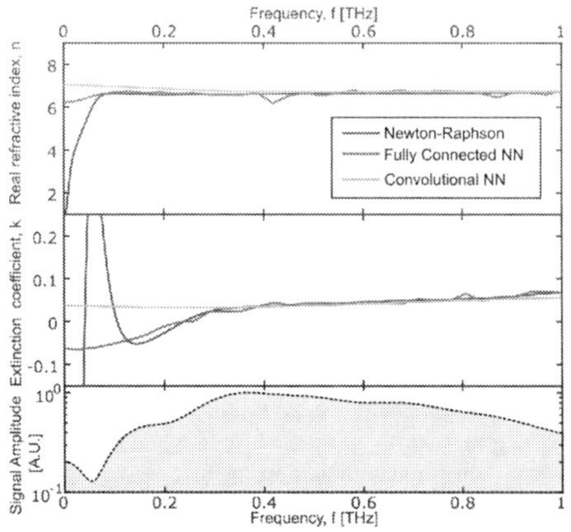

Figure 3. The first generation of DNNs and CNNs used to extract the real (top) and imaginary (middle) components of the refractive index of a lithium niobate sample. The CNN produces a more stable result, even where signal strength of the spectra of the sample measurement is low (bottom).

4 Conclusion

Terahertz time-domain data retains phase information in a way that simplifies parameter extraction as compared to other techniques. However considerable errors can arise in some steps of the extraction algorithms used by the community. Artificial neural networks have been shown to be used in place of traditional fitting functions with excellent accuracy, whilst also benefiting from fast runtime and straightforward implementation. Furthermore, deeper, more generalised networks can replace more steps of the extraction process, automating data analysis and increasing consistency. Advancing the sophistication of the networks will lead to more generalised models that can potentially improve on current methods, such as dynamically optimising the windowing process, all while retaining rapid run-times.

References

[1] M. Naftaly, *2016 41st International Conference on Infrared, Millimeter, and Terahertz waves (IRMMW-THz)*, pp. 1–2, (2016)

[2] K. Hornik, M. Stinchcombe, and H. White, *Neural Networks*, **vol. 2**, no. 5, (1989).

[3] N. Klokkou, J. Gorecki, J. S. Wilkinson, and V. Apostolopoulos, *Opt. Express* **30**, 15583-15595 (2022)

Spectral scaling transformations of nonstationary light

Jyrki Laatikainen[1,*], *Matias* Koivurova[2,3], *Jari* Turunen[1], *Tero* Setälä[1], and *Ari T.* Friberg[1]

[1]Institute of Photonics, University of Eastern Finland, P. O. Box 111, FI-80101 Joensuu, Finland
[2]Faculty of Engineering and Natural Sciences, Tampere University, FI-33720 Tampere, Finland
[3]Tampere Institute for Advanced Study, Tampere University, FI-33100 Tampere, Finland

Abstract. We present optical systems, which transform isodiffracting nonstationary beams into fields obeying either cross-spectral purity or spectral invariance. The designs are hybrid refractive-diffractive imaging systems, which are able to perform the desired transformations over a broad spectral bandwidth and irrespective of the state of spatial coherence of the input beam.

1 Introduction

Cross-spectral purity [1] and spectral invariance [2] are two cornerstone concepts of optical coherence theory, which enlighten the role of correlation-induced spectral changes on propagation and interference of light. Cross-spectral purity, introduced by Mandel in 1961, amounts to interference of two light fields, where the normalized spectrum of the superposition is the same as that of the input fields. Spectral invariance, on the other hand, was proposed by Wolf in 1986, and it refers to light fields for which the normalized far-field spectrum is independent of propagation direction.

While these concepts were originally introduced for stationary light, theoretical framework in the context of nonstationary fields was recently established for both cross-spectral purity [3] and spectral invariance [4]. An especially important class of nonstationary light is formed by the isodiffracting pulsed beams [5], which are encountered, e.g, as superpositions of Hermite-Gaussian modes in spherical-mirror resonators. These modes all have the same, frequency-independent Rayleigh range. Consequently, the transversal beam width satisfies the characteristic spectral scaling of isodiffracting beams.

In this work, we present optical systems which are capable of transforming isodiffracting beams into nonstationary fields obeying either cross-spectral purity or spectral invariance [6]. The systems are chromatically compensating hybrid lens designs consisting of one refractive and one diffractive lens, and they perform the transformations accurately over a broad spectral range.

2 Transformation systems

The transformation systems are based on the theory of paraxial, afocal imaging lenses. In particular, the transformations are achieved perfectly, within the paraxial approximation, when the spectral magnifications of the systems are [6]

$$m(\lambda) = \sqrt{\frac{\lambda_0}{\lambda}}m_0, \quad (1) \qquad m(\lambda) = \sqrt{\frac{\lambda}{\lambda_0}}m_0, \quad (2)$$

for cross-spectrally pure and spectrally invariant output beams, respectively. In these expressions, λ is the wavelength of light, λ_0 is chosen as the design wavelength of the system, and $m_0 = m(\lambda_0)$.

We employ hybrid lens systems, which consist of one achromatic refractive lens and one diffractive lens. An illustrative sketch of the thin-lens arrangement is presented in Fig. 1. Both transformation systems are similar designs, where the order of the two lenses and their separations specify the nature of the transformation. For cross-spectrally pure output beam, the refractive lens is placed before the diffractive lens, and if the order of the lenses is flipped, a spectrally invariant secondary source is achieved at the image plane of the system in Fig. 1. The systems perform the desired transforms approximately, with the ideal magnifications accomplished perfectly at λ_0. Figure 2 illustrates the accuracy of these approximations.

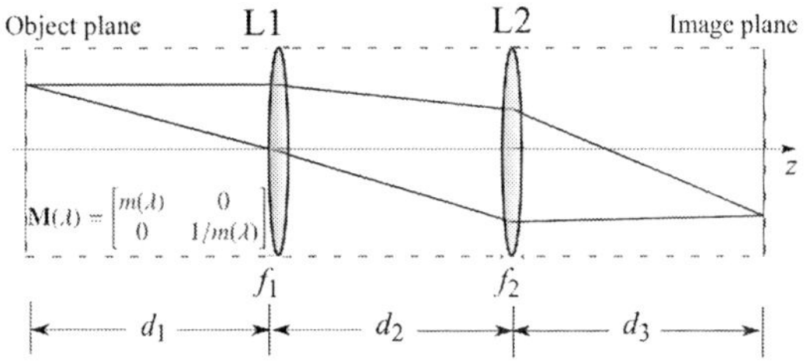

Figure 1. The transformation system, which is described within the blue dashed lines by the ray-transfer matrix **M**. The lenses L1 and L2 are refractive or diffractive thin lenses with respective focal lengths f_1 and f_2. The lenses are separated by distances d_1, d_2, and d_3, which, together with the order of the lenses, specify the nature of the transformation. Adapted from [6].

*e-mail: jyrki.laatikainen@uef.fi

 https://doi.org/10.1051/epjconf/202226613020

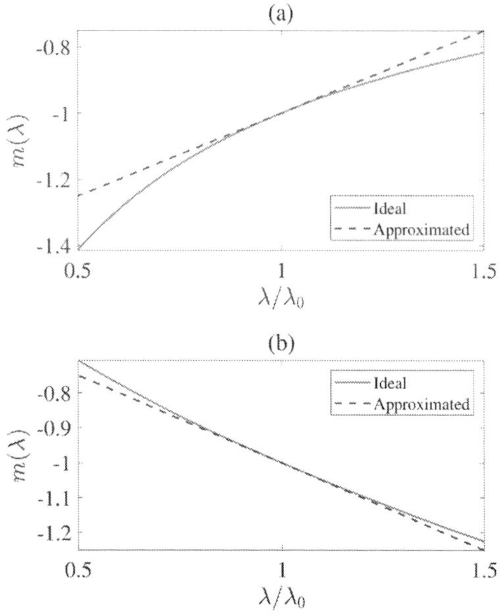

Figure 2. The magnifications for (a) cross-spectrally pure, and (b) spectrally invariant output fields. The solid lines show the ideal magnifications in Eqs. (1) and (2), together with the respective approximations (dashed line) accomplished with the thin-element designs in Fig. 1. Adapted from [6].

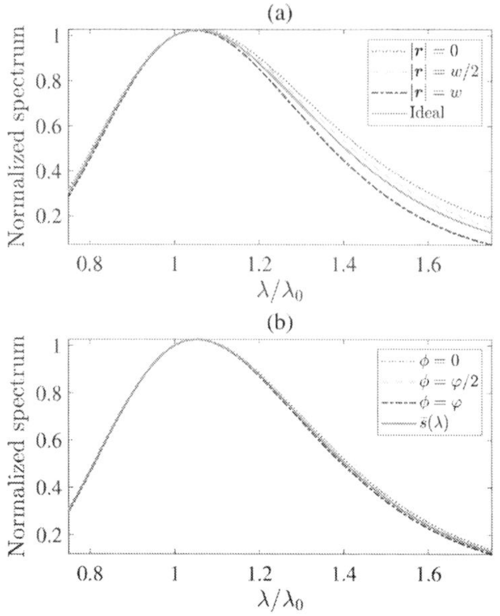

Figure 3. Numerically simulated performance of the transformation systems for pursuing a (a) cross-spectrally pure (b) spectrally invariant output beam. Adapted from [6].

3 Results

Figure 3 shows the numerically simulated performance of the presented thin-lens designs, where the input beam corresponds to a pulsed isodiffracting source with a broad spectral bandwidth. Figure 3 (a) illustrates the accuracy of the system designed for generating a cross-spectrally pure beam. The figure shows the normalized spectrum at three transversal positions $|\mathbf{r}|$ along the beam width on the image plane of the system. In addition, the ideal normalized superposition spectrum is shown. We notice that the isodiffracting beam is transformed into a nearly cross-spectrally pure field, apart from long wavelengths.

Figure 3 (b) presents the performance of the other system, which was designed to create a spectrally invariant secondary source on its image plane. The figure represents the far-field spectrum of the source for three values of the direction angle ϕ, which are relative to the $1/e^2$ beam divergence angle φ. In addition, the figure shows the normalized source-averaged spectral density, which is equal to the ideal, directionally invariant far-field spectrum obtained with the exact magnification in Eq. (2). We see that the results are even closer to the ideal situation over a wide range of wavelengths and the curves shown in Fig. 3 (b) are nearly indistinguishable.

4 Conclusions

We presented refractive-diffractive imaging systems, which transform nonstationary isodiffracting beams into cross-spectrally pure or spectrally invariant fields accurately over a broad spectral range. Finally, we note that the systems produce inverse transformations if the setup in Fig. 1 is reversed, i.e., the order of the object and image planes is flipped. An incoming cross-spectrally pure or spectrally invariant field is then transformed into an isodiffracting field, and this holds for both stationary or nonstationary beams.

References

[1] L. Mandel, J. Opt. Soc. Am. **51**, 1342 (1961).
[2] E. Wolf, Phys. Rev. Lett. **56**, 1370 (1986).
[3] M. Koivurova, C. Ding, J. Turunen, and A. T. Friberg, Phys. Rev. A **99**, 043842 (2019).
[4] C. Ding, M. Koivurova, T. Setälä, J. Turunen, and A. T. Friberg, Phys. Rev. A **101**, 033808 (2020).
[5] M. Koivurova, C. Ding, J. Turunen, and L. Pan, Phys. Rev. A **97**, 023825 (2018).
[6] J. Laatikainen, M. Koivurova, J. Turunen, T. Setälä, and A. T. Friberg, submitted (2022).

https://doi.org/10.1051/epjconf/202226613021

Applied Optics in the Development of Smart Asphalt Mixtures

Iran Rocha Segundo[*,1,2], *Salmon* Landi Jr.[3], *Cátia* Afonso[2], *Orlando* Lima Jr.[1], *Elisabete* Freitas[*,1], *Verônica* Castelo Branco[4], *Manuel F. M.* Costa[*,5] *and Joaquim* Carneiro[*,2]

[1]ISISE, Department of Civil Engineering, University of Minho, Azurém Campus, Guimarães, Portugal
[2]Centre of Physics of Minho and Porto Universities (CF-UM-UP), University of Minho, Azurém Campus, Guimarães, Portugal
[3]Federal Institute of Education, Science and Technology Goiano, Rio Verde – GO, Brazil
[4]Transportation Engineering Department, Federal University of Ceará, Fortaleza, Brazil
[5]Centre of Physics of Minho and Porto Universities (CF-UM-UP), University of Minho, Gualtar Campus, Braga, Portugal

Abstract. The functionalization of asphalt mixtures is carried out in order to provide new capabilities to the road pavements, with major social, environmental and financial benefits. Optical characterization techniques as well as optical processes like photocatalysis play a major role in the development of new asphalt mixtures with smart functions. These advanced capabilities which are being developed in asphalt mixtures are: photocatalytic, superhydrophobic, self-cleaning, de-icing/anti-ice, self-healing, thermochromic, and latent heat thermal energy storage. The main objective of this research work is to stress the importance of optics and photonics technologies giving an overview of advanced functionalized smart asphalt mixtures.

1 Introduction

The road pavements must resist to vehicle traffic and climate actions, and also assure the rolling conditions with comfort, economy, and safety with low impact on the environment. Recently, the functionalization of road pavements and their multifunctional capabilities have become an important subject due to the large area covered by these roadways cover. Optical techniques are extensively used in order to design these smart asphalt mixtures. The literature has already shown important results in the functionalization of the asphalt road pavements and also the development of smart and multifunctional asphalt mixtures, mostly for the top layer. The main important smart and multifunctional capabilities applied to asphalt mixtures addressed in this work are: photocatalytic, superhydrophobic, self-cleaning, de-icing/anti-ice, self-healing, thermochromic, and latent heat thermal energy storage [1].

2 Photocatalytic Capability

Air pollution is one of the biggest environmental problems in urban centers. Road traffic is one of the most important sources. For this reason, the photocatalytic capability can be provided in asphalt mixtures. Semiconductor oxides, such as TiO_2, have the ability to oxidize pollutants, such as NO_x, SO_2, and also to degrade organic compounds in the presence of ultraviolet (UV) light, oxygen and humidity [2]. Spectrophotometric and gas chromatography techniques are used to analyze this capability.

3 Superhydrophobic Capability

On road pavements, the worst weather condition occurs when there is water or ice on the surface. Thus, it is important to drain or repel the surface water quickly. If the water droplets are repelled from the surface, the ice/snow formation is avoided [3]. This approach provides safer roads during rain and low temperatures by the application of superhydrophobic coatings containing Polytetrafluoroethylene (PTFE), TiO_2, SiO_2, among other particles. The surface roughness is a critical aspect that is analyzed by optical methods, including by optical triangulation based-microtopography [4].

4 Self-cleaning Capability

Self-cleaning is promoted in order to remove dirt particles or degrade organic compounds, i.e., oil and greases, over the surface, contributing to the reduction of accidents in oil spilled areas and also in dusty areas. The self-cleaning capability is achieved in three types of surface: i) superhydrophobic: water droplets roll on the surfaces and carry the dirt previously deposited, this phenomenon is called Lotus Flower effect, ii) superhydrophilic: water droplets are spread over the surface, which is more easily washed by removing adsorbed dirt on its surface during rainy periods, and iii) photocatalytic: the photocatalytic materials can degrade organic pollutants (like oils and greases), cleaning the surface.

* Corresponding author: iran_gomes@hotmail.com; efreitas@civil.uminho.pt; mfcosta@fisica.uminho.pt; carneiro@fisica.uminho.pt

5 De-icing and Anti-ice Capabilities

As mentioned before, snow and ice in winter reduce road safety due to the reduced friction between tires and pavement. In order to mitigate this problem, numerous researches point to the use of de-icing agents and/or conductive materials in the asphalt mixtures [5,6]. The first one can melt the ice and snow from a chemical process. The other is associated with a microwave or an induction-heating process to melt the ice/snow.

6 Self-healing Capability

Cracking of the asphalt mixtures is one of the most important degradation processes. By using conductive materials, self-healing agents or nanomaterials, it is possible to develop the self-healing capability in order to close microcracks. This capability provides a longer lifetime to pavements, reducing the rehabilitation needs and, consequently, requiring less materials for paving, leading to less emission of CO_2, and traffic disruption [7]. Optical inspection is used in order to observe the cracking healing. Also, asphalt binder tests regarding rheology and mechanical tests concerning fatigue are carried out to evaluate the self-healing.

7 Thermochromic Capability

Thermochromic materials have been incorporated in Civil Engineering materials due to its change of optical and thermal properties. The Urban Heat Island (UHI) is an urban area that is significantly warmer than its adjacent rural areas due to human actions. By the introduction of thermochromic materials into the asphalt binder, the conventional black color is changed to another lighter color in order to reduce the energy absorption from sunlight, reducing the surface temperature. In this sense, thermochromic asphalt mixtures can mitigate the Urban Heat Island (UHI) and also improve mechanical and aging properties [8]. Spectrophotometric techniques are used to assess this capability.

8 Latent Heat Thermal Energy Storage (LHTS) Capability

A Phase Change Material (PCM) is a substance that presents a high fusion heat, melts and solidifies at a certain temperature, and is able to store/release large amounts of thermal energy. With the use of PCM, the Latent Heat Thermal Energy Storage (LHTS) capability is developed. When this technology is applied on asphalt mixtures, the UHI is mitigated, and their mechanical properties are improved by the reduction of the magnitude of temperature fluctuations [9]. By the increase of the thermal comfort in urban areas, there will exist social benefits besides the reduction of the energy consumption from heaters and air conditioners (lower use), consequently, decreased emissions.

9 Conclusions

This research work aimed to present a literature overview of smart and multifunctional capabilities applied to asphalt mixtures. A smart and multifunctional asphalt mixture is an intelligent material with different properties that differ from the conventional abilities or reacts upon an external stimulus, such as light irradiation, stress, and temperature. Optical characterization and analysis is used for its design. The roads have a large area that can provide benefits to the society and the environment, besides having a good mechanical structure and good superficial properties by the application of new capabilities. Moreover, the functionalization process of Civil Engineering materials can enhance the dynamism of the nanomaterials and nanotechnology industry due to the production and use in large scale. These sustainable surfaces can contribute to the "Green Recovery" model by developing smart materials that contribute to social, economic and environmental aspects.

Acknowledgments

This work was supported by the Portuguese Foundation for Science and Technology (FCT) in the framework of the Strategic Funding UIDB/04650/2020.

This work was partially financed by FCT - Fundação para a Ciência e a Tecnologia - under the projects of the Strategic Funding UIDB/04650/2020, MicroCoolPav project EXPL/EQU-EQU/1110/2021, and NanoAir project PTDC/FISMAC/6606/2020.

References

1. I. Rocha Segundo, E. Freitas, V.T.F.C. Branco, S. Landi, M.F. Costa, J.O. Carneiro, Renew. Sustain. Energy Rev. **151** (2021) 111552.
2. I. Rocha Segundo, E. Freitas, S. Landi Jr, M.F.M. Costa, J.O. Carneiro, Coatings **9** (2019).
3. A. Arabzadeh, H. Ceylan, S. Kim, K. Gopalakrishnan, A. Sassani, Transp. Res. Rec. J. Transp. Res. Board **2551** (2016) 10–17.
4. Costa, M.F.M. "Optical triangulation-based microtopographic inspection of surfaces". Sensors 12 (4): 4399-4420 (2012).
5. J. Gao, A. Sha, Z. Wang, Z. Tong, Z. Liu, J. Clean. Prod. **152** (2017) 429–442.
6. M. Zheng, J. Zhou, S. Wu, H. Yuan, J. Meng, Constr. Build. Mater. **84** (2015) 277–283.
7. J. Norambuena-Contreras, A. Garcia, Mater. Des. **106** (2016) 404–414.
8. H. Zhang, Z. Chen, G. Xu, C. Shi, Fuel **211** (2018) 850–858.
9. B.J. Manning, P.R. Bender, S.A. Cote, R.A. Lewis, A.R. Sakulich, R.B. Mallick, Sustain. Cities Soc. **19** (2015) 11–16.

https://doi.org/10.1051/epjconf/202226613022

Ultraestable spatiotemporal characterization of optical vortices in the visible and near infrared

Miguel López-Ripa*, *Íñigo J.* Sola, and *Benjamín* Alonso

Grupo de Aplicaciones del Láser y Fotónica (ALF), Departamento de Física Aplicada, Universidad de Salamanca, Plaza de la Merced s/n 37008 Salamanca, Spain

Abstract. We show the versatility of the bulk lateral shearing interferometer characterizing complex spatiotemporal structures in different spectral ranges. Specifically, we have characterized constant and time-varying optical vortices in the visible and near infrared spectral ranges respectively. The high stability of the system combined with its spectral versatility will ease the spatiotemporal characterization of ultrafast phenomena.

1 Introduction

Recently the evolution of laser technology has enabled the possibility of obtaining complex ultrashort laser beams with multiple applications, for example [1,2]. As the beam complexity has increased, steadily new spatiotemporal characterization techniques have been proposed to analyse them [3]. However, these techniques usually involve complex systems with high dependency to external perturbations. In our recent work [4], we have implemented an in-line and ultraestable spatiotemporal characterization technique, which solves these issues. Furthermore, the technique has the additional advantage of being able to operate in different spectral ranges without modifications.

In this contribution, we present the application of the spatiotemporal technique to characterize a type of complex ultrashort laser beams called optical vortices in different spectral ranges. In particular, we characterize constant and time-varying vortices in the visible and near infrared spectral ranges, respectively.

2 Spatiotemporal technique

The spatiotemporal characterization technique [4] is based on the properties of uniaxial birefringent crystals to implement a bulk lateral shearing interferometer.

Essentially, the technique splits the scalar beam under test into two replicas and introduces a lateral displacement (i.e., walk-off) and a temporal delay between them. Thus, it combines the principles of spectral and lateral interferometry to measure the spatiospectral spatial phase gradient of the beam in the walk-off direction. Then, knowing the spectral phase at a reference point, which can be measured with a conventional temporal characterization technique, the gradient is integrated obtaining the complete spatiospectral or spatiotemporal phase profile. In our case we used an in-line and robust temporal characterization technique called amplitude swing [5] to obtain the reference phase, so the whole system is robust against external perturbation.

In order to perform 2D characterizations it is required to measure the spatiospectral gradient for two perpendicular directions of the walk-off and this is done by rotating the crystal that introduces the lateral displacement (birefringent crystal with the optical axis not contained in the crystal face, hereafter walk-off crystal). Furthermore, if the temporal delay introduced by the walk-off crystal is not suitable, it can be adapted using a retarder plate (birefringent crystal with the optical axis contained in the crystal face), which introduces a certain delay but not lateral displacement. In Fig. 1 it is shown the scheme of the spatiotemporal technique for the configuration with horizontal walk-off.

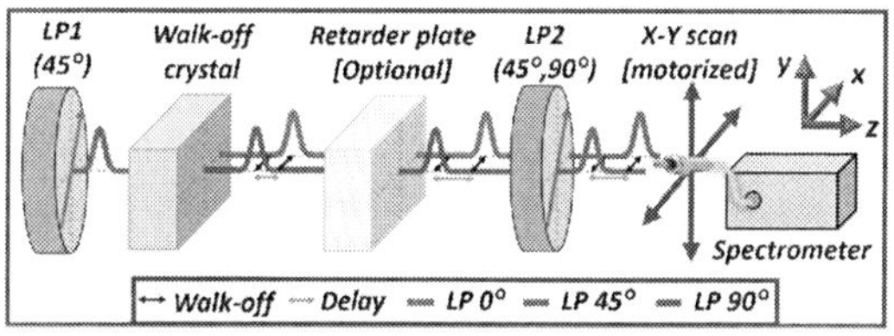

Fig. 1. Scheme of the spatiotemporal characterization technique introducing the walk-off in the horizontal direction.

3 Application to optical vortices

We used a Ti:Sapphire ultrashort laser (central wavelength around 800 nm) as light source and the optical vortices were generated using a nanostructured plate known as s-waveplate [4,6].

* Corresponding author: miguellr@usal.es

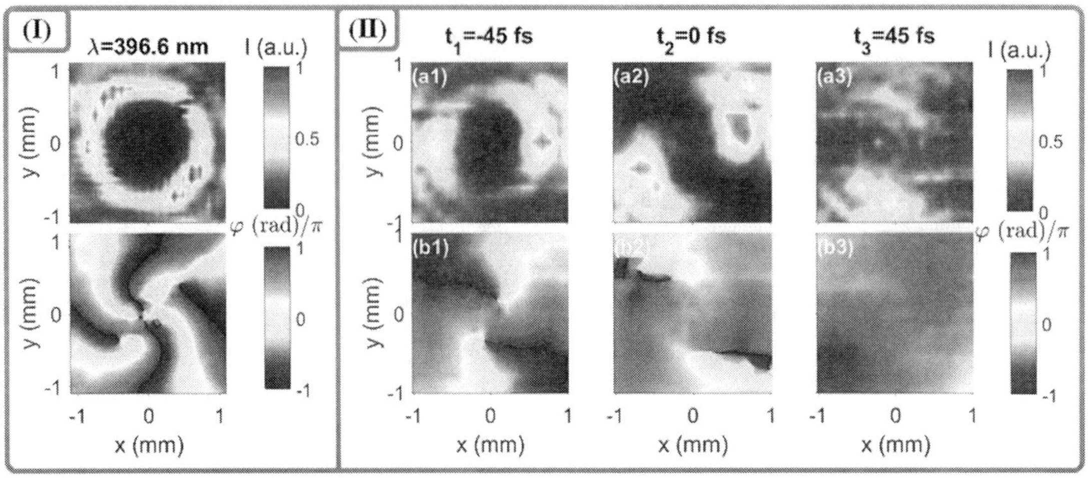

Fig. 2. Intensity (first row) and phase (second row) profiles for: (I) Frame for the central wavelength (396.6 nm) of the complete spatiospectral characterization of an ℓ=+4 optical vortex in the visible range. (II) Three temporal frames of the complete spatiotemporal characterization of the combination of ℓ=+2 and ℓ=0 optical vortices with a 65-fs delay between them.

As explained in [6], depending on the incident beam properties, the s-waveplates can generate two circularly polarized optical vortices of different orbital angular momentum (OAM or ℓ). Then, using linear polarizers and zero order waveplates we can manipulate them and select only one of the vortices or a combination of both [4].

3.1. Constant optical vortices in visible range

Although our s-waveplates are designed to generate optical vortices in the Ti:Sapphire spectral range (around 800 nm), we can generate vortices in the visible range by doubling the frequency using a second harmonic crystal.

Following this idea, we transformed a constant ℓ=+2 optical vortex in the near infrared into an ℓ=+4 optical vortex in the visible. In Fig. 2(I) it is shown the frame corresponding to the central wavelength (396.6 nm) of the complete spatiospectral characterization obtained for the ℓ=+4 optical vortex.

3.2. Time-varying optical vortex in near infrared

Another possibility as explained in [4] is to introduce a delay of the order of the FWHM of the initial beam to obtain two delayed optical vortices. The average spatiotemporal phase profile of that combination corresponds to a time-varying OAM.

Fig. 2(II) shows three frames of the complete spatiotemporal characterization of the combination of ℓ=+2 and ℓ=0 optical vortices with a delay of 65 fs between them. On the one hand, t_1 and t_3 corresponds to the tails of the average pulse in which we only have one optical vortex of ℓ=+2 and ℓ=0, respectively. On the other hand, t_2 shows the combination of both for the central time of the average pulse. This evolution agrees with the theoretical simulations as shown in [4].

4 Conclusions

In conclusion, the robust and ultraestable spatiotemporal characterization technique can ease the characterization of complex spatiotemporal structures in different spectral ranges without modifications. Due to its versatility and robustness, we believe the system can become a keystone in the characterization of ultrafast phenomena in different applications.

References

1. C. Jing, Z. Wang, Y. Cheng, Appl. Sci. **6**, 428 (2016)
2. Y. Shen, X. Wang, Z. Xie, C. Min, X. Fu, Q. Liu, M. Gong, X. Yuan, Light Sci. Appl. **8**, 90 (2019)
3. C. Dorrer, IEEE J. Sel. Top. Quantum Electron. **25**, 3100216 (2019)
4. M. López-Ripa, I. J. Sola, B. Alonso, Photon. Res. **10**, 922-931 (2022)
5. B. Alonso, W. Holgado, I. J. Sola, Opt. Express **28**, 15625-15640 (2020)
6. I. Lopez-Quintas, W. Holgado, R. Drevinskas, P. G. Kazansky, I. J. Sola, B. Alonso, J. Opt. **22**, 095402 (2020)

https://doi.org/10.1051/epjconf/202226613023

Plug and Play Colorimetric Carbon Dioxide Sensor

Nuno Mendes[1,2*], *João Mendes*[2,3], *Pedro Jorge*[1,2] and *Luís C. C. Coelho*[1,2]

[1]Department of Physics and Astronomy, Faculty of Sciences of University of Porto, 4169-007 Porto, Portugal
[2]INESC TEC – Institute for Systems and Computer Engineering, Technology and Science, and Faculty of Sciences of University of Porto, 4169-007 Porto, Portugal
3Chemistry Research Unit – Chemistry and Biochemistry Department, Faculty of Sciences of University of Porto, 4169-007, Porto, Portugal

Abstract. Carbon dioxide measurement is an important endeavor in many industries such as food packaging, grain storage and health industry. This work presents a reversible, plug and play and low-cost colorimetric CO_2 sensor calibrated in a proper concentration ranging from 1% to 3% of CO_2. The sensor showed potential for improvement to increase resolution, for measuring lower CO_2 concentrations and for more accurate readings.

1. Introduction

The Earth's atmosphere is made up mostly of Nitrogen (N_2 - 78.08%), Oxygen (O_2 - 20.95%) and Argon (Ar - 0.93%), the remainder being formed by other gases, led by Carbon Dioxide (CO_2- 0.04%) [1]. The monitoring of atmospheric CO_2 is currently one of the main topics on the table of several governments and environmental associations. For the last 200 years, the atmospheric concentration has increased from 280 parts per million(ppm) to 400 ppm.
In 2019, the market size for gas sensors was valued at around 823.1 million$[2]. This means that gas sensors are sought out from both private and public perspectives. For carbon dioxide sensors in specific, there are many industries that need accurate, small, and robust sensors for real time monitoring in a wide variety of applications.

Nowadays, there is a high demand for safe, healthy, and fresh food to be delivered at command. In modified atmosphere packaging, food is packaged, and sealed, inside an altered atmosphere, where carbon dioxide is used has a control factor to protect the food from spoils, pathogens and to extend shelf life.[3]

In this work, an optical sensor for CO_2 monitoring is presented based on the absorption of a sensitive membrane with an accurate reading obtained through optoelectronic components in a specially designed configuration.

2 Materials and Methods

2.1. Sensing Membrane

A CO_2 sensor was built around a colorimetric membrane manufactured with the following chemical reagents: p-nitrophenol (pNPh) (Sigmal-Aldrich ReagentPlus, >99%), the pNPh was used as a starting monomer for the synthesis of the poly pNPh(NP), tetraethylammonium hydroxide solution (TOA-OH; 20% in methanol, Sigma-Aldrich) was used as a quaternary ammonium (TOA-OH), Hydrogel D4 (Advance Source Biochemicals) was used as polymer to produce a homogeneous membrane, and Sylgard 184(Dow Corning, 10:1, Midland, MI, USA) was used to encapsulate and protect the chemical membrane following the procedure from[4].

For the colorimetric membrane development, a chemical cocktail was prepared by dissolving 5 mg of poly p-nitrophenol (NP) in 0.05 ml of a MeOH:H2O(1.5:1) mixture followed by the addition of 0.5 ml of a 0.5 M TOA-OH solution. After total dissolution, it was added 0.1 ml of Hydrogel D4 solution (10% in Ethanol,96%) and the resulted solution was stirred for a few minutes. The final cocktail was then spread on a mylar foil by spin-coating (900rpm for 60 seconds) and then left to dry for about two hours at room temperature and stored in the fridge prior to use.

The sensing dye chemistry is based on an ion-pair formation resulted from the interaction of the quaternary ammonium and the colorimetric polymer (TOA^+NP^-), keeping the sensitive dye in its anionic form. The interaction with the gaseous CO_2 is linked to the association of a few molecules of water ($TOA^+NP^-.xH_2O$) that come from air humidity. Equation 1 describes this process.

$$TOA^+ + NP^-.\chi H_2O + CO_2 \leftrightarrows TOA^+HCO_3^-.(\chi-1)H_2O.HNP \quad (1)$$

2.2. Sensing prototype

The prototype presented in figure 1 is based on the absorbance measurements of the sensing membrane through the transmission of a light signal from a white LED (model c503d, Cree, United Kingdom) towards to an RGB photodetector (model s9032-02, Hamamatsu, Japan) using two collimating lenses. This light signal is then acquired in the photodetector which is wavelength selective in three distinct bands with embed spectral filters corresponding to red (630 nm), green (555 nm) and blue

* NM: nuno.a.mendes@inesctec.pt; JM: joao.p.mendes@inesctec.pt; PJ: pedro.jorge@inesctec.pt; LC: luis.c.coelho@inesctec.pt

https://doi.org/10.1051/epjconf/202226613023

(450 nm). The absorption band of the sensing membrane due to the CO_2 absorbance is at 470 nm so the blue band is used for the sensing signal while the red band is used for the reference signal. Each light signal is converted into a current, amplified and digitalized locally. The digital information is processed in real time in a computer using a LabVIEW software.

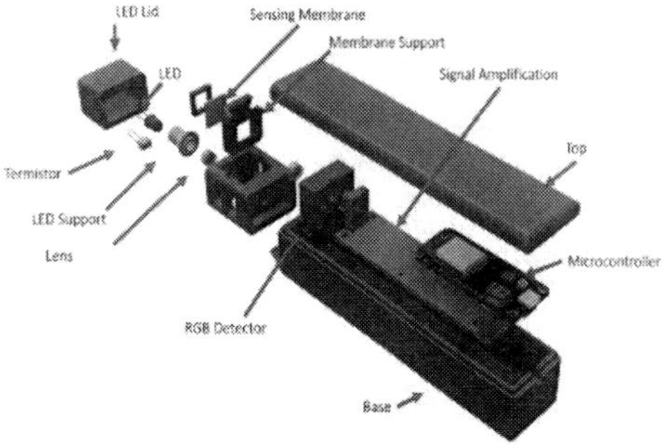

Fig. 1. Exploded view of the sensing prototype showing all the internal components

2.3. Measurements

The sensor works by detecting specific absorption variations of the chemical membrane in reaction with CO_2. As the gas concentration of CO_2 increases, an internal reaction occurs in the membrane, decreasing its pH. Depending on its internal pH levels the membrane, with a strong yellow color, becomes colorless as the pH increases and recovers its color as the pH decreases.

The sensor was tested in the setup illustrated in Figure 2, by injecting CO_2 and nitrogen (N_2) controlled by mass-flow controllers (Brooks, SLA5800 Series; Hatfield, PA, USA), in a gas mixture vessel before performing the measurements in the measurement chamber. Between controlled concentrations the gas inside the chamber was removed by injecting N_2 which pushes the remaining gas out of the chamber, as a base line atmospheric CO_2 was used. This process was repeated for each injected concentration.

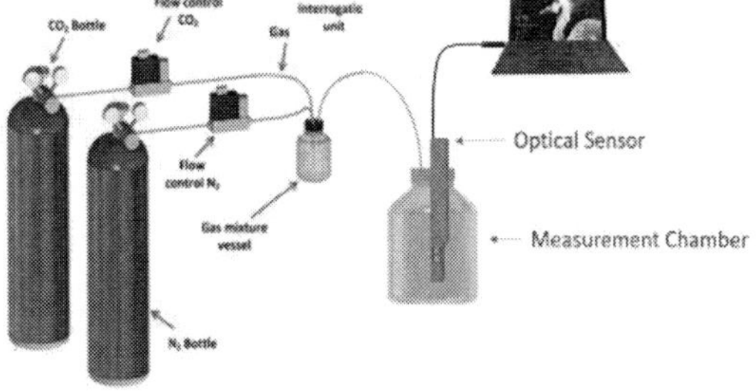

Fig. 2. Measurement setup (Adapted from[4])

3 Experimental Results and Discussion

The sensor was evaluated in controlled atmosphere and a preliminary calibration was done for 1, 2 and 3% of CO_2 in pure N_2. The signal from the white LED crosses the sensing membrane and is acquired by the RGB. Figure 3 (a) shows the time response of the sensor placed in the measurement setup for different concentrations of CO_2 and repeated in time. After each cycle the measurement chamber was filled with atmospheric gas.

The Blue/Experimental signal is used to perform the CO_2 measurement, while the red signal is only used as a reference signal to eliminate external fluctuations in real-time. The reference signal shows the stability of the LED, while correcting any external factors that could affect the whole signal. All signals have a logarithmic behavior for the three CO_2 concentrations, so a logarithm of base 10 was applied to the X axis to enable linear response.

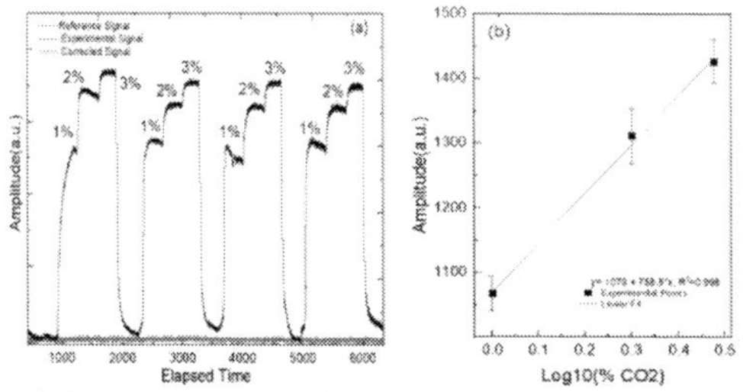

Fig. 3. (a) Sensor reaction with different CO_2 concentrations (1%, 2% and 3%). Experimental signal is used as a measurement signal, and Reference signal was used to correct of any external fluctuations (b) Sensor calibration curve plotted as a function of the logarithm percentage of CO_2 concentration.

The preliminary results from figure 3 have shown a potential for higher resolution measurement, with smaller concentration steps in between, more accurate readings and with a potential to measure lower levels of CO_2.

The gaseous carbon dioxide sensor was developed in a plug and play configuration to be used in real-time monitoring. With more improvements the sensor can be used in more challenging environments.

References

1. W. H. Schlesinger e E. S. Bernhardt, «The Atmosphere», em Biogeochemistry, Elsevier, 2020, pp. 51–97.
2. Gas sensor market size to reach $1.3 bn by 2027 at CAGR 6.4%», Focus on Catalysts, vol. 2021, n. 5, p. 2, mai. 2021.
3. N. White e D. Jayas, «Controlled Atmosphere Storage of Grain», em Handbook of Postharvest Technology, H. Ramaswamy, G. Vijaya Raghavan, A. Chakraverty, e A. Mujumdar, Eds. CRC Press, 2003, pp. 235–251.
4. J. P. Mendes et al., «Dissolved Carbon Dioxide Sensing Platform for Freshwater and Saline Water Applications: Characterization and Validation in Aquaculture Environments», Sensors, vol. 19, n. 24, p. 5513, dez. 2019

Experimental evaluation of Fourier transform holograms by a self-interferometric technique

Esther Nabadda[1], *Pascuala* Garcia-Martínez[2], *María de Mar* Sánchez-López[1,3] and *Ignacio* Moreno[1,4*]

[1]Instituto de Bioingeniería, Universidad Miguel Hernández de Elche, 03202 Elche, Spain
[2]Departamento de Óptica y Optometría y Ciencias de la Visión, Universitat de València, 46100 Burjassot, Spain
[3]Departamento de Física Aplicada, Universidad Miguel Hernández de Elche, 03202 Elche, Spain
[4]Departamento de Ciencia de Materiales, Óptica y Tecnología Electrónica, Universidad Miguel Hernández de Elche, 03202 Elche, Spain

Abstract. We present a technique that combines an encoding method to display complex-valued holograms onto a phase-only spatial light modulator (SLM) with a phase-shifting interferometric (PSI) technique for experimentally evaluating the generated complex-valued optical fields. We demonstrate an efficient common-path polarization interferometer based on the SLM itself, not requiring any external additional element. The same setup can be used to simultaneously display the complex hologram and to apply the phase-shifting values required to retrieve the phase distribution of the optical field. A simple rotation of a polarizer allows to change from the intensity configuration to the interferometer configuration.

1 Introduction

Parallel-aligned liquid-crystal on silicon (LCOS) displays are pixelated linear retarders that, under illumination with linearly polarized light parallel to the liquid-crystal (LC) director, produce a phase-only modulation [1]. They have become very popular spatial light modulator (SLM) devices for diffractive optics and structured light.

On the other hand, liquid crystal retarders are very useful devices in interferometry. In phase-shifting interferometry (PSI) they are employed to impart controlled phase bias to the interferograms to retrieve the target phase function [2]. LC-SLMs can be used to display spatially retardance functions that provide a common-path interferometric arrangement [3].

In this communication we present an extremely simple technique that makes the most of these properties to retrieve the complex information of light beams generated with an LCOS-SLM. We first apply a technique to encode complex-valued holograms based on displaying a modified checkerboard grating, useful to provide an on-axis hologram reconstruction [4-6]. The technique is demonstrated by producing different superpositions of Gaussian modes. The successful generation of these structured light beams is verified by capturing the intensity pattern in the far field. Then, the same optical system employed to generate the structured light beams is also applied to retrieve the far field phase distribution through a PSI algorithm. The change in the optical setup consists simply in rotating a polarizer. The application of PSI algorithms provides quantitative evaluation of the phase distribution that complements the traditional intensity pattern evaluation

2 Methods and techniques

2.1 Experimental setup

Figure 1 shows the experimental arrangement. A linearly polarized He-Ne laser is spatially filtered and collimated. A quarter-wave plate (QWP) converts it into circularly polarized light, so a linear polarizer (P1) can rotate arbitrarily without changing the light intensity. The beam illuminates an LCOS-SLM (Hamamatsu X10468-01, with 800×600 pixels of 20 μm pitch). The LCOS-SLM has its director axis oriented horizontally with respect to the laboratory framework and is addressed with a hologram that includes the phase of a converging lens. A camera is placed on its back focal plane, thus capturing the Fourier transform pattern generated by the hologram.

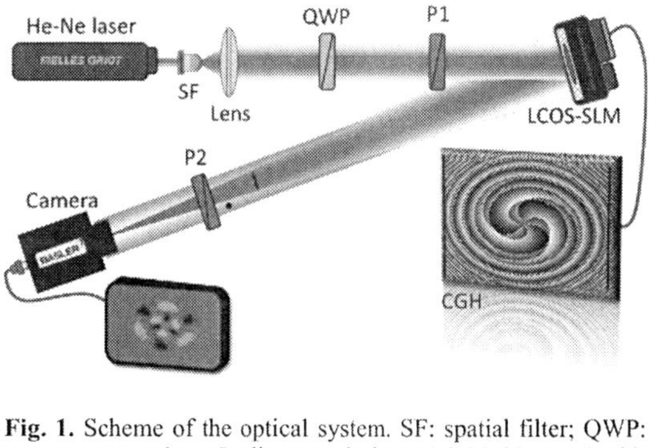

Fig. 1. Scheme of the optical system. SF: spatial filter; QWP: quarter-wave plate; P: linear polarizer; LCOS-SLM: Liquid-crystal on silicon spatial light modulator; CGH: displayed phase-only computer-generated hologram.

* Corresponding author: i.moreno@umh.es

2.2 Hologram design

A complex function $F(\mathbf{r}) = M(\mathbf{r})\exp[i\theta(\mathbf{r})]$ must be encoded on the SLM. Here $\mathbf{r} = (x, y)$ are the spatial coordinates. Since the LCOS device is a phase-only modulator, a new phase-only hologram is calculated as the following phase function

$$\phi_H(\mathbf{r}) = \theta(\mathbf{r}) + (-1)^{m+n}\arccos[M(\mathbf{r})]. \quad (1)$$

A converging lens phase function $\phi_L(\mathbf{r}) = -\pi r^2/\lambda f$ is added, where $r = |\mathbf{r}|$ is the radial coordinate, λ is the operating wavelength and f is the lens focal length. The phase function $\phi(\mathbf{r}) = \phi_H(\mathbf{r}) + \phi_L(\mathbf{r}) + \varphi_n$ is then displayed on the SLM. The phase φ_n is a constant phase bias, which does not change the spatial pattern generated in the far field, but is used to apply the PSI algorithm.

2.3. Interferometer configuration

If polarizer P1 in Fig. 1 is aligned parallel to the SLM LC director, all the light is modulated and focused on the camera, which captures the intensity of the hologram reconstruction. However, if P1 rotates to another angle, a portion of the input light is not modulated and remains collimated and can be used as the reference beam. Polarizer P2 is oriented at 45° so the components parallel and perpendicular to the LC director interfere, in a common path arrangement. The relative intensity of the reference and the test beams can be adjusted by rotating P1, while the PSI algorithms can be applied by changing the bias phase φ_n. Note that an arbitrary number of phase shifts can be applied, so different PSI algorithms can be tested. In our case we chose a synchronous detection algorithm with eight steps shifted by $\pi/4$ radians, leading to the following phase retrieving formula [7]:

$$\alpha(\mathbf{r}) = -\arctan\left\{\frac{2(I_0-I_4)+\sqrt{2}(I_1-I_3-I_5+I_7)}{2(I_2-I_6)+\sqrt{2}(I_1+I_3-I_5-I_7)}\right\}. \quad (2)$$

3 Results

Figure 2 shows some experiments where a Hermite-Gauss HG_{33} beam and a Laguerre-Gauss LG_{13} beam are encoded. Figures 2(a)-2(b) and 2(e)-2(f) show their theoretical intensity and phase patterns. Figures 2(c) and 2(g) show the experimental intensity distributions captured on the camera, while Figs. 2(d) and 2(h) show the phase patterns $\alpha(\mathbf{r})$ retrieved from the eight interferograms, which are shown in the two rows below each case. Results bear very good agreement with the theory, thus confirming the successful generation of the HG and LG beams, not only in its intensity but also in its phase distribution, which probes the accuracy and advantage of this technique. Other cases obtained by the superposition of different modes have been succesfully generated and will be presented.

Acknowledgements: This work received financial support from Ministerio de Ciencia, Innovación y Universidades, Spain (grant RTI2018-097107-B-C33). E.N. acknowledges a grant from Generalitat Valenciana, Santiago Grisolia Program (ref. GRISOLIAP/2020/004).

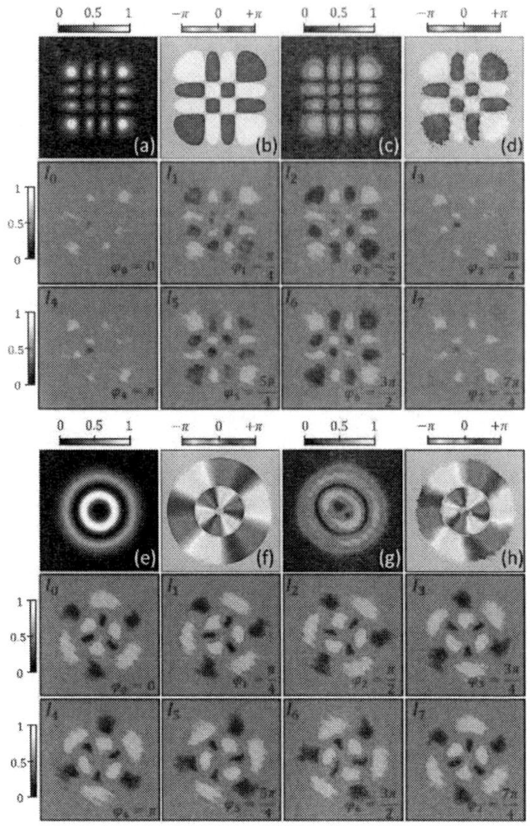

Fig. 2. Realization of a HG_{33} (a-d) and LG_{13} (e-h) beam. Theoretical intensity (a,e) and phase (b,f) distributions. Experimental intensity (c,g) and phase (d,h) distributions. The phase distribution of each beam is retrieved from the interferograms shown below each case with eight consecutive phase shifts of step $\Delta\varphi = \pi/4$.

References

1. Z. Zhang, Z. You, D. Chu, Light: Sci. Appl. **3**, 213 (2014).
2. Y.B. Machuca-Bautista, M. Strojnik, J.L. Flores, D.I. Serrano-García, G. García-Torales, Results Opt. **5**, 100197 (2021).
3. I. Moreno, J.A. Davis, F.A. Klein, M.J. Mitry, Appl. Opt. **47**(11), 1797-1801 (2008).
4. V. Arrizón, Opt. Lett. **28**(24), 2521-2523 (2003).
5. O. Mendoza-Yero, G. Mínguez-Vega, J. Lancis, Opt. Lett. **39**(7), 1740-1743 (2014).
6. J.A. Davis, E.D. Wolfe, I. Moreno, D.M. Cottrell, OSA Continuum **4**(3), 896-910 (2021).
7. K. Creath, Prog. Optics **26**, 349-393 (1988).

 https://doi.org/10.1051/epjconf/202226613025

Automation strategies and machine learning algorithms towards real-time identification of optically trapped particles

João Oliveira[1,2,*], *Vicente* Rocha[1,2], *Nuno* A. Silva[1,2], and *Pedro* A. S. Jorge[1,2]

[1]Faculdade de Ciências da Universidade do Porto, Rua do Campo Alegre 687, 4169-007 Porto, Portugal
[2]INESC TEC, Centre of Applied Photonics, Rua do Campo Alegre 687, 4169-007 Porto, Portugal

Abstract.
To automatically trap, manipulate and probe physical properties of micron-sized particles is a step of paramount importance for the development of intelligent and integrated optomicrofluidic devices. In this work, we aim at implementing an automatic classifier of micro-particles immersed in a fluid based on the concept of optical tweezers. We describe the automation steps of an experimental setup together with the implemented classification models using the forward scattered signal. The results show satisfactory accuracy around 80% for the identification of the type and size of particles using signals of 250 milliseconds of duration, which paves the path for future improvements towards real-time analysis of the trapped specimens.

1 Introduction

Ever since its introduction in 1970, the optical tweezers setup (OT) has proven itself to be a reliable tool to manipulate micron-sized particles, which eventually earned Arthur Ashkin, the Nobel Prize in Physics in 2018. More recently, OT has been integrated into microfluidic devices not only for flux control but also for the interrogation of dynamical properties of the trapped particle through dynamic light scattering[2], allowing to perform in depth analysis of a target specimen at individual level. However, despite the aiming at autonomous and intelligent optomicrofluidic devices, fully automated approaches to OT are still scarce in the literature.

In this manuscript we report our efforts in the development of a versatile and full stack approach to the automation of an optical tweezers experimental setup, integrating all the features and control within a single computational toolkit. Furthermore, introducing a strategy for identification of particles based on their dynamical properties via the analysis the forward scattered signal, we also pave the path towards intelligent devices capable of performing classification tasks at rates close to real-time.

2 Experimental System and automation strategies

The optical tweezers system used in this work consists of a conventional inverted microscope configuration for which a schematic is provided in Figure 1. The trapping laser is a fiber-coupled laser diode (Lumentum s27-7602-460) working at wavelength around 976 *nm*. Using a 100× oil immersion objective, the laser is focused to a spot size of approximately 1.1 μm, creating an optical trap. Live control of the trapping procedure is possible using a standard 1280 × 1024 pixel color CMOS camera.

For the analysis of the dynamical properties of the particle, the forward scattered signal of the trapping beam is collimated with a 10x condenser lens and subsequently directed towards a quadrant photodetector (PDQ80A-Thorlabs) placed at a conjugate back focal plane of the condenser. The detector responds in the range of 400-1050 *nm*, and the signal is acquired using a DAQ at a rate of 10KHz. After the acquisition of the signal, a set of the three signals (X, Y, SUM) is related with the transverse x and y position of the particle, while SUM corresponds to a total integral of what reaches the detector.

For a complete automation of the setup, we implemented a full-stack approach using Python, which allows us to control the system as well as perform the acquisition and analysis of the signal in a single computational tool and in real-time. For the hardware part, including laser, stages and camera, we adapted existent driver libraries and developed new ones. Then a series of routines that support typical OT experiments were created in intermediate scripts, before implementing a graphical user interface on top of these libraries using PyQt5. The implementation follows a ladder structure that covers distinct levels of expertise with distinct degrees of versatility: i) a driver layer, directed towards advanced users and developers; ii) a script layer, where an advanced user can program specific instructions for particular purposes; and iii) a graphical user interface layer, to be used by a regular user at beginner or intermediate expertise level. Furthermore, the modular approach to the organization of the code eases future scheduled improvements, such as the development of

*e-mail: jmfoliveira110@gmail.com

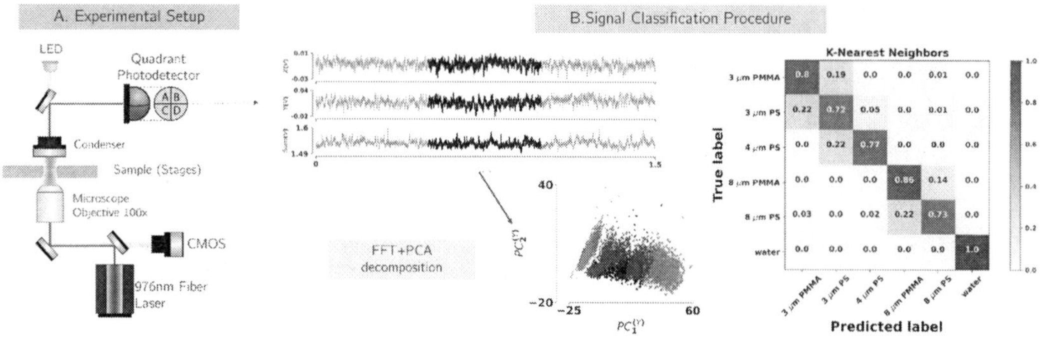

Figure 1. A. Visual summary of the Optical Tweezer system and signal classification procedure as described in the main text. B. Visual summary of the classification procedure and confusion matrix results for the particle dataset described in the main text.

an image analysis library and the addition of a spectrometer to collect the Raman spectrum of the trapped specimen.

3 Classification of particles using PCA decomposition

Being sensitive to the displacement of the trapped particle in the transverse directions, the acquired signal from the quadrant photodetector allows to probe the Brownian motion of the trapped beads which contains information regarding the physical properties of the particle[1]. In the recent years, our team explored a few strategies for single particle/cell classification using optical trapping systems and scattered radiation, both using the back-scattered[2] and forward scattered signal[3]. Focusing on the latter for the purpose of this work, we reported the classification of trapped particles by introducing an algorithm that consists on performing a dimensional reduction with Principal Component Analysis performed in the Fourier space before the application of common classification strategies such as K-nearest neighbors.

To test the classification capabilities of the system, we implemented a classification mode into the software solution described before. For the training procedure, we acquire distinct signals of 30 seconds of total duration at an acquisition rate of 10kHz for each particle type. The signals are then divided in segments of 1 second each in order to increase the dataset size, and employed to train a classification algorithm, which can then be tested in unseen particles.

The results obtained for a dataset of 5 types of particles (total of 45 particles tested, 40 used for training phase and 10 used for test) is presented in Figure 1. As observed, the overall accuracy around 80% suggests that this implementation was able to distinguish and classify accurately the solutions prepared.

4 Concluding remarks

Automation strategies are a crucial step for a seamless integration of optical tweezers (OT) tools in future intelligent optomicrofluidic solutions. In this work we presented an ab-initio approach to the automation of an experimental OT setup, developing a modular structure that allows for future improvements and addition of specialized hardware. Furthermore, by investigating the possibility of classification in living cells, we are researching the capabilities of the setup as a biomedical diagnosis tool, that performs accurately and is cheap to manufacture, which is of natural importance in the current state of medicine.

Overall, the findings enclosed pave the way for a diversity of technological solutions of OT from micro to nanoscale, establishing an integrated and acessible tool for label-free and reagentless biosensing, single cell diagnostic amongst many other applications.

References

[1] Gieseler, Jan *et. al.*, Optical tweezers — from calibration to applications: a tutorial, Advances in Optics and Photonics **13**, pp.74-241 (2021)

[2] Paiva, J. S. and Jorge, P. A. S. and Ribeiro, R. S. R. and Balmaña, M. and Campos, D. and Mereiter, S. and Jin, C. and Karlsson, N. G. and Sampaio, P. and Reis, C. A. and et al. *iLoF: An intelligent Lab on Fiber Approach for Human Cancer Single-Cell Type Identification*, Scientific Reports,10,1, 2020

[3] Carvalho, Inês Alves and Silva, Nuno Azevedo and Rosa, Carla C and Coelho, Luís CC and Jorge, Pedro AS, Sensors **21**, 6181 (2021)

Wavelength-switchable L-band fiber laser assisted by ultrafast laser fabricated random reflectors

Rosa Ana Perez-Herrera[1,2,*], *Pablo* Roldan-Varona[3,4,5], *Arturo* Sanchez-Gonzalez[1,2], *Luis* Rodriguez-Cobo[4], *Jose Miguel* Lopez-Higuera[3,4,5], and *Manuel* Lopez-Amo[1,2]

[1]Department of Electrical, Electronic and Communication Engineering, Public University of Navarra, 31006 Pamplona, Spain
[2] Institute of Smart Cities (ISC), Public University of Navarra, 31006 Pamplona, Spain
[3] Photonics Engineering Group, University of Cantabria, 39005 Santander, Spain
[4] CIBER-bbn, Instituto de Salud Carlos III, 28029 Madrid, Spain
[5] Instituto de Investigacion Sanitaria Valdecilla (IDIVAL), 39005 Cantabria, Spain

Abstract. A wavelength-switchable L-band erbium-doped fiber laser (EDFL) assisted by an artificially controlled backscattering (ACB) fiber reflector is here presented. This random reflector was inscribed by femtosecond (fs) laser direct writing on the axial axis of a multimode fiber with 50 µm core and 125 µm cladding with a length of 17 mm. This microstructure was placed inside a surgical syringe to be positioned in the center of a high-precision rotation mount to accurately control its angle of rotation. Only by rotating this mount, three different output spectra were obtained: a single wavelength lasing centered at 1574.75 nm, a dual wavelength lasing centered at 1574.75 nm and 1575.75 nm, and a single wavelength lasing centered at 1575.5 nm. All of them showed an optical signal-to-noise ratio (OSNR) of around 60 dB when pumped at 300 mW.

1 Introduction

Femtosecond (fs) laser writing has become an effective way to process any type of transparent optical material, such as silica optical fibers [1]. This technology enables the microfabrication of numerous fiber structures with excellent properties for a wide range of practical applications [2]. In particular, the development of fiber-optic microstructures based on refractive index (RI) modification under fs-laser irradiation has resulted in different implementations in the field of optical fiber sensors: surrounding refractive index sensors, strain sensors, curvature sensors, or multiparameter sensors [3]. These artificially controlled backscattering (ACB) fiber reflectors have been shown to have similar temperature sensitivity to traditional FBGs. However, the strain sensitivity can be improved by more than an order of magnitude when comparing these quasi-randomly distributed reflective microstructures with FBG-based sensors [4]. Here, an ACB fiber reflector inscribed by fs laser direct-write technique is used into an L-band erbium doped fiber laser (EDFL). This random reflector was inscribed on the axial axis of a 50/125 multimode fiber (MMF), with a length of 17 mm, and located into a high-precision rotation mount to control its angle of twist rotation. Only by rotating in a range of 8° the MMF sample where the RFG was inscribed, this wavelength-switchable EDFL can be switched among single and dual wavelength lasing.

2 Fabrication

The optical fiber inscription was carried out using a Cazadero fiber laser (Calmar Laser) that delivers 370 fs laser pulses at a central wavelength of 1030 nm. However, the light absorbed non-linearly by the fiber has a wavelength of 515 nm, due to the second harmonic generation (SHG) introduced in the setup, as shown in [5]. A pulse repetition rate of 150 Hz and a pulse energy of 0.75 µJ were used during the inscription. The laser pulses were tightly focused into the Ø50 µm core of a MMF using a 0.42 NA, 50× objective lens from Mitutoyo. The fiber was translated through the laser focus using a motorized nano-resolution XYZ stage from Aerotech.

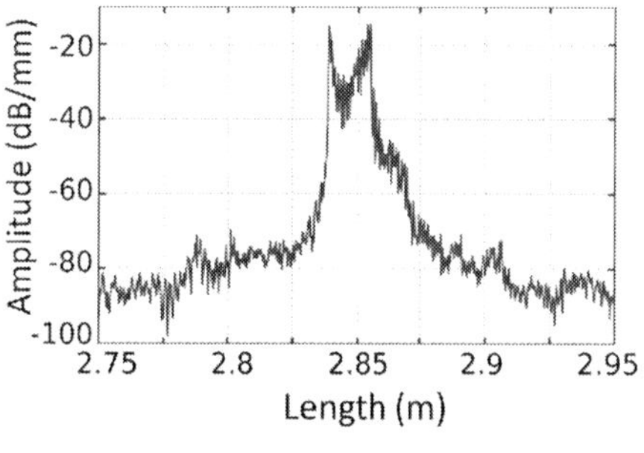

Fig. 1. Backscattered optical power as a function of fiber length for the MMF-RFG, located about 1.88 m from the connector of the OBR.

* Corresponding author: rosa.perez@unavarra.es

This ACB-MMF reflector was written on the axial axis of the fiber with a random period between Λ_{min}=1.61085 μm and Λ_{max}=1.64230 μm and a total length of 17 mm. Although this is a random fiber grating (RFG) [6], the almost total periodicity of the optical structure gives rise to Bragg resonances which, in the third order (m=3), present reflection in the spectral band centered at around 1575.4 nm. Figure 1 presents the backscattered optical power as a function of fiber length for the inscribed ACB-MMF reflector characterized with an ultra-high resolution optical backscattered reflectometer (OBR).

3 Experimental setup

Figure 2 illustrates a schematic diagram of the L-band linear-cavity fiber laser experimental setup. A 976-nm pump power (Fig. 2(a)) was injected into the linear-cavity EDFL by means of a 980/1550 nm wavelength division multiplexer (WDM) (Fig. 2(b)). The gain medium was 5 meters of highly erbium-doped fiber (EDF) I25 (980/125, Fibercore Inc.) (Fig. 2(c)), with a peak core absorption ranges from 7.7 to 9.4 dB/m at 1531 nm [7]. This EDF was connected to the common port of the WDM and followed by a 3-ports optical circulator (Fig. 2(d)) in which ports 3 and 1 were connected to conform a fiber loop mirror (FLM), as in [7]. After passing through the highly EDF section again, the reflected signal from the FLM arrived to the 1550 nm-port of the WDM up to the optical coupler (Fig. 2(e)). Then, the signal is divided into two branches where 90% of the signal reached the MMF-RFG-based reflector and the other 10% was visualized with an OSA (Fig. 2(f)). The MMF-RFG was placed inside a surgical syringe so that it could be positioned in the center of the high-precision rotation mount (Fig. 2(g)) to accurately control its angle of rotation. A photograph of the RFG reflector inscribed by femtosecond laser writing is shown in Fig. 2(h).

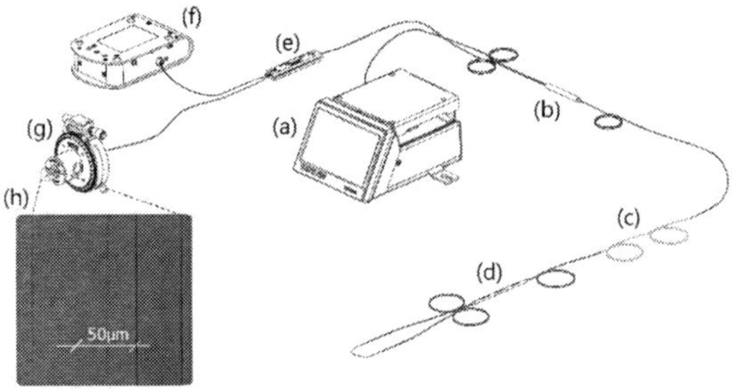

Fig. 2. Schematic diagram of the experimental L-band linear-cavity fiber laser setup

4 Results and discussion

Figure 3 shows the output spectra of the linear cavity EDF pumped at 300 mW when (a) single-wavelength laser emission centered at 1574.75 nm, (b) dual-wavelength laser emission centered at 1574.75 nm and 1575.53 nm, or (c) single-wavelength laser emission line at 1575.53 nm were obtained. By rotating the high-precision rotation mount where the multimode fiber-based reflector was located into a syringe, these three configurations can be easily reached. Output power levels of -1.02 dBm (Fig. 3(a)) and -1.85 dBm (Fig. 3(c)) were measured when single-wavelength laser emission was attained, both presenting an OSNR of 60 dB. On the other hand, when dual-wavelength laser emission was achieved, an output power level around -5 dBm and an OSNR of 58 dB was measured (Fig. 3(b)). As expected, by increasing the number of lasing wavelengths the output power values of the lasing emission lines decrease [8].

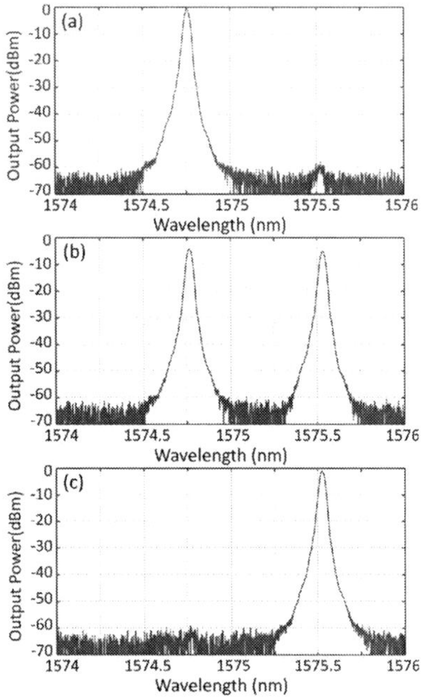

Fig. 3. Output spectra of the linear-cavity fiber laser when (a) single wavelength lasing centered at 1574.75 nm, (b) dual wavelength lasing centered at 1574.75 nm and 1575.75 nm, and (c) single wavelength lasing centered at 1575.5 nm are obtained.

This work is part of the projects PID2019-107270RB-C02, funded by MCIN/AEI/10.13039/501100011033 and FEDER "A way to make Europe", and PDC2021-121172-C01 funded by MCIN/ AEI/10.13039/501100011033 and European Union "Next generationEU"/PTR.

References

1. J. Zhao, Y. Zhao, Y. Peng, R. Lv, Q. Zhao, Optics & Laser Techn. **146**, 107473 (2022)
2. M. Chen, Y. Zhao, Y. Peng, R. Lv, Q. Zhao, Optics & Laser Techn. **147**, 107628 (2022)
3. A.G. Leal-Junior, et al. J. Lightwave Technol. **37**, (3) 971–980 (2019)
4. R.A. Perez-Herrera, et al. Scientific Reports. **11**, 20408 (2021)
5. L. Lei, et al. Nanomaterials. **11**, 2505 (2021)
6. Y. Kondo, et al. Opt. Lett. **24**, (10) 646–648 (1999)
7. R.A. Perez-Herrera, et al. Opt. Lett. **44**, 2669-2672, (2019)
8. R.A. Perez-Herrera, et al. Optical Fiber Technology. **16**, (4), 205-211 (2010)

EPJ Web of Conferences **266**, 13027 (2022)
EOSAM 2022
https://doi.org/10.1051/epjconf/202226613027

The effect of frequency modulation on the FSR of a Fabry-Perot cavity using an Optical Spectrum Analyser

João Reis[1,3*], *António V.Rodrigues*[1,2], *Paulo Robalinho*[1,2], *Susana Novais* [1], *João Maia*[1], *Paulo Marques*[1,3], *D. Roma*[4,5], *J. Salvans*[4,5], *M. Canal*[4,5], *J. Ramos*[4,5], *V. Gualani*[4,5], *S. Sisteré*[4,5], *V. Martín*[4,5], *M. Nofrarias*[4,5], *Susana Silva*[1], *Orlando Frazão*[1]

[1]INESC TEC, Rua do Campo Alegre s/n, 4169-007 Porto, Portugal
[2]Faculdade de Engenharia da Universidade do Porto, Rua Dr. Roberto Frias, s/n, 4200-465 Porto, Portugal
[3]Faculdade de Ciências da Universidade do Porto, Rua do Campo Alegre s/n, 4169-007 Porto, Portugal
[4]Institut de Ciències de l'Espai (ICE, CSIC), Campus UAB, Carrer de Can Magrans s/n, 08193 Cerdanyola del Vallès, Spain
[5]Institut d'Estudis Espacials de Catalunya (IEEC), Gran Capità, 2-4, Ed. Nexus, 08034 Barcelona, Spain

Abstract. It is presented a study of the dependence between the free spectral range (FSR) and the cavity length in Fabry-Perot interferometers. Furthermore, the effect of frequency modulation on the FSR is studied when an optical spectrum analyser (OSA) is used as an interrogator. For low frequency range it is possible to observe this behaviour in the OSA and using an appropriate processing signal it is possible to use the white light interferometry technique.

1 Introduction

The first Fabry-Perot interferometer was accomplished by the work of Charles Fabry and Alfred Perot in 1887 [1]. Essentially, a traditional Fabry-Perot (FP) interferometer consists in a cavity enclosed by two parallel reflecting surfaces. Since then, there has been a significant development of this interferometer and in 1993 after the invention of the optical fiber, the first in-fiber FP interferometer was demonstrated [2]. The importance and subsequent investigation of this type of structure is due to several features such as low cost, easy manufacturing, and immunity to electromagnetic interference.

The FP sensor can be used for many applications namely to measure strain [2-3], magnetic field [3-4], salinity or temperature [5]. Thus, in order to quantify the variation of the optical path length of the interferometer, which is defined as the product of the refractive index and the physical length of the cavity, the typical instrument used is an Optical Spectrum Analyzer (OSA).

In the present work, the study of the dependence of the FSR of an FP cavity with the frequency modulation is demonstrated. Also, the effect of frequency modulation observed in an OSA was explored using white light interferometry technique.

2 Free spectral range dependence with cavity length

The first test consists in the measurement of the optical spectrum of the FP cavity. An optical circulator was used to read the signal in reflection of the FP, using a broadband source (Optronics Tech ASE CL) with a bandwidth of 75 nm centred around 1565 nm to illuminate the fibre device.

Initially, the distance of the FP cavity was increased in the range of 50 to 600 µm by means of a micro-translation stage. An OSA (Yokogawa AQ6370C) was used to read the reflection spectrum of the increasing FP cavity and the correspondent FSR was measured. In Fig. 1 it is presented the optical spectrum of the FP cavity, with a cavity length of ~ 250 µm and high visibility.

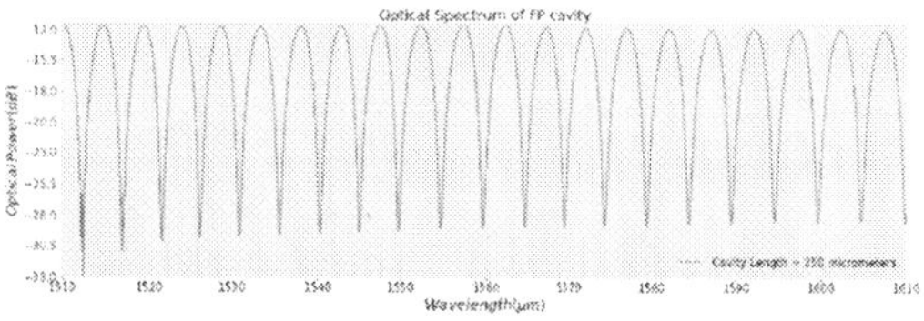

Fig. 1. Optical spectrum of the FP cavity.

In Fig. 2 it is observed the linear relationship between the FSR and the inverse of the cavity length, as predicted by the following equation [6]:

$$FSR = \frac{\lambda_0^2}{2\,n}\frac{1}{d} \qquad (1)$$

Where d is the cavity length, n is the refractive index of the air cavity ($n \approx 1$) and λ_0 is the operation wavelength. The slope obtained is of 1167.06±4.81 nm·µm^{-1}.

---
* Corresponding author: up201604696@edu.fc.up.pt

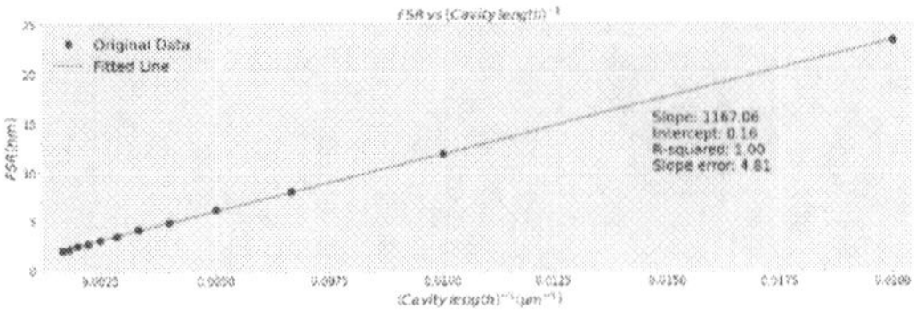

Fig. 2. FSR dependence with the inverse of the FP cavity length.

3 Frequency modulation of the correlation signal between two cascaded FP cavities

In this experiment, the setup (Fig. 3) relies on a broadband optical source with a bandwidth of 75 nm centered at 1565 nm and a four-port optical circulator. Two cascaded FP cavities are connected in independent ports of the optical circulator.

A piezoelectric stack (Thorlabs PK4GA7P1) was attached to one of the FP cavities and a high voltage amplifier (TEGAM Model 2350) was used to promote the mechanical variation of the PZT.

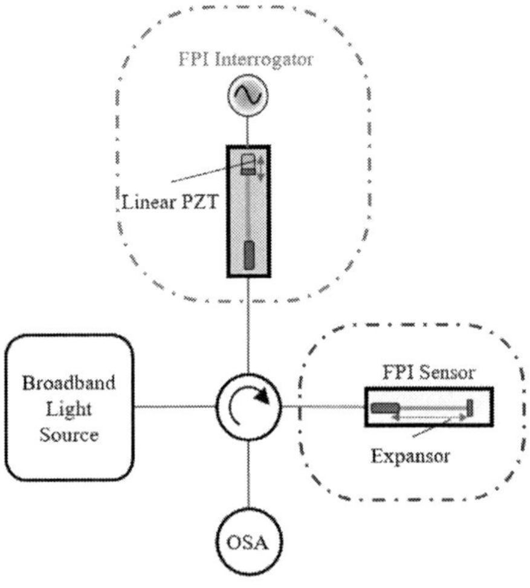

Fig. 3. Experimental setup for the characterization in frequency of the optical signal of two cascaded FP cavities.

An arbitrary wave function generator (Hewlett Packard 33120A) was used to apply a triangular electric signal to the linear PZT. The length of one cavity was continuously varied, while the other cavity length was kept constant. Therefore, the first cavity acted as an interrogator of the second cavity. The range of frequencies used was 0.5 Hz to 100 Hz and the reflected spectrum of the cascaded cavities was registered, for each frequency, in the OSA.

In Fig.4 it was observed the beat phenomenon, predicted by [7], for 0.5 Hz. When using low frequencies, the beat is generated between two cavities and the resulting modulation of the cavity variation is observed. In this case, the OSA has enough speed to keep track of the FP cavity length variation and a beat is observed (Fig.4a). However, when applying high frequency signals (above 5 Hz), the cavity length changes rapidly and the OSA is no longer able to keep track of this variation. In this case, the OSA reads the spectral response of the cavity with a background noise along the spectrum. Finally, with an appropriate signal processing it is possible to simulate the white light interferometry technique (see Fig. 4b).

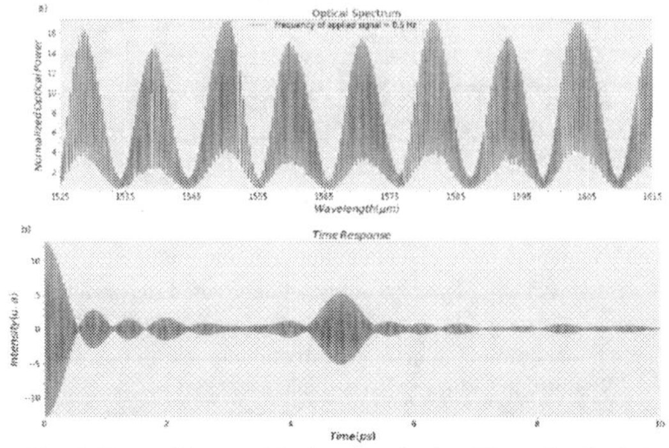

Fig. 4. Beat of the modulation signal of an FP cavity for 0.5 Hz: a) optical spectrum and b) time response.

4 Conclusion

In conclusion, it was determined that for low frequencies the OSA can be used to observe signal modulation, and with a suitable signal processing, it can been read by means of white light interferometry technique. Thus, in order to overcome this challenge for high frequency, an alternative solution involves the use of a fast detection system such as a single photodetector connected to an oscilloscope.

This work was realized with the financial support of the Lira Project in accordance with the terms of ESA Contract No. 4000135481/21/NL/AR.

References

1. M.R. Islam, M.M Ali, M. Lai, K.S. Lim, MDPI. Sens. **14**, 7451 – 7488 (2014).
2. J.S. Sirkis, D.D. Brennan, M.A. Putman, T.A. Berkoff, A.D. Kersey, E.J. Friebele, Opt. Lett. **18**, 1973 – 1975, (1993).
3. G.K.B. Costa, P.M.P. Gouvêa, L.M.B. Soares, J.M.B Pereira, F. Favero, A.M.B Braga, P. Palffy-Muhoray, A.C. Bruno, I.C.S Carvalho, Opt. Lett. **24**, 14690 – 14696, (2016).
4. K.D Oh, J. Ranade, V. Arya, A. Wang, R.O Claus, EPL, IEEE Photo. Tech. Lett. **9**, 797 – 799 (1997)
5. R. Flores, R. Janeiro, J. Viegas Scient. Repo. **9**, 9556 (2019).
6. M. Born, E. Wolf, *Principles of Optics Eletromagnetic Theory of Propagation, Interference and Diffraction of Light* (1999)
7. P. Robalinho, A. Gomes, O. Frazão, IEEE Sens. **21**, 4623 – 4627 (2021).

https://doi.org/10.1051/epjconf/202226613028

Measurement of ultrafast carrier dynamics in multilayer MPCVD graphene

Tânia M. Ribeiro[1,*], *Tiago E. C.* Magalhães[1], *Bohdan* Kulyk[2], *Alexandre F.* Carvalho[2], *António J. S.* Fernandes[2], *Florinda M.* Costa[2], and *Helder* Crespo[1]

[1]Instituto de Física de Materiais Avançados, Nanotecnologia e Fotónica (IFIMUP), Departamento de Física e Astronomia, Faculdade de Ciências, Universidade do Porto, Rua do Campo Alegre s/n, 4169-007 Porto, Portugal
[2]I3N, Departamento de Física, Universidade de Aveiro, Campus Universitário de Santiago, 3810-193 Aveiro, Portugal

Abstract. Graphene presents unique opto-electronic properties which makes it useful for a wide range of applications and devices, such as high-speed photodetectors, that rely on the relaxation dynamics of photoexcited charge carriers. These demand reliable and reproducible methods for synthesis of high quality graphene. Here we present ultrafast degenerate pump-probe measurements of multilayer graphene coatings grown by microwave plasma chemical vapour deposition (MPCVD) and analyse the impact of the synthesis growth time on the material's nonequilibrium optical response.

1 Introduction

Ultrafast time-resolved spectroscopy techniques, such as pump-probe, can be used to study photoinduced dynamical processes in materials and in nanostructures [1, 2]. In pump-probe experiments, a first "strong" light pulse called pump promotes an electronically excited state. The temporal evolution of the photoexcited carriers can be observed by measuring the transmission of a second "weak" and delayed pulse called probe. The experiment holds information on the dynamic processes that occur in the system under study and has the advantage of probing the evolution of non-emissive or dark states. Understanding the carrier dynamics of graphene on a substrate is fundamental for the development of graphene-based optoelectronic devices [3]. The study of the dynamics can also reveal the quality of graphene, since, for example, the amount of defects in graphene affects its cooling properties [1]. In this work, we present preliminary ultrafast degenerate pump-probe measurements of graphene coatings synthesized by MPCVD. A range of graphene samples with an increased number of layers was studied. We will present the sample's characterization, the constructed pump-probe setup as well as the results for the relaxation dynamics of each sample.

2 Experimental Details

Graphene samples were grown by MPCVD on copper substrates using a procedure previously reported in [4]. Three samples, labelled as A, B and C, were submitted to different growth times: 0.5, 1 and 4 minutes, respectively. The laser used for time-resolved pump-probe spectroscopy is a commercial Ti:sapphire laser amplifier (Femtolasers Compact Pro CE-phase) with 1 kHz repetition rate, 800 nm center wavelength (~ 40 nm bandwidth), delivering sub-30 fs laser pulses with 1 mJ energy. Only a small percentage (~ 5%) of the laser power was fed to our home-built pump-probe system. Pump pulses, with an energy of 220 nJ were used to photoexcite carriers while probe pulses, with an energy of 7 nJ, were used to measure the differential transmission for different delays between the pump and probe pulses. Both beams were focused onto the samples with a spot diameter of ~ 80 μm. The polarization of the probe beam was rotated by 90 degrees with respect to the pump. An optical chopper was used to modulate the pump pulses at half the laser repetition rate. Besides blocking the pump beam, an additional spatial filter was used to block scattered pump light. The differential transmission in the

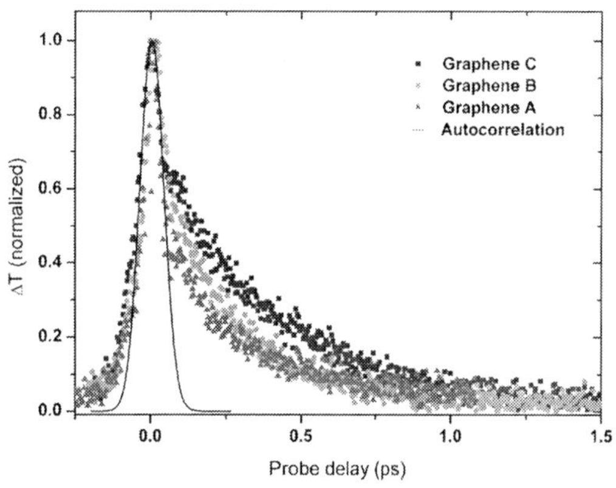

Figure 1. Measured differential transmission, ΔT, plotted as function of the probe delay for graphene samples A, B and C. ΔT is normalized at 1 for each graphene sample. The solid line represents the autocorrelation of the pump and probe pulses (full width at half maximum of ~ 90 fs).

*e-mail: tania.ribeiro@fc.up.pt

https://doi.org/10.1051/epjconf/202226613028

probe was measured with silicon photodiodes and a lock-in amplifier.

3 Results

Raman and UV-visible transmission spectra of the graphene samples indicated a variation in the number of layers proportional to the growth time. Therefore, the number of layers is proportional to the growth time and increases from A to C. Moreover, sample A presents an enhanced D peak in its Raman spectrum when compared to samples B and C, suggesting a larger degree of crystalline disorder [5]. The time-resolved response is shown in figure 1 for the three selected graphene samples (A-C). A bi-exponential decay model was used to fit the experimental data. The measured carrier relaxation time is on the order of 0.3 - 0.4 ps for the slower relaxation time (τ_2). The latter is attributed to interband scattering processes and was found to be proportional to the number of layers, i.e. decreasing from sample C to A. However, since defects can also accelerate the cooling process [1], a careful analysis with other complementary characterization techniques should also be performed.

References

[1] E. A. Pogna, A. Tomadin, O. Balci, G. Soavi, I. Paradisanos, M. Guizzardi, P. Pedrinazzi, S. Mignuzzi, K.-J. Tielrooij, M. Polini, A. C. Ferrari, G. Cerullo, ACS nano **16**, 3613–3624 (2022)

[2] R. Gatamov, A. Baydin, H. Krzyzanowska, N. Tolk, Materials Research Express **7**, 095601 (2020)

[3] M. Massicotte, G. Soavi, A. Principi, K.-J. Tielrooij, Nanoscale **13**, 8376–8411 (2021)

[4] A. F. Carvalho, T. Holz, N. F. Santos, M. C. Ferro, M. A. Martins, A. J. Fernandes, R. F. Silva, F. M. Costa, Carbon **98**, 99–105 (2016)

[5] J. M. Dawlaty, S. Shivaraman, M. Chandrashekhar, F. Rana, Michael G. Spencer, Appl. Phys. Lett. **92**, 042116 (2008).

https://doi.org/10.1051/epjconf/202226613029

Intelligent Optical Tweezers with deep neural network classifiers

Vicente Rocha[1,2,*], *João* Oliveira[1,2], *A.* Guerreiro[1,2], *Pedro* A. S. Jorge[1,2], and *Nuno* A. Silva[1,2]

[1]Faculdade de Ciências da Universidade do Porto, Rua do Campo Alegre 687, 4169-007 Porto, Portugal
[2]INESC TEC, Centre of Applied Photonics, Rua do Campo Alegre 687, 4169-007 Porto, Portugal

Abstract.
Optical tweezers use light to trap and manipulate mesoscopic scaled particles with high precision making them a useful tool in a plethora of natural sciences, with emphasis on biological applications. In principle, the Brownian-like dynamics reflect trapped particle properties making it a robust source of information. In this work, we exploit this information by plotting histogram based images of 250*ms* of position or displacement used as input to a Convolution Neural Network. Results of 2-fold stratified cross-validation show satisfying classifications between sizes or types of particles: Polystyrene and Polymethilmethacrylate thus highlighting the potential of CNN approaches in faster and non-invasive applications in intelligent opto and microfluidic devices using optical trapping tools.

1 Introduction

Optical Tweezers (OT) makes use of a fine balance between radiation pressure and dipolar forces arising in intense field gradients to manipulate particles at the mesoscopic scale. Furthermore, when integrated with a quadrant-photodetector, see figure 1, OT can also probe the dynamical properties of a trapped particle by tracking the position using the forward scattered light. Recently, it has been shown that these dynamics can be used to identify and classify particles through a careful analysis of the position timeseries and machine learning algorithms, which can be fundamental for future integration into intelligent microfluidic devices[3].

In this work, we present an alternative approach that makes use of deep learning, in particular Convolution Neural Networks (CNN), to classify particles based on information contained in the histogram plots constructed from the position timeseries. More precisely, we explore the capability to differentiate type and size of particles using position based histograms or displacement based histograms.

2 Theoretical framework

Considering that the optical trap is generated by an optical beam of Gaussian intensity profile, the trapping force can be approximated at first order by Hook-type law of harmonic oscillator type. Alone, the trap is the result of a balance between radiation pressure and electromagnetic gradient forces. As the particles are usually immersed in a medium, we should consider also the viscosity, which is fundamental for the trapping process, as well as additional collisions which subject the particle to a confined Brownian-like motion. As a whole, the stochastic dynamics can be described by a Langevin differential equation as

$$m\frac{d^2\mathbf{r}}{dt^2} + \gamma\frac{d\mathbf{r}}{dt} + \boldsymbol{k} \odot \mathbf{r} = \gamma\sqrt{2D}\chi(t), \qquad (1)$$

where m is the mass of the particle, γ the friction term between the particle and surrounding medium, D the diffusion coefficient, $\chi(t)$ the stochastic term representing Brownian motion, and k the stiffness coefficient from the Hook law approximation for the Gaussian beam profile.

The physical properties of the Brownian motion reflect both the surrounding medium properties and the trapped particles, making the position timeseries a rich source of information about the system. Also, by subtracting the position at a previous time we obtain the displacement timeseries resulting in another way of investigating the dynamics of the particle. Indeed, despite being dependent on the same physical quantities, position and displacement follow distributions of distinct momentum resulting in two distinct histograms as depicted in figure 1. The trapped particle is in equilibrium with the medium, therefore the position follows the Boltzmann distribution,

$$\rho(x) = \sqrt{\frac{k}{2\pi k_b T}}e^{-\frac{k(x-x_{eq})^2}{2k_b T}}, \qquad (2)$$

where k_B is the Boltzmann constant and T the temperature of the system. The displacement follows a zero centered normal distribution with variance given by the mean squared displacement (MSD) which considering the overdamped regime of typical experimental settings results in[1]

$$\mathrm{MSD}(\Delta t) = 2D\tau_{ot}\left(1 - e^{-\Delta t/\tau_{ot}}\right). \qquad (3)$$

where Δt is the time interval between consecutive times and $\tau_{ot} = \gamma/k$ is the optical trapping time. If we take

*e-mail: vicentevrocha.vvr@gmail.com

https://doi.org/10.1051/epjconf/202226613029

the time interval between consecutive measurements to be much smaller than the optical trapping characteristic time ($\Delta t << \tau_{ot}$) we have that MSD $\approx 2D\Delta t$, which corresponds to the variance of the displacement for a free particle under Brownian motion.

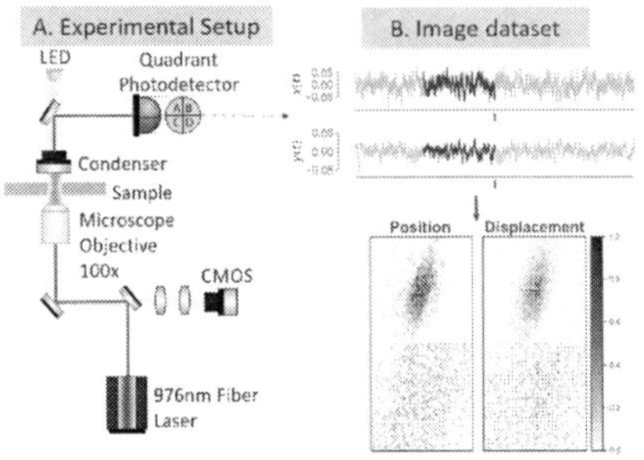

Figure 1. On the left, scheme of OT setup with quadrant photodetector for particle tracking and on the right image plotting procedure.

3 Results

In this work we investigated the capacity of CNN for classification using the forward scattered signal and in particular by looking at the displacement and position histograms retrieved for each particle. With this objective, we used 5 classes of particles: Polystyrene (PS) particles of sizes $3, 4$ and $8\mu m$ and Polymethilmethacrylate (PMMA) particles of 3 and $8\mu m$.

For a total of 50 distinct particles (10 for each class) we acquired a signal around $25s$ of duration from the quadrant photodetector, subsequently dividing them into $250ms$ segments which were used to create the images of the histograms that are the training input for the CNN, as illustrated in figure 1. The training and test procedure followed a stratified 2-fold cross-validation.

First, by fixing the size of the particle, we performed a classification based on the material (figure 2). As observed, the displacement based histograms allow better accuracy for the classification (81% average) opposed to the position-based ones (70%).

In a second test, and restricting to same particle type and the $3\mu m$ and $8\mu m$ sizes, we classified the particles based on their size. Here, excellent classification results are obtained for both position and displacement-based histograms, with 86% and 98% average accuracy for PMMA and 99% for PS(see figure 3).

Finally and in conclusion, satisfactory to excellent type and size classification results were obtained, demonstrating the capacity of CNN to classify optically trapped particles of distinct physical properties. Leveraging on the small acquistion time of the technique together with the computational power of CNN, the findings enclosed in this work can motivate future approaches towards faster and of intelligent opto and microfluidic devices involving optical trapping tools.

References

[1] Gieseler, Jan *et al.*, Advances in Optics and Photonics **13**, pp.74-241 (2021)

[2] Jones, Philip H. and Maragò, Onofrio M. and Volpe, Giovanni, *Optical Tweezers: Principles and Applications* (Cambridge University Press, 2015) pp.385-393

[3] Carvalho, Inês Alves and Silva, Nuno Azevedo and Rosa, Carla C and Coelho, Luís CC and Jorge, Pedro AS, Sensors **21**, 6181 (2021)

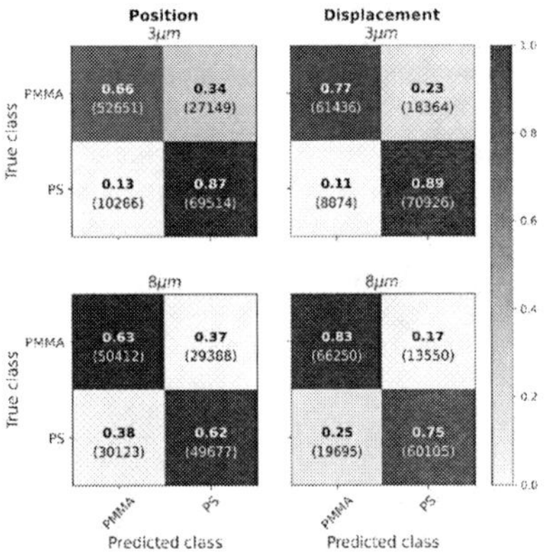

Figure 2. Confusion matrix results of classification between types of particles. Left side results using the position based histograms and right side using displacement based histograms. Top (bottom) row subset of $3\mu m$ ($8\mu m$).

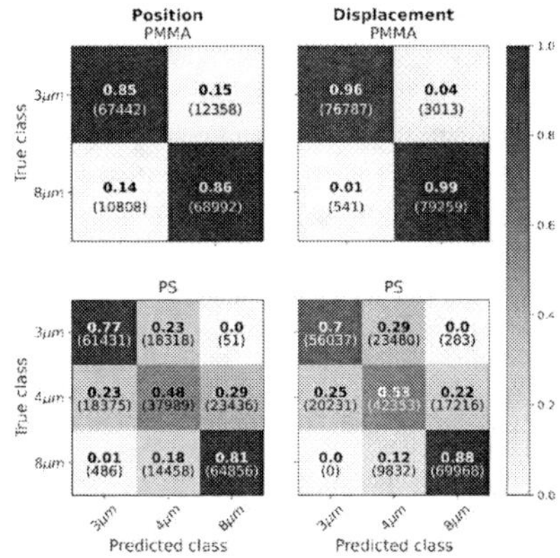

Figure 3. Size classification results between $3\mu m$ and $8\mu m$ of PMMA particles (top row) and $3\mu m$, $4\mu m$ and $8\mu m$ PS particles (bottom row). Left column (right column) is position (displacement) based classification results.

Controlling water-window high-harmonic generation with sub-cycle synthesized waveforms

Giulio Maria Rossi[1,2,*], *Roland E.* Mainz[1,2], *Fabian* Scheiba[1,2], *Miguel A.* Silva-Toledo[1,2], *Maximilian* Kubullek[1,2], and *Franz X.* Kärtner[1,2]

[1]Center for Free-Electron Laser Science, Deutsches Elektronen-Synchrotron DESY, Notkestraße 85, 22607 Hamburg, Germany
[2]Physics Department and The Hamburg Centre for Ultrafast Imaging, University of Hamburg, Luruper Chaussee 149, 22761 Hamburg, Germany

Abstract. We present the first results concerning synthesizer-driven high-harmonic generation that reach the water-window region. This approach holds the promise of offering greater spectral tunability in the generation of isolated attosecond pulses and at the same time of achieving higher photon-flux, required for attosecond-resolved soft X-ray transient absorption experiments.

1 Introduction

Understanding the ultrafast mechanisms that govern the interaction of light and matter in complex systems, such as polyatomic molecules or crystalline solids, is expected to play a fundamental role in the development of light-driven green technologies in fields such as energy generation and storage[1] or chemistry[2]. In fact, the efficiency of converting light into other forms of energy (electrical, chemical, thermal, etc.) often depends to a large extent on dynamical electronic processes on attosecond (as) and femtosecond (fs) time scales [3]. Time-resolved soft X-ray absorption spectroscopy (XAS) is one of the most powerful experimental methods for observing electron dynamics in complex systems since it returns information on the electronic structure (such as chemical bonds) and local geometric structure for the different atomic species. However, the extension of XAS to extreme time scales is not trivial. One of the most promising approach for the development of XAS setups with as/fs resolution consists in the upconversion of ultrashort laser pulses via high-harmonic generation (HHG). HHG allows for the production of as/few-fs soft X-ray pulses with excellent spectrotemporal stability and intrinsic synchronization with widely tunable pump pulses. The main challenge of soft X-ray HHG is the limited conversion efficiency of the process, which results in low photon fluxes. Moreover, HHG often spans a very large spectral region, for instance the full water-window (280-520 eV). This, although an advantage when measuring the absorption edges of different elements simultaneously, often results in a very low flux integrated over each individual absorption edge region. Therefore, the major challenge for soft X-ray HHG sources is to achieve higher efficiencies and/or spectral tunability in order to increase the number of photons in the spectral region of interest. In our group we have recently developed the technology of coherent synthesis of multiple ultrashort pulses obtained via OPAs, the parametric waveform synthesis (PWS)[5], in order to generate pulses with sub-cycle duration, controllable electric field with non-sinusoidal waveforms and energies in the order of 0.5 mJ. These waveforms, given the central wavelength of about 1200-1400 nm (depending on synthesis parameters) and the almost 2-octave spanning bandwidth, are a novel tool to try to circumvent the scaling laws of HHG driven by sinusoidal fields[4], that limit its conversion efficiency in the water-window region.

2 Tunable attosecond pulses in the water-window

Here we report the first results concerning the generation of water-window HHG via sub-cycle waveforms, probably corresponding to isolated attosecond pulses (IAPs). In previous works we already demonstrated that via the PWS it is possible to generate IAPs without gating[5] and to control the spectro-temporal characteristics of the IAPs in the < 200 eV range[6]. Our PWS setup features two spectral channels, spanning the 650-1000 nm range (NIR) and the 1200-2200 nm range (IR). By controlling the relative-phase (RP) between NIR and IR and the overall carrier-envelope phase (CEP) a manifold of non-sinusoidal waveforms can be created. We observed the HHG spectra while scanning the CEP and RP values. The generating medium is helium in a ~ 0.5 mm long gas cell with up to 10 bar of backing pressure. The measurement were taken following this procedure: after optimizing spatiotemporal overlap of NIR and IR pulses, the NIR beam is blocked and the HHG parameters (gas cell position, gas pressure, fine adjustment of iris aperture) are optimized for the IR channel pulse (0.5 mJ of energy at OPA-channel output, 8 fs duration) only in order to achieve the highest spectral intensity

*e-mail: giulio.maria.rossi@cfel.de

 https://doi.org/10.1051/epjconf/202226613030

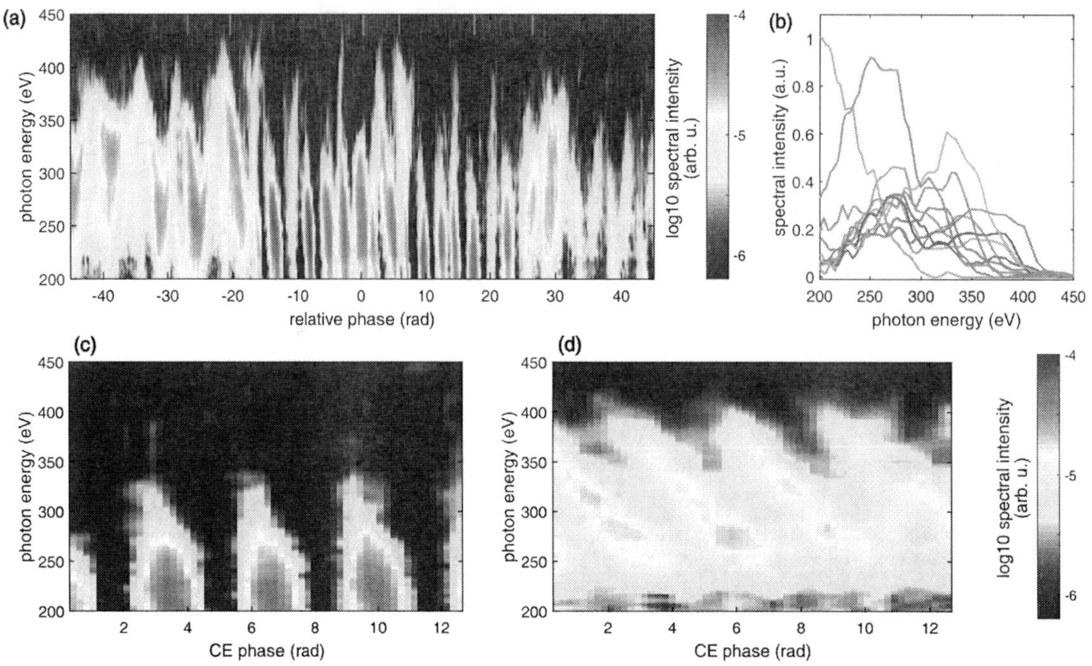

Figure 1. High-Harmonic spectra generated in helium with 10 bar of backing pressure and 2 seconds integration time. The spectra are Jacobi-corrected, transfer functions of grating and CCD are not applied. A 200 nm thick aluminum filter is used to block the IR. (a) Scan of the relative phase (RP) for a fixed CEP value, showing continua of various central photon-energies and bandwidths. (b) In red colour scale: selected HHG spectra corresponding to different RP values, marked with red ticks on the top of (a); in blue scale: IR-only HHG spectra corresponding to different CEPs covering a π range. (c) CEP scan for one particular RP value. (d) CEP scan of IR-only.

and cutoff; afterwards the NIR pulse (0.1 mJ of energy at OPA-channel output, 7 fs duration) is released and the RP scan is performed for while the CEP is actively stabilized. In the scan trace, shown in Fig. 1(a), it is clearly visible the effect of the RP on the HHG spectrum that spans the 200-420 eV region. The HHG intensity is suppressed or enhanced depending on the RP value if compared with the IR-only spectrum. The HHG spectrum is as well reshaped by the precise RP value, resulting in a tunable peak emission between 200 and 350 eV (the lower limit is dictated by the grating used in this measurement) and in maximum photon energies up to > 400 eV. In this scan, it appears that particular values of CEP and RP deliver a higher photon flux with respect to the IR-only case (Fig. 1(b)). This experiment is currently being repeated for a broad range of different macroscopic conditions (gas type/pressure, gas-cell position, focusing condition), in order to obtain a comprehensive picture. We also performed CEP scans for particular values of RP and for the IR pulse alone. As shown in Fig. 1(c,d), while the IR pulse alone produces HHG for any CEP value, the synthesized waveform completely suppresses HHG for certain CEPs, suggesting the presence of IAPs. This is the first time that synthesizer-driven HHG is shown to reach the water-window region.

References

[1] Bräm, O. et al., Phys. Chem. Chem. Phys. **14**, 7934–7937 (2012).
[2] Baldini, E. et al., ACS Photonics **5**, 4, 1241–1249 (2018).
[3] Valentini, A. et al., Phys. Chem. Chem. Phys. **22**, 22302-22313 (2020).
[4] Jin, C. et al., Nat Comm. **5**, 4003 (2014).
[5] Rossi, G.M. et al., Nat. Phot. **14**, 629–635 (2020).
[6] Yang, Y. et al., Nat. Comm. **12**, 6641 (2021).

Imaging of water samples for the detection and identification of microplastics

Matthieu Roussey[1,*], *Boniphace* Kanyathare[1], *Blaž* Hrovat[2], *Nikolaos* Papamatthaiakis[3], *Joni* Hattuniemi[4], *Benjamin O.* Asamoah[1], *Antti* Haapala[3], *Arto* Koistinen[2], and *Kai-Erik* Peiponen[1]

[1]Department of Physics and Mathematics, Institute of Photonics, University of Eastern Finland, P.O. Box 111, 80101 Joensuu, Finland
[2] SIB Labs, University of Eastern Finland, P.O. Box 1627, 70211 Kuopio, Finland.
[3] School of Forest Sciences, University of Eastern Finland, PO Box 111, 80101 Joensuu, Finland.
[4] Valmet Automation Inc., Kehräämöntie 3, 87400 Kajaani, Finland.

Abstract. We demonstrate that direct imaging is a powerful tool for the detection and recognition of microplastics in water, even in case of complex matrices. We use a commercial high-resolution imaging device (Valmet FS5, Valmet Oy.) originally developed for the observation of wood fibres and particles in papermill pulp samples. We show how to discriminate microplastics from other particles in suspension in real water samples. We show differences between several common plastic types in homemade samples.

1 Introduction

Water is the main carrier for vital resources for human and is the most polluted environment. Among the numerous chemical, biological, and physical pollutants, there are micro- and nano-plastics (MP and NP). These particles are of different sizes (MP < 5 mm, NP < 1 µm), of different shapes (spheres, fibers, flakes, arbitrary), of different compositions, and different origin. All types of waters are concerns: open and fresh water, industrial water, wastewater, and tap water. MPs and NPs are vectors for viruses, heavy metal particles, and other bacteria strongly affecting flora and fauna. An abrupt stop of the production of plastics would limit the future number of MPs in NPs but not decrease it rapidly[1]. For 70 years, plastic wastes are discarded in oceans, and they slowly degrade forming the vicious particles studied here. No efficient method currently exists to monitor MPs coming out of our industrial world or coming in the pipes feeding the water we drink. Therefore, there is a crucial and urgent need for developing techniques to detect and eventually remove MPs from water.

Water can be a complex matrix. It can contain all kind of particles and materials, which makes the study of MPs in such an environment is extremely challenging. Currently, water samples containing MPs are characterized in laboratory with techniques such as Raman spectroscopy and Fourier Transform Infrared (FTIR) spectroscopy, mainly[2]. These methods are improved by being robust and reliable, providing accurate results. However, they are expensive in cost and time, and they require a high level of expertise. In addition, the sample preparation and sorting are crucial and time demanding, which prevents the analysis of large water volumes, which is an industrially driven need. We recently demonstrated that using a combination of several optical techniques[3], i.e., spectroscopy, speckle, interferometry, and transmission measurements, one can determine many of the key properties of plastics and microplastics in water, even in complex water matrices such as ones coming from sludge samples[4]. These demonstrated photonic-based techniques are fast and relatively easy to implement. However, they still request some sample preparation.

2 Result and discussion

Fig. 1. Valmet Fiber Image Analyzer (Valmet FS5).

In this study, we investigate the possibility of using direct imaging system for the monitoring and identification of microplastics in water. We use the Valmet Fiber Image Analyzer (Valmet FS5, see Fig. 1), which is a device originally developed for the imaging of wood fibers and particles in half a litre water samples. Ultra-high-resolution images can be captured and analysed at high-

* Corresponding author: matthieu.roussey@uef.fi

speed providing an accurate figure of the particles in suspension in the considered water volume.

Besides some minor technical challenges such as the contamination of the sample container with MPs, using such a device is not difficult and provide quickly numerous images of high quality thanks to a transmission imaging system, a precise and stable illumination. The challenges come from the analysis of the collected data, and we expose here our method to recognize plastic particles.

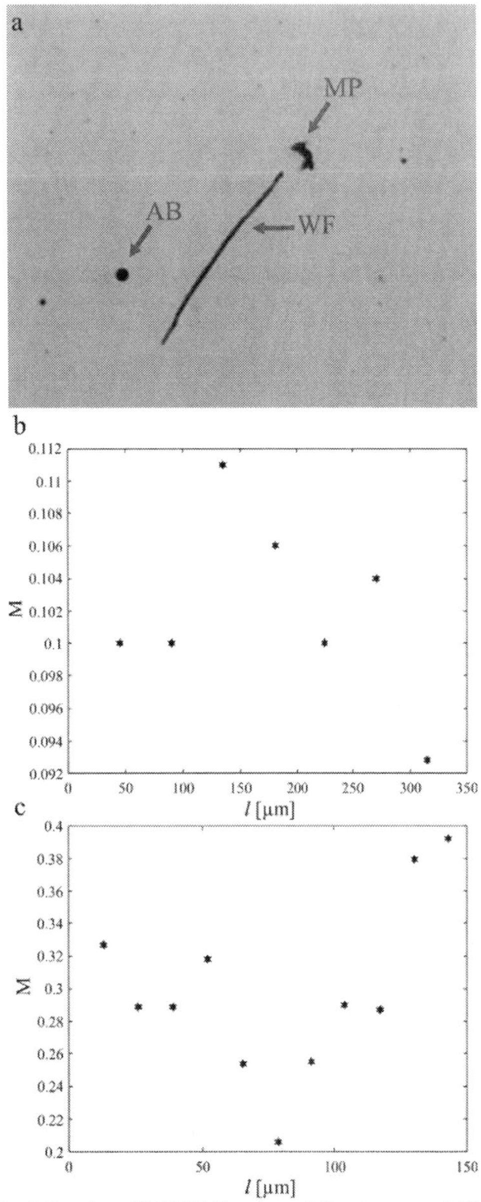

Fig. 2. Imaging of LDPE MP and wood fiber in water. a) High resolution initial image acquired with Valmet FS5 device. Intensity ratios calculated from a) for b) a MP, and c) a wood fiber (WF). AB stands for air bubble.

An example of acquired image is shown in Fig. 2a. One can see several typical compounds of a complex water sample, namely air bubbles coming from the stirring of the water sample during the measurement, microplastics, and wood fiber.

A first visual analysis allows us to discriminate between these different particles. Indeed, the air bubble is perfectly spherical and appear as a black round spot on the image, the wood fiber is elongated and not regular, the microplastics are of complete arbitrary shape. Note that primary plastics used in industry are often microbeads of very regular spherical shape that could be confused with air bubbles, preventing us to use the automatic function implemented in the device to identify the air bubbles and *remove* them from the figure.

To analyse further the image, we use a fast data processing consisting in calculating a light intensity ratio M defined as the intensity of the image normalized by a reference background image. Although most of the particles appear dark on the image due to scattering mainly, they transmit light. This transmission depends directly on the type and shape of the particles. This affects the ratio M. Some values of M are given in Figs. 2b and 2c for a microplastic particle and a wood fiber, respectively. Using this ratio, it becomes easy to distinguish between both particle since most of the plastics will have an M-value close to 0.1, while wood fibers have M-values ranging from 0.2 to 0.4. Air bubbles have a much lower value of M than MPs.

The analysis software implemented in the FS5 device could be further taught and trained to recognize plastic particles using such a method. An extensive analysis work has been already done on various water samples with known and unknown MP compositions. We will present these results.

Authors acknowledge the financial support of the Academy of Finland through the flagship on "Photonics research and innovation" (PREIN, 320166) and of Business Finland through the project on "Circulating economy of water in industrial processes" (CEIWA, #541/31/2021).

References

1. R. Geyer J. R. Jambeckand K. Lavender Law, Sci. Adv. **3**(7) (2017)
2. B. O. Asamoah, E. Uurasjärvi, J. Räty, A. Koistinen, M. Roussey, K.-E. Peiponen, Polymers **13**, 730 (2021).
3. B. O. Asamoah, M. Roussey, and K-E Peiponen, Chemosphere **254**, 126789 (2020).
4. B. O. Asamoah, P. Salmi, J. Räty, K. Ryymin, J. Talvitie, A. K. Karjalainen, J. V. K Kukkonen, M. Roussey, K.-E. Peiponen, Polymers **13**, 871 (2021).

Chip integrated photonics for ion based quantum computing

Steffen Sauer[1,2,*], *Anastasiia* Sorokina[1,2], *Carl-Frederik* Grimpe[3], *Guochun* Du[3], *Pascal* Gehrmann[1,2], *Elena* Jordan[3,5], *Tanja* Mehlstäubler[3,4], and *Stefanie* Kroker[1,2,3]

[1]Institut für Halbleitertechnik, Technische Universität Braunschweig, Hans-Sommer-Str. 66, 38106 Braunschweig, Germany
[2]Laboratory for Emerging Nanometrology, Langer Kamp 6a/b, 38106 Braunschweig, Germany
[3]Physikalisch-Technische Bundesanstalt, Bundesallee 100, 38116 Braunschweig, Germany
[4]Institut für Quantenoptik, Leibniz Universität Hannover, Welfengarten 1, 30167 Hannover, Germany
[5]DLR-Institut für Satellitengeodäsie und Inertialsensorik, Leibniz Universität Hannover, Welfengarten 1, 30167 Hannover, Germany

Abstract. Ion traps are a promising platform for the realisation of high-performance quantum computers. To enable the future scalability of these systems, integrated photonic solutions for guiding and manipulating the laser light at chip level are a major step. Such passive optical components offer the great advantage of providing beam radii in the μm range at the location of the ions without increasing the number of bulk optics. Different wavelengths, from UV to NIR, as well as laser beam properties, such as angle or polarisation, are required for different cooling and readout processes of ions. We present simulation results for different optical photonic components, such as grating outcouplers or waveguide splitters and their applications on ion trap chips. Furthermore, we will introduce the experimental setup for the optical characterisation of the fabricated structures.

1 Introduction

Today's demand for powerful computers is increasing in order to solve complex future problems. One promising approach is the quantum computer based on trapped ions on a surface chip with nanostructured components [1]. The qubits are realised by the internal states of individual trapped ions and will be controlled and transported between the loading, storage and interaction zones on future chips (see figure 1). The advantages of trapped ions as qubits are their very long coherence times, which are limited by the dephasing of technical sources, like vacuum pressure. Furthermore, the very high reliability and the existing knowledge about the preparation of the states and the readout of ions.

For the new generation of computers, growing scalability is of paramount importance: the more qubits, in our case ions, a quantum computer has, the more computing power it will provide. A big game-changer for scalability of ion-based quantum computers are the integrated passive optical components. Their integrability on chips and their small size allows for a variety of implementations compared to bulk optics. In this vein, the number of qubits can be drastically increased while complying with the DiVincenzo criteria for building a scalable, fault-tolerant quantum computing [2].

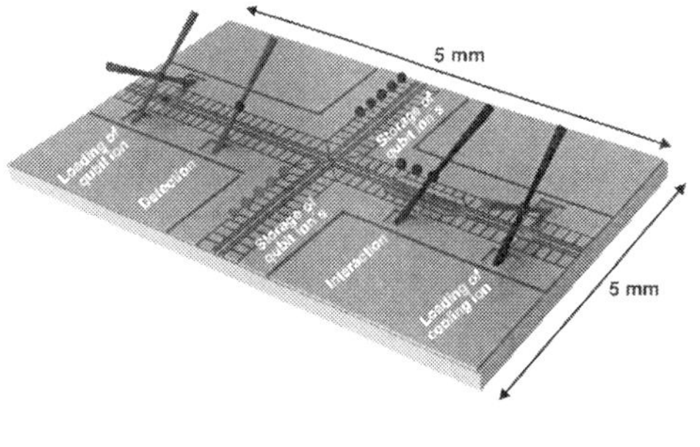

Figure 1. Possible ion trap chip design with different application zones. The ions act as qubits of the quantum computer. To cool or to detect ions, laser beams are sent via grating couplers. In the interaction zone, the qubits are used for calculations.

2 Optical componentes

Building an ion trap chip with integrated optics requires different components. The laser light has to get into the chip, it has to be distributed via waveguide splitters and extracted from the surface with grating couplers (see figure 2) so that the light can interact with the ions.

2.1 Properties and challenges

In general, you do not need a high light output power for the control of ions, in addition, there is the μm beam radius, which also has a positive effect on the required intensity. That´s why, the losses within the components are not the most significant challenge; rather, the coverage of all required wavelengths from UV to NIR should be accessi-

*e-mail: steffen.sauer@tu-braunschweig.de

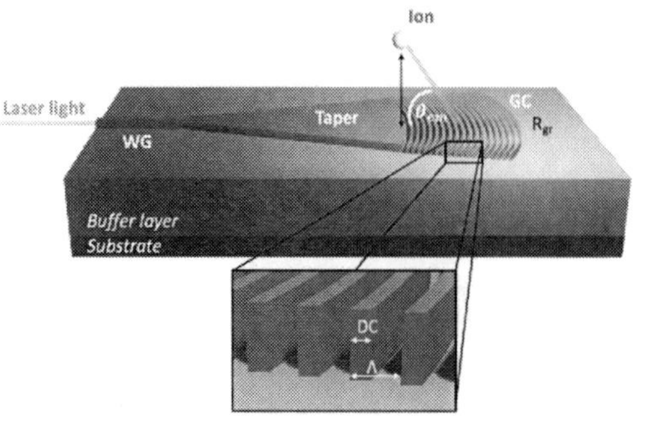

Figure 2. Sketch of a backward grating coupler. Laser light (red) is guided to the grating coupler (GC) via a waveguide (WG). A GC usually consists of a taper and a grating structure. The laser beam leaves the chip with the emission angle (θ_{em}) via the grating to interact with the ions. The properties of the emitted laser beam are defined by the period (Λ), the duty cycle (DC) and the radius of curvature R_{gr} of the grating.

ble (see figure 3). Therefore, *different materials* must be used to cover all wavelenghts of interest (see table 1). Nevertheless, it is worth bearing in mind that the outcoupled laser beams should not affect other ions. Furthermore, in addition to the state-of-the-art linear polarization for integrated grating couplers [3–5], *circular polarization* is important for cooling ions, for example.

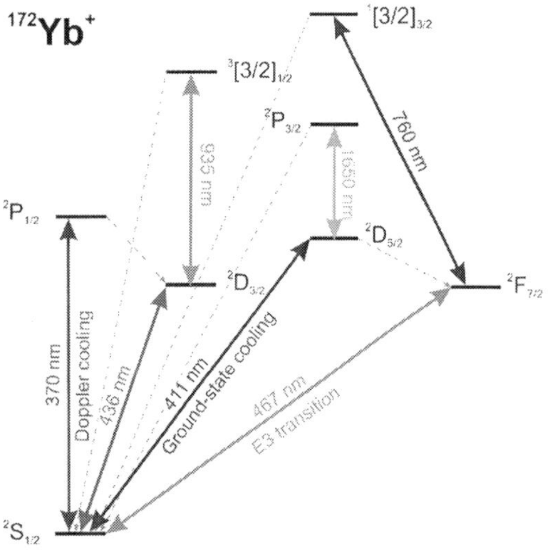

Figure 3. Level scheme of ^{172}Yb^{+} ions with all relevant transition wavelengths for our integrated optics. We use ^{172}Yb as a testbed for other future ion elements.

2.2 Simulations

Our numerical simulations consist of a combination of 2D and 3D calculations of waveguides (WG) and grating couplers (GC), calculated with COMSOL Multiphysics FEM and Lumerical FDTD solvers. We optimize the WG geometry to ensure a single-mode regime, at the same time with a high confinement of the fundamental quasi-TE mode. The design of the GCs involves optimizing the grating periods, duty cycles and curvature of the grating to ensure Gaussian beam and µm-focusing at the ion.

Table 1. Material platforms of waveguides for varying wavelengths.

Core material	Wavelength range [nm]
AlN/Al_2O_3	UV: 370 - 500
Si_3N_4	VIS-NIR: >500

Moreover, we verify the laser beam properties with Near to Far-Field Transformation, so that conclusions can be drawn about the laser beam at a typical ion height of 100 µm. We also calculate the effect of the GC and possible use of an ITO coating on top of the GC to minimize the influence of the micromotion on the ion.

3 Optical characterization

We are in the process of building optical setups to check the laser beam properties of our chips. To this end, we use microscopes with different optical components, CCD cameras and set up seven different laser systems for wavelengths between 370 and 1650 nm.

Acknowledgements

The authors gratefully acknowledge support by the Braunschweig International Graduate School of Metrology B-IGSM and the DFG Research Training Group GrK1952/1 "Metrology for Complex Nanosystems". Furthermore, we thank the Quantum Valley Lower Saxony, ATIQ and cluster of Excellence QuantumFrontiers (EXC-2123–390837967).

References

[1] C. D. Bruzewicz et al., App. Phys. Rev., **6**, 021314 (2019)
[2] D. P. DiVincenzo, Fortschr. Phys. **48**, 771 (2000)
[3] R. J. Niffenegger et al., Nat., **586**, 538 (2020)
[4] K. K. Mehta et al., Nat., **586**, 533 (2020)
[5] K. K. Mehta et al., Sci. Rep., **7**, 2019 (2017)

 https://doi.org/10.1051/epjconf/202226613033

Dispersion-managed Monolithic All Polarization-Maintaining Ultrafast Thulium-doped Fiber Oscillator

Benedikt Schuhbauer[1,*], *Andreas* Wienke[1,2], *Frithjof* Haxsen[1], *Jörg* Neumann[1,2], and *Dietmar* Kracht[1,2]

[1]Laser Zentrum Hannover e.V., Hollerithalle 8, Hannover 30419, Germany
[2]Cluster of Excellence PhoenixD (Photonics, Optics, and Engineering Innovation Across Disciplines)

Abstract. We report on our first results of a dispersion-managed monolithic polarization-maintaining ultrafast Thulium-doped fiber oscillator. The design incorporates only commercially available polarization-maintaining components and emits linearly polarized light with a polarization extinction ratio of over 17 dB . The observed optical pulse spectra with up to 10 nm width centered at 1961.5 nm beyond the water absorption lines support sub-1 ps pulse duration. The simple linear configuration and the wide spread availability of the used low cost polarization-maintaining fiber components makes the presented oscillator a valuable concept for an environmental stable seed source.

1 Introduction

Ongoing efforts on developing ultrafast Thulium-doped fiber lasers are driven by applications like nonlinear microscopy, driving of frequency conversion stages and material processing. For generating suitable output parameters in terms of average power, pulse energy and pulse duration, advanced chirped pulse amplification (CPA) schemes with multi-GW peak power have been developed [1]. Those systems require a stable seed source to avoid catastrophic damage and guarantee a reliable long term operation. As potential seed source for a CPA, various mode-locked oscillators scheme based on real saturable absorbers [2–5] or different types of artificial saturable absorbers [6–8] have been presented in the past.

Even though various configurations of mode-locked oscillator schemes have been proposed, only a small number of oscillators incorporated a polarization-maintaining (PM) architecture, which is highly desirable for environmental stable pulsed operation. Owing to the anomalous material dispersion, using only standard silica PM single mode step-index fibers is limiting oscillators to the soliton regime, which deliver rather long pulses and show the characteristic Kelly sidebands. Those sidebands corresponds to the interference of the soliton with a longer dispersive pedestal [9]. The present pedestal hinders a good contrast between main pulse and background during the amplification and at the application. Furthermore, the efficiency of the amplification can be reduced by a double-digit percentage, owing to undesired amplification of the Kelly sidebands [10]. This is mainly due to the lack of a proper PM fiber-based dispersion management, which is necessary to operate in the stretched pulse [11] or dissipative dispersion-managed soliton [3] regime without Kelly sidebands and shorter pulse durations. Michalska addressed this issue by reporting on a hybrid system of polarization-maintaining fibers and a non-PM fiber for dispersion-management, mode-locked by a nonlinear optical loop mirror [7]. The environmental stability and the polarization extinction ratio (PER) with about 7 dB remained limited. Another report on a monolithic PM Thulium-doped fiber oscillator with dispersion-management by Sotor et. al. included a PM home-built normal dispersive passive fiber and saturable absorber made out of graphene [5]. However, the authors stated no value for the achieved PER and the restrain to special home-built components is still limiting the construction of completely polarization-maintaining ultrafast oscillators with dispersion-management.

Here, we report on a monolithic ultrafast Thulium-doped fiber oscillator, which incorporates only commercially available components. This allows the wide-spread use as seed source and serves as starting point for further developments towards an environmental stable fiber front-end for CPA systems.

2 Experimental Setup and Results

The oscillator was set up in a monolithic all PM linear cavity configuration, as presented in Figure 1a. Pump light at 1550 nm was provided by an Ytterbium-Erbium-doped fiber amplifier (PE) with up to 4 W of power. In order to couple the light into the oscillator, a commercially available fused wavelength-division multiplexer (WDM) optimized for 1970 nm was used. As gain medium, a 16 cm long highly doped fiber (TDF, Nuferm PM-TSF-5/125) was used to enhance the re-absorption inside the oscillator to support operation at longer wavelengths beyond the water absorption lines. A chirped fiber Bragg

*e-mail: B.Schuhbauer@lzh.de

grating (CFBG, Teraxion) provided the wavelength selection, output coupling as well as dispersion-management. It was inscribed into the core of a PM1950 fiber and had a peak reflectivity of about 15 % with a full width at half maximum (FWHM) of 35 nm and a group delay dispersion (GDD) of 1 ps^2 at 1970 nm by design. By adjusting the length of an additional piece of PM1550 fiber, the net cavity GDD was set to 0.01 ps^2. For initiating and sustaining the mode-locked operation, a broadband semiconductor saturable absorber mirror (SAM, Batop) was used. The SAM had a saturation depth of 14 %, a non-saturable loss of 8 % and a saturation fluence of 60 $\mu J/cm^2$ at 2000 nm. For the first experiments, the SAM was butt-coupled, but the same SAM is also commercially available as mechanically stable version, directly glued on to the faced of a PM fiber connector. This glued version will be incorporated in the next steps.

With a pump power of 820 mW, the oscillator emitted mode-locked pulses with a rectangular spectrum centered at 1961.5 nm with a FWHM of about 10 nm, as depicted in Figure 1b, supporting a transform limited pulse duration of about 800 fs. The long wavelength operation beyond 1950 nm was chosen to rule out pulse and beam distortion due to water absorption in the later CPA system [12].

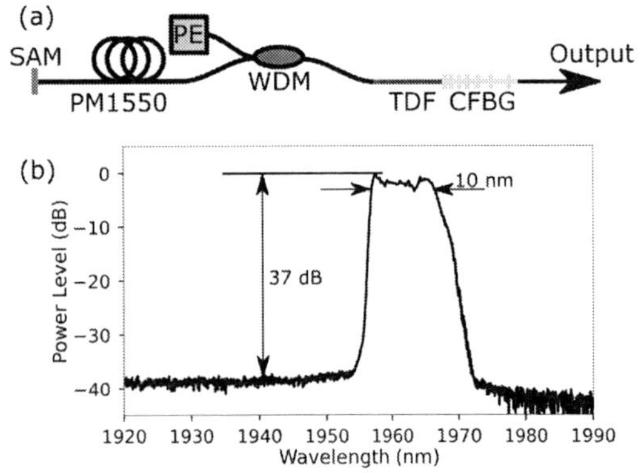

Figure 1. Experimental setup of the linear cavity (a) and output spectrum (b).

Further characterization of the oscillator in the temporal domain allowed to observe a stable pulse train with an average power of 14 mW consisting of single pulses with a temporal spacing of 57 ns, which corresponds to a fundamental repetition rate of 17.4 MHz. Operating the oscillator below 20 MHz is suitable for pulse picking in a later fiber front-end with stretcher unit and fiber-based acousto-optic modulators as pulse picker. A PER of 17.2 dB was determined, even though the cavity does not include any designated polarization selective elements. In order to further improve the PER and to ensure always operating in the slow axis, a fiber-based inline polarizer will be included into the cavity.

3 Conclusion

We presented our first results of a monolithic PM ultrafast Thulium-doped fiber oscillator, which supports sub-1 ps pulse duration. Work in the near future will include the replacement of the current SAM to the mechanically stable glued version and the further integration into an environmental stable fiber front-end with stretcher and pulse picker for use as seed for a high power Thulium-doped fiber CPA system.

Acknowledgment

This work was supported by the Deutsche Forschungsgemeinschaft (DFG, German Research Foundation) under Germany's Excellence Strategy within the Cluster of Excellence PhoenixD (EXC 2122, Project ID 390833453) and the Federal Ministry for Economic Affairs and Energy, Germany (ZIM grant No. 16KN053068).

References

[1] C. Gaida, M. Gebhardt, F. Stutzki, C. Jauregui, J. Limpert, A. Tünnermann, Opt. Lett. **41**, 4130 (2016)
[2] R. C. Sharp, D. E. Spock, N. Pan, J. Elliot, Opt. Lett. **21**, 881 (1996)
[3] Regina Gumenyuk, Ismo Vartiainen, Hemmo Tuovinen, Oleg G. Okhotnikov, Opt. Lett. **36**, 609 (2011)
[4] J. Wang, X. Liang, G. Hu, Z. Zheng, S. Lin, D. Ouyang, X. Wu, P. Yan, S. Ruan, Z. Sun et al., Scientific Reports **6**, 28885 (2016)
[5] Jarosław Sotor, Jakub Bogusławski, Tadeusz Martynkien, Paweł Mergo, Aleksandra Krajewska, Aleksandra Przewłoka, Włodek Strupiński, Grzegorz Soboń, Opt. Lett. **42**, 1592 (2017)
[6] M. A. Chernysheva, A. A. Krylov, P. G. Kryukov, N. R. Arutyunyan, A. S. Pozharov, E. D. Obraztsova, E. M. Dianov, Opt. Express **20**, B124 (2012)
[7] Maria Michalska, Optics & Laser Technology **138**, 106923 (2021)
[8] Lynn E. Nelson, Erich P. Ippen, Hermann A. Haus, Applied Physics Letters **67**, 19 (1995)
[9] R. Weill, A. Bekker, V. Smulakovsky, B. Fischer, O. Gat, Phys. Rev. A **83**, 043831 (2011)
[10] Charles W. Rudy, Karel E. Urbanek, Michel J. F. Digonnet, Robert L. Byer, J. Lightwave Technol. **31**, 1809 (2013)
[11] Frithjof Haxsen, Axel Ruehl, Martin Engelbrecht, Dieter Wandt, Uwe Morgner, Dietmar Kracht, Opt. Express **16**, 20471 (2008)
[12] Martin Gebhardt, Christian Gaida, Fabian Stutzki, Steffen Hädrich, Cesar Jauregui, Jens Limpert, Andreas Tünnermann, Opt. Express **23**, 13776 (2015)

Unravelling an optical extreme learning machine

Duarte Silva[1,2*], *Nuno A. Silva*[1,2], *Tiago D. Ferreira*[1,2], *Carla C. Rosa*[1,2], *and Ariel Guerreiro*[1,2]

[1]Departamento de Física e Astronomia, Faculdade de Ciências, Universidade do Porto, Rua do Campo Alegre s/n, 4169-007 Porto, Portugal
[2]INESC TEC, Centre of Applied Photonics, Rua do Campo Alegre 687, 4169-007 Porto, Portugal

Abstract. Extreme learning machines (ELMs) are a versatile machine learning technique that can be seamlessly implemented with optical systems. In short, they can be described as a network of hidden neurons with random fixed weights and biases, that generate a complex behaviour in response to an input. Yet, despite the success of the physical implementations of ELMs, there is still a lack of fundamental understanding about their optical implementations. This work makes use of an optical complex media to implement an ELM and introduce an ab-initio theoretical framework to support the experimental implementation. We validate the proposed framework, in particular, by exploring the correlation between the rank of the outputs, $\boldsymbol{H}$, and its generalization capability, thus shedding new light into the inner workings of optical ELMs and opening paths towards future technological implementations of similar principles.

1 Introduction

Deep neural networks are becoming a ubiquitous tool across many scientific domains and areas of knowledge. Yet, the performance of neural networks is intrinsically tied to its scalability, and with the impending plateau of Moore's law, emerges a need to continue performance scaling in the absence of electronics miniaturization. A promising route lies within hardware specialization, especially for machine learning applications. In particular, significant performance gains can be achieved by leveraging optics as hardware accelerators due to its intrinsic parallelism, high bandwidth, the easy implementations of linear operations through propagation and low energy consumption.

1.1 ELM basics

Given a target function $f(\boldsymbol{x}): R^d \to R^m$, an extreme learning machine (ELM) predicts $f_{ELM}(\boldsymbol{x}) = \sum_{i=1}^{L} \beta_i h_i(\boldsymbol{x})$, where $h_i(\boldsymbol{x}) = G(\boldsymbol{a}_i, b_i, \boldsymbol{x})$ is a nonlinear piecewise continuous function, with $\boldsymbol{a}_i \in R^d$ and $b_i \in R$, sampled from a continuous distribution function, and satisfying ELM universal approximation capability theorems. The vector β can be trained through ridge regression:

$$\min_{\beta \in R^{L \times m}} ||\boldsymbol{H}\beta - \boldsymbol{T}||_2^2 + \lambda ||\beta||_2^2 \qquad (1)$$

Where $H_{ij} = h_j(\boldsymbol{x}_i)$, $T_{ij} = (y_i)_j$ is the target data given a training set $\{\boldsymbol{x}_i, \boldsymbol{y}_i\}_{i=1}^N$, and $||\cdot||_2$ denotes the Frobenius norm.

A common approach to implement an optical version of this framework is to employ either non-linear light propagation, or ensure a non-linear detection. While effective, such works have not explored the effect of the random projection of their systems. In this work, we aim to shed some light on the mathematical intricacies of a common implementation of an optical ELM based on complex optical media.

2. Results and discussion

In Fig.1 we present an illustration of the concept and experimental setup. In short, we use a digital micromirror device (DMD) as the optical encoder capable of both amplitude and phase modulation enabled by Lee holography.

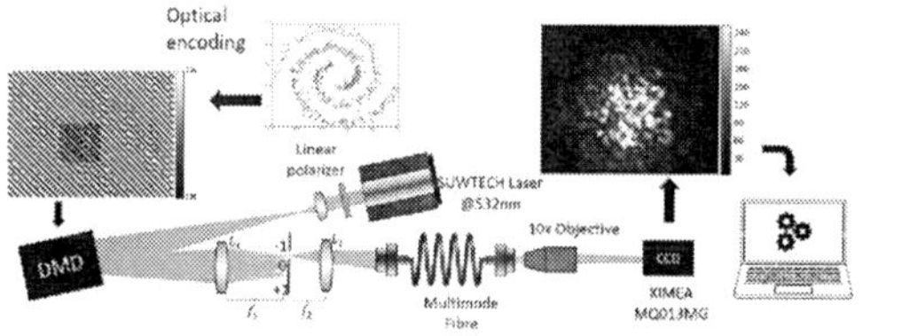

Fig. 1. Illustration of the optical set-up and methods for an optical extreme learning machine.

The light is then coupled to a multi-mode fiber where the information is mixed. The output field is a speckle pattern which is known to possess gaussian circular statistics that guarantees the randomness required by an ELM. Then, this pattern is measured on a high-speed CCD camera both in the linear and non-linear regime. We have developed a theoretical framework which demonstrates

* Corresponding author: duartejfs@hotmail.com

https://doi.org/10.1051/epjconf/202226613034

that simple linear dynamics followed by intensity measurements yield an effective activation function of polynomial and/or sinusoidal nature. Furthermore, this analysis allows us to infer on the amount of information that is carried out to $\boldsymbol{H}$ (i.e. its rank) based on the optical encoding, as well as its generalization capability, which has been experimentally confirmed, as seen in tables 1 and 2. The unmatched values of rank($\mathbf{H}$) predictions' are due to experimental noise, which is out of the scope of this manuscript.

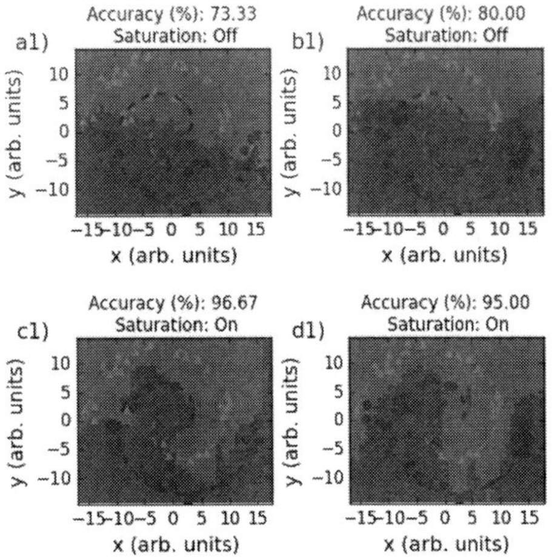

Fig. 2. Classification performance under amplitude modulation of the wavefront. a1) and b1) rely on a simple direct linear encoding mechanism according to table 1, and linear detection on the camera. c1) and d1) employ the same encodings, but with non-linear detection (i.e. camera saturation). The two classes are represented as red triangles and blue circles.

	Rank **H** (pred.)	Rank **H** (exp.)	Encoding	Accuracy (%)
a1)	6	6	$\{a_1, a_2\}$	73.33
b1)	15	7	$\{a_1, a_1^3, a_2, a_2^3\}$	80.00
c1)	N\A	12	$\{a_1, a_2\}$	96.67
d1)	N\A	13	$\{a_1, a_1^3, a_2, a_2^3\}$	95.00

Table 1. Summary of the results from figure 2. The encoding variables (a_1, a_2) is a one-to-one mapping of the (x, y) coordinates of each point to a linear amplitude modulation of the wavefront in distinct regions of the DMD screen.

	Rank **H** (pred.)	Rank **H** (exp.)	Encoding	MSE (%)
a1)	3	3	$\{b_1\}$	1.625
b1)	5	5	$\{b_1, 2\pi - b_1\}$	1.052
c1)	33	8	$\{b_1, b_1^{0.5}, b_1^2, b_1^4\}$	2.601
d1)	N\A	9	$\{b_1, 2\pi - b_1\}$	0.002

Table 2. Summary of results from figure 3. The encoding variable b_1 is a one-to-one mapping of the x coordinate of each point to a linear phase modulation of the wavefront from 0 to 2π in distinct regions of the DMD screen.

In addition, we also studied the effect of an intense physical non-linearity through saturation in the camera and its effect on the rank of the output. At the same time, the machine is benchmarked on standard regression and classification tasks, and we have found that it can yield remarkable performance under certain conditions, as seen in figures 2 and 3.

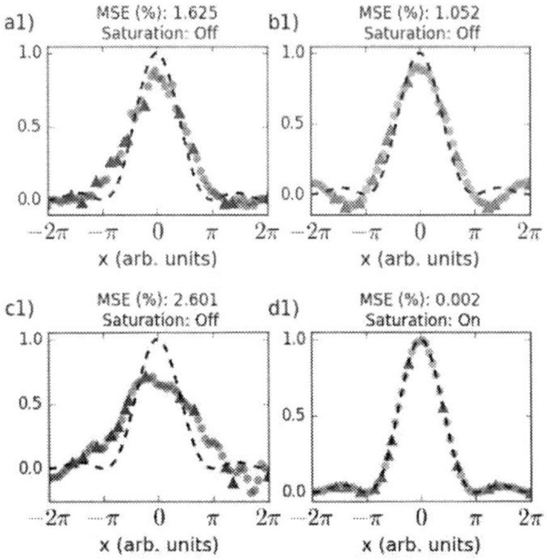

Fig. 3. Regression task performance under phase modulation of the wavefront. The target function is represented by the dashed line. a1) and b1) rely on a simple linear encoding mechanism, while c1) bears a non-linear encoding (see table 1). All with linear detection on the camera. d1) employs the same encoding as b1), but with non-linear detection (i.e. camera saturation). Semi-transparent circles represent training data and opaque triangles test data.

In conclusion, we have found a connection between the rank of the outputs, $\boldsymbol{H}$, and its generalization capability which aligns with the theoretical framework developed. However, the relationship is not unique, thus establishing the rank of the outputs only as a good rule-of-thumb when predicting the performance of an ELM, yet not the ideal metric. As for the effect of a strong non-linearity, our results are consistent with the ones present in the literature which makes a stronger case for the need of physical non-linearities for good performance. Optical ELMs are a promising platform for hardware specialization, and our work paves the way for a better understanding of such devices, allowing for a better design and task selection.

References

1. A. Saade. "*Random projections through multiple optical scattering: Approximating Kernels at the speed of light*". IEEE International Conference on Acoustics, Speech and Signal Processing. (2016)
2. G.Huang, Q. Zhu, and C. Siew. "*Extreme learning machine: a new learning scheme of feedforward neural networks*". IEEE International Joint Conference on Neural Networks. (2016)

https://doi.org/10.1051/epjconf/202226613035

Padé resummation of divergent Born series and its motivation by analysis of poles

Thomas van der Sijs[1,*], *Omar* El Gawhary[1], and *Paul* Urbach[1]

[1]Optics Research Group, Delft University of Technology, Lorentzweg 1, 2628 CJ Delft, The Netherlands

Abstract. The Born series is in principle a powerful way to solve electromagnetic scattering problems. Higher-order terms can be computed recurrently until the desired accuracy is obtained. In practice, however, the series solution often diverges, which severely limits its use. We discuss how Padé approximation can be applied to the Born series to tame its divergence. We apply it to the scalar problem of scattering by a cylinder, which has an analytical solution that we use for comparison. Furthermore, we improve our understanding of the divergence problem by analyzing the poles in the analytical solution. This helps build the case for the use of Padé approximation in electromagnetic scattering problems. Additionally, the poles reveal the region of convergence of the Born series for this problem, which agrees with actual calculations of the Born series.

1 Introduction

Forward electromagnetic scattering problems can be solved by expanding the electromagnetic field as a perturbation series: the Born series. Its terms are computed recurrently, starting from the zeroth order solution. So one goes from solving a difficult problem to solving a series of easy problems. The higher-order corrections can be added to the zeroth order field until the desired accuracy is achieved. That is, if the series converges. However, the Born series diverges in many cases of practical interest.

We discuss the approach presented in Ref. [1], where Padé approximation is used to tame the divergence of the Born series and retrieve a useful solution. As an example, we apply it to the problem of scattering by an infinitely long cylinder. We analyze the poles in the exact solution to understand the divergence problem. This allows us to determine for which parameters the Born series will converge.

2 Born-Padé method

We consider scattering problems for which the scalar formulation is rigorous. Given a polarization, either the electric or magnetic field has only one nonzero component. That component $U(\mathbf{r})$ then satisfies the inhomogeneous Helmholtz equation:

$$\left[\nabla^2 + k_0^2\right] U(\mathbf{r}) = -k_0^2 \Delta\varepsilon_r(\mathbf{r}) U(\mathbf{r}), \tag{1}$$

where k_0 is the wave number in vacuum and $\Delta\varepsilon_r(\mathbf{r}) = \varepsilon_r(\mathbf{r}) - 1$ the relative permittivity contrast relative to the background (chosen to be vacuum/air). We can consider $-k_0^2\Delta\varepsilon_r(\mathbf{r})U(\mathbf{r})$ as an induced source that acts as a perturbation superposed on the background.

*e-mail: T.A.vanderSijs@tudelft.nl

The perturbation solution is obtained by writing $U(\mathbf{r})$ as a power series in a perturbation parameter σ, i.e.,

$$U(\mathbf{r}, \sigma) = \sum_{\ell=0}^{\infty} U_\ell(\mathbf{r})\sigma^\ell. \tag{2}$$

We also add σ in front of the source term to vary the perturbation strength. The terms are computed recurrently, starting from the zeroth order field, an incident plane wave. Then summing the terms one by one is only possible if the series converges. Alternatively, we can recast the terms into a different representation that does not suffer from divergence. We use Padé approximants [2]:

$$P_N^M(\sigma) = \frac{\sum_{m=0}^{M} A_m \sigma^m}{1 + \sum_{n=1}^{N} B_n \sigma^n}, \tag{3}$$

with $M + N + 1$ unknown Padé coefficients A_m and B_n. By equating the Born and Padé representations, one can derive a linear system of $M + N + 1$ equations which yield A_m and B_n when solved. This procedure is repeated for each point in space. The power of the approximants lies in their ability to represent poles, contrary to power series.

3 Application to an infinitely long cylinder

For a normally incident plane wave scattering off an infinitely long cylinder, the solution is invariant along the cylinder's axis. The problem is effectively 2D and Maxwell's equations decouple into TE and TM polarization. Here, we consider a TE-polarized plane wave, so E_z is the only nonzero component of the electric field and it satisfies eq. (1). Figure 1 shows the geometry of the problem, with the incident plane wave propagating in the $+x$-direction. The problem has an analytical solution in terms

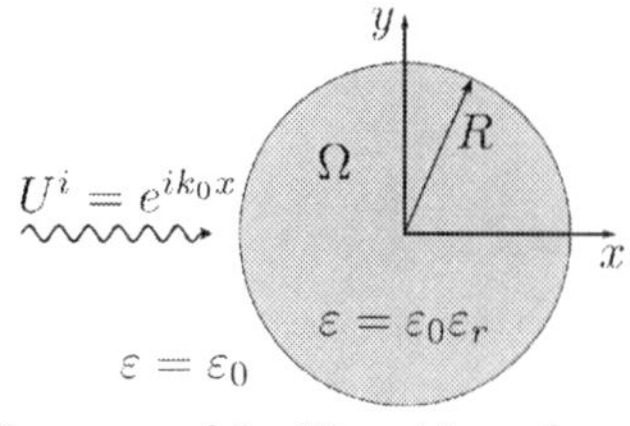

Figure 1: Geometry of the 2D problem of a cylinder. The incident plane wave U_i represents E_z, as we consider TE polarization.

of cylindrical modes, i.e., as an infinite sum of Bessel and Hankel functions [1, 3].

We showcase the Born-Padé method for these values: $\lambda = 400$ nm, radius $R = 400$ nm, and the relative permittivity of silver ($\varepsilon_r = 30.8 + 4.30i$). We compare the most accurate Padé approximant P_{12}^{12} to the analytical solution in fig. 2. We are able to retrieve the solution from a strongly diverging Born series. Namely, the magnitude of the series' terms grows with a factor of 5.67.

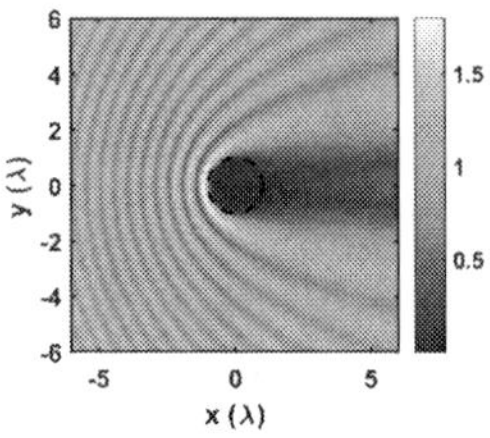

(a) Analytical solution.

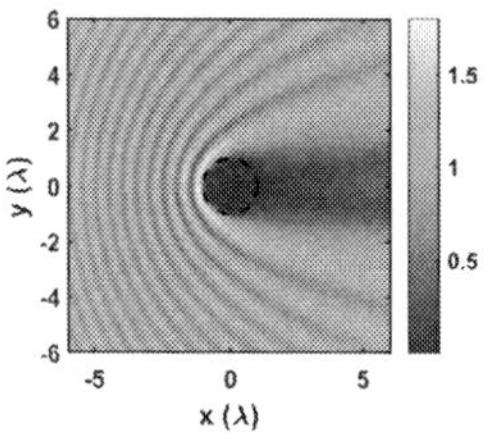

(b) Padé approximant P_{12}^{12}.

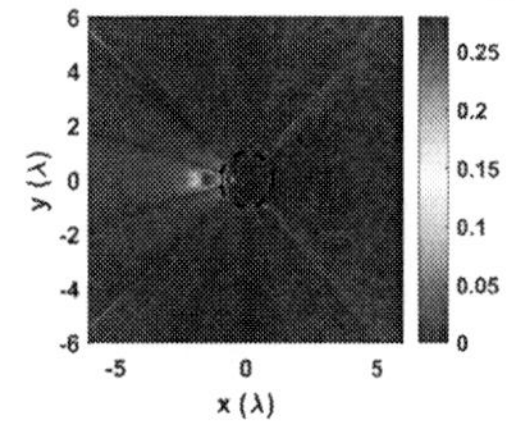

(c) Absolute error.

Figure 2: Born-Padé solution for a silver cylinder of radius 400 nm at a wavelength of 400 nm. The plots display the absolute value. The axes are relative to the wavelength.

4 Poles and convergence

Since the Born series is a Taylor series, it converges inside a disk in the complex plane of σ. The radius of the disk is determined by the pole closest to the point about which the series is expanded. To find the poles we follow the derivation of the analytical solution, by writing the general solution in- and outside the scatterer and requiring their continuity on the cylinder's boundary. This results in a 2x2 linear system of equations for each cylindrical wave mode. The determinant of the system ends up as the denominator in the analytical solution, so the poles are where the determinant vanishes. Hence, for each cylindrical mode m, we set the determinant to 0:

$$H_m^{(1)\prime}(k_0R)J_m(k_0\sqrt{\varepsilon_r}R) - \sqrt{\varepsilon_r}H_m^{(1)}(k_0R)J_m'(k_0\sqrt{\varepsilon_r}R) = 0, \tag{4}$$

where H and J are Hankel and Bessel functions, respectively. Figure 3 shows eq. (4) as function of $\Delta\varepsilon_r = \varepsilon_r - 1$ for $R = 400$ nm and $\lambda = 400$ nm. The dark spots indicate where the determinant approaches 0, and thus where the poles are. The closest pole, out of all modes m, is

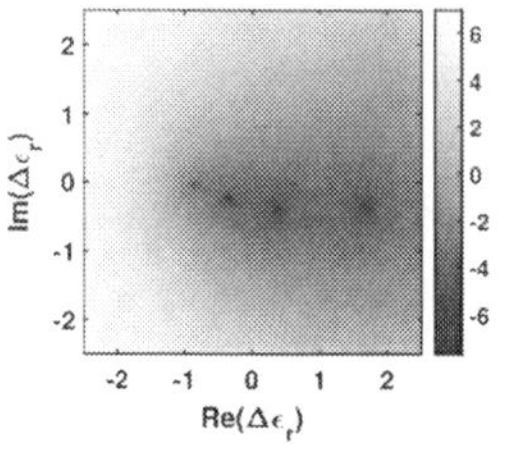

(a) Mode $m = 0$.

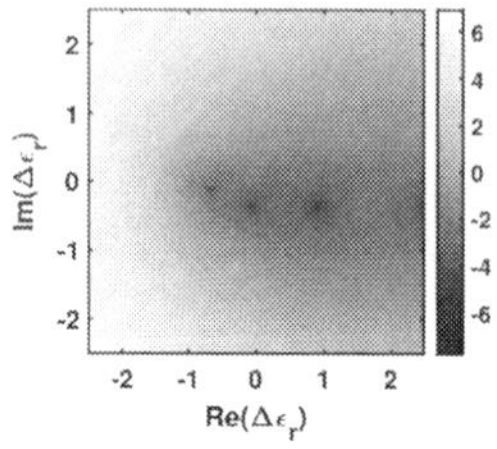

(b) Mode $m = \pm 1$.

Figure 3: Determinant for a cylinder with radius 400 nm at $\lambda = 400$ nm plotted in the complex plane of $\Delta\varepsilon_r$ (logarithm of absolute value).

$\Delta\varepsilon_r = -0.060 - 0.373i$ in the $m = \pm 1$ modes. Its distance to the origin is $|\Delta\varepsilon_r| = 0.377$, which thus is the radius of convergence. When actually computing the Born series, convergence is indeed seen only for values of $\Delta\varepsilon_r$ inside a disk with that radius.

Why do we look at the convergence in terms of $\Delta\varepsilon_r$ instead of σ? Since σ always appears in the Born series as a product $\sigma\Delta\varepsilon_r$, we can equally well set $\sigma = 1$–which we do anyways because that value corresponds to our problem–and consider the convergence as function of $\Delta\varepsilon_r$.

Furthermore, we see all poles are in the lower half plane and thus have negative imaginary part, corresponding to materials with gain. Also, setting the determinant to 0 is equivalent to solving eq. (1) in the absence of an incident field. We thus interpret the poles as a nonphysical cylinder radiating in the absence of a source. Interestingly, the divergence in physical cases is thus due to poles at nonphysical values. For obliquely incident waves, the poles could have physical values of $\Delta\varepsilon_r$, but that would still be nonphysical due to the infinity of the cylinder [4].

References

[1] T. A. van der Sijs, O. El Gawhary, and H. P. Urbach, Electromagnetic scattering beyond the weak regime: Solving the problem of divergent Born perturbation series by Padé approximants, Phys. Rev. Research **2**, 013308 (2020)

[2] C. M. Bender, and S. A. Orszag, *Advanced Mathematical Methods for Scientists and Engineers I: Asymptotic Methods for Scientists and Engineers* (Springer, New York, 1999)

[3] F. Frezza, F. Mangini, and N. Tedeschi, Introduction to electromagnetic scattering: tutorial, J. Opt. Soc. Am. A **35**, 163-173 (2018)

[4] J. A. Stratton, *Electromagnetic Theory* (John Wiley & Sons, Hoboken, 1999)

AUTHOR INDEX

AUTHOR INDEX

AUTHOR INDEX

AUTHOR INDEX

AUTHOR INDEX

AUTHOR INDEX

AUTHOR INDEX

AUTHOR INDEX

AUTHOR INDEX

EDP Sciences – Web of Conferences
17, Avenue du Hoggar
Parc d'Activité de Courtabouf
BP 112
F-91944 Les Ulis Cedex A
France

ISBN 978-1-7138-6266-6

CURRAN ASSOCIATES INC.
proceedings.com